Aijun An Stan Matwin
Zbigniew W. Raś Dominik Ślęzak (Eds.)

Foundations of Intelligent Systems

17th International Symposium, ISMIS 2008
Toronto, Canada, May 20-23, 2008
Proceedings

Springer

Series Editors

Jaime G. Carbonell, Carnegie Mellon University, Pittsburgh, PA, USA
Jörg Siekmann, University of Saarland, Saarbrücken, Germany

Volume Editors

Aijun An
York University, Department of Computer Science and Engineering
Toronto, Ontario, Canada
E-mail: aan@cse.yorku.ca

Stan Matwin
University of Ottawa, School of Information Technology and Engineering
Ottawa, Ontario, Canada
E-mail: stan@site.uottawa.ca

Zbigniew W. Raś
University of North Carolina, Department of Computer Science
Charlotte, NC, USA
E-mail: ras@uncc.edu
and
Polish Academy of Sciences, Institute of Computer Science
01-237 Warsaw, Poland

Dominik Ślęzak
Infobright Inc.
Toronto, Ontario, Canada
E-mail: dominik.slezak@infobright.com

Library of Congress Control Number: 2008926504

CR Subject Classification (1998): I.2, H.3, H.2.8, H.4, H.5, F.1, I.5

LNCS Sublibrary: SL 7 – Artificial Intelligence

ISSN 0302-9743
ISBN-10 3-540-68122-1 Springer Berlin Heidelberg New York
ISBN-13 978-3-540-68122-9 Springer Berlin Heidelberg New York

Springer is a part of Springer Science+Business Media

springer.com

Printed in Germany

Typesetting: Camera-ready by author, data conversion by Scientific Publishing Services, Chennai, India
Printed on acid-free paper SPIN: 12267781 06/3180 5 4 3 2 1 0

Preface

This volume contains the papers selected for presentation at the 17th International Symposium on Methodologies for Intelligent Systems (ISMIS 2008), held in York University, Toronto, Canada, May 21–23, 2008. ISMIS is a conference series started in 1986. Held twice every three years, ISMIS provides an international forum for exchanging scientific research and technological achievements in building intelligent systems. Its goal is to achieve a vibrant interchange between researchers and practitioners on fundamental and advanced issues related to intelligent systems.

ISMIS 2008 featured a selection of latest research work and applications from the following areas related to intelligent systems: active media human–computer interaction, autonomic and evolutionary computation, digital libraries, intelligent agent technology, intelligent information retrieval, intelligent information systems, intelligent language processing, knowledge representation and integration, knowledge discovery and data mining, knowledge visualization, logic for artificial intelligence, soft computing, Web intelligence, and Web services. Researchers and developers from 29 countries submitted more than 100 full papers to the conference. Each paper was rigorously reviewed by three committee members and external reviewers. Out of these submissions, 40% were selected as regular papers and 22% as short papers.

ISMIS 2008 also featured three plenary talks given by John Mylopoulos, Jiawei Han and Michael Lowry. They spoke on their recent research in agent-oriented software engineering, information network mining, and intelligent software engineering tools, respectively.

ISMIS 2008 could not be successful without a team effort. We would like to thank all the authors who contributed to this volume. We also thank the Program Committee members and external reviewers for their contribution in the paper selection process. We are grateful to the Chairs and Organizing Committee members for their contribution to the organization of the conference. In particular, we would like to acknowledge the generous help received from Tokuyo Mizuhara, Jiye Li, Vicki Caparello, Clara Masaro, and Ellis Lau.

We appreciate the support and sponsorship from the following institutions and organizations: York University, Natural Sciences and Engineering Research Council of Canada (NSERC), Ontario Centres of Excellence (OCE), and Springer.

May 2008

Aijun An
Stan Matwin
Zbigniew W. Raś
Dominik Ślęzak

Organization

ISMIS 2008 was organized by the Department of Computer Science and Engineering and the School of Information Technology, York Univeristy.

Executive Committee

General Chair	Zbigniew W. Raś (UNC at Charlotte, USA)
Conference Chair	Nick Cercone (York University, Canada)
Program Co-chairs	Aijun An (York University, Canada)
	Dominik Ślęzak (Infobright Inc., Canada)
	Stan Matwin (University of Ottawa, Canada)
Organizing Chair	Jimmy Huang (York University, Canada)
Workshop Chair	Parke Godfrey (York University, Canada)
Organizing Committee	Ellis Lau (York University, Canada)
	Jiye Li (York University, Canada)
	Clara Masaro (York University, Canada)

Steering Committee

Zbigniew W. Raś	UNC at Charlotte, USA, Chair
Jaime Carbonell	Carnegie Mellon University, USA
Floriana Esposito	University of Bari, Italy
Mohand-Said Hacid	University Lyon 1, France
Donato Malerba	University of Bari, Italy
Neil Murray	SUNY at Albany, USA
John Mylopoulos	University of Toronto, Canada
Setsuo Ohsuga	Waseda University, Japan
Lorenza Saitta	University of Piemonte Orientale, Italy
Giovanni Semeraro	University of Bari, Italy
Shusaku Tsumoto	Shimane Medical University, Japan
Maria Zemankova	NSF, USA
Ning Zhong	Maebashi Institute of Technology, Japan

Program Committee

Luigia Carlucci Aiello, Italy
Troels Andreasen, Denmark
Salima Benbernou, France
Petr Berka, Czech Republic
Elisa Bertino, USA
Ivan Bratko, Slovenia
Cory Butz, Canada
Sandra Carberry, USA
Juan Carlos Cubero, Spain
Michelangelo Ceci, Italy

Bruno Cremilleux, France
Alfredo Cuzzocrea, Italy
Tapio Elomaa, Finland
Ronen Feldman, Israel
Edward Fox, USA
Attilio Giordana, Italy
Parke Godfrey, Canada
Jarek Gryz, Canada
Jerzy Grzymala-Busse, USA
Mirsad Hadzikadic, USA
Howard Hamilton, Canada
Perfecto Herrera, Spain
Jimmy Huang, Canada
Seunghyun Im, USA
Nathalie Japkowicz, Canada
Janusz Kacprzyk, Poland
Mieczyslaw Klopotek, Poland
Joost N. Kok, The Netherlands
Patrick Lambrix, Sweden
Nada Lavrac, Slovenia
Charles Ling, Canada
Pawan Lingras, Canada
Jiming Liu, Canada
Ramon López de Mántaras, Spain
Michael Lowry, USA
David Maluf, USA
Nicola Di Mauro, Italy
Paola Mello, Italy
Zbigniew Michalewicz, Australia
Mitsunori Ogihara, USA
Bryan Pardo, USA
Witold Pedrycz, Canada
James Peters, Canada
Jean-Marc Petit, France
Vijay Raghavan, USA
Jan Rauch, Czech Republic
William Ribarsky, USA
Gilbert Ritschard, Switzerland
Henryk Rybinski, Poland
Nahid Shahmehri, Sweden
Luo Si, USA
Arno Siebes, The Netherlands
Andrzej Skowron, Poland
Roman Slowinski, Poland
Jerzy Stefanowski, Poland
Einoshin Suzuki, Japan
Krishnaprasad Thirunarayan, USA
Li-Shiang Tsay, USA
Athena Vakali, Greece
Christel Vrain, France
Alicja Wieczorkowska, Poland
Xindong Wu, USA
Yiyu Yao, Canada
Djamel Zighed, France

External Referees

Xiangdong An
Annalisa Appice
Teresa M.A. Basile
Henrik Bulskov
Costantina Caruso
Federico Chesani
Guillaume Cleuziou
Tomaz Curk
Chenyun Dai
Wenwen Dou
Frederic Flouvat
Francesco Folino
Matej Guid
Daan He
Chun Jin
Stamos Konstantinos
Jussi Kujala
Jiye Li
Yang Liu
Simon Marcellin
Brandeis Marshall
Elio Masciari
Marco Montali
Martin Mozina
Eileen Ni
Federica Paci
George Pallis
Andrea Pugliese
Taimur Qureshi
Fabrizio Riguzzi
Francois Rioult
Aleksander Sadikov
Marian Scuturici
Davide Sottara
Lena Strömbäck
He Tan
Xiaoyu Wang
Jure Zabkar
Sine Zambach

Sponsoring Organizations

Office of the Vice-President Academic, York University
Department of Computer Science and Engineering, York University
Atkinson Faculty of Liberal & Professional Studies, York University
School of Information Technology, York University
Natural Sciences and Engineering Research Council of Canada (NSERC)
Ontario Centres of Excellence (OCE)
Springer, Heidelberg, Germany

Table of Contents

Invited Papers

Knowledge Discovery and Data Mining - Foundations

Knowledge Discovery and Data Mining - Advances

Knowledge Discovery and Data Mining - Mining Changes and Actionable Patterns

Logic for Artificial Intelligence

Autonomic and Evolutionary Computation

Soft Computing

Databases and Data Warehouses

Digital Library

Intelligent Information Retrieval

Text Mining

Intelligent Information Systems

Knowledge Representation and Integration

Web Service and Intelligence

Applications

From Goals to High-Variability Software Design

Yijun Yu[1], Alexei Lapouchnian[2], Sotirios Liaskos[3], John Mylopoulos[2], and Julio C.S.P. Leite[4]

[1] Department of Computing, The Open University, United Kingdom
y.yu@open.ac.uk
[2] Department of Computer Science, University of Toronto, Canada
alexei@cs.toronto.edu, jm@cs.toronto.edu
[3] School of IT, York University, Canada
liaskos@yorku.ca
[4] Department of Informatics, PUC-Rio, Brazil
julio@inf.puc-rio.br

Abstract. Software requirements consist of functionalities and qualities to be accommodated during design. Through goal-oriented requirements engineering, stakeholder goals are refined into a space of alternative functionalities. We adopt this framework and propose a decision-making process to generate a generic software design that can accommodate the full space of alternatives each of which can fulfill stakeholder goals. Specifically, we present a process for generating complementary design views from a goal model with high variability in configurations, behavioral specifications, architectures and business processes.

1 Introduction

Traditionally, requirements consist of functions and qualities the system-to-be should support [8, 4]. In goal-oriented approaches [26, 27, 22], requirements are derived from the list of stakeholder goals to be fulfilled by the system-to-be, and the list of quality criteria for selecting a solution to fulfill the goals [22]. In goal models, root-level goals model stakeholder intentions, while leaf-level goals model functional system requirements. [26] offers a nice overview of Goal-Oriented Requirements Engineering, while the KAOS [8] and the i*/Tropos approaches [30, 2] represent the state-of-the-art for research on the topic.

We are interested in using goal models to develop *generic* software solutions that can accommodate many/all possible functionalities that fulfill stakeholder goals. This is possible because our goals models are extensions of AND/OR graphs, with OR decompositions introducing alternatives into the model. The space of alternatives defined by a goal model can be used as a basis for designing fine-grained variability for highly customizable or adaptable software. Through customizations, particular alternatives can be selected by using *softgoals* as criteria. Softgoals model stakeholder preferences, and may represent qualities that lead to non-functional requirements.

The main objective of this paper is to propose a process that generates a high variability software design from a goal model. The process we propose is supported by heuristic rules that can guide the design.

A. An et al. (Eds.): ISMIS 2008, LNAI 4994, pp. 1–16, 2008.

Our approach to the problem is to accommodate the variability discovered in the problem space by a variability model in the solution space. To this end, we employ three complementary design views: a feature model, a statechart and a component model. The feature model describes the system-to-be as a combination variable set of features. The statechart provides a view of the behavior alternatives in the system. Finally, the component model reveals the view of alternatives as variable structural bindings of the software components.

The goal model is used as the logical view at the requirements stage, similar to the global view in the 4+1 views [17] of the Rational Unified Process. This goal model transcends and circumscribes design views. On the other hand, a goal model is missing useful information that will guide decisions regarding the structure and behavior of the system-to-be. Our proposed process supports lightweight annotations for goal models, through which the designer can introduce some of this missing information.

The remainder of the paper is organized as follows: Section 2 introduces requirements goal models where variability is initially captured. Section 3 talks about generating high-variability design views in feature models, statecharts, component-connector architectures and business processes. Section 4 discusses tool support and maintenance of traceability. Section 5 explains a case study. Finally, Section 6 presents related work and summarizes the contributions of the paper.

2 Variability in Requirements Goal Models

We adopt the formal goal modeling framework proposed in [9, 23]. According to this framework, a goal model consists of one or more root goals, representing stakeholder objectives. Each of these is AND/OR decomposed into subgoals to form a forest. In addition, a goal model includes zero or more softgoals that represent stakeholder preferences. These can also be AND/OR decomposed. Moreover, there can be positive and negative contribution relationships from a goal/softgoal to a softgoal indicating that fulfillment of one goal/softgoal leads to partial fulfillment or denial of another softgoal. The semantics of AND/OR decompositions is adopted from AI planning. [9] and [23] provide a formal semantics for different goal/softgoal relationships and propose reasoning algorithms which make it possible to check (a) if root goals are (partially) satisfied/denied assuming that some leaf-level goals are (partially) satisfied/denied; (b) search for minimal sets of leaf goals which (partially) satisfy/deny all root goals/softgoals.

Figure 1a shows an example goal model describing the requirements for "schedule a meeting". The goal is AND/OR decomposed repeatedly into leaf-level goals. An OR-decomposition of a goal introduces a *variation point*, which defines alternative ways of fulfilling the goal.

Variability is defined as all possible combinations of the choices in the variation points. For example, the four variation points of the goal model in Figure 1a are marked VP1-VP4. VP1 contributes two alternatives, VP2 and VP4 combined contribute 3, while VP3 contributes 2. Then, the total space of alternatives for this goal model includes 2*3*2 = 12 solutions. Accordingly, we would like to have a systematic process for producing a generic design that can potentially accommodate all 12 solutions.

(a):

(b):

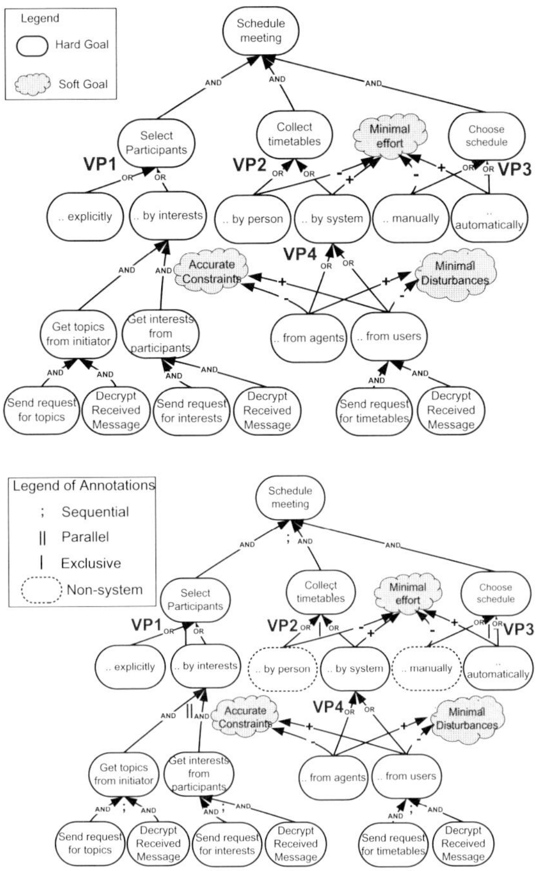

Fig. 1. An example generic goal model of the meeting scheduler: (a) Variation points by OR decompositions are indicated as VP1-4; (b) the goal model annotated with enriched labels

It is desirable to preserve requirements variability in the design that follows. We call a design parameterized with the variation point choices a *generic* design, which can implement a product-line, an adaptive product and/or a component-based architecture. A *product-line* can be configured into various products that share/reuse the implementation of their commonality. An *adaptive* product can adapt its behavior at run-time to accommodate anticipated changes in the environment as it is armed with alternative solutions. A *component-based* design makes use of interface-binding to flexibly orchestrate components to fulfill the goal. In the next section, we will detail these target design views as feature models, statecharts and component-connectors.

In this work traceability between requirements and design is achieved by an explicit transformation from goal models to design views. Goal models elicited from requirements are more abstract than any of the design views; therefore such generation is only possible after identifying a mapping between goals and design elements. To save designer's effort, we enrich the goal models with minimal information such that the generative patterns are sufficient to create preliminary design-level views.

Figure 1b shows an annotated goal model for feature model and statecharts generation, where the semantics of the goal decompositions are further analyzed: (1) VP1, VP2 and VP3 are exclusive (|) and VP4 is inclusive; (2) based on the temporal relationships of the subgoals as required by the data/control dependencies, AND-decompositions are annotated as sequential (;) or parallel (||) and (3) goals to be delegated to non-system agents are also indicated. We will explain the detailed annotations for deriving a component-connector view in the next section. As the preliminary design evolves, design elements and their relationships can change dramatically. However, the traceability to the required variability must be maintained so that one can navigate between the generated design views using the derived traceability links among them.

3 Generating Preliminary Design Views

This section presents three preliminary design views that can be generated from a high-variability goal model. For each view, we use generative patterns to simplify the task.

3.1 Feature Models

Feature modeling [14] is a domain analysis technique that is part of an encompassing process for developing software for reuse (referred to as Domain Engineering [7]). As such, it can directly help in generating domain-oriented architectures such as product-line families [15].

There are four main types of features in feature modeling: *Mandatory*, *Optional*, *Alternative*, and *OR* features [7]. A Mandatory feature must be included in every member of a product line family provided that its parent feature is included; an Optional feature may be included if its parent is included; exactly one feature from a set of Alternative features must be included if a parent of the set is included; any non-empty subset of an OR-feature set can be included if a parent feature is included.

There are fundamental differences between goals and features. Goals represent stakeholder intentions, which are manifestations of intent which may or may not be realized. Goal models are thus a space of intentions which may or may not be fulfilled. Features, on the other hand, represent properties of concepts or artifacts [7]. Goals will use the services of the system-to-be as well as those of external actors for their fulfillment. Features in product families represent *system* functions or properties. Goals may be partially fulfilled in a qualitative or quantitative sense [9], while features are either elements of an allowable configuration or they are not. Goals come with a modality: achieve, maintain, avoid, cease [8], while features have none. Likewise, AND decomposition of goals may introduce temporal constraints (e.g., fulfill subgoal A before subgoal B) while features do not.

As noted in [7], feature models must include the semantic description and the rationale for each feature (why it is in the model). Also, variable (OR/Optional/Alternative) features should be annotated with conditions describing when to select them. Since goal models already capture the rationale (stakeholder goals) and the quality criteria driving the selection of alternatives, we posit that they are proper candidates for the generation of feature models.

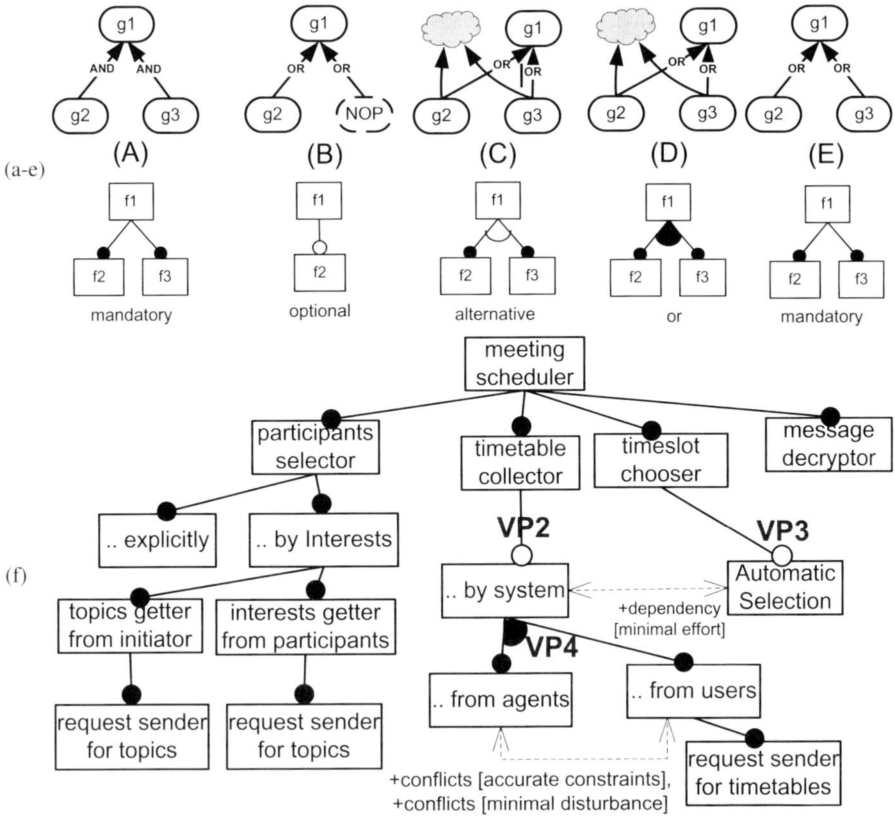

Fig. 2. Generating Features: (a-e) a set of patterns; (f) the feature model generated from Figure 1b

Feature models represent the variability in the system-to-be. Therefore, in order to generate them, we need to identify the subset of a goal model that is intended for the system-to-be. Annotations can be applied to generate the corresponding preliminary feature model. AND decompositions of goals generally correspond to sets of Mandatory features (see Figure 2a). For OR decompositions, it is important to distinguish two types of variability in goal models: *design-time* and *runtime*. Design-time variability is high-level and independent of input. It can be usually be bound at design-time with the help of user preferences or other quality criteria. On the other hand, runtime variability depends on the runtime input and must be preserved at runtime. For example, meeting participants can be selected explicitly (by name) or chosen by matching the topic of the meeting with their interests (see Figure 2). The choice will depend on the meeting type and thus both alternatives must be implemented. When subgoals cannot be selected based on some quality criteria (softgoals), they are considered runtime variability, thus, runtime variability in goal models will generate mandatory features (Figure 2e). Otherwise, as design-time variability, other OR decompositions can be mapped into sets of OR-features (Figure 2d). However, Alternative and Optional feature sets do not have counterparts in the AND/OR goal models. So, in order to generate these types of features we need to analyze whether some of the OR decompositions are, in fact, XOR decompositions (where exactly one subgoal must be achieved) and then annotate these decompositions with the symbol "|" (Figure 2c). The inclusive OR decomposition corresponds to a feature refined into a set of OR features (Figure 2d). Finally, when a goal is OR-decomposed into at least one non-system subgoal (specified by a goal annotation NOP), the sibling system subgoals will be mapped into optional features (Figure 2b). As a result, Figure 2f shows the feature model generated from Figure 1b. The implemented procedure has been reported in [31].

In a more complex design, the system may need to facilitate the environmental actors in achieving their goals or monitor the achievement of these goals. Here, the goals delegated to the environment can be replaced with user interfaces, monitoring or other appropriate features. In general, there is no one-to-one correspondence between goals delegated to the system and features. While high-level goals may be mapped directly into grouping features in an initial feature model, a leaf-level goal may be mapped into a single feature or multiple features, and several leaf goals may be mapped into a feature, by means of factoring. For example, a number of goals requiring the decryption of received messages in a secure meeting scheduling system may be mapped into a single feature "Message Decryptor" (see Figure 2[1]).

3.2 Statecharts

Statecharts, as proposed by David Harel [11], are a visual formalism for describing the behavior of complex systems. On top of states and transitions of a state machine, a statechart introduces nested super-/sub-state structure for abstraction (from a state to its super-state) or decomposition (from a state to its substates). In addition, a state can be decomposed into a set of *AND* states (visually separated by swim-lanes) or a set of *XOR* states [11]. A transition can also be decomposed into transitions among the substates.

This hierarchical notation allows the description of a system's behavior at different levels of abstraction. This property of statecharts makes them much more concise and usable than, for example, plain state machines. Thus, they constitute a popular choice for representing the behavioral view of a system.

Figure 3a0 shows a mapping from a goal in a requirements goal model to a state in a statechart. There are four basic patterns of goals in linear temporal logic formula, here we show mappings for *achieve* and *maintain* goals, whereas *cease* and *avoid* goals are similar.

An *achieve* goal is expressed as a temporal formula with P being its precondition, and Q being its post-condition. In the corresponding statechart, one entry state and one exit state are created: P describes the condition triggering the transition from the entry to the exit state; Q prescribes the condition that must be satisfied at the exit state. The transition is associated with an activity to reach the goal's desired state. The *cease* goal is mapped to a similar statechart (not shown) by replacing the condition at the exit state with $\neg Q$. For mapping a *maintain* goal into a statechart, the transition restores the state

[1] One can systematically derive feature names from the hard goal descriptions by, for example, changing the action verb into the corresponding noun (e.g., "schedule meeting" becomes "meeting scheduler").

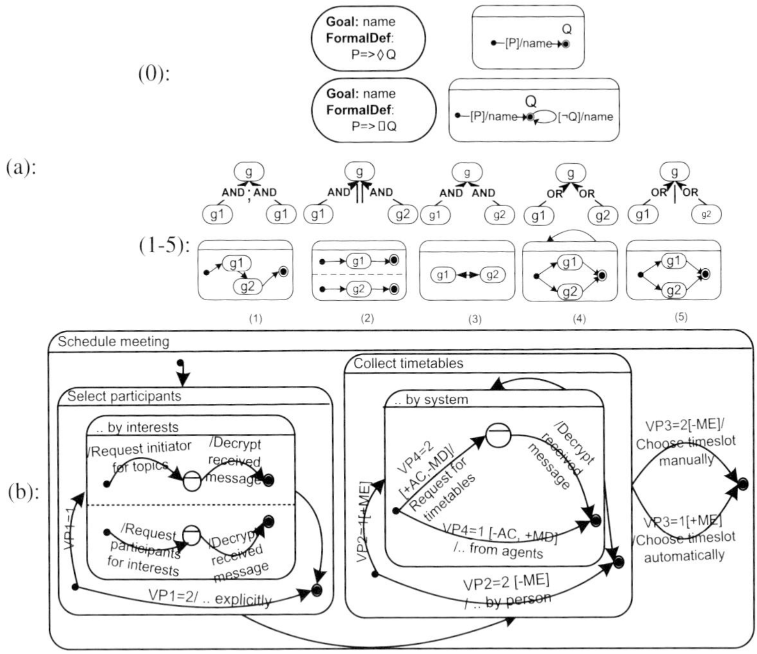

Fig. 3. Generating Statecharts: (a0-a5) a set of patterns; (b) the generated statecharts from Figure 1b

back to the one that satisfies Q whenever it is violated while P is satisfied. Similar to the *maintain* goal's statechart, the statechart for an *avoid* goal swaps Q with its negation. These conditions can be used symbolically to generate an initial statechart view, i.e., they do not need to be explicit temporal logic predicates. At the detailed design stage, the designer may provide solution-specific information to specify the predicates for a simulation or an execution of the refined statechart model.

Then applying the patterns in Figure 3a1–5, a goal hierarchy will be mapped into an isomorphic state hierarchy in a statechart. That is, the state corresponding to a goal becomes a super-state of the states associated with its subgoals. The runtime variability will be preserved in the statecharts as alternative transition paths.

The transformation from a goal model to an initial statechart can be automated even when the temporal formulae are not given: we first associate each leaf goal with a state that contains an entry substate and an exit substate. A default transition from the entry substate to the exit substate is labeled with the action corresponding to the leaf goal. Then, the AND/OR goal decompositions are mapped into compositions of the statecharts. In order to know how to connect the substates generated from the corresponding AND-decomposed subgoals, temporal constraints are introduced as goal model annotations, e.g., for an OR-decomposition, one has to consider whether it is inclusive or exclusive.

Given root goals, our statechart generation procedure descends along the goal refinement hierarchy recursively. For each leaf goal, a state is created according to Figure 3a0. The created state has an entry and an exit substates. Next, annotations that represent the temporal constraints with the AND/OR goal decompositions are considered. Composition patterns can then be used to combine the statecharts of subgoals into one statechart (Figure 3a1-5). The upper bound of number of states generated from the above patterns is $2N$ where N is the number of goals.

For the Schedule Meeting goal model in Figure 1, we first identify the sequential/parallel control patterns for AND-decompositions through an analysis of the data dependencies. For example, there is a data dependency from "Send Request for Timetable" to "Decrypt Received Message" because the time table needs to be requested first, then received and decrypted. Secondly, we identify the inclusive/exclusive patterns for the OR decompositions. For example, "Choose Time Slot" is done either "Manually" or "Automatically". Then we add transitions according to the patterns in Figure 3a. As a result, we obtain a statechart with hierarchical state decompositions (see Figure 3b). It describes an initial behavioral view of the system.

The preliminary statechart can be further modified by the designer. For example, the abstract "send requests for timetable" state can be further decomposed into a set of substates such as "send individual request for timetable" for each of the participants. Since the variability is kept by the guard conditions on the transitions, changes of the statecharts can still be traced back to the corresponding annotated goal models. Moreover, these transitions specified with entry/exit states can be used to derive test cases for the design.

3.3 Component-Connector View

A component-connector architectural view is typically formalized via an architecture description language (ADL) [21]. We adapt the Darwin ADL [20] with elements borrowed from one of its extensions, namely Koala [28]. Our component-connector view is defined by components and their bindings through interface types. An *interface type* is a collection of message signatures by which a component can interact with its environment. A *component* can be connected to other components through instances of interface types (i.e. interfaces). A **provides** interface shows how the environment can access the component's functionality, whereas a **requires** interface shows how the component can access the functionality provided by the environment. In a component configuration, a **requires** interface of a component in the system must be *bound* to exactly one **provides** interface of another component. However, as in [28], we allow alternative bindings of interfaces through the use of a special connection component, the *switch*. A switch allows the association of one **requires** interface with many alternative **provides** interfaces of the same type, and is placed within a compound component which contains the corresponding alternative components.

A preliminary component-connector view can be generated from a goal model by creating an interface type and a component for each goal. The interface type contains the signature of an operation. The operation name is directly derived from the goal description, the **IN/OUT** parameters of the operation signature must be specified for the input and output of the component. Therefore in order to generate such a preliminary component-connector view, each goal needs to be annotated with specification of input/output data. For example, the interface type generated from the goal "Collect Timetables from Users" is shown as follows:

```
interface type ICollectTimetablesFromUsers {
   CollectTimetables( IN Users, Interval,
                      OUT Constraints);
}
```

The generated component implements the interface type through a **provides** interface.

The **requires** interfaces of the component depend on how the goal is decomposed. If the goal is AND-decomposed, the component has as many **requires** interfaces as the subgoals. In our example, the initial component of the goal "Collect timetables from Users" is generated as follows:

```
component TimetableCollectorFromUsers {
      provides ICollectTimetables;
      requires IGetTimetable, IDecryptMessage;
}
```

The **requires** interfaces are bound to the **provides** interfaces of the subgoals.

If the goal is OR-decomposed, the interface types of the subgoals are first replaced with the interface type of the parent goal such that the **provides** interface of the parent goal is of the same type as the **provides** interfaces of the subgoals. Then, inside the generated component, a switch is introduced to bind these interfaces. The **provides** interface of the compound component generated for the parent goal can be bound to any of the subgoal's **provides** interfaces. Both the switch and the components of the subgoals are placed inside the component of the parent goal, and are *hidden* behind its interface.

In Figure 4, the graphical notation is directly adopted from Koala/Darwin. The boxes are components and the arrows attached to them represent **provides** and **requires** interfaces, depending on whether the arrow points inwards or outwards respectively. The lines show how interfaces are bound for the particular configuration and are annotated with the name of the respective interface type; the shape of the overlapping parallelograms represents a switch. Patterns show how AND/OR decompositions of system goals are mapped into the component-connector architecture.

In order to accommodate a scheme for *event* propagation among components, we follow an approach inspired by the C2 architectural style [21] where requests and notifications are propagated in opposite directions. As requests flow from high level components to low-level ones, notifications (events) originated from low-level components will propagate to high-level ones. Such events are generated from components associated with goals that are delegated to the environment (non-system goals). These components are responsible for supporting external actors' activity to attain the associated goal, to sense the progress of this attainment and to communicate it to the components of higher level by generating the appropriate events. We name such components *interface components* to signify that they lay at the border of the system. Interface components have an additional **requires** interface that channels the events. This interface is optionally bound to an additional **provides** interface at the parent component, its event handler. In Java, such a binding becomes a Listener interface in the parent component for receiving events from the interface component.

```xml
<view xmlns="urn:goals" xmlns:e="urn:enrichments"
  xmlns:f="urn:features" xmlns:s="urn:statecharts"
  xmlns:c="urn:components">
  < goal name="Schedule Meeting">
    <refinement type="goal" value="AND"/>
    <e:refinement type="state" value="SEQ"/>
    <e:annotation type="feature" value="SYS"/>
    <e:annotation type="component"
         input="meeting" output="schedule"/>
    <e:annotation type="state" pre="meeting!=null"
      post="!schedulable OR schedule!=null"/>
    <f:feature name="Meeting Scheduler"
      src="generated: renamed"/>
    <s:statechart name="Schedule Meeting"
      src="generated"/>
    <c:component name="Meeting Scheduler"
      src="generated: renamed"/>
    < goal name="Select Participants"> ... </goal>
    < goal name="Collect Timetables"> ... </goal>
    < goal name="Choose Schedule"> ... </goal>
  </goal>
</view>
```

(a): high-variability goal model annotated

```xml
<view xmlns:g="urn:goals" xmlns="urn:features"...>
 < g:goal name="Schedule Meeting">
  <g:refinement type="goal" value="AND"/>...
  <e:annotation type="feature" value="SYS"/>...
  < feature name="meeting scheduler"
   src="renamed"> <refinement type="feature"
    value="AND" src="generated"/>
   <cardinality value="1" src="generated"/>
   < g:goal name="Select Participants">
    < feature name="participants selector"
    src="renamed"> ...</feature>...</g:goal>
    < g:goal name="Collect Timetables">s
     < feature name="timetable collector"
      src="renamed">...</feature>...</g:goal>
    < g:goal name="Choose Schedule">
     < feature name="timetable collector"
      src="renamed">...</feature>...</g:goal>
    < feature name="message decryptor"
    src="refactored"
    origin="Decrypt Received Message">
    ...<feature/> </feature> ...</g:goal> </view>
```

(b): generated features

Fig. 5. Traceability support in representations of annotated goal models

features (e.g. "message decryptor") can correspond to a goal that is not originally decomposed from that corresponding to its parent. In this case, the original goal with which it was associated is indicated. During design, features not corresponding to any goal can be added/removed. However, such design changes cannot remove existing traceability links including the ones implied by the sibling XML elements. Deleted design elements are still kept in the model with a "deleted" `src` attribute. As the `src` attribute maintains the traceability between a goal and a design element, the link between two design views can be derived. For example, in the above document, one can obtain the traceability link between a "Meeting Scheduler" component and a "meeting scheduler" feature since they are both mapped from the "Schedule Meeting" goal.

An XMI representation of the above metamodel represents the elements of a unified model in a single name space. For each design view, the generation procedure is implemented as an XSLT stylesheet that transforms the unified model into an XML design document that can be imported by the corresponding visual tool. A generated document separates the concerns of various views into different namespaces. The default name space concerns the primary design view, whereas the relationships between adjacent tags of different namespaces capture the traceability among corresponding design views.

5 A Case Study

A case study was conducted to validate our approach. First we developed a goal model by decomposing the user's goal for *Prepare an Electronic Message* into 48 goals with 10 AND- and 11 OR-decompositions [13]. Instead of developing from scratch we considered reusing components from an existing email client. To support the *external* validity of our study, we chose a public-domain Java email client Columba[3], which had been studied for reverse engineering of its goal models [34]. This poses a threat to the *internal* validity since none of the authors is involved in the development of Columba, thus

[3] http://www.columbamail.org

our understanding of its absent design alternatives is limited. However, we relied on program understanding to analyze its implemented requirements. Due to the absence of early decisions, the purpose of this study differs from [34] which was to recover as-is requirements from the code. Rather, we compared the implementation with respect to the goal model that appears in [13] in order to check (1) whether we can reuse any components in Columba to implement our generated design views; (2) whether Columba system can cover all the variability needed by our requirements goal model; (3) whether our final design can be traced back to the original requirements goal models.

The study was done as follows (for details see technical report [33]). First, we analyzed the system to reconstruct design views including feature models, component-connector views and statecharts. Starting with the component-connector view: each component corresponds to at least one JAR archive. The system includes 3 basic components (`Core`, `AddressBook`, `Mail`), 23 third party components and 16 plug-in components. All of them become features in the system's feature model view, including 26 mandatory features and 22 optional features. Further, we found that 8 mandatory components are actually related to process-oriented goals for maintainability, such as *Testing* (`junit`, `jcoverage`), *Coding convention check* (`checkstyle`), *Command line options* (`common-cli`), *XML documentations* (`jdom`) and *Build automation* (`ant`, `maven`, `jreleaseinfo`). The other 40 components relate to product-oriented user goals. Some AND-decomposed features correspond to OR-decomposed goals. For example, the *Look and Feel* usability goal was implemented by 6 plugin components (`Hippo`, `Kunststoff`, `Liquid`, `Metoulia`, `Win`, `Thin`). At run-time, only one of these components will be enabled for the look and feel. Similar observation applies to the `IMAP` and `POP3` server features. Statecharts were obtained from the dynamic messaging trace rather than from static call graphs. Since we are interested in abstract statecharts that are closer to goals, the intra-procedural messages were ignored as long as they were structural and goals could thus be extracted and partially validated by the existing 285 test cases [34].

Since the feature models and the component-connector views were recovered from static call graphs, whereas the statecharts were recovered from dynamic messaging traces, we found it hard to establish traceability among them. For example, a non-functional maintenance goal (e.g. test cases, plugin support) often crosscuts multiple components [32, 34]. In short, there is no one-to-one mapping between goals and the tangled design elements.

Then we looked at reusing Columba to fulfill our high-variability goal, "Prepare an electronic message" [13]. We found that Columba does have reusable components to fulfill some goals of our user. For example, the above hard goal is AND decomposed into *Addressbook* and *Mail folder* management, *Composing email*, *Checking Emails*, *Sending email*, *Spell Checking*, *Printing*, etc. They are implemented by the 3 basic components. The 3rd party components are called by these components to fulfill hard goals such as *Java Mail delivery* (`Ristretto`), *Spam filtering* (`Macchiato`, `SpamAssassin`), *Full Text search* (`Lucene`), *Profile Management* (`jpim` and Berkeley DB `je`). The plug-ins help with functional goals such as *Chatting* (`AlturaMess-enger`), *mail importing*(`Addressbook`, `mbox`, etc.) and *notification* (`PlaySound-Filter`). For some non-functional softgoals, *Usability* is implemented with 7 *GUI* components (`Swing`, `JGoodies`, `Frapuccino`, `jwizz`, `JDic`) and *Online Help* (`jhall`, `usermanual`); *Security* is fulfilled by component GNU Privacy guard (`JSCF`).

There is still missing functionality or variability for our users: *Composing email* does not allow *Compose by voice* subgoal, which we implemented by switching Columba's text-based composer to the speech-based composer using `JavaMedia`; the mandatory *Auto Completion for Recipients* feature may not satisfy some of our users who wanted *Low Mistake Probability* and *Maximum Performance* when the address book is large. Thus, it was turned into optional in our implementation.

Applying our approach, we obtained preliminary design views that were traceable to each other. The derived feature view consists of 39 features where 9 non-system goals were discarded as NOP. Among the 11 variation points, 2 were turned into mandatory feature decompositions because there was no softgoal associated with them as a selection criterion. The derived statecharts view consists of 21 abstract leaf states where the transitions were turned into test cases. The component-connector view consists of 33 concrete components controlled by 9 switches. We have implemented such generic design based on Columba and additional libraries, among which a reasoning component [9, 23] was used. Thus the resulting system can be reconfigured using quality attributes derived from requirements goals, user skills and preferences [13].

6 Related Work and Conclusion

There is growing interest on the topic of mapping goal-oriented requirements to software architectures. Brandozzi et al. [1] recognized that requirements and design were respectively in problem and solution domains, thereby a mapping between a goal and a component was proposed for increasing reusability. A more recent work by van Lamsweerde et al. [25] derives software architectures from the formal specifications of a system goal model using heuristics. Specifically, the heuristics discover design elements such as classes, states and agents directly from the temporal logic formulae that define the goals. Complementary to their formal work, we apply light-weight annotations to the goal model in order to derive design views: if one has the formal specifications for each goal, some heuristics provided in [25] can be used to find the annotations we need, such as system/non-system goals, inputs/outputs and dependencies among the subgoals. Generally, this line of research has not addressed variability at the design level.

Variability within a product-line is another topic receiving considerable attention [5, 7], which has not addressed the problem of linking product family variability to stakeholder goals (and the alternative ways these can be achieved). Closer to our work, [10] proposes an extension of use case notation to allow for variability in the use of the system-to-be. More recently [3], the same group tackled the problem of capturing and characterizing variability across product families that share common requirements.

We maintain traceability between requirements and the generated design using concepts from literate programming and model-driven development. In *literate programming* [16], any form of document and program can be put together in a unified form, which can then be tangled into an executable program with documented comments or into a document with illustrative code segments. In this regards, our annotated goal model is a unified model that can derive the preliminary views with traceability to other views. In *model-driven development* (MDD), code can be generated from design models and a change from the model can propagate into the generated code [24]. To prevent a manual change to the code from being overridden by the code generation, a special *not generated* annotation can be manually added to the comments of the initially generated code. We use MDD in a broader sense – allow designers to change the generated preliminary design views and to manually change the *not generated* attribute of a generated design element. Comparing to deriving traceability links through information retrieval [12], our work proproses a generative process where such links are produced by the process itself and recorded accordingly.

In [6], softgoal models that represent non-functional requirements were associated with the UML design views. On the other hand, this paper focuses on establishing traceability between functional requirements and the preliminary designs. Since variability is explicitly encoded as alternative designs, the non-functional requirements in goal models are kept in the generic design to support larger user bases.

In summary, this paper proposes a systematic process for generating complementary design views from a goal model while preserving variability (i.e., the set of alternative ways stakeholder goals can be fulfilled). The process is supported by heuristic rules and mapping patterns. To illustrate, we report a case study reusing public domain software. The generated preliminary design is comparable in size to the goal model.

References

[1] Brandozzi, M., Perry, D.E.: Transforming goal oriented requirements specifications into architectural prescriptions. In: STRAW at ICSE 2001 (2001)

[2] Bresciani, P., Perini, A., Giorgini, P., Giunchiglia, F., Mylopoulos, J.: Tropos: An Agent-Oriented Software Development Methodology. Autonomous Agents and Multi-Agent Systems 8(3), 203–236 (2004)

[3] Bühne, S., Lauenroth, K., Pohl, K.: Modelling requirements variability across product lines. In: RE 2005, pp. 41–50 (2005)

[4] Chung, L., Nixon, B.A., Yu, E., Mylopoulos, J.: Non-Functional Requirements in Software Engineering. Kluwer Academic Publishing, Dordrecht (2000)

[5] Clements, P., Northrop, L.: Software Product Lines: Practices and Patterns. Addison-Wesley, Boston (2001)

[6] Cysneiros, L.M., Leite, J.C.S.P.: Non-functional requirements: from elicitation to conceptual models. IEEE Trans. on Softw. Eng. 30(5), 328–350 (2004)

[7] Czarnecki, K., Eisenecker, U.: Generative Programming: Methods, Tools, and Applications. Addison-Wesley, Reading (2000)

[8] Dardenne, A., van Lamsweerde, A., Fickas, S.: Goal-directed requirements acquisition. Science of Computer Programming 20(1–2), 3–50 (1993)

[9] Giorgini, P., Mylopoulos, J., Nicchiarelli, E., Sebastiani, R.: Reasoning with goal models. In: Spaccapietra, S., March, S.T., Kambayashi, Y. (eds.) ER 2002. LNCS, vol. 2503, pp. 167–181. Springer, Heidelberg (2002)

[10] Halmans, G., Pohl, K.: Communicating the variability of a software-product family to customers. Software and Systems Modeling 2, 15–36 (2003)

[11] Harel, D., Naamad, A.: The statemate semantics of statecharts. ACM Trans. on Software Engineering and Methodology 5(4), 293–333 (1996)

[12] Hayes, J.H., Dekhtyar, A., Sundaram, S.K.: Advancing candidate link generation for requirements tracing: the study of methods. IEEE Trans. on Softw. Eng. 32(1), 4–19 (2006)
[13] Hui, B., Liaskos, S., Mylopoulos, J.: Goal skills and preference framework. In: International Conference on Requirements Engineering (2003)
[14] Kang, K.C., Cohen, S.G., Hess, J.A., Novak, W.E., Peterson, A.S.: Feature-Oriented Domain Analysis (FODA) feasibility study (cmu/sei-90-tr-21, ada235785). Technical report (1990)
[15] Kang, K.C., Kim, S., Lee, J., Lee, K.: Feature-oriented engineering of PBX software for adaptability and reuseability. SPE 29(10), 875–896 (1999)
[16] Knuth, D.: Literate programming. Comput. J. 27(2), 97–111 (1984)
[17] Kruntchen, P.: Architectural blueprints – the "4+1" view model of software architecture. IEEE Software 12(6), 42–50 (1995)
[18] Lapouchnian, A., Liaskos, S., Mylopoulos, J., Yu, Y.: Towards requirements-driven autonomic systems design. In: DEAS 2005, pp. 1–7. ACM Press, New York (2005)
[19] Lapouchnian, A., Yu, Y., Mylopoulos, J.: Requirements-driven design and configuration management of business processes. In: Alonso, G., Dadam, P., Rosemann, M. (eds.) BPM 2007. LNCS, vol. 4714, pp. 246–261. Springer, Heidelberg (2007)
[20] Magee, J., Kramer, J.: Dynamic structure in software architectures. In: The 4th ACM SIGSOFT symposium on Foundations of software engineering, pp. 3–14 (1996)
[21] Medvidovic, N., Taylor, R.N.: A framework for classifying and comparing architecture description languages. SIGSOFT Softw. Eng. Notes 22(6), 60–76 (1997)
[22] Mylopoulos, J., Chung, L., Nixon, B.: Representing and using nonfunctional requirements: A process-oriented approach. IEEE Trans. on Softw. Eng. 18(6), 483–497 (1992)
[23] Sebastiani, R., Giorgini, P., Mylopoulos, J.: Simple and minimum-cost satisfiability for goal models. In: Persson, A., Stirna, J. (eds.) CAiSE 2004. LNCS, vol. 3084, pp. 20–35. Springer, Heidelberg (2004)
[24] Selic, B.: The pragmatics of model-driven development. IEEE Softw. 20(5), 19–25 (2003)
[25] van Lamsweerde, A.: From system goals to software architecture. In: Bernardo, M., Inverardi, P. (eds.) SFM 2003. LNCS, vol. 2804, Springer, Heidelberg (2003)
[26] van Lamsweerde, A.: Goal-oriented requirements engineering: From system objectives to UML models to precise software specifications. In: ICSE 2003, pp. 744–745 (2003)
[27] van Lamsweerde, A., Willemet, L.: Inferring declarative requirements from operational scenarios. IEEE Trans. Software Engineering 24(12), 1089–1114 (1998)
[28] van Ommering, R.C., van der Linden, F., Kramer, J., Magee, J.: The Koala component model for consumer electronics software. IEEE Computer 33(3), 78–85 (2000)
[29] Wang, Y., McIlraith, S.A., Yu, Y., Mylopoulos, J.: An automated approach to monitoring and diagnosing requirements. In: ASE, pp. 293–302 (2007)
[30] Yu, E.S.K.: Towards modelling and reasoning support for early-phase requirements engineering. In: RE 1997, pp. 226–235 (1997)
[31] Yu, Y., Lapouchnian, A., Leite, J., Mylopoulos, J.: Configuring features with stakeholder goals. In: ACM SAC RETrack 2008 (2008)
[32] Yu, Y., Leite, J., Mylopoulos, J.: From requirements goal models to goal aspects. In: International Conference on Requirements Engineering (2004)
[33] Yu, Y., Mylopoulos, J., Lapouchnian, A., Liaskos, S., Leite, J.C.: From stakeholder goals to high-variability software design, ftp.cs.toronto.edu/csrg-technical-reports/509. Technical report, University of Toronto (2005)
[34] Yu, Y., Wang, Y., Mylopoulos, J., Liaskos, S., Lapouchnian, A., do Prado Leite, J.C.S.: Reverse engineering goal models from legacy code. In: RE 2005, pp. 363–372 (2005)

Exploring the Power of Heuristics and Links in Multi-relational Data Mining*

Xiaoxin Yin[1] and Jiawei Han[2]

[1] Microsoft Research, One Microsoft Way, Redmond, WA, 98052, USA
xyin@microsoft.com
[2] University of Illinois at Urbana-Champaign, Urbana, IL 61801, USA
hanj@cs.uiuc.edu

Abstract. Relational databases are the most popular repository for structured data, and are thus one of the richest sources of knowledge in the world. Because of the complexity of relational data, it is a challenging task to design efficient and scalable data mining approaches in relational databases. In this paper we discuss two methodologies to address this issue. The first methodology is to use heuristics to guide the data mining procedure, in order to avoid aimless, exhaustive search in relational databases. The second methodology is to assign certain property to each object in the database, and let different objects interact with each other along the links. Experiments show that both approaches achieve high efficiency and accuracy in real applications.

1 Introduction

Most existing data mining algorithms work on single tables or a set of transactions. Although well-formatted tables or transaction sets are easy to model and analyze, most information in the world can hardly be represented by such independent tables. In a real-world dataset there are usually many types of objects, which are linked together through different types of linkages, and are usually stored in relational databases or in XML files.

Compared with individual tables, a relational database often provides much richer information for data mining. For example, the database of a computer science department often contains information about professors, students, courses, research groups, *etc.* (as shown in Figure 1). When performing data mining tasks on students, we can utilize various types of information in different relations linked with the students, such as their advisors, research groups, and publications. We can classify students according to their academic performances, cluster students based on their research, find patterns/correlations of course registrations and publications, or detect duplicate entries among authors of publications.

* The work was supported in part by the U.S. National Science Foundation NSF IIS-05-13678 and NSF BDI-05-15813. Any opinions, findings, and conclusions or recommendations expressed here are those of the authors and do not necessarily reflect the views of the funding agencies.

A. An et al. (Eds.): ISMIS 2008, LNAI 4994, pp. 17–27, 2008.

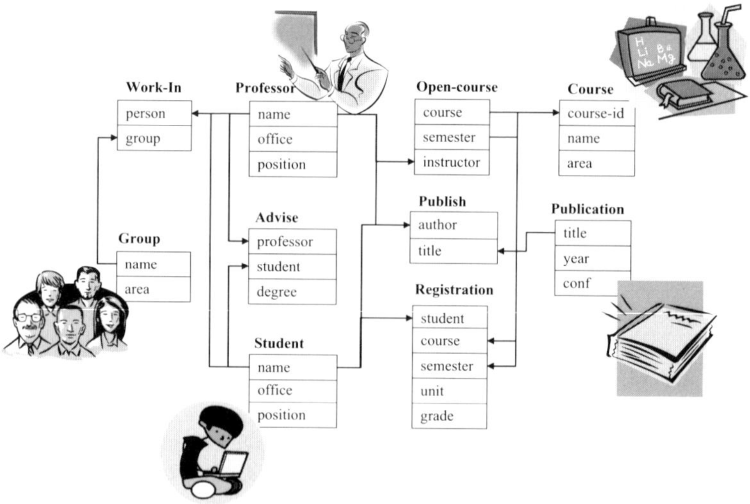

Fig. 1. The schema of the database of a computer science department

Relational data mining [4] has received much attention in recent years, such as relational decision trees [2] and relational clustering [8]. Although relational data provides richer information, it is also more complicated to model, because the hypothesis space in a relational database is usually much larger than in a single table. For example, in rule-based relational classification, each rule involves multiple relations that can be joined in a chain. If each relation is joinable with two other relations, there are 2^k join paths of length k starting from a certain relation, and many rules can be built based on each join path. It is impossible to explore all these join paths and evaluate all these rules.

In this paper we will introduce two methodologies to perform efficient data mining in relational databases or other linked environments. The first methodology is to *use certain heuristics to confine the search for good hypothesis in promising directions*. For example, it is prohibitively expensive to build a rule-based classifier by finding out all possible rules in a relational database. But it is much more affordable to explore certain join paths that are more likely to generate good rules. The second methodology is to *assign properties to objects and let different linked objects interact with each other*. Compared with the first methodology, it only explores the directly linked objects. Because many types of relationships are represented by chains of links, we usually need to compute the properties of objects iteratively. For example, when clustering a large number of inter-linked objects, we can compute the probability of each object belonging to each cluster from objects linked to it. By repeating such computation for many rounds, reasonable clustering is likely to be generated.

The rest of this paper is organized as follows. First, two approaches are introduced based on the first methodology in Section 2, and then two introduced based on the second methodology in Section 3. Finally, the study is concluded in Section 4.

2 Confining Search in Promising Directions with Heuristics

In many relational data mining approaches the hypotheses involve multiple relations (or multiple types of linked objects). For example, most relational classifiers are based on rules, and each rule involves a chain of joinable relations (*i.e.*, a join path). As mentioned above, an exhaustive search for all possible rules is usually unaffordable, and we need to confine the search procedure in promising directions. In this section we will introduce two approaches using this methodology: one for relational classification and the other for relational clustering.

2.1 CrossMine: Efficient Relational Classification

A database for relational classification consists of a set of relations, one of which is the *target relation* R_t, whose tuples are called *target tuples* and are associated with class labels. (We will use "target tuple" and "target object" exchangeably.) The other relations are *non-target relations*.

Inductive Logic Programming (ILP) [3] is the most popular approach for relational classification, and most ILP relational classifiers are either rule-based [11] or tree-based [2]. We follow the rule-based approach because it is more flexible and often achieves better accuracy. A rule is based on a join path, and there could be one or more predicates on certain attribute of each relation on the join path. Suppose we are classifying students in the database in Figure 1 according to their research areas. A rule may look like "($Student \bowtie Publish \bowtie Publication, Publication.conf =$'SIGMOD') $\Rightarrow$ $Student.area =$ 'database' ". This rule says if a student is joinable with a tuple in the relation *Publish* and then a tuple in the relation *Publication*, whose attribute *conf* has value "SIGMOD", then this student's area is "database".

In most rule-based relational classification approaches, the following heuristic is employed to generate rules: *Start from the target relation and keep expanding the range of search to relations containing good predicates, and the relations joinable with them.* Let us take FOIL [11] as an example. When generating a rule, FOIL first searches for good predicates from the target relation and the relations joinable with the target relation. In each of the following steps, FOIL searches in relations joinable with any relation containing good predicates. FOIL also has a constraint that it does not explore one-to-many joins, which means that it will not explore a relation like *Publish* in Figure 1 since one student may have multiple publications.

We propose an approach named CrossMine [16], which performs more exhaustive search for good rules in a more efficient way. We invent a technique called

Tuple ID Propagation for virtually joining different relations, which can find good predicates in different relations without doing physical joins. Compared to FOIL, CrossMine searches a broader range by removing the constraint of not exploring one-to-many joins, and by performing *look-one-ahead* (*i.e.*, searching across one more join) when reaching a *relationship* relation, such as *Publish*, which links together two *entity* relations: *Student* and *Publication*.

We test CrossMine, FOIL, and TILDE on the financial database used in PKDD CUP 1999, and the task is to predict whether a loan can be paid on time. The accuracies and running time of each approach is shown in Table 1.

Table 1. Performances on the financial database of PKDD CUP'99

Approach	Accuracy	Runtime
CrossMine w/o sampling	89.5%	20.8 sec
CrossMine	88.3%	16.8 sec
FOIL	74.0%	3338 sec
TILDE	81.3%	2429 sec

2.2 CrossClus: Relational Clustering with User's Guidance

In clustering process, objects are grouped according to their similarities among them, and thus the most essential factor for generating reasonable clusters is a good measure for object similarity. Most existing clustering approaches [6] work on a single table, and the similarity measure is pre-defined.

Clustering in multi-relational environments is a different story. A relational database usually contains information of many aspects. For example, a database of computer science department usually contains the following types of information about students: demographic information, courses, grades, research groups, advisors, publications, *etc.*. If all these relations and attributes are used indiscriminately, it is unlikely that reasonable clusters can be generated.

On the other hand, a user usually has some expectation for clustering, which is of crucial importance in multi-relational clustering. We call such expectation as "the goal of clustering". For example, the user may want to cluster students based on their research interests. It may not be meaningful to cluster them based on their attributes like phone number, ssn, and residence address, even they reside in the same table. The crucial information, such as a student's advisor and publications, are stored in some attributes in a few relations.

We propose CrossClus [14], an approach that performs relational clustering based on user's guidance. We adopt a simple form of user hint: *one or a small set of pertinent attributes*, which is easy for a user to provide.

The user hint is fundamentally different from class labels in classification, because it is used for indicating similarities between objects, instead of specifying the class labels of objects. CrossClus will learn a similarity metric from the user hints, instead of a model that predicts for each class. On the other hand, the user hint usually provides very limited information, because most users are only capable or willing to provide a very limited amount of hints, and CrossClusneeds to find other pertinent attributes for generating reasonable clusters. For example, a user may want to cluster students by research topics and specify the *area* attribute in *Group* relation, which represents very broad research areas such as database and operating systems. CrossClus will need to find highly pertinent features such as conferences of publications, advisors, and projects, and then performs clustering.

One major challenge is how to measure the pertinence of a feature. We propose a novel method for measuring whether two features cluster objects in similar ways, by comparing the inter-object similarities indicated by each feature. For a feature f, we use the similarity between each pair of objects indicated by f (a vector of $N \times N$ dimensions for N objects) to represent f. When comparing two features f and g, the cosine similarity of the two vectors for f and g is used. This measure captures the most essential information for clustering: inter-object similarities indicated by features. It treats categorical and numerical features uniformly as both types of features can indicate similarities between objects. Moreover, we design an efficient algorithm to compute similarities between features, which never materializes the $N \times N$ dimensional vectors and can compute similarity between features in linear space and almost linear time.

Another major challenge is how to search for pertinent features. We use the same heuristic as used in relational classification. Because of the large number of possible features, an exhaustive search is infeasible. CrossClus starts from the relations specified in user hint, and gradually expands the search scope to other relations. Like CrossMine, in each step CrossClus searches in the relations containing good features and in the relations joinable with them. This heuristic helps confine the search in promising directions and avoid fruitless search.

We tested the accuracies of CrossClus with different clustering algorithms (k-Medoids, k-Means, and agglomerative clustering), and the accuracies of some existing approaches (PROCLUS [1] and RDBC [8]), and a baseline approach

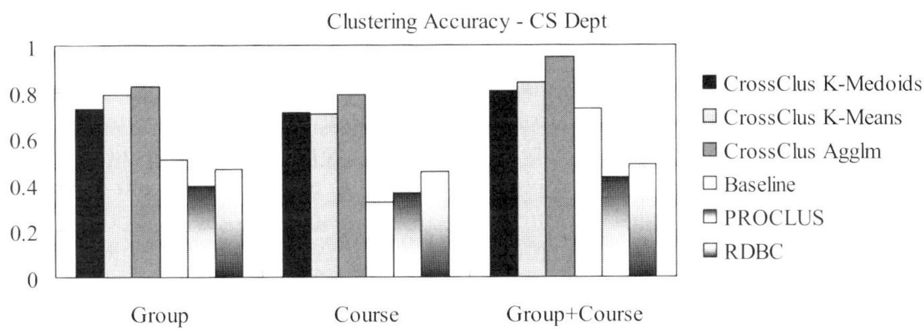

Fig. 2. Clustering accuracy on a dataset about the CS Department of UIUC

the auto-coder and then re-generating the flight software from the models for the new platform. For this to be effective, the intent is for all software (except a fixed reuse core) to be generated 'pristinely': no hand modification after auto-coding. This redirection through changing the auto-coder will also enable the same models to be used for simulation and training software. Compared to previous baselines this approach is also expected to decrease cost and schedule, reduce coding defects, simplify integration issues, and support rapid development through model simulation and debugging.

NASA research centers are developing intelligent analysis tools that can be used for NASA oversight, and are adapting them to this innovative software development process. One objective is to verify that the software works correctly over a wide range of both nominal and off-nominal scenarios. Doing this analysis at the model level has advantages in both finding defects early and in scaling the analysis to the large software systems expected for Orion. There are two related objectives to this model analysis. The first is to thoroughly exercise the different execution paths through the software. This objective has been extended to white-box testing: automating the generation of suites of test cases that provide coverage for exercising the execution paths through the software. The second related objective extends this to black-box testing technology: clustering the behavior of the system under simulation according to nominal and off-nominal behavior, and automatically testing the system over a wide range of mission parameters in order to determine governing factors between nominal and off-nominal behavior. The final objective described in this paper is to independently verify the output of the auto-coder for compliance with safety policies, and to generate documentation suitable for humans doing technical reviews.

In order to explain the expected use of these tools, the following sub-sections provide a synopsis of the UML/MDA software development approach. A key observation is that this approach naturally results in system software structured as a set of interacting finite state machines. This structuring facilitates the scaling and use of model-checking technology for model analysis. A second observation is that the adaptable auto-coding approach itself provides a means for interfacing between UML models and model-checking technology: the auto-coder is adapted to the model-checking platform. The core model-checking platform is based on a virtual machine technology that further facilitates this adaptation.

2.1 Model-Driven Architecture

The Object Modeling Group (OMG) defines model-driven architecture as the separation of system application logic from the underlying platform technology. This enables platform-independent models of a system's function and behavior, built using UML, to be implemented on many different target platforms. The platform-independent models represent the function and behavior of a software system independent of the avionics platform (encompassing hardware, operating system, programming language, etc.) used to implement it. Platform dependence has been a perennial challenge for long-lived aerospace systems, such as the International Space Station (ISS). For example, the upgrade from Ada 83 to Ada 95 for the ISS required a substantial re-engineering effort. Orion is expected to be used through the Lunar Sortie and Lunar Habitat phases of NASA's exploration program, and likely into Martian exploration.

2.2 Software Development with UML

This subsection simplifies the relevant aspects of UML-based software development from the viewpoint of the NASA analysis tools. The interested reader can find more details in the extensive literature on UML and object-oriented design including the UML variant chosen by the Orion project – executable UML (xUML) [1].

The UML software development process begins by grouping concepts into semi-independent domains that are largely self-cohesive. A domain reflects subject matter expertise, for example Guidance, Navigation, and Control (GN&C). Within each domain, the *classes* of objects and their static relationships are then defined in a class diagram. Each object class then has attributes defined, for example in GN&C a vehicle class would have attributes attitude and velocity. The static relationships between object classes are also defined, for example spacecraft are a specialization of vehicle.

Once the static class diagrams are defined, then the dynamic aspects of the domain are defined, such as *operations*, which are procedures that operate on objects and related objects. Of particular interest are *statecharts* that define the transition of an object through a succession of states. For example, an Orion capsule could be modeled as an object of class *spacecraft* that has states including pre-launch, launch, various phases of ascent, orbit insertion, docking with the space station, etc. Most objects transition through their states through interaction with other objects; such as the launch pad infrastructure signaling Orion that the countdown has reached zero, and Orion in turn signaling the rocket booster to ignite. These interactions are modeled through formal definitions of signals between state machines. UML allows a rich variety of actions that can be taken when signals are sent and received, and the objects enter, remain in, and exit states. Some variants of UML allow these actions to be defined in a conventional programming language. The xUML variant chosen by Lockheed defines these actions through an Action Semantics Language (ASL), which is then auto-coded to a programming language.

Verification of interacting state machines against logical and temporal properties is done through model-checking. For Orion software, the operations and actions that define the behavior of the xUML state machines will be auto-coded to an object-oriented programming language. Thus the model-checker needs to handle object-oriented software in a manner that permits scaling to large software systems.

3 NASA Analysis Tools and their Interface to the Orion Software Development Process

NASA's intelligent analysis tools for model-based software development draw upon core intelligent system methods including automated symbolic reasoning, constraint solving, intelligent search, automated theorem proving, and automated data analysis. The tools interface directly to Orion's model-based software development process (MDA/UML development process), providing the following capabilities:

- Early analysis at the model level to find defects when they are inexpensive to fix. The model-based analysis does a sophisticated graph search through the state space of a software system formulated in UML to automatically find defects against required properties.

- Automated testing and test suite generation that ensures coverage. This includes verification testing based upon white-box coverage criteria of paths through the software, and validation testing for black-box determination of the robustness of the system under varying mission parameters. The white-box testing relies on symbolic reasoning and constraint solving, while the black-box testing relies on intelligent search and machine data analysis.
- Innovative methods for verifying auto-generated code. The auto-code verification independently checks the C++ code against safety policies, and also develops detailed documentation that can be used during a code review. This mitigates the trust that needs to be put into the auto-coder.

On the left of figure 1 below is a fragment in the middle of Lockheed's Orion tool chain: an auto-coder that maps from UML models and Matlab models to C++ flight software. This fragment is preceded by requirements and design steps, and succeeded by compilation to partitions in the target ARINC 653 platform, and many stages of testing. The later stages of testing on the target hardware with a simulated mock-up of the Orion environment are expensive and require unique facilities, so it is important that defects are found as early as possible in the process to avoid costly rework downstream. All of the steps in the tool chain are embedded in a configuration management system. On the right of figure 1 are the NASA intelligent analysis tools, described next in the following sub-sections.

3.1 Model Analysis

Model Analysis compares the UML and Matlab models against properties specifying verification conditions and then generates design error reports in the form of counter-examples. The models are translated to the model-checking target environment through an alternative auto-coder in the same family of commercial auto-coders that are adapted to generate the flight, training, and simulation software. The model-checking target environment is Java Pathfinder [2,3]: a Java virtual machine that is optimized for state-space exploration through dynamic execution of software encoded as Java byte code instructions. Java Pathfinder dynamically checks a software system against both generic properties (such as consistency and completeness of transitions) and specified verification properties. The verification properties are derived from requirements and design specifications.

Java Pathfinder is an explicit state model checker with some similarity to SPIN [4]. Explicit state model checkers use an explicit representation of individual states. They run a backtracking search over the graph of the state space, saving state information for matching on subsequent nodes in the search. If a new node matches a previously generated node, then the search from the new node is terminated. For software, the transitions correspond to instruction executions, and the states correspond to the value of program variables across all threads and processes. Explicit state model checkers are particularly good for asynchronous concurrent systems and reactive embedded systems, where the system interacts with the environment. The environment is itself modeled as a finite state machine, e.g., a software simulation, operating in parallel with the system.

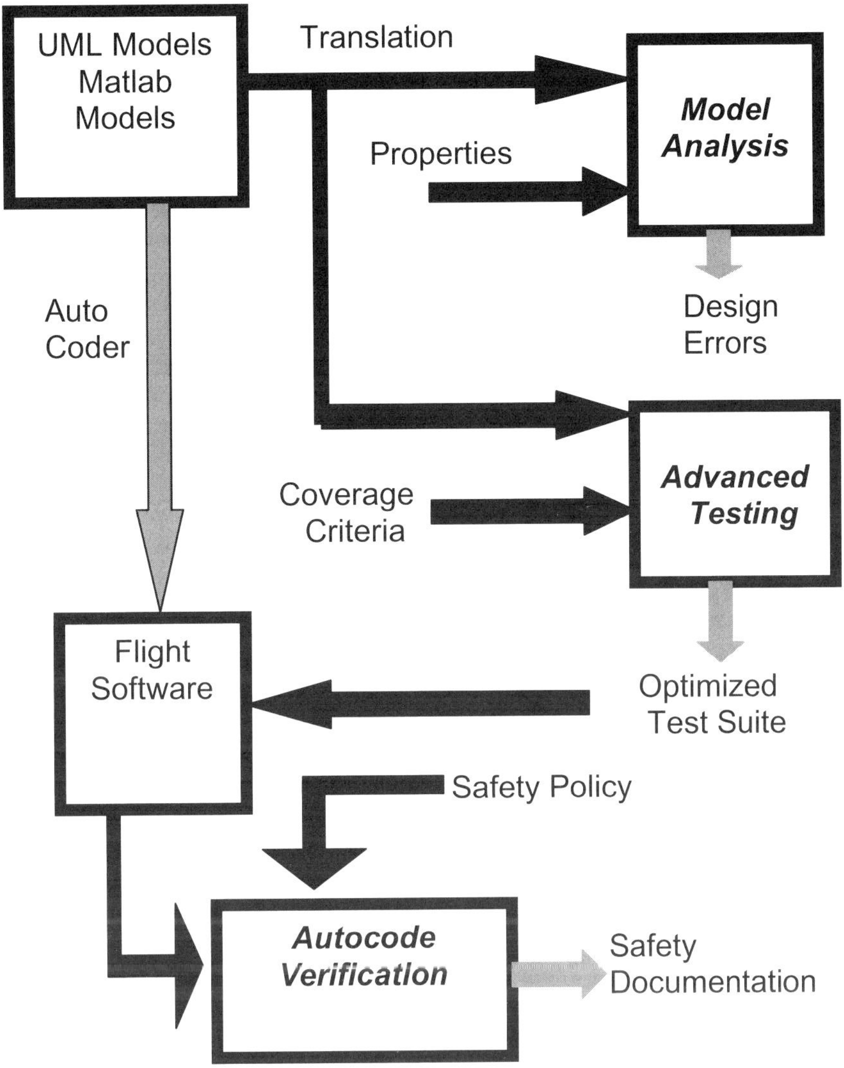

Fig. 1. On the left, a portion of Lockheed's Orion software development process. On the right, NASA's intelligent software analysis tools.

Orion software falls into the category of a reactive embedded system with rich constructs (ASL) for the transition relations, and complex states corresponding to Object-oriented (OO) software. Java Pathfinder was built from the ground up to handle object-oriented (OO) software; it has highly effective means of compactly representing state information for OO software executions. In addition, UML's use of state machines for modeling the dynamics of a system provide an abstraction layer that has been modeled as an extension of the Java virtual machine at the core of Java Pathfinder; providing further computational savings in analyzing Orion software. There is a considerable literature on methods for handling the computational complexity of model-checking, many of these methods are built-in to Java Pathfinder.

The architecture of Java Pathfinder also enables changing the operational semantics of the byte-code instructions on the fly; the default semantics are the same as standard Java virtual machines. This flexibility enables plugging-in operational semantics for other analysis tools, which is used for verification testing.

3.2 Advanced Verification Testing – Test Suite Generation

There are two components to advanced testing: verification testing and validation testing. Verification testing is done through an alternate operational semantics with an extension of Java Pathfinder [5]. Instead of executing the byte codes on concrete inputs, the byte codes are executed on symbolic inputs. The result is to accumulate symbolic expressions denoting the value of variables at successive points in the program execution. When a conditional statement is encountered, the condition becomes a symbolic constraint on variables. For a simple if-then-else statement, a path condition is generated on the variables of the conditional expression for the then branch, and the negation of the path condition is generated for the else branch. The symbolic execution accumulates path conditions through nested conditionals and guarded iterative constructs; the constraints in the accumulated path conditions are then solved to generate test vectors. This is done for the different paths and conditions in order to ensure that the generated test vectors provide sufficient coverage, as described below.

Verification testing takes as input the same translated models as model analysis, and in addition a specification of desired coverage criteria. Examples of coverage criteria include statement coverage (execute each program statement at least once), and branch coverage (execute all combinations of conditional branches, taking into account nesting structure). Symbolic execution [6,7] then determines the cumulative path conditions for the different points in the code required for the specified coverage. For each such point in the code, a vector of input values are then generated that satisfy the constraints. The result is a set of vectors that comprise a test suite that provides the specified coverage criteria. This set of vectors can be optimized to eliminate duplication. The test suite can be applied at many different stages of testing from the model-level through high-fidelity hardware testing. Verification testing as described here has mainly been applied to automatically generating test suites for individual units. Methods for extending it to larger portions of software are under development, including innovative approaches to integrating with system-level validation testing.

3.3 Advanced Validation Testing

The software validation testing technology performs large validation test runs (optionally on computer clusters or supercomputers) of a system under simulation, followed by automated machine analysis of the test logs to cluster the results, identify anomalous runs, and identify predicates on the inputs that separate nominal from off-nominal runs. It largely automates the laborious and expensive traditional process of human validation testing and test-log analysis, filtering the large amount of data to something which is much more manageable by human engineers. It thus enables validating a simulated system over a much wider range of scenarios, defined by variations in mission parameters, thereby providing assurance that the system does what is needed and more sharply defining under what range of off-nominal parameters the system is no longer robust. Conceptually, it is an extension of Monte-Carlo analysis – where a simulated system is tested under a statistical distribution of parameters, leading to a scatter-plot of results.

The innovation over standard Monte-Carlo analysis is two-fold. First, combinatorial testing techniques are used to factor pairs or triples of parameters into varying sets in order to identify critical parameters. This mitigates the combinatorics of the number of tests, which then allows searching over a much larger range of statistical variation than the two or three sigma that is standard for Monte Carlo testing. The objective of this extended variation is to determine system robustness through explicit identification of failure boundaries. The coverage criteria for advanced validation testing are specified through settings for the combinatorial and Monte-Carlo test generation. The second innovation is the automated machine analysis of test runs. Clustering algorithms based on expectation-maximization are generated automatically through AutoBayes [8] – a program synthesis system for data analysis. Treatment learning provides learning predicates on parameters that separate nominal from off-nominal behavior.

This technology has already been applied to analysis of ascent and re-entry simulations for Orion [9]. At present, several iterations with manual input are required to identify factors that determine off-nominal behavior – e.g., atmospheric re-entry points that lead to off-range landings. The manual input is to adjust test generation ranges and factoring. Under development are methods to automate the adjustments for successive iterations based on machine data analysis.

3.4 Autocode Verification

Auto-coders are a critical component of Orion's Model-Driven Architecture, enabling avionics platform re-targeting over the decades-long expected lifetime of Orion. While auto-coders have historically been used for rapid prototyping and design exploration, their use in safety-critical domains is more recent. One approach in safety-critical domains is to treat the auto-generated code as if it were manually developed for purposes of verification and test, however this approach provides only limited immediate productivity gains over manual development. Another approach is to *qualify* the code generator, which requires the same certification standards as the production flight code but enables analysis activities on the model-level to receive formal verification credit. However, code generator qualification is expensive and needs to be completely redone for every upgrade to the auto-coder and every adaptation and reconfiguration for a project.

A third approach is to exploit information about the auto-coder to automate portions of the analysis and documentation needed for certifying the generated code. This is the approach being taken with the NASA autocode verification tool. Certain aspects of the safety certification for flight-readiness are being automated, through a Hoare-style verification of auto-generated code against safety policies, with automated generation of detailed documentation that can be used during a code review. The technology has already been demonstrated on a number of safety policies relevant to Orion, including programming-language safety conditions (e.g., variable initialization before use, and memory access safety) to domain-specific safety conditions (e.g., correct handling of physical units and co-ordinate frames [10]. The documentation of conformance with safety policies is generated after an automated, formal proof that the code meets the safety requirements, and provides an understandable hyperlinked explanation suitable for code reviews.

The algorithm synopsis [11,12] is to work backwards through the generated code from a safety postcondition, first heuristically generating annotations that exploit information about the autocoder (such as stylized code fragments that encode co-ordinate transformations), and then collecting verification conditions. These are then submitted to an automated theorem prover, which typically succeeds in discharging the verification conditions. The approach is fail-safe, in that the annotations are not trusted and if incorrect will lead to failure to prove the verification condition. The approach is loosely motivated by proof-carrying code.

4 Summary

NASA's Orion project is using an MDA software development approach for NASA's new crew exploration vehicle. NASA research centers have adapted intelligent software engineering analysis technology to this model-based approach in order to provide tools for insight and oversight of Orion software development. The tools include model-checking to find design errors in UML and Matlab models, verification testing technology that automatically generates test suites providing white-box testing coverage for units and subsystems, validation testing to find mission parameter ranges that distinguish nominal versus off-nominal behavior, and autocode verification technology that automates aspects of the safety certification of the MDA auto-generated code.

Acknowledgement. The author wishes to acknowledge the many years of work by members of NASA's Intelligent Software Design project, whose present members at NASA Ames include Ewen Denney, Karen Gundy-Burlet, Masoud Mansouri-Samani, Peter Mehlitz, Corina Pasareanu, Thomas Pressburger, and Johan Schumann.

References

1. Raistrick, C., Francis, P., Wright, J., Carter, C., Wilke, I.: Model Driven Architecture with Executable UML. Cambridge University Press, Cambridge (2004)
2. Java PathFinder; http://javapathfinder.sourceforge.net
3. Visser, W., Havelund, K., Brat, G., Park, S., Lerda, F.: Model Checking Programs. Automated Software Engineering Journal 10(2) (2003)
4. Holzmann, G.: The Spin Model Checker: Primer and Reference Manual. Addison-Wesley, Menlo Park (2003)
5. Anand, S., Pasareanu, C.S., Visser, W.: JPF-SE: A symbolic execution extension to Java PathFinder. In: Grumberg, O., Huth, M. (eds.) TACAS 2007. LNCS, vol. 4424, Springer, Heidelberg (2007)
6. Kurshid, K., Pasareanu, C.S., Visser, W.: Generalized symbolic execution for model checking and testing. In: Garavel, H., Hatcliff, J. (eds.) ETAPS 2003 and TACAS 2003. LNCS, vol. 2619, Springer, Heidelberg (2003)

7. Pasareanu, C., Visser, W.: Symbolic Execution and Model Checking for Testing. In: Invited talk for Haifa Verification Conference (2007)
8. Fischer, B., Schumann, J.: AutoBayes: A System for Generating Data Analysis Programs from Statistical Model. J. Functional Programming 13(3), 483–508 (2003)
9. Gundy-Burlet, K., Schumann, J., Barrett, T., Menzies, T.: Parametric Analysis of Antares Re-Entry Guidance Algorithms using Advanced Test Generation and Data Analysis. In: Proc. 9th International Symposium on Artificial Intelligence, Robotics, and Automation in Space (2008)
10. Denney, E., Trac, S.: A Software Safety Certification Tool for Automatically Generated Guidance, Navigation and Control Code. In: Proc. IEEE Aerospace Conference (2008)
11. Denney, E., Fischer, B.: A generic annotation inference algorithm for the safety certification of automatically generated code. In: Proc. GPCE 2006: 5th International Conference on Generative Programming and Component Engineering (2006)
12. Denney, E., Fischer, B.: Extending Source Code Generators for Evidence-based Software Certification. In: Second International Symposium on Leveraging Applications of Formal Methods, Verification and Validation (2006)

Boosting Support Vector Machines for Imbalanced Data Sets

Benjamin X. Wang and Nathalie Japkowicz

School of information Technology and Engineering,
University of Ottawa, 800 King Edward Ave., P.O.Box 450 Stn.A,
Ottawa, Ontario, K1N 6N5, Canada
{bxwang,nat}@site.uottawa.ca

Abstract. Real world data mining applications must address the issue of learning from imbalanced data sets. The problem occurs when the number of instances in one class greatly outnumbers the number of instances in the other class. Such data sets often cause a default classifier to be built due to skewed vector spaces or lack of information. Common approaches for dealing with the class imbalance problem involve modifying the data distribution or modifying the classifier. In this work, we choose to use a combination of both approaches. We use support vector machines with soft margins as the base classifier to solve the skewed vector spaces problem. Then we use a boosting algorithm to get an ensemble classifier that has lower error than a single classifier. We found that this ensemble of SVMs makes an impressive improvement in prediction performance, not only for the majority class, but also for the minority class.

1 Introduction

A data set is imbalanced if the number of instances in one class greatly outnumbers the number of instances in the other class. Some examples of domains presenting a class imbalance are: fraudulent telephone calls, telecommunications management, text and image classification, and disease detection. For reasons of simplicity, and with no loss in generality, only binary class data sets are considered in this paper.

Recently, the class imbalance problem has received a lot of attention in the Machine Learning community by virtue of the fact that the performance of the algorithms used degrades significantly if the data set is imbalanced (Japkowicz and Stephen, 2002). Indeed, in very imbalanced domains, most standard classifiers will tend to learn how to predict the majority class. While these classifiers can obtain higher predictive accuracies than those that also try to consider the minority class, this seemingly good performance can be argued as being meaningless.

The next section will discuss some of the approaches previously applied to deal with the class imbalance problem. Section 3 introduces the performance measures we use to evaluate our research. Section 4 discusses the motivations for our approach. Next, Section 5 describes our approach while Section 6 presents the results we obtained. Section 7 is the conclusion.

A. An et al. (Eds.): ISMIS 2008, LNAI 4994, pp. 38–47, 2008.

2 Previous Work

The machine learning community has addressed the issue of class imbalance in two different ways in order to solve the skewed vector spaces problem. The first method, which is classifier-independent, is to balance the original dataset. The second way involves modifying the classifiers in order to adapt them to the data sets. Here we will talk about the most effective approaches that have been proposed. We will discuss these approaches in terms of both their benefits and their limitations.

Balancing the data set. The simplest way to balance a data set is by under-sampling (randomly or selectively) the majority class while keeping the original population of the minority class (Kubat & Matwin, 1997)

Obviously this method results in information loss for the majority class. Over-sampling (Japkowicz & Stephen, 2002; Chawla et al., 2000) is the opposite of under-sampling approach. It duplicates or interpolates minority instances in the hope of reducing class imbalance. With over-sampling, the neighborhood of a positive instance is assumed to be also positive as are the instances between two positive instances. Assumptions like these, however, are data dependent and do not apply in all cases. Experimental results show that under-sampling produces better results than over-sampling in many cases. The belief is that although over-sampling does not lose any information about the majority class, it introduces an unnatural bias in favour of the minority class. Using synthetic examples to augment the minority class is believed be better than over-sampling with replacement (Chawla et al., 2000). It does not cause any information loss and could potentially find "hidden" minority regions. The disadvantage of this method is that it creates noise for the classifiers which could result in a loss of performance. Nonetheless, a method such as this one has the potential of being better than the other approaches discussed since it used a non-skewed mechanism to solve the problem of skewed data.

Modifying the classifiers. Working with classifiers to adapt data sets could be another way to deal with the imbalanced data problem. Assigning distinct costs to the training examples seems to be the best approach of this kind. Various experimental studies of this type have been performed using different kinds of classifiers (Chen et al., 2004; Guo & Viktor, 2004). In terms of SVMs, several attempts have been made to improve their class prediction accuracy (Akbani et al., 2004; Morik et al., 1999). We will discuss them in detail in Section 4. These experiments show that SVMs may be able to solve the problem of skewed vector spaces without introducing noise. However, the resulting classifiers may over-fit the data, as we will discuss later.

In this paper, we present a system that combines the two general methods described for solving the problem of data set imbalance. The system works by modifying the classifier using cost assignation, but counters the modification by using a combination scheme, which is in effect similar to modifying the data distribution. We choose to use boosting as our combination scheme since it is works very well in terms of being able to produce very accurate prediction rules without causing over-fitting. Boosting has the added advantage of working with any type of classifier. In this paper we will focus on support vector machines, which have demonstrated remarkable success in many different applications. Our experiments show that boosting methods can be combined with

SVMs very effectively in the presence of imbalanced data. Our results show that this method is not only able to solve the skewed vector spaces problem, but also the overfitting problem caused by support vector machines.

3 Motivation for Our Approach

In this section, we begin by explaining why SVM with soft margins is not sufficient for solving the class imbalance problem. We then discuss methods that have been previously devised in order to improve on this scheme and explain how our methods compares to them. Our scheme will be described in detail in Section 5.

3.1 SVMs and the Skewed Boundary

Support vector machines are based on the principle of Structural Risk Minimization from statistical learning theory. The idea of structural risk minimization is to find a hypothesis h for which we can guarantee the lowest true error. In the presence of noise, the idea of using a soft margin was introduced by Vapnik (1995).

As noted earlier, data imbalance causes a default classifier to be learned which always predicts the "negative" class. Wu and Chang (2003) observed two potential causes for the problem of a skewed boundary: (1) the imbalanced training data ratio and (2) the imbalanced support-vector ratio. For the first cause, we note that on the minority side of the boundary, the positive examples may not always reside as close to the "ideal boundary" as the negative examples. In terms of the second cause, consider the following: according to the KKT conditions, the values for α_i must satisfy $\Sigma_{i=1}^{n} \alpha_i y_i = 0$. Since the values for the minority class tend to be much larger than those for the majority class and the number of positive support vectors substantially smaller, the nearest neighborhood of a test point is likely to be dominated by negative support vectors. In other words, the decision function is more likely to classify a boundary point as negative.

3.2 Analysis of Strategies for the Imbalanced Problem for SVMs

To deal with the imbalanced boundary problem, several approaches were given for adjusting the skewed boundary. We first present three approaches, then, in the next section, our strategy for handling this problem.

3.2.1 Kernel Transformation Method

Adaptively modifying the kernel function K based on the training data distribution is an effective method for improving SVMs. Amari and Wu (1999) propose a method of modifying a kernel function to improve the performance of a support vector machine classifier. This method is based on the structure of the Riemannian geometry induced by the kernel function. The idea is to increase the separability between classes by enlarging the space around the separating boundary surface.

Improving upon Amari and Wu's method, Wu and Chang (2003) propose a class-boundary-alignment algorithm, which also modifies the kernel matrix K based on the distribution of the training data. Instead of using an input space, they conduct the kernel transformation based on the spatial distribution of the support vectors in feature space. A new kernel function is defined as: $\tilde{K}(x, x') = D(x)D(x')K(x, x')$ Where an RBF distance function $D(x) = \sum_{k \in SV} \exp(-\frac{|x - x_k|}{\tau_k^2})$ is used as a positive conformal function in this equation.

This method takes advantage of the new information learned in every iteration of the SVM algorithm while leaving the input-space distance unchanged. The class boundary alignment algorithm can be applied directly to adjust the pair-wise object distance in the kernel matrix K in cases where the input space may not physically exist. Theoretical justifications and empirical studies show that kernel transformation method is effective on imbalanced classification, but this technique is not sufficiently simple to be implemented efficiently.

3.2.2 Biased Penalties Method

Shawe-Taylor & Cristianini (1999) show that the distance of a test point from the boundary is related to its probability of misclassification. This observation has motivated a related technique which is used in his paper. The technique is to provide a more severe penalty if an error is made on a positive example than if it is made on a negative example. By using the cost factors and adjusting the cost of false positives and false negatives, such penalties can be directly incorporated into the SVM algorithm.

Morik et al. (1999) and Shawe-Taylor & Cristianini (1999) propose an algorithm to use the L_1 norm (k=1). Two cost-factors are chosen so that the potential total cost of the false positives equals the potential total cost of the false negatives. This means that the parameters of the SVM are selected such that they obey the ratio: $C_+/C_- = n_-/n_+$. By increasing the margin on the side of the smaller class, this method provides a way to induce a decision boundary which is much more distant from the "critical" class than it is from the other. But in this model, the balance between sensitivity and specificity cannot be controlled adaptively resulting in over-fitting.

Instead of using the L_1 norm for the loss measure, Veropoulos et al. (1999) use the square of the L_2 norm (k=2). This method enables the algorithm to control the balance between sensitivity and specificity, not adding any information. Experimental results (Veropoulos et al., 1999) show that this method has the power to effectively control the sensitivity and not the specificity of the learning machine.

From this analysis we can see that what is really required is a method that is able to introduce some information to the problem in order to increase both sensitivity and specificity.

3.2.3 Boosting Method (Our Approach)

Our approach seeks to improve upon Morik et al. (1999)'s method. Instead of increasing C_+ or C_- to get the balance between sensitivity and specificity, we provide another solution that modifies the training data sets x_i in order to adjust some α_i on both the positive and negative side. The respective adjustments are based on the contribution of each. We choose to use boosting, a general method which combines several simple classifiers, to modify the training data sets. The details of this technique are given in the next section.

4 Boosting SVM with Asymmetric Misclassification Cost

Boosting and other ensemble learning methods have been recently used with great success in many applications (Chawla et al., 2003; Guo & Viktor, 2004). In our algorithm, we choose to use the L_1 norm (k=1) SVM (as described in Section 4) with asymmetric misclassification cost for the component classifiers in a boosting scheme. Our method will now be presented formally: Given a set of labeled instances $\{x_i, y_i\}_{i=1}^{n}$, the class prediction function of our base classifier is formulated in terms of the kernel function K:

$$sign(f(x) = \sum_{i=1}^{n} y_i \alpha_i K(x, x_i) + b)$$

where b is the bias and the optimal coefficients are found by maximizing the primal Lagrangian:

$$L_p = \frac{\| \omega \|^2}{2} + C_+ \sum_{\{i|y_i=+1\}}^{n_+} \xi_i^2 + C_- \sum_{\{j|y_j=-1\}}^{n_-} \xi_j^2$$

$$- \sum_{i=1}^{n} \alpha_i [y_i(\omega \cdot x_i + b) - 1 + \xi_i] - \sum_{i=1}^{n} \mu_i \xi_i$$

where $C_+ \geq \alpha_i \geq 0$, $C_- \geq \alpha_i \geq 0$, $\frac{C_+}{C_-} = \frac{n_-}{n_+}$ *and* $\mu_i \geq 0$. Using this component classifier, we find that the points labeled ξ_i^*, where since $\xi_i^* = \xi_i / \|\beta\|$, are said to be on the *wrong side of the margin*, as shown in figure 1. In terms of the L_1 norm margin slack vector optimization, the feasibility gap can be computed since the ξ_i are not specified when moving to the dual. The values for ξ_i can therefore be chosen in order to ensure that the primary problem is feasible. The values are calculated using the following equation:

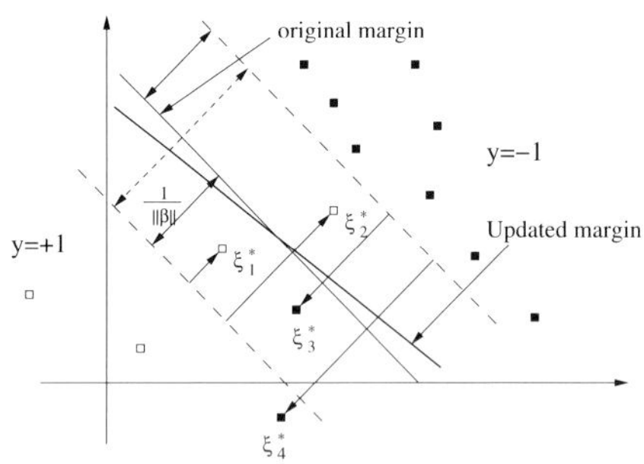

Fig. 1. Support vector machines with Asymmetric Misclassification Cost in the imbalanced non-separable case. The points labeled ξ_i^* are on the wrong side of the margin by an amount $\xi_i^* = \xi_i / \|\beta\|$; points on the correct side have $\xi_i^* = 0$. The margin shown results after the update to the points labeled ξ_i^*.

Algorithm Boosting-SVM:
Given: Sequence of N examples X_{Train}, $X_{Validation}$
M; /* *the maximum running iterations* */
Output: G; /* *output ensemble classifier* */
Variables:
ω_i; /* *weights to training observations* (x_i, y_i), i=1,2,...,N */
T; /* *the selected running iterations* */
ρ /* G-mean value */
Function Calls:
S; /* *single classifier* */
SVMTrain(X); /* *training the single classifier S using SVMs with Asymmetric Cost* */
SVMClassify(X,S); /* *classify X by the classifier S* */
Gmean(G); /* *obtain the G-mean value from G* */
Begin
Initialize
$\omega_i = 1$, i=1,2,...N
$\rho = 0$; $\rho_{best} = 0$
T=1.
Do for m=1, 2,, M
(a) $X_{train}(x) \leftarrow X_{train}(x)$ using weights ω_i.
(b) $S_m \leftarrow SVMTrain(X_{train})$.
(c) Compute $\varepsilon_m = \frac{\Sigma_{i=1}^{N} \omega_i I(y_i \neq SVMClassify(X_{train}, S_m))}{\Sigma_{i=1}^{N} \omega_i}$
(d) Compute $\alpha_m = \lambda \log(1 - \varepsilon_m)\varepsilon_m \;\; (0 < \lambda \leq 1)$
(e) Set $\omega_i \leftarrow \omega_i \cdot \exp[\alpha_m \cdot I(y_i \neq SVMClassify(X_{train}, S_m))]$, i=1,2,...,N.
(f) $G_m = sign[\Sigma_{j=1}^{m} \alpha_j S_j]$
(g) $\rho_m = Gmean[G_m(X_{validation})]$
(g) if $\rho_m > \rho_{best}$, then T=m and $\rho_{best} = \rho_m$.
Return G_t.
End

Algorithm 1. Boosting-SVM with Asymmetric Cost algorithm

$$\xi_i = \max(0, 1 - y_i(\sum_{j=1}^{n} y_j \alpha_j K(x_j, x_i) + b))$$

Here, the task consists of modifying the weights ω_i of the training observations x_i in the input space in order to modify the point labeled ξ_i^*. The advantage of this technique is that we are able to easily build a modified version of the training data and improve the class prediction function of the boosting procedure.

Our purpose is to sequentially apply the component classification algorithm to the modified versions of the data, thereby producing a sequence of component classifiers $G_m(x)$, m=1, 2,..., M.

The predictions from all of the component classifiers are then combined by a weighted majority vote to produce the final prediction: $G(x) = sign(\Sigma_{m=1}^{M} \alpha_m G_(m)(x))$. Here the values for α_1,α_2,...α_M are computed by the boosting algorithm and are used to weight the contribution of each respective $G_m(x)$. The resulting effect is to give greater influence to the more accurate classifiers in the sequence.

Figure 2 shows the details of our boosting-SVM algorithm. In this algorithm, the classifier S is induced from the current weight observation. The resulting weighted error rate ε is computed as shown at line (c). The weight α_m is then found by calculating $\alpha_m = \lambda \log(1-\varepsilon)/\varepsilon$. Here the λ is an empirical parameter used to tune the magnitude of the penalization for each iteration. We use G-mean instead of prediction accuracy to evaluate the classifier since it combines the values of both sensitivity and specificity. We apply our algorithm on the training data set X_{train} until the G-mean value on the test set $X_{validation}$ cannot be improved.

The final classification is given following a vote by the sequence of component classifiers. Figure 2 provides an example of a final classifier built from three component classifiers. The ensemble classifier will have lower training error on the full training set than any other single component classifier. The ensemble classifier will also have lower error than a single linear classifier trained on the entire data set.

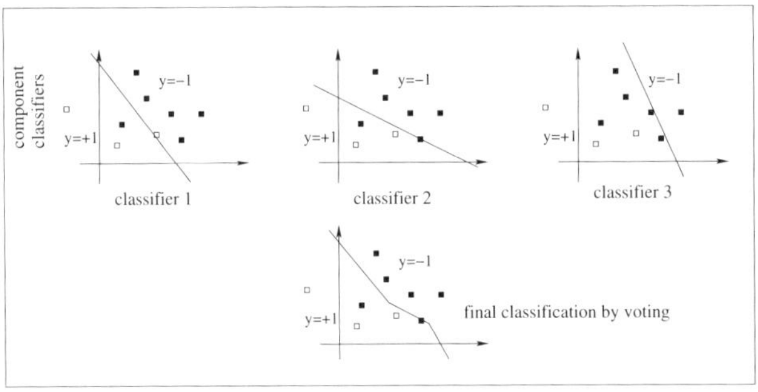

Fig. 2. The final classification is given by the voting of the component classifiers and yields a nonlinear decision boundary. The three component classifiers trained by SVM are shown on the top and the resulting classifier is given on the bottom. The ensemble classifier has lower error than a single classifier trained on the entire data set.

5 Experiments and Discussion

In our experiments, we compare the performance of our classifier with eight other popular methods: (I)Adacost (Fan et al, 1999), (II)SMOTEboost (Chawla et al., 2003), (III)WRF (Chen et al., 2004), (IV)Databoost-IM (Guo & Viktor, 2004), (V)Under-sampling with SVMs, (VI)SMOTE (Chawla et al., 2000) with SVMs, (VII)SVMs with Asymmetric Cost (Morik et al., 1999), (VIII)SMOTE combined with VII (Akbani et al., 2004). For under-sampling we used a random sampling method. For both under-sampling and SMOTE, the minority class was over-sampled at 200%, 300%, 400% and 500%. We use the same component classifier for all methods. The results obtained were then averaged. For our method and the SVMs with Asymmetric Cost and L_1 norm, we set the cost ratio by: $\frac{C_+}{C_-} = \frac{n_-}{n_+}$. In our experiment we use 10-fold cross-validation to

Table 1. Kubat's G-mean for each algorithm from 10-fold cross validation

DATASET	I	II	III	IV
ABALONE	56.14	56.95	57.39	61.09
B-CANCER	55.64	58.74	58.03	60.01
CAR	91.95	89.13	90.93	91.89
GLASS	**94.41**	91.07	90.54	92.34
H-DISEASE	47.34	47.09	46.60	48.76
LETTER	86.03	87.23	86.22	87.99
LUPUS-I	74.68	74.61	74.56	77.54
LUPUS-II	64.07	63.13	66.00	67.41
SEGMENT	96.10	96.23	95.76	**97.29**
STROKE-I	63.10	63.14	62.88	65.25
STROKE-II	62.04	61.42	62.05	63.30
YEAST	66.00	67.57	69.52	66.94
MEAN	71.46	71.36	71.70	73.32

Table 2. Kubat's G-mean for each algorithm from 10-fold cross validation

DATASET	V	VI	VII	VIII	B-SVM
ABALONE	56.27	61.53	78.39	76.92	**79.52**
B-CANCER	58.05	60.99	58.63	59.83	**61.89**
CAR	76.15	79.50	91.67	91.60	**92.49**
GLASS	85.81	89.27	88.22	82.50	91.33
H-DISEASE	47.21	47.86	38.74	47.85	**54.36**
LETTER	53.54	70.73	88.90	87.99	**89.66**
LUPUS-I	46.80	74.18	75.38	77.17	**84.54**
LUPUS-II	55.14	57.41	68.19	67.02	**71.05**
SEGMENT	94.05	95.49	94.31	94.78	96.36
STROKE-I	64.58	64.58	64.23	63.69	**66.70**
STROKE-II	62.29	62.10	62.10	61.86	**64.74**
YEAST	67.08	69.61	66.31	66.87	**71.42**
MEAN	63.91	69.44	72.92	73.17	**77.01**

train our classifier since it provides more realistic results than the holdout method. In the boosting schemes we use 70% of the data set for training, 20% to set the threshold for each boosting iteration. The remaining 10% of the data is used as normal in the 10-fold cross validation testing. All training, validation, and test subsets were sampled in a stratified manner that ensured each of them had the same ratio of negative to positive examples (Morik et al., 1999). For all SVM classifiers, we used a linear kernel function to prevent the choice of kernel function from affecting our results.

We chose to experiment on 12 different imbalanced data sets. Abalone19, B-cancer, Car3, Glass7, Heart-disease1, Letter4, Segment and Yeast are from UCI datasets. Lupus-I, Lupus-II, Stroke-I, Stroke-II, are health related data sets.

The next table lists Kubat's G-mean (as a percentage) measure (Kubat & Matwin, 1997) obtained for each algorithm. This measure is more representative of an algorithm's performance.

As a result, when comparing the four approaches we can see that Boosting-SVM with Asymmetric Cost C_+ and C_- yields the best average performance. The result demonstrates that our approach has the power to effectively control both sensitivity and specificity without adding noise. Our approach is always better than SVMs with Asymmetric Cost and L_1 norm since we use it as the component classifier. The improvement in terms of both sensitivity and specificity means that this method is able to avoid overfitting the data.

6 Conclusion

We have proposed the boosting-SVMs with Asymmetric Cost algorithm for tackling the problem associated with imbalanced data sets. Through theoretical justifications and empirical studies, we demonstrated this method to be effective. We find that our boosted SVM classifiers are robust in two ways: (1) they improve the performance of the SVM classifier trained with training set; and (2) they are sufficiently simple to be immediately applicable.

In future work, we hope to test more effective boosting methods on our algorithm. We will test this framework on different kernel functions and we will use more efficient measures to evaluate performance in our experiments.

References

[1] Akbani, R., Kwek, S., Japkowicz, N.: Applying Support Vector Machines to Imbalanced Datasets. In: Boulicaut, J.-F., Esposito, F., Giannotti, F., Pedreschi, D. (eds.) ECML 2004. LNCS (LNAI), vol. 3201, Springer, Heidelberg (2004)

[2] Amari, S., Wu, S.: Improving support vector machine classifiers by modifying kernel functions. Neural Networks 12, 783–789 (1999)

[3] Chawla, N., Bowyer, K., Hall, L., Kegelmeyer, W.P.: SMOTE: synthetic minority oversampling technique. In: International Conference on Knowledge Based Computer Systems (2000)

[4] Chawla, N., Lazarevic, A., Hall, L., Bowyer, K.: SMOTEBoost: Improving Prediction of the Minority Class in Boosting. In: 7th European Conference on Principles and Practice of Knowledge Discovery in Databases, Cavtat-Dubrovnik, Croatia, pp. 107–119 (2003)

[5] Chen, C., Liaw, A., Breiman, L.: Using random forest to learn unbalanced data. Technical Report 666, Statistics Department, University of California at Berkeley (2004)

[6] Fan, W., Stolfo, S., Zhang, J., Chan, P.: AdaCost: Misclassification Cost-Sensitive Boosting, In: Proceedings of 16th International Conference on Machine Learning, Slovenia (1999)

[7] Guo, H., Viktor, H.L.: Learning from Imbalanced Data Sets with Boosting and Data Generation: The DataBoost-IM Approach. ACM SIGKDD Explorations 6(1), 30–39 (2004)

[8] Japkowicz, N., Stephen, S.: The Class Imbalance Problem: A Systematic Study. Intelligent Data Analysis 6(5), 429–450 (2002)

[9] Kubat, M., Matwin, S.: Addressing the curse of imbalanced training sets: One-sided selection. In: Proceddings of the Fourteenth International Conference on Machine Learning, pp. 179–186 (1997)

[10] Morik, K., Brockhausen, P., Joachims, T.: Combining Statistical Learning with a Knowledge-Based Approach - A Case Study in Intensive Care Monitoring. In: ICML, pp. 268–277 (1999)

[11] Shawe-Taylor, J., Cristianini, N.: Further results on the margin distribution. In: Proceedings of the 12th Conference on Computational Learning Theory (1999)

[12] Vapnik, V.: The nature of statistical learning theory. Springer, New York (1995)

[13] Veropoulos, K., Campbell, C., Cristianini, N.: Controlling the sensitivity of support vector machines. In: Proceedings of the International Joint Conference on Artificial Intelligence, pp. 55–60 (1999)

[14] Wu, G., Chang, E.: Adaptive feature-space conformal transformation for imbalanced data learning. In: Proceedings of the 20th International Conference on Machine Learning (2003)

Class-Oriented Reduction of Decision Tree Complexity

José-Luis Polo, Fernando Berzal, and Juan-Carlos Cubero

Department of Computer Sciences and Artificial Intelligence
University of Granada. 10871 Granada, Spain
{jlpolo,fberzal,jc.cubero}@decsai.ugr.es

Abstract. In some classification problems, apart from a good model, we might be interested in obtaining succinct explanations for particular classes. Our goal is to provide simpler classification models for these classes without a significant accuracy loss. In this paper, we propose some modifications to the splitting criteria and the pruning heuristics used by standard top-down decision tree induction algorithms. This modifications allow us to take each particular class importance into account and lead us to simpler models for the most important classes while, at the same time, the overall classifier accuracy is preserved.[1]

1 Introduction

Traditional classification techniques treat all problem classes equally. This means that classification models are built without focusing on particular problem classes. In practice, however, not all problem classes are equally important. Obtaining a simpler model for the important classes, even when it might be slightly less accurate, might be of interest. In this paper, we introduce *class-complexity-sensitive* classification techniques, CCS classifiers for short, to address this issue.

The aim of CCS algorithms is to build classification models that, being as simple as possible for the most important classes, preserve the global classification accuracy their traditional counterparts provide. It should be noted that, given their goal, CCS algorithms will not treat all class values equally during the classifier construction process.

CCS classification models can be useful in the following scenarios:

- **Extreme class values**, when experts are specially interested in succinctly describing classes whose importance is premium within the decision making process (a common situation when classes are ranked).
- **Ontologies relating different problem classes**, when classes can be organized somehow (i.e. focusing on a subset of related classes that can be specially meaningful for the expert).
- **Typical binary classification problems**, when obtaining a proper description of one of the classes is more important for the user than having a typical classification model where both problem classes are equally treated.

[1] Work partially supported by grant TIN2006-07262.

A. An et al. (Eds.): ISMIS 2008, LNAI 4994, pp. 48–57, 2008.

It should also be noted that the class values we are interested in might change depending on our personal goals, even for the same problem. We incorporate the relative importance of each problem class into the classification model building process. We consider *relative weights* representing each class relative importance. When all the problem classes are equally important, we assign a relative weight of 1 to each one of them. When a class is 50%, 100%, 150%, 200%, 400%, and 800% more important than other, its relative weight should be 1.5, 2, 2.5, 3, 5, and 9 times the weight of the less important class. Afterwards, the resulting values can always be normalized to fall within the [0,1] interval while preserving the relative values they define in their ratio scale. The relative importance for the class values must be set empirically, according to each problem.

The rest of our paper is organized as follows. Section 2 describes existing work where special class features are taken into account. Section 3 briefly describes standard decision trees and how to modify them in order to build CCS classification models. Section 4 proposes how to evaluate the resultant models from a CCS perspective. Experimental results are provided in Section 5. Finally, Section 6 includes some conclusions.

2 Related Work

All classes should not be equally treated in all classification problems. Class differences make traditional classification models ineffective when class features are ignored. Some techniques have been proposed to deal with particular class features:

- **Imbalanced learning methods** [1] do not ignore the less frequent classes when there are very frequent classes that would lead traditional methods astray. Unlike CCS models, which are focused on classifier complexity, imbalanced learning methods are focused on the resulting classifier accuracy.
- **Cost-sensitive classification** [4] techniques take into account that the cost of misclassification is not the same for all the problem classes. Unlike CCS models, these learners are mainly focused on classifier accuracy (to avoid high-cost classification mistakes).
- **Subgroup discovery** [5] tries to find interesting subgroups within the training data set with respect to a target class value from a statistical point of view. They only provide a model for the target class, whereas CCS learners build complete classification models.

3 CCS Decision Trees

Decision trees [8] are one of the most widely used classification models. The availability of efficient and scalable decision tree learners [9,6] makes them useful for data mining tasks. Moreover, their interpretability makes them specially suitable for CCS classification.

We have modified the standard top-down induction of decision trees algorithm (TDIDT) to address CCS classification by modifying both its heuristic splitting criterion and the tree pruning strategy.

3.1 Splitting Criteria

A heuristic splitting criterion is used to decide how to branch the tree [2]. Quinlan's *Gain Ratio* criterion [8], for instance, chooses the attribute maximizing the information gain ratio with respect to the class attribute:

$$GainRatio(A) = \frac{H(C) - \sum_{j=1}^{J_A} p(a_j) \cdot H(C|a_j)}{-\sum_{j=1}^{J_A} p(a_j) log_2(p(a_j))}$$

where $H(C)$ is the entropy of the class, J_A corresponds to the number of different values for the A attribute, and $p(a_j)$ is the relative frequency of value a_j. $H(C|a_j)$ represents the class entropy for the a_j value of the A attribute:

$$H(C|a_j) = -\sum_{k=1}^{K} p(c_k|a_j) \cdot log_2(p(c_k|a_j))$$

where K is the number of classes and $p(c_k|a_j)$ is the relative frequency of the k-th value of the class given the j-th value of the attribute A.

Standard splitting criteria measure how good an attribute is for separating the examples belonging to different classes, but they do not take the relative importance of each class into account. In a CCS classification context, however, we should bias the heuristic criteria towards nodes with a better representation of examples belonging to the most important classes.

Criteria based on class entropy average the contribution of each class. We could include CCS information in this averaging process. The resulting criterion family, or CCS evaluation criteria, E_f, consider the relative importance of each class and can be formalized as follows:

$$E_f(C|a_j) = \sum_{k=1}^{K} f(w_k) \cdot E(c_k|a_j)$$

where $f(w_k)$ is the function used to aggregate the contributions from each class according to its relative importance, which is uniquely determined by its weight.

$E(c_k|a_j)$ is the function we use to compute how good the j-th value of the attribute A is in determining the k-th value of the class. Here, we can resort to the value given by the standard entropy-based splitting criteria, that is

$$E(c_k|a_j) = -p(c_k|a_j) \cdot log_2(p(c_k|a_j))$$

Please note that, when $f(w_k) = 1$, $E_f \equiv H$. We propose two alternative splitting criteria using two different aggregation functions:

Simple CCS Evaluation (E_I). We could directly resort to the class weights to aggregate the contributions from each class, i.e., using the identity function, $f(w_k) \equiv I(w_k) = w_k$. The resulting criterion is then:

$$E_I(C|a_j) = \sum_{k=1}^{K} w_k \cdot E(c_k|a_j)$$

Weighted Premium CCS Evaluation (E_{WP}). In some cases, the previous criterion could lead to classifiers that would tend to ignore the least important classes as we increase the relative importance of some classes. In some sense, this is similar to the problem imbalanced learning methods try to address and it should be avoided. Hence we propose a different criterion that uses a softened aggregation function: *Weighted Premium* (WP). For a given class weight w_k, its weighted premium is

$$\mathrm{WP}(w_k) = 1 + \frac{w_k - min\_weight}{max\_weight} \quad (1)$$

where min_weight corresponds to the weight of the least important class and max_weight represents the most important class weight. Therefore, the weighted premium is 1 for the least important classes and it is greater than one for more important classes. It favors the most important classes without ignoring the least important ones.

The normalized version of weighted premiums can then be used as the $f(w_k)$ aggregation function to define a Weighted Premium Evaluation criterion, E_{WP}:

$$E_{\mathrm{WP}}(C|a_j) = \sum_{k=1}^{K} \left[\frac{\mathrm{WP}(w_k)}{\sum_{i=1}^{K} \mathrm{WP}(w_i)} \right] \cdot E(c_k|a_j)$$

For instance, in a 2-class problem where one class is nine times more important than the other, the relative weights for the two classes would be $\frac{9}{10}$ and $\frac{1}{10}$. After the WP transformation, the resulting values would be softened: 0.65 and 0.35, respectively.

3.2 Tree Pruning

Tree pruning is necessary in TDIDT algorithms to avoid overfitting. Quinlan's pessimistic pruning [8], for instance, performs a postorder traversal of the tree internal nodes in order to decide, for each subtree, if it should be replaced for a single leaf node, which would then be labeled with the most common class in that subtree.

From a CCS classification perspective, a subtree should be pruned if the errors that tree pruning introduces correspond to examples belonging to the less important classes. In order to perform CCS tree pruning, we define a CCS error rate that takes class weights into account:

$$CCSError = \frac{\sum_{k=1}^{K} w_k \cdot e_k}{\sum_{k=1}^{K} w_k \cdot n_k}$$

where e_k is the number of misclassified training examples from the k-th class and n_k is the support of the k-th class in the training set. In other words, a misclassified example is taken into account according to its relative class weight. Please note that no smoothing function, such as WP, is needed when defining the CCSError rate because it is a direct measure of the CCS model quality. For splitting criteria, however, the smoothing function was needed for the algorithm not to ignore the least important classes.

We propose two pruning strategies that take CCSError into account:

- The first pruning strategy, ***CCS Pruning***, adapts Quinlan's pessimistic error estimate by replacing *Error* with *CCSError*. However, there are some situations in which *CCS pruning* is not effective enough. Let us imagine a node whose examples mainly belong to unimportant classes, maybe with some occasional examples belonging to a very important class. When the relative importance of the important class is very high, pruning will not be performed. However, it is clear that pruning might be beneficial in this case, since complexity would be reduced while accuracy would not be severely affected. This leads us to a second pruning strategy:
- ***Double Pruning*** addresses the aforementioned situation by allowing the use of the *CCSError* rate to perform *additional* pruning, apart from Quinlan's standard pruning. In other words, we will prune a subtree if the pessimistic estimation of the CCS error for the subtree is higher than the CCS error for a single leaf node, even when the standard estimation of the leaf node error is greater than the standard estimation of the subtree error.

4 CCS Classifier Evaluation

In order to evaluate CCS models, we have used three different metrics corresponding to three different aspects we would like a CCS classifier to address: simplicity with respect to important classes (that is, the main driver behind CCS classification), a good overall accuracy (since the resultant model should still be useful in practice), and a small false positive rate (to check that the complexity reduction does not come at the cost of too many false positives for the important classes).

4.1 Classifier Weighted Complexity

AvgCSDepth is defined as the CCS version of the average tree depth typically used to measure decision tree complexity. The average tree depth, without taking class weights into account, can be computed as

$$AvgDepth = \frac{\sum_{k=1}^{K} \sum_{l=1}^{L} n_{kl} \cdot l}{\sum_{k=1}^{K} n_k}$$

where K is the number of classes, L is the number of tree levels (i.e. its overall depth), n_k is the number of examples belonging to the k-th class, and n_{kl} is the number of examples belonging to the k-th class that end up in a leaf at the l-th tree level.

The average class-sensitive depth, $AvgCSDepth$, is computed as a weighted average by taking class weights into account, so that each class importance determines that class contribution to the CCS depth:

$$AvgCSDepth = \frac{\sum_{k=1}^{K} w_k \sum_{l=1}^{L} n_{kl} \cdot l}{\sum_{k=1}^{K} w_k n_k}$$

This way, a decision tree with important class leaves nearer to the root will have a lower CS depth. When all weights are the same, the classifier CCS depth measure is equivalent to the standard average tree depth.

4.2 Classifier Accuracy

Even though our goal is to achieve simpler models for the most important classes, we must still consider the resulting classifier accuracy. The simplest classification model, from a CCS classification perspective, would be useless if it misclassified too many examples. Hence, we will also include the overall classifier accuracy in our experiment results:

$$Accuracy = \frac{\sum_{k=1}^{K} TP_k}{N}$$

where TP_k is the number of true positives belonging to the k-th class and N is the total number of examples.

4.3 F Measure: False Positives and False Negatives

Finally, we also check that the complexity reduction we achieve does not come at the cost of an inordinate number of false positives. For this, we resort to the F measure typically used in information retrieval. This measure computes the harmonic mean of the resulting classifier precision and recall. In particular, we resort to the macro-averaging F measure [10] that is defined as follows:

$$MacroF = \frac{2 \cdot Pr^M \cdot Re^M}{Pr^M + Re^M}$$

where Pr^M and Re^M represent the macro-averaged precision and recall measures (i.e. the average of the individual measurements performed for each one of the problem classes).

5 Experimental Results

We have tested CCS versions of the standard C4.5 TDIDT algorithm [8] with some well-known classification problems from the UCI Machine Learning Repository [7]. Table 1 shows the data sets we have used in our experiments.

Table 1. Data sets used to evaluate CCS decision trees

Dataset	Records	#Attr.	Classes
ADULT	48842	15	2
AUSTRALIAN	690	15	2
AUTOS	205	25	6
BALANCE-SCALE	625	4	3
BREAST	699	9	2
CAR	1728	7	4
CHESS	3196	36	2
CMC	1473	9	3
FLAGS	194	29	8
GLASS	214	9	6
HAYESROTH	160	5	3
HEART	270	14	2
IMAGE	2310	18	7

Dataset	Records	#Attr.	Classes
IRIS	150	5	3
MAGIC	19020	10	2
MUSHROOM	8124	23	2
NURSERY	12960	9	5
PIMA	768	9	2
SPAMBASE	4601	57	2
SPLICE	3175	61	3
TICTACTOE	958	10	2
TITANIC	2201	4	2
VOTES	435	17	2
WAVEFORM	5000	22	3
WINE	178	14	3

For each classification problem, we have performed an experiment suite for each class value. In each suite, a different class is chosen to be the most important one whereas the others are equally important among them. Each suite, itself, includes experiments with seven different relative weights. We have tested the algorithms performance when the higher weight is 1, 1.5, 2, 2.5, 3, 5, and 9 times the lower weight, where 1 corresponds to the non-weighted case, 1.5 corresponds to a 50% premium, and so on. Since each particular weight assignment is evaluated using a 10-folded cross validation, that leads to $10 \cdot 7 \cdot K$ experiments for each algorithm tested on a particular data set, where K is the number of different classes in the data set. Average results for each problem will be used in order to draw conclusions.

In addition, further statistical analysis [3] have been performed in order to ensure the validity of the conclusions drawn from our experiments. In our results, the number of wins-ties-losses and the average value each measure will be deemed as significant according to the Sign Test [11] and Wilcoxon's test [12], respectively.

Figures 1-3 compare the results we have obtained with CCS decision trees with the results standard decision trees would achieve. The charts show the changes that the use of CCS heuristics cause for the three metrics described in the previous section: *average class-sensitive tree depth* (`AvgCSDepth`), *classifier accuracy* (`Accuracy`), and *macro-averaged F measure* (`MacroF`). In these charts, the results for different class weights are shown along the X axis, while the Y axis corresponds to the percentage variation CCS techniques introduce for each metric. The X axis corresponds to the results we would achieve by using a traditional decision tree classifier, which are the results we will always obtain with equal class weights ($w = 1$).

Splitting criteria. Figure 1 shows the results we obtain when we use the E_I and E_{WP} splitting criteria instead of the C4.5 gain ratio criterion, always using the standard C4.5 pessimistic pruning strategy [8].

Regarding E_I, no effective CCS depth reduction is observed. Moreover, neither `Accuracy` nor `MacroF` are reduced. In fact, only the `MacroF` result for $w = 9$ is

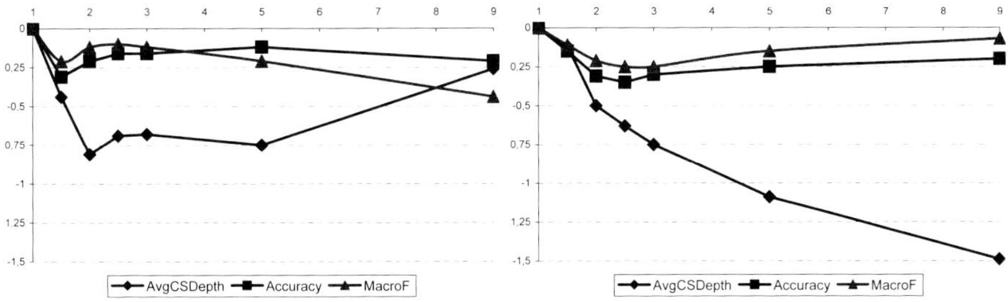

E_I + C4.5 Pessimistic Pruning E_{WP} + C4.5 Pessimistic Pruning

Fig. 1. Results obtained using CCS splitting criteria

significant according to Wilcoxon's test. The remaining differences are never significant according to this statistical test.

With respect to E_{WP}, the average CCS depth is progressively reduced when w is increased, while both `Accuracy` and `MacroF` are hardly reduced. However, the differences are never significant according to Wilcoxon's test.

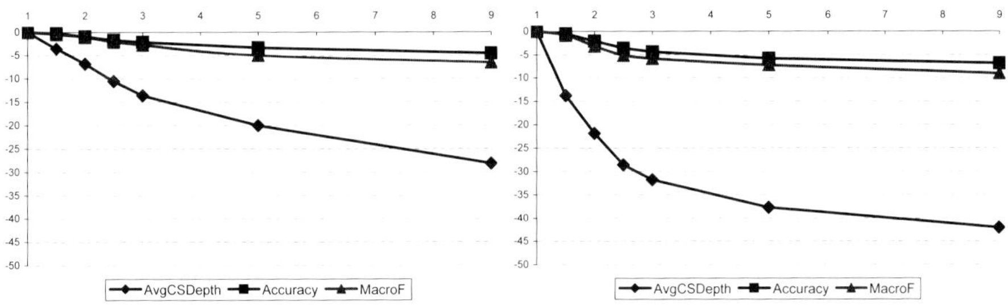

Gain Ratio + CCS Pruning Gain Ratio + Double Pruning

Fig. 2. Results obtained using CCS pruning techniques

Pruning techniques. Figure 2 depicts the summarized results we have obtained when using CCS pruning and double pruning. Both pruning techniques lead to much simpler models from a CCS classification point of view, with relatively small accuracy and precision losses.

Double pruning has proved to be more effective than simple CCS pruning, since greater reductions are achieved even for smaller relative weights. The average CCS depth obtained by using double pruning is obviously never worse than what we would achieve using the standard pessimistic pruning and the differences are always statistically significant, even though they are less pronounced for accuracy and MacroF than for tree depth.

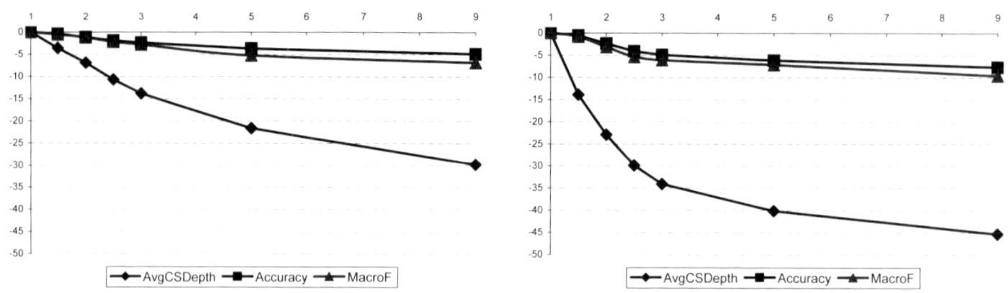

E_{WP} + CCS Pruning E_{WP} + Double Pruning

Fig. 3. Results obtained when combining WP Criterium and CCS pruning techniques

Combined methods. Figure 3 summarizes the results we have obtained by combining WP criterium and CSS pruning techniques (similar results are obtained by considering E_I instead of E_{WP}). In all cases, CCS depth is significantly reduced (according to Wilcoxon's test) with minor reductions on classifier accuracy and precision. Both splitting criteria (E_I and E_{WP}) always achieve, in combination with CCS pruning, simpler models than the standard splitting criterion, but CCS pruning is the main driver behind the observed differences.

If we consider the number of times that the average CCS tree depth is reduced, we consistently obtain significant improvements according to the Sign Test [11]. Out of the 81 experiments we performed, simple CCS pruning reduces the classifier complexity in 51 to 72 situations depending on the relative weights we use (the higher w, the higher the number of wins). Double pruning achieves even better results: from 71 to 76 wins out of 81 tests (consistently reducing classifier complexity even for small values of w).

6 Conclusions

Classification model complexity is very important since it is closely related to its interpretability. In many real-life problems, some class values are, somehow, more important than others. In such situations, experts might be interested in building succinct models for the mnost important classes. Class-complexity-sensitive (CCS) classification techniques are designed to address this issue and they build simpler models for the most important classes without incurring into high accuracy losses.

In this paper, we have introduced several heuristics that let us adapt standard decision tree learning algorithm in order to take class importance into account. Both splitting criteria and pruning strategies have been devised to deal with CCS classification, thus providing the mechanisms needed to build CCS decision trees using the standard TDIDT algorithm.

Our experimental results show that CCS splitting criteria do not provide significant improvements with respect to their traditional counterparts, a result that is consistent with prior research on splitting criteria [2].

However, CCS pruning techniques help us achieve considerable reductions in CCS model complexity within a reasonable accuracy loss. Depth and accuracy/precision are traded off, as expected, when weights are introduced into the standard TDIDT model. Combining both CCS splitting criteria and CCS pruning techniques leads us to even smaller CCS classification models.

References

1. Batista, G.E.A.P.A., Prati, R.C., Monard, M.C.: A study of the behavior of several methods for balancing machine learning training data. SIGKDD Explorations 6(1), 20–29 (2004)
2. Berzal, F., Cubero, J.C., Cuenca, F., Martín-Bautista, M.J.: On the quest for easy-to-understand splitting rules. Data and Knowledge Engineering 44(1), 31–48 (2003)
3. Demsar, J.: Statistical comparisons of classifiers over multiple data sets. Journal of Machine Learning Research 7, 1–30 (2006)
4. Domingos, P.: Metacost: A general method for making classifiers cost-sensitive. In: 5th ACM SIGKDD International Conference on Knowledge Discovery and Data Mining, pp. 155–164 (1999)
5. Gamberger, D., Lavrac, N.: Expert-guided subgroup discovery: Methodology and application. Journal of Artificial Intelligence Research 17, 501–527 (2002)
6. Gehrke, J., Ramakrishnan, R., Ganti, V.: Rainforest - a framework for fast decision tree construction of large datasets. Data Mining and Knowledge Discovery 4(2/3), 127–162 (2000)
7. Blake, C.L., Newman, D.J., Merz, C.J.: UCI repository of machine learning databases (1998)
8. Quinlan, J.R.: C4.5: Programs for Machine Learning. Morgan Kaufmann, San Francisco (1993)
9. Rastogi, R., Shim, K.: PUBLIC: A decision tree classifier that integrates building and pruning. Data Mining and Knowledge Discovery 4(4), 315–344 (2000)
10. Sebastiani, F.: Machine learning in automated text categorization. ACM Comput. Surv. 34(1), 1–47 (2002)
11. Sheskin, D.J.: Handbook of parametric and nonparametric statistical procedures. Chapman and Hall/CRC, Boca Raton (2000)
12. Wilcoxon, F.: Individual comparisons by ranking methods. Biometrics Bulletin 1(6), 80–83 (1945)

Evaluating Decision Trees Grown with Asymmetric Entropies

Simon Marcellin[1], Djamel A. Zighed[1], and Gilbert Ritschard[2]

[1] Université Lumière Lyon 2 Laboratoire ERIC, Bât L, Campus Porte des Alpes
5, av. Pierre Mendès-France, F-69600 Bron, France
{abdelkader.zighed,simon.marcellin}@univ-lyon2.fr
http://eric.univ-lyon2.fr

[2] Université de Genève, Département d'économétrie, 40 bd du Pont-d'Arve
CH-1211 Geneva 4, Switzerland
gilbert.ritschard@unige.ch

Abstract. We propose to evaluate the quality of decision trees grown on imbalanced datasets with a splitting criterion based on an asymmetric entropy measure. To deal with the class imbalance problem in machine learning, especially with decision trees, different authors proposed such asymmetric splitting criteria. After the tree is grown a decision rule has to be assigned to each leaf. The classical Bayesian rule that selects the more frequent class is irrelevant when the dataset is strongly imbalanced. A best suited assignment rule taking asymmetry into account must be adopted. But how can we then evaluate the resulting prediction model? Indeed the usual error rate is irrelevant when the classes are strongly imbalanced. Appropriate evaluation measures are required in such cases. We consider ROC curves and recall/precision graphs for evaluating the performance of decision trees grown from imbalanced datasets. These evaluation criteria are used for comparing trees obtained with an asymmetric splitting criterion with those grown with a symmetric one. In this paper we only consider the cases involving 2 classes.

1 Introduction

Learning from imbalanced datasets is an important issue in datamining [1,2]. A dataset is imbalanced when the distribution of the modalities of the class variable is far away from the uniform distribution. This happens in a lot of real world applications: in the medical field, to predict a rare illness; in the industry to predict a device breakdown; or in the bank field, to predict insolvent costumers or frauds in transactions. In these cases, there is one rare state of the class variable (ill, breakdown, insolvent, fraud) that should be detected in priority. Standard methods do not take such specificities into account and just optimize a global criterion with the consequence that all the examples would be classified into the majority class, i.e. that which minimizes the global error rate. This kind of prediction models is useless because it does not carry any information. In decision trees, this problem appears at two levels: during the generation of the

A. An et al. (Eds.): ISMIS 2008, LNAI 4994, pp. 58–67, 2008.

tree with the splitting criterion, and during the prediction with the assignment rule of a class in each leaf.

First, to choose the best feature and the best split point to create a new partition, classical algorithms use an entropy measure, like the Shannon entropy or quadratic entropy. Entropy measures evaluate the quantity of information about the outcome provided by the distribution of the class variable. They consider the uniform distribution, i.e that for which we have the same number of examples in each class, as the most entropic situation. So the worst situation according to these measures is the balanced distribution. However, if in the real world for example 1% of the people are sick, ending with a leaf in which 50% of the members are sick would be very interesting and would carry a lot of information for the user of the model. Thus, using a classical entropy measure precludes obtaining such branches and hence the relevant associated rules for predicting the rare class. The second important aspect of decision trees is the assignment rule. Once the decision tree is grown, each branch defines the condition of a rule. The conclusion of the rule depends on the distribution of the leaf. Classical algorithms conclude to the majority class, i.e the most frequent modality in the leaf. But this is not efficient: In the previous example where 1% of the people are sick, a rule leading to a leaf with a frequency of the 'sick' class of 30% would conclude to 'not sick'. According to the importance of predicting correctly the minority class, it may be better however in that case to conclude to 'sick'. This will lead to a higher total number of errors, but a lower number of errors on the rare class and hence a better model.

In decision trees, the imbalance of the prediction class influences the learning process during these two steps. This paper focuses on the first issue. Asymmetric criterion were proposed to deal with this imbalance aspect in decision trees. How do such criteria influence the learning? If we use an asymmetric criterion, what performance measure should be used to evaluate the gain of using this criterion? Our proposition is to consider ROC curves and recall/precision graphs and apply them for evaluating the gain brought by using splitting criteria based on asymmetric measures over those based on symmetrical measures. In section 2 we present the decision trees and the asymmetric criterion. In section 3 we propose evaluation methods to compare trees built with a symmetric criterion versus those grown with an asymmetric one. Section 4 presents the methodology of our evaluation and exposes our results. We finish by the section 5 that concludes and proposes future works.

2 Asymmetric Criteria for Decision Trees

2.1 Notations and Basic Concepts

We note Ω the population concerned by the learning problem. The profile of any example ω in Ω is described by p explicative or exogenous features $X_1, \ldots, X_p$. Those features may be qualitative or quantitative ones. We also consider a variable C to be predicted called either endogenous, class or response variable. The values taken by this variable within the population are discrete and form a finite set $\mathcal{C}$. Letting m_j be the number of different values taken by X_j and n the number of modalities of C, we have $\mathcal{C} = \{c_1, \ldots, c_n\}$. And when it is not ambiguous, we denote the class c_i simply by i. Algorithms of trees induction generate a model $\phi(X_1, \ldots, X_p)$ for the prediction of C represented by a decision tree [3,4] or an induction graph [5]. Each branch of the tree represents a rule. The set of these rules is the prediction model that permits to determine the predicted value of the endogenous variable for any new example for which we know only the exogenous features. The development of the tree is made as follows: The learning set Ω_a is iteratively segmented, each time on one of the exogenous features $X_j; j = 1, ...p$ so as to get the partition with the smallest entropy for the distribution of C. The nodes obtained at each iteration define a partition on Ω_a. Each node s of a partition S is described by a probability distribution of the modalities of the endogenous features C: $p(i/s); i = 1, \ldots, n$. Finally, these methods generate decision rules in the form **If** *condition* **then** *Conclusion*. Splitting criteria are often based on entropies. The notion of entropy is defined mathematically by axioms out of the context of machine learning. See for instance [6] and [7] for details. The entropy H on the partition S to minimize is generally a mean entropy such as $H(S) = \sum_{s \in S} p(s) h(p(1|s), \ldots, p(i|s), \ldots, p(n|s))$ where $p(s)$ is the proportion of cases in the node s and $h(p(1|s), \ldots, p(n|s))$ an entropy function such as Shannon's entropy for instance $H_n(p_1, p_2, \ldots, p_n) = -\sum_{i=1}^{n} p_i \log_2 p_i$ .. By continuity we set $0 \log_2 0 = 0$. There are other entropy measures [8] such as the quadratic entropy $H_n(p_1, p_2, \ldots, p_n) = \sum_{i=1}^{n} p_i(1 - p_i)$ for instance.

2.2 Asymmetric Criteria

The properties of classical entropy measures such as those cited above (Shannon, quadratic) are not suited to inductive learning for reasons exposed in [5]. First, the uniform distribution is not necessarily the most uncertain. Second, the computation of the entropy being based on estimates of the probabilities, it should account for the precision of those estimates, i.e. account for the sample size. That is why we proposed in [5] a new axiomatic leading to a new family of more general measures allowing for a user defined maximal entropy reference and sensitive to the sample size. Let $\lambda_i = \frac{Nf_i + 1}{N + n}$ be the Laplace estimator of p_i, $W = (w_1, w_2, \ldots, w_n)$ the vector with maximal entropy and N the sample size. The asymmetric entropy we proposed reads:

$$h_W(N, f_1, f_2, \ldots, f_n) = \sum_{i=1}^{n} \frac{\lambda_i(1 - \lambda_i)}{(-2w_i + 1)\lambda_i + w_i 2}$$

An other non-centered entropy has been proposed in [9]. It results from a different approach that transforms the frequencies p_i's of the relevant node by means of a transformation that turns W into a uniform distribution. In the two class case, the transformation function is composed of two affine functions: $\pi = \frac{p}{2w}$ if $0 \leq p \leq w$ and $\pi = \frac{p + 1 - 2w}{2(1 - w)}$ if $w \leq p \leq 1$. The resulting non-centered entropy is then defined as the classical entropy of the transformed distribution. Though this method can be used with any kind of entropy measure, it is hardly extensible to the more than two class problem.

3 Evaluation Criteria of Trees in the Imbalanced Case

3.1 Performance Measures

There exist different measures for evaluating a prediction model. Most of them are based on the confusion matrix (see Table 1). Some measures are designed for the prediction of a specific modality: the recall rate ($\frac{TP}{TP+FN}$)and the precision rate ($\frac{TP}{TP+FP}$) The F-Measure is the harmonic mean of recall and precision. Other measures do not distinguish among outcome classes. We may cite here overall error rate, and the sensibility and specificity (mean of recall and precision on each class). The latter measures are less interesting for us, since by construction they favor accuracy on the majority class. (Still, we may cite the PRAGMA measure [10] that allows the user to specify the importance granted for each class as well as its preferences in terms of recall and precision). It follows that recall and precision are the best suited measures when the concern is the prediction of a specific (rare) class as in our setting.

Table 1. Confusion matrix for the two classes case

	Class +	Class -
Class +	True positives (TP)	False negatives (FN)
Class -	False positives (FP)	True negatives (TN)

The confusion matrix depicted in Table 1 is obtained for a decision tree by applying the relevant decision rule to each leaf. This is not a problem when the assigned class is the majority one. But with an asymmetric criterion this rule is not longer suited [11]: If we consider that the worst situation is a distribution W, meaning that the probability of class i is w_i in the most uncertain case, then no decision can be taken for leaves having this distribution. Hence, leaves where the class of interest is better represented than in this worst reference case ($f_i > w_i$) should be assigned to the class i. This simple and intuitive rule could be replaced by a statistical test, has we proposed it with the implication intensity [12] for instance. In this paper, we consider however the following simple decision rule: $C = i$ if $f_i > w_i$. This rule is adapted to the 2-class case. With k classes, the condition can indeed be satisfied for more than one modality and should then be reinforced. To avoid the rule's limitation, we also move the decision threshold between 0 and 1 to observe the recall / precision graph. This allows us to see if a method dominates an other one for different thresholds of decision, and can also help us to choose the most appropriate decision rule.

3.2 ROC Curve

A ROC curve (Receiver operating characteristics) is a well suited tool for visualizing the performances of a classifier regarding results for a specific outcome class. Several works present its principles [13,14]. First, a score is computed for each example. For decision trees, it is the probability to classify this example as positive. This probability is estimated by the proportion of positive examples in the leaf. Then, all examples are plotted in a false positive rate / true positive rate space, cumulatively from the best scored to the last scored. A ROC curve close to the main diagonal means that the model provides no useful additional information about the class. *A contrario* a ROC curve with a point in [0,1] means that the model separates perfectly positive and negative examples. The area under the ROC curve (AUC) summarizes the whole curve. We now examine how the ROC curve and the AUC may be affected when an asymmetric measure is used instead of a symmetric one.

4 Evaluations

4.1 Compared Models and Datasets

Our study is based on decision trees evaluated in 10 cross-validation to avoid the problems of over-fitting on the majority class. For each dataset we consider the quadratic entropy and the asymmetric entropy. The chosen stopping criterion, required to avoid over-fitting, is a minimal information gain of 3%. Other classical stopping criteria such as the minimal support of a leaf, or the maximal depth of the tree, would preterite the minority class. We selected the 11 datasets listed in Table 2. For each of them we have a two class outcome variable. We consider predicting the overall last frequent class. A first group of datasets is formed by strongly imbalanced datasets of the UCI repository [15]. In the dataset *letter* (recognition of hand-writing letters) we consider predicting the letter 'a' vs all the others (*letter_a*) and the vowels vs the consonants (*letter_vowels*). The classes of the dataset *Satimage* were merged into two classes as proposed by Chen and Liu [16]. The datasets *Mammo1* and *Mammo2* are real data from the breast cancer screening and diagnosis collected within an industrial partnership. The goal is to predict from a set of predictive features whether some regions of interest on digital mammograms

Table 2. Datasets

Dataset	# of examples	# of features	Imbalance
Breast	699	9	34%
Letter_a	2000	16	4%
Letter_vowels	2000	16	23%
Pima	768	8	35%
Satimage	6435	36	10%
Segment_path	2310	19	14%
Waveform_merged	5000	40	34%
Sick	3772	29	6%
Hepatisis	155	19	21%
Mammo1	6329	1038	8%
Mammo2	3297	1038	15%

are cancers or not. This last example provides a good illustration of learning on a imbalanced dataset: Missing a cancer could lead to death, which renders the prediction of this class very important. A high precision is also requested since the cost of a false alarm is psychologically and monetary high.

4.2 Results and Interpretation

Table 3 shows the AUC values obtained for each dataset. Figures 1,2,3,4 and 5 exhibit the ROC curves and the recall / precision graphs respectively for the datasets *Mammo1*, *Mammo2*, *Letter_a*, *Waveform_merged* and *Satimage*.

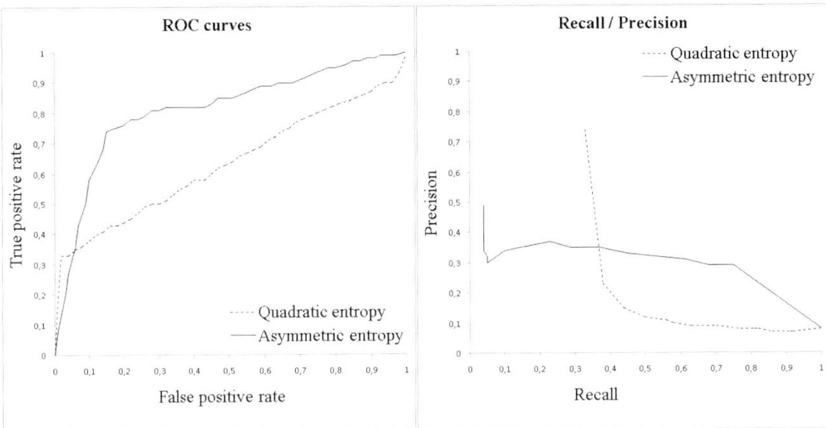

Fig. 1. Results for Mammo1

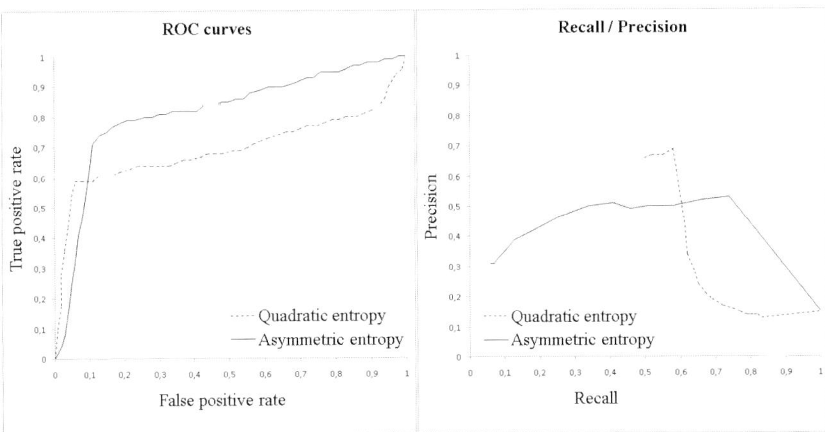

Fig. 2. Results for Mammo2

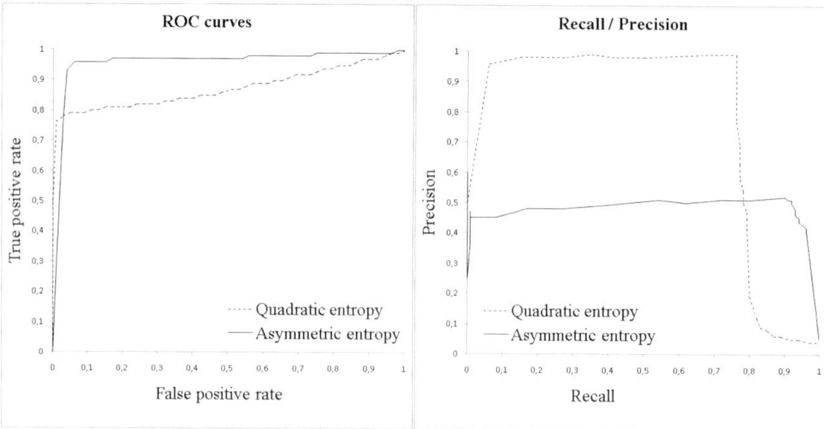

Fig. 3. Results for Letter_a

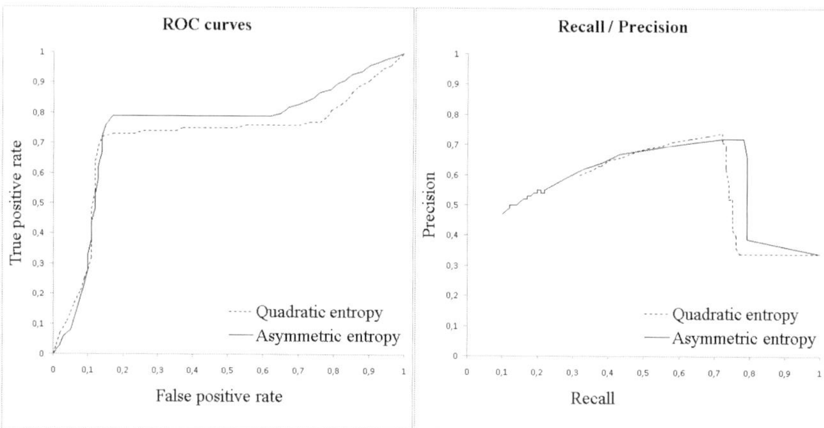

Fig. 4. Results for Waveform_merged

The recall / precision graphs show that when recall is high, the asymmetric criterion ends up with a better precision. This means that decision rules derived from a tree grown with an asymmetrical entropy are more accurate for predicting the rare class. On both real datasets (Figures 1 and 2) we see that if we try to maximize the recall (or to minimize the number of 'missed' cancers, or false negatives), we obtain fewer false positives with the asymmetric entropy. This is exactly the desired effect.

The ROC curves analysis shows that using the asymmetric entropy improves the AUC criterion (Table 3). More importantly is however the form of the curves. The ROC curves of the quadratic entropy are globally higher on the left side of

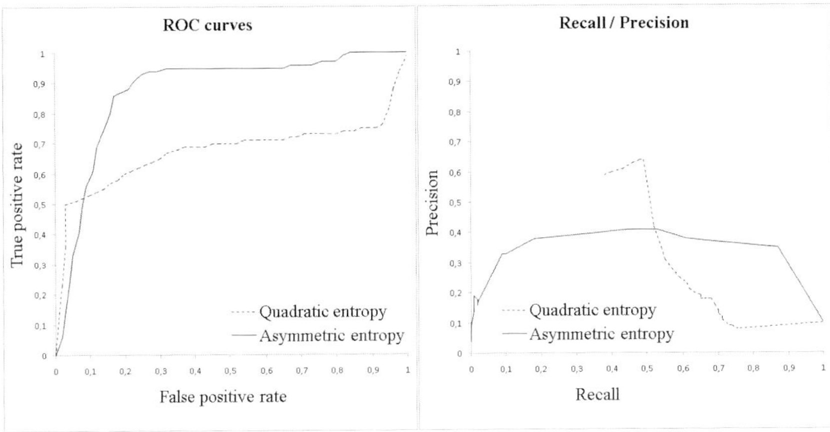

Fig. 5. Results for Satimage

Table 3. Obtained AUC

Dataset	AUC with quadratic entropy	AUC with asymmetric entropy
Breast	0.9288	0.9359
Letter_a	0.8744	0.9576
letter_voyelles	0.8709	0.8818
pima	0.6315	0.6376
satimage	0.6715	0.8746
segment_path	0.9969	0.9985
Waveform_merged	0.713	0.749
sick	0.8965	0.9572
hepatisis	0.5554	0.6338
mammo1	0.6312	0.8103
mammo2	0.6927	0.8126

the graph, i.e. for high scores. Then the two ROC curves cross each other, and on the right side the asymmetric criterion is almost always dominating. We can thus conclude that the lower the score, the more suited the use of an asymmetric entropy. We saw in section 2 through several examples that when predicting rare events, we have to use small acceptation threshold (we accept a leaf when the observed frequency of the minority class exceeds the corresponding probability in the more uncertain distribution). Thus, ROC curves clearly highlight the usefulness of asymmetric entropies for predicting rare classes.

The two previous remarks mean that for seeking 'nuggets' of the minority class, we always get better recall and precision rates with an asymmetric criterion. In other words, if we accept predicting the class of interest with a score below 50%, then the smaller the score, the better the recall and precision rates when compared with those obtained with a symmetric criterion.

5 Conclusion

We evaluated how using a splitting criterion based on an asymmetrical entropy to grow decision trees for imbalanced datasets influences the quality of the prediction of the rare class. If the proposed models are as expected less efficient in terms of global measures such as the error rate, ROC curves as well as the behavior of recall and precision as function of the acceptation threshold reveals that models based on asymmetric entropy outperform those built with a symmetric entropy, at least for low decision threshold.

An important issue with asymmetric criterion is how can we determine the "most" uncertain reference distribution W? When the probability of each class is known, it is consistent to use these probabilities. Otherwise, we could estimate them from the overall class frequencies in the learning dataset. For our empirical experimentation, we set this distribution once and for all. A different approach would be to use at each node the distribution in the parent node as reference W. The criterion would in that case adapt itself at each node. A similar approach is to use Bayesian trees [17], where in each node we try to get rid of the parent node distribution. Finally, we noticed during our experimentations that the choice of the stopping criterion is very important when we work on imbalanced datasets. Therefore, we plan to elaborate a stopping criterion suited for imbalanced data, that would, for instance, take into account the number of examples at each leaf, but allow for a lower threshold for leaves where the relevant class is better represented. In a more general way, various measures of the quality of association rules should help us to build decision trees.

We did not decide about the question of the decision rule to assign a class to each leaf. Since an intuitive rule is the one proposed in section 2, consisting in accepting the leaves where the class of interest is better represented than in the original distribution, we propose two alternative approaches: the first is to use statistical rules, or quality measures of association rules. The second is to use the graphs we proposed in this article, by searching optimal points on the recall / precision graph and on the ROC curve. We should consider the break-even Point (BEP, [18]) to find the best rate, or the Pragma criterion [10]. The choice of a rule will allow us to precisely quantify the use of an asymmetric criterion.

The extension of the concepts exposed in this article to the case of more than two modalities raises several problems. First, even if the asymmetric entropy applies to the multiclass case, some other measures are not. The problem of the decision rule is very complex with several classes. Indeed, setting a threshold on each class is not efficient, because this rule can be satisfied for several classes simultaneously. A solution is to choose the class with the frequency that departures the most from its associated threshold, or that with the smallest contribution to the entropy of the node. The methods of evaluation proposed in this paper (ROC curves and recall / precision graphs) are adapted for a class vs all the others, i.e. in the case with more than 2 classes, for the case where one modality among the others is the class of interest. It would be more difficult evaluating the model when two or more rare classes should be considered as equally relevant. The evaluation of multiclass asymmetric criteria will be the topic of future works.

References

1. Provost, F.: Learning with imbalanced data sets. In: AAAI 2000 Workshop on Imbalanced Data Sets (2000)
2. Barandela, R., Sánchez, J.S., García, V., Rangel, E.: Strategies for learning in class imbalance problems. Pattern Recognition 36(3), 849–851 (2003)
3. Breiman, L., Friedman, J.H., Olshen, R.A., Stone, C.J.: Classification And Regression Trees. Chapman and Hall, New York (1984)
4. Quinlan, J.R.: C4.5: Programs for Machine Learning. Morgan Kaufmann, San Mateo (1993)
5. Zighed, D.A., Marcellin, S., Ritschard, G.: Mesure d'entropie asymétrique et consistante. In: EGC, pp. 81–86 (2007)
6. Rényi, A.: On measures of entropy and information. In: 4th Berkely Symp. Math. Statist. Probability, vol. 1, pp. 547–561 (1960)
7. Aczel, J., Daroczy, Z.: On measures of information and their characterizations (1975)
8. Zighed, D., Rakotomalala, R.: Graphe d'induction Apprentissage et Data Mining. Hermés, Paris (2000)
9. Lallich, S., Lenca, P., Vaillant, B.: Construction d'une entropie décentrée pour l'apprentissage supervisé. In: 3éme Atelier Qualité des Connaissances á partir des Données (QDC-EGC 2007), Namur, Belgique, pp. 45–54 (2007)
10. Thomas, J., Jouve, P.E., Nicoloyannis, N.: Mesure non symétrique pour l'évaluation de modéles, utilisation pour les jeux de données déséquilibrés. In: 3éme Atelier Qualité des Connaissances á partir des Données (QDC-EGC 2007), Namur, Belgique (2007)
11. Marcellin, S., Zighed, D., Ritschard, G.: An asymmetric entropy measure for decision trees. In: 11th Information Processing and Management of Uncertainty in knowledge-based systems (IPMU 2006), Paris, France, pp. 1292–1299 (2006)
12. Ritschard, G., Zighed, D., Marcellin, S.: Données déséquilibrées, entropie décentrée et indice d'implication. In: Gras, R., Orús, P., Pinaud, B., Gregori, P. (eds.) Nouveaux apports théoriques á l'analyse statistique implicative et applications (actes des 4émes rencontres ASI4, Castellón de la Plana (Espaïia), Departament de Matemàtiques, Universitat Jaume, October 18-21, 2007, vol. I, pp. 315–327 (2007)
13. Egan, J.: Signal detection theory and roc analysis. Series in Cognition and Perception (1975)
14. Fawcett, T.: An introduction to roc analysis. Pattern Recognition Letter 27(8), 861–874 (2006)
15. Hettich, S., Bay, S.D.: The uci kdd archive (1999)
16. Chen, C., Liaw, A., Breiman, L.: Using random forest to learn imbalanced data (2004)
17. Chai, X., Deng, L., Yang, Q.: Ling: Test-cost sensitive naive bayes classification. In: ICDM (2005)
18. Sebastiani, F.: Machine learning in automated text categorization. ACM Comput. Surv. 34(1), 1–47 (2002)

Stepwise Induction of Logistic Model Trees

Annalisa Appice, Michelangelo Ceci, Donato Malerba, and Savino Saponara

Dipartimento di Informatica, Università degli Studi di Bari
via Orabona, 4 - 70126 Bari, Italy
{appice,ceci,malerba}@di.uniba.it, savinos@email.it

Abstract. In statistics, logistic regression is a regression model to predict a binomially distributed response variable. Recent research has investigated the opportunity of combining logistic regression with decision tree learners. Following this idea, we propose a novel Logistic Model Tree induction system, SILoRT, which induces trees with two types of nodes: regression nodes, which perform only univariate logistic regression, and splitting nodes, which partition the feature space. The multiple regression model associated with a leaf is then built stepwise by combining univariate logistic regressions along the path from the root to the leaf. Internal regression nodes contribute to the definition of multiple models and have a global effect, while univariate regressions at leaves have only local effects. Experimental results are reported.

1 Introduction

In its original definition, logistic regression is a regression model for predicting the value of a binomially distributed response variable $Y = \{C_1, C_2\}$. Given a training set $D = \{(\boldsymbol{x}, y) \in \boldsymbol{X} \times Y \mid y = g(\boldsymbol{x})\}$ where $\boldsymbol{X}$ represents the search space spanned by m independent (or predictor) continuous variables X_i, a logistic regression model M is induced by generalizing observations in D in order to estimate the posterior class probability $P(C_i|x)$ that any unlabeled example $x \in \boldsymbol{X}$ belongs to C_i. Differently from the classical regression setting where the value of a (continuous) response variable is directly predicted, in logistic regression the response to be predicted is the probability that an example belongs to a given class. This probability can then be used for classification purposes.

Logistic regression was widely investigated in the literature for the classification task [3] [8] [5] [4]. The results of an empirical and formal comparison between logistic regression and decision trees [7] motivated the attempt of combining tree induction procedures with logistic regression. For example, Landwehr et al. [4] proposed a top-down fitting process to construct Logistic Model Trees. In this case, coefficients of a logistic regression model at leaves are constructed by exploiting the coefficients of logistic regression models constructed in the highest levels of the tree. Although correct, this approach considers only full regression models (i.e., regression models at leaves include all predictor variables) by ignoring that a regression model based on a subset of predictor variables may give more precise predictions than the model based on more variables [2]. This

A. An et al. (Eds.): ISMIS 2008, LNAI 4994, pp. 68–77, 2008.

depends on the fact that the variable subset selection avoids to poorly estimate regression coefficients in presence of two or more predictor variables linearly related to each other (collinearity) [6]. However, finding the best subset of variables while choosing the best split becomes too costly when applied to large datasets since the search procedure may require the computation of a high number of multiple regression models. Chan et al. [1] proposed to recursively partitioning the data and fitting a piecewise (multiple or univariate) logistic regression function in each partition. A user-defined parameter allows the user to choose between learning either multiple or univariate functions when constructing the logistic model to be associated with a node of the tree. In the case of a univariate logistic model, the collinearity problem is easily faced. In fact, the leaf model involves just a single predictor variable (univariate regression), but ignores all the others. Although its advantages in facing collinearity, the method proposed by Chan et al., as well as all the methods cited in this work, do not permit to capture the possible global effect of some predictor variable, that is, the case that the contribution of a predictor variable is equally shared by several models. Finally, Zeleis et al. [10] have recently proposed a model-based recursive partitioning which can be applied to various regression problems (linear regression and logistic regression), but, as in other cited methods, this partitioning is not able to discriminate between global and local effect of variables.

In this paper we propose a new top-down logistic model tree induction method, called SILoRT, which integrates the predictive phase and the splitting phase in the induction of a logistic model tree. Such logistic model trees include two types of nodes: regression nodes and split nodes [6]. The former are associated with univariate logistic regressions involving one continuous predictor variable, while the latter are associated with split tests. The multiple logistic model associated with each leaf is then built stepwise by combining all univariate logistic regressions reported along the path from the root to the leaf. This stepwise construction has several advantages. Firstly, it overcomes the computational problem of testing a large number of multiple linear regression models. Secondly, differently from original logistic regression formulation problem, it permits to consider both continuous and discrete variables. Thirdly, it solves the problem of collinearity since only the subset of variables selected with the regression nodes along the path from to root to the leaf is practically used to construct the logistic model to be associated with the leaf itself. Fourthly, it allows modeling phenomena where some variables have a global effect while others have only a local effect. Modeling such phenomena permits to obtain simpler model that can be easily understood by humans. A variable selected with a regression node at higher level of the tree has a global effect. In fact, the effect of the univariate logistic regression with this regression node is shared by all multiple models associated with the leaves of the sub-tree rooted in the regression node itself.

The paper is organized as follows. In the next Section, we briefly describe the stepwise construction of logistic model trees, while in Section 3 we describe the construction of the logistic regression model in SILoRT. Experimental results are commented in Section 4. Finally, conclusions and future works are drawn.

2 Stepwise Construction of Logistic Model Trees

The development of a tree structure is not only determined by a recursive partitioning procedure, but also by some intermediate prediction functions. This means that there are two types of nodes in the tree: regression nodes and splitting nodes. They pass down observations to their children in two different ways. For a splitting node t, only a subgroup of the $N(t)$ observations in t is passed to each child, and no change is made on the variables. For a regression node t, all the observations are passed down to its only child, but both the values of the response variable and the values of the (continuous) predictor variables not yet included in the model are transformed. The value of the response variable is transformed in order to take into account the error performed by the logistic regression function. Predictor variables not selected in the regression nodes along the path from the root to the current node are transformed in order to remove the linear effect of those variables already included in the model. Hence, each continuous predictor variable X_j not selected in a regression node is transformed in X'_j with $X'_j = X_j - (\alpha_0 + \alpha_1 X_i)$. X_i is the regression variable, α_0 and α_1 are intercept and slope estimated to model the straight-line regression between X_j and X_i ($\widehat{X_j} = \alpha_0 + \alpha_1 X_i$). α_0 and α_1 are computed according to the procedure in [2]. Thus, descendants of a regression node do operate on a modified dataset. This transformation is coherent with the stepwise procedure that is adopted in statistics to construct incrementally a multiple linear regression model: each time a new continuous predictor variable is added to the model its linear effect on remaining continuous variables has to be removed [2].

The validity of either a regression step or a splitting test on a variable X_i is based on two distinct evaluation measures, $\rho(X_i, Y)$ and $\sigma(X_i, Y)$ respectively. The variable X_i is of a continuous type in the former case, and of any type in the latter case. Both $\rho(X_i, Y)$ and $\sigma(X_i, Y)$ are error rates measures (i.e., percentage of misclassified cases), therefore they can be actually compared to choose between three different possibilities: (i) growing the model tree by adding a regression node t, (ii) growing the model tree by adding a split node t, (iii) stopping the tree growth at node t. The evaluation measure $\sigma(X_i, Y)$ is coherently defined on the basis of the multiple logistic regression model to be associated with each leaf. In the case of a split node, it is sufficient to consider the best univariate logistic regression associated to each leaf t_R (t_L), since regression nodes along the path from the root to t_R (t_L) already partially define a multiple logistic regression model. If X_i is continuous and α is a threshold value for X_i then $\sigma(X_i, Y)$ is defined as:

$$\sigma(X_i, Y) = \frac{N(t_L)}{N(t)} E(t_L) + \frac{N(t_R)}{N(t)} E(t_R) \tag{1}$$

where $N(t)$ is the number of cases reaching t, $N(t_L)$ ($N(t_R)$) is the number of cases passed down to the left (right) child, and $E(t_L)$ ($E(t_R)$) is the error rate of the left (right) child.

The error rate $E(t_L)(E(t_R))$ is computed by considering labeled (training) cases falling in t_L (t_R) and counting the cases whose class is different from the class predicted by combining all logistic regression functions associated to regression nodes along the path from the root to t_L (t_R). More precisely:

$$E(t) = (1/N(t)) \sum_{\boldsymbol{x} \in t} d(y, \hat{y}). \tag{2}$$

where $\hat{y}$ is the class predicted for $\boldsymbol{x}$ and $d(y, \hat{y}) = 0$ if $y = \hat{y}$, 1 otherwise. The details on the construction of the logistic regression function at a node t are provided in Section 3.

Possible values of α are found by sorting the distinct values of X_i in the training set associated to t, then identifying one threshold between each pair of adjacent values. Therefore, if the cases in t have k distinct values for X_i, $k-1$ thresholds are considered. Obviously, the lower $\sigma(X_i, Y)$, the better the split.

If X_i is discrete, SILoRT partitions attribute values into two sets, the system starts with an empty set $LeftX = \emptyset$ and a full set $RightX = Sx$. It moves one element from $RightX$ to $LeftX$ such that the move results in a better split. The evaluation measure $\sigma(X_i, Y)$ is computed as in the case of continuous variables, therefore, a better split decreases $\sigma(X_i, Y)$. The process is iterated until there is no improvement in the splits.

The split selection criterion explained above can be improved to consider the special case of identical logistic regression model associated to both children (left and right). When this occurs, the straight-line regression associated to t is the same as that associated to both t_L and t_R, up to some statistically insignificant difference. In other words, the split is useless and can be filtered out from the set of alternatives. To check this special case, SILoRT compares the two regression lines associated to the children according to a statistical test for coincident regression lines [9].

Similarly to the splitting case, the evaluation of a regression step on a regression variable X_i is based on the error rate at node t: $\rho(X_i, Y) = E(t)$. $E(t)$ is computed as reported in (2). The choice between the best regression and the best split is performed by considering the following function $max\{\gamma \times max_i\{\sigma(X_i, Y)\}, max_i\{\rho(X_i, Y)\}\}$ where γ is a user defined parameter.

Three different stopping criteria are implemented in SILoRT. The first requires the number of cases in each node to be greater than a minimum value ($\sqrt{n}$). The second stops the induction process when all continuous predictor variables along the path from the root to the current node are selected in regression steps. The third stops the induction process when the predictive accuracy at the current node is 1 (i.e., error rate is 0).

3 Logistic Regression in SILoRT

SILoRT associates each node with a multiple logistic regression model that is constructed by combining all univariate logistic regressions associated with regression nodes along the path from the root to the node itself. Details on both the univariate logistic regression construction and the stepwise construction of multiple logistic regression functions in SILort are reported below.

3.1 Computing Univariate Logistic Regression Functions

A univariate logistic model on a (continuous) predictor variable X_i is the estimate of the posterior probability $P(Y|x_i)$ by means of the logit function:

$$P(C_1|x_i) = \frac{e^{\beta_0+\beta_1 x_i}}{1+e^{\beta_0+\beta_1 x_i}}$$
$$P(C_0|x_i) = 1 - P(C_1|x_i) = \frac{1}{1+e^{\beta_0+\beta_1 x_i}}$$

Parameters $\beta = [\beta_0, \beta_1]^T$ are computed by maximizing the conditional log-likelihood of the class labels in the training dataset:

$$L(\beta) = \sum_{j=1..n} \widetilde{y}_i ln(P(C_1|x_{ij}, \beta)) + (1-\widetilde{y}_i) ln(P(C_0|x_{ij}, \beta)) \quad (3)$$

where $\widetilde{y}_i = 1$ if $y_i = C_1$, 0 otherwise and n is the number of examples.

The values β_0 and β_1 which maximize $L(\beta)$ are found by computing the first order partial derivative of $L(\beta)$ with respect to β and solving the system of equations:

$$\begin{cases} \frac{\partial L(\beta)}{\partial \beta_0} = 0 \\ \frac{\partial L(\beta)}{\partial \beta_1} = 0 \end{cases} \quad (4)$$

This system of equations can be solved by means of the Newton-Raphson algorithm. In particular, β is iteratively modified in order to reach the $L(\beta)$ zero. In the matrix representation,

$$\beta_{new} = \beta_{old} - \left(\frac{\partial^2 L(\beta)}{\partial \beta \partial \beta^T}\right)^{-1} \frac{\partial L(\beta)}{\partial \beta} \quad (5)$$

Operatively, β is computed according to Algorithm 1, where *minDist* and *numIters* are user defined thresholds used to stop the iteration.

3.2 Combining Univariate Logistic Regressions in a Stepwise Fashion

In order the combination of logistic regression functions, let us consider a simple example where training cases are described by the continuous predictor variables X_1 and X_2, while the response variable Y assumes values 0 and 1. A logistic model of Y on X_1 and X_2 is built stepwise by combining univariate logistic regressions with the regression nodes on X_1 and X_2, respectively.

We firstly derive parameters β_0 and β_1 of the univariate logistic regression of Y on X_1, such that:

$$\hat{Y} = \begin{cases} 1 & if \;\; \frac{e^{\beta_0+\beta_1 X_1}}{1+e^{\beta_0+\beta_1 X_1}} > th \\ 0 & otherwise \end{cases} \quad (6)$$

β_0 and β_1 are computed according to the Algorithm 1. The fitted logistic model reported in Equation 6 may not predict Y exactly, but the error in predicting Y

Algorithm 1. Newton-Raphson application

$\beta \leftarrow [0,0]^T$; $numIters \leftarrow 0$;

$Q \leftarrow \begin{bmatrix} 1 & 1 & \dots & 1 \\ x_{i1} & x_{i2} & \dots & x_{in} \end{bmatrix}^T$;

$T \leftarrow [\widetilde{y}_1, \widetilde{y}_2, \dots, \widetilde{y}_n]^T$;

repeat

$\beta_{new} \leftarrow \beta$; $numIters++$;

$P \leftarrow [P(C_1|x_{i1},\beta), P(C_1|x_{i2},\beta), \dots, P(C_1|x_{in},\beta)]^T$;

$\widetilde{X} \leftarrow \begin{bmatrix} P(C_1|x_{i1})P(C_0|x_{i1})[1, x_{i1}] \\ P(C_1|x_{i2})P(C_0|x_{i2})[1, x_{i2}] \\ \dots \\ P(C_1|x_{in})P(C_0|x_{in})[1, x_{in}] \end{bmatrix}$;

$\beta_{new} \leftarrow \beta + (Q\widetilde{X})^{-1}Q(T-P)$;

until $||\beta_{new} - \beta||_2 > minDist$ OR $numIters > maxIters$

return β_{new}

may be reduced by adding the new variable X_2. Instead of starting from scratch and building a model with both X_1 and X_2, we exploit the stepwise procedure and now derive the slope and intercept of the straight-line regression to predict X_2 from X_1, that is:

$$\hat{X}_2 = \alpha_{2_0} + \alpha_{2_1} X_1. \tag{7}$$

Then we compute the residuals of both X_2 and Y as reported in (8)-(10).

$$X_2' = X_2 - (\alpha_{2_0} + \alpha_{2_l} X_i) \tag{8}$$

$$Y_{FN}' = \begin{cases} 1 & if \quad Y - \frac{e^{\beta_0+\beta_1 X_1}}{1+e^{\beta_0+\beta_1 X_1}} - th > 0 \\ 0 & otherwise \end{cases} \tag{9}$$

$$Y_{FP}' = \begin{cases} 1 & if \quad \frac{e^{\beta_0+\beta_1 X_1}}{1+e^{\beta_0+\beta_1 X_1}} - Y - th > 0 \\ 0 & otherwise \end{cases} \tag{10}$$

where th is a user defined threshold (default value 0.5) that identifies the "change of class" probability. Y_{FN}' are residuals for False Negative errors and Y_{FP}' are residuals for False Positive errors.

Both Y_{FN}' and Y_{FP}' are heuristically compared in order to choose the residual variable that minimizes the error rate on training cases. We denote this minimum with Y' and now compute the univariate logistic regression between Y' on X_2'. We obtain the logistic regression model:

$$\widehat{Y'} = \begin{cases} 1 & if \quad \frac{e^{\beta_0'+\beta_1' X_2'}}{1+e^{\beta_0'+\beta_1' X_2'}} > th \\ 0 & otherwise \end{cases} \tag{11}$$

and combining residuals, we have that

$$\hat{Y} = \begin{cases} 1 & if \quad \frac{e^{\beta_0+\beta_1 X_1}}{1+e^{\beta_0+\beta_1 X_1}} + s' \frac{e^{\beta_0'+\beta_1' X_2'}}{1+e^{\beta_0'+\beta_1' X_2'}} > th \\ 0 & otherwise \end{cases} \tag{12}$$

where s'=1 if we are performing a logistic regression on Y'_{FN}, s'=-1 if we are performing a logistic regression on Y'_{FP}.

A generalization of this procedure is that adopted by SILoRT to construct the multiple logistic regression model at a node t in a stepwise fashion. Logistic univariate regressions to be combined are those constructed in regression nodes along the path from the root to the node t. Parameters of both logistic and straight-line regressions at each node of the tree are estimated by processing only the training cases falling in the node itself. Intuitively, the stepwise construction of a logistic model tree is motivated by the need of reducing the classification error performed in regression nodes at higher levels of the tree by adding new predictor variables to the model by means of the regression nodes (or leaves) at lower levels of the tree.

4 Evaluation

In order to evaluate performances of SILoRT, we have commented an illustrative example and reported an empirical evaluation.

4.1 An Illustrative Example

Let us consider training data described by two continuous predictor variables X_1 and X_2 and a binary response variable Y, that is:

$X1$	1	2	3	3	3	4	5	6	7	8	20
$X2$	1	2	9	10	11	8	7	6	5	4	3
Y	1	0	0	0	0	0	0	0	0	0	1

SILoRT ($\gamma = 0.9$, $th = 0.5$) chooses the regression node on X_1 as root of the tree (Figure 1). This means that a global effect of X_1 on Y has been observed. The univariate logistic regression on $X1$ is the following:

$$\widehat{Y} = \begin{cases} 1 & if \ \frac{e^{-2.797+0.191*X1}}{1+e^{-2.797+0.191*X1}} > th \\ 0 & otherwise \end{cases}$$

and the linear regression of $X2$ on $X1$ is: $\widehat{X2} = (6.97 - 0.17 * X1)$.

The parameters of the logistic regression function are determined as reported in Algorithm 1. Training data are then transformed in order to remove the effect of performed (logistic) regression and they are passed down to the unique child. In particular, $X2$ is replaced with residual $X2'$ while response variable Y is replaced by a pair of variables including the residuals of False Positive and False Negative errors.

$X2'$	-5.8	-4.6	2.5	3.5	4.5	1.7	0.9	0.1	-0.8	-1.6	-0.5
Y'_{FN}	1	0	0	0	0	0	0	0	0	0	0
Y'_{FP}	0	0	0	0	0	0	0	0	0	0	0

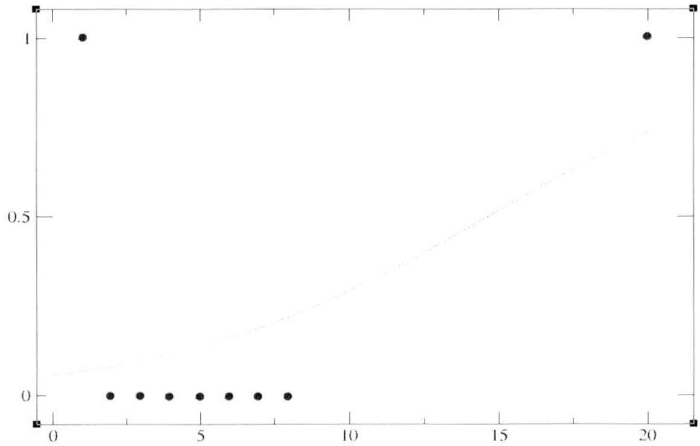

Fig. 1. Regression on $X1$

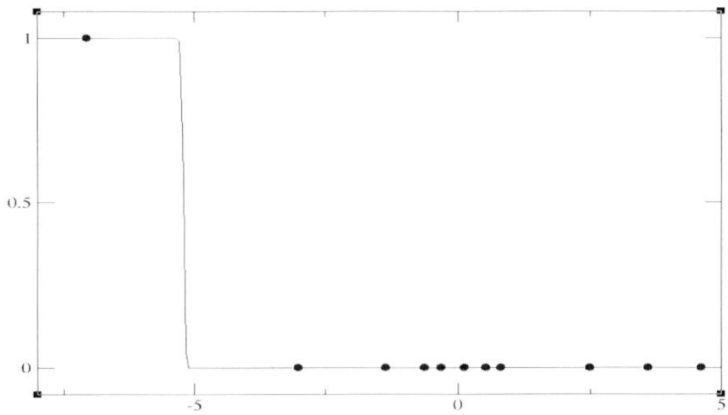

Fig. 2. Regression on $X2'$

SILoRT chooses to compute the logistic regression of Y'_{FN} from $X2'$ and obtains the logistic regression model (Figure 2):

$$\widehat{Y'_{FN}} = \begin{cases} 1 & if \quad \frac{e^{-330.4-63.4*X2'}}{1+e^{-330.4-63.4*X2'}} > th \\ 0 & otherwise \end{cases}$$

This logistic regression correctly predicts all training cases (error rate=0). If SILoRT chose to regress Y'_{FP}, error rate on the training set would be 9%.

Since there are no other continuous variables, SILoRT stops the search and returns a regression tree with two regression nodes (one of which is a leaf). This model is relatively simple and shows the global effect of $X1$ and $X2$ on Y.

4.2 Evaluation on Real World Datasets

SILoRT has been empirically evaluated on six datasets taken from the UCI Machine Learning Repository (http://www.ics.uci.edu/ mlearn/MLRepository.html) Each

Table 1. Datasets used in SILoRT evaluation

Dataset	#Cases	#Attributes	#Cont. Attributes
Breast-cancer-wisconsin	683	10	10
Ionosphere	351	34	34
Iris	100	4	4
Pima-indians-diabetes	768	8	8
Storm	3664	4	3
Wdbc	569	31	31

Table 2. SILoRT evaluation (average accuracy)

Dataset	SILoRT	Baseline
Breast-cancer-wisconsin	93.82	92.22
Ionosphere	85.09	81.76
Iris	91	91
Pima-indians-diabetes	69.02	74.35
Storm	79.29	69.38
Wdbc	93.15	88.74

Table 3. SILoRT: model complexity

Dataset	Avg. #Leaves	Avg. max tree depth
Breast-cancer-wisconsin	13.5	6.5
Ionosphere	12.7	6.56
Iris	4.3	3.76
Pima-indians-diabetes	40.9	12.44
Storm	133.1	35.64
Wdbc	10.5	8.46

data dataset is analyzed by means of a 10-fold cross-validation that is, the dataset was divided into ten *folds* and then, for every fold, SILoRT was trained on the remaining folds and tested on it. Main characteristics of used datasets are reported in Table 1. Results are obtained with the following parameters values $\gamma = 1, th = 0.5$, $minDist = 0.001, numIters = 400$.

As baseline of our experiments, we considered the simple classifier whose classification model is a univariate logistic regression model (see section 3.1). Results reported in Table 2 show a clear advantage of SILoRT with respect to the baseline in most of cases. This result is not so clear in case of few continuous attributes when SILoRt seems to suffer from overfitting problems. This is confirmed by the relative complexity of induced models (see Table 3). On the contrary, the system shows good results in case of discrete attributes. This result was somehow expected since the baseline algorithm, as original approaches for logistic regression, does not consider discrete attributes.

5 Conclusions

In this paper, we propose a new Logistic Model Trres induction method, SILoRT, whose induced model is characterized by two types of nodes: logistic regression nodes and splitting nodes. Main peculiarities of SILoRT are in the capability to solve collinearity problems without additional computational costs and in the capability of facing the problem of modeling phenomena, where some variables have a global effect while others have only a local effect. This permits to obtain models that can be easily understood by humans.

Similarly to many decision tree induction algorithms, SILoRT may generate model trees that overfit training data. Therefore, a future research direction is the a posteriori simplification of model trees with both regression nodes and splitting nodes. For future work, we also intend to evaluate the opportunity of considering discrete variables in regression models, empirically compare SILoRT with other classifiers and extend experiments by considering artificial data.

Acknowledgments

This work is supported in partial fulfillment of the research objectives of "FAR" project "Laboratorio di bioinformatica per la biodiversità molecolare".

References

1. Chan, K.Y., Loh, W.Y.: Lotus: An algorithm for building accurate and comprehensible logistic regression trees. Journal of Computational and Graphical Statistic 13(4), 826–852 (2004)
2. Draper, N.R., Smith, H.: Applied regression analysis. John Wiley & Sons, Chichester (1982)
3. Hosmer, D.W., Lemeshow, S.: Applied Logistic Regression. Wiley, New York (1989)
4. Landwehr, N., Hall, M., Frank, E.: Logistic model trees 95(1-2), 161–205 (2005)
5. le Cessie, S., van Houwelingen, J.: Ridge estimators in logistic regression. Applied Statistics 41(1), 191–201 (1992)
6. Malerba, D., Esposito, F., Ceci, M., Appice, A.: Top down induction of model trees with regression and splitting nodes. IEEE Transactions on Pattern Analysis and Machine Intelligence 26(5), 612–625 (2004)
7. Perlich, C., Provost, F., Simonoff, J.S.: Tree induction vs. logistic regression: a learning-curve analysis. J. Mach. Learn. Res. 4, 211–255 (2003)
8. Venables, W.N., Ripley, B.D.: Modern Applied Statistics with S-PLUS, 3rd edn. Springer, New York (1999)
9. Weisberg, S.: Applied regression analysis. Wiley, New York (1985)
10. Zeileis, A., Hothorn, T., Hornik, K.: A model-based recursive partitioning. Journal of Computational and Graphical Statistics (2008)

Stochastic Propositionalization for Efficient Multi-relational Learning

N. Di Mauro, T.M.A. Basile, S. Ferilli, and F. Esposito

Department of Computer Science, University of Bari, Italy
{ndm,basile,ferilli,esposito}@di.uniba.it

Abstract. The efficiency of multi-relational data mining algorithms, addressing the problem of learning First Order Logic (FOL) theories, strongly depends on the search method used for exploring the hypotheses space and on the coverage test assessing the validity of the learned theory against the training examples. A way of tackling the complexity of this kind of learning systems is to use a propositional method that reformulates a multi-relational learning problem into an attribute-value one. We propose a population based algorithm that using a stochastic propositional method efficiently learns complete FOL definitions.

1 Introduction

The efficacy of induction algorithms has been shown on a wide variety of benchmark domains. However, current machine learning techniques are inadequate for learning in more difficult real-world domains. The nature of the problem can be of different type as noise in the descriptions and lack of data, but also the representation language exploited to describe the examples of the target concept.

It is well known that the choice of the right representation for a learning problem has a significant impact on the performance of the learning systems. In traditional supervised learning problems, the learner is provided with training examples typically represented as vectors of attribute-value. In most complex real-world domains, however, it is necessary to adopt a more powerful representation language, such as a FOL language, a natural extension of the propositional representation, that is able to describe any kind of relation between objects. Inductive Logic Programming (ILP) systems are able to learn hypotheses expressed with this more powerful language. However, this representation language allows a potentially great number of mappings between descriptions (relational learning), differently from the feature vector representation in which only one mapping is possible (propositional learning). The obvious consequence of such a representation is that both the hypotheses space to search and the coverage test to assess the validity of the learned hypotheses against positive (all covered) and negative (all rejected) examples result more costly than in the case of propositional one.

A possible solution is a reformulation of the original multi-relational learning problem in a propositional one [4,3,2,5]. During the reformulation, a fixed set of structural features is built from relational background knowledge and the structural properties of the individuals occurring in the examples. In such a process,

A. An et al. (Eds.): ISMIS 2008, LNAI 4994, pp. 78–83, 2008.

each feature is defined in terms of a corresponding program clause whose body is made up of a set of literals derived from the relational background knowledge. When the clause defining the feature is called for a particular individual (i.e., if its argument is bound to some example identifier) and this call succeeds at least once, the corresponding boolean feature is defined to be true.

Alternative approaches avoid the reformulation process and apply propositionalization directly on the original FOL context: the relational examples are *flattened* by substituting them with all (or a subset of) their matchings with a pattern. Following this idea, [7,9,1] proposed a multi-instance propositionalization in which each relational example is reformulated in its multiple matchings with a pattern. After that, to each initial observation correspond many feature vectors and the search for hypotheses may be recasted in this propositional representation as the search for rules that cover at least one instance per positive observation and no instance of negative observations.

This work proposes a propositionalization technique in which the transposition of the relational data is performed by an *online* flattening of the examples. The proposed method is a population based algorithm that stochastically propositionalizes the training examples in which the learning phase may be viewed as a bottom-up search in the hypotheses space. The method is based on a stochastic reformulation of examples that, differently from other proposed propositionalization techniques, does not use the classical subsumption relation used in ILP. For instance, in PROPAL [1], each example E, described in FOL, is reformulated into a set of matchings of a propositional pattern P with E by using the classical θ-subsumption procedure, being in this way still bond to FOL context. On the contrary, in our approach the reformulation is based on an rewriting of the training examples on a fixed set of domain constants.

2 The Proposed Technique

In this section the underlying idea of the propositionalization technique along with its implementation in the Sprol system will be given.

Let e an example of the training set, represented as a Datalog[1] ground clause, and let $consts(e)$ the set of the constants appearing in e. One can write a new example e' from e by changing one or more constants in e, i.e. by renaming.

In particular, e' may be obtained by applying an antisubstitution[2] and a substitution[3] under Object Identity to e, $e' = e\sigma^{-1}\theta_{OI}$. In the Object Identity framework, within a clause, terms that are denoted with different symbols must be distinct, i.e. they must represent different objects of the domain.

Definition 1 (Renaming of an example). *A ground renaming of an example E, is obtained by applying a substitution $\theta = \{V_1/t_1, V_2/t_2, \dots V_n/t_n\}$ to $E\sigma^{-1}$,*

[1] Horn clause language without function symbols with arity greater than 0.

[2] An antisubstitution is a mapping from terms into variables.

[3] A substitution θ is a finite set of the form $\{V_1/t_1, V_2/t_2, \dots V_n/t_n\}$ ($n \geq 0$) where in each *binding* V_i/t_i the V_i is a variable ($V_i \neq V_j$) and each t_i is a term. θ is a *ground substitution* when each t_i is a ground term.

such that $\{V_1, V_2, \dots V_n\} \subseteq vars(E\sigma^{-1})$, $\{t_1, t_2, \dots t_n\}$ *are distinct constants of* $consts(E)$, *and* σ^{-1} *is an antisubstitution.*

In this way, we do not need to use the θ-subsumption test to compute the renamings of an example E, we just have to rewrite it considering the permutations of the constants in $consts(E)$.

2.1 Generalizing Examples

In the general framework of ILP the generalization of clauses is based on the concept of *least general generalization* originally introduced by Plotkin. Given two clauses C_1 and C_2, C_1 generalizes C_2 (denoted by $C_1 \leq C_2$) if C_1 subsumes C_2, i.e. there exists a substitution θ such that $C_1\theta \subseteq C_2$.

In our propositionalization framework, a generalization C (a non-ground clause) of two positive examples E_1 and E_2 may be calculated by turning constants into variables in the intersection between a renaming of E_1 and a renaming of E_2. In order to obtain consistent intersections, it is important to note that all the renamings, for both E_1 and E_2, must be calculated on the same fixed set of constants. Hence, given $E_1, E_2, \dots, E_n$ examples, the set C of the constants useful to build the renamings may be chosen equal to $C = \arg\max_{E_i}(|consts(E_i)|)$.

Furthermore, to avoid empty generalizations, the constants appearing in the head literal of the renamings must be taken fixed. More formally, let $ren(E, C)$ a generic renaming of an example E onto the set of constants C, a generalization G such that subsumes both E_1 and E_2 is $(ren(E_1, C) \cap ren(E_2, C))\sigma^{-1}$.

Example 1. Given two positive examples $E_1 : h(a) \leftarrow q(a, b),\ c(b),\ t(b, c),\ p(c, d)$ and $E_2 : h(d) \leftarrow q(d, e),\ c(d),\ t(e, f)$, let $C = \arg\max_{E_i}(|consts(E_i)|) = consts(E_1) = \{a, b, c, d\}$. A generalization G of E_1 and E_2 is $G = (ren(E_1, C) \cap ren(E_2, C))\sigma^{-1} = (\{h(a), \neg q(a, b), \neg c(b), \neg t(b, c), \neg p(c, d)\} \cap \{h(a), \neg q(a, b), \neg c(a), \neg t(b, c)\})\sigma^{-1} = \{h(a), \neg q(a, b), \neg t(b, c)\}\sigma^{-1} = (h(a) \leftarrow q(a, b), t(b, c))\sigma^{-1} = h(X) \leftarrow q(X, Y), t(Y, Z)$ with $\sigma^{-1} = \{a/X, b/Y, c/Z\}$.

2.2 Covering Examples

In the classical ILP setting, generalizations are evaluated on the training examples using the θ-subsumption as a covering procedure. Here, the covering test is based on a syntactic lazy matching more efficient than the θ-subsumption.

Given a generalization G and an example E, it is possible to prove that G subsumes E under OI iff exists a permutation $P(n, r) = (c_1, c_2, \dots, c_r)$ of size r of the constants $consts(E)$, with $r = |vars(G)|$, $n = |consts(E)|$ and $r \leq n$, such that $G\theta \cap E = G$ with $\theta = \{V_1/c_1, V_2/c_2, \dots, V_r/c_r\}$, $V_i \in vars(G), V_i \neq V_j$. In order to be complete, the procedure must prove the test $G\theta \cap E = G$ for all the permutations $P(n, r)$. However, we can make the test stochastic by randomly choosing a number α of all the possible permutations.

Algorithm 1. Sprol

Input: $\underline{E^+}$: pos exs; $\underline{E^-}$: neg exs; $\underline{\alpha}$: the parameter for neg coverage; $\underline{\beta}$: the parameter for pos coverage; $\underline{k}$: the dimension of the population; $\underline{r}$: number of restarts;
Output: the hypotheses h

$C = \arg\max_{E_i \in E^+}(|consts(E_i)|)$;
while $E^+ \neq \emptyset$ **do**
  select a seed e from E^+
  Population $\leftarrow ren(k, e, C)$; /* *select k renamings of e* */
  PopPrec $\leftarrow$ Population; $i \leftarrow 0$;
  while $i < r$ **do**
    P $\leftarrow \emptyset$;
    for each element $v \in$ Population **do**
      for each positive example $e^+ \in E^+$ **do**
        $V_{e^+} \leftarrow ren(t, e^+, C)$; /* *select t renamings of e^+* */
        P $\leftarrow$ P $\cup \{u | u = v \cap w_i, w_i \in V_{e^+}\}$; /* *generalization* */
    Population $\leftarrow$ P;
    /* *Consistency check* */
    for each negative example $e^- \in E^-$ **do**
      $V_{e^-} \leftarrow ren(\alpha, e^-, C)$; /* *select α renamings of e^-* */
      for each element $v \in$ Population **do**
        if v covers an element of V_{e^-} **then** remove v from Population
    /* *Completeness check* */
    for each element $v \in$ Population **do** completeness$_v \leftarrow 0$;
    for each positive example $e^+ \in E^+$ **do**
      $V_{e^+} \leftarrow ren(\beta, e^+, C)$; /* *select β renamings of e^+* */
      for each element $v \in$ Population **do**
        if $\exists u \in V_{e^+}$ s.t. $u \cap v = v$ **then** completeness$_v \leftarrow$ completeness$_v + 1$;
    $i \leftarrow i + 1$;
    if $|\text{Population}| = 0$ **then**
      Population $\leftarrow$ PopPrec; /* *restart with the previous population* */
    else
      leave in Population the best k generalizations only; PopPrec $\leftarrow$ Population;
  add the best element $b \in$ Population to h;
  remove from E^+ the positive examples covered by b

To reduce the set of possible permutations we can fix the associations for the variables in the head of the generalization G. In particular if

$$G : h(V_1, V_2, \ldots, V_d) \leftarrow \ldots \text{ and } E : h(c_1, c_2, \ldots, c_d) \leftarrow \ldots$$

then we can fix the associations $\{V_1/c_1, V_2/c_2, \ldots, V_d/c_d\}$, $d \leq r, n$ in all the generated permutations. Furthermore, we can ulteriorly reduce the set of permutations by taking into account the positions of the costants in the literals. Supposing $p(V_1, V_2, \ldots, V_k)$ be a literal of the generalization G. Then, all the constants that may be associated to V_i, $1 \leq i \leq k$, are all those appearing in position i in the literals p/k of the example E.

2.3 The Algorithm

Algorithm 1 reports the sketch of the Sprol system, implemented in Yap Prolog 5.1.1, that incorporates ideas of the propositional framework we proposed. Sprol is a population based algorithm where several individual candidate solutions are simultaneously maintained using a constant size population. The population of candidate solutions provides a straightforward means for achieving search diversification and hence for increasing the exploration capabilities of the search process. In our case, the population is made up of candidate generalizations over the training positive examples. In many cases, local minima are quite common in search algorithms and the corresponding candidate solutions are typically not of sufficiently high quality. The strategy we used to escape from local minima is a *restart strategy* that simply reinitializes the search process whenever a local minimum is encountered.

Sprol takes as input the set of positive and negative examples of the training set and some user-defined parameters characterizing its stochastic behavior. In particular, α and β represent the number of renamings of a negative, respectively positive, example to use for the covering test; k is the size of the population; and, r is the number of restarts.

As reported in Algorithm 1, Sprol tries to find a set of clauses that cover all the positive examples and no negative one, by using an iterative population based covering mechanism. It sets the initial population made up of k randomly chosen renamings of a positive example (lines 3-4). Then, the elements of the population are iteratively generalized on the positive examples of the training set (lines 9-11). All the generalizations that cover at least one negative example are taken out (lines 14-17), and the quality of each generalization, based on the number of covered positive examples, is calculated (lines 18-24). Finally, best k generalizations are considered for the next iteration (line 28). In case of an empty population a restart is generated with the previous population (line 26).

Renamings of an example are generate randomly choosing a number of k renamings of the example E onto the set of its constants C.

It is important to note that our approach constructs hypotheses that are only approximately consistent. Indeed, in the consistency check it is possible that there exists a matching between an hypothesis and a negative example. The number α of allowed permutations is responsible of the induction cost as well as the consistency of the produced hypotheses. An obvious consequence is that the more permutations allowed, the more consistent the hypotheses found and, perhaps, the more learning time.

3 Discussion and Conclusion

In order to evaluate the system Sprol, we performed experiments on the classical mutagenesis dataset [8],consisting of structural descriptions of molecules to be classified into mutagenic and non-mutagenic ones, for which we considered the *atom bond* descriptions, the *indicator* variables, *logp* and *lumo*. The size k of the population has been set to 50, at the same way of α and β, and making 5 restarts.

As measures of performance, we use predictive accuracy and execution time. Results have been compared to that obtained by Progol [6]. The experiments were performed exploiting a 10-fold cross-validation. The Sprol results, averaged over the 10-folds, show an improvement of the execution time with respect to Progol (56.35 sec. of Sprol vs 546.25 sec. of Prolog) obtaining a good predictive accuracy of the learned theory (80.4% of Sprol vs 79.81% of Prolog).

As a concluding remark, the proposed population based algorithm is able to efficiently solve multi-relational problems by using a stochastic propositional method. The result of an empirical evaluation on the mutagenesis dataset of the proposed technique is very promising and proves the validity of the method.

It is important to note that α and β parameters used in the algorithm for checking consistency and completeness lead to different behaviors of the induction process. In particular, in complex domains with a lot of failure derivations between hypotheses and negative examples, low values for α may lead to inconsistent hypotheses. On the other way, low values for β may produce a theory with many hypotheses. An extensive study of this problem is needed and it represents an important future work. Furthermore, we plan to automatically discover, in an online manner, the correct input parameters of Sprol for a given learning task.

Finally, more experiments, and comparisons with other classical ILP systems, are needed to better evaluate the methodology, especially using synthetic data well suited for a parameter setting study.

References

1. Alphonse, E., Rouveirol, C.: Lazy propositionalization for relational learning. In: Horn, W. (ed.) Proceedings of ECAI 2000, pp. 256–260. IOS Press, Amsterdam (2000)
2. Krogel, M.-A., Rawles, S., Zelezny, F., Flach, P., Lavrac, N., Wrobel, S.: Comparative evaluation of approaches to propositionalization. In: Horváth, T., Yamamoto, A. (eds.) ILP 2003. LNCS (LNAI), vol. 2835, pp. 194–217. Springer, Heidelberg (2003)
3. Lavrac, N., Dzeroski, S.: Inductive Logic Programming: Techniques and Applications. Ellis Horwood, Chichester (1994)
4. Lavrac, N., Dzeroski, S., Grobelnik, M.: Learning nonrecursive definitions of relations with LINUS. In: Kodratoff, Y. (ed.) EWSL 1991. LNCS, vol. 482, pp. 265–281. Springer, Heidelberg (1991)
5. Lavrač, N., Železný, F., Flach, P.A.: RSD: Relational subgroup discovery through first-order feature construction. In: Matwin, S., Sammut, C. (eds.) ILP 2002. LNCS (LNAI), vol. 2583, pp. 149–165. Springer, Heidelberg (2003)
6. Muggleton, S.: Inverse Entailment and Progol. New Generation Computing, Special issue on Inductive Logic Programming 13(3-4), 245–286 (1995)
7. Sebag, M., Rouveirol, C.: Tractable induction and classification in first order logic via stochastic matching. In: Proceedings of IJCAI 1997, pp. 888–893 (1997)
8. Srinivasan, A., Muggleton, S., King, R.D.: Comparing the use of background knowledge by inductive logic programming systems. In: De Raedt, L. (ed.) Proceedings of ILP 1995, pp. 199–230. Springer, Heidelberg (1995)
9. Zucker, J.-D., Ganascia, J.-G.: Representation changes for efficient learning in structural domains. In: Proceedings of ECML 1996, pp. 543–551 (1996)

Analyzing Behavior of Objective Rule Evaluation Indices Based on Pearson Product-Moment Correlation Coefficient

Hidenao Abe and Shusaku Tsumoto

Shimane University
89-1 Enya-cho Izumo Shimane, 6938501, Japan
abe@med.shimane-u.ac.jp, tsumoto@computer.org

Abstract. In this paper, we present an analysis of behavior of objective rule evaluation indices on classification rule sets using Pearson product-moment correlation coefficients between each index. To support data mining post-processing, which is one of important procedures in a data mining process, at least 40 indices are proposed to find out valuable knowledge. However, their behavior have never been clearly articulated. Therefore, we carried out a correlation analysis between each objective rule evaluation indices. In this analysis, we calculated average values of each index using bootstrap method on 32 classification rule sets learned with information gain ratio. Then, we found the following relationships based on the correlation coefficient values: similar pairs, discrepant pairs, and independent indices. With regarding to this result, we discuss about relative functional relationships between each group of objective indices.

1 Introduction

In recent years, enormous amounts of data have been stored on information systems in natural science, social science, and business domains. People have been able to obtain valuable knowledge due to the development of information technology. Besides, data mining techniques combine different kinds of technologies such as database technologies, statistical methods, and machine learning methods. Then, data mining has been well known for utilizing data stored on database systems. In particular, if-then rules, which are produced by rule induction algorithms, are considered as one of the highly usable and readable outputs of data mining. However, to large datasets with hundreds of attributes including noise, the process often obtains many thousands of rules. From such a large rule set, it is difficult for human experts to find out valuable knowledge, which are rarely included in the rule set.

To support such a rule selection, many studies have done using objective rule evaluation indices such as recall, precision, and other interestingness measurements [1,2,3] (Hereafter, we refer to these indices as "objective indices"). Although their properties are identified with their definitions, their behavior on rule sets are not investigated with any promising method.

A. An et al. (Eds.): ISMIS 2008, LNAI 4994, pp. 84–89, 2008.

With regard to the above-mentioned issues, we present a correlation analysis method to identify the functional properties of objective indices in Section 2. Then, with the 39 objective indices and classification rule sets from 32 UCI datasets, we identified the following relationships based on the correlation analysis method: similar pairs of indices, contradict pairs of indices, and independent indices. Based on the result in Section 3, we discuss about these relationships and differences between functional properties and original definitions.

2 Correlation Analysis for the Objective Rule Evaluation Indices

In this section, we describe a correlation analysis method to identify behavior of objective indices. To analyze functional relationships between objective indices, we should gather the following materials: values of objective indices of each classification rule set learned from each dataset, and correlation coefficients between objective indices with the values. The process of the analysis is shown in Figure 1.

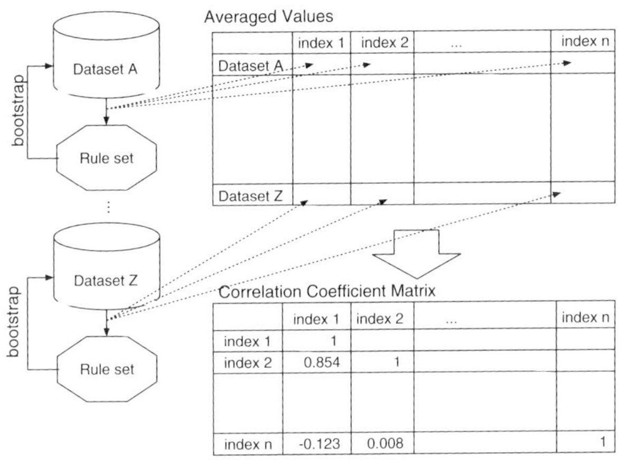

Fig. 1. An overview of the correlation analysis method

First, we obtain multiple rule sets from some datasets to get values of objective indices. When gathering these values, we should care the statistical correctness of each value. Therefore, the values are averaged adequately large number (> 100) of values from bootstrap samples.

Then, Pearson product-moment correlation coefficients r between indices, x and y, are calculated for n datasets.

$$r = \frac{\frac{1}{n}\sum_{i=1}^{n}(x_i - \overline{x})(y_i - \overline{y})}{\sqrt{\frac{1}{n}\sum_{i=1}^{n}(x_i - \overline{x})^2}\sqrt{\frac{1}{n}\sum_{i=1}^{n}(y_i - \overline{y})^2}}$$

With these coefficient values, we identified similar pairs, contradict pairs, and independent indices.

3 Analyzing the Objective Rule Evaluation Indices on UCI Datasets

In this section, we describe the correlation analysis of the 39 objective indices with 32 UCI datasets. Table 1 shows the 39 objective indices investigated and reformulated for classification rules by Ohsaki et al. [4].

Table 1. Objective rule evaluation indices for classification rules used in this research. **P:** Probability of the antecedent and/or consequent of a rule. **S:** Statistical variable based on P. **I:** Information of the antecedent and/or consequent of a rule. **N:** Number of instances included in the antecedent and/or consequent of a rule. **D:** Distance of a rule from the others based on rule attributes.

Theory	Index Name (**Abbreviation**) [Reference Number of Literature]
P	Coverage (**Coverage**), Prevalence (**Prevalence**) Precision (**Precision**), Recall (**Recall**) Support (**Support**), Specificity (**Specificity**) Accuracy (**Accuracy**), Lift (**Lift**) Leverage (**Leverage**), Added Value (**Added Value**)[2] Klösgen's Interestingness (**KI**)[5], Relative Risk (**RR**)[6] Brin's Interest (**BI**)[7], Brin's Conviction (**BC**)[7] Certainty Factor (**CF**)[2], Jaccard Coefficient (**Jaccard**)[2] F-Measure (**F-M**)[8], Odds Ratio (**OR**)[2] Yule's Q (**YuleQ**)[2], Yule's Y (**YuleY**)[2] Kappa (**Kappa**)[2], Collective Strength (**CST**)[2] Gray and Orlowska's Interestingness weighting Dependency (**GOI**)[9] Gini Gain (**Gini**)[2], Credibility (**Credibility**)[10]
S	χ^2 Measure for One Quadrant (χ^2**-M1**)[11] χ^2 Measure for Four Quadrant (χ^2**-M4**)[11]
I	J-Measure (**J-M**)[12], K-Measure (**K-M**)[13] Mutual Information (**MI**)[2] Yao and Liu's Interestingness 1 based on one-way support (**YLI1**)[3] Yao and Liu's Interestingness 2 based on two-way support (**YLI2**)[3] Yao and Zhong's Interestingness (**YZI**)[3]
N	Cosine Similarity (**CSI**)[2], Laplace Correction (**LC**)[2] ϕ Coefficient (ϕ)[2], Piatetsky-Shapiro's Interestingness (**PSI**)[14]
D	Gago and Bento's Interestingness (**GBI**)[15] Peculiarity (**Peculiarity**)[16]

As for datasets, we have taken the 32 datasets from UCI machine learning repository[17], which are distributed with Weka [18].

For the above datasets, we obtained rule sets with PART [19] implemented in Weka. PART constructs a rule set based on information gain ratio. This means the obtained rule sets are biased with the correctness of classification.

3.1 Constructing a Correlation Coefficient Matrix of the 39 Objective Indices

For the 32 datasets, we obtained the rule sets using PART. This procedure is repeated 1000 times with bootstrap re-sampling for each dataset. As a representative value for each bootstrap iteration, the average for a rule set has been calculated. Then, we averaged the average values from 1000 times iterations.

With the average values for each dataset, we calculated correlation coefficients between each objective index.

3.2 Identifying Characteristic Relationships between Objective Indices Based on Correlation Coefficient Matrix Analysis

Based on the correlation coefficients, we identify characteristic relationship between each objective index. We defined the three characteristic relation ship as follows:

- Similar pair: two indices has strong positive correlation $r > 0.8$.
- Discrepant pair: two indices has strong negative correlation $r < -0.8$.
- Independent index: a index has only weak correlations $-0.8 \leq r \leq 0.8$ for the other indices.

Figure 2 shows similar and discrepant pairs of objective indices on the correlation analysis. There are several groups having mutual correlations. The largest group, which has correlation to Cosine Similarity and F-Measure, includes 23 indices. Relative Risk and Odds Ratio make another group. χ^2-M1, χ^2-M4 and PSI also make different functional group. These pairs indicate distinct functional property for the rule sets.

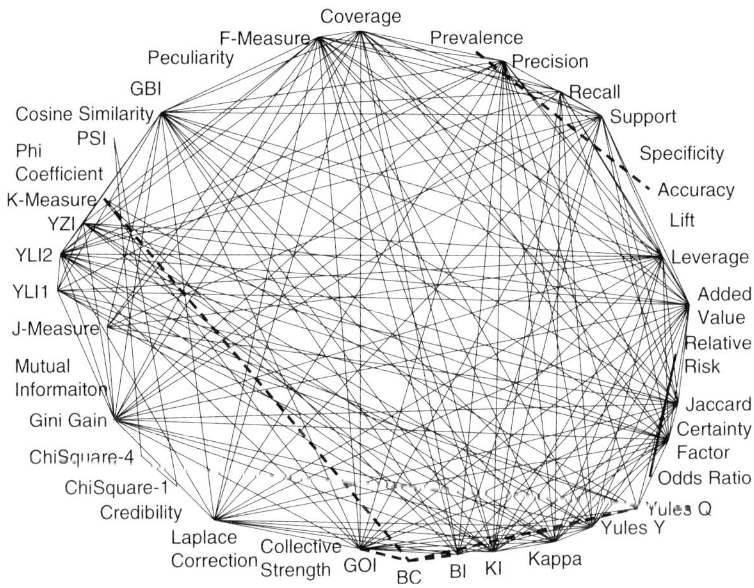

Fig. 2. Similar pairs (with solid line) and discrepant (with dotted line) of objective indices on the correlation analysis

As for discrepant pairs, there are smaller number of groups. Accuracy and Prevalence has discrepant property each other. Likewise, BI and BC also indicate discrepant property because of negative correlation between them. BC shows discrepant property to several indices, which belong to the biggest group of similar pairs.

4 Discussion

With regarding to Figure 2, we can say that the following objective indices indicate similar property: Coverage, Precision, Recall, Support, Leverage, Added Value, Jaccard, Certainty Factor, YulesQ, YulesY, Kappa, KI, BI, GOI, Laplace Correction, Gini Gain, J-Measure, YLI1, YLI2, YZI, K-Measure, Cosine Similarity, and F-Measure. The other groups also show similar functional property to the classification rule sets based on information gain ratio. Considering their definitions, although they have different theoretical backgrounds, their functional property is to represent correctness of rules. This indicates that these indices evaluate given rules optimistically.

On the other hand, BC indicates opposite functional property comparing with the former indices. Therefore, the result indicates that BC evaluate given rules not so optimistically. As for Accuracy and Prevalence, although Accuracy measures ratio of both of true positive and false negative for each rule, Prevalence only measures ratio of mentioned class value of each rule. It is reasonable to indicate discrepant property, because accurate rules have high Accuracy values irrespective of their mentioned class value.

As for the independent indices, GBI and Peculiarity suggested with the different theoretical background comparing with the other indices. Therefore, what they have different functional properties is reasonable. However, Corrective Strength, Credibility, Mutual Information and ϕ Coefficient indicate the different functional property comparing with the other indices which have the same theoretical backgrounds (**P**,**S** and **N**). These indices evaluate given rules from each different viewpoint.

5 Conclusion

In this paper, we described the method to analyze functional properties of objective rule evaluation indices.

We investigated functional properties of objective indices with 32 UCI datasets and their rule sets as an actual example. With regarding to the result, several groups are found as functional similarity groups in cross-sectional manner for their theoretical backgrounds.

In the future, we will investigate functional properties of objective indices to other kind of rule sets obtained from the other rule mining algorithms. At the same time, we will investigate not only Pearson product-moment correlation coefficient but also rank correlation coefficients and other correlations.

References

1. Hilderman, R.J., Hamilton, H.J.: Knowledge Discovery and Measure of Interest. Kluwer Academic Publishers, Dordrecht (2001)
2. Tan, P.N., Kumar, V., Srivastava, J.: Selecting the right interestingness measure for association patterns. In: Proceedings of International Conference on Knowledge Discovery and Data Mining KDD 2002, pp. 32–41 (2002)

3. Yao, Y.Y., Zhong, N.: An analysis of quantitative measures associated with rules. In: Zhong, N., Zhou, L. (eds.) PAKDD 1999. LNCS (LNAI), vol. 1574, pp. 479–488. Springer, Heidelberg (1999)
4. Ohsaki, M., Abe, H., Yokoi, H., Tsumoto, S., Yamaguchi, T.: Evaluation of rule interestingness measures in medical knowledge discovery in databases. Artificial Intelligence in Medicine 41(3), 177–196 (2007)
5. Fayyad, U.M., Piatetsky-Shapiro, G., Smyth, P., Uthurusamy, R. (eds.): Explora: A Multipattern and Multistrategy Discovery Assistant. Advances in Knowledge Discovery and Data Mining, pp. 249–271. AAAI/MIT Press, California (1996)
6. Ali, K., Manganaris, S., Srikant, R.: Partial classification using association rules. In: Proceedings of the International Conference on Knowledge Discovery and Data Mining KDD 1997, pp. 115–118 (1997)
7. Brin, S., Motwani, R., Ullman, J., Tsur, S.: Dynamic itemset counting and implication rules for market basket data. In: Proceedings of the ACM SIGMOD International Conference on Management of Data, pp. 255–264 (1997)
8. Rijsbergen, C.: Information retrieval, ch. 7 (1979), `http://www.dcs.gla.ac.uk/Keith/Chapter.7/Ch.7.html`
9. Gray, B., Orlowska, M.E.: CCAIIA: Clustering categorical attributes into interesting association rules. In: Wu, X., Kotagiri, R., Korb, K.B. (eds.) PAKDD 1998. LNCS, vol. 1394, pp. 132–143. Springer, Heidelberg (1998)
10. Hamilton, H.J., Shan, N., Ziarko, W.: Machine learning of credible classifications. In: Australian Conf. on Artificial Intelligence AI 1997, pp. 330–339 (1997)
11. Goodman, L.A., Kruskal, W.H.: Measures of association for cross classifications. Springer Series in Statistics, vol. 1. Springer, Heidelberg (1979)
12. Smyth, P., Goodman, R.M.: Rule induction using information theory. In: Piatetsky-Shapiro, G., Frawley, W.J. (eds.) Knowledge Discovery in Databases, pp. 159–176. AAAI/MIT Press, Cambridge (1991)
13. Ohsaki, M., Kitaguchi, S., Kume, S., Yokoi, H., Yamaguchi, T.: Evaluation of rule interestingness measures with a clinical dataset on hepatitis. In: Boulicaut, J.-F., Esposito, F., Giannotti, F., Pedreschi, D. (eds.) PKDD 2004. LNCS (LNAI), vol. 3202, pp. 362–373. Springer, Heidelberg (2004)
14. Piatetsky-Shapiro, G.: Discovery, analysis and presentation of strong rules. In: Piatetsky-Shapiro, G., Frawley, W.J. (eds.) Knowledge Discovery in Databases, pp. 229–248. AAAI/MIT Press (1991)
15. Gago, P., Bento, C.: A metric for selection of the most promising rules. In: European Conference on the Principles of Data Mining and Knowledge Discovery PKDD 1998, pp. 19–27 (1998)
16. Zhong, N., Yao, Y.Y., Ohshima, M.: Peculiarity oriented multi-database mining. IEEE Transactions on Knowledge and Data Engineering 15(4), 952–960 (2003)
17. Hettich, S., Blake, C.L., Merz, C.J.: UCI repository of machine learning databases. University of California, Department of Information and Computer Science, Irvine, CA (1998), `http://www.ics.uci.edu/~mlearn/MLRepository.html`
18. Witten, I.H., Frank, E.: Data Mining: Practical Machine Learning Tools and Techniques with Java Implementations. Morgan Kaufmann, San Francisco (2000)
19. Frank, E., Witten, I.H.: Generating accurate rule sets without global optimization. In: The Fifteenth International Conference on Machine Learning, pp. 144–151 (1998)

Obtaining Low-Arity Discretizations from Online Data Streams

Tapio Elomaa, Petri Lehtinen, and Matti Saarela

Department of Software Systems, Tampere University of Technology
P. O. Box 553 (Korkeakoulunkatu 1), FI-33101 Tampere, Finland
{tapio.elomaa,petri.lehtinen,matti.saarela}@tut.fi

Abstract. Cut point analysis for discretization of numerical attributes has shown, for many commonly-used attribute evaluation functions, that adjacent value range intervals with an equal relative class distribution may be merged together without risking to find the optimal partition of the range. A natural idea is to relax this requirement and rely on a statistical test to decide whether the intervals are probably generated from the same distribution. ChiMerge is a classical algorithm for numerical interval processing operating just in this manner.

ChiMerge handles the interval mergings in the order of their statistical probability. However, in online processing of the data the required $n \log n$ time is too much. In this paper we propose to do the mergings during a left-to-right scan of the intervals. Thus, we reduce the time requirement of merging down to more reasonable linear time. Such linear time operations are not necessary in connection of every example. Our empirical evaluation shows that intervals get effectively combined, their growth rate remains very moderate even when the number of examples grows excessive, and that the substantial reduction of interval numbers can even benefit prediction accuracy.

1 Introduction

Continuous-valued (numerical) attributes need to be discretized somehow in, e.g., decision tree learning. The number of techniques proposed for discretization is overwhelming and many of the available algorithms work well in practice [1,2]. Probably the best-known discretization approaches are unsupervised equal-width and equal-frequency binning, which overlook the class labels provided for examples. A well-known robust supervised discretization algorithm — taking class labels into account — is Kerber's [3] ChiMerge. There are also several subsequent improvements of it [4,5,6,7].

For example standard decision tree learning is an *univariate* approach in which the attributes are handled separately; one at a time without regard to the other attributes. In other words, only the value of the attribute under consideration and the class label are examined. Examples are first sorted by the value of the numerical attribute in question. This yields a sequence of initial intervals. ChiMerge algorithms then repeatedly merge together two adjacent intervals if a

A. An et al. (Eds.): ISMIS 2008, LNAI 4994, pp. 90–99, 2008.

statistical test (χ^2 in the original algorithm [3]) implies that their relative class distribution (RCD) for the instances is the same. Equal RCD indicates that the numerical attribute at hand does not affect the class label of instances (at this point). Therefore, we have no reason to divide the value range in the cut point candidate being examined.

Another line of research is *cut point analysis* [8,9,10,11,12]. In it one analytically determines at which locations in the example sequence do the split thresholds fall for a given attribute evaluation function. Prior work has shown that when applying many of the commonly-used attribute evaluation functions one can significantly prune the potential cut points of a continuous-valued attribute without losing the possibility to come up with the optimal partition. We have recently extended the applicability of cut point analysis to the *data stream* setting [13]. In particular, we showed that the sufficient statistics for optimal multi-way splits can be maintained online so that only constant time per example is required for processing. The shortcoming of our proposal is that it only applies to the case in which no bound on the number of intervals (arity) is enforced for the resulting discretization. Hence, the arity can remain quite high for some attributes, which may hurt the interpretability of the classifier. Our goal in this work is to extend the approach and be able to come up with low-arity partitions.

On one hand one can see that cut point analysis leads to actually fitting the hypothesis too closely to the training data present, because exactly the same RCD is required from two adjacent intervals before they can be merged. This requirement, obviously, is too tight in practice. When the number of intervals in the resulting partitioning is restricted, the fitting to training data naturally loosens, since intervals with different RCDs necessarily need to be merged. Actually, we are interested in whether the two intervals come from the same underlying distribution in which case they still may slightly vary in their observed RCD.

We will propose and evaluate an approach to online discretization that does not overfit in this one aspect of learning. The main difference to cut point analysis is that we combine adjacent intervals together if a statistical test indicates that their RCDs are realizations of the same underlying distribution. Thus, we end up proposing an approach that resembles ChiMerge. For efficiency reasons, however, we need to evaluate the combination candidates during a left-to-right scan. Our empirical evaluation demonstrates that typically some 90% of the potential cut point candidates get discarded. The approach copes well with increasing data stream size. Our initial evaluation also demonstrates that the mergings do not reduce classification accuracy; on the contrary, the accuracy may even increase.

The remainder of this paper is organized as follows. In Section 2 we briefly recapitulate cut point analysis for optimal splits in numerical attribute ranges. Section 3, then, discusses the possibility of applying cut point analysis to obtain low-arity partitions from data streams. We propose to apply χ^2 merging during a left-to-right scan over (a particular grouping of) the data. In Section 4 basics of the statistical test are recapitulated and possibilities for its efficient computation online are charted. Section 5 presents an empirical evaluation of the proposed approach. Finally, Section 6 gives the concluding remarks of this paper.

2 Cut Point Analysis

Recursive binary splitting of the value range of a numerical attribute is the most common supervised discretization technique for a continuous range. It has obvious shortcomings that could potentially be avoided if one-shot multi-way splits were used instead [14,15]. On the other hand, multi-way splitting of a numerical value range can be computationally expensive. Hence, the most popular approaches are based on heuristic multi-way splitting through successive binary splits [14,15]. Such partitions cannot, though, be guaranteed to be optimal with respect to the attribute evaluation function being used.

Without loss of generality, let us consider only one real-valued attribute X. A training sample $S = \{ (x_1, y_1), \ldots, (x_n, y_n) \}$ consists of n labeled examples. For each $(x, y) \in S$, $x \in \mathbb{R}$ and y is the label of x from the set of classes $C = \{ c_1, \ldots, c_m \}$. A k-interval discretization of the sample is generated by picking $k - 1$ *interval thresholds* or *cut points* $T_1 < T_2 < \cdots < T_{k-1}$. Empty intervals are not allowed. The set of $k - 1$ thresholds defines a partition $\biguplus_{i=1}^{k} S_i$ of the set S as follows:

$$S_i = \begin{cases} \{ (x, y) \in S \mid x \le T_1 \} & \text{if } i = 1, \\ \{ (x, y) \in S \mid T_{i-1} < x \le T_i \} & \text{if } 1 < i < k, \\ \{ (x, y) \in S \mid x > T_{k-1} \} & \text{if } i = k. \end{cases}$$

The *global* discretization problem is to find a partition $\biguplus_{i=1}^{k} S_i$ that has the minimum attribute evaluation function value over all partitions of S. The maximum number of intervals k may also be given as a parameter. Then the problem is to find the optimal partition among those that have at most k intervals. This is called *bounded-arity* discretization.

If one could make its own partition interval out of each data point in the data set, this discretization would have zero training error. However, one cannot — nor wants to — discern between all data points. Only those that differ in their value of X can be separated from each other. Consider, e.g., the data set shown in Fig. 1. There are 27 integer-valued examples. They are instances of two classes; α and β. Interval thresholds can only be set in between those points where the attribute value changes. Therefore, one can process the data into *bins*. There is one bin for

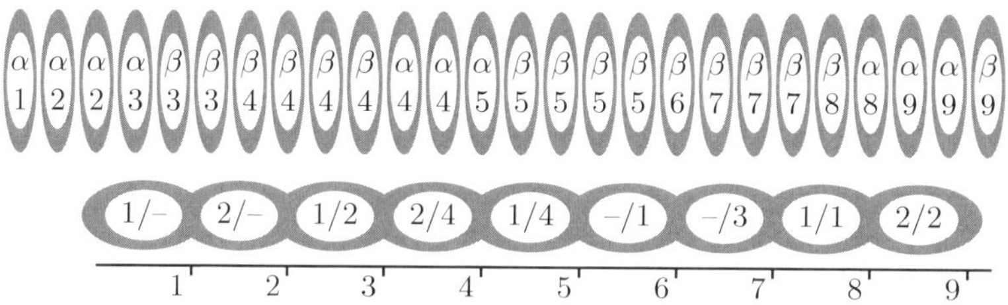

Fig. 1. A sequence of examples sorted according to their numerical values (above). The class labels (α and β) of the instances are also shown. The sequence of data bins with their respective class distributions (below).

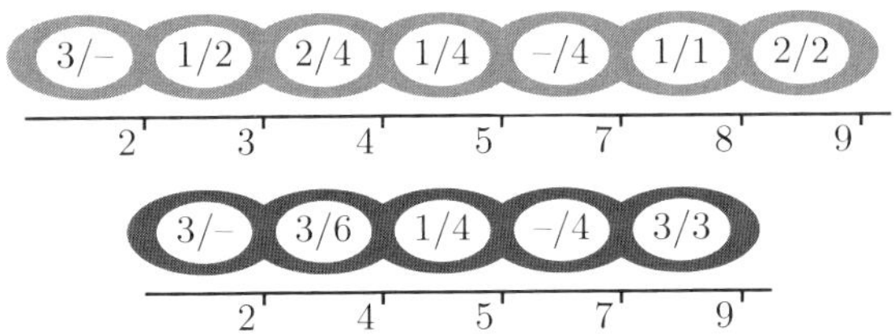

Fig. 2. The blocks (above) and segments (below) in the sample of Fig. 1. Block borders are the boundary points of the numerical range, segment borders are a subset of them.

each existing attribute value. Within each bin we record the class distribution of the instances that belong to it. This information suffices to evaluate the goodness of the partition; the actual data set does not need to be maintained.

The sequence of bins has the minimal *attainable* misclassification rate for some, but not all, attribute evaluation functions (see [11,12]). However, the same rate can usually be obtained with a smaller number of intervals. Fayyad and Irani's [8] analysis of the entropy function has shown that cut points embedded into class-uniform intervals need not be taken into account, only the end points of such intervals — the *boundary points* — need to be considered to find the optimal discretization. Thus, optimal binary splits of *Average Class Entropy* and *Information Gain* fall on boundary points. Hence, only they need to be examined in optimal binary partitioning of the value range of a numerical attribute.

Elomaa and Rousu [11] showed that the same is true for many commonly-used evaluation functions. By this analysis we can merge together adjacent class uniform bins with the same class label to obtain example *blocks* (see Fig. 2). The boundary points of the value range are the borders of its blocks. Block construction still leaves all bins with a mixed class distribution as their own blocks. A dynamic programming algorithm lets one find optimal arity-restricted multi-way partitions efficiently in these cases [9,10,11].

Subsequently, a more general property was also proved for some evaluation functions [12]: *segment borders* — points that lie in between two adjacent bins with different RCDs — are the only points that need to be taken into account. It is easy to see that segment borders are a subset of boundary points. Example *segments* are easily obtained from bins by merging together adjacent bins with the same RCD (see Fig. 2).

3 Cut Point Analysis in Online Data Streams

We have recently shown [13] that as long as a balanced binary search tree (BST) on the observed values for each numerical attribute is maintained, only local changes happen to bins, blocks, and segments as a result of incorporating a new example from the data stream into our data structures. Sending the new example down the BST costs $O(\lg V)$ time per numerical attribute, where V is the number of different values (bins) in the value range. After that only constant

store other information for convenience. Bins are the building blocks of extended intervals and, hence, we can also access their sufficient statistics.

Let us study the linear-time process of merging intervals in one left-to-right scan. Does this process have better locality properties than Kerber's procedure? At first it might appear that this is purely local as the new example only affects one interval and it has two neighbor intervals the merging with which needs to be checked. Even though the intervals and their changing counts are local (cf. [13]), the statistical test is global by nature.

When a new example from the stream falls in interval i we would need to recompute the deviations $D_{i-1,i}$ and $D_{i,i+1}$. If the example is an instance of the class j, the counts N_{ij}, N_i, C_j, and N change, which also affects the expectation E_{ij}. The deviations $D_{i-1,i}$ and $D_{i,i+1}$ can be easily updated online and compared to the χ^2 critical value in constant time. If the statistical test indicates that there is no reason to merge interval i with one of its neighbors, then no further processing is needed. On the other hand, if the intervals get merged, then the new interval may have a deviation value below the critical value when combined with one of its own neighbors. These changes can in the worst case propagate through all intervals. Hence, the changes induced by one new example from the data stream are not always limited to its own and the two adjacent intervals. As it is, exact computation of all deviation values is not feasible online.

Of course the changes are the largest for the interval of the new example. The skewness caused to other counts is not large and can be corrected later by periodic linear-time recomputation of the deviations for all interval pairs. In summary, we propose to do the constant-time local changes that concern the interval into which the new example falls. Then to repair the skewness that is caused by not updating all changed information online, we propose to periodically recompute the statistics for the whole data (in linear time).

5 Empirical Evaluation

Let us now experiment with some larger data sets from the UCI repository which contain truly numerical values. In our evaluation we disregard attributes labeled as numerical if they have less than ten different values. We also overlook those examples that have missing values for numerical attributes.

Fig. 3 shows for ten UCI domains the reduction in cut points obtained by moving from bins to segments and that of using the χ^2 independence test to merge intervals during a left-to-right scan (with significance level 5%). The figures on right are the bin counts for the attributes and the black bars represent the relative fraction of resulting segments. White bars are the relative numbers of intervals when χ^2 merging is executed. Truly continuous attributes are rare. Usually the number of bins is (clearly) less than 10% of the number of examples. The exceptions are found from domains `Abalone`, `German`, and `Segmentation`.

The segment reduction percentage varies from 80% to 0%. Usually for attributes with a high number bins large reductions are obtained by moving to operate on segments (see e.g., `Adult` and `Segmentation`). However, in domains

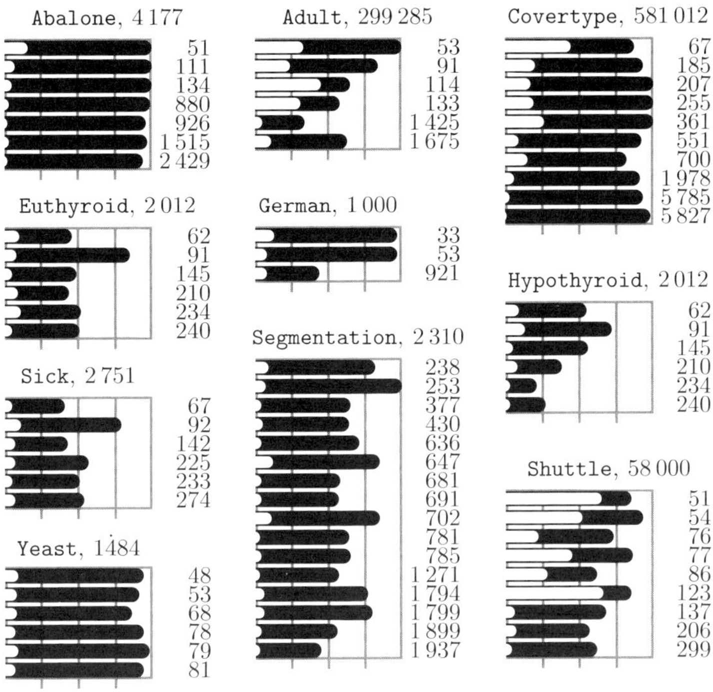

Fig. 3. The relative fraction of segments (black bars) and χ^2 merged intervals (white bars) out of the bins observed for numerical attributes (the figures on right) for ten UCI domains. The figures on top denote the number of training examples (without missing numerical values) in the domain.

Abalone, Covertype, and Yeast only quite small reductions are recorded for all attributes. The χ^2 independence test leads most often to intervals counts of only few percentages of the original bins. There are only a few exceptions (in domains Adult, Covertype, and Shuttle) in which clearly smaller reductions are recorded. We can, thus, conclude that we have achieved our goal of being able to come up with truly low-arity partitions.

In our second test we monitor for the change of interval counts in a data stream using the Waveform data generator [17] from the UCI repository. Fig. 4 displays for three attributes the evolution of the number the intervals when the number of examples grows from one thousand to ten million. The curves eventually stabilize to be more or less linear. Because the x-axis is in logarithmic scale, the true growth rate for all examined intervals is also logarithmic. The number of intervals present after χ^2 merging can be seen to grow for all three attributes, but the growth rate is very moderate. Thus, we are able to reach our goal of being able to operate on low-arity partitions even when the data stream grows to be excessive. Segment reduction cannot offer us the same benefit: as more and more examples are received, the relative fraction of segments grows closer to the number of bins.

Finally, we briefly tested, using Naïve Bayes, whether the intervals have an effect on classification accuracy. Two of our test domains — Adult and Sick —

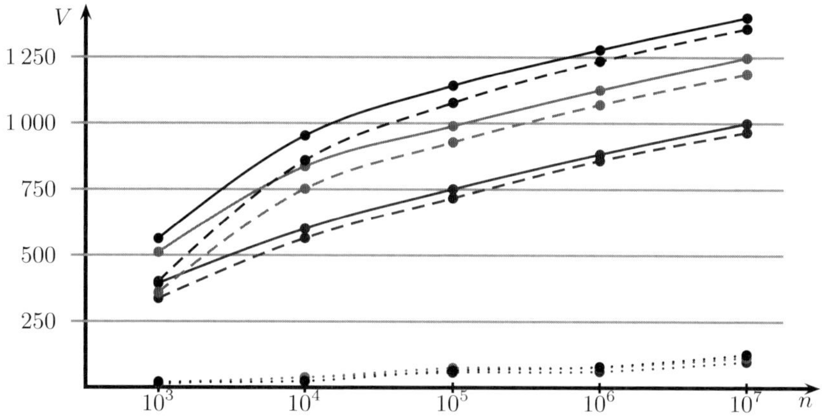

Fig. 4. Evolution of interval counts for three attributes of `Waveform` as more data is generated. Bin counts are denoted by the solid curves, the dashed ones under them are the related segment counts, and the dotted curves stand for intervals after χ^2 merging. Note the logarithmic scale of the x-axis.

come divided into training and test sets. We trained Naïve Bayes with training data in which numerical attribute ranges were segmented and obtained classification accuracies 94.2% and 90.9%, respectively, on the test sets. For `Adult` all reasonable significance levels lead to the same classification accuracy after χ^2 merging of numerical ranges. For the domain `Sick` the at most 92.3% classification accuracy was recorded using most commonly used significance levels. However, preventing intervals get easily merged (after significance level 30%) reduces the accuracy down to the same level as when using segments.

6 Conclusion

Motivated by data stream applications and the obvious overfitting, we have proposed to relax the requirement of two adjacent intervals having exactly equal RCD in cut point analysis. Moreover, because of efficiency reasons, we put forward interval merging during one (left-to-right) scan of the data instead of ChiMerge's recursive search for the best combination candidates. Our empirical evaluation demonstrated that the proposed approach is efficient and effective.

The solution proposed can be seen to be only partially satisfying. Periodic recomputation of the χ^2 test is still a relatively heavy operation to perform in the online setting. Our future work will consider whether ChiMerge can be implemented efficiently in the online setting.

Acknowledgments

This work has been financially supported by Academy of Finland. projects Intents (206280), Alea (210795), and "Machine learning and online data structures" (119699).

References

1. Dougherty, J., Kohavi, R., Sahami, M.: Supervised and unsupervised discretization of continuous features. In: Proc. 12th International Conf. on Machine Learning, pp. 194–202. Morgan Kaufmann, San Francisco (1995)
2. Liu, H., Hussain, F., Tan, C.L., Dash, M.: Discretization: An enabling technique. Data Mining and Knowledge Discovery 6(4), 393–423 (2002)
3. Kerber, R.: ChiMerge: Discretization of numeric attributes. In: Proc. 10th National Conference on Artificial Intelligence, pp. 123–128. AAAI Press, Menlo Park (1992)
4. Pfahringer, B.: Compression-based discretization of continuous attributes. In: Proc. 12th International Conference on Machine Learning, pp. 456–463. Morgan Kaufmann, San Francisco (1995)
5. Richeldi, M., Rossotto, M.: Class-driven statistical discretization of continuous attributes. In: Lavrač, N., Wrobel, S. (eds.) ECML 1995. LNCS, vol. 912, pp. 335–338. Springer, Heidelberg (1995)
6. Liu, H., Setiono, R.: Feature selection via discretization. IEEE Transactions on Knowledge and Data Engineering 9(4), 642–645 (1997)
7. Tay, F.E.H., Shen, L.: A modified Chi2 algorithm for discretization. IEEE Transactions on Knowledge and Data Engineering 14(3), 666–670 (2002)
8. Fayyad, U.M., Irani, K.B.: On the handling of continuous-valued attributes in decision tree generation. Machine Learning 8(1), 87–102 (1992)
9. Fulton, T., Kasif, S., Salzberg, S.: Efficient algorithms for finding multi-way splits for decision trees. In: Proc. 12th International Conference on Machine Learning, pp. 244–251. Morgan Kaufmann, San Francisco (1995)
10. Zighed, D., Rakotomalala, R., Feschet, F.: Optimal multiple intervals discretization of continuous attributes for supervised learning. In: Proc. 3rd International Conference on Knowledge Discovery and Data Mining, pp. 295–298. AAAI Press, Menlo Park (1997)
11. Elomaa, T., Rousu, J.: General and efficient multisplitting of numerical attributes. Machine Learning 36(3), 201–244 (1999)
12. Elomaa, T., Rousu, J.: Efficient multisplitting revisited: Optima-preserving elimination of partition candidates. Data Mining and Knowl. Discovery 8(2), 97–126 (2004)
13. Elomaa, T., Lehtinen, P.: Maintaining optimal multi-way splits for numerical attributes in data streams. In: Washio, T., Suzuki, E., Ting, K.M., Inokuchi, A. (eds.) PAKDD 2008. LNCS (LNAI), vol. 5012, pp. 544–553. Springer, Heidelberg (2008)
14. Catlett, J.: On changing continuous attributes into ordered discrete attributes. In: Kodratoff, Y. (ed.) EWSL 1991. LNCS, vol. 482, pp. 164–178. Springer, Heidelberg (1991)
15. Fayyad, U.M., Irani, K.B.: Multi-interval discretization of continuous-valued attributes for classification learning. In: Proc. 13th International Joint Conference on Artificial Intelligence, pp. 1022–1027. Morgan Kaufmann, San Francisco (1993)
16. Gama, J., Rocha, R., Medas, P.: Accurate decision trees for mining high-speed data streams. In: Proc. 9th ACM SIGKDD Conference on Data Mining and Knowledge Discovery, pp. 523–528. ACM Press, New York (2003)
17. Breiman, L., Friedman, J.H., Olshen, R.A., Stone, C.J.: Classification and Regression Trees, Wadsworth, Pacific Grove, Calif. (1984)

Maps Ensemble for Semi-Supervised Learning of Large High Dimensional Datasets

Elie Prudhomme and Stéphane Lallich

Université Lumière Lyon 2, Laboratoire ERIC,
5 avenue Pierre Mendès-France
69676 Bron
eprudhomme@eric.univ-lyon2.fr,
stephane.lallich@univ-lyon2.fr

Abstract. In many practical cases, only few labels are available on the data. Algorithms must then take advantage of the unlabeled data to ensure an efficient learning. This type of learning is called semi-supervised learning (SSL). In this article, we propose a methodology adapted to both the representation and the prediction of large datasets in that situation. For that purpose, groups of non-correlated attributes are created in order to overcome problems related to high dimensional spaces. An ensemble is then set up to learn each group with a self-organizing map (SOM). Beside the prediction, these maps also aim at providing a relevant representation of the data which could be used in semi-supervised learning. Finally, the prediction is achieved by a vote of the different maps. Experimentations are performed both in supervised and semi-supervised learning. They show the relevance of this approach.

1 Motivations

In supervised learning, classification algorithms deduce a predictive model from a set of labeled data. Nevertheless, obtaining a label for each example in the training data is often an expensive process which needs a human expert. On the other hand, automatic data-gathering is no more subjected to storage constraints. In practice, this leads to large data that has only a small subset of examples labeled by an expert, while the rest are unlabeled. For this reason, it would be interesting if unlabeled training data could be used by the learning algorithm. For example, in content-based image retrieval, user would have to label only a few number of images when mining a huge image database. To achieve this goal, Semi-Supervised Learning (SSL) uses unlabeled data to enhance the predictive model. Particularly, unlabeled data allows to better define the data topology.

Data concerned by SSL are often large. First, SSL arises from automatic acquisition processes which leads, beyond labeling problems, to an important number of examples (e.g. web pages classification, [3]). Moreover, SSL is often applied to complex objects such as text or image, described by an important number of features. This two-fold complexity examples/features must be taken

A. An et al. (Eds.): ISMIS 2008, LNAI 4994, pp. 100–110, 2008.

into account by the learning algorithm. In fact, time spent by learning process is directly linked to the quantity of data being analysed. As a result, methods with a complexity that is not linear to the number of examples and features would not ensure scalability with bulky data. In this case, using these methods needs either heuristics or sampling strategies which both may degrade learning. Furthermore, a growing number of features leads to theoretical problems for learning algorithms. First, the number of points needed to describe a space grows exponentially with its dimensionality (curse of dimensionality, [2]). The learning algorithm is thus less precise in high dimensional spaces. In addition, distances between points tend to be constant at space infinity (concentration of measures, [8]). This last point compromises navigation through examples and data representation. Yet, navigation and representation are important for the learning process. On one hand, navigation allows to provide the end-users with existing examples similar to a new one during the labeling process. In SSL, user could then refine the model with these unlabeled examples. On the other hand, representation is useful for SSL as well as for preprocessing, because it is used to choose the examples to give to the expert. As for postprocessing, representation is used to validate the model. Indeed, a major problem in SSL is the model selection, especially for a low number of labeled data and whatever the used strategy, as reported by [7]. In fact, in that case, cross-validation is likely to be unreliable. To avoid that problem, a solution is to have recourse to statistical validation of learning. A practical way to do this could be the validation using a cross-product statistic [19].

Thus, navigation and representation capacity can improve performance and usability of SSL methods, especially in the context of high dimensional datasets. For that purpose, we propose to develop an ensemble of Self-Organizing Maps (SOM, [14]). The rest of this paper is organized as follows. Next section presents SSL and different solutions using the concept of multiple classifiers (Sect. 2). Ensembles are then introduced before describing our methodology that is based on maps ensemble (Sect. 3). At last, experiments conducted with this strategy on high dimensional datasets are presented for supervised (Sect. 4.2) and semi-supervised datasets (Sect. 4.2).

2 Semi-Supervised Learning

SSL supposes that unlabeled data could help in supervised learning. Unfortunately, this is not always true. To benefit from unlabeled data, some assumptions on the data will have to hold. Three of these assumptions were distinguished by [7]:

- semi-supervised smoothness assumption: "if two points $x1, x2$ in a high-density region are closed, then should be the corresponding outputs $y1, y2$",
- cluster assumption: "if points are on the same cluster, they are likely to be of the same class",
- manifold assumption: "the (high-dimensional) data lie (roughly) on a low-dimensional manifold".

The first assumption is more likely a condition to learn from data while the other two specify the way to learn. The main difference between cluster and manifold assumption concerns the space where the distance is calculated. This space represents respectively the input space and the high-density manifold. If one of these assumptions holds in data, unlabeled data could be used to delimit either a cluster or a manifold. Nevertheless, an algorithm exploiting the cluster assumption on manifold data risks to fail.

Methods that were sucessfully applied to SSL could be brought together into several families, among which generative models, graph-based methods and co-training (see [29] for a review). This last family of methods trains multiple classifiers. Same as an ensemble approach in supervised learning, the diversity of these classifiers helps to enhance prediction, as explained in Sect. 3.1. In fact, also co-training uses these base classifiers to manage unlabeled data. In their former work, [3] trained two classifiers on two different features spaces during an iterative process. First, each classifier was trained on the labeled examples, then it classified two different subsets of unlabeled examples. The classifiers were thus trained again with both labeled and newly labeled examples. These steps were repeated until each classifier was trained on the whole dataset. Because of the low number of classifiers, features subspaces must be sufficiently independent to enhance the performance. [28] avoided that difficulty by setting up an ensemble of different learning algorithms using the same dataset. Training this classifier ensemble follows globally the strategy proposed by [3]. However, [28] control the quality of newly labeled examples before adding them in the learning process.

These two approaches (and others like [1; 16]) show the advantage of using multiple learners in SSL. Nevertheless, these approaches lie on supervised learning algorithms which need to predict unlabeled examples before using them. In contrast, when the prediction takes advantage of an unsupervised learning, unlabeled examples can contribute to the final prediction without being labeled. Following this idea, we propose to apply an ensemble of Self-Organizing Maps in SSL. SOMs, learned on the whole set of examples (label or not), will endow our ensemble of representation capacity whereas the classifier ensemble will improve learning on a single map. The next section presents our approach in further details.

3 Maps Ensemble

3.1 Ensembles

Ensembles are sets of learning machines whose decisions are combined to improve performance of overall system. This implies a three-step learning. First, information (carried by examples, features or classes) is shared out between several classifiers. Then, this information is learned by each classifier. Finally, their predictions are combined to get the class of a new example. That approach have been applied to a wide range of real problems with theoretical and empirical evidences of its effectiveness. For example, bagging (*Boostrap AGGregatING*) builds K classifiers by making bootstrap replicates of the learning set [4]. In this way, bagging reduces error variance, mainly due to learning set. In bagging, samples are drawn with replacement using an uniform probability distribution, while in boosting methods [10] weighted samples are drawn from training data. That technique places the highest weight on examples most often misclassified by previous base classifiers. Thus, boosting enlarges margins of the training set and then reduces the generalization error. These two methods sample the training set to learn different classifiers but this is not the only strategy to build ensembles. For example, in mixture of experts, each classifier are competing each others to learn an example [12]. Or, when data are stemming from several sensors, each classifier learns data belonging to a specific one [9]. A more complete description of ensembles could be found in [24].

One of the major reasons of the ensembles success is summarized by the diversity concept ([6] for a review). Although there is no formal definition, diversity is the capacity of different classifiers to make their mistakes on different examples. Indeed, more the classifiers are mistaken on different examples of the training set, more it is probable that a majority of them find the correct response for a particular example. That intuition on diversity is confirmed by several theoretical results [11; 15; 23]. In classification context, [27] express the added error of an ensemble given the a posteriori probability of each class predicted by base classifier. That error could be broken into two terms. The first is the base classifiers mean error and the second – named ambiguity term – contains both variance and covariance of the ensemble. By measuring correlations between base classifiers error, covariance is an expression of diversity. Moreover, contrary to bias and variance included in the first term, covariance can be negative. Thus, it can reduce the average quadratic error of an ensemble.

Finally, setting up an ensemble approach requires to consider three points: (1) base classifiers: which algorithm is the most suitable for each dataset, (2) diversity: how to induce diverse classifiers from initial data and (3) aggregation: how to aggregate different classifiers prediction to obtain the class of a new example. Concerning base classifiers, SOM has been chosen for their representation capability and their linear complexity according to features and examples (Sect. 3.2). To manage their diversity, different features subspaces are generated as described in Sect. 3.3. In this paper, we do not studied aggregation process and simply use a majority voting procedure.

3.2 Self-Organizing Map

SOM allows both a fast unsupervised learning of examples and their representation. Because of those properties (fast algorithm and topology preservation), they have been adapted to supervised learning in several ways. For one of them, the class of examples is only used after a classical learning of SOM on features. During that second step, neurons take the class of the examples they represent. The reverse happens during prediction: the class of a new example is determined by class of the neuron which best matches that example with the function Kohonen-Opt proposed by [19].

Furthermore, supervised learning of SOM can be validated by assessing the strength of the link between proximity in the sense of the map and proximity in the sense of the classes. In [19], we propose statistics narrowly correlated with the error in generalization in order to evaluate the level of quality of the learning. In SSL, that statistic can be useful for model selection.

3.3 Features Subspaces

Using an ensemble is very interesting when working with high-dimensional spaces. As the problem of high-dimensionality resides in the important number of features to learn, reducing that number using the ensemble approach is a well suited solution. By setting up several features subspaces, each one learned by a base classifier, we aim several objectives. First is, of course, to circumvent the theoretical problems involved in high dimensional spaces by carrying out the learning of each map on a reduced number of dimensions. However, because different maps learn different subspaces, an important number of features are used, unlike feature selection that could be used during the preprocessing. The advantage, as underlined by [25], is that the redundancy of information takes part in the noise reduction on each feature. The second objective is the improvement in prediction which results from the diversity of base classifiers built on each group. For high dimensional data, the ensemble based on features subspaces handles information in its globality while avoiding problems arising from high dimensional space.

Even if it's possible to randomly generate several subspaces (as in [5]), we propose to control features gathering to ensure both a low mean error of the maps and a high level of diversity. The strategy used to select the subspace is often controled by an error measure [21]. This kind of strategy requires labels, and thus is inconsistent with our choice to incorporate unlabeled examples in the primary clustering. To achieve our goal, feature is considered as an information support and the correlation among features is considered as information sharing. Therefore, as the quantity of information received by a map increases, the prediction accuracy increases too and so does the risk to share that information with other maps which results in poor diversity. In order to optimize the prediction, the quantity of information allowed to each map have to be balanced to perform the trade-off between the prediction accuracy and the risk to share feature information. Features correlation helps us to resolve that trade-off, allowing to distribute features that share the same information (i.e. correlated features) on different maps. Thus, maps could share information while trying to keep their diversity by being trained on different information.

To achieve this, a feature clustering is firstly applied to generate k clusters of correlated features. We choose Varclus [22] among several classical algorithms, like hierarchical clustering [26] or k-means [17] performed on the transposed matrix. Varclus is a top-down method which is specific to feature clustering. It presents the main advantage to fix automatically the cluster number. For each iteration, the first two eigenvectors of the features correlation matrix are computed. Clusters are then generated from these vectors, assigning each feature to the closest vector according to their correlation (r^2 maximum). Nevertheless,

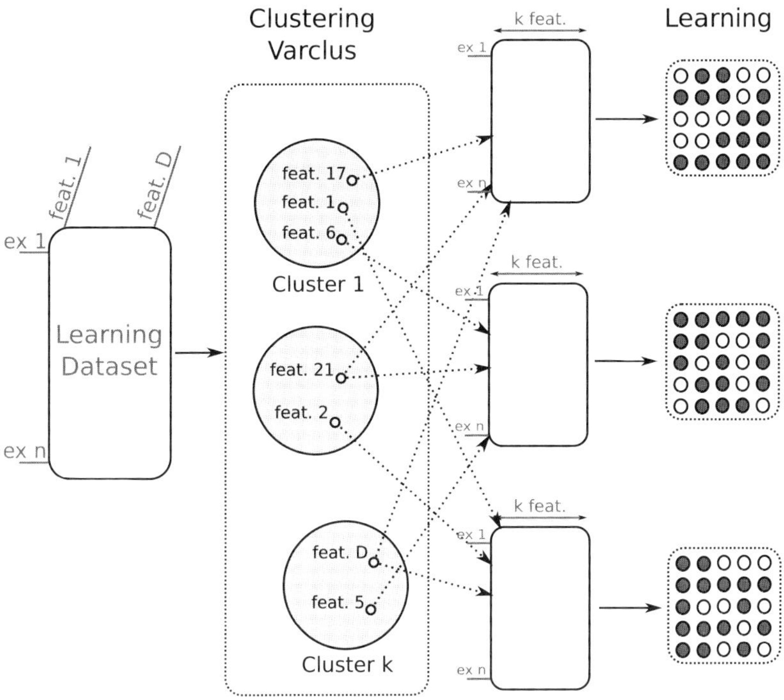

Fig. 1. Creation of different datasets from feature clustering

to overcome a massive assignation of features to the first eigenvector, eigenvectors are rotated by the varimax method [13]. Iterations on each cluster are done until the second eigenvalue is less than 1.0 . The k clusters obtained are then processed to constitute k groups of uncorrelated features (see Fig. 1). For that purpose, a feature is randomly picked in each cluster resulting in k groups of k features. Because a cluster could have less or more than k features, each feature could be picked several times or missed out.

3.4 Complexity

The computational complexity of our approach can be otained by adding the complexities of the Varclus and the SOMs approaches. Varclus needs first to compute the correlation matrix between the d features. This computation has a complexity of d^2. Then, for each $\log_2 k$ iterations of Varclus (with k the numbers of features groups and $k < d$), we compare the correlation between features and the first two eigenvectors of the correlation matrix, which implies $2 \times d \times \log_2 k$ comparisons. Concerning the SOMs, k maps are learned (one for each group) on k features arising a total complexity of $c \times n \times k^2$ (c is the neurons number) [14]. Therefore, the overall complexity of our approach is on the order of $d^2 + (2 \times d \times \log_2 k) + (c \times n \times k^2)$.

3.5 Application on Semi-Supervised Learning

In semi-supervised learning, information lying under unlabeled examples must be used to improve the prediction. In our approach, prediction relies on the unsupervised learning of the self-organizing maps. By considering unlabeled examples, we can improve maps quality, and thus reach a better prediction. To perform this, the overall examples (labeled or not) are used to build features subspaces and learn the SOMs in each one. Finally, maps are labeled on the basis of the labeled examples in order to predict new examples.

4 Experimentations

In general, the relative quality of an approach is not data independant, whether under a supervised or a semi-supervised learning paradigm. The purpose of our experiments is to show to which extend our approach – which has the inherent advantage of allowing both visualization and data exploration – can still be competitive to other existing approaches, especially in terms of error-rate.

4.1 Supervised Datasets

We first test our approach in supervised learning. For that, we compared it to the ensemble methods of boosting and random forests [5], those methods being standard. For those algorithms, we used Tanagra's implementation [20]. Datasets come from UCI site [18]. Table 1 shows their characteristics. On these datasets, parameters of our approach don't change: maps of size 20×20 with rectangular lattice of the grid, 17 learning cycles and a learning rate and a neighbourhod kernel which decrease linearly. Error rate obtained by 10-cross-validation are shown in Table 2.

Table 1. Datasets

Data	Features	Classes	Examples
(1) Ionosphere	34	2	351
(2) Multi-features (Profile correlations)	76	10	2000
(3) Multi-features (Fourier coefficients)	216	10	2000

Results of our approach are always better than $ID3$ or a single SOM, as excepted for an ensemble approach. For datasets (1) and (2), our approach is between random forests and boosting. Dataset (3) is interesting. On that dataset, boosting gives worst results than $ID3$ (several parameters have been tested). This kind of result occurs when boosting overfits which generally happen with noisy data. In that case, our approach still gives interesting results, better than those obtained by random forests. These preliminary results in supervised

Table 2. Methods error rate on three datasets

Data	Maps Ensemble	SOM	ID_3	Boosting ID_3	Random Forests
(1)	8.4	11.4	10.1	8.5	**6.0**
(2)	**19.8**	26.0	33.3	37.6	21.6
(3)	7.8	9.5	23.6	8.1	**5.6**

learning validate the predictive capability of our approach. In the next subsection, we look at its behavior in SSL.

4.2 Semi-Supervised Datasets

In SSL, we follow the validation process of [7]. So, datasets and results directly come from – and are more detailed in – their book. Table 3 shortly describes those datasets. For each one, 12 folds of labeled data are provided. Error rate is then estimated by the mean error rate obtained on these folds.

To validate our SSL approach, we first control that the 1400 unlabeled examples enhance the prediction. To perform this, we compare the learning of the entire example set (labeled or not) to the learning of the labeled examples only. As expected, unlabeled examples have a real impact on the prediction (Fig. 2)

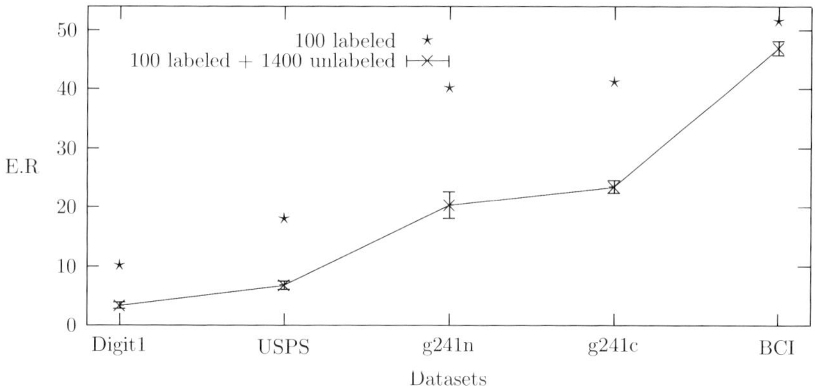

Fig. 2. The impact of unlabeled data on map ensemble prediction in SSL context

Next, our approach is compared to other SSL approaches. In [7], two standard supervised learning methods were applied (1-NN and SVM, their results are reported in Table 3) as well as 11 SSL algorithms. For short, we simply present the rank of our approach according to the error rate (where 1 stand for best and 12 for worst) and the error rate range (Table 3). Results are given for both 10 and 100 labeled examples. Parameters of our approach remaind unchanged: maps of size 10×10 (resp. 6×6) and 10 learning cycles with 100 (resp. 10) labeled examples.

Table 3. SSL datasets and results

	g241c	**g241d**	**Digits1**	**USPS**	**BCI**
Features	241	241	241	241	117
Examples	1500	1500	1500	1500	400
Classes	2	2	2	2	2
Comparison with 100 labeled examples					
1NN	43.9	42.4	3.9	5.8	48.7
SVM	23.1	24.6	5.5	9.7	34.3
Maps Ens.	23.5	**20.3**	**3.4**	6.8	47.0
Rank (/12)	7	7	3	7	9
[min; max]	[13.5;43.6]	[4.9;41.6]	[2.8;6.1]	[4.7;9.8]	[31.4;47.4]
Comparison with 10 labeled examples					
1NN	47.9	46.7	13.6	16.7	49
SVM	47.3	46.7	30.6	20	49.8
Maps Ens.	**40.5**	**39.2**	16.2	18.9	49.9
Rank (/12)	6	2	9	5	10
[min; max]	[22.8;47.1]	[18.6;42]	[5.4;23.5]	[13.6;25.4]	[46.9;50.4]

In comparison with supervised learning methods, our approach improves results for dataset *g241c* and *g241d* (except for dataset *g241c* with 100 labeled examples for which we suppose that the bad results are due to parameters problem). As expected by [7], those datasets are cluster-like and correspond to the assumption made by our approach. Compared to others SSL approaches, our results are often close to those obtained by most of the algorithms. Regarding error rates, maps ensemble also provide good results on datasets *digits1* and *USPS*. However, on the dataset BCI, we failed – as many other methods – to learn anything. Finally, when the number of labeled examples decrease, our approach performance seems to be less decreasing that performance of the other SSL algorithms.

5 Conclusion and Future Work

To conclude, we propose an ensemble method adapted to semi-supervised learning and based on the cluster assumption. This method is designed for large datasets while allowing a data representation. Experiments performed in supervised learning show its efficiency, particularly with noisy data. In SSL, experiments provide creditable results, particularly with datasets holding the cluster assumption. Nevertheless, our approach can be enhanced in three ways. First, model selection can be achieved through iterative validation of the map representation. Next, representation is able to manage the labeling process by supplying well distributed examples to expert. This procedure will optimize both labeling cost and its use. Finally, the learning algorithm could be improve. In fact, we do not exploit the co-learning strategy and based the SSL only on the ensemble approach.

Bibliography

[1] Becker, S.: JPMAX: Learning to recognize moving objects as a model-fitting problem. In: Advances in Neural Information Processing Systems, vol. 7, pp. 933–940. MIT Press, Cambridge (1995)
[2] Bellmann, R.: Adaptive Control Processes: A Guided Tour. Princeton University Press, Princeton (1975)
[3] Blum, A., Mitchell, T.: Combining labeled and unlabeled data with co-training. In: COLT: Proceedings of the Workshop on Computational Learning Theory, pp. 92–100. Morgan Kaufmann, San Francisco (1998)
[4] Breiman, L.: Bagging predictors. Machine Learning 24(2), 123–140 (1996)
[5] Breiman, L.: Random forests. Machine Learning 45(1), 5–32 (2001)
[6] Brown, G., Wyatt, J., Harris, R., Yao, X.: Diversity creation methods: a survey and categorisation. Information Fusion 6(1), 5–20 (2005)
[7] Chapelle, O., Schölkopf, B., Zien, A. (eds.): Semi-Supervised Learning. MIT Press, Cambridge (2006)
[8] Demartines, P.: Analyse de données par réseaux de neurones auto-organisés. Ph.d. dissertation, Institut National Polytechnique de Grenoble, France (1994)
[9] Duin, R., Tax, D.: Experiments with classifier combining rules. In: Kittler, J., Roli, F. (eds.) MCS 2000. LNCS, vol. 1857, pp. 16–29. Springer, Heidelberg (2000)
[10] Freund, Y.: Boosting a weak learning algorithm by majority. In: Proceedings of the Workshop on Computational Learning Theory, Morgan Kaufmann, San Francisco (1990)
[11] Geman, S., Bienenstock, E., Doursat, R.: Neural networks and the bias/variance dilemma. Neural Computing 4(1), 1–58 (1992)
[12] Jacobs, R., Jordan, M., Barto, A.: Task decomposition through competition in a modular connectionist architecture: The what and where vision tasks. Cognitive Science 15, 219–250 (1991)
[13] Kaiser, H.: The varimax criterion for analytic rotation in factor analysis. Psychometrika 23, 187–200 (1958)
[14] Kohonen, T.: Self-Organizing Maps, vol. 30. Springer, Heidelberg (2001)
[15] Krogh, A., Vedelsby, J.: Neural network ensembles, cross validation, and active learning. Advances in NIPS 7, 231–238 (1995)
[16] Leskes, B.: The Value of Agreement, a New Boosting Algorithm. Springer, Heidelberg (2005)
[17] McQueen, J.: Some methods for classification and analysis of multivariate observations. In: Proc. of the Fifth Berkeley Symposium on Mathematical Statistics and Probability, vol. 1, pp. 281–297 (1967)
[18] Newman, D., Hettich, S., Blake, C., Merz, C.: UCI repository of machine learning databases (1998)
[19] Prudhomme, E., Lallich, S.: Quality measure based on Kohonen maps for supervised learning of large high dimensional data. In: Proc. of ASMDA 2005, pp. 246–255 (2005)

[20] Rakotomalala, R.: Tanagra: un logiciel gratuit pour l'enseignement et la recherche. In: Sloot, P.M.A., Hoekstra, A.G., Priol, T., Reinefeld, A., Bubak, M. (eds.) EGC 2005. LNCS, vol. 3470, pp. 697–702. Springer, Heidelberg (2005)
[21] Ruta, D., Gabrys, B.: Classifier selection for majority voting. Information Fusion 6, 63–81 (2005)
[22] SAS, SAS/STAT user's guide, vol. 2. SAS Institute Inc. (1989)
[23] Tumer, K., Ghosh, J.: Theoretical foundations of linear and order statistics combiners for neural pattern classifiers. Technical report, Computer and Vision Research Center, University of Texas, Austin (1995)
[24] Valentini, G., Masulli, F.: Ensembles of learning machines. In: Marinaro, M., Tagliaferri, R. (eds.) WIRN 2002. LNCS, vol. 2486, pp. 3–20. Springer, Heidelberg (2002)
[25] Verleysen, M., François, D., Simon, G., Wertz, V.: On the effects of dimensionality on data analysis with neural networks. In: International Work-Conference on ANNN: Computational Methods in Neural Modeling, vol. II, pp. 105–112. Springer, Heidelberg (2003)
[26] Ward, J.H.: Hierarchical grouping to optimize an objective function. Journal of American Statistical Association 58(301), 236–244 (1963)
[27] Zanda, M., Brown, G., Fumera, G., Roli, F.: Ensemble learning in linearly combined classifiers via negative correlation. In: International Workshop on Multiple Classifier Systems (2007)
[28] Zhou, Y., Goldman, S.: Democratic co-learning. In: ICTAI, pp. 594–202 (2004)
[29] Zhu, X.: Semi-supervised learning literature survey. Technical report (2005)

Mining Induced and Embedded Subtrees in Ordered, Unordered, and Partially-Ordered Trees

Aída Jiménez, Fernando Berzal, and Juan-Carlos Cubero

Dept. Computer Science and Artificial Intelligence,
ETSIIT - University of Granada, 18071, Granada, Spain
{aidajm,jc.cubero,fberzal}@decsai.ugr.es

Abstract. Many data mining problems can be represented with non-linear data structures like trees. In this paper, we introduce a scalable algorithm to mine partially-ordered trees. Our algorithm, POTMiner, is able to identify both induced and embedded subtrees and, as special cases, it can handle both completely ordered and completely unordered trees (i.e. the particular situations existing algorithms address).

1 Introduction

Non-linear data structures are becoming more and more common in data mining problems. Graphs, for instance, are commonly used to represent data and their relationships in different problem domains, ranging from web mining and XML documents to bioinformatics and computer networks. Trees, in particular, are amenable to efficient mining techniques and they have recently attracted the attention of the research community.

The aim of this paper is to present a new algorithm, POTMiner, to identify frequent patterns in partially-ordered trees, a particular kind of trees that is present in several problems domains. However, existing tree mining algorithms cannot be directly applied to this important kind of trees.

Our paper is organized as follows. We introduce the idea of partially-ordered trees as well as some standard terms in Section 2. Section 3 describes the state of the art in tree mining algorithms. Our algorithm is presented in Section 4. Section 5 shows some experimental results. Finally, in Section 6, we present some conclusions and pointers to future work in this area.

2 Background

We will first review some basic concepts related to labeled trees using the notation from [1].

A **tree** is a connected and acyclic graph. A tree is rooted if its edges are directed and a special node, called root, can be identified. The root is the node from which it is possible to reach all the other vertices in the tree. In contrast, a

A. An et al. (Eds.): ISMIS 2008, LNAI 4994, pp. 111–120, 2008.

tree is free if its edges have no direction, that is, when it is an undirected graph. Therefore, a free tree has no predefined root.

Rooted trees can be classified as:

- **Ordered trees**, when there is a predefined order within each set of siblings.
- **Unordered trees**, when there is not such a predefined order among siblings.

In this paper, we consider **partially-ordered trees**, witch can be defined as trees that have both ordered and unordered sets of siblings. They can be useful when the order within some sets of siblings is important but it is not necessary to establish an order relationship among all the tree nodes.

In Figure 1, we show a dataset example with different kinds of rooted trees. In this figure, circled nodes represent ordered nodes, while squared nodes represent unordered ones.

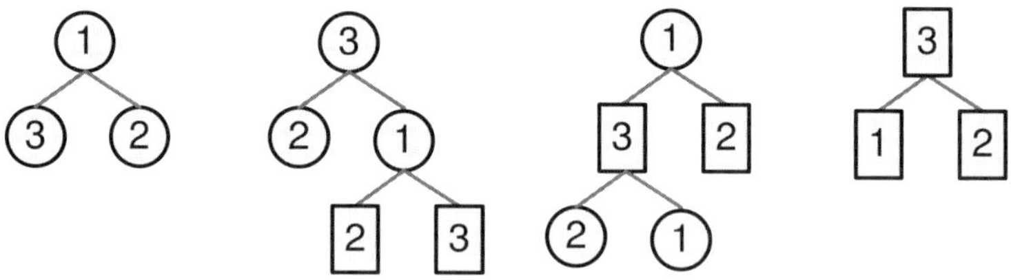

Fig. 1. Example dataset with different kinds of rooted trees (from left to right): *(a)* completely ordered tree, *(b) and (c)* partially-ordered trees, *(d)* completely unordered tree

Different kinds of subtrees can be mined from a set of trees depending on the way we define the matching function between a pattern and a tree. We can obtain **induced subtrees** from a tree T by repeatedly removing leaf nodes from a bottom-up subtree of T (i.e. the subtree obtained by taking one vertex from T and all its descendants). We can also obtain **embedded subtrees** by removing nodes from a bottom-up subtree provided that we do not to break the ancestor relationship between the vertices of T.

Figure 2 shows the induced subtrees of size 3 that appear at least 3 times in the example dataset from Figure 1. In this example, this would also be the set of frequent embedded subtrees of size 3.

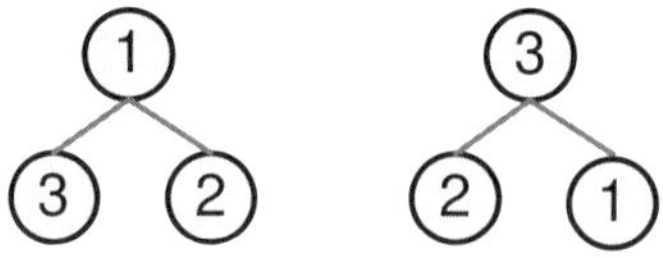

Fig. 2. Frequent subtrees of size 3 found in the dataset from Figure 1

3 Tree Pattern Mining

Once we have introduced some basic terminology, we will survey the state of the art in tree pattern mining.

The goal of frequent tree mining is the discovery of all the frequent subtrees in a large database of trees D, also referred to as *forest*, or in an unique large tree.

Let δ_T(S) be the occurrence count of a subtree S in a tree T and d_T a variable such that d_T(S)=0 if $\delta_T(S) = 0$ and d_T(S)=1 if $\delta_T(S) > 0$. We define the **support** of a subtree as $\sigma(S) = \sum_{T \in D} d_T(S)$, i.e., the number of trees in D that include at least one occurrence of the subtree S. Analogously, the **weighted support** of a subtree is $\sigma_w(S) = \sum_{T \in D} \delta_T(S)$, i.e., the total number of occurrences of S within all the trees in D.

We say that a subtree S is **frequent** if its support is greater than or equal to a predefined minimum support threshold. We define F_k as the set of all frequent subtrees of size k.

A frequent tree t is **maximal** if is not a subtree of other frequent tree in T and is **closed** if is not a subtree of another tree with the same support in D.

Before we proceed to introduce the algorithms proposed in the literature, we will describe the most common tree representations used by such algorithms.

3.1 Tree Representation

A canonical tree representation is an unique way of representing a labeled tree. This representation makes the problems of tree comparison and subtree enumeration easier.

Three alternatives have been proposed in the literature to represent rooted ordered trees as strings:

- **Depth-first codification:** The string representing the tree is built by adding the label of the tree nodes in a depth-first order. A special symbol $\uparrow$, which is not in the label alphabet, is used when the sequence comes back from a child to his parent. In Figure 1, the depth-first codification of tree *(b)* is $32\uparrow12\uparrow3\uparrow\uparrow$ while the codification of tree *(c)* is $132\uparrow1\uparrow\uparrow2\uparrow$.
- **Breadth-first codification:** Using this codification scheme, the string is obtained by traversing the tree in a breadth-first order, i.e., level by level. Again, we need an additional symbol $, which is not in the label alphabet, to separate sibling families. The breadth-first codification for trees (b) and (c) in the previous example is 3$21$$23 and 1$32$21, respectively.
- **Depth-sequence-based codification:** This codification scheme is also based on a depth-first traversal of the tree, but it explicitly stores the depth of each node within the tree. The resulting string is built with pairs (l, d) where the first element (l) is the node label and the second one (d) is the depth of the node. The depth sequence codifications of the previous examples is (3,0)(2,1)(1,1)(2,2)(3,2) for tree (b) and (1,0)(3,1)(2,2)(1,2)(2,1) for tree (c).

The canonical representation for unordered trees can be defined as the minimum codification, in lexicographical order, of all the ordered trees that can be derived from it. You can use any of the codification schemes described above.

Free trees have no predefined root, but it is possible to select one node as the root in order to get an unique canonical representation of the tree, as described in [2].

3.2 Tree Mining Algorithms

Several frequent tree pattern mining algorithms have been proposed in the literature. These algorithms are mainly derived from two well-known frequent pattern mining algorithms: Apriori [3] and FP-Growth [4].

Many algorithms follow the well-known Apriori [3] iterative pattern mining strategy, where each iteration is broken up into two distinct phases:

- *Candidate Generation*: Potentially frequent candidates are generated from the frequent patterns discovered in the previous iteration. Most algorithms generate candidates of size $k+1$ by merging two patterns of size k having $k-1$ elements in common. There are several strategies for candidate subtree generation:
  - The **rightmost expansion** strategy generates subtrees of size $k+1$ from frequent subtrees of size k by adding nodes only to the rightmost branch of the tree. This technique is used in algorithms like FREQT [5], uFreqt [6], and UNOT [7].
  - The **equivalence class-based extension** technique is based on the depth-first canonical representation of trees. It generates a candidate $(k+1)$-subtree through joining two frequent k-subtrees with $(k-1)$ nodes in common. Zaki used this extension method in his TreeMiner [8] and SLEUTH [9] algorithms.
  - The **right-and-left tree join** method was proposed with the AMIOT algorithm [10]. In candidate generation, this algorithm considers the rightmost and the leftmost leaves of a tree.
  - The **extension and join** technique is based on the breadth-first codification of trees and defines two extension mechanisms and a join operation to generate candidates. This method is used by HybridTreeMiner [2].
- *Support Counting*: Given the set of potentially frequent candidates, this phase consists of determining their support and keeping only those candidates that are actually frequent.

Other algorithms are derived from the FP-Growth [4] algorithm. For instance, the PathJoin algorithm [11] uses compacted structures called FP-Trees to encode input data, while CHOPPER and XSpanner [12] use a sequence codification for trees and extract subtrees using frequent subsequences.

Table 1 summarizes some frequent tree mining algorithms that have been proposed in the literature, pointing out the kind of input trees they can be applied to (ordered, unordered, or free) and the kind of subtrees they are able to identify (induced, embedded, or maximal).

Table 1. Some frequent tree mining algorithms

	Input trees			Discovered patterns		
Algorithm	Ordered	Unordered	Free	Induced	Embedded	Maximal
FreqT[5]	•			•		
AMIOT [10]	•			•		
uFreqT [6]		•		•		
TreeMiner [8]	•				•	
CHOPPER [12]	•				•	
XSpanner [12]	•				•	
SLEUTH [9]		•			•	
Unot [7]		•			•	
TreeFinder [13]		•			•	•
PathJoin [11]		•		•		•
CMTreeMiner [14]	•	•		•		•
FreeTreeMiner [15]			•	•		
HybridTreeMiner [2]		•	•	•		

4 Mining Partially-Ordered Trees

The algorithms described in the previous section can extract frequents patterns from trees that are completely ordered or completely unordered, but none of them works with partially-ordered trees.

In this section, we describe how Zaki's TreeMiner [8] and SLEUTH [9] algorithms can be adapted to mine partially ordered trees. The algorithm we have devised, POTMiner (Partially-Ordered Tree Miner), is able to identify frequent subtrees, both induced and embedded, in ordered, unordered, and partially-ordered trees.

Our algorithm is based on Apriori [3]. Therefore, it has two phases: candidate generation and support counting.

4.1 Candidate Generation for Partially-Ordered Trees

We use a depth-first codification to represent trees and generate $(k+1)$-subtree candidates by joining two frequent k-subtrees with $k-1$ elements in common.

We use Zaki's class extension method to generate candidates. Two k-subtrees are in the same equivalence class $[P]$ if they share the same codification string until the node $k-1$. Each element of the class can then be represented by a single pair (x, p) where x is the k-th node label and p specifies the depth-first position of its parent.

TreeMiner [8] and SLEUTH [9] use the class extension method to mine ordered and unordered embedded subtrees, respectively. The main different between TreeMiner and SLEUTH is that, in SLEUTH, which works with unordered trees, only those extensions that produce canonical subtrees are allowed in order to avoid the duplicate generation of candidates.

In POTMiner, as in TreeMiner, all extensions are allowed because tree nodes might be ordered or unordered. The candidate generation method is defined as follows:

Let (x, i) and (y, j) denote two elements in the same class $[P]$, and $[P_x^i]$ be the set of candidate trees derived from the tree that is obtained by adding the element (x, i) to P. The join procedure considers two scenarios [9]:

1. *Cousin extension*, when the element (y,j) is a right relative of the element (x,i): If $j \leq i$ and $|P| = k - 1 \geq 1$, then $(y, j) \in [P_x^i]$.
2. *Child extension*, when the element (y,j) is a descendant of the element (x,i): If $j = i$ then $(y, k - 1) \in [P_x^i]$.

4.2 Support Counting for Induced and Embedded Subtrees

For the support counting phase, we define the scope of a node [9] as a pair $[l,u]$ where l is the position of the node in depth-first order and u is the position of its rightmost descendant.

Our algorithm preserves each occurrence of a pattern X in each database tree using a tuple $(t,\ m,\ s,\ d)$ where t is the tree identifier, m stores which nodes of the tree match those of the (k-1) prefix of the pattern X in depth-first order, s is the scope of the last node in the pattern X, and d is a depth-based parameter used for mining induced subtrees (it is not needed when mining embedded subtrees). The scope list of a pattern X is the list of all the tuples $(t,\ m,\ s,\ d)$ representing the occurrences of X in the database.

For patterns of size 1, m is an empty list and the element d is initialized with the depth of the pattern only node in the database tree.

The scope list for a new candidate is built by joining the scope lists of the subtrees involved in the generation of the candidate. Let $[t_x,\ m_x,\ s_x,\ d_x]$ and $[t_y,\ m_y,\ s_y,\ d_y]$ be the scope lists of the subtrees involved in the generation of the candidate. The scope list for this candidate is built by a join operation that depends on the candidate extension method used, i.e., whether it has been generated by cousin extension or child extension.

1. *In-scope test* (used in conjunction with child extension):
   If
   (a) $t_x = t_y = t$ and
   (b) $m_x = m_y = m$ and
   (c) $s_y \subset s_x$, (i.e., $l_x \leq l_y$ and $u_x \geq u_y$)
   then add [t, m $\bigcup \{l_x\}$, s_y, $d_y - d_x$] to the scope list of the generated candidate.
2. *Out-scope test* (used in conjunction with cousin extension):
   If
   (a) $t_x = t_y = t$ and
   (b) $m_x = m_y = m$ and
   (c) If the node is ordered and $s_x < s_y$ (i.e. $u_x < l_y$) or the node is unordered and either $s_x < s_y$ or $s_y < s_x$ (i.e. either $u_x < l_y$ or $u_y < l_x$)
   then add [t, m $\bigcup \{l_x\}$, s_y ,d_y] to the scope list of the generated candidate.

The weighted support of an embedded pattern is the number of elements in its scope list. The weighted support of an induced pattern is the number of elements within its scope list whose d parameter equals 1. Intuitively, d represents the distance between the last node in the pattern and its prefix m. This parameter is needed to perform the support counting phase for induced patterns, since only the occurrences with $d_x = 1$ have to be considered when joining scope lists.

5 Experimental Results

All the experiments described in this section have been performed on a 2GHz Core 2 Duo processor with 2GB of main memory running on Windows Vista. POTMiner has been implemented in Java using Sun Microsystems JDK 5.0, while Zaki's TreeMiner and SLEUTH C++ implementations were obtained from *http://www.cs.rpi.edu/~zaki/*.

The experiments were performed with 5 synthetic datasets generated by the tree generator available at *http://www.cs.rpi.edu/~zaki/software/TreeGen.tgz*. The datasets were obtained using the generator default values and varying the number of trees from 10 to 100000.

In our first experiments, we have compared POTMiner and TreeMiner / SLEUTH execution times. POTMiner and TreeMiner [8] are compared for (completely) ordered trees, while POTMiner and SLEUTH [9] are compared for unordered trees. The results we have obtained are summarized in Figure 3.

Looking at the charts in Figure 3, which use a logarithmic scale for their y axis, we can see that TreeMiner, SLEUTH, and POTMiner are efficient, scalable algorithms for mining induced and embedded subtrees in ordered (TreeMiner and POTMiner) and unordered (SLEUTH and POTMiner) trees. The observed differences are probably due to the different programming platforms used in their implementation (Java for POTMiner, C++ for TreeMiner and SLEUTH).

Figure 3 also shows that there are no significant differences between the POTMiner's execution times for ordered and unordered trees in the experiments performed with the aforementioned synthetic datasets.

Execution times of POTMiner and TreeMiner algorithm (*left*)for extracting induced (*top*) and embedded (*down*) subtrees in completely ordered trees. POTMiner and SLEUTH behaviour when mining completely unordered trees, induced (*top*) and embedded (*down*)is shown in right images.

We have also performed some experiments with partially-ordered trees. In this case, since TreeMiner [8] and SLEUTH [9] cannot be applied to partially-ordered trees, we have studied the behavior of POTMiner when dealing with this kind of trees. Starting from the same datasets used in the previous experiments, we have varied the percentage of ordered nodes in the datasets. The results we have obtained are shown in Figure 4.

As expected, we have found that execution times slightly decrease when the percentage of ordered nodes is increased, since ordered trees are easier to mine than unordered trees.

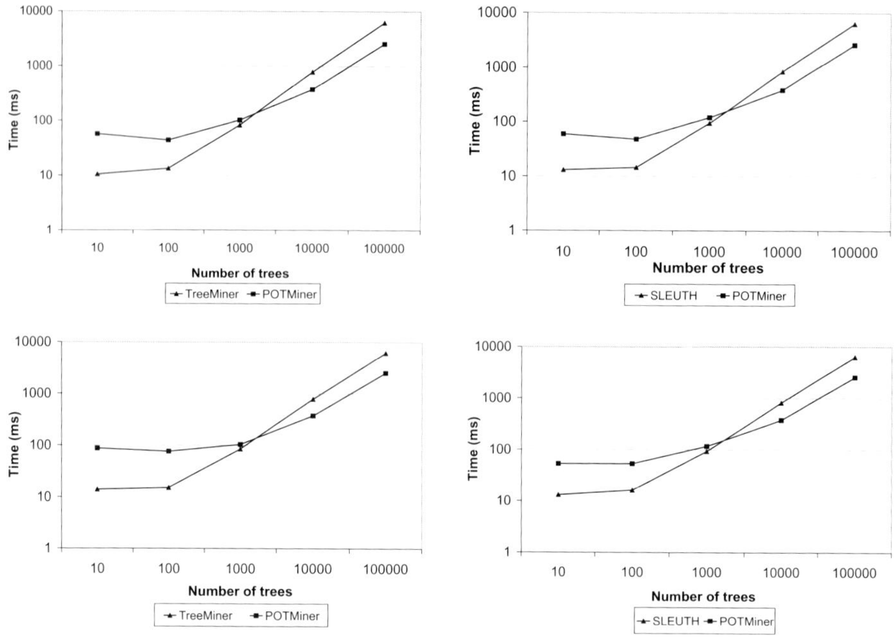

Fig. 3. Execution times of POTMiner and TreeMiner algorithms(*left images*) for extracting induced (*top*) and embedded (*down*) subtrees in completely ordered trees. POTMiner and SLEUTH execution times and completely unordered (*right*) trees.

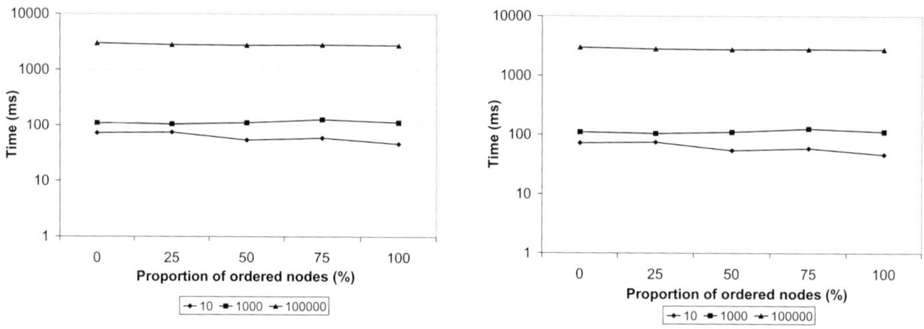

Fig. 4. POTMiner execution times varying the percentage of ordered nodes. The charts display POTMiner behavior when mining induced *(left)* and embedded *(right)* subtrees. The series show execution times for different dataset sizes (10, 1000, and 100000 trees).

6 Conclusions and Future Work

There are many different tree mining algorithms that work either on ordered or unordered trees, but none of them, to our knowledge, works with partially-ordered trees, that is, trees that have both ordered and unordered nodes. We have devised a new algorithm to address this situation that is as efficient and scalable as existing algorithms that exclusively work on either ordered or unordered trees.

Partially-ordered trees are important because they appear in different application domains. In the future, we expect to apply our tree mining algorithm to some of these domains. In particular, we believe that our algorithm for identifying frequent subtrees in partially-ordered trees can be useful in the following contexts:

– XML documents [16], due to their hierarchical structure, are directly amenable to tree mining techniques. Since XML documents can contain both ordered and unordered sets of nodes, partially-ordered trees provide a better representation model for them and they are better suited for knowledge discovery than existing ordered (or unordered) tree mining techniques.
– Multi-relational data mining [17] is another emerging research area where tree mining techniques can be useful. They might help improve existing multi-relational classification [18] and clustering [19] algorithms.
– In Software Engineering, it is usually acknowledged that mining the wealth of information stored in software repositories can "support the maintenance of software systems, improve software design/reuse, and empirically validate novel ideas and techniques" [20]. For instance, there are hierarchical program representations, such as dependence higraphs [21], which can be viewed as partially-ordered trees, hence the potential of tree mining techniques in software mining.

We also have the goal of making some refinements to our POTMiner algorithm, such as extending it for dealing with partial or approximate tree matching, a feature that would be invaluable for many real-world problems, from entity resolution in XML documents to program element matching in software mining.

Acknowledgements

Work partially supported by research project TIN2006-07262.

References

1. Chi, Y., Muntz, R.R., Nijssen, S., Kok, J.N.: Frequent subtree mining - an overview. Fundamenta Informaticae 66(1-2), 161–198 (2005)
2. Chi, Y., Yang, Y., Muntz, R.R.: HybridTreeMiner: An efficient algorithm for mining frequent rooted trees and free trees using canonical form. In: SSDBM 2004, pp. 11–20 (2004)
3. Agrawal, R., Srikant, R.: Fast algorithms for mining association rules in large databases. In: VLDB 1994, pp. 487–499 (1994)
4. Han, J., Pei, J., Yin, Y.: Mining frequent patterns without candidate generation. In: SIGMOD 2000, pp. 1–12 (2000)
5. Abe, K., Kawasoe, S., Asai, T., Arimura, H., Arikawa, S.: Efficient substructure discovery from large semi-structured data. In: SDM 2002 (2002)

6. Nijssen, S., Kok, J.N.: Efficient discovery of frequent unordered trees. In: First International Workshop on Mining Graphs, Trees and Sequences (MGTS 2003), in conjunction with ECML/PKDD 2003, pp. 55–64 (2003)
7. Asai, T., Arimura, H., Uno, T., ichi Nakano, S.: Discovering frequent substructures in large unordered trees. In: Grieser, G., Tanaka, Y., Yamamoto, A. (eds.) DS 2003. LNCS (LNAI), vol. 2843, pp. 47–61. Springer, Heidelberg (2003)
8. Zaki, M.J.: Efficiently mining frequent trees in a forest: Algorithms and applications. IEEE Transactions on Knowledge and Data Engineering 17(8), 1021–1035 (2005)
9. Zaki, M.J.: Efficiently mining frequent embedded unordered trees. Fundamenta Informaticae 66(1-2), 33–52 (2005)
10. Hido, S., Kawano, H.: AMIOT: induced ordered tree mining in tree-structured databases. In: ICDM 2005, pp. 170–177 (2005)
11. Xiao, Y., Yao, J.F., Li, Z., Dunham, M.H.: Efficient data mining for maximal frequent subtrees. In: ICDM 2003, pp. 379–386 (2003)
12. Wang, C., Hong, M., Pei, J., Zhou, H., Wang, W., Shi., B.: Efficient pattern-growth methods for frequent tree pattern mining. In: Dai, H., Srikant, R., Zhang, C. (eds.) PAKDD 2004. LNCS (LNAI), vol. 3056, pp. 441–451. Springer, Heidelberg (2004)
13. Termier, A., Rousset, M.C., Sebag, M.: TreeFinder: a first step towards xml data mining. In: ICDM 2002, pp. 450–457 (2002)
14. Chi, Y., Xia, Y., Yang, Y., Muntz, R.R.: Mining closed and maximal frequent subtrees from databases of labeled rooted trees. IEEE Transactions on Knowledge and Data Engineering 17(2), 190–202 (2005)
15. Chi, Y., Yang, Y., Muntz, R.R.: Indexing and mining free trees. In: ICDM 2003, pp. 509–512 (2003)
16. Nayak, R., Zaki, M.J. (eds.): Knowledge discovery from xml documents. In: Nayak, R., Zaki, M.J. (eds.) KDXD 2006. LNCS, vol. 3915, Springer, Heidelberg (2006)
17. Džeroski, S.: Multi-relational data mining: An introduction. SIGKDD Explorations Newsletter 5(1), 1–16 (2003)
18. Yin, X., Han, J., Yang, J., Yu, P.S.: CrossMine: efficient classification across multiple database relations. In: ICDE 2004, pp. 399–410 (2004)
19. Yin, X., Han, J., Yu, P.S.: Cross-relational clustering with user's guidance. In: KDD 2005, pp. 344–353 (2005)
20. Gall, H., Lanza, M., Zimmermann, T.: 4th International Workshop on Mining Software Repositories (MSR 2007). In: ICSE COMPANION 2007, pp. 107–108 (2007)
21. Berzal, F., Cubero, J.C., Jimenez, A.: Hierarchical program representation for program element matching. In: Yin, H., Tino, P., Corchado, E., Byrne, W., Yao, X. (eds.) IDEAL 2007. LNCS, vol. 4881, pp. 467–476. Springer, Heidelberg (2007)

A Structure-Based Clustering on LDAP Directory Information

Vassiliki Koutsonikola, Athena Vakali, Antonios Mpalasas, and Michael Valavanis

Department of Informatics
Aristotle University
54124 Thessaloniki, Greece
{vkoutson,avakali,antoniom,mvalavan}@csd.auth.gr

Abstract. LDAP directories have rapidly emerged as the essential framework for storing a wide range of heterogeneous information under various applications and services. Increasing amounts of information are being stored in LDAP directories imposing the need for efficient data organization and retrieval. In this paper, we propose the LPAIR & LMERGE (LP-LM) hierarchical agglomerative clustering algorithm for improving LDAP data organization. LP-LM merges a pair of clusters at each step, considering the *LD-vectors*, which represent the entries' structure. The clustering-based LDAP data organization enhances LDAP server's response times, under a specific query framework.

1 Introduction

Directory services provide a generic and appropriate framework for accessing a variety of information. They act as database repositories ensuring more efficient data retrieval mechanisms through the usage of Lightweight Directory Access Protocol (LDAP)[15]. LDAP is an open industry standard that gains wide acceptance due to its flexibility and the fact that it integrates with an increasing number of data retrieval and management applications [6].

To date, there are multiple applications that rely on LDAP servers. Most of the operating LDAP-based servers store information that describe user profiles and address books for messaging applications, configuration files of network devices and network security policies, under the Directory Enabled Networks (DEN) initiative [5]. Directory servers are also used to store certificates and revocation lists for PKI applications [1] as well as access control lists for authentication systems [13]. The new H.350 standard uses LDAP to provide a uniform way to store information related to video and voice over IP (VoIP) in directories [3]. Moreover, Grid computing has emerged as a very promising infrastructure for distributed computing, having its foundation and core on the distributed LDAP directories [2].

Due to the heterogeneity of LDAP data, finding appropriate data organization schemes such as clustering will ensure LDAP servers' functionality and performance. Earlier research efforts have focused either on proposing application-oriented LDAP schema definitions [12], or on introducing pure caching [7] and

A. An et al. (Eds.): ISMIS 2008, LNAI 4994, pp. 121–130, 2008.

indexing [8] approaches that can improve performance and scalability of directory based services. However, a framework that will propose a well-distributed and scalable LDAP data organization, regardless of the underlying application, enhancing at the same time system's performance, is necessary.

In this paper we propose the LPAIR & LMERGE (LP-LM) algorithm, an agglomerative structure-based clustering algorithm, for LDAP data organization. According to the authors' knowledge, LDAP and data clustering technologies have been barely combined. A clustering approach of LDAP metadata has been proposed to facilitate discovering of related directory objectclasses to better enable their reconciliation and reuse [11]. Our work applies clustering analysis on LDAP entries and uses clustering results in order to define the LDAP data arrangement. More specifically, our main contributions can be summarized as follows:

- We introduce the notion of LD-vectors to capture LDAP entries' structure
- We propose the structure-based LPAIR & LMERGE clustering algorithm which organizes LDAP data, regardless of the underlying applications.
- We carry out experiments to evaluate the LP-LM's efficiency as well as LDAP server's performance that adjusts its data organization to clustering results.

The rest of the paper is organized as follows: Section 2 discusses some basic concepts of LDAP data representation and the introduced LD-vector structures. Section 3 describes our problem formulation and the proposed LDAP clustering algorithm. Section 4 presents the experimentation while conclusions and future work insights are given in Section 5.

2 LDAP Background and Data Representation

Data is stored in LDAP directories in the form of entries arranged in hierarchical information trees. Each LDAP entry is identified by a distinguished name (DN) that declares its position in the hierarchy. The hierarchy's structure forms the directory information tree (DIT), which originates from a root (RootDN). In the basic LDAP notation, "dc" stands for domain component and "ou" for organizational unit. For example, the RootDN of the DIT that maintains clients' and products' data for a Greek company would be "dc=company-name, dc=gr", while the DN of the clients' and products' nodes would be "ou=clients, dc=company-name, dc=gr" and "ou=products, dc=company-name, dc=gr" respectively.

All information within a directory entry is stored as attribute-value pairs. The set of attributes that can appear in a given entry is determined by the objectclasses that are used to describe it. The definition of an objectclass specifies that some attributes may be mandatory while others optional. For example, the user defined objectclass "client", which can be used to describe a company's clients, would consider as mandatory the "surname" and "clientid" attributes while it would define as optional the "email" attribute. Moreover, the user defined "product" objectclass may consider as mandatory the "productNumber" and "price"

```
dn: clientid=500,ou=clients,dc=company-name,dc=gr
objectclass: client
clientid: 500
surname: Vakali
email: avakali@csd.auth.gr
```

(a) A client's entry

```
dn: productNumber=1989,ou=products,dc=company-name,dc=gr
objectclass: product
objectclass: shop
productNumber: 1989
price: 540
shopName: ACA Company
shopAddress: Tsimiski 15, Thessaloniki, Greece
```

(b) A product's entry

Fig. 1. LDAP entries LDIF

attributes, while the objectclass "shop", the "shopName" and "shopAddress" attributes that refer to the shop that the described product exists. Figures 1(a) and 1(b) present the LDIF[1] of a sample client and product entry respectively.

The set of rules that define the objectclasses and the attributes they contain constitutes the LDAP schema. LDAP allows the definition of new objectclasses and attributes while it supports objectclasses inheritance and thus new objectclasses can be created to extend existing ones. In each case, the LDAP schema defines whether there is relation between pairs of objectclasses or objectlass-attribute pairs.

In this paper, we consider a particular framework where we have as source a set $E = \{e_1, \ldots, e_f\}$ of f LDAP entries. Let $O = \{o_1, \ldots, o_m\}$ denote the set of m objectclasses and $A = \{a_1, \ldots, a_n\}$ the set on n attributes used to describe E. As discussed above, the LDAP schema defines the related objectclasses pairs (due to inheritance) or the related objectlass-attribute pairs.

Definition 1 (LD-Pair of an entry). *Given an LDAP entry $e_i \in E$, the **LD-Pair** of e_i is a set of pairs $LDP(e_i) = \{(d_x, d_y)\} : d_x \in O, d_y \in \{O \cup A\}, d_x \neq d_y$, if and only if $\forall (d_x, d_y)$ pair, d_y is an objectclass that inherits objectclass d_x or d_y is an attribute describing d_x in e_i.*

Example 1. Consider the two LDAP entries depicted in Figure 1. The LD-Pair for the client entry denoted as e_{client} is $LDP(e_{client})$={(client, clientid), (client, surname), (client, email)} while the LD-Pair for the product entry denoted as $e_{product}$ is $LDP(e_{product})$={(product, productNumber), (product,price), (shop, shopName), (shop, shopAddress)}. □

The definition of the entry's LD-Pair can also be extended to a set of entries.

Definition 2 (LD-Pair of an entries' set). *Given a set $E^* \subseteqq E$ of f^* LDAP entries where $f^* \leq f$, the $LDP(E^*)$ is defined as $LDP(E^*) = \{\cup LDP(e_i)\}$, $\forall e_i \in E^*$.*

Example 2. We consider the set $E^* = \{e_{client}, e_{product}\}$ and the $LDP(e_{client})$, $LDP(e_{product})$ as discussed in the Example 1. According to Definition 2,

[1] LDIF (LDAP Data Interchange Format) is a standard for representing LDAP entries in human readable format.

LDP(E^*)= {(client, clientid), (client, surname), (client, email), (product, productNumber), (product,price), (shop, shopName), (shop, shopAddress)}. □

Next, we use the LD-Pair concept to define a vector data structure which represents the LDAP entries' structure.

Definition 3 (LD-VECTOR). *Given the E^* set of f^* LDAP entries and its LDP(E^*), we use l to denote the number of (d_x, d_y) pairs where $(d_x, d_y) \in$ LDP(E^*). Then $\forall e_i \in E^*$ we define the binary multivariate* ***LD-vector*** *$LDV(e_i, :)$ of l values as follows:*

$$\underset{1 \leq i \leq f^*, 1 \leq r \leq l}{LDV(e_i, r)} = \begin{cases} 1 \text{ if the } r\text{-th } (d_x, d_y) \text{ pair of } LDP(E^*) \in LDP(e_i) \\ 0 \text{ otherwise} \end{cases}$$

Example 3. Given the set $E^* = \{e_{client}, e_{product}\}$ and its *LDP*(E^*), then based on Defintion 3 the LD-vectors of e_{client} and $e_{product}$ are $LDV(e_{client})$=[1 1 1 0 0 0 0] and $LDV(e_{product})$=[0 0 0 1 1 1 1], respectively. □

3 LDAP Data Structure-Based Clustering

The binary nature of LD-vectors identifies them as categorical data. According to [4],[10], pure distance-based clustering algorithms may not be as effective on categorical data as link-based ones. The proposed clustering algorithm adopts the link-based perspective which groups entries in terms of their common neighbors, as expressed by the link value.

3.1 Problem Formulation

The representation of LDAP entries as binary vectors makes dissimilarity coefficients an appropriate choice for measuring distance between them. We choose Czekanowski (or Dice) dissimilarity coefficient as distance measure, instead of popular Jaccard, to give more gravity to the elements that two entries have in common. Given two binary vectors $LDV(e_i, :)$ and $LDV(e_j, :)$ of length l where $e_i, e_j \in E, i \neq j$, their distance $D(e_i, e_j)$ in terms of Czekanowski coefficient is defined as:

$$D(e_i, e_j) = \frac{b + c}{2a + b + c}$$

where, for $1 \leq t \leq l$, $a = |t| : \{LDV(e_i, t) = LDV(e_j, t) = 1\}$, $b = |t| : \{LDV(e_i, t) = 1 \wedge LDV(e_j, t) = 0\}$, $c = |t| : \{LDV(e_i, t) = 0 \wedge LDV(e_j, t) = 1\}$.

The values of D range between 0 and 1 with higher values indicating higher dissimilarity between the involved entries. The Czekanowski dissimilarity coefficient has been used to capture distances in various clustering approaches [14],[9].

Two entries are considered to be neighbors if their distance is less than a user defined threshold θ. The set $LN(e_i)$ contains the LDAP entries that are neighbors to $e_i \in E$ and is defined as $LN(e_i) = \{\cup e_j\} : D(e_i, e_j) \leq \theta, \forall e_i, e_j \in E$. Moreover, for each of the entries belonging to one of the obtained clusters we compute the expected number of its neighbors as follows: Let C_i denote the $i-th$ of the k obtained clusters of size c_i. When $\theta = 0$, each entry belonging to a cluster C_i is expected to have only 1 neighbor, itself, while for $\theta = 1$ any other entry belonging to C_i is neighbor of e_i (their distance is always ≤ 1), resulting in c_i neighbors. For any other value of θ, where $0 < \theta < 1$, it is expected that higher values of θ will result in more neighbors of e_i. A quantity that applies to the above situation and can express the expected number of neighbors for an e_i entry is given by $c_i^{\frac{2\theta}{1+\theta}}$.

Furthermore, the link between two entries e_i and e_j expresses the number of their common neighbors and is calculated by $link(e_i, e_j) = \mid \mathrm{LN}(e_i) \cap \mathrm{LN}(e_j) \mid$ $\forall e_i, e_j \in E$. Given the expected number of neighbors for each e_i entry in cluster C_i, the e_i contributes to a link value equal to $c_i^{\frac{4\theta}{1+\theta}}$ (one for each pair of its neighbors)[2]. Then, the total number of expected links in C_i cluster caused by all c_i in number entries will be $c_i^{1+\frac{4\theta}{1+\theta}}$.

The link-based clustering approach aims at maximizing the link between each pair of entries belonging to a single cluster. According to the Definition 3, there may be a set of LDAP entries that are represented by the same LD-vector, which our structure-based clustering algorithm will assign to the same cluster (due to their common structure). This could lead to unbalanced cluster and thus, inspired by [4], we define a criterion function $J(E)$ where the total number of links between a cluster's entries is divided by the expected link value for this cluster weighed by the the number of its entries.

$$J(E) = \sum_{i=1}^{k} c_i * \sum_{e_x, e_y \in C_i} \frac{link(e_x, e_y)}{c_i^{1+\frac{4\theta}{1+\theta}}} \tag{1}$$

Our goal is to maximize the link value of entries contained in a cluster. Therefore, we define the LDAP Clustering problem as follows:

Problem 1 (LDAP CLUSTERING). Given a set E of f LDAP entries, an integer value k, and the criterion function $J(E)$, find a CL clustering of E into k clusters such that the $J(E)$ is maximized.

3.2 The LPAIR & LMERGE (LP-LM) Clustering Algorithm

The proposed LPAIR & LMERGE (LP-LM) algorithm is a hierarchical agglomerative clustering algorithm aiming at finding a solution to Problem 1. Since the goal of LP-LM is the maximization of the criterion function $J(E)$ (Equation 1),

---

[2] The total number of pairs that have e_i as neighbor is given by $2\binom{c_i^{\frac{2\theta}{1+\theta}}}{2} + c_i^{\frac{2\theta}{1+\theta}} = c_i^{\frac{4\theta}{1+\theta}}$ (each pair is measured twice, e.g. (e_x,e_y) and (e_y,e_x) while each entry e_x forms the (e_x,e_x) pair).

we need to specify the best pair of clusters to be merged at each step of the algorithm.

According to Equation 1, the maximization of $J(E)$ signifies maximization of each cluster's link value. Thus, in each iteration, the best pair of (C_i, C_j) clusters candidate for merging is the one with the highest link value defined as $link(C_i, C_j) = \sum_{e_x \in C_i, e_y \in C_j} link(e_i, e_j)$. Similarly to the definition of $J(E)$, in order to prevent the continuous merging of large-size clusters, we divide the $link(C_i, C_j)$ with an expected link value between C_i and C_j clusters. To compute the expected link value between two clusters we need to calculate the total link value of the two clusters if we considered them as one (i.e. $(c_i + c_j)^{1+\frac{4\theta}{1+\theta}}$) and subtract the link value of C_i (i.e. $c_i^{1+\frac{4\theta}{1+\theta}}$) and C_j (i.e. $c_j^{1+\frac{4\theta}{1+\theta}}$). We use this normalization factor as a heuristic to steer towards the maximization of the criterion function value. Therefore, the *merging criterion* $mc(C_i, C_j)$ of clusters C_i and C_j is defined as:

$$mc(C_i, C_j) = \frac{link(C_i, C_j)}{(c_i + c_j)^{1+\frac{4\theta}{1+\theta}} - c_i^{1+\frac{4\theta}{1+\theta}} - c_j^{1+\frac{4\theta}{1+\theta}}} \tag{2}$$

The pair of clusters that maximizes mc will be merged at each algorithm's step.

The LP-LM algorithm takes as input a set E of f LDAP entries, the number k of clusters to be created and a decimal θ (distance threshold), $0 \leq \theta \leq 1$ and results in the assignment of LDAP entries to the k clusters.

Algorithm 1. The LPAIR & LMERGE algorithm

Input: A set $E = \{e_1 \dots e_f\}$ of f LDAP entries, a threshold θ and the number of clusters k.
Output: Assignment of the LDAP entries in the k clusters, such that J is maximised.

```
1: /*Preprocessing*/
2: LDP = CreateLDPair(E)
3: LDV = CreateLDVectors(LDP)
4: D = ComputeDistance(LDV)
5: LN = ComputeNeighbors(D, θ)
6: link = ComputeLink(LN)
7: /*Clustering process*/
8: while NumClusters ≥ k do
9:    (C1, C2) = FindMergingClusters(link, mc)
10:   C* = merge(C1, C2)
11:   update(link, C*)
12: end while
```

Initially in LP-LM, a preprocessing takes place, where given the initial set E of LDAP entries the algorithm computes the entries' LD-Pairs (line 2) and then the respective LD-vectors (line 3). Using the Czekanowski coefficient (Section 3.1), the table D of distances between LDAP entries is computed (line 4) and then, based on D and θ, the algorithm calculates each entry's neighbors and stores them in table LN (line 5). The LN table is used for the computation of the link value for each pair of entries, resulting in table *link* (line 6). After the preprocessing step, an iterative process follows which constitutes the main clustering process. This process lasts until k clusters are obtained (line 8). During each iteration [3] of this step, the LP-LM algorithm finds the best pair of clusters (C_1,C_2) to me merged according to the mc values (line 9). The two clusters C_1 and C_2 are merged and a new C^* cluster is obtained (line 10). The table *link* is updated (line 11) with the new link values in terms of C^* cluster.

4 Experimentation

Our data consists of around 10000 entries that describe DBLP data[4]. The DBLP dataset contains about 2000 entries for each of the following publication categories: articles, inproceedings, masterthesis, phdthesis and www. The LDAP schema involves a set of objectclasses (e.g. article, phdthesis) and a set of attributes (e.g. author, title, pages). For the experiments we have used the OpenLDAP directory server with Berkeley DB backend, setting cache size to zero (to obtain unbiased results) and having the "publicationid" attribute indexed.

In the first section of our experimentation we study the LP-LM algorithm's performance for different values of θ and $k = 5$, given that our entries belong to 5 different categories, and we calculate the percentage of successfully clustered entries per category. The results for $\theta = 0.2, 0.3, 0.4$ and 0.5 are depicted in Figure 2(a). Lower values of θ demand higher similarity between two entries in order to be considered as neighbors, resulting in lower percentages of entries successfully assigned to clusters. Increasing the values of θ, the LP-LM manages to cluster successfully more entries while for $\theta = 0.5$ (and $\theta > 0.5$) there is no missclustered entry. The proper clustering is achieved for $\theta = 0.5$ and not for a lower value, because our dataset is described by a large number of distinct LD-vectors (e.g. there are 38 different LD-vectors capturing articles' structure). A higher number of LD-vectors indicates more dissimilar in structure LDAP entries and demands a higher value of θ to result in the proper clustering.

Moreover, as depicted in Figure 2(a), articles can not comprise a cluster for $\theta < 0.5$ because of the greater dissimilarity that exists between them compared to the other categories' entries. Furthermore, entries of both inproceedings and masterthesis categories can not be successfully clustered for low values of θ, eventhough lower percentages of their entries compared to articles, are not assigned properly. The LP-LM algorithmn assigns www and phdthesis entries correctly, even for low θ values. In all cases, the overall calculated percentage of successfully assigned entries is over 90% which proves the efficiency of the LP-LM algorithm.

In the second section of our experimentation we evaluate an LDAP server's performance, comparing its response times to a set of queries, in case of an un-

[3] In the first iteration, the pair of clusters is a pair of entries.

[4] DBLP data: http://www.sigmod.org/dblp/db/index.html The DBLP data were retrieved in XML format and converted to LDAP entries.

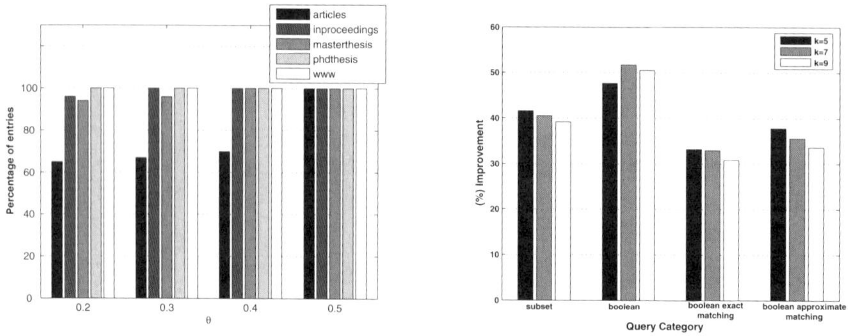

(a) Entries successfully assigned to clusters per category

(b) LDAP server's response time improvements

Fig. 2. Experimentation results

clustered data organization (flat DIT) and a clustered one, as indicated by the LP-LM clustering algorithm, for $\theta = 0.5$ and $k = 5, 7, 9$. In order to benefit from the data organization proposed, we apply a query processing which directs users' queries to specific LDAP data clusters. The idea behind the query processing operation is that, based on the users' queries keywords (expressed by the contained attributes and objectclasses) and a mapping scheme (which reveals the relations between clusters and sets of objectclasses and attributes), the responses' pathways in the LDAP DIT are determined. In the worst case scenario, the query's keywords will be found in all clusters and the search will have to start by the RootDN, resulting in response time equal to that of the unclustered scheme. In any other case, the search space would be reduced, resulting in improved response times. We have examined LDAP server's performance running a set of queries that belong to the following four categories:

– *Subset queries*: Queries that retrieve subsets of entries providing no query filter (e.g. all articles, all inproceedings).
– *Boolean queries*: Queries containing boolean expressions without involving specific attributes value (e.g. entries that have a booktitle but not an ISBN).
– *Boolean queries of exact matching*: Queries containing boolean expressions and filters specifying an exact matching (e.g. all inproceedings, www of 2002).
– *Boolean queries of approximate matching*: Boolean queries with approximate matching filters (e.g. articles, www containing in title the term "database").

It should be noted, that in all cases we executed the same set of queries for each category, and calculated the difference in response times. For all query categories, the response times were better in the clustering-based data organization. The obtained improvements were averaged and are depicted in Figure 2(b).

In case of "subset queries" we observe the same levels of improvements for the different values of k because the search space is reduced equally regarding the clustered data. For instance, in an unclustered data organization, a query looking for all phdthesis, must search all 10000 entries in order to retrieve them while in the clustering-based data organization, the query processing locates the one cluster containing the phdthesis, reducing the search space to 2000 entries. For $k = 7$, the LP-LM has created one cluster for each of the articles, inproceedings, masterthesis and www entries while the phdthesis have been distributed in 3 clusters. The search space for phdthesis remains to 2000 entries while the transmission between clusters causes negligible delay to the response time. For $k = 9$, the LP-LM creates one cluster for each of the masterthesis and www categories, 3 clusters for phdthesis, 2 clusters for articles and 2 clusters for inproceedings without affecting the search space of the "subset queries".

The improvement observed in case of "boolean queries" depends significantly on the query and the obtained clusters. For example, a query asking for "ee" values of phdthesis will be directed to the one of the 3 phdthesis clusters which is the only one containing entries with "ee" attribute, reducing significantly the search space and resulting in high improvements of about 78% (for $k = 7, 9$). On the other hand, asking for "volume" values results in 14% improvement (for all k values), since "volume" is a common attribute contained in entries that belong to all categories except for masterthesis.

Similarly, for the boolean queries of exact and approximate matching, the obtained improvement is dependent of the query's keywords and the way they restrict the search space. For example, a query requesting articles that contain in their title the word "database" resulted in 75% improvement while a query retrieving articles, phdthesis, www and inproceedings published in one of the IEEE journals led to 15% improvement.

The above indicative (due to the lack of space) discussion clearly shows that the clustering-based data organization yields enhancement of LDAP server's performance, in terms of all LDAP query types. The recorded improvements depend on the search space defined by the query keywords (e.g. www or inproceedings) as well as the distinctiveness of attributes. The response time improvements are noticeable even in case of common attributes (e.g. "title") but they are remarkable when attributes appearing in restricted publication types (e.g. "ee") are involved. Moreover, the discussed improvements refer to a rather homogeneous dataset since publications are not described by a considerable variety of attributes and objectclasses. An LDAP server storing more heterogeneous data is expected to be more benefitted from the proposed clustering approach.

5 Conclusions and Future Work

This paper presents a structure-based clustering algorithm which is used to define the organization of the LDAP Data Information Tree. The proposed LPAIR & LMERGE (LP-LM) algorithm has been proved to perform efficiently in case of LDAP data described by different sets of objectclasses and attributes. Moreover, the LDAP server that adjusts its data organization to the clustering results presents improved response times, under a specific query framework. The recorded improvements are especially high in case of queries containing keywords that correspond to distinctive clusters. Therefore, the LP-LM algorithm can be particularly beneficial for an LDAP server storing various information such as multimedia, network device configuration files and user profiles, enhancing the performance of the underlying applications.

For the future, we plan to incorporate knowledge about data's content to the overall clustering process and experiment with more distance metrics.

References

1. Chadwick, D.: Deficiencies in LDAP when used to support PKI. Communications of the ACM 46, 99–104 (2003)
2. Fan, Q., Wu, Q., He, Y., Huang, J.: Optimized Strategies of Grid Information Services. In: Proc. of the First Int. Conf. on Semantics, Knowledge, and Grid, p. 90 (2005)
3. Gemmill, J., Chatterjee, S., Miller, T., Verharen, E.: ViDe.Net Middleware for Scalable Video Services for Research and Higher Education. In: ACM Southeastern Conf. GA. ACM 1-58113-675-7/030/03, pp. 463–468 (2003)
4. Guha, S., Rastogi, R., Shim, K.: ROCK: A Robust Clustering Algorithm For Categorical Attributes. In: Proc. 15th Int. Conf. Data Eng., pp. 512–521 (1999)
5. Howes, T., Smith, M.: LDAP: Programming Directory-Enabled Applications with Lightweight Directory Access Protocol. Macmillan Technical Publishing, Basingstoke (1997)
6. Koutsonikola, V., Vakali, A.: LDAP: Framework, Practices, and Trends. IEEE Internet Computing 8, 66–72 (2004)
7. Kumar, A., Gupta, R.: Edge Caching for Directory Based Web Applications: Algorithms and Performance. In: Proc. of the 8th international workshop in Web content caching and distribution, pp. 39–56 (2004)
8. Lee, H., Mun, S.-G., Huh, E.-N., Choo, H.: Efficient Data Indexing System Based on OpenLDAP in Data Grid. In: Int. Conf. on Computational Science, vol. 1, pp. 960–964 (2006)
9. Li, T.: A Unified View on Clustering Binary Data. Machine Learning 62, 199–215 (2006)
10. Lian, W., Cheung, D., Mamoulis, N., Yiu, S.-M.: An Efficient and Scalable Algorithm for Clustering XML Documents by Structure. IEEE Trans. on Knowledge and Data Engineering 16, 82–96 (2004)
11. Liang, J., Vaishnavi, V., Vandenberg, A.: Clustering of LDAP directory schemas to facilitate information resources interoperability across organizations. IEEE Trans. on Systems, Man and Cybernetics, Part A 36, 631–642 (2006)
12. Lim, S., Choi, J., Zeilenga, K.: Design and Implementation of LDAP Component Matching for Flexible and Secure Certificate Access in PKI. In: Proc. of the 4th Annual PKI R&D Workshop, pp. 41–51 (2005)
13. Park, J., Sandhu, R., Ahn, G.-J.: Role-based access control on the web. ACM Trans. on Information and System Security (TISSEC) 4, 37–71 (2001)
14. Ponaramenko, J., Bourne, P., Shindyalov, I.: Building an Automated Classification of DNA-binding Protein Domains. Bioinformatics 18, S192–S201 (2002)
15. Whal, M., Howes, T., Kille, S.: Lightweight Directory Access Protocol (v3). IETF RFC 2251 (1997)

iZi: A New Toolkit for Pattern Mining Problems

Frédéric Flouvat, Fabien De Marchi, and Jean-Marc Petit

Université de Lyon, LIRIS, UMR5205 CNRS, F-69621, France
firstname.lastname@liris.cnrs.fr

Abstract. Pattern mining problems are useful in many applications. Due to a common theoretical background for such problems, generic concepts can be re-used to easier the development of algorithms. As a consequence, these problems can be implemented with only minimal effort, i.e. programmers do not have to be aware of low-level code, customized data structures and algorithms being available for free. A toolkit, called *iZi*, has been devised and applied to several problems such as itemset mining, constraint mining in relational databases and query rewriting in data integration systems. According to our first results, the programs obtained using our library offer a very good tradeoff between performances and development simplicity.

1 Introduction

In the last decade, many algorithms have been devised for pattern mining problems (such as for **F**requent **I**temset **M**ining). This is especially true for pattern mining problems known to be representable as set [1], as for instance, frequent itemset mining and variants [2,3], functional or inclusion dependency inference [4] or learning monotone boolean function [1]. Recently, other application domains have been identified such as discovery of schema matching [5] or query rewriting in integration systems [6]. In this setting, a common idea is to say that algorithms devised so far should be useful to answer these tasks and available open source implementations are a great source of know-how. Unfortunately, it seems rather optimistic to envision the application of most of publicly available implementations of frequent itemset mining algorithms, even for closely related problems. Indeed, sophisticated data structures specially devised for monotone predicates turn out to give very efficient algorithms but limit their application to other data mining tasks. As a consequence, low-level implementations hamper the rapid advances in the field.

Paper contribution. This paper takes advantage of the common theoretical background of problems isomorphic to boolean lattices. We provide a generic architecture and an implementation framework for this family of pattern mining problems. It encompasses efficient data structures for boolean lattice representation and several generic implementations of well known algorithms, such as a levelwise algorithm and a dualization-based algorithm. By the way, any problem can be implemented with only minimal effort, i.e. the programmers do not have to be aware of low-level code, customized data structures and algorithms being available for free. To the best of our knowledge, our contribution is the

A. An et al. (Eds.): ISMIS 2008, LNAI 4994, pp. 131–136, 2008.

only one providing a generic theoretical and implementation framework for this family of pattern mining problems. A toolkit called *iZi* has been devised and applied to several problems such as itemset mining, constraint mining in relational databases and query rewriting in data integration systems. According to our first results, the programs obtained using our toolkit have very interesting performances regarding simplicity of their development.

2 Related Work

Several packages and libraries have also been proposed for data mining. However, most of them do not focus on interesting pattern discovery problems and address more specifics data mining tasks (classification, clustering,...).

To our knowledge, only the DMTL library has objectives close to *iZi* w.r.t. code reusability and genericity. DMTL (Data Mining Template Library) is a C++ library for frequent pattern mining which supports any types of patterns representable as graphs (sets, sequences, trees and graphs). However, the motivations are quite different: while DMTL focuses on patterns genericity w.r.t. the frequency criteria only, *iZi* focuses on a different class of patterns but on a wider class of predicates. Moreover, *iZi* is based on a well established theoretical framework, whereas DMTL does not rely on such a theoretical background. However, DMTL encompasses problems that cannot be integrated into *iZi*, for instance frequent sequences or graphs mining since such problems are not isomorphic to a boolean lattice.

3 Theoretical Framework

We recall in this section the theoretical KDD framework defined in [1] for interesting pattern discovery problems. Given a database d, a finite language $\mathcal{L}$ for expressing patterns or defining subgroups of the data, and a predicate Q for evaluating whether a pattern $\varphi \in \mathcal{L}$ is true or "interesting" in d, the discovery task is to find the theory of d with respect to $\mathcal{L}$ and Q, i.e. the set $Th(\mathcal{L}, d, Q) = \{\varphi \in \mathcal{L} \mid Q(d, \varphi) \text{ is true}\}$.

Let us suppose a specialization/generalization relation between patterns of $\mathcal{L}$. Such a relation is a partial order $\preceq$ on the patterns of $\mathcal{L}$. We say that φ is more general (resp. more specific) than θ, if $\varphi \preceq \theta$ (resp. $\theta \preceq \varphi$). Let $(I, \preceq)$ be a partially ordered set of elements. A set $S \subseteq I$ is *closed downwards* (resp. *closed upwards*) if, for all $X \in S$, all subsets (resp. supersets) of X are also in S. The predicate Q is said to be monotone (resp. anti-monotone) with respect to $\preceq$ if for all $\theta, \varphi \in \mathcal{L}$ such that $\varphi \preceq \theta$, if $Q(d, \varphi)$ is true (resp. false) then $Q(d, \theta)$ is true (resp. false). As a consequence, if the predicate is monotone (resp. anti-monotone), the set $Th(\mathcal{L}, d, Q)$ is upward (resp. downward) closed, and can be represented by either his *positive border* or his *negative border*. The *positive border*, denoted by $\mathcal{B}d^+(Th(\mathcal{L}, d, Q))$, made up of the MOST SPECIALIZED true patterns when $Th(\mathcal{L}, d, Q)$ is downward closed, and the MOST SPECIALIZED false patterns when $Th(\mathcal{L}, d, Q)$ is upward closed. The *negative border*, denoted by $\mathcal{B}d^-(Th(\mathcal{L}, d, Q))$, made up of the MOST GENERALIZED false patterns when $Th(\mathcal{L}, d, Q)$ is downward closed, and the MOST GENERALIZED true patterns when $Th(\mathcal{L}, d, Q)$ is upward closed.

The last hypothesis of this framework is that the problem must be representable as sets via an isomorphism, i.e. the search space can be represented by a boolean lattice (or subset lattice). Let $(\mathcal{L}, \preceq)$ be the ordered set of all the patterns defined by the language $\mathcal{L}$. Let E be a finite set of elements. The problem is said to be *representable as sets* if a bijective function $f : (\mathcal{L}, \preceq) \rightarrow (2^E, \subseteq)$ exists, and its inverse function f^{-1} is computable, such that: $X \preceq Y \iff f(X) \subseteq f(Y)$.

Example 1: Key mining. *Let us consider the key discovery problem in a relation, which can be enounced as follows: Let r be a relation over a schema R, extract the (minimal) keys satisfied in r. The patterns are : $\{X \mid X \subseteq R\} = \mathcal{P}(R)$. X is true if X is a superkey, i.e. if $|\pi_X(r)| = |r|$, where $\pi_X(r)$ is the projection onto X over r. It is clear that any superset of a superkey is also a superkey, it justifyes that only minimal keys are really interested. One can deduce that* minimal keys constitute the positive border of superkeys, with natural set inclusion. *The transformation function here is the identity since patterns are sets.*

4 A Generic Toolkit for Pattern Discovery

Based on the theoretical framework presented in section 3, we propose a C++ library, called *iZi*, for these pattern mining problems. The basic idea is to offer a toolbox for a rapid development of efficient and robust programs. The development of this toolkit takes advantage of the past experience to solve particular problems such as frequent itemsets mining, functional dependency mining, inclusion dependency mining and query rewriting...

4.1 Generic Algorithms and Data Structures

Even if this framework has been frequently used at a theoretical level, it has never been exploited at a technical point of view. One of our goal is to factorize some technical solutions which can be common to any pattern mining problem representable as sets. We are interested in algorithms and data structures that apply directly on sets, since they can be applied without any change for any problem, exploiting the isomorphic transformation. Our solution reuse some previous works done for frequent itemset mining, which is a problem "directly" representable as sets.

Currently, many algorithms from the multitude that has been proposed for the FIM problem could be implemented into *iZi*, from classical levelwise [7] and depth-first approaches, to more sophisticated dualization-based algorithms [8,4]. Since the generic part of our library only manipulates sets, we use a data structure based on prefix-tree (also called *trie*) specially devoted to this purpose [9]. They have not only a power of compression by factorizing common prefix in a set collection, but are also very efficient for candidate generation. Moreover, prefix-trees are well adapted for inclusion and intersection tests, which are basic operations when considering sets.

4.2 Architecture

Figure 1 represents the architecture of our library. The figure 2 presents how the library solves the inclusion dependency (IND) mining problem using the levelwise strategy. The *algorithm* is initialized (*initialization* function) with patterns corresponding to singletons in the set representation, using the data (*data access* component). Then, during the execution of the algorithm, the *predicate* is used to test each pattern against the data. Before testing a element, the algorithm uses the *set transformation* function to transform each set generated into the corresponding pattern. This architecture is directly derived from the studied framework and has the main advantage of decoupling algorithms, patterns and data. Only the *predicate*, *set transformation* and *initialization* components are specifics to a given problem. Consequently, to solve a new problem, users may have to implement or reuse with light modifications some of these components.

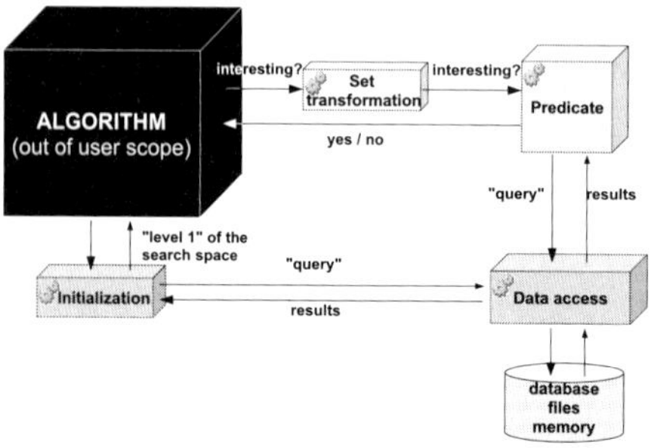

Fig. 1. *iZi* architecture

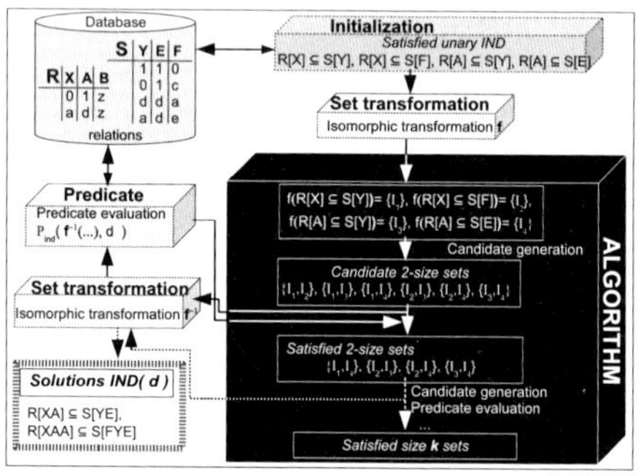

Fig. 2. IND mining example

As shown in figure 1, algorithms are **decoupled of the problems** and are **black box for users**. Each algorithm can be used directly to solve any problem fitting in the framework without modifications. This leads to the rapid construction of robust programs without having to deal with low-level details. Currently, the library offers two bottom-up algorithms (an *Apriori*-like [7,1] and *ABS* [4]), two top-down algorithms (top-down versions of *Apriori* and *ABS*) and depth-first strategies are currently being integrated.

Another important aspect of our library is that data access is totally decoupled of all other components (see figure 1). Currently, data access in most of the implementations is tightly coupled with algorithm implementations and predicates. Consequently, algorithms and "problem" components can be used with different data formats without modifications.

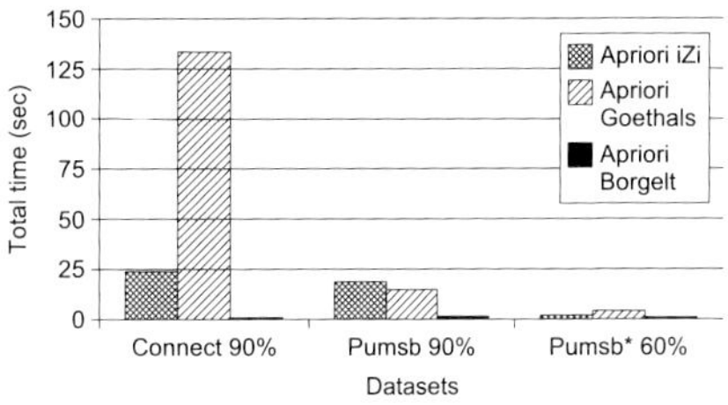

Fig. 3. Comparison of three *Apriori* implementations

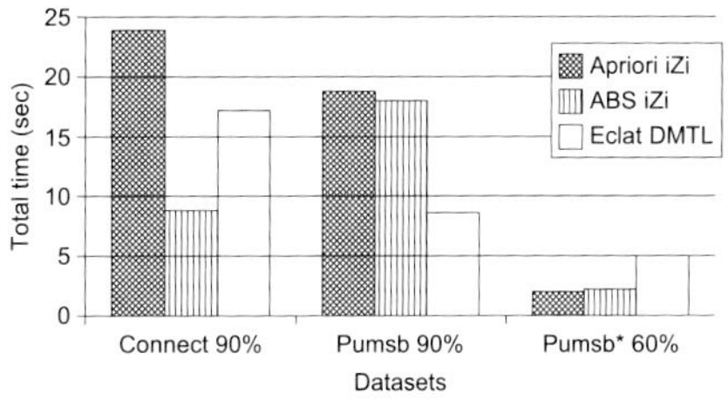

Fig. 4. Comparison of iZi and DMTL implementations

5 Experimentation

From our past experience in the development of pattern mining algorithms, we note that the adaptation of existing implementations is extremely difficult. In some cases, it implies the redevelopment of most of the implementation and could take more time than developing a new program from scratch.

Many problems have been implemented in our library along with several components. Our library was tested against 5 problems (and 5 different data formats): frequent and frequent essential itemset mining (FIMI file format [10,11]), inclusion dependency and key mining (FDEP file format [12], CSV file format or *MySQL* DBMS), and query rewriting in integration systems (specific file format). As indication, the use of our library to implement a program for the key mining problem has been done in less than one working day. Note that thanks to our library, an external module has been developed and integrated into a query rewriting prototype, allowing the scalability with respect to the number of views. From our point of view, this is a typical case where our library is very useful, providing a scalable component, almost for free, for the data-centric problems of a larger problem/application.

Performances. Implementations for FIM are very optimized, specialized, and consequently very competitive. The most performant ones are often the results of many years of research and development. In this context, our experimentations aims at proving that our generic algorithms implementations behave well compared to specialized ones. Moreover, we compare *iZi* to the DTML library, which is also optimized for frequent pattern mining. Experiments have been done on some classical benchmark datasets [10,11]. We compared our *Apriori* generic implementation to two others devoted implementations: one by B. Goethals and one by C. Borgelt [13]. The first one is a quite natural version, while the second one is, to our knowledge, the best existing *Apriori* implementation, developed in *C* and strongly optimized. Then, we compared "*iZi Apriori* and *ABS*" to the eclat implementation provided with DMTL. As shown in figures 3 and 4, our generic version has good performances with respect to other implementations. The difference between the two libraries is mainly due to the algorithm used during the experimentations. These results are very encouraging, in regards of the simplicity to obtain an operational program.

6 Discussion and Perspectives

In this paper, we have considered a classical problem in data mining: the discovery of interesting patterns for problems known to be *representable as sets*, i.e. isomorphic to a boolean lattice. As far as we know, this is the first contribution trying to bridge the gap between fundamental studies in data mining and practical aspects of pattern mining discovery. Our work concerns plenty of applications from different areas such as databases, data mining, or machine learning. Many perspectives exist for this work. First, we may try to integrate the notion of *closure* which appears under different flavors in many problems. The basic research around concept lattices could be a unifying framework. Secondly, we are interested in integrating the library as a plugin for a data mining software such as Weka. Analysts could directly use the algorithms to solve already implemented problems or new problems by dynamically loading their own components. Finally, a natural perspective of this work is to develop a declarative version for such mining problems using query optimization techniques developed in databases [14].

References

1. Mannila, H., Toivonen, H.: Levelwise search and borders of theories in knowledge discovery. Data Min. Knowl. Discov. 1(3), 241–258 (1997)
2. Agrawal, R., Imielinski, T., Swami, A.N.: Mining association rules between sets of items in large databases. In: SIGMOD Conference, pp. 207–216 (1993)
3. Bastide, Y., Taouil, R., Pasquier, N., Stumme, G., Lakhal, L.: Mining frequent patterns with counting inference. SIGKDD Explorations 2(2), 66–75 (2000)
4. De Marchi, F., Flouvat, F., Petit, J.M.: Adaptive strategies for mining the positive border of interesting patterns: Application to inclusion dependencies in databases. In: Constraint-Based Mining and Inductive Databases, pp. 81–101 (2005)
5. He, B., Chang, K.C.C., Han, J.: Mining complex matchings across web query interfaces. In: Das, G., Liu, B., Yu, P.S. (eds.) DMKD, pp. 3–10. ACM, New York (2004)
6. Jaudoin, H., Petit, J.M., Rey, C., Schneider, M., Toumani, F.: Query rewriting using views in presence of value constraints. In: Description Logics (2005)
7. Agrawal, R., Srikant, R.: Fast algorithms for mining association rules in large databases. In: VLDB, pp. 487–499 (1994)
8. Gunopulos, D., Khardon, R., Mannila, H., Saluja, S., Toivonen, H., Sharm, R.S.: Discovering all most specific sentences. ACM Trans. Database Syst. 28(2) (2003)
9. Bodon, F.: Surprising results of trie-based fim algorithms [11]
10. Bayardo Jr., R.J., Zaki, M.J. (eds.): FIMI 2003, Proceedings of the IEEE ICDM Workshop on Frequent Itemset Mining Implementations, USA (November 2003)
11. Bayardo Jr., R.J., Goethals, B., Zaki, M.J. (eds.): FIMI 2004, Proceedings of the IEEE ICDM Workshop on Frequent Itemset Mining Implementations, UK (November 2004)
12. Flach, P.A., Savnik, I.: Database dependency discovery: A machine learning approach. AI Commun. 12(3), 139–160 (1999)
13. Borgelt, C.: Efficient implementations of Apriori and Eclat [10]
14. Chaudhuri, S.: Data mining and database systems: Where is the intersection? IEEE Data Eng. Bull. 21(1), 4–8 (1998)

A Multi-relational Hierarchical Clustering Method for DATALOG Knowledge Bases

Nicola Fanizzi, Claudia d'Amato, and Floriana Esposito

Dipartimento di Informatica – Università degli Studi di Bari
Campus Universitario, Via Orabona 4, 70125 Bari, Italy
{fanizzi,claudia.damato,esposito}@di.uniba.it

Abstract. A clustering method is presented which can be applied to relational knowledge bases (e.g. DATALOG deductive databases). It can be used to discover interesting groups of resources through their (semantic) annotations expressed in the standard logic programming languages. The method exploits an effective and language-independent semi-distance measure for individuals., that is based on the resource semantics w.r.t. a number of dimensions corresponding to a committee of features represented by a group of concept descriptions (discriminating features). The algorithm is a fusion of the classic BISECTING K-MEANS with approaches based on medoids that are typically applied to relational representations. We discuss its complexity and potential applications to several tasks.

1 Unsupervised Learning with Complex Data

In this work, we investigate on unsupervised learning for knowledge bases (KBs) expressed in relational languages. In particular, we focus on the problem of conceptual clustering of semantically annotated resources. The benefits of *conceptual clustering* [10] in such a context are manifold: 1) *concept formation*: clustering annotated resources enables the definition of new emerging concepts on the grounds of the primitive concepts asserted in a KB; 2) *evolution*: supervised methods can exploit these clusters to induce new concept definitions or to refining existing ones; 3) *search and ranking*: intensionally defined groupings may speed-up the task of search upon queries; a hierarchical clustering also suggests criteria for ranking the retrieved resources.

Essentially, many existing clustering methods are based on the application of similarity (or density) measures defined over a fixed set of attributes of the domain objects. Classes of objects are taken as collections that exhibit low interclass similarity (density) and high intraclass similarity (density). Often these methods cannot take into account *background knowledge* that could characterize object configurations by means of global concepts and semantic relationship. As pointed out in related surveys [11], initially, most of the proposed similarity measures for concept descriptions focus on the similarity of atomic concepts within simple concept hierarchies or are strongly based on the structure of the terms for specific FOL fragments [4]. Alternative approaches are based on the notions of *feature* similarity or *information content*. In the perspective of exploiting similarity measures in inductive (instance-based) tasks like those mentioned above, the need for a definition of a semantic similarity measure for *instances* arises [2, 8].

A. An et al. (Eds.): ISMIS 2008, LNAI 4994, pp. 137–142, 2008.

Early conceptual clustering methods aimed at defining groups of objects using conjunctive descriptions based on selected attributes [10]. Anyway, in the conceptual clustering perspective, the expressiveness of the language adopted for describing objects and clusters (concepts) is equally important. Alternative approaches, suitable to concept languages, have pursued a different way for attacking the problem, devising logic-based methods [3]. However, these methods may suffer from noise in the data. This motivates our investigation on similarity-based clustering methods which can be more noise-tolerant. We propose a multi-relational extension of effective clustering techniques. It is intended for grouping similar resources w.r.t. a semantic dissimilarity measure in order to discover new concepts. Our relational method derives from the *Bisecting k-means* algorithm [5], a well-known partitional clustering method. Specifically, we recur to the notion of *medoids* (like in algorithm *PAM* [6]) as central individual in a cluster, rather than to the notion of means characterizing the algorithms descending from *k-means* and *EM* [5] developed for numeric (or ordinal) features. Upgrading existing algorithms to work on multi-relational representations such as clausal languages, requires novel similarity measures that are suitable for such representations. Moreover, rather than fix a given number k of clusters of interest (that may be hard when scarce domain knowledge is available), a partitional method may be employed up to reaching a minimal threshold value for cluster *quality* [6, 5] which makes any further bisections useless.

In the next section the dissimilarity measure adopted in the algorithm is defined. The clustering algorithm is presented in Sect. 3. Possible developments are examined in Sect. 4.

2 A Family of Metrics for Instances

In the following, we assume that objects (instances), concepts and relationships among them are defined in terms of a function-free (yet not constant-free) clausal language such as DATALOG, endowed with the standard semantics (see [7]). A *knowledge base* is defined as $\mathcal{K} = \langle \mathcal{P}, \mathcal{D} \rangle$, where $\mathcal{P}$ is a logic program representing the *schema*, with concepts (entities) and relationships defined through definite clauses, *database* $\mathcal{D}$ is a set of ground facts concerning the world state. Without loss of generality, we will consider concepts as described by unary atoms. *Primitive* concepts are defined in $\mathcal{D}$ extensionally by means of ground facts only, whereas *defined* concepts will be defined in $\mathcal{P}$ by means of clauses. The set of the objects occurring in $\mathcal{K}$ is denoted with $\mathsf{const}(\mathcal{D})$. As regards the necessary inference services, our measures will require performing *instance-checking*, which amounts to determining whether an object belongs (is an instance) of a concept in a certain interpretation.

Instances lack a syntactic structure that may be exploited for a comparison. However, on a semantic level, similar objects should *behave* similarly w.r.t. the same concepts, i.e. similar assertions (facts) should be shared. Conversely, dissimilar instances should likely instantiate disjoint concepts. Therefore, we introduce novel dissimilarity measures for objects, whose rationale is the comparison of their semantics w.r.t. a fixed number of dimensions represented by concept descriptions (predicate definitions). Instances are compared on the grounds of their behavior w.r.t. a reduced (yet not necessarily disjoint) committee of features (concept descriptions) $\mathsf{F} = \{F_1, F_2, \ldots, F_m\}$, expressed in the language taken into account, acting as discriminating *features*. We will consider unary predicates which have a definition in the KB. Following [9], a family of totally semantic distance measures for objects can be defined for clausal representations. In its simplest formulation, inspired by Minkowski's metrics, it is defined as:

Definition 2.1 (family of measures). *Let $\mathcal{K}$ be a KB. Given a set of concept descriptions* $\mathsf{F} = \{F_1, F_2, \ldots, F_m\}$*, a family* $\{d_p^{\mathsf{F}}\}_{p \in \mathbb{N}}$ *of functions* $d_p^{\mathsf{F}} : \mathsf{const}(\mathcal{D}) \times \mathsf{const}(\mathcal{D}) \mapsto [0, 1]$ *is defined as follows*

$$\forall a, b \in \mathsf{const}(\mathcal{D}) \qquad d_p^{\mathsf{F}}(a, b) := \frac{1}{m} \left[\sum_{i=1}^{m} (\delta_i(a, b))^p \right]^{1/p}$$

where $\forall i \in \{1, \ldots, m\}$ *the* i-th dissimilarity function δ_i *is defined:*

$$\forall a, b \in \mathsf{const}(\mathcal{D}) \qquad \delta_i(a, b) = \begin{cases} 0 & \mathcal{K} \vdash F_i(a) \text{ iff } \mathcal{K} \vdash F_i(b) \\ 1 & \text{otherwise} \end{cases}$$

The superscript F will be omitted when the set of features is fixed.

These functions are semi-distances (or pseudo-metrics) [1], namely, it cannot be proved that if $d_p(a, b) = 0$ then $a = b$. However, if the *unique names assumption* is made for the constant names, then a distance can be obtained by using a further feature set F_0 based on the equality: $\delta_0(a, b) = 1$ if $a = b$; $\delta_0(a, b) = 0$ otherwise.

Here, we make the assumption that the feature-set F represents a sufficient number of (possibly redundant) features that are able to discriminate really different objects. In [1], we propose a method for performing a randomized search of optimal feature sets.

Compared to other proposed distance (or dissimilarity) measures, the presented functions are not based on structural (syntactical) criteria.

The definition above might be further refined and extended by recurring to model theory. The set of Herbrand models $\mathcal{M}_\mathcal{K} \subseteq 2^{|\mathcal{B}_\mathcal{K}|}$ of the KB may be considered, where $\mathcal{B}_\mathcal{K}$ stands for its Herbrand base. Given two instances a and b to be compared w.r.t. a certain feature F_i, $i = 1, \ldots, m$, we might check if they can be distinguished in the world represented by a Herbrand interpretation $\mathcal{I} \in \mathcal{M}_\mathcal{K}$: $\mathcal{I} \models F_i(a)$ and $\mathcal{I} \models F_i(b)$. Hence, a distance measure should count the cases of disagreement, varying the Herbrand models of the KB. The resulting measure definition will be in this case:

$$\forall a, b \in \mathsf{const}(\mathcal{D}) \quad d_p^{\mathsf{F}}(a, b) := \frac{1}{m \cdot |\mathcal{M}_\mathcal{K}|} \left[\sum_{\mathcal{I} \in \mathcal{M}_\mathcal{K}} \sum_{i=1}^{m} \mid \delta_i^{\mathcal{I}}(a, b) \mid^p \right]^{1/p}$$

where the dissimilarity functions $\delta_i^{\mathcal{I}}$ are computed for a specific world state as encoded by a Herbrand interpretation $\mathcal{I}$:

$$\forall a \in \mathsf{const}(\mathcal{D}) \quad \delta_i^{\mathcal{I}}(a, b) = \begin{cases} 1 & F_i(a) \in \mathcal{I} \text{ and } F_i(b) \in \mathcal{I} \\ 0 & \text{otherwise} \end{cases}$$

3 Grouping Objects through Hierarchical Clustering

The conceptual clustering procedure implemented in our method works top-down, starting with one universal cluster grouping all instances. Then it iteratively finds two clusters bisecting an existing one up to the desired number of clusters is reached. Our algorithm can be ascribed to the category of the heuristic partitioning algorithms such as K-MEANS and EM [5]. Each cluster is represented by the center of the cluster. In our setting we consider the medoid [6] as a notion of cluster center. In particular our algorithm can be seen as a hierarchical extension of the PAM algorithm (*Partition Around Medoids* [6]): each cluster is represented by one of the individuals in the cluster, the medoid, that is, in our case, the one with the lowest average distance w.r.t. all the others individuals in the cluster. The bi-partition is repeated level-wise producing a dendrogram. In the following, a sketch of the algorithm is reported.

```
HBAM(allIndividuals, k, maxIterations): clusterVector;
input  allIndividuals: set of individuals; k: number of clusters;
       maxIterations: max number of inner iterations;
output clusterVector: array [1..k] of sets of clusters

level := 0;  clusterVector[1] := allIndividuals;
repeat
       ++level;
       cluster2split := selectWorstCluster(clusterVector[level]);
       iterCount := 0;
       stableConfiguration := false;
       (newMedoid1,newMedoid2) := selectMostDissimilar(cluster2split);
       repeat
           ++iterCount;
           // E step
           (medoid1,medoid2) := (newMedoid1,newMedoid2);
           (cluster1,cluster2) := distribute(cluster2split,medoid1,medoid2);
           // M step
           newMedoid1 := medoid(cluster1);
           newMedoid2 := medoid(cluster2);
           stableConfiguration := (medoid1 = newMedoid1) ∧ (medoid2 = newMedoid2);
       until stableConfiguration ∨ (iterCount = maxIterations);
       clusterVector[level+1] := replace(cluster2split,cluster1,cluster2,clusterVector[level]);
until (level = k);
```

The algorithm essentially consists of two nested loops: the outer one computes a new level of the resulting dendrogram and it is repeated until the desired number of clusters is obtained; the inner loop consists of a run of the PAM algorithm at the current level. Per each level, the next worst cluster is selected (*selectWorstCluster*() function) on the grounds of its quality, e.g. the one endowed with the least average inner similarity (or cohesiveness [10]). This cluster is candidate to being parted in two. The partition is constructed around two medoids initially chosen (*selectMostDissimilar*() function) and then iteratively adjusted in the inner loop. In the end, the candidate cluster is replaced by the newly found parts at the next level of the dendrogram. The inner loop basically resembles to a 2-means (or EM) algorithm, where medoids are considered instead of means, which can hardly be defined in symbolic computations. Then, the classical two steps are performed in an iteration: **E step:** given the current medoids, the first distributes the other individuals in one of the two partitions under construction on the grounds of their similarity w.r.t. either medoid; **M step:** given the bipartition obtained by *distribute*(), this second step computes the new medoids for either cluster. These tend to change on each iteration until eventually they converge to a stable couple (or when a maximum number of iteration have been performed). The medoid of a group of individuals is the individual that has the lowest distance w.r.t. the others. Formally. given a cluster $C = \{a_1, a_2, \ldots, a_n\}$, the medoid is defined: $m = \mathrm{medoid}(C) = \mathrm{argmin}_{a \in C} \sum_{j=1}^{n} d(a, a_j)$. The representation of centers by means of medoids has two advantages. First, it presents no limitations on attributes types, and, second, the choice of medoids is dictated by the location of a predominant fraction of points inside a cluster and, therefore, it is lesser sensitive to the presence of outliers.

Each node of the tree (a cluster) may be labeled with an intensional concept definition which characterizes the individuals in the given cluster while discriminating those in the twin cluster at the same level. Labeling the tree-nodes with concepts can be regarded as a number of supervised learning problems in the specific multi-relational representation targeted in our setting. A straightforward solution may be given by the computation of the *least general generalization* (lgg) operator [7] and (an approximation of) the *most specific concept* (msc) operator, which amounts to building a new ground clause whose body is made up of the ground literals in the set of derivable facts (from $\mathcal{K}$) that are linked through their constants:

$$\mathsf{msc}_{\mathcal{K}}(a) := \{L : \text{literal}, \mathcal{K} \models L \mid \exists a \in \mathrm{args}(L) \textit{ or } \exists L' \in \mathsf{msc}(a) \textit{ s.t. } \\ \exists b \in \mathrm{args}(L) \cap \mathrm{args}(L') \textit{ and } \exists c \in \mathrm{args}(L')\}$$

This solution involves the following steps:

- **input:** clusters of individuals C_j
- **output:** new clause
  1. **for each** individual $a_{ij} \in C_j$ **do**
     (a) compute $M_{ij} \leftarrow \mathsf{msc}_{\mathcal{K}}(a_{ij})$
     (b) **let** $\mathsf{Clause}_{ij} \leftarrow (\mathsf{newConcept}_j(a_{ij})$:- $M_{ij})$;
  2. **return** $\mathsf{lgg}(\mathsf{Clause}_{ij})$

As an alternative, algorithms for learning concept descriptions expressed in DATALOG may be employed.

4 Conclusions and Future Work

This work has presented a clustering method for DATALOG knowledge bases. The method exploits a novel dissimilarity measure, that is based on the resource semantics w.r.t. a number of dimensions corresponding to a committee of features represented by a group of concept descriptions (discriminating features). The algorithm is an adaptation of the classic bisecting k-means to complex LP representations. We have discussed its complexity and the potential applications to a variety of important tasks.

Ongoing work concerns the feature selection task. Namely, we aim at inducing an optimal set of concepts for the distance measure by means of randomized algorithms based on genetic programming and simulated annealing. Furthermore, also the clustering process itself may be carried out by means of a randomized method based on the same approaches. We are also exploiting the outcome of the clustering algorithm for performing similarity search grounded on a lazy-learning procedure and specifically based on the weighted k-nearest neighbor approach, exploiting the distance measures presented in this work. Further applications regards the tasks specified in Sec. 1.

References

[1] d'Amato, C., Fanizzi, N., Esposito, F.: Induction of optimal semantic semi-distances for clausal knowledge bases. In: Proceedings of the 17th International Conference on Inductive Logic Programming, ILP 2007. LNCS (LNAI), Springer, Heidelberg (2007)

[2] Emde, W., Wettschereck, D.: Relational instance-based learning. In: Saitta, L. (ed.) Proceedings of the 13th International Conference on Machine Learning, ICML 1996, pp. 122–130. Morgan Kaufmann, San Francisco (1996)

[3] Fanizzi, N., Iannone, L., Palmisano, I., Semeraro, G.: Concept formation in expressive description logics. In: Boulicaut, J.-F., Esposito, F., Giannotti, F., Pedreschi, D. (eds.) ECML 2004. LNCS (LNAI), vol. 3201, pp. 99–113. Springer, Heidelberg (2004)

[4] Hutchinson, A.: Metrics on terms and clauses. In: van Someren, M., Widmer, G. (eds.) ECML 1997. LNCS, vol. 1224, pp. 138–145. Springer, Heidelberg (1997)

[5] Jain, A., Murty, M., Flynn, P.: Data clustering: A review. ACM Computing Surveys 31(3), 264–323 (1999)

[6] Kaufman, L., Rousseeuw, P.: Finding Groups in Data: an Introduction to Cluster Analysis. John Wiley & Sons, Chichester (1990)

[7] Nienhuys-Cheng, S., de Wolf, R.: Foundations of Inductive Logic Programming. LNCS (LNAI), vol. 1228. Springer, Heidelberg (1997)

[8] Ramon, J., Bruynooghe, M.: A framework for defining distances between first-order logic objects. Technical Report CW 263, Department of Computer Science, Katholieke Universiteit Leuven (1998)

[9] Sebag, M.: Distance induction in first order logic. In: Džeroski, S., Lavrač, N. (eds.) ILP 1997. LNCS, vol. 1297, pp. 264–272. Springer, Heidelberg (1997)

[10] Stepp, R.E., Michalski, R.S.: Conceptual clustering of structured objects: A goal-oriented approach. Artificial Intelligence 28(1), 43–69 (1986)

[11] Zezula, P., Amato, G., Dohnal, V., Batko, M.: Similarity Search: The Metric Space Approach. Springer, Heidelberg (2007)

LAREDAM – Considerations on System of Local Analytical Reports from Data Mining

Jan Rauch and Milan Šimůnek

Faculty of Informatics and Statistics, University of Economics,
Prague, nám. W. Churchilla 4 130 67 Prague, Czech Republic
{Rauch,Simunek}@vse.cz

Abstract. LAREDAM is a research project the goal of which is to study possibilities of automatic formulation of analytical reports from data mining. Each such report presents answer to one analytical question. Lot of interesting analytical questions can be answered by GUHA procedures implemented in the LISp-Miner system. The paper presents first steps in building system of reasonable analytical questions and corresponding analytical reports.

Keywords: Data mining, GUHA method, background knowledge, analytical report, automatization.

1 Introduction

Presentation of results of data mining to a data owner is one of topics related to *10 challenging problems in data mining research*; see http://www.cs.uvm.edu/~icdm/. A reasonable way is to arrange results in an analytical report structured both according to the given analytical question and to the analyzed data. An early attempt to produce analytical reports from data mining is in [1]. It is important that the core of such analytical report is a string of formal assertions concerning analyzed data. Thus the whole analytical report can be understood as a formal object and we can try to index and retrieve particular reports using formulas expressing knowledge instead of usual key-words [2].

It is natural to see the situation also from a point of view of semantic web [3]. We can ask if analytical reports presenting results of data mining can be automatically compared, analyzed, summarized etc. Such questions led to the research project SEWEBAR the goal of which is to study possibilities of automatic creation of both "local" and "global" analytical reports [4, 5]. The local analytical report presents answer to a "local" analytical question concerning particular data set. The global analytical report presents answer to a "global" analytical question that concerns at least two local analytical reports that can be related to different datasets. Automatic creation of the global report requires detailed formal description of used local reports. The local analytical report must be considered not only as a chain of formal assertions. We have to consider also function of pieces of text connecting particular formal assertions and the way in which the local analytical report was prepared. LAREDAM (**L**ocal **A**nalytical **RE**ports from **DA**ta **M**ining) is a subsystem of the SEWEBAR project that tries to deal with local analytical reports in this way.

A. An et al. (Eds.): ISMIS 2008, LNAI 4994, pp. 143–149, 2008.

The whole SEWEBAR project is based on the LISp-Miner system the core of which is created by six GUHA procedures. The input of GUHA procedure consists of several parameters defining large set of relevant patterns. The GUHA procedure generates and verifies each relevant pattern in the given data. Its output consists of all true patterns. There are usually very fine tools to tune the set of relevant patterns to be generated and verified. The most used GUHA procedure is the 4ft-Miner procedure that mines for various forms of association rules [6]. Its implementation is not based on apriori algorithm but it uses representation of analyzed data by strings of bits. The same approach is used to implement the additional five GUHA procedures [7].

There is a large variety of local analytical questions that can be answered by these GUHA procedures. There is also a well defined structure of analytical report that presents answer to each particular analytical question. Both the structure and the way of creation of the analytical report are determined by the analytical question. It means that it is possible to automate the process of creation of analytical report using methods of knowledge engineering. It however requires lot of knowledge related both to the LISp-Miner system and to area of application. The system SEWEBAR enhances the LISp-Miner system by tools that make possible to store, maintain and apply relevant pieces of knowledge. They can be also used to formulate reasonable local analytical questions. First experience is presented in [8].

We believe that it is possible to automate also the indexing of local analytical reports in the way that will make possible to automate also the whole process of formulating global analytical question and of producing of global analytical reports. The goal of this paper is to present considerations on producing and indexing of local analytical reports. We use data set STULONG and related background knowledge; see section 2. An example of dealing with local analytical report is in section 3.

2 Data Set STULONG and Background Knowledge

We use the ENTRY data matrix of the data set STULONG describing *Longitudinal Study of Atherosclerosis Risk Factors*[1]. Each row of the data matrix corresponds to one of 1 417 patients. The data matrix has 64 columns, corresponding to particular attributes (characteristics) of the patients. These attributes are divided into 11 basic groups. We use namely three of them: *social status* (6 attributes), *physical examination* (8 attributes) and *alcohol consumption* (9 attributes).

There are various pieces of relevant background knowledge, e.g.:

- basic value limits for particular attributes (e.g. 0°C and 100°C for temperature);
- typical interval lengths when categorizing values (e.g. *5 or 10 years* for age);
- important groups of attributes (e.g. *social status* and *alcohol consumption*);
- mutual influence among attributes.

---

[1] The study (STULONG) was carried out at the 2nd Department of Medicine, of the 1st Faculty of Medicine of Charles University and the Charles University Hospital, Prague 2 under the supervision of Prof. F. Boudík, MD, ScD, with the collaboration of M. Tomečková, MD, PhD and Ass. Prof. J. Bultas, MD, PhD. The data were transferred to electronic form by the European Centre of Medical Informatics, Statistics and Epidemiology of Charles University and the Academy of Sciences. The data is available on http://euromise.vse.cz/challenge2004

As an example of background knowledge we present mutual influence table showing relations among several attributes from a medical domain, see Fig. 1. At intersection of each row and column there is a symbol representing mutual influence of corresponding attributes. We present two examples – positive influence ↑↑ and negative influence ↑↓, see circles in Fig. 1. An example of the positive influence is: "*If the beer consumption increases then the body mass index (BMI) increases too*", formally *Beer* ↑↑ *BMI*. An example of negative influence is "*If education increases then body mass index decreases*", formally we can write *Education* ↑↓ *BMI*.

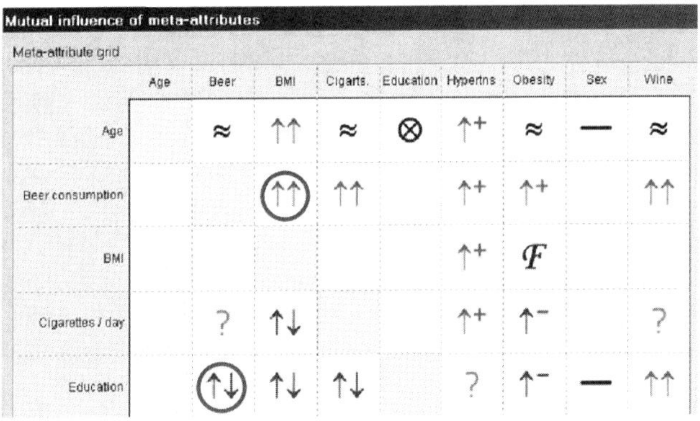

Fig. 1. Mutual influence among attributes

3 Local Analytical Report – Example

Various possibilities how to use background knowledge to automatically formulate reasonable analytical questions are mentioned in [8]. Here we use a simple variant – the scheme: *What strong relations between Boolean characteristics of two groups of attributes are observed in the STULONG data?* Example of the question according to this scheme is: *What strong relations between Boolean characteristics of the groups* ***social status + alcohol consumption*** *and* ***physical examination*** *are observed in the STULONG data*? Symbolically we write $\mathcal{B}$[Social+Alcohol] $\approx^?$ $\mathcal{B}$ [Physical].

We use the GUHA procedure 4ft-Miner. It mines for association rules $\varphi \approx \psi$. Here φ (i.e. antecedent) and ψ (i.e. succedent) are Boolean attributes derived from columns of analyzed data matrix. Symbol $\approx$ is 4ft-quantifier. A condition concerning a contingency table of φ and ψ (see Table. 1) is assigned to each 4ft-quantifier.

Table 1. Contingency table 4ft(φ,ψ,$\mathcal{M}$) of φ and ψ in data matrix $\mathcal{M}$

$\mathcal{M}$	ψ	$\neg\psi$
φ	a	b
$\neg\varphi$	c	d

Contingency table 4ft($\varphi,\psi,\mathcal{M}$) of φ and ψ in data matrix $\mathcal{M}$ is a quadruple $\langle a,b,c,d \rangle$ of integer numbers, a is number of rows of $\mathcal{M}$ satisfying both φ and ψ, b is number of rows satisfying φ and not satisfying ψ etc., see Tab. 1. The rule $\varphi \approx \psi$ is true in $\mathcal{M}$ if the condition associated to $\approx$ is satisfied in 4ft($\varphi,\psi,\mathcal{M}$).

To solve the question $\mathcal{B}$ [Social+Alcohol] $\approx^{?}$ $\mathcal{B}$ [Physical] we define the set $\mathcal{B}$ [Social+Alcohol] as the set of relevant antecedents and similarly for $\mathcal{B}$ [Physical] and set of relevant succedents. One possibility how to do it is shown in Fig. 2. These definitions are however very simple and it serves only to show principles.

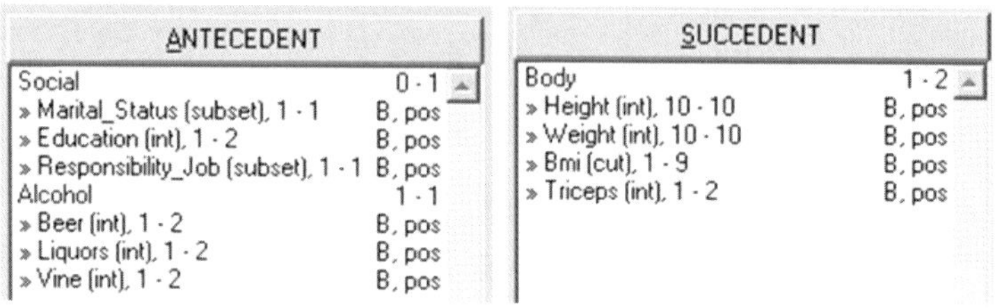

Fig. 2. Definition of the sets $\mathcal{B}$ [Social + Alcohol] and $\mathcal{B}$ [Physical]

The definition in Fig. 2 means that each Boolean attribute $\varphi \in \mathcal{B}$ [Social+Alcohol] is a conjunction $\varphi_S \wedge \varphi_A$ of Boolean attributes $\varphi_S \in \mathcal{B}$ [Social] and $\varphi_A \in \mathcal{B}$ [Alcohol]. Here $\mathcal{B}$ [Social] is a set of Boolean attributes derived from the group of three attributes *Marital_Status*, *Education*, and *Responsibility_Job* (i.e. responsibility in job). $\mathcal{B}$ [*Marital_Status*] denotes the set of Boolean attributes defined from *Marital_Status* etc. The expression Social 0 - 1 means that maximally one Boolean attribute from $\mathcal{B}$ [*Marital_Status*], $\mathcal{B}$ [*Education*] and $\mathcal{B}$ [*Responsibility_Job*] can be used as φ_S. The set $\mathcal{B}$ [*Marital_Status*] is defined by the expression Marital_Status(subset),1-1. It means that each Boolean atribute $\varphi_M \in \mathcal{B}$ [*Marital_Status*] has a form of *Marital_Status*(α) where α is a subset containing exactly one (see (subset),1-1) of possible values *married*, *divorced*, *single*, *widower*.

The set $\mathcal{B}$ [*Education*] is defined by the expression Education(int),1-2. It means that each Boolean atribute $\varphi_E \in \mathcal{B}$ [*Education*] has a form of *Education*(α) where α is an interval of 1-2 possible ordinal values of the atribute *Education*. The attribute *Education* has values *basic school*, *apprentice school*, *secondary school*, and *university*. Thus *Education*(*secondary school*, *university*) is an example of the Boolean attribute derived from *Education*. Analogously for $\mathcal{B}$ [*Responsibility_Job*].

The expression Alcohol 1-1 means that exactly one Boolean attribute from $\mathcal{B}$ [*Beer*], $\mathcal{B}$ [*Liquors*] and $\mathcal{B}$ [*Vine*] can be used as φ_A. The set $\mathcal{B}$ [*Beer*] is defined by the expression Beer(int),1-2 and the attribute *Beer* has values of *he does not drink*, *medium*, *a lot*. The sets $\mathcal{B}$ [*Liquors*] and $\mathcal{B}$ [*Vine*] are defined in a similar way.

The set $\mathcal{B}$ [Physical] of relevant succedents is defined similarly. The attribute *Triceps* describes the skinfold in mm above musculus triceps and it has values $(0;5\rangle$, …, $(20;25\rangle$, > 25. The set $\mathcal{B}$ [*Bmi*] is defined by the expression Bmi(cut),1 - 9. It means that cuts of the length 1–9 are generated. There are left cuts i.e. intervals with the left value equal to the minimal value of the atribute *Bmi*. There are also right cuts defined as intervals with the right value equal to the maximal value of the atribute *Bmi*.

We applied two different 4ft-quantifiers – *founded implication* $\Rightarrow_{p,B}$ and *above average implication* $\Rightarrow^{+}_{p,B}$ [6]. The 4ft-quantifier $\Rightarrow_{p,B}$ is defined by the condition $\frac{a}{a+b} \geq p \wedge a \geq B$ (see Table 1). Thus the rule $\varphi \Rightarrow_{p,B} \psi$ says that both at least 100p per cent of patients satisfying φ satisfy also ψ (i.e. the confidence is p) and that there are at least B of patients satisfying both φ and ψ. The 4ft-quantifier $\Rightarrow^{+}_{p,B}$ is defined by the condition $\frac{a}{a+b} \geq (1+p)\frac{a+c}{a+b+c+d} \wedge a \geq B$. The rule $\varphi \Rightarrow^{+}_{p,B} \psi$ says that both the relative frequency of patients satisfying ψ among the patients satisfying φ is at least *100p* per cent higher than the relative frequency of patients satisfying ψ in the whole analyzed data matrix and that there are at least B patients satisfying both φ and ψ.

We used two 4ft-Miner runs, the first one with the 4ft-quantifier $\Rightarrow_{0.9,20}$. The procedure 4ft-Miner generated and verified more than 535 000 of rules in 13 sec (on PC with 3.06 GHz and 512 MB SDRAM). However no rule satisfying given conditions was found. Thus we used the quantifier $\Rightarrow^{+}_{1,20}$ and 24 of rules satisfying given conditions were found. The strongest rule what concerns the difference of relative frequencies of succedent is the rule

$$Education(apprentice\ school) \wedge Beer(medium,\ a\ lot) \Rightarrow^{+}_{1,24,28} Bmi(>30) \wedge Triceps(<10)$$

with the 4ft table 4ft(E-B, B-T, ENTRY) given in Table 2 where E-B denotes Boolean attribute *Education*(apprentice school) ∧ *Beer*(medium, a lot) and B-T denotes Boolean attribute *Bmi*(>30) ∧ *Triceps*(<10).

Table 2. 4ft-table of 4ft(E-B, B-T, ENTRY)

ENTRY	ψ	$\neg\psi$
φ	28	268
$\neg\varphi$	29	1027

A lot of output rules of the 4ft-Miner procedure can be uninteresting because of they follow from background knowledge. The rule

$$Education(apprentice\ school) \wedge Beer(medium,\ a\ lot) \Rightarrow^{+}_{1,24,28} Bmi(>30) \wedge Triceps(<10)$$

is an example. It can be deduced from the item *Beer* ↑↑ *BMI* of background knowledge, see section 2. The precise formal description of the process of deduction is out of the scope of this paper, we present only its principle. The expression *Beer* ↑↑ *BMI* says that if the beer consumption increases then the body mass index increases too. It means that among the patients with high beer consumption there will be relatively more patients with high BMI than in the whole data matrix. Thus we can expect that the rules $Beer(\mathrm{HIGH}_{Beer}) \Rightarrow^{+}_{p,Base} Bmi(\mathrm{HIGH}_{Bmi})$ will be true for significantly higher values of p and *Base*. These values can be found experimentally and can be included into background knowledge. The coefficient HIGH_{Bmi} of the Boolean attribute $Bmi(\mathrm{HIGH}_{Bmi})$ is a subset consisting only of high values of the attribute *Bmi*. It is even done by standards that *Bmi* >30 is too high. Similarly we can accept that the Boolean attribute *Beer*(*medium, a lot*) has coefficient consisting of high values. This way we can deduce the rule *Beer*(*medium, a lot*) $\Rightarrow^{+}_{p,Base}$ *Bmi*(>30) from the item *Beer* ↑↑ *BMI*. Let us again emphasize that the whole process is described informally and requires much work to be specified. The exact deduction rules concerning association rules in logic of association rules will be also used [9].

This way we get a relatively simple answer to the analytical question $\mathcal{B}$[Social + Alcohol] $\approx^{?}$ $\mathcal{B}$ [Physical]. The question can be automatically generated on the basis of the knowledge of basic groups of attributes. The results obtained in the above outlined way can be arranged into an analytical report sketched in Fig. 3. The structure of the report is given by the solved analytical question.

1. Introduction
Goal of the analysis – textual formulation of $\mathcal{B}$[Social+Alcohol] $\approx^{?}$ $\mathcal{B}$[Physical], principle of the solution, structure of the report
2. Analysed data
Names and basic statistics of used attributes of data matrix ENTRY
3. Analytical question and set of relevant rules
Detailed definition and explanation of the sets B[Social+Alcohol], B[Physical] and quantifiers $\Rightarrow_{p,Base}$ and $\Rightarrow^{+}_{p,Base}$
4. Filtering results
Principles of filtering and overview of removed results
5. Results overview
Overview of found rules both for $\Rightarrow_{0.9,20}$ and for $\Rightarrow^{+}_{1,20}$ suitable statistics of particular attributes occurrences. Generalized assertions like "There is no pattern concerning Education" can be used.
6. Particular rules
Detailed description of particular found patterns.

Fig. 3. Sketch of an analytical report

To keep the analytical report readable it is necessary to present only reasonable number of rules. The reasonable number of rules depends on the solved analytical question and it is done by minimal and maximal value. Getting of specified number of rules is an iterative process that can be solved automatically using various heuristics. To get larger number of rules we have to use a weaker 4ft-quantifier and/or to define larger sets of Boolean attributes to be taken into account. Similarly, to get smaller number of rules we have to use a stronger 4ft-quantifier and/or to define smaller sets of Boolean attributes.

The sets of relevant Boolean attributes can be modified e.g. by modifications of definitions of coefficients of relevant Boolean attributes. There are various types of coefficients of Boolean attribute [6], some of them are mentioned above. There is lot of possibilities how to modify these definitions and also lot of heuristics how to do it. We can e.g. use Height(int),5-10 instead of Height(int),10-10 or Marital_Status(subset),1-2 instead of Marital_Status(subset),1-1. The heuristics how to modify the definitions of coefficients often strongly depend on the used attributes (namely on number of their possible values – there is however a danger of combinatorial explosion of coefficients of the type subset). The more detailed description of the iterative process of getting specified number of rules does not fit within the scope of this paper.

Note that lot of tasks concerning data matrices with several thousands of rows can be solved in few minutes and that we use bit-string that ensures reasonable linear dependency of the solution time on the number of rows of analyzed data matrix [6]. Thus the modifications of parameters can be done in an automatically controlled iterative process that will work in reasonable time.

4 Conclusions and Further Work

We outlined possibilities of automatic generation of analytical reports. The simple way how to index the resulting reports is to index the report by its chapters and sections where each section is indexed by the included formal assertions. It can be again done automatically. However the important goal of local analytical reports is to serve as input for global analytical reports. Thus the way how to index the local analytical reports will be determined by its role in preparing global analytical reports. Our next work concerns not only detailed implementation of the above described process of creating local analytical repots but also the investigation of global analytical reports.

Acknowledgement. The work described here has been supported by Grant No. 201/08/0802 of the Czech Science Foundation and by Grant No. ME913 of Ministry of Education, Youth and Sports, of the Czech Republic.

References

1. Matheus, J., et al.: Selecting and Reporting What is Interesting: The KEFIR Application to Healthcare Data. In: Fayyad, U.M., et al. (eds.) Advances in Knowledge Discovery and Data Mining, pp. 495–515. AAAI Press / The MIT Press (1996)
2. Rauch, J.: Logical Calculi for Knowledge Discovery in Databases. In: Komorowski, J., Zytkow, J. (eds.) Principles of Data Mining and Knowledge Discovery, pp. 47–57. Springer, Heidelberg (1997)
3. Lín, V., Rauch, J., Svátek, V.: Analytic Reports from KDD: Integration into Semantic Web. In: Horrocks, I., Hendler, J. (eds.) ISWC 2002. LNCS, vol. 2342, p. 38. Springer, Heidelberg (2002)
4. Rauch, J.: Project SEWEBAR – Considerations on Semantic Web and Data Mining. In: Proceedings of IICAI 2007, pp. 1763–1782. Florida A&M University (2007)
5. Rauch, J., Šimůnek, M.: Semantic Web Presentation of Analytical Reports from Data Mining – Preliminary Considerations. In: Lin, T.Y., et al. (eds.) Web Intelligence 2007 Proceedings, pp. 3–7 (2007)
6. Rauch, J., Šimůnek, M.: An Alternative Approach to Mining Association Rules. In: Lin, T.Y., et al. (eds.) Data Mining: Foundations, Methods, and Applications, pp. 219–238. Springer, Heidelberg (2005)
7. Rauch, J., Šimůnek, M.: GUHA Method and Granular Computing. In: Hu, X., et al. (eds.) Proceedings of IEEE conference Granular Computing, pp. 630–635 (2005)
8. Rauch, J., Šimůnek, M.: Dealing with Background Knowledge in the SEWEBAR Project. In: Berendt, B., Svátek, V., Železný, F. (eds.) Prior Conceptual Knowledge in Machine Learning and Knowledge Discovery. ECML/PKDD 2007 Workshop Proceedings, pp. 97–108 (2007)
9. Rauch, J.: Logic of Association Rules. Applied Intelligence 22, 9–28

- First, we mine the *patterns* of executions that are likely to characterize the behavior of a given log. In fact, we specialize earlier frequent pattern mining approaches to the context of process logs, by *(i)* defining a notion of pattern which effectively characterizes concurrent processes by accounting for typical routing constructs, and by *(ii)* presenting an algorithm for their identification.
- Second, we use an outlier detection approach which is *cluster-based*, i.e., it computes a clustering for the logs (where the similarity measure roughly accounts for how many patterns jointly characterize the execution of the traces) and finds outliers as those individuals that hardly belong to any of the computed clusters or that belong to clusters whose size is definitively smaller than the average cluster size.

By this way, we will discover, e.g., that traces $s_9, ..., s_{14}$ do not follow any of the frequent behaviors registered in the log. Moreover, we will reduce the risk of both false positives (traces are compared according to behavioral patterns rather than to the pure sequencing of activities) and false negatives (traces that comply with the model might be seen as outliers, if they correspond to unfrequent behavior)—cf. **(C1)** and **(C2)**.

Organization. The above techniques are illustrated in more details in Section 2, while basic algorithmic issues are discussed in Section 3. After illustrating experimental results in Section 4, we draw some concluding remarks in Section 5.

2 Formal Framework

Process-oriented commercial systems usually store information about process enactments by tracing some events related to the execution of the various activities. By abstracting from the specificity of the various systems, as commonly done in the literature, we may view a *log* L for them as a bag of *traces* over a given set of activities, where each trace t in L has the form $t[1]t[2]...t[n]$, with $t[i]$ $(1 \leq i \leq n)$ being an activity identifier. Next, these traces are assumed to be given in input and the problem of identifying anomalies among them is investigated.

Behavioral Patterns over Process Logs. The first step for detecting outliers is to characterize the "normal" behavior registered in a process log. In the literature, this is generally done by assessing the causal relationships holding between pairs of activities (e.g., [3,10]). However, this does not suffice to our aims, as abnormality may emerge not only w.r.t. the sequencing of activities, but also w.r.t. constructs such as branching and synchronization. Hence, towards a richer view of process behaviors, we next focus on the identification of those features that emerge as complex patterns of executions.

Definition 1 (S-Pattern). A *structural* pattern (short: S-pattern) over a given set A of activities is a graph $p = \langle A_p, E_p \rangle$, with $A_p = \{n, n_1, \ldots n_k\} \subseteq A$ such that either: **(a)** $E_p = \{n\} \times (\{n_1, \ldots n_k\})$ – in this case, p is called a *FORK*-pattern–, or **(b)** $E_p = (\{n_1, \ldots n_k\}) \times \{n\}$ – in this case, p is called a *JOIN*-pattern. Moreover, the *size* of p, denoted by $size(p)$, is the cardinality of E_p. □

In particular, an S-pattern with size 1 is both a *FORK*-pattern and a *JOIN*-pattern, and simply models a causal precedence between two activities. This is, for instance, the case of patterns p_3, p_4, and p_5 in Figure 1. Instead, higher size patterns account for fork and join constructs, which are typically meant to express parallel execution (cf. p_1) and synchronization (cf. p_2), respectively, within concurrent processes. The crucial question is now to formalize the way in which patterns emerge for process logs.

Definition 2 (Pattern Support). Let t be a trace and $p = \langle A_p, E_p \rangle$ be an S-pattern. We say that t *complies with* p, if **(a)** t includes all the activities in A_p and **(b)** the projection of t over A_p is a topological sorting of p, i.e., there are not two positions i, j inside t such that $i < j$ and $(t[j], t[i]) \in E_p$. Then, the support of p w.r.t. t, is defined as:

$$supp(p,t) = \begin{cases} \min_{(t[i],t[j]) \in E_p} e^{-|\{t[k] \notin A_p | i<k<j\}|}, & \text{if } t \text{ complies with } p \\ 0, & \text{otherwise.} \end{cases}$$

This measure is naturally extended to any trace bag L and pattern set P as follows: $supp(p, L) = \frac{1}{|L|} \times \sum_{t \in L} supp(p, t)$ and $supp(P, t) = \frac{1}{|P|} \times \sum_{p \in P} supp(p, t)$. □

In words, a pattern p is not supported in a trace t if some relation of precedence encoded in the edges of p is violated t. Otherwise, the support of p decreases at the growing of the minimum number of spurious activities (i.e., $\{t[k] \notin A_p \mid i < k < j\}$) that occur between any pair of activities in the endpoints of the edges in p.

While at a first sight this notions may appear similar to classical definitions from frequent pattern mining research, some crucial and substantial differences come instead into play. Indeed, the careful reader may have noticed that our notion of support is not *anti-monotonic* regarding graph containment. This happens because adding an edge of the form (x, y) to a given pattern may well lead to increase its support, since one further activity (either x or y) may be no longer viewed as a spurious one. Consequently, the space of all the possible S-patterns does not form a lattice, and classical *level-wise* approaches cannot be used to single out those patterns whose support over a log L is greater than a given threshold σ, hereinafter called σ-*frequent* patterns.

In addition, differently from many pattern mining approaches, the frequency of a pattern p does not necessarily indicate its relevance to modeling the process behavior. In particular, when comparing two σ-frequent patterns p and p' such that p is a subgraph of p', we can safely focus on p' if its frequency does not differ significantly from that of p; otherwise, i.e., if p is much more frequent than p', the subpattern p is also interesting its own, as it can help recognizing relevant behavioral clusters. This is formalized below.

Definition 3 (Interesting Patterns). Let L be a log, and σ, γ be two real numbers. Given two S-patterns p and p', we say that p' γ-*subsumes* p, denoted by $p \sqsubseteq_\gamma p'$, if p is a subgraph of p' and $supp(p, L) - supp(p', L) < \gamma \cdot supp(p', L)$. Further, an S-pattern p is (σ, γ)-*maximal* w.r.t. L if **(a)** p is σ-frequent on L and **(b)** there is no other S-pattern p' over A s.t. $size(p') = size(p) + 1$, p' is σ-frequent on L, and $p \sqsubseteq_\gamma p'$. □

Cluster-based Outliers. Once that "normality" has been modeled by means of the discovery of interesting patterns, we can then look for those individuals whose behavior deviates from the normal one. To this end, the second step of our outlier detection approach performs a "co-clustering" (see, e.g., [2]) of patterns and traces, based on their mutual correlation captured by the $supp$ measure. Intuitively, we look for associating pattern clusters with trace clusters, so that outliers emerge as those individuals that are not associated to any pattern cluster or that belong to clusters whose size is definitively smaller than the average cluster size. Abstracting from the specificity of the mining algorithm (discussed in Section 3), the output of this method is formalized below.

Definition 4 (Coclusters and Outliers). An α*-coclustering* for a log L and a set P of S-patterns is a tuple C=$\langle \widehat{P}, \widehat{L}, \mathcal{M} \rangle$ where:

- $\widehat{P}$=$\{\widehat{p}_1, ..., \widehat{p}_k\}$ is a set of non-empty P's subsets (named *pattern clusters*) s.t. $\bigcup_{j=1}^{k} \widehat{p}_j = P$;
- $\widehat{L}$=$\{\widehat{l}_1, ..., \widehat{l}_h\}$ is a set of non-empty disjoint L's subsets (named *trace clusters*) such that $\bigcup_{i=1}^{h} \widehat{l}_i$ = $\{t \in L \mid \exists \widehat{p}_i \in \widehat{P}$ s.t. $supp(\widehat{p}_i, t) \geq \alpha\}$;
- $\mathcal{M} : \widehat{P} \mapsto \widehat{L}$ is an invertible function that associates each pattern cluster $\widehat{p}_j$ with a trace cluster $\widehat{l}_i$ and vice-versa, i.e., $\widehat{l}_i = \mathcal{M}(\widehat{p}_j)$ and $\widehat{p}_j = \mathcal{M}^{-1}(\widehat{l}_i)$.

Moreover, given two real numbers α, β in $[0..1]$, a trace $t \in L$ is an (α, β)*-outlier* w.r.t. C if either **(a)** $t \notin \bigcup_{i=1}^{h} \widehat{l}_i$, or **(b)** $|\widehat{l}_i| < \beta \times \frac{1}{h} \sum_{\widehat{l}_j \in \widehat{L}} |\widehat{l}_j|$, where $t \in \widehat{l}_i$. □

In words, we define outliers according to a number of clusters, discovered for both traces and patterns based on their mutual correlations, which represent different behavioral classes. More specifically, two different kinds of outlier emerge; indeed, condition **(a)** deems as outlier any trace that is not assigned to any cluster (according to the minimum support α), while condition **(b)** estimates as outliers all the traces falling into small clusters (smaller than a fraction β of the average clusters' size).

3 An Algorithm for Detecting Outliers

In this section, we discuss the `TraceOutlierMining` algorithm that discovers a set of outliers, based on the computation scheme and the framework described so far. The algorithm is shown in Figure 2: Given a log L, a natural number $pattSize$ and four real thresholds σ,γ, α and β, it first computes a set P of (σ, γ)-maximal S-patterns via the function `FindPatterns`, while restricting the search to patterns with no more than $pattSize$ arcs. Then, an α-coclustering for L and P is extracted with the function `FindCoClusters` (Step 2). The remaining steps are just meant to build a set U of traces that are (α, β)-outliers w.r.t. this coclustering, by checking the conditions in Definition 4 on all traces. Clearly, the main computation efforts hinge on the functions `FindPatterns` and `FindCoClusters`, which are thus thoroughly discussed next.

Function `FindPatterns`. The main task when mining (σ, γ)-maximal S-patterns is the mining of σ-frequent S-patterns, as the former S-patterns directly derive from the latter ones. Unfortunately, a straightforward level-wise approach cannot be used to this end, since the support $supp$ is not anti-monotonic w.r.t. pattern containment.

To face this problem, `FindPatterns` uses a relaxed notion of support (denoted by $supp'$) which optimistically decreases the counting of spurious activities by a "bonus" that depends on the size of the pattern at hand: the lower the size the more the bonus. More precisely, within Definition 2, for each arc $(t[i], t[j])$ in p, we replace the term $|\{t[k] \notin A_p \mid i < k < j\}|$ with $min\{\ |\{t[k] \notin A_p \mid i < k < j\}|, pattSize - size(p)\}$. The

Input: A log L, an upper bound $pattSize \in \mathbb{N}^+$ for pattern size, and four real numbers σ, γ, α and β
Output: A set of (α, β)-outlier
Method: Perform the following steps:

1 $P :=$ `FindPatterns` $(L, pattSize, \sigma)$;
2 $\langle \widehat{P}, \widehat{L}, \mathcal{M} \rangle :=$ `FindCoClusters`(L, P, α);
3 $U := \emptyset$; $avgSize := \frac{1}{|\widehat{L}|} \sum_{\hat{l}_j \in \hat{L}} |\widehat{l}_j|$;
4 **for each** trace t in L **do**
5   **if** $t \notin \bigcup_{i=1}^{h} \widehat{l}_i$, **or** $|\widehat{l}_i| < \beta \times \frac{1}{h} \sum_{\hat{l}_j \in \hat{L}} |\widehat{l}_j|$, where $t \in \widehat{l}_i$ **then** $U := U \cup \{t\}$;
6 **return** U;

Function `FindPatterns`(L: log; $pattSize$: natural number; σ: real number): set of S-patterns;

P1 Compute the set $L_1 = \{p$ is an S-pattern $\mid supp'(p, L) \geq \sigma$ and $size(p) = 1\}$ in a scan of L;
P2 $k := 2$; $R := \emptyset$
P3 **repeat**
P4 $Cand_k := generateCandidates(L_{k-1}, L_1)$;
P5   Compute $supp(p, L)$ and $supp'(p, L)$ for each $p \in Cand_k$ through a scan of L;
P6 $L_k := \{p \in Cand_k \mid supp'(p, L) \geq \sigma\}$; *// filter out "unfrequent" patterns*
P7 $R := R \cup \{p \in L_{k-1} \mid \nexists p' \in L_k$ s.t. $p \sqsubseteq_\gamma p'\}$; *// select (σ, γ)-maximal patterns (cf. Def.* **??**)
P8 $k := k + 1$;
P9 **until** $L_k = \emptyset$ **or** $k + 1 = pattSize$;
P10 **return** R;

Function `FindCoClusters`(L: log; P: S-patterns; α: real number): α-coclustering;

C1 **for each** pair of patterns p_i, p_j in P **do** $M(i, j) := \frac{|\{t' \mid supp(p_i, t') \geq \alpha \wedge supp(p_j, t') \geq \alpha\}|}{|\{t' \mid supp(p_i, t') \geq \alpha \vee supp(p_j, t') \geq \alpha\}|}$
C2 Compute a partition P^* of P by applying the MCL clustering algorithm to M;
C3 $\widehat{L} := \emptyset$; $\widehat{P} := \emptyset$; $\mathcal{M} := \emptyset$;
C4 **for each** trace t in L
C5 $\widehat{p}^t := \bigcup_{\widehat{p}^* \in P^*} \{\widehat{p}^* \mid supp(\widehat{p}^*, t) \geq \alpha\}$;
C6   **if** P contains $\widehat{p}^t$ *// cluster $\widehat{p}^t$ already exists and is hence associated with some trace cluster*
C7   Let $\widehat{l}^t = \mathcal{M}(p^t)$ be the cluster currently associated with $\widehat{p}^t$, and $\widehat{l}^t_{new} = \widehat{l}^t \cup \{t\}$;
C8 $\widehat{L} := \widehat{L} - \{\widehat{l}^t\} \cup \{\widehat{l}^t_{new}\}$; $\mathcal{M}(\widehat{p}^t) := \widehat{l}^t_{new}$;
C9   **else**
C10 $\widehat{L} := \widehat{L} \cup \{\{t\}\}$; $\widehat{P} := \widehat{P} \cup \{p_t\}$; $\mathcal{M}(\widehat{p}^t) := \{t\}$;
C11   **end if**
C12 **end for**
C13 **return** $\langle \widehat{P}, \widehat{L}, \mathcal{M} \rangle$;

Fig. 2. Algorithm `TraceOutlierMining`

reason for this is that, in the best case, each of the $pattSize - size(p)$ arcs that might be added to p, along the level-wise computation of patterns, will just fall between i and j.

It can be shown that function $supp'$ is both anti-monotonic and "safe", in that it does not underestimate the real support of candidate patterns. We can hence exploit it to implement a level-wise search of patterns: After building (in Step P1) the basic set L_1 of frequent S-patterns with size 1 (i.e., frequent activity pairs), an iterative scheme is used to incrementally compute any other set L_k, for increasing pattern size k (Steps P4–P8), until either no more patterns are generated or k reaches the upper bound given as input. In more detail, for each $k > 1$, the function $generateCandidates$ is first used to produce the set $Cand_k$ of k-sized candidate patterns, by suitably extending the patterns in L_{k-1} with those in L_1 (Step P4). The set L_k is then filled with the candidate patterns in $Cand_k$ that really get an adequate support in the log (Steps P5- P6). By construction of $supp'$, L_k is guaranteed to include (at least) all σ-frequent S-patterns with size k.

Eventually, by a straightforward application of Definition 3 to the patterns in L_{k-1} and L_k, all (σ, γ)-maximal S-patterns with size $k - 1$ are correctly extracted and added to the set R, the ultimate outcome of `FindPatterns`. In fact, in Step P7 the original function $supp$ is actually used for checking $(\sigma.\gamma)$-maximality.

Function `FindCoClusters`. The function `FindCoClusters` encodes a method for simultaneously clustering a log and its associated S-patterns. Provided with a log L, a set P of S-patterns and a threshold α, the function computes, in a two-step fashion, an α-coclustering $\langle \widehat{P}, \widehat{L}, \mathcal{M} \rangle$ for L and P, where $\widehat{P}$ (resp., $\widehat{L}$) is the set of pattern (resp., trace) clusters, while $\mathcal{M}$ is a mapping from $\widehat{P}$ to $\widehat{L}$.

At start, a preliminary partition P^* of P is built by applying a clustering procedure to a similarity matrix S for P, where the similarity between two patterns roughly estimates the likelihood that they occur in the same trace. Precisely, similarity values are computed (Step C1) by regarding $supp$ as a sort of contingency table over P and L (i.e., (p, t) measures the correlation between the pattern p and the trace t), and by filtering out low correlation values according to the threshold α. Clearly, many classical clustering algorithms could be used to extract P^* out of the matrix M (Step C2). In fact, we used an enhanced implementation of the *Markov Cluster Algorithm*, which achieved good results on several large datasets [4], and choose the number of clusters autonomously.

In the second phase (Steps C3-C13), the preliminary clustering P^* of the patterns is refined, and yet used as a basis for simultaneously clustering the traces of L: new, "high order" pattern clusters are built by merging together basic pattern clusters that relate to the same traces. More precisely, each trace t in the log induces a pattern cluster $\widehat{p}^t$, which is the union of all the (basic) clusters in P^* that are correlated enough to t, still based on the function $supp$ and threshold α. It may happen that the cluster $\widehat{p}^t$ is already in $\widehat{P}$, for it was induced by some other traces; in this case we retrieve, by using the mapping $\mathcal{M}$, the cluster $\widehat{l}^t$ containing these traces (Step C7), and extend it with the insertion of t (Step C8). Otherwise, we save a new trace cluster, just consisting of t, in $\widehat{L}$, and update $\mathcal{M}$ to store the association between this new cluster and $\widehat{p}^t$, which is stored as well, in $\widehat{P}$, as a novel pattern cluster (Step C10).

We pinpoint that algorithm `TraceOutlierMining` can be implemented without importing the entire input log in main memory, as we may just scan it k times to find the patterns, plus two further times to build the matrix M and map the traces to the clusters (Steps C4-C12). Thus, main memory computation is just limited to the clustering of interesting patterns, whose overall size can be kept low by suitably setting the thresholds σ and γ (cf. Def. 3). This can ensure scalability over huge datasets.

4 Experimental Results

The proposed approach has been implemented in a Java prototype system, and thoroughly tested to assess its scalability and accuracy, on a 1800MHz/2GB Pentium IV machine running *Windows XP Pro*. To this aim, we developed a generator that randomly produces a process log of N_T traces, by enabling to control several data distribution features. Traces in the log are distributed along N_C clusters, so that $p_C^{out} \times N_T$ of them fall into clusters whose size is smaller than the average. In addition, $p^{out} \times N_T$ traces are produced that do not comply with any of the clusters. Hence, the total percentage of outliers in the dataset is $p_C^{out} + p^{out}$. In a nutshell, the generator works as follows: First, it builds a set P of disjoint subschemas whose sizes are taken from a gaussian

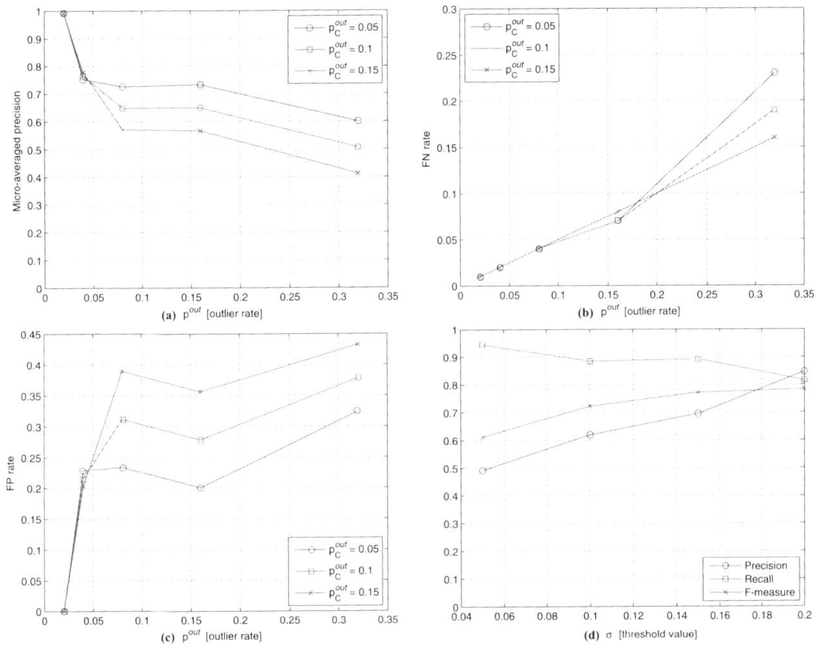

Fig. 3. Accuracy Results

distribution with mean S_P, and then combines them into a schema W_P over N_A activities, where each sub-schema is allowed to be run independently. Then, N_C subsets of P are randomly selected and enacted (according to p_C^{out}) in W_P, thereby generating the various clusters of traces over a total of $(1 - p^{out}) \times N_T$ traces. Finally, $p^{out} \times N_T$ traces are generated by simulating enactments that do not comply with W_P.

Accuracy. Firstly, we evaluated the effectiveness of the approach against various input logs, containing different percentage of outliers. To this purpose, logs were generated by varying both p^{out} (from 0.02 to 0.32) and p_C^{out} (from 0.05 to 0.15), and using fixed values for the other parameters: N_A=180, N_T=16000, N_C=4, and S_P=6. Figures 3.(a), 3.(b), and 3.(c) illustrate accuracy results obtained on these data, by applying the `TraceOutlierMining` algorithm with γ=4, α=0.4 and β=0.5, and *pattSize*=8. More precisely, Figure 3.(a) depicts the accuracy of the approach in rediscovering all the clusters in the input log, according to the standard *micro-averaged precision* measure—computed by averaging, over all the mined clusters, the frequency of the majority class in each cluster (i.e., the maximal percentage of elements assigned to that mined cluster and coming from one input "true" cluster). Instead, Figures 3.(b) and 3.(c) depict the capability of the approach to correctly recognize anomalous traces, by reporting the rates of False Negatives (FN), i.e., outliers deemed as normal, and False Positives (FP), i.e., normal traces deemed as outliers, resp.—in a sense, the outlier detection problem is regarded here as a classification problem with two classes of objects: outliers and normal individuals. These quality measures worsen when

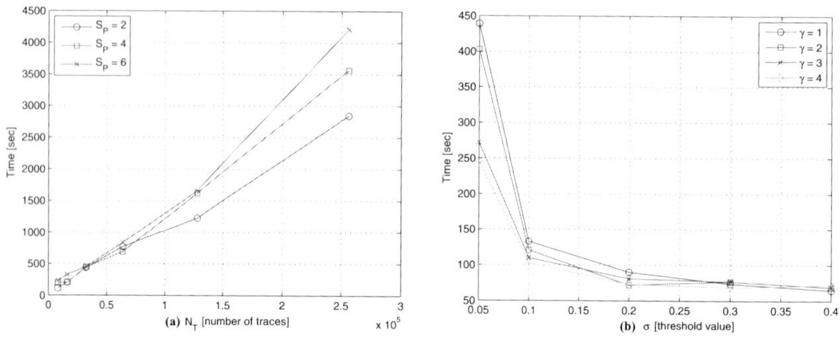

Fig. 4. Scalability Results

increasing the overall percentage of outliers, still getting quasi-optimal values when this latter is under 9%.

In order to evaluate the sensitivity of the algorithm to its parameters, we generated a log with p^{out}=0.05 and p_C^{out}=0.09, and the same value as in the previous test for any other data parameter. Figure 3.(d) reports three standard accuracy measures (namely, precision, recall, and F-measure), computed again for the binary classification problem outliers vs. normal individuals, obtained when varying σ, and for α=0.8, γ=4, β=0.5 and *pattSize*=8. The figure evidences a trade-off between precision and recall, thereby suggesting that parameters must be chosen depending on the application needs, possibly with the help of self-tuning heuristics (as in [11]).

Scalability. In another series of experiments, we assessed the scalability of the approach w.r.t. the size of the input data, by building a number of datasets with increasingly larger number of traces and activities in the schema, while fixing p^{out}=0.02, N_A=125, and N_C=4. Fixed values were taken as well for all `TraceOutlierMining`'s parameters: α=0.2, β=0.5, γ=4 and σ=0.15, and *pattSize*=10. As shown in Figure 4.(a), the total computation time linearly scales both with the number of log traces N_T and with the size of the process schema W_P used to generate the log itself (cf. S_P).

Finally, Figure 4.(b) reports the total computation time spent by the algorithm over the log for Figures 3.(a), 3.(b), and 3.(c), when keeping fixed all the parameters but σ and γ (actually, we set again α=0.8, γ=4, β=0.5 and *pattSize*=8). Note that σ and γ do impact on computation time: the lower their value the higher the time. However, a notable increase only occurs when σ passes from 0.1 to 0.05. This effect is emphasized when γ too is kept low, and any σ-frequent pattern is also (σ, γ)-maximal.

5 Discussion and Conclusion

Even though singling out anomalies in process executions can help in recognizing critical points in the process, current process mining approaches adopt very simple and pragmatical solutions in its facing. Indeed, the general idea (e.g., [10]) is to exploit user-defined thresholds to define the minimum frequency for activities below which an execution is considered noisy. Only very recently, [11] embarked on a systematic investigation of noisy environments, by focusing on the mining of conversation logs and by proposing automatic techniques for identifying the right threshold value. In this paper, we have expanded this research line, by devising an *outlier detection* approach specifically tailored for process models and logs, which, differently from [11], takes account for concurrency constructs. Moreover, due to our focus on outlier detection rather than on noise filtering, anomalies are defined not only w.r.t. statistical global features, but also w.r.t. major behavioral clusters. In fact, clustering-based outlier detection techniques have already been used in different contexts (see, e.g., [8,13,9]). Here, this methodology has been applied to process models and specialized to deal with categorical data (cf. the various behavioral patterns), by sharing the perspective of [7] where outliers are characterized in terms of infrequent patterns.

The proposed approach paves the way for further elaborations. For example, an avenue of research is to integrate it with the self-tuning techniques in [11], as to reduce as much as possible human intervention in the mining process. Also, it would be relevant to investigate on the definition of explanation techniques, i.e., on methods that, given a set of outliers, abductively formulate hypotheses for recognizing critical points in the process that can yield invalid or inappropriate behavior.

References

1. Apostolico, A., Bock, M.E., Lonardi, S., Xu, X.: Efficient detection of unusual words. Journal of Computational Biology 7(1/2), 71–94 (2000)
2. Dhillon, I.S., Mallela, S., Modha, D.S.: Information-theoretic co-clustering. In: Proc. 9th ACM SIGKDD Conf. on Knowledge Discovery and Data Mining (KDD 2003), pp. 89–98 (2003)
3. Dustdar, S., Hoffmann, T., van der Aalst, W.M.P.: Mining of ad-hoc business processes with teamlog. Data and Knowledge Engineering 55(2), 129–158 (2005)
4. Enright, A.J., Van Dongen, S., Ouzounis, C.A.: An efficient algorithm for large-scale detection of protein families. Nucleic Acids Res. 30(7), 1575–1584 (2002)
5. Fawcett, T.E., Provost, F.: Fraud detection. In: Handbook of data mining and knowledge discovery, pp. 726–731. Oxford University Press, Oxford (2002)
6. Greco, G., Guzzo, A., Pontieri, L., Saccà, D.: Discovering expressive process models by clustering log traces. IEEE Trans. on Knowledge and Data Engin. 18(8), 1010–1027 (2006)
7. He, Z., Xu, Z., Huang, J.Z., Deng, S.: Fp-outlier: Frequent pattern based outlier detection. In: Hao, Y., Liu, J., Wang, Y.-P., Cheung, Y.-m., Yin, H., Jiao, L., Ma, J., Jiao, Y.-C. (eds.) CIS 2005. LNCS (LNAI), vol. 3801, pp. 735–740. Springer, Heidelberg (2005)
8. Jaing, M.F., Tseng, S.S., Su, C.M.: Two-phase clustering process for outliers detection. Pattern Recogn. Lett. 22(6-7), 691–700 (2001)
9. Jiang, S., Song, X., Wang, H., Han, J.-J., Li, Q.-H.: A clustering-based method for unsupervised intrusion detections. Pattern Recogn. Lett. 27(7), 802–810 (2006)
10. Maruster, L., Weijters, A.J.M.M., van der Aalst, W.M.P., van den Bosch, A.: A rule-based approach for process discovery: Dealing with noise and imbalance in process logs. Data Mining and Knowledge Discovery 13(1), 67–87 (2006)
11. Motahari Nezhad, H.R., Saint-Paul, R., Benatallah, B., Casati, F.: Protocol discovery from imperfect service interaction logs. In: Proc. of ICDE 2007, pp. 1405–1409 (2007)
12. van der Aalst, W.M.P., van Dongen, B.F., Herbst, J., Maruster, L., Schimm, G., Weijters, A.: Workflow mining: a survey of issues and approaches. Data & Know. Engin. 47(2), 237–267 (2003)
13. Yu, D., Sheikholeslami, G., Zhang, A.: Findout: finding outliers in very large datasets. Knowledge Information Systems 4(4), 387–412 (2002)

Action Rule Extraction from a Decision Table: ARED

Seunghyun Im[1] and Zbigniew W. Raś[2,3]

[1] University of Pittsburgh at Johnstown, Department of Computer Science
Johnstown, PA 15904, USA
[2] University of North Carolina, Department of Computer Science
Charlotte, NC, 28223, USA
[3] Polish Academy of Sciences, Institute of Computer Science, 01-237 Warsaw, Poland
sim@pitt.edu, ras@uncc.edu

Abstract. In this paper, we present an algorithm that discovers action rules from a decision table. Action rules describe possible transitions of objects from one state to another with respect to a distinguished attribute. The previous research on action rule discovery required the extraction of classification rules before constructing any action rule. The new proposed algorithm does not require pre-existing classification rules, and it uses a bottom up approach to generate action rules having minimal attribute involvement.

1 Introduction

In this paper, we present an algorithm that discovers action rules. An action rule is a rule extracted from a database that describes a possible transition of objects from one state to another with respect to a distinguished attribute called a decision attribute [14]. Values of some attributes, used to describe objects stored in a database, can be changed. This change can be influenced and controlled by users. For example, let us assume that a number of customers have closed their bank accounts. We construct a description of this group of customers. Then, we search for another description of a new group of customers who keep the bank accounts active. If these descriptions have a form of rules, they can be seen as actionable rules. For instance, by comparing those two descriptions, we may identify the reason for closing their accounts, and formulate an action, which if undertaken by the bank, may prevent other customers from closing their accounts. In this case, an action rule may say that, if the bank lowers the interest rate by 2 percent on credit cards for certain group of customers, it is almost guaranteed that they do not close their accounts. A similar definition, but with different notation, of an action rule was given earlier in [4]. Also, interventions introduced in [5] are conceptually similar to action rules. Action rules introduced in [14] has been further investigated in [17][13][12][18][15].

Paper [7] was probably the first attempt towards formally introducing the problem of mining action rules without pre-existing classification rules. Authors explicitly formulated it as a search problem in a support-confidence-cost framework. The proposed algorithm is similar to Apriori [1]. Their definition of an action rule allows changes on stable attributes. Changing the value of an attribute, either stable or flexible, is linked with a cost [18]. In order to rule out action rules with undesired changes on stable attributes, authors have assigned very high cost to such changes. However, that way, the

A. An et al. (Eds.): ISMIS 2008, LNAI 4994, pp. 160–168, 2008.

cost of action rules discovery is getting unnecessarily increased. Also, they did not take into account the dependencies between attribute values which are naturally linked with the cost of rules used either to accept or reject a rule.

Algorithm *ARED*, presented in this paper, is based on Pawlak's model of an information system S [10] and its goal is to identify certain relationships between granules defined by the indiscernibility relation on objects in S. Some of these relationships uniquely define action rules for S.

The rest of this paper is organized as follows. Section 2 describes the background and objectives of our work. The details of the algorithm *ARED* are presented in Section 3. Experimental results are shown in Section 4. Section 5 discusses possible future work and concludes the paper.

2 Backgrounds and Objectives

Action rules are extracted from an information system. By an information system S, we mean $S = (X, A, V)$ where X is a nonempty finite set of objects, A is a nonempty finite set of attributes, and $V = \bigcup\{V_a : a \in A\}$ is set of their values. For example, Table 1 presents an information system S with $X = \{x_1, x_2, x_3, x_4, x_5, x_6, x_7, x_8\}$, $A = \{a, b, c, d\}$, and $V = \{a_1, a_2, b_1, b_2, b_3, c_1, c_2, d_1, d_2, d_3\}$.

An information system $S = (X, A, V)$ is called a decision system (or decision table), if $A = A_{St} \cup A_{Fl} \cup \{d\}$, where d is a distinguished attribute set called the decision. Attributes in A_{St} are called *stable* and attributes in A_{Fl} are called *flexible*. They jointly form the set of conditional attributes. "Date of birth" is an example of a stable attribute. "Interest rate" for each customer account is an example of a flexible attribute.

In earlier works in [14][17][13][12][15], action rules are constructed from classification rules. This means that we use pre-existing classification rules or generate them using a rule discovery algorithm, such as LERS [6] or ERID [2], then, construct action rules either from certain pairs of the rules or from a single classification rule. For instance, algorithm $ARAS$ [15] generates sets of terms (built from values of attributes) around classification rules and constructs action rules directly from them. In this study, we propose a different approach to achieve the following objectives:

1. Extract action rules directly from an information system without using pre-existing conditional rules.
2. Extract all distinct action rules that have minimal attribute involvement.

To meet these two goals, we first generate action rules using two attributes, and iteratively apply the technique to generate more specific action rules.

3 Algorithm ARED

We describe the algorithm, *ARED* (Action Rule Extraction from Decision table), by working through an example using the decision Table S shown in Table 1. We assume that the decision attribute is d, stable attributes $A_{St} = \{a\}$, and the flexible attributes $A_{Fl} = \{b, c\}$. The minimum support (λ_1) and confidence (λ_2) are given as 1 and 0.85.

Table 1. Decision Table S

	a	b	c	d
x_1	a_1	b_1	c_1	d_1
x_2	a_2	b_1	c_2	d_1
x_3	a_2		c_2	d_1
x_4	a_2	b_1	c_1	d_1
x_5	a_2	b_3	c_2	d_1
x_6	a_1	b_1		d_2
x_7	a_1	b_2	c_2	d_1
x_8	a_1	b_2	c_1	d_3

For simplicity reason, we will consider decision tables with only one decision in this paper.

The first step is to find the pessimistic interpretation in S of all attribute values in V, as shown in Table 2 (for conditional attributes) and Table 3 (for decision attribute). The resulting sets are called granules. In S, $Dom(a) = \{a_1, a_2\}$, $Dom(b) = \{b_1, b_2, b_3\}$, $Dom(c) = \{c_1, c_2\}$, and $Dom(d) = \{d_1, d_2, d_3\}$. The granule a_1^* associated with attribute value a_1 in S (see Table 1) is the set of objects having property a_1 (e.g. objects $\{x_1, x_6, x_7, x_8\}$). The set of granules associated with an attribute a in S is defined as $\{v^* : v \in V_a\}$.

Table 2. Granules associated with attributes in a, b, c

$$\begin{aligned} {a_1}^* &= \{x_1, x_6, x_7, x_8\} \\ {a_2}^* &= \{x_2, x_3, x_4, x_5\} \\ {b_1}^* &= \{x_1, x_2, x_4, x_6\} \\ {b_2}^* &= \{x_7, x_8\} \\ {b_3}^* &= \{x_5\} \\ {c_1}^* &= \{x_1, x_4, x_8\} \\ {c_2}^* &= \{x_2, x_3, x_5, x_7\} \end{aligned}$$

Table 3. Granules associated with attribute d

$$\begin{aligned} {d_1}^* &= \{x_1, x_2, x_3, x_4, x_5, x_7\} \\ {d_2}^* &= \{x_6\} \\ {d_3}^* &= \{x_8\} \end{aligned}$$

Next, we define two sets, τ and δ, to examine possible *property transitions* between objects in S. Let T be a set of proper conjuncts built from elements in $\cup\{V_i, i \neq d, i \in A\}$. By proper conjunct, we mean a conjunct which contains maximum one element from each V_i.

- $\tau = T \cdot d_1$, where $d_1 \in V_d$, and $(\forall \rho_1 \in T \cdot d_1)(sup(\rho_1) \geq \lambda_1)$.
- $\delta = T \cdot d_2$, where $d_2 \in V_d$, and $(\forall \rho_2 \in T \cdot d_2)(sup(\rho_2) \geq \lambda_1)$.

By $T \cdot d_i$, we mean $\{t \cdot d_i : t \in T\}$, i=1,2. The support of ρ_i, $sup(\rho_i)$, is the number of objects in S supporting all attribute values listed in ρ_i, $i = 1, 2$. They can be easily calculated by intersecting two granules listed in Table 2 and Table 3. For example, $(a_1 \cdot d_2)^* = \{x_1, x_6, x_7, x_8\} \cap \{x_6\} = \{x_6\}$. So, $sup((a_1 \cdot d_2)^*) = 1$.

Each set contains only terms built from one decision value and at least one value of conditional attribute. Therefore, these sets represent (1) a relationship between conditional attributes and the decision attribute, and (2) property of a set of objects. If the property transition from τ to δ is valid, τ and δ are interpreted as the *condition* and the *decision* of an *action rule*.

As mentioned, *ARED* attempts to discover the shortest action rules in terms of the number of attributes, then iteratively generates longer action rules. Therefore, we first construct τ containing two elements (which is the shortest form of τ) by combining one attribute from Table 2 and the other from Table 3. This process aims to find the meaning in S of all terms in a conjunctive form built from values in V [16]. The concatenation functor used to build these terms is interpreted as the set-theoretical intersection.

Table 4. 2-element τ and δ

τ	δ
$(a_1 \cdot d_1)$	$(a_1 \cdot d_1)$
$(a_1 \cdot d_2)$	$(a_1 \cdot d_2)$
$(a_1 \cdot d_3)$	$(a_1 \cdot d_3)$
$(a_2 \cdot d_1)$	$(a_2 \cdot d_1)$
$(b_1 \cdot d_1)$	$(b_1 \cdot d_1)$
$(b_1 \cdot d_2)$	$(b_1 \cdot d_2)$
$(b_2 \cdot d_1)$	$(b_2 \cdot d_1)$
$(b_2 \cdot d_3)$	$(b_2 \cdot d_3)$
$(b_3 \cdot d_1)$	$(b_3 \cdot d_1)$
$(c_1 \cdot d_1)$	$(c_1 \cdot d_1)$
$(c_1 \cdot d_3)$	$(c_1 \cdot d_3)$
$(c_2 \cdot d_1)$	$(c_2 \cdot d_1)$

Table 5. Action Rules

τ		δ	sup	$conf$	$rule$
$(b_1 \cdot d_2)$	$\mapsto$	$(b_2 \cdot d_1)$	1	1	y
$(b_1 \cdot d_2)$	$\mapsto$	$(b_2 \cdot d_3)$	1	1	y
$(b_1 \cdot d_2)$	$\mapsto$	$(b_3 \cdot d_1)$	1	1	y
$(b_2 \cdot d_1)$	$\mapsto$	$(b_1 \cdot d_2)$	1	1	y
$(b_2 \cdot d_3)$	$\mapsto$	$(b_1 \cdot d_1)$	1	1	y
$(b_2 \cdot d_3)$	$\mapsto$	$(b_1 \cdot d_2)$	1	1	y
$(b_2 \cdot d_3)$	$\mapsto$	$(b_3 \cdot d_1)$	1	1	y
$(b_3 \cdot d_1)$	$\mapsto$	$(b_1 \cdot d_2)$	1	1	y
$(b_3 \cdot d_1)$	$\mapsto$	$(b_2 \cdot d_3)$	1	1	y
$(c_1 \cdot d_3)$	$\mapsto$	$(c_2 \cdot d_1)$	1	1	y

Table 6. Invalid Transitions

τ		δ	sup	$conf$	$rule$
$(b_1 \cdot d_1)$	$\mapsto$	$(b_2 \cdot d_3)$	1	0.33	n
$(c_2 \cdot d_1)$	$\mapsto$	$(c_1 \cdot d_3)$	1	0.25	n
$(a_1 \cdot d_1)$		$(a_1 \cdot d_1)$			
$(a_1 \cdot d_2)$		$(a_1 \cdot d_2)$			
$(a_1 \cdot d_3)$		$(a_1 \cdot d_3)$			
$(a_2 \cdot d_1)$		$(a_2 \cdot d_1)$			

Clearly, it is required that $sup(\rho_1) \geq \lambda_1 \wedge sup(\rho_2) \geq \lambda_1$ because we need to find property transitions that involve at least λ_1 objects. Terms satisfying these criteria, related to the example in Table 1, are shown in Table 4.

Now, we construct action rules using τ to δ in Table 4 by evaluating the validity of their transitions. We use the following code to explain the evaluation method.

input : $\{\tau\}, \{\delta\}$
output : action_rule [],gen_τ[], gen_δ[]

1: for each $\tau_i = [t_1 \cdot d_1] \in \tau, \delta_j = [t_2 \cdot d_2] \in \delta$
2: if $d_1 \neq d_2$ then
3:     if $\exists(a(x_i) \in t_1, a(x_j) \in t_2)$, where $a \in A_{Fl}$ and $a(x_i) \neq a(x_j)$ then
4:         if $conf(\tau_i \mapsto \delta_j) \geq \lambda_2$
5:             action_rule [end+1] = $\tau_i \mapsto \delta_j$
6:         elseif
7:             gen_τ[end+1] = τ_i, gen_δ [end+1] = δ_j
8:         end if
9:     else
10:         gen_τ[end+1] = τ_i, gen_δ [end+1] = δ_j
11:     end if
12: end if

In line 1, we read each τ_i and δ_i listed in Table 4. Note that an information system may or may not produce two-way action rules (e.g. if b changes from b_1 to b_2 then d changes from d_1 to d_2, and vice versa). Therefore, we generate all pairs of elements from τ and δ. The condition in line 2 is clear. No action can be performed if the decision values are equal. The condition in line 3 checks if there exist different flexible attribute values between t_1 and t_2, and they are from the same domain. For example, $d_1 \neq d_2$ and $a_1 \neq a_2$ for $(a_1 \cdot d_2)$ and $(a_2 \cdot d_1)$. However, a is a stable attribute, so no action rule is constructed. If two sets do not meet this condition, we put them in two separate arrays (line 10) and use them to generate τ and δ for the next iteration. Last 4 rows in Table 6 are the sets in this category. The reasoning behind this strategy is simple. Clearly, there are at least λ_1 objects supporting $(a_1 \cdot d_2)$ and $(a_2 \cdot d_1)$. These sets may be concatenated with one or more flexible attribute values in later iterations, and produce action rules. An example of this case is $(a_1 \cdot b_1 \cdot d_1) \mapsto (a_1 \cdot b_2 \cdot d_3)$ in Table 8. The condition in line 4 checks the confidence of $\tau_i \mapsto \delta_j$. If it is greater than or equal λ_2, $\tau_i \mapsto \delta_j$ becomes an action rule. To compute the confidence, we first find the support of $ar = \tau \mapsto \delta$ which is defined as the minimum support of two sets.

$$sup(ar) = min(sup(\tau), sup(\delta))$$

$min(sup, sup)$ finds the exact number of object transitions because at least one element is different between τ and δ, and by definition of S, an object cannot support two different values of the same attribute. Note that we do not need to check the support of ar in line 4 because the support of τ and δ were checked when they were generated, and the support of the action rule is the smaller value between them. The confidence of an ar is defined as,

$$conf(ar) = \frac{sup(ar)}{sup(\tau)}$$

Two-element action rules extracted from S are shown in Table 5. For example, $(b_1 \cdot d_2) \mapsto (b_2 \cdot d_1)$ is an action rule, and it is interpreted as, "if b changes from

Table 7. 3-element τ and δ

τ	δ
$(a_2 \cdot b_1 \cdot d_1)$	$(a_1 \cdot c_2 \cdot d_3)$
$(a_2 \cdot c_3 \cdot d_1)$	$(a_1 \cdot b_2 \cdot d_3)$
$(a_1 \cdot b_1 \cdot d_1)$	$(b_2 \cdot c_2 \cdot d_3)$
$(a_1 \cdot c_3 \cdot d_1)$	
$(b_1 \cdot c_3 \cdot d_1)$	

Table 8. Action Rule

τ	δ	sup	$conf$	$rule$
$(a_1 \cdot b_1 \cdot d_1) \mapsto$	$(a_1 \cdot b_2 \cdot d_3)$	1	1	y
$(a_1 \cdot c_3 \cdot d_1) \mapsto$	$(a_1 \cdot c_2 \cdot d_3)$	1	1	y
$(b_1 \cdot c_3 \cdot d_1) \mapsto$	$(b_2 \cdot c_2 \cdot d_3)$	1	1	y

Table 9. Invalid Transition

τ	δ	sup	$conf$	$rule$
$(a_2 \cdot b_1 \cdot d_1) \mapsto$	$(a_1 \cdot b_2 \cdot d_3)$	1	0.50	n
$(a_2 \cdot c_3 \cdot d_1) \mapsto$	$(a_1 \cdot c_2 \cdot d_3)$	1	0.33	n

Table 10. 4-element τ and δ

τ	δ
$(a_2 \cdot b_1 \cdot c_3 \cdot d_1)$	$(a_1 \cdot b_2 \cdot c_2 \cdot d_3)$

Table 11. Action Rule

τ	δ
$(a_2 \cdot b_1 \cdot c_3 \cdot d_1) \mapsto$	$(a_1 \cdot b_2 \cdot c_2 \cdot d_3)$

b_1 to b_2, then d changes from d_1 to d_2". Finally, if the confidence is less than λ_2 (line 7) τ_i and δ_j are considered in the next iteration to generate 3-element candidate sets. In Table 6, we have two transitions having confidence of 0.33 and 0.25 that are less than 1 (λ_2).

Assume that $\mid t_i \mid$, for any conjunct term t_i, denotes the set of all values of attributes listed in t_i. To find the action rules of length 3, we generate τ of length 3 from τs in Table 6. Two terms $\tau_1 = t_1 \cdot d_1$ and $\tau_2 = t_2 \cdot d_2$ are concatenated if (1) $d_1 = d_2$ (2) $\mid t_1 \cup t_2 \mid - \mid t_2 \cap t_1 \mid = \{v_1 \in V_a, v_2 \in V_b\}$, where $a \neq b$. The set $\{\delta\}$ is generated from δs in Table 6 using the same method. That means they are generated independently. The reason that we build 3-element candidate sets separately in this way is that all pairs are considered in the initial step. Therefore, any super set of the candidate set identified as an action rule will not be considered. Table 7 shows those 3-element candidate sets. Corresponding action rule and invalid transitions are show in Tables 8 and 9 respectively.

In the next iteration, we construct τ and δ listed in Table 10 and build an action rule containing 4 elements as shown in Table 11. However, it is not included in the list of action rules because it's τ and δ are supersets of $(b_1 \cdot c_3 \cdot d_1) \mapsto (b_2 \cdot c_2 \cdot d_3)$ in Table 8, which is a more general action rule.

The process stops because there are no sets to be combined.

4 Experiment

We implemented the algorithm in Matlab on a Pentium M 1.6 GHz computer running Windows XP, and tested it using a sample data set (nursery database) obtained from [8]. The data set contains information about applications for nursery schools, and used to rank them. The decision attribute has five classes (not recommend, recommend,

Table 12. Experiment Result

Type	Attribute Name	Description
flexible	parents (p)	Parents' occupation
stable	has_nurs (n)	Child's nursery
stable	form (o)	Form of the family
stable	children (c)	Number of children
flexible	housing (h)	Housing conditions
flexible	finance (f)	Financial standing of the family
stable	social (s)	Social conditions
flexible	health (t)	Health conditions
flexible	decision (d)	Rank

Table 13. Experiment Result

Experiment No.	Num. of objects	Stab. attribute	Flex. attribute	Sup	Conf	Num. of action rules	Computation time (in seconds)
1	12960	4	5	5%	85%	86	2.36
2	12960	4	5	10%	85%	53	1.07
3	12960	4	5	15%	85%	12	0.88

Table 14. Action Rules

condition		decision	sup	conf
$(f)convenient \rightarrow (d)notrecom$	$\mapsto$	$(f)inconvenient \rightarrow (d)priority$	2022	0.9
$(f)convenient \rightarrow (d)notrecom$	$\mapsto$	$(f)inconvenient \rightarrow (d)spec\_prior$	2160	1
$(f)convenient \rightarrow (d)priority$	$\mapsto$	$(f)inconvenient \rightarrow (d)notrecom$	2160	1
$(f)convenient \rightarrow (d)priority$	$\mapsto$	$(f)inconvenient \rightarrow (d)spec\_prior$	2188	1
$(f)inconvenient \rightarrow (d)notrecom$	$\mapsto$	$(f)convenient \rightarrow (d)priority$	2160	1
$(f)inconvenient \rightarrow (d)priority$	$\mapsto$	$(f)convenient \rightarrow (d)notrecom$	2022	1
$(f)inconvenient \rightarrow (d)spec\_prior$	$\mapsto$	$(f)convenient \rightarrow (d)notrecom$	2160	1
$(f)inconvenient \rightarrow (d)spec\_prior$	$\mapsto$	$(f)convenient \rightarrow (d)priority$	2188	1
$(t)priority \rightarrow (d)spec\_prior$	$\mapsto$	$(t)notrecom \rightarrow (d)notrecom$	2466	1
$(t)priority \rightarrow (d)spec\_prior$	$\mapsto$	$(t)recommended \rightarrow (d)priority$	2412	1
$(t)recommended \rightarrow (d)priority$	$\mapsto$	$(t)notrecom \rightarrow (d)notrecom$	2412	1
$(t)recommended \rightarrow (d)priority$	$\mapsto$	$(t)priority \rightarrow (d)spec\_prior$	2412	1

very recommend, priority, special priority).We partitioned the attributes into stable and flexible based on the description given by the provider. For example, the number of children in a family (1, 2, 3, more) is considered as a stable attribute, while financial standing of the family (convenient, inconvenient) or parents' occupation (usual, pretentious, great pretentious) are considered as flexible attributes.

Table 12 shows the attributes names, descriptions, and the partitions. All attributes are categorical in the data set. If a data set contains continuous data, one can use a discretization method (such as Rosetta [9]) to convert them to categorical data. Table 13 shows the parameters used in our experiment, the number of action rules generated, and the time required to complete the task.

Table 14 shows the action rules generated during experiment No. 3. The first action rule can be read as, if the housing standing of the family changes from *convenient* to *inconvenient*, then the decision changes from *not recommend* to *priority* with support value of 2022 and confidence of 0.9. The overall result shows that the change of rank (decision attribute) is strongly related to the changes in financial standing of the family and health conditions in experiment 3.

5 Conclusion and Future Work

We presented an algorithm that discovers action rules from a decision table. The proposed algorithm generates a complete set of shortest action rules without using pre-existing classification rules. During the experiment with several data sets, we noticed that the flexibility of attributes are not equal. For example, the social condition is most likely less flexible than the health condition in the data set used in our experiment, and this may have to considered. Future work shall address this issue as well as further analysis of the algorithm with real world data sets.

Acknowledgment

This work was partially supported by the National Science Foundation under grant IIS-0414815.

References

1. Agrawal, R., Srikant, R.: Fast algorithm for mining association rules. In: Proceeding of the Twentieth International Conference on VLDB, pp. 487–499 (1994)
2. Dardzińska, A., Raś, Z.: Extracting rules from incomplete decision systems. In: Foundations and Novel Approaches in Data Mining, Studies in Computational Intelligence, vol. 9, pp. 143–154. Springer, Heidelberg (2006)
3. Fensel, D.: Ontologies: a silver bullet for knowledge management and electronic commerce. Springer, Heidelberg (1998)
4. Geffner, H., Wainer, J.: Modeling action, knowledge and control. In: ECAI, pp. 532–536 (1998)
5. Greco, S., Matarazzo, B., Pappalardo, N., Slowiński, R.: Measuring expected effects of interventions based on decision rules. J. Exp. Theor. Artif. Intell. 17(1-2), 103–118
6. Grzymala-Busse, J.: A new version of the rule induction system LERS. Fundamenta Informaticae 31(1), 27–39 (1997)
7. He, Z., Xu, X., Deng, S., Ma, R.: Mining action rules from scratch. Expert Systems with Applications 29(3), 691–699 (2005)
8. Hettich, S., Blake, C.L., Merz, C.J.: UCI Repository of machine learning databases, University of California, Irvine, Dept. of Information and Computer Sciences (1998), `http://www.ics.uci.edu/~mlearn/MLRepository.html`
9. Øhrn, A., Komorowski, J.: ROSETTA: A Rough Set Toolkit for Analysis of Data (1997)
10. Pawlak, Z.: Information systems - theoretical foundations. Information Systems Journal 6, 205–218 (1981)

11. Qiao, Y., Zhong, K., Wang, H.-A., Li, X.: Developing event-condition-action rules in real-time active database. In: Proceedings of the 2007 ACM symposium on Applied computing, pp. 511–516. ACM, New York (2007)
12. Raś, Z.W., Dardzińska, A.: Action rules discovery, a new simplified strategy. In: Esposito, F., Raś, Z.W., Malerba, D., Semeraro, G. (eds.) ISMIS 2006. LNCS (LNAI), vol. 4203, pp. 445–453. Springer, Heidelberg (2006)
13. Raś, Z.W., Tzacheva, A., Tsay, L.-S., Gürdal, O.: Mining for interesting action rules. In: Proceedings of IEEE/WIC/ACM International Conference on Intelligent Agent Technology (IAT 2005), Compiegne University of Technology, France, pp. 187–193 (2005)
14. Raś, Z.W., Wieczorkowska, A.: Action-Rules: How to increase profit of a company. In: Zighed, A.D.A., Komorowski, J., Żytkow, J.M. (eds.) PKDD 2000. LNCS (LNAI), vol. 1910, pp. 587–592. Springer, Heidelberg (2000)
15. Raś, Z., Wyrzykowska, E., Wasyluk, H.: ARAS: Action rules discovery based on agglomerative strategy. In: Mining Complex Data, Post-Proceedings of 2007 ECML/PKDD Third International Workshop (MCD 2007). LNCS (LNAI), vol. 4944, pp. 196–208. Springer, Heidelberg (2007)
16. Skowron, A.: Rough sets and boolean reasoning. In: Granular Computing: an Emerging Paradigm, pp. 95–124. Physica-Verlag (2001)
17. Tsay, L.-S., Raś, Z.W.: Action rules discovery system DEAR3, in Foundations of Intelligent Systems. In: Esposito, F., Raś, Z.W., Malerba, D., Semeraro, G. (eds.) ISMIS 2006. LNCS (LNAI), vol. 4203, pp. 483–492. Springer, Heidelberg (2006)
18. Tzacheva, A., Ras, Z.W.: Constraint based action rule discovery with single classification rules. In: An, A., Stefanowski, J., Ramanna, S., Butz, C.J., Pedrycz, W., Wang, G. (eds.) RSFDGrC 2007. LNCS (LNAI), vol. 4482, pp. 322–329. Springer, Heidelberg (2007)

Discovering the Concise Set of Actionable Patterns

Li-Shiang Tsay[1] and Zbigniew W. Raś[2,3]

[1] North Carolina A&T State Univ., School of Tech., Greensboro, NC 27411
[2] Univ. of North Carolina, Dept. of Comp. Science, Charlotte, NC 28223
[3] Polish-Japanese Inst. of Inf. Tech., 02-008 Warsaw, Poland
ltsay@ncat.edu, ras@uncc.edu

Abstract. It is highly expected that knowledge discovery and data mining (KDD) methods can extract useful and understandable knowledge from large amount of data. Action rule mining presents an approach to automatically construct relevantly useful and understandable strategies by comparing the profiles of two sets of targeted objects – those that are desirable and those that are undesirable. The discovered knowledge provides an insight of how relationships should be managed so that objects of low performance can be improved. Traditionally, it was constructed from one or two classification rules. The quality and quantity of such *Action Rules* depend on adopted classification methods. In this paper, we present *StrategyGenerator*, a new algorithm for constructing a complete set of *Action Rules* which satisfies specified constraints. This algorithm does not require prior extraction of classification rules. Action rules are generated directly from a database.

Keywords: Action Rules, Interestingness Measure, Reclassification Models.

1 Introduction

Knowledge Discovery and Data mining (KDD) is the process which identifies and exploits useful and understandable knowledge buried in large volumes of data. The products of KDD have been proven very effective in many fields, such as business, science, government, etc. While most of the KDD algorithms generate predictions and describe behaviors, a focus on understanding changes in object behaviors normally improves the quality of the decision making process. *Action Rule* mining constructs relatively interesting and useful strategies by comparing the profiles of two groups of targeted objects – those that are desirable and those that are undesirable. It is formed as a term $[(\omega)\wedge(\alpha\rightarrow\beta)]\Rightarrow(\phi\rightarrow\psi)$, where ω is a conjunction of fixed condition features shared by both groups, $(\alpha\rightarrow\beta)$ represents proposed changes in values of flexible features, and $(\phi\rightarrow\psi)$ is a desired effect of the action. The discovered knowledge provides an insight of how relationships should be managed so the undesirable objects can be changed to desirable objects. For example, in society, one would like to find a way to improve his or her salary from a low-income to a high-income. Another example in business area is when an owner would like to improve his or her company's profits by going from a high-cost, low-income business to a low-cost, high-income business.

A. An et al. (Eds.): ISMIS 2008, LNAI 4994, pp. 169–178, 2008.

The goal of this research is twofold: (1) making the discovered patterns actionable by providing specific action plans; (2) facilitate the decision-making process in an efficient and easy way by giving users the information they need. Making a decision implies that there are several choices to be considered in a short time frame. It is important not only to identify as many of these choices as possible but also to avoid a redundancy among them.

Action Rules algorithms exam the data in an objective way and represent the discovered information in a short and clear statement. The discovered rules can be served as choices to help a decision maker to produce better decisions. The rules presented to a decision maker should only consist of simple, understandable, and complete strategies that allow a reasonably easy identification of preferable rules. The support and confidence are used to determine which candidate pattern passes the criteria and becomes a desired *action rule*.

Conventional actionable patterns [6-15], and [3] are built on the basis of previously discovered classification rules, so the quality and quantity of the action rules strictly depend on the adopted classification methods. Because these methods may fail to discover some useful action strategies, there is a strong need to develop an algorithm which can derive a set of actionable patterns directly from a given data set. Paper [4] is probably the first attempt towards formally introducing the problem of mining action rules from the scratch. Authors explicitly formulated it as a search problem in a support-confidence-cost framework and next they presented an Apriori-like [1] algorithm for mining action rules. Their definition of an action rule is an object-oriented one and it allows changes on stable attributes. Changing the value of an attribute, either stable or flexible, is linked with a cost [15]. In order to rule out action rules with undesired changes on stable attributes, authors have assigned very high cost to such changes. However, that way, the cost of action rules discovery is getting unnecessarily increased. Also, they did not take into account the dependencies between attribute values which are naturally linked with the cost of rules used either to accept or reject a rule. In this paper, we investigate properties of action rules and present a new efficient algorithm, *StrategyGenerator*, generating a simple and complete set of action rules without using classification rules. This type of action rules is called *Object-Based Action Rules* (OBAC). Three thresholds, *Right Support*, *Left Support*, and *Confidence* of OBAC, are defined and used to identify which action rules are interesting.

2 Mining Action Rules

An information system is used for representing knowledge. Its definition, presented here, was proposed in [5]. By an information system we mean a pair $S = (U, A)$, where:

- U is a nonempty, finite set of objects,
- A is a nonempty, finite set of attributes, i.e. $a : U \rightarrow V_a$ is a function for any $a \in A$, where V_a is called the domain of a.

Elements of U are called objects. In this section, for the purpose of clarity, objects are interpreted as customers. Attributes are interpreted as features such as, offers made by a bank, characteristic conditions etc. We only consider a special type of information systems called decision tables.

A decision table consists of a set of objects where each object is described by a set of attributes. Attributes are partitioned into conditions and decisions. Additionally, we assume that the set of conditions is partitioned into stable conditions and flexible conditions. In our example, we take "profit ranking" as the decision attribute. Its domain is defined as a set of integers. The decision attribute classifies objects (customers) with respect to the profit gained by a bank. Date of birth is an example of a stable attribute. The interest rate on any customer account is an example of a flexible attribute because the bank can adjust rates. We adopt the following definition of a decision table:

By a decision table we mean any information system $S = (U, A_{St} \cup A_{Fl} \cup \{d\})$, where $d \notin (A_{St} \cup A_{Fl})$ is a distinguished attribute called a decision. The set of attributes A in S is partitioned into stable conditions A_{St} and flexible conditions A_{Fl}.

The number of elements in $d(U) = \{ k: (\exists x \in U)[d(x)=k]\}$ is called the rank of d and it is denoted by $r(d)$. Let us observe that the decision d determines the partition $Part_S(d) = \{X_1, X_2,\ldots, X_{r(d)}\}$ of the universe U, where $X_k = d^{-1}(\{k\})$ for $1 \leq k \leq r(d)$. $Part_S(d)$ is called the classification of objects in S with respect to the decision d.

As we have mentioned before, objects in U are interpreted as bank customers. Additionally, we assume that customers in $d^{-1}(\{k_2\})$ are more profitable than customers in $d^{-1}(\{k_1\})$, for any $k_1 < k_2$. The set $d^{-1}(\{r(d)\})$ contains the most profitable customers. Clearly one of the main goals of any bank is to increase its profit. One way to do that is to shift some customers from a group $d^{-1}(\{k_1\})$ to $d^{-1}(\{k_2\})$, for any $k_1 < k_2$. Action rules can be used for that purpose since they provide hints about what type of special offers can me made by a bank to guarantee that values of targeted flexible attributes will be changed in a way that a desired group of customers should move from a group of a lower profit ranking to a group of a higher profit.

The basic principle of reclassification is a process of learning a function that maps one class of objects into another class by changing values of some conditional attributes describing them. The conditional attributes are divided into stable and flexible. The goal of the learning process is to create a reclassification model, for objects in a decision system, which suggests possible changes that can be made within values of some flexible attributes to reclassify these objects the way user wants. In other words, reclassification is the process of showing what changes in values of some of the flexible attributes for a given class of objects are needed in order to shift them from one decision class into another more desired one.

A decision system S classifies a set of objects so that for each object there exists a class label assigned to it.

By *action rule* in S we mean an expression $r = [[(a_1 = \omega_1) \wedge (a_2 = \omega_2) \wedge \ldots \wedge$

$$(a_q = \omega_q)] \wedge (b_1, \alpha_1 \rightarrow \beta_1) \wedge (b_2, \alpha_2 \rightarrow \beta_2) \wedge \ldots \wedge (b_p, \alpha_p \rightarrow \beta_p)] \Rightarrow [(d, k_1 \rightarrow k_2)],$$

where $\{b_1, b_2, \ldots, b_p\}$ are flexible attributes and $\{a_1, a_2,\ldots, a_q\}$ are stable in S. Additionally, we assume that $\omega_i \in Dom(a_i)$, $i=1,2,\ldots,q$ and $\alpha_i, \beta_i \in Dom(b_i)$, $i=1,2,\ldots,p$. The term $(a_i = \omega_i)$ states that the value of the attribute a_i is equal to ω_i, and $(b_j, \alpha_j \rightarrow \beta_j)$ means that value of the attribute b_j has been changed from α_j to β_j.

We say that object $x \in U$ supports an action rule r in S, if there is an object $y \in U$ such that: $(\forall i \leq p)[\,[b_i(x) = \alpha_i] \wedge [b_i(y) = \beta_i]\,]$, $(\forall i \leq q)\,[a_i(x) = a_i(y) = \omega_i]$, $d(x) = k_1$ and $d(y) = k_2$.

An action rule is meaningful only if it contains at least one flexible attribute. If we apply the left hand side of an action rule to object x, then the rule basically says: the values ω_i of stable attributes a_i (i=1,2,…,q) have to remain unchanged in x and then if we change the value of attribute b_i in x from α_i to β_i, for i=1,2,…,p, then the object x which is in the class k_1 is expected to move to class k_2.

From the point of reclassification, we are not targeting all possible cases on the decisional part of reclassification. Since some states are more preferable than other states, we should basically ask users to specify in what direction they prefer to see the changes. On the conditional part of action rules, we have no information to verify if the rule is applicable. If the domain expert can supply prior knowledge of a given domain then some of the rules cannot be applied. For example, the size of a tumor's growth can not increase when the status of a patient is changing from sick to becoming cured. Therefore, some combinations can be ruled out automatically just by having an expert who is involved in the application domain.

Since action plans are constructed by comparing the profiles of two sets of targeted customers, we can assume that there are two patterns associated with each object-based action rule, a left hand side pattern P_L and a right hand side pattern P_R. There are three objective measures of rule interestingness including *Left Support*, *Right Support*, and *confidence*.

The *Left Support* defines the domain of an action rule which identifies objects in U on which the rule can be applied. The larger its value is, the more interesting the rule will be for a user. The left hand side pattern of action rule

$$r = [[(a_1 = \omega_1) \wedge (a_2 = \omega_2) \wedge \ldots \wedge (a_q = \omega_q)] \wedge (b_1, \alpha_1 \rightarrow \beta_1) \wedge (b_2, \alpha_2 \rightarrow \beta_2) \wedge \ldots \wedge (b_p, \alpha_p \rightarrow \beta_p)] \Rightarrow [(d, k_1 \rightarrow k_2)]$$

is defined as the set $P_L = V_L \cup \{k_1\}$, where $V_L = \{\omega_1, \omega_2, \ldots, \omega_q, \alpha_1, \alpha_2, \ldots, \alpha_p\}$. The domain $Dom_S(V_L)$ of the left pattern P_L is a set of objects in S that exactly match V_L. Card[$Dom_S(V_L)$] is the number of objects in that domain. Card[$Dom_S(P_L)$] is the number of objects in S that exactly match P_L and Card[U] is the total number of objects in the decision system S. By the left support $supL$ of an action rule r, we mean $supL(r)$ = Card[$Dom_S(P_L)$] /Card[U].

The *Right Support* shows how well is the rule supported by objects in S from the preferable decision class. The higher its value is, the stronger case of the reclassification effect will be. The pattern P_R of an action rule r is defined as $P_R = V_R \cup \{k_2\}$, where $V_R = \{\omega_1, \omega_2, \ldots, \omega_q, \beta_1, \beta_2, \ldots, \beta_p\}$.

By domain $Dom_S(V_R)$ we mean a set of objects matching V_R. Card[$Dom_S(P_R)$] is the number of objects that exactly match P_R. By the right support $supR$ of action rule r, we mean $supR(r)$ = Card[$Dom_S(P_R)$] /Card[U].

The *confidence* of rule r shows the success measure in transforming objects from a lower preference decision class to a higher one. The *support* of action rule r in S, denoted by Sup_S(r), is the same as the left support $supL(r)$ of action rule r. This is the percentage of objects that need to be reclassified into more preferable class. By the *confidence* of the action rule *r* in *S*, denoted by $Conf_S(r)$, we mean

$$Conf_S(r)=(\text{Card}[Dom_S(P_L)]/\text{Card}[Dom_S(V_L)])*(\text{Card}[Dom_S(P_R)]\text{Card}[Dom_S(V_R)]).$$

3 Algorithm and Example: *StrategyGenerator*

The Brute Force method used in [10] to construct all action rules, directly from the decision table, is expensive and inefficient because it considers all possible pair combinations of flexible attributes. Hence, we propose the *StrategyGenerator* algorithm to find the set of most concise action rules. It considers each change of value within a single flexible attribute and each value of a stable attribute as an atomic expression from which more complex expressions are built. The algorithm operates similarly to LERS [2] and the same it guarantees that all discovered action rules are the shortest. This is an agglomerative type of a strategy used for instance in [9] to construct action rules. However, the new method does not require prior extraction of classification rules.

There are two basic steps in the proposed approach. (1) Partition the decision table and select target sub-tables: The original decision table *S* is first partitioned into a number of sub-tables S_1, S_2, ..., S_p according to the decision attribute in the decision table. Two relevant sub-tables are selected based on the reclassification goal for forming workable strategies. (2) Form actionable plans: The workable strategies are formed by comparing the domains of these two chosen sub-tables. In this case, we can avoid generating unqualified candidate terms similarly to *LERS* algorithm [2]. First, single-element candidate terms are computed and checked for its relations with the reclassification goal. If the relation holds, a positive mark is placed on it and the rule is generated. By doing this, we guarantee that the discovered action rules are the most concise ones. The anti-monotonic property is applied to filter candidate terms. When one of the support values is below the threshold, a negative mark is placed on the candidate term. The algorithm recursively takes unmarked candidate terms and extends them by one new unmarked atomic term till no new candidates are found.

Now, we present a decision table used to illustrate the *StrategyGenerator* for construction of action rules step by step.

Assume that $S = (\{x_1, x_2, x_3, x_4, x_5, x_6, x_7, x_8\}, \{a\} \cup \{b, \text{c}, \text{e}\} \cup \{d\})$ is a decision table represented by Table 1. Attributes in $\{b, c, e\}$ are flexible, attribute *a* is stable, and *d* is the decision attribute. We assume that *H* denotes customers of a high profit ranking and *L* denotes customers of a low profit ranking. The direction of reclassification is from *L* to *H*. The minimum support for both *supR* and *supL* is 12.5%, and the minimum confidence for rules is 75%.

Partition the decision table. In this example, the domain of the decision attribute is *L, N,* and *H* and the reclassification direction is from *L* to *H*. That means the customers with decision value *N* are not the focus point in this case. Therefore, the decision table *S* can be divided into S_1 *and* S_2 according to the decision value *L* and *H* as represented in Figure 1. Actionable strategies will be constructed based on sub-tables S_1 and S_2 only.

Table 1. Decision table *S*

Objects	a	b	c	e	d
X_1	0	2	1	1	H
X_2	0	2	2	1	H
X_3	2	1	2	2	L
X_4	0	3	1	0	N
X_5	2	3	2	0	N
X_6	2	3	1	0	H
X_7	2	1	2	1	L
X_8	2	1	1	1	L

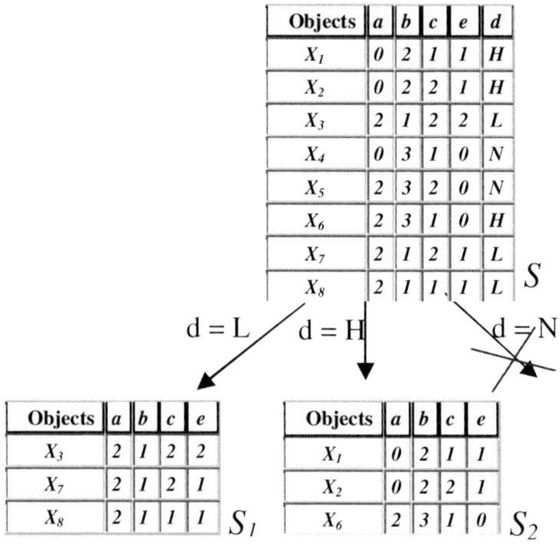

Fig. 1. Partition objects

Forming actionable strategies. The main idea of the reclassification goal is to move objects from an undesirable group into a more desirable one. Objects in *S* having property L are denoted by L_S* and objects in *S* having property R are denoted by R_S*. These two sets are also called granules. *StrategyGenerator* algorithm starts with atomic terms for *S* generated in its first loop. These terms are classified into two groups: premise-type and decision-type. Premise-type atomic terms are split into stable and flexible. As we mentioned before, an action rule without at least one flexible premise-type atomic term is meaningless. Stable atomic terms can not be solely used to construct action rules but they are important in boosting their confidence [10], [12]. In this example, one valid candidate term which is a stable atom (*a*, 2) is generated. In order to create atomic terms for a flexible attribute we check its domain in both sub-tables. Referring back to Example 1, the values of attribute *b* are "*1*" in sub-table S_1 and "2" and "*3*" in sub-table S_2. It means that the action recommendations for attribute *b* say that its value should be changed from *1* to 2 or from *1* to *3*. The corresponding atomic terms are presented as (*b*, *1*→*2*) and (*b*, *1*→*3*). Following the same procedure for attributes *c* and *e*, their corresponding atomic terms can be formed and they are listed below.

One-element term loop:

```
// Granules corresponding to values of a decision attribute
Decision-type atomic term: (d, L→H),
Granules: L* = {x3, x7, x8}, R*={x1, x2, x6}
// Granules corresponding to values of condition attributes
Premise-type stable atomic expressions:
(a,0),   L*= Ø    Marked   "-"
(a,2),   L* = {x3,x7,x8},   R* = {x6}
```

Premise-type flexible atomic expressions:

(b, 1→2) , L* = {x_3,x_7,x_8}, supL(r) = 3/8; R*= {x_1,x_2}, supR(r) = 2/8; Conf(r) = (3/3) ×(2/2) = 100% Marked "+"
(b, 1→3), L* = {x_3,x_7,x_8}, supL(r) = 3/8; R* = {x_6}, supR(r) = 1/8; Conf(r) = (3/3) ×(1/3) = 33%
(c, 2→1) , L* = {x_3,x_7}, supL(r) = 2/8; R* = {x_1,x_6}, supR(r) = 2/8; Conf(r) = (2/4)×(2/4) = 25%
(c, 1→2) , L* = {x_8}, supL(r) = 1/8; R* = {x_2}, supR(r) = 1/8; Conf(r) = (1/4)×(1/4) = 6.25%
(e, 2→1), L* = {x_3}, supL(r) = 1/8; R* = {x_1,x_2}, supR(r) = 2/8; Conf(r) = (1/1)×(2/4) = 50%
(e, 2→0), L* = {x_3}, supL(r) = 1/8; R* = {x_6}, supR(r) = 1/8; Conf(r) = (1/1)×(1/3) = 33.3%
(e, 1→0), L* = {x_7,x_8}, supL(r) = 1/8; R* = {x_6}, supR(r) = 1/8; Conf(r) = (2/4)×(1/3) = 16.7%

The action rule *r* linking each premise-type term and the decision-type term is acceptable when the values of the corresponding *supL(r)*, *supR(r)*, and *Conf(r)* meet the user specified thresholds. The primary idea of the StrategyGenerator algorithm lies in the property of anti-monotonic property of the support. It is used to prune unqualified candidates. This is achieved by placing a "-" mark when a term does not have sufficient support. Going back to the example, the support of the atomic term *(a,0)* does not satisfy the minimum support requirement, so it is marked with "-" symbol and it is not considered in later steps of the algorithm. The goal of this algorithm is to find the shortest action rules. It means when a premise-type term t_l jointly with a decision-type term form an acceptable action rule, then t_l is not investigated any further. In this example, the term *(b, 1→2)* jointly with *(d, L→H)* meet all three thresholds, so the action rule *(b, 1→2)* ⇒ *(d, L→H)* is discovered and the term *(b, 1→2)* is marked as "+".

Build two-element premise-type terms by concatenating any two unmarked premise-type terms that have different attributes. Below is the list of two-element terms. There is no action rule generated in this step, since none of the terms jointly with *(d, L→H)* satisfy all three thresholds.

Two-elements term loop:

(a,2) ∧ (b, 1→3), L* = {x_3,x_7,x_8}, supL(r) = 3/8; R* ={x_6}, supR(r) = 1/8; Conf(r) = (3/3)×(1/2) = 50%
(a,2) ∧ (c, 2→1), L* = {x_3,x_7}, supL(r) = 2/8; R* = {x_6}, supR(r) = 1/8; Conf(r) = (2/3)×(1/2) = 33.3%
(a,2) ∧ (c, 1→2), L* = {x_8}, supL(r) = 1/8; R* = ∅; Marked "-"
(a,2) ∧ (e, 2→1), L* = {x_3}, supL(r) = 1/8; R* = ∅; Marked "-"
(a,2) ∧ (e, 2→0), L* = {x_3}, supL(r) = 1/8; R* = {x_6}, supR(r) = 1/8; Conf(r) = (1/1)×(1/2) = 50%
(a,2) ∧ (e, 1→0), L* = {x_7,x_8}, supL(r) = 2/8; R* = {x_6}, supR(r) = 1/8;Conf(r) = (2/2)×(1/2) = 50%
(b, 1→3) ∧ (c, 2→1), L* = {x_3,x_7}, supL(r) = 2/8; R* = {x_6}, supR(r) = 1/8; Conf(r) = (2/2)×(1/2) = 50%
(b, 1→3) ∧ (c, 1→2), L* = {x_8}, supL(r) = 1/8; R* = ∅ Marked "-"
(b, 1→3) ∧ (e, 2→1), L* = {x_8}, supL(r) = 1/8; R* = ∅ Marked "-"

(b, 1→3) ∧ (e, 2→0), L* = {x_3}, supL(r) = 1/8; R* = {x_6},supR(r) = 1/8; Conf(r) = (1/1)×(1/3) = 33.3%
(b, 1→3) ∧ (e, 1→0), L* = {x_7, x_8}, supL(r) = 2/8; R* = {x_6}, supR(r) = 1/8; Conf(r) = (2/2)×(1/3) = 33.3%
(c, 2→1) ∧ (e, 2→1), L* = {x_3}, supL(r) = 1/8; R* = {x_1}, supR(r) = 1/8; Conf(r) = (1/1)×(1/2) = 50%
(c, 2→1) ∧ (e, 2→0), L* = {x_3}, supL(r) = 1/8; R* = {x_6}, supR(r) = 1/8; Conf(r) = (1/1)×(1/2) = 50%
(c, 2→1) ∧ (e, 1→0), L* = {x_7}, supL(r) = 1/8; R* = {x_6}, supR(r) = 1/8; Conf(r) = (1/2)×(1/2) = 25%
(c, 1→2) ∧ (e, 2→1), L* = ∅; Marked "-"
(c, 1→2) ∧ (e, 2→0), L* = ∅; Marked "-"
(c, 1→2) ∧ (e, 1→0), L* = {x_8}, supL(r) = 1/8; R* = ∅; Marked "-"

Build three-element terms by concatenating any two unmarked terms that have different attributes. Below is the list of three-element terms. There are three action rules discovered.

Three-elements term loop:

(a,2)∧(b, 1→3)∧(c, 2→1), L* = {x_3,x_7}, supL(r)=2/8; R*= {x_6}, supR(r) = 1/8; Conf(r) = (2/2)×(1/1) = 100%; Marked "+"
(a,2)∧(b, 1→3)∧(e, 2→0), L* = {x_3}, supL(r) = 1/8; R*={x_6}, supR(r) = 1/8; Conf(r) = (1/1)×(1/2) = 50%
(a,2)∧(b, 1→3)∧(e, 1→0), L* = {x_7, x_8}, supL(r)=2/8; R*= {x_6}, supR(r) = 1/8; Conf(r)=(2/2)×(1/2)=50%
(a,2)∧(c, 2→1)∧(e, 2→0), L* = {x_3}, supL(r)=1/8; R*= {x_6}, supR(r) = 1/8; Conf(r)=(1/1)×(1/1)=100%; Marked "+"
(a,2)∧(c, 2→1)∧(e, 1→0), L* = {x_7}, supL(r)= 1/8; R*={x_6}, supR(r)=1/8; Conf(r)= 1/1)×(1/1) = 100%; Marked "+"
(b, 1→3)∧(c, 2→1)∧(e, 2→1), L* = {x_3}, supL(r)=1/8; R*= ∅ Marked "-"
(b, 1→3)∧(c, 2→1)∧(e, 2→0), L* = {x_3}, supL(r)=1/8; R*={x_6}, supR(r)=1/8; Conf(r)=(1/1)×(1/2) = 50%
(b, 1→3)∧(c, 2→1)∧(e, 1→0), L* = {x_7}, supL(r)=1/8; R*={x_6}, supR(r)=1/8; Conf(r)=(1/1)×(1/2)=50%

In Example 1, we have the following four action rules:

(b, 1→2)⇒(d, L→H), supL(r)=3/8, supR(r)=2/8, Conf(r)=100%
((a,2)∧(b, 1→3)∧(c, 2→1))⇒(d, L→H), supL(r)=2/8, supR(r)=1/8, Conf(r)=100%
((a,2)∧(c, 2→1)∧(e, 2→0))⇒(d, L→H), supL(r)=1/8, supR(r)=1/8, Conf(r)=100%
((a,2) ∧(c, 2→1)∧(e, 1→0))⇒(d, L→H), supL(r)=1/8, supR(r)=1/8, Conf(r)=100%

We claim that the new method guarantees that the actionable patterns are concise, general, and reliable. As we can see the discovered action rules contain relatively few attribute-value pairs on the classification side and the number of these rules is also relatively small. Such rules are more readable, easier to understand and apply later on.

The algorithm, StrategyGenerator, was implemented under Windows XP. It was tested on several public domain databases and on the medical database HEPAR prepared in the Medical Center of Postgraduate Education (Warsaw, Poland) by Dr. med. Hanna Wasyluk. In all cases the recall of the new algorithm was higher than DEAR [11][12].

Finally, let us compare the action rules generated by StrategyGenerator with action rules constructed by the tree-based algorithms DEAR [7], [14], [10], [11], [6], [12], [13], [9]. For the same Example 1, thirteen classification rules have been generated by LERS algorithm and they are listed below:

$(b,2)\rightarrow(d,H)$ $(b,1)\rightarrow(d,L)$ $(e,2)\rightarrow(d,L)$
$(a,0)\wedge(b,3)\rightarrow(d,N)$ $(a,0)\wedge(c,2)\rightarrow(d,H)$ $(a,0)\wedge(e,1)\rightarrow(d,H)$
$(a,0)\wedge(e,0)\rightarrow(d,N)$ $(a,2)\wedge(e,1)\rightarrow(d,L)$ $(b,3)\wedge(c,2)\rightarrow(d,N)$
$(a,2)\wedge(b,3)\wedge(c,1)\rightarrow(d,H)$ $(a,2)\wedge(b,3)\wedge(c,2)\rightarrow(d,N)$
$(a,2)\wedge(c,1)\wedge(e,0)\rightarrow(d,H)$ $(a,2)\wedge(c,2)\wedge(e,0)\rightarrow(d,N)$.

Five classification rules have been generated by C4.5 algorithm and they are listed below:

$(b,2)\rightarrow(d,H)$ $(b,1)\rightarrow(d,L)$ $(a,0)\wedge(b,3)\rightarrow(d,N)$
$(a,2)\wedge(b,3)\wedge(c,1)\rightarrow(d,H)$ $(a,2)\wedge(b,3)\wedge(c,2)\rightarrow(d,N)$.

DEAR algorithms generated from them only one action rule: $(b,1\rightarrow2)\Rightarrow(d,L\rightarrow H)$. The new method generates more action rules than DEAR as we have seen in the above example.

4 Conclusion

The ability to discover useful knowledge hidden in large volumes of data and to act on that knowledge is becoming increasingly important in today's competitive world. The knowledge extracted from data can provide a competitive advantage in support of decision-making. In this paper, we focus on analyzing a complete information system and obtaining a set of concise workable strategies. Any action rule provides a brief and clear hint to a user about required changes within flexible attributes that are needed to re-classify some objects from a lower ranked class to a higher one. This knowledge can be turned into action and this action may help to achieve user's goal. StrategyGenerator is a novel method of a reclassification strategy which extracts higher level actionable knowledge from large volumes of data.

Acknowledgements

This work was supported by the National Science Foundation under grant IIS-0414815.

References

1. Agrawal, R., Srikant, R.: Fast algorithm for mining association rules. In: Proceeding of the Twentieth International Conference on VLDB, pp. 487–499 (1994)
2. Chmielewski, M.R., Grzymala-Busse, J.W., Peterson, N.W., Than, S.: The rule induction system LERS - a version for personal computers. Foundations of Computing and Decision Sciences 18(3-4), 181–212 (1993)
3. Greco, S., Matarazzo, B., Pappalardo, N., Slowinski, R.: Measuring expected effects of interventions based on decision rules. Journal of Experimental and Theoretical Artificial Intelligence 17(1-2), 103–118 (2005)
4. He, Z., Xu, X., Deng, S., Ma, R.: Mining action rules from scratch. Expert Systems with Applications 29(3), 691–699 (2005)
5. Pawlak, Z.: Information systems - theoretical foundations. Information Systems Journal 6, 205–218 (1981)
6. Raś, Z.W., Tzacheva, A., Tsay, L.-S., Gurdal, O.: Mining for interesting action rules. In: Proceedings of IEEE/WIC/ACM International Conference on Intelligent Agent Technology (IAT 2005), Compiegne University of Technology, France, pp. 187–193 (2005)
7. Raś, Z.W., Tsay, L.-S.: Discovering extended action-rules (System DEAR), in Intelligent Information Systems. In: Proceedings of the IIS 2003 Symposium, Advances in Soft Computing, pp. 293–300. Springer, Heidelberg (2003)
8. Raś, Z., Wieczorkowska, A.: Action rules: how to increase profit of a company. In: Zighed, A.D.A., Komorowski, J., Żytkow, J.M. (eds.) PKDD 2000. LNCS (LNAI), vol. 1910, pp. 587–592. Springer, Heidelberg (2000)
9. Raś, Z.W., Wyrzykowska, E.: ARAS: Action rules discovery based on agglomerative strategy. In: Mining Complex Data, Post-Proceedings of the ECML/PKDD 2007 Third International Workshop, MCD 2007. LNCS (LNAI), vol. 4944, pp. 196–208. Springer, Heidelberg (2008)
10. Tsay, L.-S.: Discovery of extended action rules, Ph.D. Dissertation, Department of Computer Science, University of North Carolina, Charlotte (2005)
11. Tsay, L.-S., Raś, Z.W.: Action rules discovery: system DEAR2, method and experiments. Journal of Experimental and Theoretical Artificial Intelligence 17(1-2), 119–128 (2005)
12. Tsay, L.-S., Raś, Z.W.: Action rules discovery systems DEAR3. In: Esposito, F., Raś, Z.W., Malerba, D., Semeraro, G. (eds.) ISMIS 2006. LNCS (LNAI), vol. 4203, pp. 483–492. Springer, Heidelberg (2006)
13. Tsay, L.-S., Raś, Z.W.: E-Action Rules. In: Foundations of Data Mining, Studies in Computational Intelligence, Springer, Heidelberg (will appear, 2007)
14. Tsay, L.-S., Raś, Z., Wieczorkowska, A.: Tree-based algorithm for discovering extended action-rules (System DEAR2). In: Intelligent Information Processing and Web Mining, Advances in Soft Computing, Proceedings of the IIS 2004 Symposium, pp. 459–464. Springer, Heidelberg (2004)
15. Tzacheva, A., Raś, Z.W.: Action rules mining. International Journal of Intelligent Systems 20(7), 719–736 (2005)

Discovering Emerging Patterns for Anomaly Detection in Network Connection Data

Michelangelo Ceci, Annalisa Appice, Costantina Caruso, and Donato Malerba

Dipartimento di Informatica, Università degli Studi di Bari
via Orabona, 4 - 70126 Bari, Italy
{ceci,appice,caruso,malerba}@di.uniba.it

Abstract. Most intrusion detection approaches rely on the analysis of the packet logs recording each noticeable event happening in the network system. Network connections are then constructed on the basis of these packet logs. Searching for abnormal connections is where the application of data mining techniques for anomaly detection promise great potential benefits. Anyway, mining packet logs poses additional challenges. In fact, a connection is composed of a sequence of packets, but classical approaches to anomaly detection loose information on the possible relations (e.g., following) between the packets forming one connection. This depends on the fact that the attribute-value data representation adopted by classical anomaly detection methods does not allow either the distinction between connections and packets or the discovery of the interaction between packets in a connection. In order to face this issue, we resort to a Multi-Relational Data Mining approach which makes possible to mine data scattered in multiple relational tables (typically one for each object type). Our goal is to analyse packet logs of consecutive days and discover multivariate relational patterns whose support significantly changes from one day to another. Discovered patterns provide a human-interpretable description of the change in the network connections occurring in consecutive days. Experimental results on real traffic data collected from the firewall logs of our University Department are reported.

1 Introduction

In the last years, an increasing number of organizations are becoming vulnerable to a wide variety of cyber threats, which come from hardware failures, software flaws, tentative probing and malicious attacks. Intrusion detection (ID) is the process of analyzing the events occurring in a network system in order to detect the set of malicious actions that may compromise the integrity, confidentiality, and availability of information resources (security violations) [3]. Traditional methods for intrusion detection are classified into two broad categories: misuse detection and anomaly detection [10]. Misuse detection works by searching for the traces or patterns of well-known attacks while anomaly detection uses a model of normal user or system behavior and flags significant deviations from this model as potentially malicious.

A. An et al. (Eds.): ISMIS 2008, LNAI 4994, pp. 179–188, 2008.

In this paper we are interested in analyzing the firewall logs of a network system in several consecutive days and discovering significant deviations (or changes) in daily network traffic. Hence, we concentrate on detecting anomalies from network connection data. Although anomaly detection has been deeply investigated in the literature [7,8], all proposed methods assume that data are stored in a single table of a relational database (attribute-value representation). This representation allows efficient algorithmic solution but it does not allow to represent the packet-based structure of a single connection.

To overcome limitation posed by single table representation, we exploit findings of research conducted in Multi-Relational Data Mining [6] in order to distinguish between connections (i.e., reference objects of analysis) and packets (i.e., task-relevant objects of analysis) and to mine their interactions: a connection is constructed from one or more packets and the packets are timely related to define a sequence. Coherently with the goals posed by the anomaly discovery task, we propose investigate the opportunity of discovering descriptions of abnormal connections. Such descriptions are in form of relational patterns whose support significantly decreases from one day (target day) to another (background day). Such patterns, known as relational emerging patterns [2], may be employed to capture the "possible" deviation in the traffic network from a day to another: the larger the difference of pattern support, the more interesting the patterns to detect a deviation in network traffic. The interpretation of emerging patterns would add additional depth to the administrators defenses, and allow them to better determine what are the threats against the network they manage.

The paper is organized as follows. In the next Section, we formally present the faced problem. A method to discover relational emerging pattern is described in Section 3. Anomalies detected by mining relational emerging patterns from four successive weeks of firewall logs of a network system are described in Section 4. Lastly, some conclusions are drawn.

2 Problem Definition

Network connection data can be constructed from packet logs recorded by means of packet capturing utilities [4]. The basic premise is that when audit mechanisms are enabled, distinct evidence of anomalies in daily network connections (i.e., differences in connections recorded in consecutive days) will be manifested in the recorded audit data.

Definition 1 (Anomaly detection in Network Connection Data). *Let us consider the connection data constructed from the packet logs* $L_1, \ldots, L_n$ *such that each pair* $\langle L_i, L_{i+1} \rangle$ *describes the network traffic recorded in two consecutive days. Anomaly detection aims at identifying significant deviations (anomalies) in connections recorded one day with respect to connections recorded the day before (or after).*

Such deviations may involve features which describe the connections (e.g., the machine that was contacted, the service that was adopted, the duration of connection) or features which describe one or more packets within each connection (e.g., number of bytes or duration) or features which describe the interaction between two consecutive packets within the same connection (e.g., distance). Hence, the anomalies of a connection may depend on anomalous values of related features of different object type. Therefore, anomaly detection needs distinguishing between connections and packets and mining their inherent interaction. In fact, connection data are naturally stored in "separate" tables of a relational database D according to a schema S: one table for each object type (connections and packets). The relation between connections and packets is expressed by means of foreign key constraints, while the interaction between packets (e.g., the packet P is consecutive to the packet Q) is stored in a separate table of S. By this mapping of packet logs into a relational database, it is then possible to take into account attributes of related task relevant objects (i.e., packets) when investigating properties of the reference objects (i.e., connections) which are the main subject of analysis. By taking into account the multi-relational structure of data, anomalous connections are described by means of relational patterns. A formal definition of relational pattern is provided in the following.

Definition 2 (Relational pattern). *Let S be a database schema. A relational pattern P over S is a conjunction of predicates:*

$$p_0(t0_1), p_1(t1_1, t1_2), p_2(t2_1, t2_2), \ldots, p_m(tm_1, tm_2)$$

where $p_0(t0_1)$ is the key predicate associated with the target table of the task at hand (i.e., table that contains reference objects) and $\forall i = 1, \ldots, m \;\; p_i(ti_1, ti_2)$ is either a structural predicate [1] *or a property predicate*[2] *over S.*

Henceforth, we will also use the set notation for relational patterns, that is, a relational pattern is considered a set of atoms.

The change in network traffic can be properly modeled by means of relational emerging patterns [2], that is, multi-variate features whose support significantly decreases from one data class (target class) to another class (background class). The class feature is associated with the reference objects stored in the target table, while explanatory features refer to either the reference objects or the task-relevant objects which are someway related to the reference objects. The structural information required to mine relational emerging patterns can be automatically obtained from the database schema by navigating foreign key constraints.

Definition 3 (Relational Emerging Patterns). *Let D_t and D_b be two instances of a database schema S such that each D_i ($i = t, b$) contains a set of*

[1] A structural predicate is a binary predicate $p(t, s)$ associated with a pair of tables T_i and T_j with T_i and T_j related by a foreign key FK in S. The name p denotes FK, while the term t (s) is a variable that represents the primary key of T_i (T_j).

[2] A property predicate is a binary predicate $p(t, s)$ associated with the attribute ATT of the table T_i. The name p denotes the attribute ATT, the term t is a variable representing the primary key of T_i and s is a constant which represents a value belonging to the range of ATT in T_i.

Table 1. Number of relational emerging patterns of daily connections from the day after (or before). Emerging patterns are grouped with respect to the grow-rate value.

Day	Grow Rate Range					Day	Grow Rate Range				
	[1,1.5]	]1.5,4]	]4,8]	]8,∞]	∞		[1,1.5]	]1.5,4]	]4,8]	]8,∞]	∞
1 from 2	43	104	81	49	8	2 from 1	1	36	271	20	2
2 from 3	231	15	0	0	22	3 from 2	203	37	0	0	0
3 from 4	11	308	0	0	0	4 from 3	38	63	30	35	0
4 from 5	25	96	1	0	0	5 from 4	68	63	33	26	0
5 from 6	143	85	0	4	0	6 from 5	10	19	0	0	0
6 from 7	23	30	66	51	0	7 from 6	7	113	287	3	0
7 from 8	392	0	0	0	0	8 from 7	62	10	0	0	0
8 from 9	73	24	0	0	0	9 from 8	382	0	0	0	0
9 from 10	272	70	0	0	0	10 from 9	128	7	0	0	22
10 from 11	166	5	0	0	2	11 from 10	184	16	0	0	0
11 from 12	236	113	0	0	29	12 from 11	66	53	4	0	0
12 from 13	258	24	0	0	0	13 from 12	47	34	0	0	0
13 from 14	55	40	4	0	0	14 from 13	186	116	0	0	0
14 from 15	83	34	0	0	0	15 from 14	287	42	0	0	0
15 from 16	147	18	0	0	0	16 from 15	250	1	0	0	0
16 from 17	359	0	0	0	0	17 from 16	79	20	5	6	0
17 from 18	151	157	0	0	0	18 from 17	57	125	108	291	62
18 from 19	67	71	88	275	153	19 from 18	10	333	0	0	0
19 from 20	133	73	4	0	0	20 from 19	326	1	0	0	66
20 from 21	242	93	0	0	3	21 from 20	112	139	2	0	0
21 from 22	2	290	35	0	32	22 from 21	61	56	56	65	36
22 from 23	16	41	0	4	0	23 from 22	134	38	2	19	2
23 from 24	145	63	21	29	0	24 from 23	5	17	2	5	1
24 from 25	48	36	29	70	0	25 from 24	18	183	132	0	0
25 from 26	259	42	0	0	0	26 from 25	84	4	0	0	0
26 from 27	89	39	0	0	0	27 from 26	313	27	0	0	0
27 from 28	95	19	0	0	19	28 from 27	186	124	72	4	0

The emerging patterns whose growth rate is close to 1 ($GR <= 1.5$) capture the profile of connections ingoing the firewall which have approximately the same frequency (support) on consecutive days. Hence, emerging patterns with relatively low value of growth rate ($GR \approx 1$) capture some behavior in daily connection data, and this behavior is maintained on at least two consecutive days. Differently, the larger the growth rate, the more interesting the emerging patterns to detect change in network traffic.

According to such considerations, we can explore the distribution of emerging patterns with respect to the growth rate range and then observe that there is a high number of emerging patterns with growth rate greater than 8 ($GR >= 8$) on June 2nd. These patterns capture a significant deviation in the network traffic on June 2nd from the traffic on June 1st (3rd). In fact, a deeper analysis of these patterns reveals some intersting anomalies. For example, the pattern $P1$:

$P1$: $connection(C), service(C, unknown), packet(C, P), next(I, P, Q), timeDistance(I, [0..100])$

describes the connections C reconstructed from at least two consecutive packets, denoted by P and Q, such that the time distance between P and Q is between 0 and 100 and the service of the connection is unknown. The support of $P1$ on June 2nd is 0.13 with $GR(P1) = 3.42$ from June 1st and $GR(P1) = \infty$ from June 3rd. This means that only few connections satisfying $P1$ incomes firewalls on June 1st, while no connection satisfying $P1$ incomes the firewalls on June 3th. In addition, we verify that $P1$ is unfrequent on all days observed after June 2nd, hence, $P1$ describes the profile of isolated connections (outliers) incoming on June 2nd. Furthermore, the profile of these connections is described by fully exploiting the relational nature of data: $P1$ involves some properties of connections (i.e., service is unknown) and describes the interaction between consecutive packets incoming within the same connection.

Similarly, the analysis of emerging patterns discovered on June 18th reveals some new anomalies. For example, the pattern $P2$:

$P2$: $connection(C), packet(C, P), nationTimeZone(C, 1), time(C, [10h, 12h]), sourceExtIP\_0(C, 193), destination(C, 151)$

has $support = 0.11$ on June 18th with $GR(P2) = \infty$ from June 19th and $GR(P2) = 187.87$ from June 17th. Furthermore, $P2$ is unfrequent ($support < 0.1$) on all observed days after (and before) June 19th (17th). Also in this case, $P2$ identifies some outlier connections incoming only on June 18th. The pattern also includes a human interpretable profile of such connections.

Differently, by analyzing emerging patterns on June 22nd we discover $P3$:

$P3$: $connection(C), packet(C, P), service(C, 4671), destination(C, 153)$,

such that $support(P3) = 0.69$ on June 22nd with $GR(P3) = \infty$ from June 21st and $GR(P3) = 1.15$ from June 23rd. $P3$ is unfrequent on all observed days before June 21st, while $support(P3) = 0.60$ on June 23rd. This suggests the idea that $P3$ is describing a change in the traffic behavior that is persistent for at least two consecutive days.

5 Conclusions

The problem of detecting anomalies in network connection data can be formalized in the multi-relational framework. In fact, network connections are typically reconstructed from the packet logs daily recorded from firewalls of a network system. Connections and packets are naturally stored in separate tables of a relational database. This allows distinguishing between objects of different types (connections and packets), which are naturally described by a different set of properties, and representing their interactions. Relational emerging patters, that is, multivariate features involving properties of the connection or properties of the packets inside the connection or the interaction between packets within the same connection (a packet P incomes after a packet Q), are then discovered to capture significant change from one day to the day after (or before): the larger the difference of pattern support, the more interesting the patterns to detect a deviation in the network traffic. Such patterns are employed to detect abnormal activities in the logs without too much human inputs.

As future work, we plan to use emerging patterns to define profiles useful to detect anomalies in run-time. We are interested in extending the emerging pattern discovery in order to discover patterns discriminating the network traffic of one day from the network traffic in a "sequence" of days after (or before). This new kind of emerging pattern will make possible to automatically distinguish outliers and change points [12]. An isolated change not preserved in several days may identify the presence of outlier connections, while a change whose effect is observed for several consecutive days may identify some changing pattern.

Acknowledgments

This work is supported by the Strategic Project: "Telecommunication Facilities and Wireless Sensor Networks in Emergency Management".

References

1. Agrawal, R., Imielinski, T., Swami, A.N.: Mining association rules between sets of items in large databases. In: Buneman, P., Jajodia, S. (eds.) International Conference on Management of Data, pp. 207–216 (1993)
2. Appice, A., Ceci, M., Malgieri, C., Malerba, D.: Discovering relational emerging patterns. In: Basili, R., Pazienza, M. (eds.) AI*IA 2007: Artificial Intelligence and Human-Oriented Computing. LNCS (LNAI), pp. 206–217. Springer, Heidelberg (2007)
3. Bace, R.: Intrusion Detection. Macmillan Technical Publishing, Basingstoke (2000)
4. Caruso, C., Malerba, D., Papagni, D.: Learning the daily model of network traffic. In: Hacid, M.-S., Murray, N.V., Raś, Z.W., Tsumoto, S. (eds.) ISMIS 2005. LNCS (LNAI), vol. 3488, pp. 131–141. Springer, Heidelberg (2005)
5. Dong, G., Li, J.: Efficient mining of emerging patterns: Discovering trends and differences. In: International Conference on Knowledge Discovery and Data Mining, pp. 43–52. ACM Press, New York (1999)
6. Džeroski, S., Lavrač, N.: Relational Data Mining. Springer, Heidelberg (2001)
7. Knorr, E.M., Ng, R.T.: Algorithms for mining distance-based outliers in large datasets. In: Gupta, A., Shmueli, O., Widom, J. (eds.) VLDB, pp. 392–403. Morgan Kaufmann, San Francisco (1998)
8. Mahoney, M.V., Chan, P.K.: Learning nonstationary models of normal network traffic for detecting novel attacks. In: KDD, pp. 376–385. ACM Press, New York (2002)
9. Mannila, H., Toivonen, H.: Levelwise search and borders of theories in knowledge discovery. Data Mining and Knowledge Discovery 1(3), 241–258 (1997)
10. Mounji, A.: Languages and Tools for Rule-Based Distributed Intrusion Detection. PhD thesis, Facultes Universitaires Notre-Dame de la Paix Namur, Belgium (1997)
11. Plotkin, G.D.: A note on inductive generalization. Machine Intelligence 5, 153–163 (1970)
12. Takeuchi, J., Yamanashi, K.: A unifying framework for identifying changing points and outliers. IEEE Transactions on Knowledge and Data Engineering 18(4) (2006)
13. Zhang, X., Dong, G., Ramamohanarao, K.: Exploring constraints to efficiently mine emerging patterns from large high-dimensional datasets. In: Knowledge Discovery and Data Mining, pp. 310–314 (2000)

Temporal Extrapolation within a Static Clustering

Tim K. Cocx, Walter A. Kosters, and Jeroen F.J. Laros

Leiden Institute of Advanced Computer Science, Leiden University, The Netherlands
tcocx@liacs.nl

Abstract. Predicting the behaviour of individuals is a core business of policy makers. This paper discusses a new way of predicting the "movement in time" of items through pre-defined classes by analysing their changing placement within a static, preconstructed 2-dimensional clustering. It employs the visualization realized in previous steps within item analysis, rather than performing complex calculations on each attribute of each item. For this purpose we adopt a range of well-known mathematical extrapolation methods that we adapt to fit our need for 2-dimensional extrapolation. Usage of the approach on a criminal record database to predict evolvement of criminal careers, shows some promising results.

1 Introduction

The ability to predict (customer) behaviour or market trends plays a pivotal role in the formation of any policy, both in the commercial and public sector. Ever since the coming of the information age, the procurement of such prognoses is becoming more and more an automated process, extracting and aggregating knowledge from data sources, that are often very large.

Mathematical computer models are frequently used to both describe current and predict future behaviour. In many cases these models are chosen on the basis of *detection theory* [2]. They employ algorithms like *Naive Bayes*, *K-Nearest Neighbour* or concepts like *Support Vector Machines*. Next to the prediction of certain unknown attributes by analysing the available attributes, it might also be of interest to predict behaviour based upon past activities alone, thus predicting the continuation of a certain sequence of already realized behaviour.

Sequences play an important role in classical studies of instrumental conditioning [3], in human skill learning [10], and in human high-level problem solving and reasoning [3]. It is logical that sequence learning is an important component of learning in many task domains of intelligent systems. Our approach aims to augment the currently existing set of mathematical constructs by analysing the "movement in time" of a certain item through a static clustering of other items. The proposed model can be added seamlessly to already performed steps in item analysis, like clustering and classification, using their outcome as direct input.

Section 2 mentions extrapolation methods. The main contribution of this paper is in Section 3, where the new insights into temporal sequence prediction are discussed. Section 4 shows experiments, and Section 5 concludes.

A. An et al. (Eds.): ISMIS 2008, LNAI 4994, pp. 189–195, 2008.

2 Background

A lot of work has been done in the development of good clustering and strong extrapolation methods that we can resort to within our approach.

2.1 Clustering

It is common practice to visualize a clustering within the 2-dimensional plane, utilizing some form of *Multi Dimensional Scaling* (MDS) [6] to approximately represent the correct, multi-dimensional distances. These methods include, e.g., "associative array" clustering techniques [9] and systems guided by human experience [5]. An obvious choice for our approach would be to select the method yielding the smallest error.

2.2 Extrapolation

Extrapolation is the process of constructing new data points outside a discrete set of known data points, i.e., predicting some outcome on a yet unavailable moment. It is closely related to the process of interpolation, which constructs new points between known points and therefore utilizes many of its concepts, although its results are less reliable.

Interpolation. Interpolation is the method of constructing a function which closely fits a number of known data points and is sometimes referred to as *curve fitting* or *regression analysis*. There are a number of techniques available to interpolate such a function, most of the time resulting in a polynomial of a predefined degree n. Such a polynomial can always exactly fit $n+1$ data points, but if more than $n+1$ points are available one needs to resort to approximation measures like the *least squares error* method [1]. The two main interpolation methods that are suitable to be incorporated in our approach are *polynomial interpolation* and *splines*. Polynomial interpolation uses linear algebra to solve a system of linear equations in order to find one polynomial that best approximates the given data points (see Figure 1).

Data points can also be interpolated by specifying a separate low degree polynomial (e.g., degree 2 or 3) between each couple of data points or *knots*. This

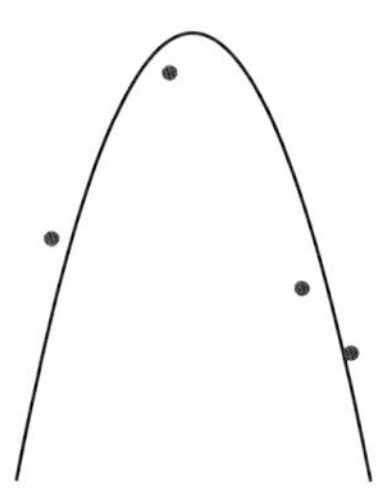

Fig. 1. A function of degree 2 that best fits 4 data points

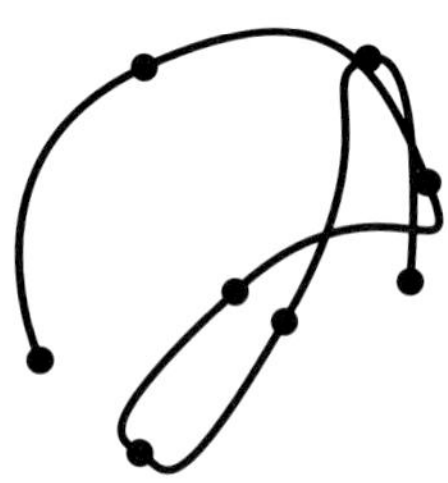

Fig. 2. An example of a spline

interpolation scheme, called a *spline*, exactly fits the derivative of both polynomials ending in the same knot. Demanding that the second derivatives also match and specifying the requested derivative in both end points yields $4n$ equations for $4n$ unknowns. Following Bartels et al. [4] one can specify a third degree polynomial for both the x and y coordinates between two separate knots, resulting in an interpolation like the graph in Figure 2. Due to the liberty this method allows in the placement of the existing data points, it seems well suited for the task of 2-dimensional extrapolation, see below.

Extrapolation. All interpolation schemes are suitable starting points for the process of extrapolation. It should, however, be noted that higher level polynomials can lead to larger extrapolation errors: the *Runge phenomenon*. Polynomials of degrees higher than 3 are often discouraged here.

In most cases, it is sufficient to simply continue the fabricated interpolation function after the last existing data point. In the case of the spline, however, a choice can be made to continue the polynomial constructed for the last interval (which can lead to strange artifacts), or extrapolate with a straight line, constructed with the last known derivative of that polynomial. The difference between the two methods is displayed in Figure 3.

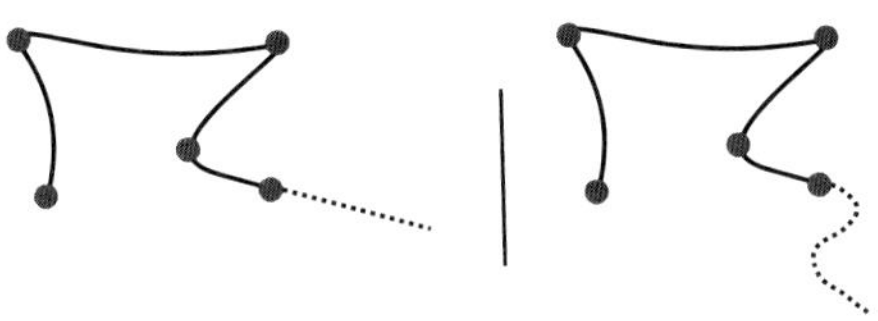

Fig. 3. Straight line extrapolation (left) and polynomial continuation (right)

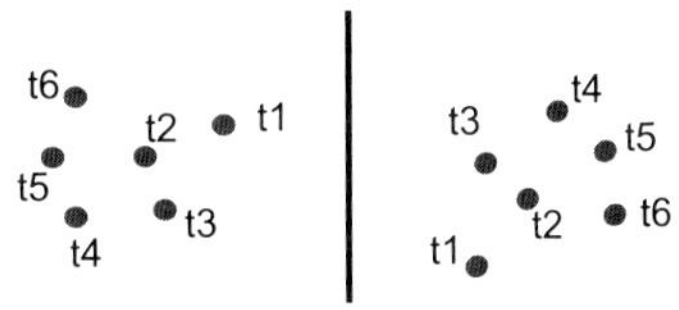

Fig. 4. Rotation with the best left-right ordering on the x-axis. Note that 2 and 3 remain in the wrong order.

2-Dimensional Extrapolation. In our approach both x and y are coordinates and therefore inherently independent variables. They depend on the current visualization alone. Within our model, they do however depend on the time variable t. Because our methods aims to extrapolate x, y out of one other variable t, we need a form of 2-dimensional extrapolation. After rotation and under the assumption that x is in fact the independent variable guiding y, standard methods can be used. For this scenario we need to establish a rotation that best fits the time ordering to a left-right ordering on the x-axis as displayed in Figure 4.

It is also possible to use the polynomial extrapolation for the x and y variables separately and combine them into a linear system, much like spline interpolation, only for the entire domain (referred to as x,y system): $x = p_1(t)$, $y = p_2(t)$. Naturally, the dependence of x and y on t within the spline interpolation scheme makes that method very well suited for the task of 2-dimensional extrapolation.

This leaves six methods that are reasonably suited for our approach:

1. Second degree polynomial extrapolation
2. Third degree polynomial extrapolation

3. x,y system with second degree polynomial extrapolation
4. x,y system with third degree polynomial extrapolation
5. Spline extrapolation with straight line continuation
6. Spline extrapolation with polynomial continuation

3 Approach

The number of attributes describing each item in a database can be quite large. Taking all this information into account when extrapolating can therefore be quite a hassle. Since this information is inherently present in an already performed visualization of a clustering, we can theoretically narrow the information load down to two attributes (x and y) per item whilst retaining the same accuracy. The stepwise strategy is illustrated in Figure 5.

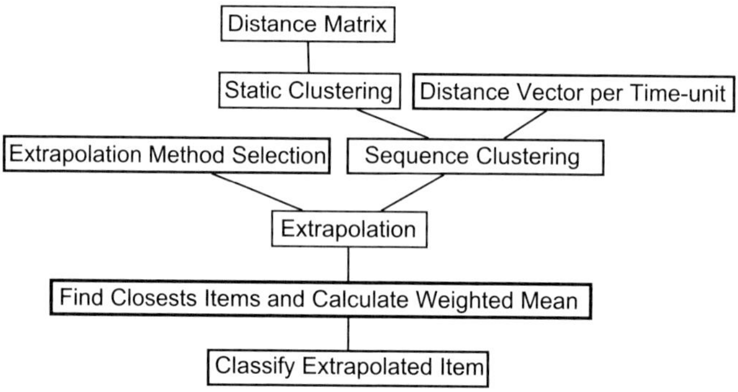

Fig. 5. Stepwise approach

3.1 Distance Matrix and Static Clustering

The data used as reference within our approach is represented by a square $q \times q$ distance matrix describing the proximity between all q items. These items are such that their data is fully known beforehand. The number of items q should be large enough to at least provide enough reference material on which to base the extrapolation. These items are clustered and visualized according to some MDS technique resulting in a 2-dimensional plane with dots representing our reference items. This step in the approach is done only once so the focus should be on the quality of the clustering instead of the computational complexity. From this point on this clustering is considered to be fixed or static.

3.2 Distance Vector Time-Unit and Sequence Clustering

Analysis of the behaviour of new items should start with the calculation of the values for each time-unit t. These units are supposed to be cumulative, meaning that they contain all the item's *baggage*, i.e., its whole history, up to the specified moment. Using the same distance measure that was used to create the initial distance matrix, the *distance vector per time-unit* can now be calculated. This should be done for all t time-units, resulting in t vectors of size q. These vectors can now be visualized as before. The chosen visualization method should naturally allow for incremental placement of individual items, e.g., as in [9]. These new data points within the clustering will be used to extrapolate the items behaviour through the static clustering.

3.3 Extrapolation

After selecting the best extrapolation scheme for our type of data our method creates a function that extrapolates item behaviour. For the same data the different schemes can yield different results as illustrated in Figure 6, so care should be taken to select the right type of extrapolation for the data under consideration.

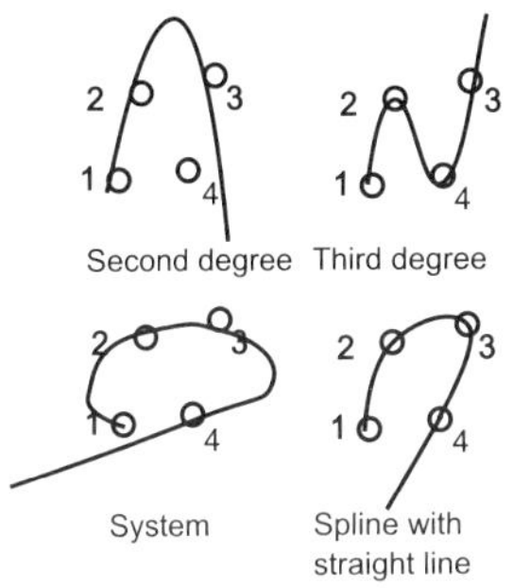

Fig. 6. Different extrapolation methods yield very different results

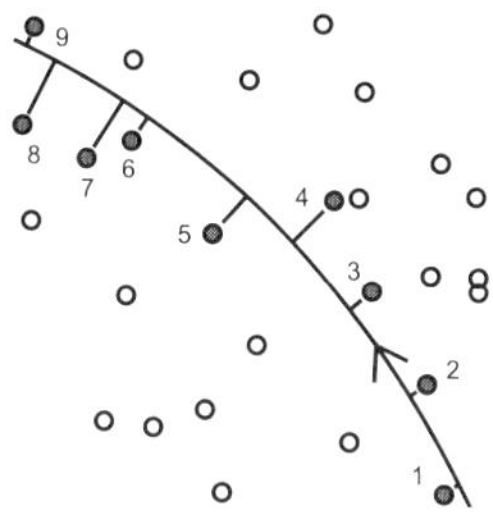

Fig. 7. Selecting points with the shortest distance to the extrapolation line

One advantage of this approach is that the extrapolation or prediction is immediately visualized to the end-user rather than presenting him or her with a large amount of numerical data. If the user is familiar with the data under consideration, he/she can analyse the prediction in an eye blink. Augmenting the system with a click and point interface would enable the end user to use the prediction as a starting point for further research.

3.4 Final Steps

In most cases it is desirable to predict which class the item under consideration might belong to in the future. In that case it is important to retrieve further information from some of the reference items and assign future attribute values and a future class to the item.

A first step would be to select r reference items closest to the extrapolation line. This can easily be done by evaluating the geometric distance of all reference points to the line and selecting those with the smallest distance, see Figure 7.

We order these points by their respective distance to the last known data point of the extrapolated item: the confidence of the prediction declines with this distance. We estimate the value for the future attribute j:

$$Attrib_j(future) = \frac{2}{r(r+1)} \cdot \sum_{i=1}^{r} (r - i + 1)\ Attrib_j(point\ i)$$

The extrapolated item can now be visualized into the clustering according to its future attributes and be classified accordingly.

4 Experiments

The detection, analysis, progression and prediction of criminal careers is an important part of automated law enforcement analysis [7, 8]. Our approach of temporal extrapolation was tested on the national criminal record database of The Netherlands. This database contains approximately one million offenders and their respective crimes (approximately 50 types).

We clustered 1,000 criminals on their criminal careers, i.e., all the crimes they committed throughout their careers. In this test-case r will be set to 30. We employed a ten-fold cross validation technique within this group using all of the different extrapolation methods in this static clustering, and standard extrapolation on each of the attributes (methods 7 and 8). For each item (i.e., person) in the test set we only consider the first 4 time periods. The accuracy is described by the mean similarity between the calculated and the expected values of the attributes. The results are presented in Table 1, where *time factor* represents runtime slowdown with respect to the fastest method under consideration.

Although the runtime needed for visual extrapolation is much less than that of regular methods, the accuracy is comparable. For this database the best result is still a regular second degree extrapolation but its accuracy is just marginally higher than that of the spline extrapolation with a straight line, where its runtime is much larger. The simpler x,y system with third degree extrapolation is very fast but still reaches an acceptable accuracy.

Table 1. Results of Static Clustering Extrapolation for the analysis of Criminal Careers

	method	*time factor*	*accuracy*
1	Second degree polynomial extrapolation	1.0	79.1%
2	Third degree polynomial extrapolation	1.1	79.3%
3	x,y system with second degree polynomial extrapolation	1.9	81.5%
4	x,y system with third degree polynomial extrapolation	2.1	87.5%
5	Spline extrapolation with straight line continuation	13.4	88.7%
6	Spline extrapolation with polynomial continuation	13.4	79.6%
7	Regular second degree attribute extrapolation	314.8	89.0%
8	Regular third degree attribute extrapolation	344.6	82.3%

5 Conclusion and Future Directions

In this paper we demonstrated the applicability of temporal extrapolation by using the prefabricated visualization of a clustering of reference items. We demonstrated a number of extrapolation techniques and employed them to predict the future development of item behaviour. Our methods were tested within the arena of criminal career analysis, predicting the future of unfolding criminal careers.

We showed that our novel approach largely outperforms standard prediction methods in the sense of computational complexity, without a loss in accuracy larger than 1 percentage point. The visual nature of our method enables the analyst of the data to immediately continue his/her research since the prediction results are easily displayed within a simple graphical interface.

Future research will aim at reaching even higher accuracy values by improving the selection of reference items close to the extrapolation line. Different types of data might well be more susceptible to errors, providing another research issue.

Acknowledgment. The authors would like to thank Kees Vuik and Robert Brijder. This research is part of the DALE (Data Assistance for Law Enforcement) project as financed in the ToKeN program from the Netherlands Organization for Scientific Research (NWO) under grant number 634.000.430.

References

1. Abdi, H.: Least squares. In: Lewis-Beck, M., Bryman, A., Futing, T. (eds.) Encyclopedia for Research Methods for the Social Sciences, pp. 792–795. Sage, Thousand Oaks (2003)
2. Abdi, H.: Signal detection theory. In: Salkind, N.J. (ed.) Encyclopedia of Measurement and Statistics, Sage, Thousand Oaks (2007)
3. Anderson, J.: Learning and Memory. Wiley and Sons, New York (1995)
4. Bartels, R.H., Beatty, J.C., Barsky, B.A.: An Introduction to Splines for Use in Computer Graphics and Geometric Modelling. Morgan Kaufmann, San Francisco (1987)
5. Broekens, J., Cocx, T., Kosters, W.A.: Object-centered interactive multidimensional scaling: Let's ask the expert. In: Proceedings of the Eighteenth Belgium-Netherlands Conference on Artificial Intelligence (BNAIC 2006), pp. 59–66 (2006)
6. Davison, M.L.: Multidimensional Scaling. John Wiley and Sons, New York (1983)
7. de Bruin, J.S., Cocx, T.K., Kosters, W.A., Laros, J.F.J., Kok, J.N.: Data mining approaches to criminal career analysis. In: Proceedings of the Sixth IEEE International Conference on Data Mining (ICDM 2006), pp. 171–177 (2006)
8. de Bruin, J.S., Cocx, T.K., Kosters, W.A., Laros, J.F.J., Kok, J.N.: Onto clustering criminal careers. In: Proceedings of the ECML/PKDD 2006 Workshop on Practical Data Mining: Applications, Experiences and Challenges, pp. 92–95 (2006)
9. Kosters, W.A., van Wezel, M.C.: Competitive neural networks for customer choice models. In: E-Commerce and Intelligent Methods, Studies in Fuzziness and Soft Computing 105, pp. 41–60. Physica-Verlag, Springer (2002)
10. Sun, R., Merrill, E., Peterson, T.: From implicit skills to explicit knowledge: A bottom-up model of skill learning. Cognitive Science 25(2), 203–244 (2001)

Discovering Explanations from Longitudinal Data

Corrado Loglisci and Donato Malerba

Dipartimento di Informatica - Universita' degli Studi di Bari, Italy
{loglisci,malerba}@di.uniba.it

Abstract. The inference of Explanations is a problem typically studied in the field of *Temporal Reasoning* by means of approaches related to the *reasoning about action and change*, which aim usually to infer statements that explain a given change. Most of proposed works are based on inferential logic mechanisms that assume the existence of a general domain knowledge. Unfortunately, the hypothesis to have a domain theory is a requirement not ever guaranteed. In this paper we face the problem from a data-driven perspective where the aim is to discover the events that can plausibly explain the change from a state to another one of the observed domain. Our approach investigates the problem by splitting it in two issues: *extraction of temporal states* and *finding out the events*. We applied the approach to the scenarios of Industrial Process Supervision and Medical Diagnosis in order to support the task of domain experts: the experimental results show interesting aspects of our proposal.

Keywords: Explanation, longitudinal data, temporal state, event, data mining.

1 Introduction

Longitudinal data can be defined as data resulting from the repeated observations of subjects described by a number of variables over time. They are useful to investigate the evolution of a phenomenon which is characterized by at least two states, initial and final, and by a number of events. Two classes of inference problems are particularly important when studying the evolution of a phenomenon: *prediction* and *explanation*. A prediction problem is defined when the initial state of the phenomenon under study, the nature and timing of events is known, and the final state (after the last event) is of interest. In an explanation problem, information is provided about both the final state and previous events, and questions are asked about the initial state or more generally about earlier states. This kind of problems is particularly investigated in the literature on *reasoning about action and change* [3], where prediction, also called *temporal forward projection*, is based on deductive inference, while explanation, also called *backward temporal projection*, is based on abductive inference. More precisely, as pointed out in [8], if T represents the general knowledge of the phenomenon , S the set of observed facts (events or properties) concerning both the initial state and intermediate

A. An et al. (Eds.): ISMIS 2008, LNAI 4994, pp. 196–202, 2008.

events, and G the set of facts related to the final state, then the prediction can be formalized as the logical entailment of G from T and S ($T, S \models G$). In the case of the explanation, T and G are given and the problem is that of finding S such that $T, S \models G$ and S is consistent with T.

2 Motivation and Contribution

Unfortunately the possibility to have the use of the general knowledge T of the phenomenon is not ever guaranteed. The present work is addressed to fill just this lack: it investigates the Problem of Explanation with a data-driven approach that is based only on the set S of observed facts, and that does consider no one domain theory T.

More precisely, our aim is that to discover backwards the *events* that can have led the phenomenon to a given *state* from the past state, where the nature and timing of the past state are known, and the events and their timing are of interest. The notion of state roughly follows the notion of *situation* well-known in Situation Calculus[6]: a state consists of the set of all facts of phenomenon, called *fluents*, which are true over a certain time-period during which there are no changes. The meaning of event rather corresponds to the usual notion of whatever action [9] that occurs within a certain time-period and that leads the phenomenon to evolve: one or more events can initiate a state or terminate another one (Figure 1).

The rest of this paper is organized as follows. In next section we provide the formal statement of the faced problem. In section 4 we describe the methodology for the discovery of explanations by detailing the procedures that resolve two particular sub-problems: the *extraction of temporal states* and the *finding out the events*. Discovering explanations in several real scenarios it can turn out to be useful and to give a lot of support to the domain experts. To this end we applied our approach to the real cases of *Industrial Process Supervision* and *Medical*

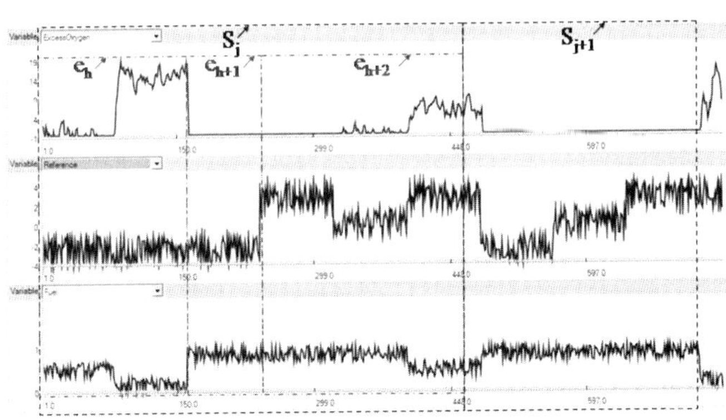

Fig. 1. One or more events $\langle e_h, e_{h+1}, e_{h+2} \rangle$ can occur within a time-period (e.g., $[ts^j ... te^j]$) during which the phenomenon remains in a state (e.g., S_j): they trigger the evolution of the phenomenon from a state (e.g., S_j) to another one (e.g., S_{j+1})

Diagnosis: we present the experiments by showing the results and discussing some interesting aspects as well.

3 The Problem Definition

In order to define the problem, we first report some useful definitions, then the formal setting.

Let F:$\{d_1,\ldots,d_m\}$ be the finite set of attributes or descriptors of the phenomenon with ranges $\{D_1,\ldots,D_m\}$ and Dp:$<Dp_{t_1},Dp_{t_2},\ldots,Dp_{t_n}>$ be the input finite sequence (i.e. longitudinal data) of data-points, where each Dp_{t_i}:$\langle v_{i1}, v_{i2}, \ldots, v_{im}, t_i \rangle \in D_1 \times D_2 \ldots \times D_m \times T^P$, $i \in \mathbf{Z}^+$, v_{ik} value of d_k at the time-point t_i, $t_i \in T^P$.

We say that a **temporal state S_j** is represented by the signature $\langle ts^j,te^j,C_j\rangle$, where: ts^j, $te^j \in T^P$, $ts^j \leq te^j$[1] and C_j:$\{r_1,r_2,\ldots\}$ is a finite set of fluents represented in a language *L*.

We say that an **event e** is represented by the followings:

- $\langle ed_1,\ldots,ed_r,\ldots,ed_{m'}\rangle$ which corresponds to a subset of descriptors F;
- w_k:$[ts^k,te^k]$ and w_{k+1}:$[ts^{k+1},te^{k+1}]$, which are time-windows on Dp. We say that the event **e** lasts over the period $[ts^k \ldots te^k]$;
- $\langle cv_1, \ldots, cv_r,\ldots, cv_{m'}\rangle$, which consists in the set of categorical values for each ed_r in w_k. Roughly speaking each r-th categorical value is a qualitative representation of values of ed_r;
- $\langle[\inf_{ed_1}\ldots\sup_{ed_1}], \ldots, [\inf_{ed_r}\ldots\sup_{ed_r}],\ldots, [\inf_{ed_{m'}},\sup_{ed_{m'}}]\rangle$, which consists in the set of numerical ranges for each ed_r in w_k. Roughly speaking, each r-th numerical range is a quantitative representation of values of ed_r;

In doing so, the problem can be formulated as follows:

given: longitudinal data Dp,
find: 1) the finite set of temporal states S:$\{S_1,S_2,\ldots,S_s\}$ induced from Dp, and 2) for each pair (S_j,S_{j+1}), S_j:$\langle ts^j,te^j,C_j\rangle$, S_{j+1}:$\langle ts^{j+1},te^{j+1},C_{j+1}\rangle$, j=1,…,s-1, the *explanation* $L_{j,j+1}$:$\langle e_1,e_2,\ldots,e_h,\ldots,e_p\rangle$ where $ts^j \leq ts^{k_{e_1}}$, $ts^{k_{e_{h+1}}} = te^{k_{e_h}}+1$, $te^{k_{e_p}} \leq te^j$. Thus, discovering the explanation $L_{j,j+1}$ can be seen as finding out the sequence of events $\langle e_1,e_2,\ldots,e_h,\ldots,e_p\rangle$.

4 Discovering Explanations

In this section we describe the details of the procedures that solve respectively the sub-problems of *extraction of temporal states* and *finding out the events*.

4.1 Extraction of Temporal States

In accordance with the concept of state previously provided, a state S_j can be seen as a sort of snapshot of the phenomenon which spans a certain time-period.

[1] The partial order relation '$\leq$' holds for the set T^P.

Thus, extracting the set S of states means identifying the several snapshots occurring in Dp. At this end we resort to the approach proposed in [4] which is shortly sketched. First, a process of *temporal segmentation* performed on Dp allows to identify the time-periods $[ts^j...te^j]$ for each state. Next, for each of resulting segment, an *inductive learning process* generates the fluents C_j (in a propositional logic language L) such that the following holds: given two states $S_j:\langle ts^j, te^j, C_j\rangle$, $S_{j+1}:\langle ts^{j+1}, te^{j+1}, C_{j+1}\rangle$, the fluents C_j do hold in $[ts^j...te^j]$ but do not in $[ts^{j+1}...\ te^{j+1}]$, conversely, C_{j+1} do hold in $[ts^{j+1}...\ te^{j+1}]$ but do not in $[ts^j...te^j]$, $te^j<ts^{j+1}$.

4.2 Finding Out the Events

Here we describe the procedure to find out the sequence of events that can explain the evolution of the phenomenon to a given state $S_{j+1}:\langle ts^{j+1}, te^{j+1}, C_{j+1}\rangle$ from the previous one $S_j:\langle ts^j, te^j, C_j\rangle$. The approach originates in the fact that such events can be seen as *significative variations* occurring in Dp. Since generally whatever variation in the data is reflected in the models generated from ([1]), detecting the events that can trigger the evolution from S_j to S_{j+1} means detecting variations or differences between the models induced respectively from $\{Dp_{ts^j},\ldots,Dp_{te^j}\}$ and $\{Dp_{ts^{j+1}},\ldots,Dp_{te^{j+1}}\}$. At this aim we propose a two-step procedure (i) *mining candidate events* and ii) *generation of the sequence of events*) in the following reported.

In the first phase, a sliding-window technique iteratively scans Dp and, at each scan, generates a single sequence of candidate events. A single candidate event is found by detecting significative differences[2] among the coefficients of two regression models[3] induced from w_k and w_{k+1} respectively ($ts^j \le ts^k$, $te^k<ts^{k+1}$, $te^{k+1} \le te^{j+1}$). In particular, in the case a candidate is found the searching for the next candidate continues with two slided time-windows, otherwise the process is repeated with a wider w_k. Once a candidate event is mined, it is represented as follows:

– $\langle ed_1,\ldots,ed_r,\ldots,ed_{m'}\rangle$ constitued by the set of descriptors whose coefficients make the significative differences;

– $\langle cv_1,\ldots, cv_r,\ldots, cv_{m'}\rangle$ constitued by the qualitative descriptions of the trend of each ed_r during w_k. These are generated by resorting to the temporal abstraction techniques [7];

– $\langle [inf_{ed_1}...sup_{ed_1}],\ldots, [inf_{ed_r}...sup_{ed_r}],\ldots, [inf\_ed_{ed_{m'}}, sup_{ed_{m'}}]\rangle$ constituted by the quantitative descriptions of each ed_r during w_k. These are produced by means the same inductive learning process [5] used in section 4.1, which generalizes the value intervals assumed by each ed_r in w_k and discriminates the ones assumed in w_{k+1}.

---

[2] The notion of significative variation exploits an heuristic function which is here not reported because of space limitations.

[3] The linear regression models $f(\mathbf{x}) = \beta_0 + \beta_1 x_1 + \ldots + \beta_m x_m$ are induced by solving problems of recursive least squares. Here the independent and dependent variables x_i correspond to the descriptors F.

The first phase returns a set of sequences of candidate events $\{\langle e'_1, e'_2, \ldots \rangle, \langle e''_1, e''_2, \ldots \rangle, \ldots\}$ which constitutes the input to the phase of generation of the explanation. The latter will be produced as the sequence of mostly informative events by exploiting the following:

Definition 1. A candidate event e_u is called **mostly informative** with *support* $f(\lambda)$, $f(\lambda)>0$, iff:

- $\exists\ \{e_1, e_2, \ldots, e_\lambda\} \ni'$ for each e_q, $q=1 \ldots \lambda$, $e_q \neq e_u$: $\langle ed_1, \ldots, ed_r, \ldots, ed_{m'} \rangle^{e_q} \subseteq \langle ed_1, \ldots, ed_r, \ldots, ed_{m'} \rangle^{e_u}$ and $ts^{k_{e_u}} < ts^{k_{e_q}} \wedge te^{k_{e_u}} > te^{k_{e_q}}$ and for each ed_r: $cv_r^{e_u} = cv_r^{e_q} \wedge [inf_{ed_r} \ldots sup_{ed_r}]^{e_u} \supseteq [inf_{ed_r} \ldots sup_{ed_r}]^{e_q}$;
- $\nexists\ e_v$ with support $f'(\lambda)$, $e_v \neq e_u$, $f'(\lambda) > f(\lambda) \ni' \langle ed_1, \ldots, ed_r, \ldots, ed_{m'} \rangle^{e_u} \subseteq \langle ed_1, \ldots, ed_r, \ldots, ed_{m'} \rangle^{e_v}$ and $ts^{k_u} > ts^{k_v} \wedge te^{k_u} < te^{k_v}$ and for each ed_r: $cv_r^{e_u} = cv_r^{e_v} \wedge [inf_{ed_r} \ldots sup_{ed_r}]^{e_u} \supseteq [inf_{ed_r} \ldots sup_{ed_r}]^{e_v}$.

Informally speaking, Definition 1 allows to identify the events that generalize the information contained in all the found candidate events.

5 Applications

In this section we discuss the application to some real world scenarios by deepening the fact that the explanations can generally provide information of 'when an event occurs' and 'what is the associated change'. We performed experiments with datasets coming from different applicative contexts, namely *Industrial Process Supervision* and *Medical Diagnosis* which can claim a different support each other. Since, at our knowledge, competiting works have not been proposed yet, we compare our results and conclusions with the ones obtained from other works which use the same datasets. Because of space limitations we discuss only the context to the Industrial Process Supervision.

Industrial Process Supervision. In this scenario our approach could be applied for the typical tasks of process monitoring and/or of quality control of the final product. In particular, the aim is to provide information of the historic evolution of the process parameters, which and when particular operations can have lead to failure conditions and how certain actions affect the final product of the process. The dataset has been produced from a system of *Steam Generator* modelled [2] at Abbott Power Plant in Champaign IL and is consisting of 9600 measurements sampled at 3 seconds, where each measurement is described in terms of 8 numerical variables (i.e. descriptors): *fuel, air, reference level, disturbance, drum pressure, excess oxygen, water level, steam flow.*

In accordance with [4], by varying the input parameters of the algorithm reported in Section 4.1 different state sets can be extracted: each set represents differently the whole process. For instance, one of these consists of 18 states each of which spans at least 5 hours (namely 100 data-points). We discuss the

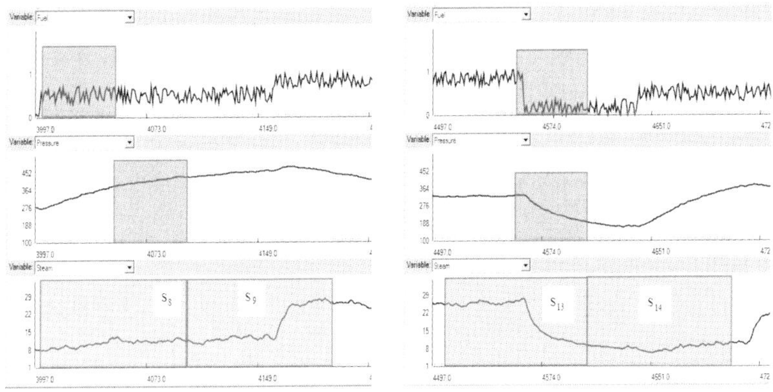

(a) (b)

Fig. 2. Representations (from top to bottom) of fuel, drum pressure and steam flow during the process of steam generation from S_8 to S_9 (a) and from S_{13} to S_{14} (b). For clarity, the time-windows of states are only drawn on the axis of steam flow, while the time-windows of the events are drawn on the axis of fuel and drum pressure.

explanations (reported in the table below) of the evolution from $\langle t_{4001}, t_{4101}, C_8\rangle$ to $\langle t_{4102}, t_{4202}, C_9\rangle$, from $\langle t_{4507}, t_{4606}, C_{13}\rangle$ to $\langle t_{4607}, t_{4708}, C_{14}\rangle$ and from $\langle t_{4202}, t_{4303}, C_{10}\rangle$ to $\langle t_{4304}, t_{4404}, C_{11}\rangle$ in comparison with the results obtained in [2].

$\langle\ldots,ed_r,\ldots\rangle$	$\langle\ldots,[inf_{ed_r}\ldots sup_{ed_r}],\ldots\rangle$	w_k	$\langle\ldots,cv_r,\ldots\rangle$	w_{k+1}	$f(\lambda)$
drum pressure	[386.45053 .. 435.63766]	$[t_{4051}\cdots t_{4101}]$	VERY_INCREASE	$[t_{4102}\cdots t_{4202}]$	1
fuel	[0.33408 .. 0.73257]	$[t_{4001}\cdots t_{4050}]$	VERY_INCREASE	$[t_{4051}\cdots t_{4101}]$	
drum pressure	[196.07344 .. 334.885]	$[t_{4556}\cdots t_{4606}]$	DECREASE	$[t_{4607}\cdots t_{4708}]$	1
fuel	[0.00368 .. 1.01325]	$[t_{4556}\cdots t_{4606}]$	VERY_DECREASE	$[t_{4607}\cdots t_{4708}]$	
fuel	[0.67145 .. 1.06162]	$[t_{4253}\cdots t_{4303}]$	VERY_INCREASE	$[t_{4304}\cdots t_{4404}]$	1
air	[0.0048 .. 1.0651]	$[t_{4202}\cdots t_{4252}]$	VERY_INCREASE	$[t_{4253}\cdots t_{4303}]$	

In particular, in their model the steam flow depends on drum pressure and, the latter, in turn, is related to previous variations on fuel quantity. Our results are consistent with this study, in fact the difference showed by *steam flow* from S_8 to S_9[4] can be explained with a strong increasing of *fuel* during the period $[t_{4001}\ldots t_{4050}]$ which, in turn, triggers a strong increase of *drum pressure* in $[t_{4051}\ldots t_{4101}]$ (see Fig.2 a)). Analogously, the reduction of the quantity of generated steam from S_{13} to S_{14}[5] can be explained (with maximum statistical support, $f(\lambda)=1$) with the fact that the *fuel*, which is decreasing in $[t_{4556}\ldots t_{4606}]$, can cause the decrease of the *drum pressure* in $[t_{4556}\ldots t_{4606}]$(see Fig.2 b)).

Further findings confirm the influence of *fuel* and *air* on the excessive generation of oxygen [2]. In particular, the sequence of events 'air steeply increasing

[4] C_8:{...∧ steam flow in [7.98298..13.10828]∧ steam flow in [11.70881..12.86494]...}, C_9:{...∧steam flow in [11.69014..28.06678]... steam flow in [10.04345..23.16669]...}.

[5] C_{13}:{...∧ steam flow in [24.70965..27.47846]∧ steam flow in [10.70601..26.94199]∧ steam flow in [25.79022..27.63655]...}, C_{14}:{...∧ steam flow in [6.4073..9.54047]∧ steam flow in [7.16914..10.4356]...}.

during [t_{4202}...t_{4252}] and fuel also increasing during [t_{4253}...t_{4303}]' seems to trigger the evolution from S_{10} to S_{11}[6].

6 Conclusions

The Problem of Explanation has attracted interest in the field of *Temporal Reasoning* particularly and also in the field of *Uncertain Reasoning*, while a little bit of effort has been spent in other disciplines. In this work we offer an inductive perspective to the problem and provide a data-mining approach to discover events that potentially explain the evolution of an observed domain. Our added value consists in fact that the explations express information about 'what are the associated evolutions' and 'when these occur'. Although the human generally is interested in 'what will it happen' than 'what did happen', the explanation of the past can seemingly be useless and not interesting at all. The reported experiments show that this kind of information can turn out to be important and helpful in several contexts.

Acknowledgments

This work is partial fulfillment of the research objective of DDTA project "Centro Servizi per il Distretto Digitale Tessile-Abbigliamento".

References

1. Baron, S., Spiliopoulou, M.: Monitoring Change in Mining Results. In: Proc. of the Third International Conference on Data Warehousing and Knowledge Discovery (2001)
2. Espinosa, J., Vandewalle, J.: Predictive Control Using Fuzzy Models Applied to a Steam Generating Unit. In: International Workshop FLAIRS (1998)
3. Fisher, M., Gabbay, D., Vila, L.: Handbook of Temporal Reasoning in Artificial Intelligence. Foundations of Artificial Intelligence. Elsevier Science Inc., Amsterdam (2005)
4. Loglisci, C., Berardi, M.: Segmentation of Evolving Complex Data and Generation of Models. In: Proc. of the 6-th ICDM - Workshops (2006)
5. Malerba, D.: Learning Recursive Theories in the Normal ILP Setting. Fundamenta Informaticae 57(1), 39–77 (2003)
6. McCarthy, J., Hayes, P.J.: Some philosophical problem from standpoint of Artificial Intelligence. Machine Learning 4, 463–502 (1969)
7. Shahar, Y.: A Framework for Knowledge-Based Temporal Abstraction. Artif. Intell. 90(1-2), 79–133 (1997)
8. Shanahan, M.: Prediction is Deduction but Explanation is Abduction. In: Proc. of IJCAI conference, pp. 1055–1060 (1989)
9. Van Belleghem, K., Denecker, M., De Schreye, D.: Representing Continuous Change in the Abductive Event Calculus. In: ICLP (1994)

[6] $C_{10}\{\ldots \wedge$ excess oxygen in [-0.04801..2.88602]$\wedge \ldots\}$, C_{11}:$\{\ldots \wedge$ excess oxygen in [-0.04304..18.3013]$\ldots\}$.

Reduced Implicate/Implicant Tries*

Neil V. Murray[1] and Erik Rosenthal[2]

[1] Department of Computer Science, State University of New York, Albany, NY 12222, USA
nvm@cs.albany.edu
[2] Department of Mathematics, University of New Haven, West Haven, CT 06516, USA
erosenthal@newhaven.edu

Abstract. The *reduced implicate trie*, introduced in [10], is a data structure that may be used as a target language for knowledge compilation. It has the property that a query can be processed in time *linear in the size of the query*, regardless of the size of the compiled knowledge base. This data structure can be used with propositional databases, where a query amounts to asking whether a clause is an implicate of a logical formula. In this paper, *reduced implicant tries* are investigated, and the dual question is addressed: determining the implicants of a formula. The main result is that a single trie — the *reduced implicate/implicant trie*, with a structure that is similar to that of reduced implicate tries — can serve dual roles, representing both implicates and implicants. As a result, there can be significant savings in both time and space.

1 Introduction

Several investigators have represented knowledge bases as propositional theories. A query of such a propositional theory typically has the form, *Is a CNF clause an implicate of the theory?* However, since the question, *Does* $\mathcal{NP} = \mathcal{P}$*?* remains open — i.e., there are no known polynomial algorithms for problems in the class $\mathcal{NP}$ — the time to answer queries is (in the worst case) exponential. The *reduced implicate trie* was developed [10,11] as a solution to a problem posed by Kautz and Selman [8] in 1991. Their idea, known as *knowledge compilation*, was to pay the exponential penalty once by compiling the knowledge base into a *target language* that would guarantee fast response to queries. A number of languages — for example, *Horn sets*, *ordered binary decision diagrams*, sets of *prime implicates/implicants*, *decomposable negation normal form*, *factored negation normal form*, and *pairwise-linked formulas* — have been proposed as targets for knowledge compilation. (See, for example, [1,2,3,4,5,6,7,9,13,15].)

The reader is assumed to be familiar with the basic terminology of propositional logic. Consequences expressed as minimal clauses that are implied by a formula are its *prime implicates*; minimal conjunctions of literals that imply a formula are its *prime implicants*. Asking whether a given formula is entailed by a DNF clause is equivalent to asking whether the clause is an implicant of the formula. Throughout the paper, this question is what is meant by DNF query.

* This research was supported in part by the National Science Foundation under grants IIS-0712849 and IIS-0712752.

A. An et al. (Eds.): ISMIS 2008, LNAI 4994, pp. 203–214, 2008.

A *trie* is a simply labeled tree. They can be used to represent logical formulas: The nodes along each branch represent the literals of a clause, and the conjunction of all such clauses is a CNF equivalent of the formula represented by the trie. A *reduced implicate trie* (ri-trie) is a trie in which each branch represents an implicate, and every implicate is represented, possibly implicitly, by a branch. A data structure called a *reduced implicate trie* was Introduced in [10,11]; they have the property that response time is *linear in the size of the query*. The dual query is considered in this paper: *Is a DNF clause an implicant of the theory?* The *reduced implicant trie* is described in Section 2.1, and *reduced implicate/implicant tries* are introduced in Section 3.

2 A Data Structure That Enables Fast DNF Query Processing

Determining whether a DNF clause — i.e., a conjunction of literals — is an implicant of a logical formula is considered in this section. Tries can be used to represent prime implicants as well prime implicates [14]: The nodes along each branch represent the literals of a DNF clause, and the disjunction of all such clauses is a DNF equivalent of the formula represented by the trie. Tries that represent DNF formulas can be interpreted directly as formulas in negation normal form (NNF): A trie consisting of a single node represents the label of that node. Otherwise, the trie represents the conjunction of the label of the root with the disjunction of the formulas represented by its children.

A trie that stores all (non-contradictory) implicants of a formula is called a *complete implicant trie*. To define it formally, first select an ordering $\{p_1, p_2, ..., p_n\}$ of the variables that appear in the (propositional) formula $\mathcal{F}$. Let q_i be either the literal p_i or the literal $\neg p_i$, and order the literals by $q_i \prec q_j$ if $i < j$. (This can be extended to a total order by defining $\neg p_i \prec p_i, 1 \leq i \leq n$. But neither queries nor branches in the trie will contain such complementary pairs.) A *prefix* of a DNF clause $\{q_1, q_2, ..., q_k\}$ is defined to be a clause of the form $\{q_1, q_2, ..., q_i\}$, where $0 \leq i \leq k$. If $i = 0$, then the prefix is the empty clause. This is extended in the obvious way to the nodes/labels of a branch.

The *complete implicant trie* for $\mathcal{F}$, with respect to a given variable ordering, is a tree defined as follows: If $\mathcal{F}$ is truth constant, the tree consists of a root labeled with that constant. Otherwise, the complete implicate trie is a tree in which the root is labeled 1, all other nodes are labeled with literals, and the tree satisfies these properties:

1. No node has distinct children with the same label.
2. The set of labels of any prefix of any branch is a prefix of an implicant of $\mathcal{F}$.
3. If a prefix of the labels of a branch represent an implicant, then the last node in that prefix is marked with the end symbol; all leaves are marked with the end symbol.
4. Every implicant is the (not necessarily proper) prefix of some branch.

Observations

1. Every leaf is labeled with q_n (i.e., with p_n or with $\neg p_n$).
2. Whether a clause is an implicant of $\mathcal{D}$ can be determined in time linear in the size of the clause simply by traversing the corresponding branch.
3. Since any superset of an implicant is an implicant, if a node labeled q_k is marked with the end symbol, and if $k < n$, then the node will have as children nodes labeled $\neg q_{k+1}$ and q_{k+1}, and these children will be marked with the end symbol.

4. Part 4 of the definition slightly abused the term *branch* by using it to mean the clause represented by its labels. Branch will be used in this way throughout this paper, typically assuming implicitly that any constants along the branch have been simplified away. (A ternary representation of reduced implicant tries in which branches may have multiple constants will be described below.)

2.1 Reduced Implicant Tries

The following simplification rules are useful (even if trivial).

SR1.	$\mathcal{F} \longrightarrow \mathcal{F}[\mathcal{G}/\mathcal{G} \vee 0]$	$\mathcal{F} \longrightarrow \mathcal{F}[\mathcal{G}/\mathcal{G} \wedge 1]$
SR2.	$\mathcal{F} \longrightarrow \mathcal{F}[0/\mathcal{G} \wedge 0]$	$\mathcal{F} \longrightarrow \mathcal{F}[1/\mathcal{G} \vee 1]$
SR3.	$\mathcal{F} \longrightarrow \mathcal{F}[0/p \wedge \neg p]$	$\mathcal{F} \longrightarrow \mathcal{F}[1/p \vee \neg p]$

Applications of **SR1** and **SR2** will be restricted to leaves of implicant tries. Any trie that is produced by a sequence of these rules and that has no leaves labeled with constants (other than the root) is called an *implicant trie*. Observe that some applications of the simplification rules will produce a trie that has nodes other than the root — necessarily leaves — labeled with constants. These are not implicant tries. Of course, such a trie can be simplified to an implicant trie. An implicant trie that cannot be simplified with these rules is called a *reduced implicant trie* or, more simply, an ri^c-trie.

Observations

1. Suppose q is (the label of) a node with two leaf children, $\neg p$ and p. Then **SR3** replaces the two children with a single leaf labeled 1, and **SR1** deletes the new leaf.
2. If a node of an implicant trie is marked with the end symbol, then, since any superset of an implicant is an implicant, all extensions of that branch are implicants. As a result, if a node labeled q_k is marked with the end symbol, $k < n$, it will have children labeled $\neg q_j$ and q_j, $k < j \leq n$, and they will be also be marked with the end symbol. Thus, repeated applications of the simplification rules will delete the entire subtrie below q_k.
3. If the formula $\mathcal{F}$ is a contradiction, then repeated applications of the rules will produce a trie in which the root has a single child labeled 0; **SR1** then produces the reduced implicant trie consisting of a root labeled 0.
4. The only nodes in an ri^c-trie with the end marker are leaves. In particular, no proper prefix of a branch in a reduced implicant trie represents an implicant of the trie.
5. Any implicant with no proper prefix as an implicant is a branch in the ri^c-trie.

The last two observations lead naturally to a definition: If $\mathcal{F}$ is a logical formula, and if the variables of $\mathcal{F}$ are ordered, then a *relatively prime implicant* is one for which no proper prefix is also an implicant. The next theorem is now immediate.

Theorem 1. Given a logical formula $\mathcal{F}$ and an ordering of the variables of $\mathcal{F}$, then the branches of the corresponding reduced implicant trie represent precisely the relatively prime implicants. In particular, the prime implicants are relatively prime, and each is represented by a branch in the trie. □

2.2 Computing Reduced Implicant Tries

In this section, techniques for computing reduced implicant tries are developed. Let $\mathrm{Imp}^c(\mathcal{F})$ denote the set of all implicants of $\mathcal{F}$.

Lemma 1. Let $\mathcal{F}$ and $\mathcal{G}$ be formulas. Then $\mathrm{Imp}^c(\mathcal{F}) \cap \mathrm{Imp}^c(\mathcal{G}) = \mathrm{Imp}^c(\mathcal{F} \wedge \mathcal{G})$. □

Let $\mathcal{F}$ be a logical formula with variable set $V = \{p_1, p_2, ..., p_n\}$, in that order. Then the RIT^c operator[1] is defined by

$$\mathrm{RIT}^c(\mathcal{F}, V) = \begin{cases} \mathcal{F} & V = \emptyset \\ \\ \neg p_i \wedge \mathrm{RIT}^c(\mathcal{F}[0/p_i], V - \{p_i\}) & \\ \quad\vee & \\ p_i \wedge \mathrm{RIT}^c(\mathcal{F}[1/p_i], V - \{p_i\}) & p_i \in V \\ \quad\vee & \\ \mathrm{RIT}^c((\mathcal{F}[0/p_i] \wedge \mathcal{F}[1/p_i]), V - \{p_i\}) & \end{cases}$$

where p_i is the variable of lowest index in V.

Implicit in this definition is the use of simplification rules **SR1-3**. The RIT^c operator produces a forest; define $ri^{cn}(\mathcal{F}, V) = 1 \vee \mathrm{RIT}(\mathcal{F}, V)$ to produce the n-ary trie.

Theorem 2. If $\mathcal{F}$ is any logical formula with variable set V, then $\mathrm{RIT}^c(\mathcal{F}, V)$ is logically equivalent to $\mathcal{F}$, and each branch of $\mathrm{RIT}^c(\mathcal{F}, V)$ is an implicant of $\mathcal{F}$. □

The theorem below says, in essence, that reduced implicant tries have the desired property that determining whether a clause $C = \{q_1, q_2, ..., q_k\}$ is an implicant can be done by traversing a single branch. Clause C will be an implicant iff the labels of some branch form a prefix of C.

Theorem 3. Let $\mathcal{F}$ be a logical formula with variable set V, and let C be an implicant of $\mathcal{F}$. Then there is a unique branch of $\mathrm{RIT}(\mathcal{F}, V)$ that is a prefix of C, and every branch is a relatively prime implicant. □

Corollary. Every prime implicant of $\mathcal{F}$ is a branch in $\mathrm{RIT}^c(\mathcal{F}, V)$and every subsuming implicant (including any prime implicant) of a branch in $\mathrm{RIT}^c(\mathcal{F}, V)$ contains the literal labeling the leaf of that branch. □

2.3 Intersecting Reduced Implicant Tries

Observe that the RIT^c operator essentially produces a disjunction of three tries. It is therefore natural to represent a reduced implicant trie as a ternary trie. In the ternary representation, the root of the third (right-most) subtrie is labeled 1. One advantage of this representation is that the ith variable appears only at level i. Another is that any subtrie (including the entire trie) is easily expressed as a four-tuple consisting of its root and the three subtries. For example, for a subtrie $\mathcal{T}$ we might write $\langle r, \mathcal{T}^+, \mathcal{T}^-, \mathcal{T}^0 \rangle$,

[1] There is a slight abuse of notation in that $\mathcal{F}$ and V are used for both the initial formula and variable set, respectively, and for the parameters passed by the recursive calls.

where r is the root label of $\mathcal{T}$, and $\mathcal{T}^+$, $\mathcal{T}^-$, and $\mathcal{T}^0$ are the three subtries. Obtaining the ternary representation with the RIT^c operator requires only a minor change, namely conjoining 1 to the third disjunct.

A trivial technical difficulty arises with the ternary representation: The ones along branches interfere with the prefix property of Theorem 3. But this is easily dealt with by interpreting the statement, *A branch B is a prefix of a clause C*, to mean *The clause represented by B* **with ones simplified away** *is a prefix of C*.

The notation $ri^c(\mathcal{F}, V) = 1 \wedge \mathrm{RIT}(\mathcal{F}, V)$ will be used for the ternary reduced implicant trie of $\mathcal{F}$ with variable ordering V. For the remainder of this paper, we will generally assume this ternary representation. As a result, the forest denoted by $\mathrm{RIT}^c(\mathcal{F}, V)$ will contain three tries whose roots are labeled by a variable, its complement, and one.

Theorem 4. Let $\mathcal{F}$ and $\mathcal{G}$ be logically equivalent formulas. Then, with respect to a fixed variable ordering, $\mathrm{RIT}^c(\mathcal{F})$ is isomorphic to $\mathrm{RIT}^c(\mathcal{G})$.

Given any two formulas $\mathcal{F}$ and $\mathcal{G}$, fix an ordering of the union of their variable sets, and let $\mathcal{T}_{\mathcal{F}}$ and $\mathcal{T}_{\mathcal{G}}$ be the corresponding reduced implicant tries. The *intersection* of $\mathcal{T}_{\mathcal{F}}$ and $\mathcal{T}_{\mathcal{G}}$, is defined to be the trie that represents the intersection of the implicant sets with respect to the given variable ordering. By Theorems 2 and 4 and Lemma 1, this is the reduced implicant trie for $\mathcal{F} \wedge \mathcal{G}$.

The INT operator is defined for ri-tries in [11] and produces an ri-trie representing the intersection of the implicate sets of its arguments. The same operator when applied to ri^c-tries produces an ri^c-trie representing the intersection of the implicant sets of its arguments in a completely analogous way – see [12]. Therefore, due to space limitations, the details of the INT operator are omitted.

Theorem 5 provides a formal basis for a definition of the RIT^c operator that produces ri^c-tries using intersection and structure sharing: The third disjunct is replaced by

$$1 \wedge \mathrm{INT}(\mathrm{RIT}^c(\mathcal{F}[0/p_i], V - \{p_i\}), \mathrm{RIT}^c(\mathcal{F}[1/p_i], V - \{p_i\}).$$

Theorem 5. Let $\mathcal{T}_{\mathcal{F}}$ and $\mathcal{T}_{\mathcal{G}}$ be the reduced implicant tries for $\mathcal{F}$ and $\mathcal{G}$ having the same variable ordering. Then $\mathrm{INT}(\mathcal{T}_{\mathcal{F}}, \mathcal{T}_{\mathcal{G}})$ is the reduced implicant trie that is the intersection of $\mathcal{T}_{\mathcal{F}}$ and $\mathcal{T}_{\mathcal{G}}$ and, as a result, is the reduced implicant trie for $\mathcal{F} \wedge \mathcal{G}$ with respect to the given variable ordering. □

3 Reduced Implicate/Implicant Tries

The goal in this section is to build a trie in which both the implicates and implicants of a formula are represented. Let $V = \{p_1, \ldots, p_n\}$ be the variable set of a logical formula $\mathcal{F}$, and consider $\mathcal{T}_{\mathcal{F}} = ri(\mathcal{F}, V)$ and $\mathcal{T}^c_{\mathcal{F}} = ri^c(\mathcal{F}, V)$. Each is a ternary trie in which the ith variable appears at the ith level. Any node in either trie can be uniquely specified by its position.[2] Note that, from the definitions of RIT — see [11] — and RIT^c, if N has label q, then N^c, the node in the corresponding position in the ri^c-trie, has the complementary label, which is equivalent to $\neg q$.

[2] The term *position* is precisely defined in [12] in the obvious way.

Theorem 6. Let $\mathcal{F}_2 \models \mathcal{F}_1$, and let $\mathcal{T}_1$ and $\mathcal{T}_2^c$ be the ri-trie and ri^c-trie, respectively, of $\mathcal{F}_1$ and $\mathcal{F}_2$, under a given variable ordering. Then $\mathcal{T}_1$ and $\mathcal{T}_2^c$ have no leaf positions in common. Also, if P is a position in both tries with no common extension, then the node at P in one trie has a first child but not a second, and the corresponding node in the other trie has a second child but not a first; neither has a third child. □

Theorem 6 makes it easy to build a trie whose set of branches is the union of the branch set of $\mathcal{T}$ and the branch set of $\mathcal{T}^c$. In a simple recursive process, the tries can be traversed in parallel. A trie can be constructed that is isomorphic to the common parts of the two tries. When a node is encountered where the tries diverge, its children can be assigned the first child from one trie and the second child from the other. Note that the third child will be empty since it is empty in both tries being traversed.

Let $\mathcal{T}_\mathcal{F} = \langle 0,\ \mathcal{T}_\mathcal{F}^+, \mathcal{T}_\mathcal{F}^-,\ \mathcal{T}_\mathcal{F}^0 \rangle$ be the ri-trie for $\mathcal{F}$, and let $\mathcal{T}_\mathcal{G}^c = \langle 1,\ {}^c\mathcal{T}_\mathcal{G}^+, {}^c\mathcal{T}_\mathcal{G}^-,\ {}^c\mathcal{T}_\mathcal{G}^0 \rangle$ be the ri^c-trie for $\mathcal{G}$ under a fixed variable ordering, where $\mathcal{G} \models \mathcal{F}$. The *merge* of $\mathcal{T}_\mathcal{F}$ and $\mathcal{T}_\mathcal{G}^c$ is defined to be the trie whose branches appear in either $\mathcal{T}_\mathcal{F}$ or $\mathcal{T}_\mathcal{G}^c$. To distinguish the branches in the merge according to their trie of origin, the leaf nodes will be marked as type-d or type-c, depending on whether they come from the ri-trie or from the ri^c-trie, respectively. Branches leading to type-d (type-c) leaves are called type-d (type-c) branches. Since nodes in positions common to both tries have complementary labels, we will use a toggling symbol $\propto$ to indicate the label that a node at a given position would have if it were at that position in $\mathcal{T}_\mathcal{F}$ or in $\mathcal{T}_\mathcal{G}^c$. When considering implicates, $\propto$ is interpreted as identity, but for implicants it is interpreted as complement.

The tries $\mathcal{T}_\mathcal{F}$ and $\mathcal{T}_\mathcal{G}^c$ can be interpreted as logical formulas using conjunction and disjunction. Their merge, in effect, by Theorem 6, represents both logical formulas: The type-d branches represent $\mathcal{T}_\mathcal{F}$, and the type-c branches represent $\mathcal{T}_\mathcal{G}^c$.

Given an ri-trie $\mathcal{T}$ and an ri^c-trie $\mathcal{T}^c$, $d(\mathcal{T})$ denotes the trie produced by marking the leaves of $\mathcal{T}$ as type-d and prepending the symbol $\propto$ to all its labels, and $c(\mathcal{T}^c)$ denotes the trie produced by marking the leaves of $\mathcal{T}^c$ as type-c, complementing all nodes, and prepending $\propto$ to its labels.

The merge of $\mathcal{T}_\mathcal{F}$ and $\mathcal{T}_\mathcal{G}^c$ is defined as follows.

$$\mathrm{MERGE}(\mathcal{T}_\mathcal{F},\ \mathcal{T}_\mathcal{G}^c) = \begin{cases} \emptyset & \mathcal{T}_\mathcal{F} = \emptyset \text{ and } \mathcal{T}_\mathcal{G}^c = \emptyset \\ d(\mathcal{T}_\mathcal{F}) & \mathcal{T}_\mathcal{G}^c = \emptyset \\ c(\mathcal{T}_\mathcal{G}^c) & \mathcal{T}_\mathcal{F} = \emptyset \\ \langle \propto r,\ \mathrm{MERGE}(\mathcal{T}_\mathcal{F}^+,\ {}^c\mathcal{T}_\mathcal{G}^+), & \\ \qquad \mathrm{MERGE}(\mathcal{T}_\mathcal{F}^-,\ {}^c\mathcal{T}_\mathcal{G}^-), & \text{otherwise} \\ \qquad \mathrm{MERGE}(\mathcal{T}_\mathcal{F}^0,\ {}^c\mathcal{T}_\mathcal{G}^0)\ \rangle & \end{cases}$$

Note that in the recursive call to MERGE (fourth case), the root labels of the two arguments are complementary. The root r of the constructed trie is defined to be the label of the ri-trie but with $\propto$ prepended. This yields exactly the correct label when these nodes are viewed for implicates, dually for implicants. For uniformity, $\propto$ is prepended to all labels in the base cases as well. For the second case, the construction requires only that the leaves be marked as type-d and that $\propto$ be added to the labels, which is precisely what $d(\mathcal{T}_\mathcal{F})$ does. For the third case, $\mathcal{T}_\mathcal{G}^c$ is a correctly labeled ri^c-trie. When searching for implicants, $\propto$ means complement, so, as a result, for this case the labels must be complemented and then prepended with $\propto$. In essence, evaluating $c(\mathcal{T}_\mathcal{G}^c)$ complements the correct labels twice. This proves Lemma 2 below; for details, see [12].

Lemma 2. If $\mathcal{G} \models \mathcal{F}$, the d-branches and c-branches of $\mathrm{MERGE}(\mathcal{T}_\mathcal{F}, \mathcal{T}_\mathcal{G}^c)$ are disjoint and are exactly the branches of $\mathrm{RIT}(\mathcal{T}_\mathcal{F})$ and of $\mathrm{RIT}^c(\mathcal{T}_\mathcal{G})$, respectively. The sub-trie consisting of the type-d branches is identical to $\mathrm{RIT}(\mathcal{T}_\mathcal{F})$ if $\propto$ and leaf marks are removed, and the subtrie consisting of the type-c branches is identical to $\mathrm{RIT}^c(\mathcal{T}_\mathcal{G})$ if labels are complemented and $\propto$ and leaf marks are removed. □

As a result, the rii-trie for $\mathcal{F}$ is defined to be $\mathrm{MERGE}(\mathcal{T}_\mathcal{F}, \mathcal{T}_\mathcal{F}^c)$. The goal, however, is to define the RIIT operator to compute this trie without first computing the ri-trie and the reduced implicant trie.

3.1 Intersecting rii-Tries

If an rii-trie is viewed as the MERGE of an ri-trie and a ri^c-trie, then the notion of intersection need not be addressed explicitly. Intersections of implicates and implicants have already been computed as necessary in forming the tries to be merged. However, it is necessary to address intersection directly in order to compute rii-tries directly.

Given any two formulas $\mathcal{F}$ and $\mathcal{G}$, fix an ordering of the union V of their variable sets, let $\mathcal{T}_\mathcal{F}$ and $\mathcal{T}_\mathcal{G}$ be the corresponding ri-tries, and let $\mathcal{T}_\mathcal{F}^c$ and $\mathcal{T}_\mathcal{G}^c$ be the corresponding ri^c-tries. Denote by $\mathcal{T}_\mathcal{F}^{ii}$ and $\mathcal{T}_\mathcal{G}^{ii}$ the rii-tries for $\mathcal{F}$ and for $\mathcal{G}$. The *intersection* of $\mathcal{T}_\mathcal{F}^{ii}$ and $\mathcal{T}_\mathcal{G}^{ii}$ is defined to be $\mathrm{MERGE}(\mathrm{RIT}(\mathcal{F} \vee \mathcal{G}, V),\ \mathrm{RIT}^c(\mathcal{F} \wedge \mathcal{G}, V))$, which, by Lemma 1 and its dual (see [11]), and Theorem 6, is

$$\mathrm{MERGE}(\mathrm{INT}(\mathcal{T}_\mathcal{F}, \mathcal{T}_\mathcal{G}),\ \mathrm{INT}(\mathcal{T}_\mathcal{F}^c, \mathcal{T}_\mathcal{G}^c)).$$

This trie, while not necessarily the rii-trie of any formula, is the trie whose d-branches correspond precisely to the branches of $\mathrm{RIT}(\mathcal{F} \vee \mathcal{G}, V)$, and whose c-branches correspond precisely to the branches of $\mathrm{RIT}^c(\mathcal{F} \wedge \mathcal{G}, V)$. Note that this merge of the ri-trie and ri^c-tries of, respectively, the different formulas $(\mathcal{F} \vee \mathcal{G})$ and $(\mathcal{F} \wedge \mathcal{G})$ is well-defined because the latter entails the former – see [12]. Our goal here is to define IINT, the rii-trie intersection operator, directly without the use of MERGE, RIT, or RIT^c.

The IINT operator is a recursion that traverses its trie arguments. The base cases that end the recursion involve either the empty trie or a leaf node. If one argument is empty, then so is the intersection. When one argument is a type-d leaf, then the intersection is all branches in the other argument that end in type-d leaves. Dually, when one argument is a type-c leaf, then the intersection is all branches in the other argument that end in type-c leaves.

Let $\mathcal{T}_\mathcal{F}$ and $\mathcal{T}_\mathcal{G}$ be the rii-tries for $\mathcal{F}$ and $\mathcal{G}$, respectively. [3] Let $\langle r, \mathcal{T}_\mathcal{F}^+, \mathcal{T}_\mathcal{F}^-, \mathcal{T}_\mathcal{F}^0 \rangle$ and $\langle r, \mathcal{T}_\mathcal{G}^+, \mathcal{T}_\mathcal{G}^-, \mathcal{T}_\mathcal{G}^0 \rangle$ be the 4-tuples denoting $\mathcal{T}_\mathcal{F}$ and $\mathcal{T}_\mathcal{G}$, respectively. Let $\mathcal{T}^\gamma$ be the sub-trie of $\mathcal{T}$ whose branches end in γ-*leaves*, $\gamma = d, c$, and let $\mathcal{T}_\mathcal{F} \bigoplus_{ii} \mathcal{T}_\mathcal{G}$ be the four-tuple $\langle r,\ \mathrm{IINT}(\mathcal{T}_\mathcal{F}^+, \mathcal{T}_\mathcal{G}^+),\ \mathrm{IINT}(\mathcal{T}_\mathcal{F}^-, \mathcal{T}_\mathcal{G}^-),\ \mathrm{IINT}(\mathcal{T}_\mathcal{F}^0, \mathcal{T}_\mathcal{G}^0) \rangle$. Then

[3] There is a slight abuse of notation in that $\mathcal{F}_\mathcal{F}$ and $\mathcal{T}_\mathcal{G}$ are used for the initial formulas here and for the parameters passed by the recursive calls in the definition of IINT.

$$\text{IINT}(\mathcal{T}_{\mathcal{F}}, \mathcal{T}_{\mathcal{G}}) = \begin{cases} \emptyset & \mathcal{T}_{\mathcal{F}} = \emptyset \text{ or } \mathcal{T}_{\mathcal{G}} = \emptyset \\ \mathcal{T}_{\mathcal{F}}^{\gamma} & \gamma\text{-}leaf(\mathcal{T}_{\mathcal{G}}) \\ \mathcal{T}_{\mathcal{G}}^{\gamma} & \gamma\text{-}leaf(\mathcal{T}_{\mathcal{F}}) \\ \emptyset & leaf(\mathcal{T}_{\mathcal{F}} \bigoplus_{ii} \mathcal{T}_{\mathcal{G}}) \\ \mathcal{T}_{\mathcal{F}} \bigoplus_{ii} \mathcal{T}_{\mathcal{G}} & otherwise \end{cases}$$

Lemma 3. Let $\mathcal{T}_{\mathcal{F}}$ and $\mathcal{T}_{\mathcal{G}}$ be the rii-tries of $\mathcal{F}$ and of $\mathcal{G}$ with the same variable ordering. Let $C_{\mathcal{F}}$ be a non-empty prefix of $C_{\mathcal{G}}$, where $C_{\mathcal{F}}$ is a d-branch (c-branch) in $\mathcal{T}_{\mathcal{F}}$ and $C_{\mathcal{G}}$ is a d-branch (c-branch) in $\mathcal{T}_{\mathcal{G}}$. Then $C_{\mathcal{G}}$ is a d-branch (c-branch) in $\text{IINT}(\mathcal{T}_{\mathcal{F}}, \mathcal{T}_{\mathcal{G}})$. □

Essentially, the converse of Lemma 3 also holds.

Lemma 4. Let $\mathcal{T}_{\mathcal{F}}$ and $\mathcal{T}_{\mathcal{G}}$ be rii-tries, and let C be a branch in $\text{IINT}(\mathcal{T}_{\mathcal{F}}, \mathcal{T}_{\mathcal{G}})$. Then C is a branch in one of $\mathcal{T}_{\mathcal{F}}$ or $\mathcal{T}_{\mathcal{G}}$, and a prefix of C is a branch in the other. □

Theorem 7. Let $\mathcal{T}_{\mathcal{F}}$ and $\mathcal{T}_{\mathcal{G}}$ be the rii-tries for $\mathcal{F}$ and $\mathcal{G}$ with the same variable ordering. Then $\text{IINT}(\mathcal{T}_{\mathcal{F}}, \mathcal{T}_{\mathcal{G}})$ is the trie that represents the intersection of $\mathcal{T}_{\mathcal{F}}$ and $\mathcal{T}_{\mathcal{G}}$ with respect to the given variable ordering. □

Theorem 7 guarantees that when applied to rii-tries, IINT produces precisely the trie whose d-branches represent (uniquely) the implicates (CNF clauses) represented in both sub-tries and whose c-branches represent (uniquely) the implicants (DNF clauses) represented in both sub-tries. They coexist peacefully in one trie because branches may end in one type or the other, but not both.

3.2 The RIIT Operator

If $\mathcal{T}$ is the rii-trie for a logical formula $\mathcal{F}$, and if L is an ordered set of literals, then, since L can be interpreted as either a disjunctive or a conjunctive clause, one can ask, respectively, whether L is an implicate or an implicant of $\mathcal{F}$. The rii-trie will have the property that in either case, the answer is yes if and only if a unique prefix of L is a branch in $\mathcal{T}$. The notation *d-clause* and *c-clause* will be used to indicate, respectively, disjunctive and conjunctive clauses. Similarly, *d-search* and *c-search* will be used to indicate that the trie is being searched for, respectively, d-clauses or c-clauses. The choice of interpretation will determine the connectives along branches and those between branches, node labels, and the truth constants labeling interior nodes. There is a straightforward duality of the logical connectives in these tries. For ri-tries, branches are disjunctions that are conjoined to each other; for reduced implicant tries, branches are conjunctions that are disjoined. In the rii-trie, the connectives must be interpreted; i.e., whether each connective is a disjunction or a conjunction depends upon whether implicates or implicants are being sought.

The root of an ri-trie of a non-contradictory, non-tautological formula is 0; for the ri^c-trie its 1. With the ternary structure, the third sub-trie is rooted at 0 for an ri-trie; 1 for an ri^c-trie. In ri-tries, interior zeros are disjoined to the conjunction of their children; in ri^c-tries, interior ones are conjoined to the disjunction of their children. In each case, the ternary structure is maintained by *not* applying simplification rules to such interior nodes. This convention is the same in both types of trie because the simplification rules, if applied, would have the same effect on interior constants. In particular, as with connectives in an rii-trie, the value of constants labeling interior nodes depends on the search (d or c), but their behavior with respect to the structure of the trie is independent of the search.

The RIIT operator is defined using $\otimes$ and $\odot$ to represent, respectively, the connective between sub-tries and the connective along a branch. For example, with a d-search, $\otimes = \wedge$ and $\odot = \vee$. It will also be convenient to regard the label of an interior node as either unnegated or negated, depending on the search. The symbol $\propto$ will be used as a unary operator representing the identity — i.e., unnegated — for a d-search, and negation for a c-search. In the ternary structure, the third child of a node is always labeled with the constant $\propto 0$, producing 0 for an implicate search and 1 for implicants.

The definition of the RIIT operator requires the IINT operator and the simplification rules. The latter are more complicated than they are for ri-tries, because they must allow for the dual connectives between and along branches. The definitions of the required simplification rules follow the definition of RIIT.

$$\mathrm{RIIT}(\mathcal{F}, V) = \begin{cases} \mathcal{F} & V = \emptyset \\ (\propto p_i \odot \mathcal{B}_1) \otimes (\propto \neg p_i \odot \mathcal{B}_2) \otimes (\propto 0 \odot \mathcal{B}_3) & p_i \in V \end{cases}$$

where p_i is the variable of lowest index in V, $\mathcal{B}_1 = \mathrm{RIIT}(\mathcal{F}[0/p_i], V - \{p_i\}), \mathcal{B}_2 = \mathrm{RIIT}(\mathcal{F}[1/p_i], V - \{p_i\})$, and $\mathcal{B}_3 = \mathrm{IINT}(\mathcal{B}_1, \mathcal{B}_2)$. Notice that the truth constants substituted for variables in $\mathcal{F}$ are not preceded by the symbol $\propto$. The RIIT operator produces a forest and does not include the root labeled $\propto 0$, which is similar to RIT and RIT^c. The solution is also similar: define $rii(\mathcal{F}, V) = \propto 0 \otimes \mathrm{RIIT}(\mathcal{F}, V)$.

Additional simplification rules are required; the asterisks indicate the presence of the interpretation-dependent operators $\otimes$ and $\odot$.

$$\begin{array}{llll}
\mathbf{SR^*4}. & \mathcal{F} \longrightarrow \mathcal{F}[\propto q^d/(\propto q \odot 0)] & \mathcal{F} \longrightarrow \mathcal{F}[\propto q^c/(\propto q \odot 1)] \\
\mathbf{SR^*5}. & \mathcal{F} \longrightarrow \mathcal{F}[0/(\propto p^d) \otimes (\propto \neg p^d)] & \mathcal{F} \longrightarrow \mathcal{F}[1/(\propto p^c) \otimes (\propto \neg p^c)] \\
\mathbf{SR^*6}. & \mathcal{F} \longrightarrow \mathcal{F}[\mathcal{G}/\mathcal{G} \otimes (\propto 0 \odot \propto 1)] &
\end{array}$$

Rules **SR1 – SR3** are unchanged but are now applicable only in simplifying $\mathcal{F}$ after substituting truth values for variables and never along or between the trie branches — the trie itself has no occurrences of $\vee$ or $\wedge$. As a result, **SR1 – SR3** cannot be extended to $\odot$ and $\otimes$. The result of such analogous operations is unknown unless the interpretation of the operator is known. The result is determined only after the choice of implicate or implicant search has been made: In one case the proposition (i.e., sub-trie) vanishes, and in the other the constant vanishes. To allow for both cases, these simplifications cannot be performed on the rii-trie but instead must be accounted for during each search.

Observe that nodes labeled with 0 or 1 can occur only as leaves[4] without siblings and can be simplified away with **SR*4**. A 0-leaf occurs exactly when the branch represents an implicate, and a 1-leaf occurs with an implicant. This distinction must be maintained when the leaves are simplified away, so **SR*4** makes the parent into a *type-d* or *type-c* leaf by giving the label the appropriate superscript.

Constants also arise when the IINT operation produces the empty trie. In this case, the third sub-trie, whose root is labeled $\propto 0$, can simply be removed. To see why, note that IINT produces a trie whose branches represent both the intersection of implicate sets and the intersection of implicant sets. Consider the implicate interpretation. An empty intersection of implicates is an empty conjunction and hence is equal to the constant 1; $\propto 0 \odot \propto 1 = 0 \vee 1 = 1$, and $1 \wedge \mathcal{G} = \mathcal{G}$. Then **SR*6** applies, and there is no third branch. Similarly, in the implicant interpretation, an empty intersection of implicants is an empty disjunction and hence is equal to 0; $\propto 0 \odot \propto 1 = 1 \wedge 0 = 0$, $0 \vee \mathcal{G} = \mathcal{G}$, and again **SR*6** applies.

Rule **SR*5** reduces complementary sibling type-d leaves to 0 and complementary type-c leaves to 1. (It is easy to verify that this is correct under both implicate and implicant interpretations.) In essence, **SR*5** eliminates sections of the trie that are implicit in one interpretation and ignored in the other. Note that complementary sibling leaves of opposite type cannot be reduced: One, but not the other, is treated as if it were not there once the type of search has been specified.

Lemma 2 can be reconsidered in the context of the simplification rules. If attention is restricted to implicates — i.e., if $\odot$, $\otimes$, and $\propto$ are specified as $\vee$, $\wedge$, and the identity — then the c-branches can be removed from the rii-trie for $\mathcal{F}$; the result is (essentially) the ri-trie for $\mathcal{F}$. In essence, the second half of **SR*4** is replaced by the second half of **SR2** since branches are disjunctions. Type-c leaves represent their label disjoined with 1 and are simplified away up to the first ancestor having another child. Similarly, if the d-branches are removed, the result is the reduced implicant trie for $\mathcal{F}$.

Theorem 8 below states that the object built by the RIIT operator is exactly the rii-trie of its first argument.

Theorem 8. If formula $\mathcal{F}$ has variable set V, then

$$\mathrm{RIIT}(\mathcal{F}, V) = \mathrm{MERGE}(\mathrm{RIT}(\mathcal{F}, V), \mathrm{RIT}^c(\mathcal{F}, V)). \qquad \square$$

Searching an rii-trie for implicates and implicates is straightforward. To determine whether a given d-clause is an implicate, traverse the branch consisting of the literals in the clause. If the search fails or if a c-leaf is encountered, the clause is not an implicate. If the search ends in a d-leaf, then the path to the leaf must be a prefix of the clause, and the clause is an implicate. The search for implicants is the same, except that a yes answer requires termination at a c-leaf. A simple example is the rii-trie for $((p \leftrightarrow q) \vee r)$, displayed below.

[4] Internal nodes labeled with constants are labeled $\propto 0$.

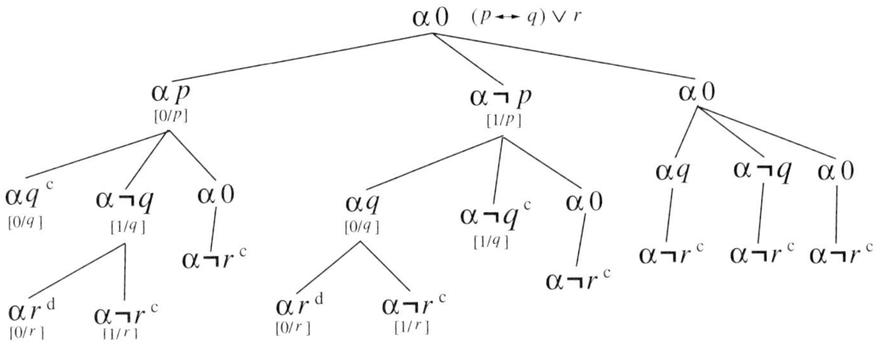

The assignments of truth values to variables determined by the definition of RIIT are indicated below the corresponding nodes. Third sub-tries of interior nodes are not associated with particular truth assignments, because they are computed as the intersection (IINT) of the first two. In this example, the implicants $(\neg p \wedge r)$, $(p \wedge \neg r)$, $(\neg q \wedge r)$, $(q \wedge r)$, and (r) are all computed as the result of the IINT operator. This formula has only two relatively prime implicates: $(p \vee \neg q \vee r)$ and $(\neg p \vee q \vee r)$. They appear as branches ending in type-d leaves in the rii-trie. The nine relatively prime implicants appear as the branches ending in type-c leaves. For a sample search, consider determining whether the clause $(q \vee \neg r)$ is an implicate. The $\propto$ operator is interpreted as the identity, and the leftmost branch of the third child of the root would be traced. The leaf is $\neg r^c$, and so the answer is no.

References

1. Bryant, R.E.: Symbolic Boolean manipulation with ordered binary decision diagrams. ACM Comput. Surv. 24(3), 293–318 (1992)
2. Cadoli, M., Donini, F.M.: A survey on knowledge compilation. AI Commun. 10, 137–150 (1997)
3. Darwiche, A.: Compiling devices: A structure-based approach. In: Proc. Int'l Conf. on Principles of Knowledge Representation and Reasoning (KR 1998), pp. 156–166. Morgan-Kaufmann, San Francisco (1998)
4. Darwiche, A.: Decomposable negation normal form. J. ACM 48(4), 608–647 (2001)
5. Forbus, K.D., de Kleer, J.: Building Problem Solvers. MIT Press, Cambridge, Mass (1993)
6. Hähnle, R., Murray, N.V., Rosenthal, E.: Normal Forms for Knowledge Compilation. In: Ras, Z. (ed.) Proceedings of the International Symposium on Methodologies for Intelligent Systems, ISMIS 2005. LNCS, Springer, Heidelberg (to appear, 2005)
7. Hai, L., Jigui, S.: Knowledge compilation using the extension rule. J. Automated Reasoning 32(2), 93–102 (2004)
8. Kautz, H., Selman, B.: A general framework for knowledge compilation. In: Boley, H., Richter, M.M. (eds.) PDK 1991. LNCS, vol. 567, Springer, Heidelberg (1991)
9. Marquis, P.: Knowledge compilation using theory prime implicates. In: Proc. Int'l Joint Conf. on Artificial Intelligence (IJCAI), pp. 837–843. Morgan-Kaufmann, San Mateo, Calif (1995)
10. Murray, N.V., Rosenthal, E.: Efficient query processing with compiled knowledge bases. In: Beckert, B. (ed.) TABLEAUX 2005. LNCS (LNAI), vol. 3702, pp. 231–244. Springer, Heidelberg (2005)

11. Murray, N.V., Rosenthal, E.: Efficient query processing with reduced implicate tries. Journal of Automated Reasoning 38(1-3) (2207), 155–172
12. Murray, N.V., Rosenthal, E.: Linear response time for implicate and implicant queries. Technical Report SUNYA-CS-08-01, Department of Computer Science, University at Albany - SUNY (January 2008), Download in pdf at: `http://www.cs.albany.edu/~nvm/ritries/papers.html`
13. Ramesh, A., Murray, N.V.: An application of non-clausal deduction in diagnosis. Expert Systems with Applications 12(1), 119–126 (1997)
14. Reiter, R., de Kleer, J.: Foundations of assumption-based truth maintenance systems: preliminary report. In: Proceedings of the 6th National Conference on Artificial Intelligence, Seattle, WA, July 12-17, pp. 183–188 (1987)
15. Selman, B., Kautz, H.: Knowledge compilation and theory approximation. J. ACM 43(2), 193–224 (1996)

Golden Ratio Annealing for Satisfiability Problems Using Dynamically Cooling Schemes

Juan Frausto-Solis[1] and Felix Martinez-Rios[2]

[1] Tecnológico de Monterrey, Campus Cuernavaca, Autopista del Sol Km 104, Colonia Real del Puente, 62790, Xochitepec, Morelos, México
juan.frausto@itesm.mx
[2] Universidad Panamericana, Campus Ciudad de México, Augusto Rodín 498, Col. Insurgentes Mixcoac, 03920, Distrito Federal, México
fmartin@up.edu.mx

Abstract. Satisfiability (SAT) Problem is an *NP-Complete* problem which means no deterministic algorithm is able to solve it in a polynomial time. Simulated Annealing (SA) can find very good solutions of SAT instances if its control parameters are correctly tuned. SA can be tuned experimentally or by using a Markov approach; the latter has been shown to be the most efficient one. Moreover Golden Ratio (GR) is an unconventional technique used to solve many problems. In this paper a new algorithm named Golden Ratio for Simulated Annealing (GRSA) is presented; it is tuned for three different cooling schemes. GRSA uses GR to dynamically decrease the SA temperature and a Markov Model to tune its parameters. Two SA tuned versions are compared in this paper: GRSA and a classical SA. Experimentation shows that the former is much more efficient than the latter.

Keywords: Simulated Annealing, Golden Ratio, Satisfiability Problem.

1 Introduction

One of the principal challenges of computer science is to solve *NP-Complete* problems [1]. SAT was the first problem classified as *NP–Complete* [2] and it is fundamental to the analysis of computational complexity theory [3]. Satisfiability is widely studied in different areas such as: planning, circuit testing, temporal reasoning, complexity theory, scheduling, cryptology, constraint satisfaction problems, computer network design, and many others [4]. Besides, any instance of an *NP* problem can be transformed to a SAT instance by using a polynomial transformation [2, 5]. Therefore, if SAT can be solved efficiently with a particular algorithm, then a similar result could probably be obtained for other *NP–Complete* problems using the same algorithm. Polynomial and exponential time algorithms are frequently referred to as "good algorithms" and "bad algorithms", respectively [6]; however this classification is not always true, as can be noticed in the following example: Cook [2] gave us the example of a particular polynomial algorithm which is not a good one because it requires n^{100} steps (it is impractical even for values around *1000*). Another example is the simplex method which is an exponential algorithm; however, it is known that for many practical problems it is the best one [8].

A. An et al. (Eds.): ISMIS 2008, LNAI 4994, pp. 215–224, 2008.

Golden Ratio (GR), a nature inspired search technique, has been shown to be very efficient in many areas [9, 10]. It is in fact an old idea, for instance it is supposed that Leonardo Da Vinci and Leonardo Da Pisa (Fibonacci) used GR to solve some problems [11]. GR is now used as a searching strategy, dividing the search space of a problem into two subspaces; the way of dividing it follows a rule called "Golden Ratio" (GR), which works as follows:

GR (Search Space S)
1) If the stop criterion is fulfilled then STOP GR;
2) Otherwise divide S using GR.

The stop criterion is fulfilled when the optimal (or the item searched) is found or when a certain number g of GR's is applied. The longer subdivision obtained with GR is named "golden section" (GS). GS has a particular proportion of the total space; in ancient times one supposed that esthetic issues and vital features of animals and plants followed this proportion [11]. Let us suppose a line segment which is divided by GR in two segments a (the longest) and b. GR establishes that, the total length $a+b$ is the longest segment a as a is to the shortest segment b; *GR* also establishes that the golden section φ is determined algebraically by solving:

$$\frac{a+b}{a}=\frac{a}{b}=\varphi \tag{1}$$

Therefore,

$$\varphi=\frac{1+\sqrt{5}}{2} \tag{2}$$

This relation is an important parameter in the GR search strategy [11]. In the paper GR is introduced into an optimization method.

On the other hand, let us briefly review that for *NP-Complete* problems there are two main types of methods [12]: Complete and incomplete. Complete methods always discover if an instance is satisfiable or not, while incomplete methods usually use random algorithms, and not always are able to find the solution of a given SAT instance [13]. Moreover, most of the time, complete algorithms use enormous periods of execution time, while incomplete methods can achieve a good approximation to the optimal solution (or probably the optimal) in a smaller execution time.

Because SAT is *NP-Complete*, random methods are probably the most efficient for it. Since the publication of the random method known as Simulated Annealing (SA) [14, 15, 16], it is known that it is able to achieve very good solutions (the optimal or very close to the optimal) only if its cooling scheme's parameters are correctly tuned. One of the most efficient tuning approaches is to use Markov Models (MM) [17]. Tuning SA algorithms with MM (TMSA algorithms or TMSA in short) are among the most efficient SA algorithms with very suitable solution quality [18].

There are several cooling schemes for SA, but all of them have some problems. Let us explain them by using the most common cooling scheme named as Geometric Cooling Scheme which uses a geometric function, that is $c_{k+1}=\alpha c_k$, where $0<\alpha<1$ and

$k \in N$ identifies the temperature number used in SA. Then the next tuning strategies are possible:

1) To set the initial ($k=0$) and final ($k=k_{max}$) temperatures (c_i and c_f respectively) with an analytical approach, merely by using a Markov approach [17]; but the c_i and c_f values founded are usually extremely high and extremely low respectively. Therefore, the execution time for hard SAT instances can be extremely big.
2) To set experimentally c_i and c_f; usually a low range of temperatures is chosen and a small execution time is required; however the solutions quality may be very poor.

Another set of problems in SA is related with the way of decreasing the temperature (i.e. the α value):

1) If the temperature is slowly reduced (i.e. big α values for geometric cooling scheme Equation 6), then very good solutions are achieved [17], but the execution time can be very high as well.
2) If the temperature is reduced in big steps (i.e. small α values for geometric cooling scheme Equation 6), (this is a method known as Simulated Quenching (SQ)) [18], then the SQ execution time can be reduced considerably, but the solution quality may be very poor [19].

In order to obtain a good solution with a short execution time, an algorithm named Golden Ratio Simulation Annealing (GRSA) is proposed in this paper. It is a hybrid algorithm among the later approaches. Briefly GRSA is devised as follows:

1) As a classical SA, with initial and final temperatures analytically tuned using Markov Models.
2) As SQ, decreasing its temperature very fast only at extremely high or very high temperatures.

In GRSA, the temperatures are obtained as in a classical SA (i.e. by an analytical model), but the temperature range is divided by using GR again and again until the stop criterion is reached. The g number of GR rules is given as an input data, and the performance obtained with different g values were experimentally measured. In the paper it is shown that the GRSA performance can be considerably increased even with a small g value. GRSA was also applied to solve SAT instances with three cooling schemes; then its results are compared versus those obtained with a classical SA algorithm (TMSA algorithm was used), showing that GRSA is more efficient, while the solution quality obtained is very similar.

2 Tuning Cooling Scheme

In SA the maximum Markov chain L_{max} occurs at the final temperature, and $L_{max}=C\,|V_{Si}|$, where $|V_{Si}|$ is the neighborhood size, $C=-Ln(P_R(S_i))$ and $P_R(S_i)$ is the rejection probability for a solution S_i. C ranges from 1 to 4.6 guaranteeing a good exploration level of the neighborhood at the final temperature. Different exploration levels P_R can be applied, for instance if *99%* of the solution space is going to be explored, then *C=4.6.* Let ΔZ_{Vmax} and ΔZ_{Vmin} be the maximum and minimum cost deteriorations of the objective function through the neighborhood set *V*. Then the initial and final temperatures c_i and c_f are [18]:

$$c_i = \frac{-\Delta Z_{V\max}}{Ln(P_A(\Delta Z_{V\max}))} \tag{4}$$

$$c_f = \frac{-\Delta Z_{V\min}}{Ln(P_A(\Delta Z_{V\min}))} \tag{5}$$

```
Initialization (c=c_i, c_f, S_i, ß)
Repeat
   Repeat L times
       Generate S_j
       if E(S_j)<E(S_i) then
              S_i=S_j
       else
          if random<Boltzman then
                   S_i=S_j
   End Repeat
   new c, L
Until (c <c_f)
```

Fig. 1. TMSA (Tuning Markov Simulated Annealing) pseudocode

```
Initialization (c=c_i, c_f, S_i, ß)
Repeat GR applied times
      Calculate c_fp with GR of c_f
      Repeat
         Repeat L times
            Generate S_j
            if E(S_j)<E(S_i) then
                  S_i=S_j
            else
               if random<Boltzman then
                        S_i=S_j
         End Repeat
         new c, L
      Until (c < c_fp)
End Repeat
```

Fig. 2. GRSA (Golden Ratio for Simulated Annealing) pseudocode

At the beginning of the annealing process, the probability to accept any proposed new solution $P_A(\Delta Z_{Vmax})$ is high, and close to one in order to accept any solution at the beginning of the process. The acceptance probability of new solutions is reduced as the temperature is decreased and at very low temperature $P_A(\Delta Z_{Vmin})$ must be near to zero in order to reject most of the bad solutions [18]. In Fig. 1, the SA algorithm with this Markov tuning method (or TMSA) is presented. Other parameters depend on the last temperature parameters and were derived in [18].

As can be noticed in Fig. 2, GRSA uses a temperature (or external) loop (lines *2-12*) which changes the temperature parameter following the GR rule (i.e. faster at the beginning and lower at the ending).

GRSA calculates a partial final temperature (c_{fp}) using the GR rule with the initial temperature (c_i) but in the next cycle this c_{fp} becomes the new initial temperature (c_i). In line *13*, three popular cooling schemes are implemented:

$$\text{Geometric: } T_{k+1} = \alpha T_k \tag{6}$$

$$\text{Exponential: } T_{k+1} = \exp(-\alpha) T_k \tag{7}$$

$$\text{Logarithmic: } T_{k+1} = T_k / \ln(\alpha) \tag{8}$$

Where the α parameter is determined experimentally as is explained in section 5.

3 Experimentation Results

As mentioned before a previous work has implemented dynamically the length of the Markov Chain in SA for SAT problem [18]. This implementation is referred here as TMSA. Also a temperature parameters model was tuned analytically for TMSA in this reference, showing that TMSA execution time is lower than a classical SA with constant Markov chains, while the quality solution remains similar. GRSA uses the same tuning method as TMSA but the cooling scheme is changed by modifying the α values according to the golden ratio rule. Therefore, in order to analyze the impact of this modification it is necessary to compare GRSA with similar implementation (i.e. TMSA) where the only variation is to maintain constant the α value.

GRSA was tested with the SAT instances shown in Table 1; these instances have different σ relation of clauses/variables [12, 20] and were taken from SATLIB [21] or generated with Hories algorithm [22]. The performance of GRSA was based on the time of execution and the solution quality. The quality solution was measured for any instance at the end of the execution of any instance as the percentage of "true" clauses with respect to their total. Both algorithms were implemented in a Dell Intel Core Duo with *2 GB* of RAM and Pentium *4* processor running at *2.40* GHz. Each instance was executed *40* times and the average execution time and the average solution quality were obtained. The quality of the solution was established by the next expression:

$$Q = \frac{clauses \quad true}{total \quad clauses} \times 100 \tag{9}$$

Table 1. SAT instances tested

Sat Instance	σ	Sat Instance	σ
aim-50-1_6-yes1-3	1.60	par8-5-c	3.97
aim-50-1_6-no-2	1.60	G2_V100_C400_P4_I1	4.00
aim-100-1_6-yes1-1	1.60	hole8	4.13
aim-200-1_6-no-1	1.60	uuf225-045	4.27
aim-50-2_0-no-4	2.00	RTI_k3_n100_m429_150	4.29
aim-50-2_0-yes1-1	2.00	uuf100-0789	4.30
aim-50-2_0-no-3	2.00	uf175-023	4.30
G2_V100_C200_P2_I1	2.00	uf50-01	4.36
dubois21	2.67	uuf50-01	4.36
dubois26	2.67	ii8a2	4.44
dubois27	2.67	G2_V50_C250_P5_I1	5.00
BMS_k3_n100_m429_161	2.83	hole10	5.10
G2_V50_C150_P3_I1	3.00	ii32e1	5.34
G2_V300_C900_P3_I1	3.00	anomaly	5.44
BMS_k3_n100_m429_368	3.08	aim-50-6_0-yes1-1	6.00
hole6	3.17	G2_V50_C300_P6_I1	6.00
par8-1	3.28	jnh201	8.00
aim-50-3_4-yes1-2	3.40	jnh215	8.00
hole7	3.64	medium	8.22
par8-3-c	3.97	jnh301	9.00

For TMSA algorithm, the best α values were obtained experimentally for each cooling scheme (Equations 6, 7 and 8) as follows into the next intervals: a) Geometric: $0.7 \leq \alpha \leq 0.99$, b) Exponential: $0.01 \leq \alpha \leq 0.4$ and c) Logarithmic: $2.745 \leq \alpha \leq 4.2$.

The best performance with TMSA was obtained as follows: a) Geometric, $\alpha=0.99$, b) Exponential: $\alpha=0.01$ and c) Logarithmic: $\alpha=2.745$. For these same cooling schemes, GRSA used alpha's values in the same intervals previously defined. In each case the temperature was decreased using GR rule and modifying it's α value.

To compare GRSA and TMSA results, the solution quality and the execution time were standardized respectively, as follows (these quotients are shown in Figure 3 and Figure 4):

$$Q_q = \frac{Q_{GRSA}}{Q_{TMSA}} \times 100 \tag{10}$$

$$Q_t = \frac{Time_{GRSA}}{Time_{TMSA}} \times 100 \tag{11}$$

In Figure 3, it can be observed that GRSA maintains the quality of solution (Equation 10) with very close values to those obtained by TMSA. Nevertheless, as can be observed in Figure 4, the GRSA execution time (Equation 11) is lower than the TMSA execution time. Notice that the execution time is reduced as the number g is increased.

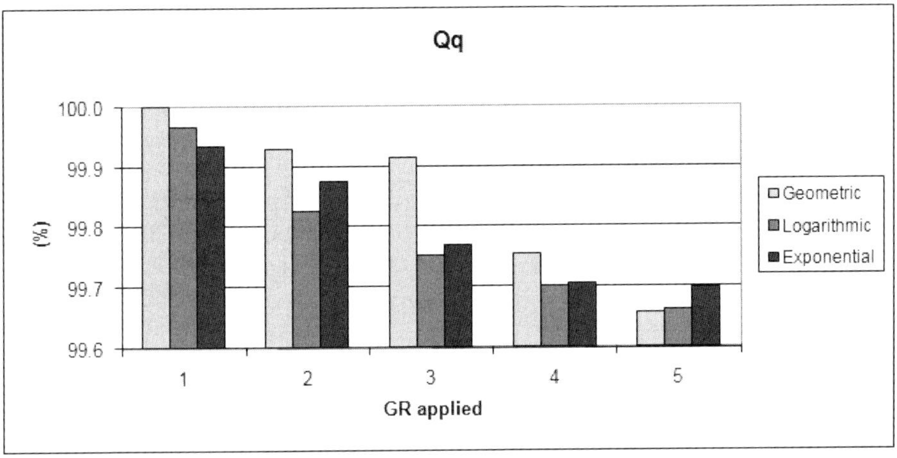

Fig. 3. Q_q solution quality of GRSA Vs TMSA (Equation 10)

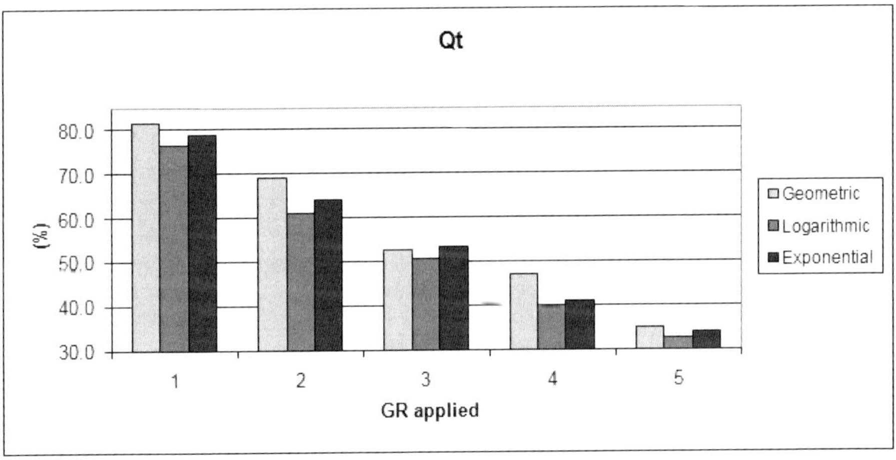

Fig. 4. Q_t with execution time in GRSA versus execution time in TMSA (Equation 11)

In Figure 5, the results by applying *20* golden ratios can be observed. In this figure, quality improvement (Equation 10) and time improvement (Equation 11) were placed together. Notice that the quality improvement of the solution remains constant while the time improvement is considerably reduced. The behaviour of these curves helps us to observe that the improved time is decreased in terms of a polynomial (i.e. $T_{GRSA} = T_{TMSA}/g^2$), whereas the quality solution remains very similar.

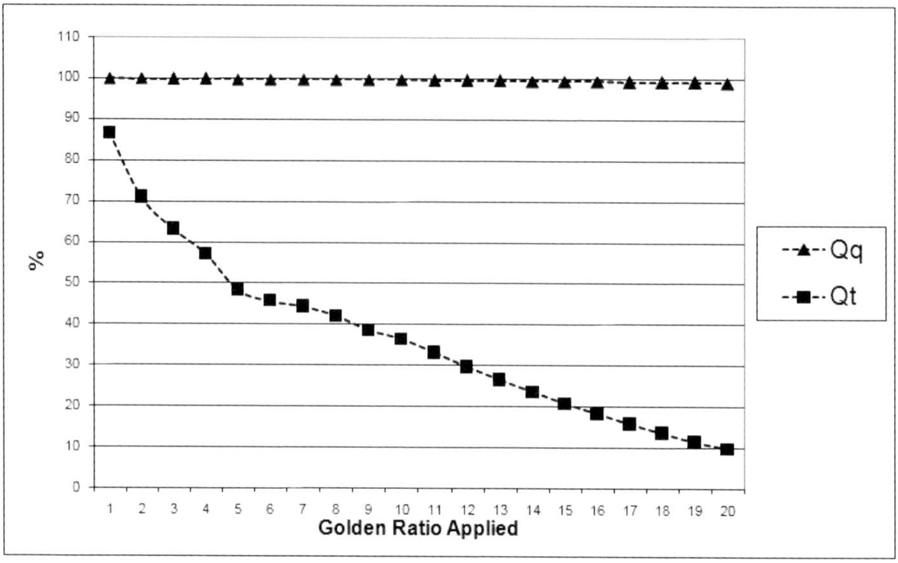

Fig. 5. GRSA performance

4 Conclusions

In this paper, a new SA algorithm for the satisfiability problem named GRSA is presented. This new algorithm is based on Markov Models and a heuristic rule known as Golden Ratio (GR). Even though a formal demonstration is not presented here, this paper shows that including this heuristic in a classical Simulated Annealing (TMSA) its performance is improved. This was tested using hard instances of the SATLIB, with three cooling schemes and starting with the same temperature in both GRSA and TMSA. To change the temperature parameter GRSA uses the GR rule a certain g number of times, in the three cooling schemes. According to the experimentation presented in this paper, the execution time was proportionally improved with g^2. In addition, GRSA quality remains very similar to that obtained with TMSA.

References

1. Garey, M.R., Johnson, D.S.: Computers and Intractability: A guide to the theory of NP-Completeness. W.H. Freeman, New York (1979)

2. Cook, S.A.: The complexity of theorem proving procedures. In: Proceedings of 3rd Annual ACM symposium on the Theory of Computing, pp. 151–158. ACM, New York (1971)
3. Papadimitriou, C.H.: Computational Complexity. Addison Wesley Longman, Redwood City (1995)
4. Gu, J.: Multispace search for satisfiability and NP-hard problems. In: Satisfiability Problem: Theory and Applications: Proceedings of a DIMACS Workshop. DIMACS Series in Discrete Mathematics and Theoretical Computer Science, vol. 35, pp. 407–517 (1996)
5. Creignou, N.: The class of problems that are linearly equivalent to satisfiability or a uniform method for proving NP-completeness. In: Martini, S., Börger, E., Kleine Büning, H., Jäger, G., Richter, M.M. (eds.) CSL 1992. LNCS, vol. 702, pp. 115–133. Springer, Heidelberg (1993)
6. Edmons, J.: Minimum partition of a matroid into independent subset. J. Res. Nat. Bur. Standards Sect. B 69, 67–72 (1965)
7. Cook, S.: Computational Complexity of Higher Type Functions. In: Proc. International Congress of Mathematicians, Kyoto, Japan, pp. 51–69. Springer, Heidelberg (1991)
8. Bertzekas, D.P.: Network Optimization: Continuos and Discrete Models. Athena Scientific (1998)
9. Bazaraa, M.S., Sherali, H.D., Shetty, C.M.: Non Linear Programming: Theory and Algorithms, 2nd edn. Wiley, Chichester (1993)
10. Olivetti the Fianca, F., Von Zuben, F.J., Nunes de Castro, L.: An Artificial Immune Network for Multimodal Function Optimization on Dynamic Environments. In: GECCO 2005, Washington DC, USA, June 25-29 (2005)
11. Livio, M.: The Golden Ratio: The Story of Phi, The World's Most Astonishing Number. Broadway Books, New York (2002)
12. Cook, S., Mithchel, D.G.: Finding Hard Instances of the Satisfiability Problem: A Survey. In: Satisfibility Problem Theory and Applications. Dimacs Series Discrete Mathematics and Theoretical Computer Sciences, pp. 1–17 (1997)
13. Gu, J., Purdom, P., Franco, J., Wah, B.: Algorithms for the satisfaibility (SAT) problem: A Survey. In: Satisfibility Problem Theory and Applications. Dimacs Series Discrete Mathematics and Theoretical Computer Sciences, pp. 19–151 (1997)
14. Cerny, V.: A Thermodynamical Approach to the Traveling Salesman Problem: An efficient Simulation Algorithm, Report, Comenius University, Bratislava, Czechoslovakia (1982)
15. Cerny, V.: Thermodynamical approach to the traveling salesman problem: An efficient simulation algorithm. Journal of Optimization Theory and Applications 45, 41–51 (1985)
16. Kirkpatrick, S., Gelatt, C.D., Vecchi, M.P.: Optimization by Simulated Annealing. Science (4598) 220, 671–680 (1983)
17. Aarts, E., y Korst, E.: Simulated annealing and Boltzman machines: A stochastic approach to combinatorial optimization and neural computing. John Wiley & Sons, Great Bretain (1989)
18. Frausto, J., Sanvicente, H., Imperial, F.: ANDYMARK: An analytical Method to Establish Dynamically the Length of the Markov Chian in Simulated Annealing for the Satisfiablity Problem. Springer, Heidelberg (2006)
19. Ingber, L.: Simulated Annealing; Practice Versus Theory. J. MATHL. Comput. Modeling 18(11), 29–57 (1993)

20. Mezard, M., Parisi, G., Zecchina, R.: Analytic and algorithmic solution of random satisfiability problems. Science, June 27 (2002)
21. SATLIB - The Satisfiability Library, http://www.cs.ubc.ca/~hoos/SATLIB/index-ubc.html
22. Horie, S., Watanabe, O.: Hard instance generation for SAT, Technical Report TR97-0007, Dept. of Computer Science, Tokyo Inst. of Tech. (Available from CS Dept. TR Archive); the extended abstract appeared in Proc. ISAAC 1997. LNCS, vol. 1350. Springer, Heidelberg (1997)

Modeling Cooperation in P2P Data Management Systems

Luciano Caroprese and Ester Zumpano

DEIS, Univ. della Calabria, 87030 Rende, Italy
{lcaroprese,zumpano}@deis.unical.it

Abstract. This paper investigates the data exchange problem among distributed independent sources. It is based on previous works in [9,10] in which a (declarative) semantics for P2P systems. In this semantics only facts not making the local databases inconsistent are imported *Weak Models*, and the *Preferred Weak Models* are those in which peers import maximal sets of facts not violating integrity constraints. The framework proposed in [9,10] does not provide any mechanism to set priorities among mapping rules. Anyhow, while collecting data it is quite natural for a source peer to associate different degrees of reliability to the portion of data provided by its neighbor peers. Starting from this observation, this paper enhances previous semantics by using priority levels among mapping rules in order to select the weak models containing a maximum number of mapping atoms according to their importance. We will call these weak models, *Trusted Weak Models* and we will show they can be computed as stable models of a logic program with weak constraints.

1 Introduction

Modeling cooperation between independent sources is a very old problem that emerges in all the areas in which data must be exchanged between a source and a target. In this paper, we consider *Peer-to-Peer* (P2P) setting and provide a contribution to the problem of mapping data in peer-to-peer data sharing systems. Each peer, joining a P2P system, can both provide or consume data and has information about its neighbors, that is about the peers that are reachable and can provide data of interest. In a P2P system each peer exhibits a set of *mapping rules*, that is a set of semantic correspondences to its neighbor peers. A mapping rule transfer data from a source peer to a target peer. The entry of a new peer in the system is extremely simple as it just requires the definition of its mapping rules. By using mapping rules as soon as it enters the system, a peer can participate and access data available in its neighborhood, and through its neighborhood it becomes accessible to all the other peers in the system. Therefore, due to the specified mechanism for collecting data, a query, that can be posed to any peer in the system, is answered by using locally stored data and all the information that can be imported, without corrupting the local database, from its neighbors.

This paper is based on previous work in [9,10] in which a different interpretation of mapping rules, that allows importing from other peers only atoms not

A. An et al. (Eds.): ISMIS 2008, LNAI 4994, pp. 225–235, 2008.

violating integrity constraints, has been proposed. This has led to the proposal of a new semantics for P2P systems. Under this semantics only facts not making the local databases inconsistent can be imported - *Weak Models* - and the *Preferred Weak Models* are those in which peers import maximal sets of facts not violating integrity constraints. This paper extends previous proposal and enriches the semantics of cooperation among peers by associating different levels of priority to mapping rules, that is different degree of reliability to data they could import.

In the rest of this section we will intuitively introduce the proposed semantics.

Example 1. Consider the P2P system consisting of the following three peers $\mathcal{P}_1$, $\mathcal{P}_2$ and $\mathcal{P}_3$ where: (i) $\mathcal{P}_2$ contains the atom $r(a)$ and $(ii)\mathcal{P}_3$ contains the atom $s(a)$, (iii) $\mathcal{P}_1$ imports data from $\mathcal{P}_2$ using the mapping rule $m_1:\ p(X) \hookleftarrow r(X)$[1] and imports data from $\mathcal{P}_3$ using the mapping rule $m_2:\ q(X) \hookleftarrow s(X)$. Moreover, it also contains the constraint $\leftarrow p(X), q(X)$ stating that it is not possible to import both $p(a)$ and $q(a)$. The intuition is that, with $r(a)$ and$s(a)$ being *true* respectively in $\mathcal{P}_2$ and $\mathcal{P}_3$, either $p(a)$ or $q(a)$ can be imported in $\mathcal{P}_1$ (but not both, otherwise the integrity constraint is violated). Therefore, the choice of import either from $\mathcal{P}_2$ or from $\mathcal{P}_3$ produces, the following preferred weak models $M_1 = \{r(a), s(a), p(a)\}$ and $M_2 = \{r(a), s(a), q(a)\}$. □

The framework in [9,10] does not provide any mechanism to set priorities among mapping rules. Anyhow, while collecting data it is quite natural for a source peer to associate different degree of reliability to the portion of data provided by its neighbor peers, so that allowing us to model concepts like *"it is more important to import the atom A with respect to the atom B if they conflict"* or *"the data imported from the peer P_1 are more reliable than data imported from the peer P_2"*.

This paper, stems from above observations and extends the definition of peer and P2P system, given in [9,10], with a framework that allows us to model previous concepts by associating different levels of priority to mapping rules, that is different degrees of reliability to data they import. This facility is then used to drive the cooperation among peers, and let to the proposal of a new semantics for P2P systems. More specifically, this paper enhances the semantics introduced in [9] by using priority levels among mapping rules in order to select the weak models containing a maximum number of mapping atoms according to their importance. We will call these weak models, *Trusted Weak Models.*

Example 2. Let's consider again previous example and suppose $\mathcal{P}_1$ *trust more data coming from* $\mathcal{P}_2$ *w.r.t data coming from* $\mathcal{P}_3$, therefore the portion of data coming from $\mathcal{P}_2$ *is more important* than that coming from $\mathcal{P}_3$. In order to model this concept an higher priority is assigned to the mapping rule m_1 w.r.t. m_2. From the introduction of this priority condition it results M_1 to be the unique trusted weak model.

The paper will provide a rewriting technique that allows to compute the trusted weak models of a peer by evaluating the stable models of a logic program with weak constraints.

---

[1] Please, note the special syntax we use for mapping rules introduced in [9].

2 Background

Familiarity is assumed with deductive database [1], logic programming and computational complexity [14,17,18]. Now we briefly recall some notation used in the rest of the paper. A *normal rule* or ($\vee - free$ - disjunction free) is of the form $H \leftarrow \mathcal{B}$, where H is an atom (*head* of the rule) and $\mathcal{B}$ is a conjunction of literals (*body* of the rule). A *normal program* is a finite set of normal rules. Given a program $\mathcal{P}$, $ground(\mathcal{P}) = \bigcup_{r \in \mathcal{P}} ground(r)$, where $ground(r)$ denotes the set of all rules obtained from the rule r by replacing variables with constants in all possible ways. An *interpretation* is a set of ground atoms. The *semantics* of a program $\mathcal{P}$ is given by the set of its *stable models*, denoted as $\mathcal{SM}(\mathcal{P})$. A normal positive program $\mathcal{P}$ has one stable model, that coincides with the *least model*.

Prioritized logic programs. Various forms of priorities have been investigated in logic languages [5,19]. We refer to the extension proposed in [19]. Given two atoms e_1 and e_2, the statement $e_2 \succeq e_1$ is a *priority* stating that for each a_2 instance of e_2 and for each a_1 instance of e_1, a_2 has higher priority than a_1. If $e_2 \succeq e_1$ and $e_1 \not\succeq e_2$ we write $e_2 \succ e_1$. If $e_2 \succ e_1$ the sets of instantiations of e_1 and e_2 must have empty intersection. The relation $\succeq$ is transitive and reflexive. A *prioritized logic program* $\mathcal{P}_\Phi$ is a pair $(\mathcal{P}, \Phi)$ where $\mathcal{P}$ is a program and Φ is a set of priorities. Φ^* denotes the set of priorities which can be reflexively or transitively derived from Φ. Given a PLP $(\mathcal{P}, \Phi)$, the relation $\sqsupseteq$ is defined over the stable models of $\mathcal{P}$. For any stable models M_1, M_2 and M_3 of $\mathcal{P}$: (i) $M_1 \sqsupseteq M_1$; (ii)$M_2 \sqsupseteq M_1$ if a) $\exists\ e_2 \in M_2 - M_1$, $\exists\ e_1 \in M_1 - M_2$ such that $(e_2 \succeq e_1) \in \Phi^*$ and b) $\nexists\ e_3 \in M_1 - M_2$ such that $(e_3 \succ e_2) \in \Phi^*$; (iii) if $M_2 \sqsupseteq M_1$ and $M_1 \sqsupseteq M_3$, then $M_2 \sqsupseteq M_3$. If $M_2 \sqsupseteq M_1$ holds, then we say that M_2 is *preferable* to M_1 w.r.t. Φ. Moreover, we write $M_2 \sqsupset M_1$ if $M_2 \sqsupseteq M_1$ and $M_1 \not\sqsupseteq M_2$. An interpretation M is a *preferred* stable model of $(\mathcal{P}, \Phi)$ if M is a stable model of $\mathcal{P}$ and there is no stable model N s.t. $N \sqsupset M$. $\mathcal{PSM}(\mathcal{P}_\Phi)$ denotes the set of preferred stable models of $\mathcal{P}_\Phi$.

Integrity Constraints. A *(strong) integrity constraint ic* expresses a relationship among data that every database instance is required to satisfy, and it is expressed by a rule of the form $\leftarrow \beta$, where $\mathcal{B}$ is a conjunction of literals. A database $\mathcal{DB}$ satisfies *ic* if $\mathcal{DB} \not\models \beta$. A *weak integrity constraint* [6,16] is a variant of an integrity constraint and is expressed by a rule of the form: $\Leftarrow \mathcal{B}\ [l]$, where l is a positive integer, called *level*, expressing the priority of the constraint. Weak constraints express desiderata, that is *conditions to be possibly satisfied* and their informal semantics is to minimize the number of violated instances. Given a program $\mathcal{P}$, a set of strong constraint $\mathcal{SC}$ and a set weak constraints $\mathcal{WC}$, an interpretation M is a stable model of $\mathcal{P} \cup \mathcal{SC} \cup \mathcal{WC}$ if M is a stable model of $\mathcal{P}$ that (i) satisfies $\mathcal{SC}$ and (ii) minimizes the number of unsatisfied instances of weak constraints in $\mathcal{WC}$ according to their importance, that is it minimizes the number of violated weak constraints in the highest priority level[2], then minimizes the number of the violated weak constraints in the next lower level, and so on and so

[2] Higher values for priority levels mark weak constraints of higher importance.

forth [3]. Formally, we associate to each priority level $i \in [1..n]$ a positive weight $f(i)$ inductively defined as follows: $f(1) = 1$, $f(i) = f(i-1) * |ground(\mathcal{WC})| + 1$, $1 < i \leq n$, where $|ground(\mathcal{WC})|$ is the total number of ground instances of weak constraints in $\mathcal{WC}$. Further, given a stable model M for a program $\mathcal{P}$, the *objective function* $H_{\mathcal{P},\mathcal{M}}$ is defined as: $H_{\mathcal{P},\mathcal{M}} = f(1) * N_1^M + \ldots + f(n) * N_n^M$, where N_i^M $1 \leq i \leq n$ is the number of ground instances of weak constraints whose priority level is i which are violated in M. Therefore, among the stable models of $\mathcal{P}$ that satisfies $\mathcal{SC}$ we select those minimizing the objective function. Observe that, the objective function guarantees that the violation of a single constraint of priority level i is more *expensive* than the violation of *all* weak constraints of the lower levels (i.e. all levels $< i$).

3 A New Semantics for P2P Systems

A *(peer) predicate symbol* is a pair $i : p$, where i is a *peer identifier* and p is a predicate symbol. A *(peer) atom* is of the form $i : A$, where i is a *peer identifier* and A is a standard atom. A *(peer) literal* is a peer atom $i : A$ or its negation $not\ i : A$. A conjunction $i : A_1, \ldots, i : A_m, not\ i : A_{m+1}, \ldots, not\ i : A_n, \phi$, where ϕ is a conjunction of built-in atoms[4], will be also denoted as $i : \mathcal{B}$, with $\mathcal{B}$ equals to $A_1, \ldots, A_m, not\ A_{m+1}, \ldots, not\ A_n, \phi$. A *(peer) rule* can be of one of the following three types:

1. STANDARD RULE. It is of the form $i : H \leftarrow i : \mathcal{B}$, where $i : H$ is an atom and $i : \mathcal{B}$ is a conjunction of atoms and built-in atoms.
2. INTEGRITY CONSTRAINT. It is of the form $\leftarrow i : \mathcal{B}$, where $i : \mathcal{B}$ is a conjunction of literals and built-in atoms.
3. MAPPING RULE. It is of the form $i : H \hookleftarrow j : \mathcal{B}$, where $i : H$ is an atom, $j : \mathcal{B}$ is a conjunction of atoms and built-in atoms and $i \neq j$.

In the previous rules $i : H$ is called *head*, while $i : \mathcal{B}$ (resp. $j : \mathcal{B}$) is called *body*. Negation is allowed just in the body of integrity constraints. The operator $ground(\cdot)$ is defined is standard way. Whenever the peer is understood, the peer identifier can be omitted. The definition of a predicate $i{:}p$ consists of the set of rules in whose head the predicate symbol $i{:}p$ occurs. A predicate can be of three different kinds: *base predicate*, *derived predicate* and *mapping predicate*. A base predicate is defined by a set of ground facts; a derived predicate is defined by a set of standard rules and a mapping predicate is defined by a set of mapping rules. An atom $i : p(X)$ is a *base atom* (resp. *derived atom*, *mapping atom*) if $i : p$ is a base predicate (resp. standard predicate, mapping predicate). Given an interpretation M, $\mathcal{DB}(M)$ (resp. $\mathcal{LP}(M)$, $\mathcal{MP}(M)$) denotes the subset of base atoms (resp. derived atoms, mapping atoms) in M.

In [9,10] a P2P system is defined as a set of peers. A peer $\mathcal{P}_i$ is defined as a tuple $\langle \mathcal{DB}_i, \mathcal{LP}_i, \mathcal{MP}_i, \mathcal{IC}_i \rangle$ where $\mathcal{DB}_i$ is a set of atoms (the *local database*),

[3] A program with a weak constraint $\Leftarrow p(X)$ can be regarded as modeling a minimization problem whose objective function is the cardinality of p.

[4] A *built-in atom* is of the form $\theta(X, Y)$, where X and Y are terms and θ is a comparison predicate.

$\mathcal{LP}_i$ is a set of standard rules, $\mathcal{MP}_i$ is a set of mapping rules and $\mathcal{IC}_i$ is a set of integrity constraints. $\mathcal{P}_i$ is assumed to be *locally consistent* that is $\mathcal{LP}_i \cup \mathcal{DB}_i \models \mathcal{IC}_i$. However, inconsistencies could be introduced when mapping atoms are imported into $\mathcal{P}_i$. Therefore, the mapping rules in $\mathcal{MP}_i$ are used to import into $\mathcal{P}_i$ maximal sets S of mapping atoms that do not generate violations of the integrity constraints in $\mathcal{IC}_i$, i.e. $\mathcal{LP}_i \cup \mathcal{DB}_i \cup S \models \mathcal{IC}_i$.

This framework does not provide any mechanism to set priorities among mapping rules disallowing us to model concept as *"it is more important to import the atom A with respect to the atom B if they conflict"* or *"the data imported from the peer P_1 are more reliable than data imported from the peer P_2"*. Here we extend the definition of *peer* and a *P2P system*, given in [9,10], with a framework that allows us to model such concepts setting priorities among mapping rules.

Definition 1. P2P SYSTEM. A *peer* $\mathcal{P}_i$ is a tuple $\langle \mathcal{DB}_i, \mathcal{LP}_i, \langle \mathcal{MP}_{i,1}, \dots, \mathcal{MP}_{i,k_i} \rangle, \mathcal{IC}_i \rangle$, where (i) $\mathcal{DB}_i$ is a set of atoms; (ii) $\mathcal{LP}_i$ is a set of standard rules; (iii) $\mathcal{MP}_{i,j}$ - with $j \in [1..k_i]$ - are sets of mapping rules and (iv) $\mathcal{IC}_i$ is a set of integrity constraints over predicates defined by $\mathcal{DB}_i$, $\mathcal{LP}_i$ and $\cup_{j \in [1..k_i]} \mathcal{MP}_{i,j}$. A *P2P system* $\mathcal{PS}$ is a set of peers $\{\mathcal{P}_1, \dots, \mathcal{P}_n\}$. □

The tuple $\langle \mathcal{MP}_{i,1}, \dots, \mathcal{MP}_{i,k_i} \rangle$ of a peer $\mathcal{P}_i$ establishes priorities among mapping rules. The intuitive meaning of this structure is that the data imported by means of a mapping rule r belonging to $\mathcal{MP}_{i,u}$ are 'more important' to data imported by means of a mapping rule s belonging to $\mathcal{MP}_{i,v}$ for each $u < v$.

Without loss of generality, we assume that every mapping predicate is defined by only one mapping rule of the form $i : p(X) \hookleftarrow j : q(X)$[5]. Given a mapping rule $r = H \hookleftarrow \mathcal{B}$, $St(r)$ denotes the corresponding standard logic rule $H \leftarrow \mathcal{B}$. Analogously, given a set of mapping rules $\mathcal{MP}$, $St(\mathcal{MP}) = \{St(r) \mid r \in \mathcal{MP}\}$. Given a P2P system $\mathcal{PS} = \{\mathcal{P}_1, \dots, \mathcal{P}_n\}$, where $\mathcal{P}_i = \langle \mathcal{DB}_i, \mathcal{LP}_i, \langle \mathcal{MP}_{i,1}, \dots, \mathcal{MP}_{i,k_i} \rangle, \mathcal{IC}_i \rangle$, we define $\mathcal{DB}(\mathcal{PS}) = \bigcup_{i \in [1..n]} \mathcal{DB}_i$, $\mathcal{LP}\ (\mathcal{PS}) = \bigcup_{i \in [1..n]} \mathcal{LP}_i$, $\mathcal{MP}(\mathcal{PS}) = \bigcup_{i \in [1..n], j \in [1..k_i]} \mathcal{MP}_{i,j}$ and $\mathcal{IC}(\mathcal{PS}) = \bigcup_{i \in [1..n]} \mathcal{IC}_i$.

We first recall the concept of satisfaction of mapping rules in [9,10]. Given an interpretation M, an atom H and a conjunction of atoms $\mathcal{B}$: (i) $val_M(H \leftarrow \mathcal{B}) = val_M(H) \geq val_M(\mathcal{B})$, (ii) $val_M(H \hookleftarrow \mathcal{B}) = val_M(H) \leq val_M(\mathcal{B})$. Thus, if the body is *true*, the head of a standard rule *must* be *true*, whereas the head of a mapping rule *could* be *true*. The idea in [9,10] is that if the source peer satisfies the body of a mapping rule, the head can be inferred in the target peer only if it does not generate inconsistencies.

Definition 2. WEAK MODEL [9,10]. Given a P2P system $\mathcal{PS}$, an interpretation M is a *weak model* for $\mathcal{PS}$ if M is the least model of $\mathcal{LP}(\mathcal{PS}) \cup \mathcal{DB}(\mathcal{PS}) \cup St(\mathcal{MP}(\mathcal{PS})^M) \cup \mathcal{IC}(\mathcal{PS})$, where $\mathcal{MP}(\mathcal{PS})^M$ is obtained from *ground*($\mathcal{MP}$ ($\mathcal{PS}$)) by removing all mapping rules whose head is *false* with respect to M.

The set of weak models for $\mathcal{PS}$ is denoted as $\mathcal{WM}(\mathcal{PS})$. □

---

[5] The definition of a mapping predicate $i{:}p$ consisting of n rules of the form $i{:}p(X) \hookleftarrow \mathcal{B}_k$, with $k \in [1..n]$, can be rewritten into $2 * n$ rules of the form $i{:}p_k(X) \hookleftarrow \mathcal{B}_k$ and $i{:}p(X) \leftarrow i{:}p_k(X)$, with $k \in [1..n]$.

Given a P2P system $\mathcal{PS}$, the set $\mathcal{WM}(\mathcal{PS})$ models all the ways the peers can import atoms from their neighbors without violating integrity constraints.

Example 3. Consider the P2P system $\mathcal{PS}$ containing the three peers: $\mathcal{P}_1 = \langle\emptyset, \emptyset, \langle\{a \hookleftarrow\}, \{b \hookleftarrow d\}\rangle, \{\leftarrow a, b\}\rangle$, $\mathcal{P}_2 = \langle\{c\}, \emptyset, \emptyset, \emptyset\rangle$ and $\mathcal{P}_3 = \langle\{d\}, \emptyset, \emptyset, \emptyset\rangle$.

The weak models of $\mathcal{PS}$ are: $M_1 = \{a, c, d\}$, $M_2 = \{b, c, d\}$ and $M_3 = \{c, d\}$. Note that M_1 and M_2 import respectively a and b, whereas no mapping atom is imported by M_3. □

We now recall the proposal of *Preferred Weak Models* given in [9,10] that following the reasonable principle of *"importing as much knowledge as possible"* selects among weak models those that contain maximal sets of mapping atoms.

Definition 3. Preferred Weak Model [9,10]. Let $\mathcal{PS}$ be a P2P system and M and N weak models of $\mathcal{PS}$. Then M is *preferable over* to N with respect to $\mathcal{PS}$, denoted as $M \sqsupset N$, if $\mathcal{MP}(M) \supset \mathcal{MP}(N)$. A weak model M is said to be *preferred* if there is no weak model N such that $N \sqsupset M$. The set of *preferred weak models* for $\mathcal{PS}$ is denoted as $\mathcal{PWM}(\mathcal{PS})$. □

A preferred weak model is a weak model with a maximal set of mapping atoms.

Example 4. (Example 3 continued). The preferred weak models of $\mathcal{PS}$ are: $M_1 = \{a, c, d\}$ and $M_2 = \{b, c, d\}$. In fact, there are two different ways to import maximal sets of mapping atoms into $\mathcal{P}_1$: importing a using the mapping rule $a \hookleftarrow c$ or importing b using the mapping rule $b \hookleftarrow d$. □

Here we introduce a new semantics for P2P systems that enhances the one modeled by Definition 3 with priority levels among mapping rules in order to select the weak models containing a maximum number of mapping atoms according to their importance. We will call these weak models, *Trusted Weak Models.*

Definition 4. Mapping Atoms with Trusted Level. Given a weak model M of a P2P system $\mathcal{PS} = \{\mathcal{P}_1, \ldots, \mathcal{P}_n\}$, we denote as $\mathcal{MP}(M, k)$ the set of mapping atoms belonging to M and inferred by means of mapping rules of level k that is $\mathcal{MP}(M, k) = \{i : a(t) \mid i : a(t) \in M \wedge\ i : a(t) \hookleftarrow \mathcal{B} \in \mathcal{MP}_{i,k}$ for $i \in [1..n]\}$. □

As each mapping predicate is defined by means of just one mapping rule, in the previous definition $\mathcal{MP}(M, k) \cap \mathcal{MP}(M, h) = \emptyset$ for each $k \neq h$.

Definition 5. Trusted Weak Model. Let $\mathcal{PS}$ be a P2P system and M and N weak models of $\mathcal{PS}$. We say that M *is trustable over* N *with respect to* $\mathcal{PS}$, denoted as $M \succ N$, if there exists $j \geq 1$ such that for each $1 \leq i < j$ we have $|\mathcal{MP}(M, i)| = |\mathcal{MP}(N, i)|$ and $|\mathcal{MP}(M, j)| > |\mathcal{MP}(N, j)|$. We say that M is a *trusted weak model* of $\mathcal{PS}$ if there is no weak model N of $\mathcal{PS}$ such that $N \succ M$. The set of trusted weak models for $\mathcal{PS}$ is denoted as $\mathcal{TWM}(\mathcal{PS})$. □

Therefore, the new semantics first selects the weak models containing a maximum number of atoms inferred by means of mapping rule of level 1 (the highest level), then among that models those containing a maximum number of atoms inferred by means of mapping rule of level 2 and so on.

By definition, for each P2P systems $\mathcal{PS}$ the following relations hold: $\mathcal{PWM}(\mathcal{PS}) \subseteq \mathcal{WM}(\mathcal{PS})$ and $\mathcal{TWM}(\mathcal{PS}) \subseteq \mathcal{WM}(\mathcal{PS})$. Moreover, the interesting property that each trusted weak model is a preferred weak model holds.

Theorem 1. *Let $\mathcal{PS}$ be a P2P system. Then $\mathcal{TWM}(\mathcal{PS}) \subseteq \mathcal{PWM}(\mathcal{PS})$.* □

Summarizing, for each P2P $\mathcal{PS}$, the relations between weak, preferred weak and trusted weak models are: $\mathcal{TWM}(\mathcal{PS}) \subseteq \mathcal{PWM}(\mathcal{PS}) \subseteq \mathcal{WM}(\mathcal{PS})$.

Example 5. (Example 3 and Example 4 continued). As stated before, the preferred weak models of $\mathcal{PS}$ are: $M_1 = \{a, c, d\}$ and $M_2 = \{b, c, d\}$. We observe that $|\mathcal{MP}(M_1, 1)| = 1$, $|\mathcal{MP}(M_2, 1)| = 0$. Therefore, as the first mapping rule has priority w.r.t. to the second one (as it belongs to the level 1), M_1 is the unique trusted weak model of $\mathcal{PS}$. □

4 An Alternative Characterization of the New Semantics

This section describes two alternative characterizations of previously presented semantics. The first one has been proposed in [9,10] and captures the preferred weak model semantics by a logic program with priorities. Here, we present the second one that captures the trusted weak model semantics by means of a logic program with weak constraints. The interesting result is that both characterizations use the same logic program that basically generates all weak models. Preferred weak models are then obtained by adding to this program a st of priorities and trusted weak models are obtained by adding to it a set of weak constraints. We recall that, given a mapping rule $H \hookleftarrow \mathcal{B}$, if $\mathcal{B}$ is *true* in the source peer then two mutually exclusive actions are possible in the target peer: *either importing H or not importing H*. This behavior can be modeled by a logic rule of the form: $H \oplus H' \leftarrow \mathcal{B}$[6], where if $H = i : p(X)$ then $H' = i : p'(X)$. The meaning of this rule is that if $\mathcal{B}$ is *true* then exactly one atom between H and H' must be *true*. Therefore, while the head atom H models the *import* action, H' models the *not-import* action.

Definition 6. *Given a P2P system $\mathcal{PS}$,*

- $Rew(\mathcal{MP}(\mathcal{PS})) = \{H \oplus H' \leftarrow \mathcal{B} | H \hookleftarrow \mathcal{B} \in \mathcal{MP}(\mathcal{PS})\}$
- $Rew(\mathcal{PS}) = \mathcal{DB}(\mathcal{PS}) \cup \mathcal{LP}(\mathcal{PS}) \cup Rew(\mathcal{MP}(\mathcal{PS})) \cup \mathcal{IC}(\mathcal{PS})$ □

Given a stable model M for $Rew(\mathcal{PS})$, the subset of non-primed atoms of M is denoted by $\Omega(M)$. The operator $\Omega(\cdot)$ is extended to sets of models. The weak models of a P2P system $\mathcal{PS}$ can be obtained from the stable models of $Rew(\mathcal{PS})$.

Theorem 2. Given a P2P system $\mathcal{PS}$, $\mathcal{WM}(\mathcal{PS}) = \Omega(\mathcal{SM}(Rew(\mathcal{PS})))$. □

[6] $H \oplus H' \leftarrow \mathcal{B}$ is just shorthand for $H \leftarrow \mathcal{B}, not\ H'$ and $H' \leftarrow \mathcal{B}, not\ H$.

Prioritized programs and preferred stable models. The preferred weak model semantics can be computed by adding to the logic program in Definition 6 a set of priorities so that obtaining a prioritized logic program [5,19].

Definition 7. Given a P2P system $\mathcal{PS}$, (i) $\Phi(\mathcal{PS}) = \{H \succeq H' | H \hookleftarrow \mathcal{B} \in \mathcal{MP}(\mathcal{PS})\}$, (ii) $Rew^P(\mathcal{PS}) = (Rew(\mathcal{PS}), \Phi(\mathcal{PS}))$ □

The priority statement $H \succeq H'$ in the previous definition intuitively stands that *it is preferable to import* H *w.r.t. not import it.*

The following theorem shows the equivalence between the preferred weak models of a P2P system $\mathcal{PS}$ and the preferred stable models of its rewriting $Rew^P(\mathcal{PS})$.

Theorem 3. Given a P2P system $\mathcal{PS}$, $\mathcal{PWM}(\mathcal{PS}) = \Omega(\mathcal{PSM}(Rew^P(\mathcal{PS})))$. □

Trusted weak models with stable model semantics. Trusted weak model semantics can be computed by adding to the logic program in Definition 6 a set of weak constraints [6,16].

Definition 8. Given a P2P system $\mathcal{PS} = \{\mathcal{P}_1, \ldots, \mathcal{P}_n\}$, where $\mathcal{P}_i = \langle \mathcal{DB}_i, \mathcal{LP}_i, \langle \mathcal{MP}_{i,1}, \ldots, \mathcal{MP}_{i,k_i} \rangle, \mathcal{IC}_i \rangle$, for $i \in [1..n]$, and l is the maximum priority level occurring in $\mathcal{PS}$, that is $l = max(\{k_i | i \in [i..n]\})$. (i) $\mathcal{G}(\mathcal{PS}) = \{i : p^g(t) | i : p(t) \hookleftarrow \mathcal{B} \in ground(\mathcal{MP}(\mathcal{PS}))\}$, (ii) $\mathcal{WC}(\mathcal{PS}) = \{\Leftarrow H^g, not\ H\ [l-k+1] | H \hookleftarrow \mathcal{B} \in \mathcal{MP}_{i,k}$ for $i \in [1..n]\}$ where if $H = i : p(X)$ then $H^g = i : p^g(X)$, (iii) $Rew^T(\mathcal{PS}) = Rew(\mathcal{PS}) \cup \mathcal{G}(\mathcal{PS}) \cup \mathcal{WC}(\mathcal{PS})$ □

The weak constraint $\Leftarrow H^g, not\ H\ [l-k+1]$ in the previous definition intuitively expresses that the *desirable action of importing* H *is in the level* $l - k + 1$. We recall that, as pointed out in the Section 2, the violation of a single weak constraint of priority level i is more expensive than the violation of all weak constraints of the lower levels. Therefore, our objective function asks for the minimization of violations starting from the highest priority level of weak constraints that correspond to the lowest level of mapping rules.

The following theorem shows the equivalence between the trusted weak models of a P2P system $\mathcal{PS}$ and the stable models of its rewriting $Rew^T(\mathcal{PS})$.

Theorem 4. Given a P2P system $\mathcal{PS}$, $\mathcal{TWM}(\mathcal{PS}) = \Omega(\mathcal{SM}(Rew^T(\mathcal{PS})))$. □

Example 6. Consider the P2P system $\mathcal{PS}$ of Example 3. $Rew(\mathcal{PS})$ contains the following rules: $a \oplus a' \leftarrow c$; $b \oplus b' \leftarrow d$; $\leftarrow a, b$; c; d. The stable models of $Rew(\mathcal{PS})$ are: $M_1 = \{a, b', c, d\}$, $M_2 = \{a', b, c, d\}$ and $M_3 = \{a', b', c, d\}$. The set of weak models obtained by removing the primed atoms, by using the operator $\Omega(\cdot)$, are $W_1 = \{a, c, d\}$, $W_2 = \{b, c, d\}$ and $W_3 = \{c, d\}$.

$Rew^P(\mathcal{PS}) = (Rew(\mathcal{PS}), \Phi(\mathcal{PS}))$, where $\Phi(\mathcal{PS}) = \{a \succeq a',\ b \succeq b'\}$. The preferred stable models of $Rew^P(\mathcal{PS})$ are M_1 and M_2. Therefore, the preferred weak models are W_1 and W_2.

$Rew^T(\mathcal{PS}) = Rew(\mathcal{PS}) \cup \mathcal{G}(\mathcal{PS}) \cup \mathcal{WC}(\mathcal{PS})$, where $\mathcal{G}(\mathcal{PS}) = \{a^g, b^g\}$ and $\mathcal{WC}(\mathcal{PS}) = \{\Leftarrow a^g, not\ a\ [2], \Leftarrow b^g, not\ b\ [1]\}$. The unique stable model of $Rew^T(\mathcal{PS})$ is M_1. Therefore, the unique trusted weak model is W_1. □

Observe that, the prioritized program $Rew^P(\mathcal{PS})$ cannot be processed directly using ASP solvers such as DLV [16] or Smodels [20]. Therefore, a more complex rewriting technique embedding priorities into a unique logic program has been presented in [11]. On the other hand, as the mechanism of weak constraints is implemented in ASP solvers, the program $Rew^T(\mathcal{PS})$ can be directly evaluated.

5 Discussion

Complexity Results. Complexity results for computing the *preferred weak models* can be immediately determined by considering analogous results on stable model semantics for prioritized logic program [5,19]. More specifically, for ($\vee$-free) prioritized program deciding whether an atom is *true* in some preferred weak model is Σ_2^p-complete, whereas deciding whether an atom is *true* in every preferred weak model is Π_2^p-complete. Complexity results for computing the *trusted weak models* can be immediately fixed by considering analogous results on stable model semantics for the fragment of DATALOG allowing negation ($\neg$), strong (s) and weak constraints with priorities ($w^{\succ}$), denoted as DATALOG$^{\neg,s,w^{\succ}}$ [6]. It is rather clear that strong constraints do not affect at all the computational complexity [7]. Additionally, in [6] it has been shown that adding weak constraints to a logic program do not cause a tremendous increase of the complexity as it always remains in the same level of polynomial hierarchy. Our framework models trusted weak models by using DATALOG$^{\neg,s,w^{\succ}}$, that is Datalog plus strong and weak constraints with priorities. In this case deciding whether an atom is *true* in some trusted weak model increases from $\mathcal{NP}$ to Δ_2^P.

Related Works. The problem of integrating and querying databases in a P2P system has been investigated in [3,7,8,12,13]. In [7,8] a new semantics for a P2P system, based on epistemic logic, is proposed. A peer collects data and constructs its epistemic theory. Epistemic logic ensures that each peer only exports the data it really knows, that is its *certain data*. In [12,13] a characterization of P2P database systems and a model-theoretic semantics dealing with inconsistent peers is proposed. The basic idea is that if a peer does not have models all (ground) queries submitted to the peer are *true*. Thus, if some database is inconsistent it does not mean that the entire system is inconsistent. The semantics in [12,13] coincides with the epistemic semantics in [7,8]. None of the previous proposals take into account the possibility of modeling some preference criteria while performing the data integration process. A new interesting semantics for data exchange systems that goes in this direction has been recently proposed in [3,4]. This semantics allows for a cooperation among pairwise peers that related each other by means of data exchange constraints (i.e. mapping rules) and trust relationships. The decision by a peer on what other data to consider (besides its local data) does not depend only on its data exchange constraints, but also on the trust relationship that it has with other peers. For example, if peer $\mathcal{P}_1$ trusts

[7] In fact, under stable model semantics a strong constraint of the form $\leftarrow \mathcal{B}$ is actually a shorthand for $p \leftarrow \mathcal{B}, \neg p$.

peer $\mathcal{P}_2$'s data more than its own, then $\mathcal{P}_1$ will accommodate its data to $\mathcal{P}_2$ and will keep the data exchange constraints between them and its local integrity constraints satisfied while respecting its trust relationship (that is it will give priority to $\mathcal{P}_2$'s data with respect to its own data). The main difference between our proposal and the recent proposal by Bertossi and Bravo is related to the different levels allowed for modeling priorities among peers. More specifically, in [4] trust relationships allows just modeling two different reliability levels w.r.t the database of a neighbor peer: $\mathcal{P}_1$ trusts itself *less* than $\mathcal{P}_2$ or $\mathcal{P}_1$ trusts itself the *same* as $\mathcal{P}_2$, whereas we can associate different priority levels to different portions of the database exported by the same peer. Moreover, our approach allows to import mapping atoms only if they do not produce a local violation of integrity constraints (a peer trusts its own data *more* than any other imported data), whereas in [4] a peer collects data from its neighbors using data exchange constraints and then applies a local repair semantics [2,15].

References

1. Abiteboul, S., Hull, R., Vianu, V.: Foundations of Databases. Addison-Wesley, Reading (1994)
2. Arenas, M., Bertossi, L., Chomicki, J.: Consistent Query Answers in Inconsistent Databases. In: PODS, pp. 68–79 (1999)
3. Bertossi, L., Bravo, L.: Query Answering in Peer-to-Peer Data Exchange Systems. In: Extending Database Technology Workshops (2004)
4. Bertossi, L., Bravo, L.: Query The semantics of consistency and trust in peer data exchange systems. In: LPAR, pp. 107–122 (2007)
5. Brewka, G., Eiter, T.: Preferred Answer Sets for Extended Logic Programs. AI 109(1-2), 297–356 (1999)
6. Buccafurri, F., Leone, N., Rullo, P.: Enhancing Disjunctive Datalog by Constraints. TKDE (12)5, 845–860
7. Calvanese, D., De Giacomo, G., Lembo, D., Lenzerini, M., Rosati, R.: Inconsistency Tolerance in P2P Data Integration: an Epistemic Logic Approach. In: IS (2007)
8. Calvanese, D., De Giacomo, G., Lenzerini, M., Rosati, R.: Logical foundations of peer-to-peer data integration. In: PODS, pp. 241–251 (2004)
9. Caropresе, L., Greco, S., Zumpano, E.: A Logic Programming Approach to Querying and Integrating P2P Deductive Databases. In: FLAIRS, pp. 31–36 (2006)
10. Caropresе, L., Molinaro, C., Zumpano, E.: Integrating and Querying P2P Deductive Databases. In: IDEAS, pp. 285–290 (2006)
11. Caropresе, L., Zumpano, E.: Consistent Data Integration in P2P Deductive. In: SUM, pp. 230–243 (2007)
12. Franconi, E., Kuper, G.M., Lopatenko, A., Zaihrayeu, I.: Queries and Updates in the coDB Peer to Peer Database System. In: VLDB, pp. 1277–1280 (2004)
13. Franconi, E., Kuper, G.M., Lopatenko, A., Zaihrayeu, I.: A robust logical and computational characterisation of Peer-to-Peer database systems. In: DBISP2P, pp. 64–76 (2003)
14. Gelfond, M., Lifschitz, V.: The Stable Model Semantics for Logic Programming. In: ICLP/SLP, pp. 1070–1080 (1988)
15. Greco, G., Greco, S., Zumpano, E.: Repairing and Querying Inconsistent Databases. TKDE 15(6), 1389–1408 (2003)

16. Leone, N., Pfeifer, G., Faber, W., Eiter, T., Gottlob, G., Perri, S., Scarcello, F.: The DLV system for knowledge representation and reasoning. ACM Trans. Comput. Log. 7(3), 499–562 (2006)
17. Lloyd, J.W.: Foundations of Logic Programming. Springer, Heidelberg (1987)
18. Papadimitriou, C.H.: Computational Complexity. Addison-Wesley, Reading (1994)
19. Sakama, C., Inoue, K.: Prioritized logic programming and its application to commonsense reasoning. AI 123(1-2), 185–222 (2000)
20. Syrjanen, T., Niemela, I.: The Smodels System. In: LPNMR, pp. 434–438 (2001)

Interactions between Rough Parts in Object Perception

Andrzej W. Przybyszewski

Department of Psychology, McGill University Montreal, Canada
Dept of Neurology, University of Massachusetts Medical Center, Worcester MA USA
przy@ego.psych.mcgill.ca

Abstract. The visual systems of humans and primates outperform the best artificial vision systems by almost any measure. Humans can easily recognize as complex objects as faces even if they haven't seen them in such conditions before. However, experiments with the inverted faces (Thatcher illusion) show strong dependences between parts and their configuration. We propose pattern recognition rules similar to the primate visual brain on the basis of the simple shape classification in the intermediate area of the visual cortex (V4). In the present work we have described interactions between parts and their configurations using single cell responses as the brain expertise (decision attribute). Experimental data as the set of condition (stimulus) and decision (cell responses) attributes were placed into a decision table. Applying the rough set theory (Pawlak, 1992) we have divided our stimuli into equivalent classes determined by evoked cell responses. On the basis of the decision table, we found the decision rules. Comparing decision rules for responses to object and to its parts, we have found the interaction rules in the receptive fields of the area V4. We have proposed the interaction rules for objects that are simpler than faces but we expect that such rules can give us neurophysiological basis for the Gestalt perception of the complex objects. By comparing responses of different cells we have found equivalent concept classes. However, many different cells show inconsistency between their decision rules, which may suggest that the brain uses several different decision logics in order to make object perception insensitive to changes in properties of its parts (rough parts).

Keywords: Visual brain, imprecise computation, bottom-up, top-down processes, neuronal activity.

1 Introduction

After Pawlak [1], we define an information system as $S = (U, A)$, where U is a set of objects and A is a set of attributes. If $a \in A$ *and* $u \in U$, the value $a(u)$ is a unique element of V (a value set). The *indiscernibility* relation of any subset B of A, or $IND(B)$, is defined [1] as the equivalence relation whose elements are the sets $\{u: b(u) = v\}$ as v varies in V, and $[u]_B$ - the equivalence class of u form *B-elementary granule*. The concept $X \subseteq U$ is *B-definable* if for each $u \in U$ either $[u]_B \subseteq X$ or $[u]_B \subseteq U \backslash X$. $\underline{B}X = \{u \in U: [u]_B \subseteq X\}$ is a lower approximation of X. The concept $X \subseteq U$ is *B-indefinable* if exists such $u \in U$ such that $[u]_B \cap X \neq \phi$ }. $\overline{B}X = \{u \in$

A. An et al. (Eds.): ISMIS 2008, LNAI 4994, pp. 236–245, 2008.

U: $[u]_B \cap X \neq \phi\}$ is an upper approximation of X. The set $BN_B(X) = \overline{B}X - \underline{B}X$ will be referred to as the *B-boundary region* of. If the boundary region of X is the empty set then X is *exact* (*crisp*) with respect to B; otherwise if $BN_B(X) \neq \phi$ X is not *exact* (*rough*) with respect to B.

In this paper the universe U is a set of simple visual patterns that were used in our experiments [2], which can be divided into equivalent indiscernibility classes or *B-elementary granules*, where $B \subseteq A$. The purpose of our research is to find how these objects are classified in the brain. Therefore we will modify definition of the information system as $S = (U, C, D)$ where C and D are condition and decision attributes. Decision attributes will classify elementary granules in agreement with neurological responses from the specific visual brain area. In this work we are looking into single cell responses only in one area - V4 that will divide all patterns into equivalent indiscernibility classes of *V4-elementary granules.* Neurons in V4 are sensitive only to the certain attributes of the stimulus, like for example space localization – the pattern must be in the receptive field, and most of them are insensitive to contrast changes. Different V4 cells have different receptive field properties, which means that one *B-elementary granule* can be classified in many ways by different *V4-elementary granules.*

2 Method

We will represent experimental data ([2]) in the following table. In the first column are neural measurements. Neurons are identified using numbers related to a collection of figures in the previous paper [2]. Different measurements of the same cell are denoted by additional letters (a, b,...). For example, 11a denotes the first measurement of a neuron numbered 1 Fig. 1 of [2], 11b the second measurement, etc. Stimuli typically used in neuroscience have the following properties (see Fig 1):

1. orientation in degrees appears in the column labeled *o*, and orientation bandwidth is labeled by *ob*.
2. spatial frequency is denoted as *sf*, and spatial frequency bandwidth is *sfb*
3. x-axis position is denoted by *xp* and the range of x-positions is *xpr*
4. y-axis position is denoted by *yp* and the range of y-positions is *ypr*
5. x-axis stimulus size is denoted by *xs*
6. y-axis stimulus size is denoted by *ys*
7. stimulus shape is denoted by *s*, values of s are following: for grating $s=1$, for vertical bar $s=2$, for horizontal bar $s=3$, for disc $s=4$, for annulus $s=5$

Decision attributes are divided into several classes determined by the strength of the neural responses. Small cell responses are classified as *class 0*, medium to strong responses are classified as *classes 1 to n-1* ($min(n)=2$), and the strongest cell responses are classified as class n. Therefore each cell divides stimuli into its own family of equivalent objects.

Cell responses (r) are divided into $n+1$ ranges:

class 0 : activity below the threshold (e.g. 10 sp/s) labeled by r_0;

class 1: activity above the threshold labeled by r_1; ...

class n: maximum response of the cell (e.g. 100-200 sp/s) labeled by r_n.

Thus the full set of stimulus attributes is expressed as $B = \{o, ob, sf, sfb, xp, xpr, yp, ypr, xs, ys, s\}$.

3 Results

3.1 Analysis of the Interactions between Parts

We have analyzed the experimental data from several neurons recorded in the monkey's V4 [2]. One example of V4 cell responses to thin (0.25 deg) vertical bars in different horizontal - x positions is shown in the upper left part of Fig. 1 (Fig. 1E). Cell responses show a maximum for the middle ($XPos$ = 0) bar position along the x-axis. Cell responses are not symmetrical around 0. In Fig. 1F the same cell (cell 61 in table 1) is tested with two bars. The first bar stays at the 0 position, while the second bar changes its position along the x-axis. Cell responses show several maxima dividing the receptive field into four areas. However, this is not always the case as responses to two bars in another cell (cell 62 in table 1) show only three maxima (Fig. 1G). Horizontal lines in plots of both figures divide cell responses into the three classes: r_0, r_1, r_2, which are related to the response strength (see Methods). Stimuli attributes and cell responses divided into two: r_1 and r_2 classes are shown in table 1 for cells from Fig. 1 and in table 2 for cells from Fig. 2.

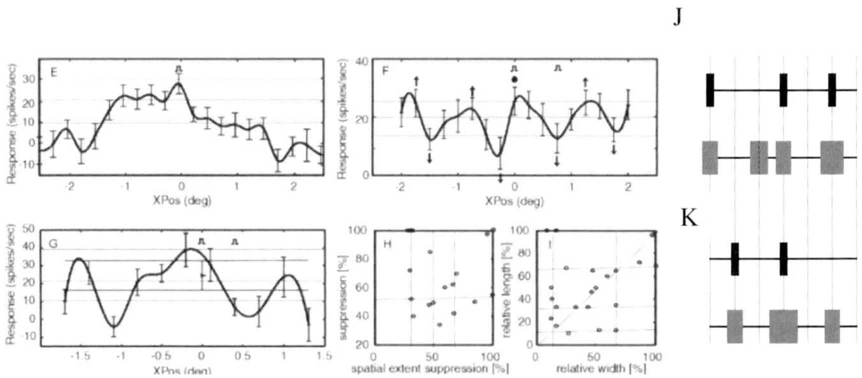

Fig. 1. Modified plots from [2]. Curves represent responses of several cells from area V4 to small single (**E**) and double (**F, G**) vertical bars. Bars change their position along x-axis (Xpos). Responses are measured in spikes/sec. Mean cell responses $\pm$ SE are marked in E, F, and G. Thick horizontal lines represent a 95% confidence interval for the response to single patch in position 0. Cell responses are divided into three ranges by thin horizontal lines. (**H**) A scatter plot showing peak percentage reduction of response to the central bar when a second bar is simultaneously presented. Assuming that the single bar give response r_2 50% suppression means that the second bar reduce cell response to r_1 (horizontal line). Cell properties can be divided into approximately three types (vertical lines): with a maximum suppression every 30, 50, or 70% of the receptive field extension. (**I**) A similar scatter plot as H but on x-axis is the ratio of optimal length and width, on y-axis is the spatial extent of the stimulus. (**J**) "Window sharpening": The schematic for the cell from part (F) and table 1 (rows 61f*) showing bar positions giving r_2 (upper part in black), and r_1 (lower part in gray) cell response. (**K**) The same as in (J) but for cell plotted in (G) and in table 1 rows 61g1 to 61g5.

Table 1. Decision table for cells from Fig. 1. Attributes *o, ob, sf, sfb* were constant and are not presented in the table. In experiments where two stimuli were used, the shape value was following**:** for two bars s=22, for two discs s=44.

Cell	xp	xpr	xs	ys	s	r
61e	-0.7	1.4	0.25	4	2	1
61f1	-1.9	0.2	0.25	4	22	2
61f2	0.1	0.2	0.25	4	22	2
61f3	1.5	0.1	0.25	4	22	2
61f4	-1.8	0.6	0.25	4	22	1
61f5	-0.8	0.8	0.25	4	22	1
61f6	0.4	0.8	0.25	4	22	1
61f7	1.2	0.8	0.25	4	22	1
62g1	-1.5	0.1	0.25	4	22	2
62g2	-0.15	0.5	0.25	4	22	2
62g3	-1.5	0.6	0.25	4	22	1
62g4	-0.25	1.3	0.25	4	22	1
62g5	1	0.6	0.25	4	22	1
63h1	-0.5	0.5	1	1	44	2
63h2	1	1	1	1	44	1
63h3	0.2	0.1	0.25	4	22	2

We assign the narrow (xpr_n), medium (xpr_m), and wide (xpr_w) x position ranges as follows: xpr_n *if* $(xpr\text{: } 0<xpr\leq 0.6)$, medium xpr_m *if* $(xpr\text{: } 0.6 <xpr\leq 1.2)$, wide xpr_w *if* $(xpr\text{: } xpr>1.2)$.

On the basis of Fig. 1 and Tab.1 the **two-bar** horizontal interaction study for cell 61f can be presented as the following **two-bar decision rules:**

DRT1: $(o_{90} \wedge xpr_n \wedge (xp_{-1.9} \vee xp_{0.1} \vee xp_{1.5}) \wedge xs_{0.25} \wedge ys_4)_1 \wedge$
$(o_{90} \wedge xp_0 \wedge xs_{0.25} \wedge ys_4)_0 \text{->} r_2$

DRT2: $(o_{90} \wedge xpr_m \wedge (xp_{-1.8} \vee xp_{-0.8} \vee xp_{0.4} \vee xp_{1.2}) \wedge xs_{0.25} \wedge ys_4)_1 \wedge$
$(o_{90} \wedge xp_0 \wedge xs_{0.25} \wedge ys_4)_0 \text{->} r_1$

One-bar decision rules [3] can be interpreted as follows: the narrow vertical bar evokes a strong response in certain positions, medium size bars evoke medium responses in certain positions, and wide horizontal or vertical bars evoke no responses. *Two-bar decision rules* claim that: the cell responses to two bars are strong if one bar is in the middle of the receptive field (RF) (bar with index 0 in decision rules) and the second narrow bar (bar with index 1 in decision rules) is in the certain positions of the RF (DRT1). But when the second bar is in medium position range, the max cell responses became weaker (DRT2). Responses of other cells are sensitive to other bar positions (Fig. 1G, H).

The decision table (Table 2) based on Fig. 2 describes cell responses to two patches placed in different positions along x-axis in the receptive field (RF). Figure 2 shows that adding the second patch reduced single patch cell responses. We have assumed that cell response to a single patch places in the middle of the RF is r_2. The second patch suppresses cell responses stronger when is more similar to the first patch (Fig. 2D).

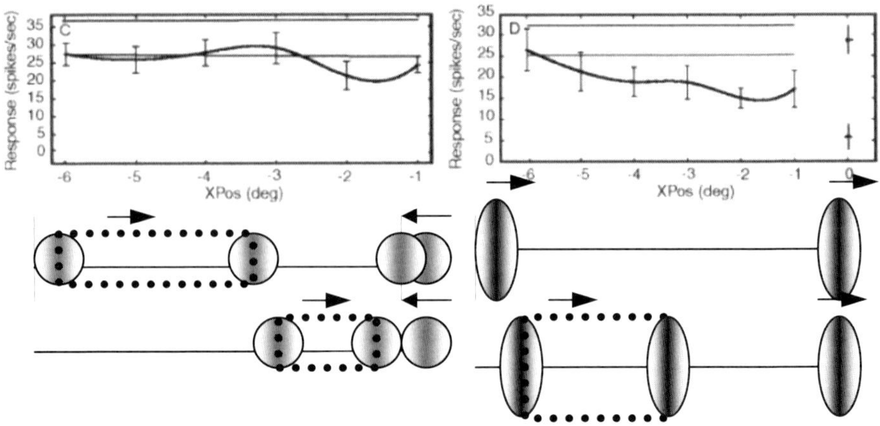

Fig. 2. Modified plots from [2]. Curves represent V4 cell responses to two 1 deg patches with gratings moving in opposite (**C**) and in the same (**D**) directions. One patch is always at *x*-axis position 0 and the second patch changes its position as it is marked in XPos coordinates. The horizontal lines represent 95% confidence intervals for the response to single patch in position 0. Below C and D schematics showing positions of the patches for *class 2* (upper parts) and *class 1* (lower parts) responses. Arrows are showing directions of moving gratings. Double dotted lines mark range of the possible positions of the second patch that give the same response.

Table 2. Decision table for one cell shown in Fig. 2. Attributes *xpr, ypr, s* = 44 are constant and are not presented in the table**.** We introduce another parameter of the stimulus, difference in the direction of drifting grating of two patches: *ddg* = 0 when drifting are in the same directions, and *ddg* = 1 if drifting in two patches are in opposite directions.

Cell	xp	xpr	xs	ys	ddg	r
64c	-4.5	3	1	1	1	2
64c1	-1.75	1.5	1	1	1	1
64c2	-0.5	1	1	1	1	2
64d	-6	0	1	8	0	2
64d1	-5.5	3	1	8	0	1

Two-patch horizontal interaction **decision rules** are as follows:

DRT3: $ddg_1 \wedge (o_0 \wedge xpr_3 \wedge xp_{-4.5} \wedge xs_1 \wedge ys_1)_1 \wedge (o_0 \wedge xp_0 \wedge xs_1 \wedge ys_1)_0 \rightarrow r_2$,
DRT4: $ddg_1 \wedge (o_0 \wedge xpr_1 \wedge xp_{-0.5} \wedge xs_1 \wedge ys_1)_1 \wedge (o_0 \wedge xp_0 \wedge xs_1 \wedge ys_1)_0 \rightarrow r_2$,
DRT5: $ddg_0 \wedge (o_0 \wedge xpr_3 \wedge xp_{-5.5} \wedge xs_1 \wedge ys_8)_1 \wedge (o_0 \wedge xp_0 \wedge xs_1 \wedge ys_1)_0 \rightarrow r_1$,

These decision rules can be interpreted as follows: patches with drifting in opposite directions gratings give strong responses when positioned very near (overlapping) or 150% of their width apart one from the other (DRT3, DRT4). Interaction of patches with a similar gratings evoked small responses in large extend of the RF (DRT5). Generally, interactions between similar stimuli evoke stronger and more extended inhibition than between different stimuli. These and other examples can be generalized to other classes of objects.

We propose following classes of the Stimuli Interaction Rules

SIR1: facilitation when stimulus consists of multiple similar thin bars with small distances (about 0.5 deg) between them, and suppression when distance between bars is larger than 0.5 deg. Suppression/facilitation can be periodic along the receptive field with dominating periods of about 30, 50, or 70% of the RF width.

SIR2: inhibition when stimulus consists of multiple similar discs with distance between their edges ranging from 0 deg (touching) to 3 deg through the RF width.

SIR3: if bars or patches have different attributes like polarity or drifting directions than suppression is smaller and localized facilitation at the small distance between stimuli is present.

SIR4: center-surround interaction, described below in detail.

We will concentrate on the center-surround interaction described above as **SIR4**. We make a decision table for nine different cells tested with discs or annuli (Pollen et al. [2] Fig. 10). If the center is stimulated with a stimulus different from that in the surround then the surround inhibitory mechanism is weak (Fig. 9B in [2]). In order to compare different cells, we have normalized their optimal orientation, denoted it as 1, and removed orientation and its values from the table.

Table 3. Decision table for eight cells comparing the center-surround interaction. All stimuli were concentric discs or annuli with *xo* – outer diameter, *xi* – inner diameter. All stimuli were localized around the middle of the receptive field, so that $xp = yp = xpr = ypr = 0$ were constant and we did not put them in the table.

Cell	sf	sfb	xo	xi	s	r
101	0.5	0	7	0	4	0
101a	0.5	0	7	2	5	1
102	0.5	0	8	0	4	0
102a	0.5	0	8	3	5	0
103	0.5	0	6	0	4	0
103a	0.5	0	6	2	5	1
104	0.5	0	8	0	4	0
104a	0.5	0	8	3	5	2
105	0.5	0	7	0	4	0
105a	0.5	0	7	2	5	1
106	0.5	0	6	0	4	1
106a	0.5	0	6	2	5	2
107	0.5	0.25	6	0	4	2
107a	2.1	3.8	6	2	5	2
107b	2	0	4	0	4	1
108	0.5	0	6	0	4	1
108a	0.9	0.9	4	0	4	2
108b	5	9	6	2	5	2
20a	0.5	0	6	0	4	1
20b	0.5	0	6	0	4	2

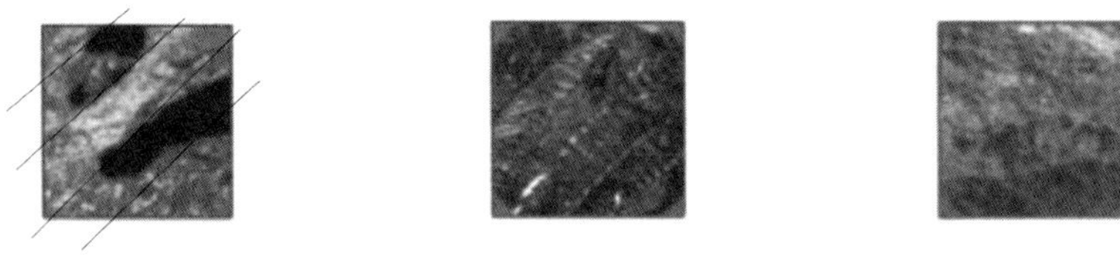

Fig. 3. In their paper David et al. [4] stimulated V4 neurons (medium size of their receptive fields was 10.2 deg) with natural images. Several examples of their images are shown above. We have divided responses of these cells into three classes. The image on the left represents cell, which gives strong response related to stimulus concept 2. The image in the middle evokes response above 20 spikes/s; that is related to stimulus concept 1. The image on the right gives very weak response; it is related to the stimulus concept 0.

This image can be seen as a group of medium x position range bars (bars of medium width), which means using the *DR3* decision rule. Even if this image shows differences between its central and surround parts, they have also many similar features like orientation or spatial frequencies. Therefore even if the center and surround alone would give strong cell responses, their interactions will be inhibitory (rule *SIR4*). In consequence, the middle image will give class 1 responses in V4 and it is related to stimulus concept 1. In the image on the right there is no significant difference between the stimulus in the center and the surround. Therefore the response will be similar to that obtained when a single disc covers the whole receptive field: *DR11, DR12*. In most cells such stimuli class will be equivalent to a stimulus concept 0.

4 Discussion

In this work we have concentrated on the pre-attentive processes. These so-called early processes extract and integrate into many parallel channels the basic features of the environment. These processes are related to the human perceptions property of objects with unsharp boundaries of values of attributes put together by similarities [5]. These similarities may be related to synchronizations of the multi-resolution parallel computations that are difficult to simulate in the digital computer [6]. It seems that it is relatively straightforward task to classify different objects on the basis of their physical properties, which define values of their attributes. Generally problem appears when the same object in different conditions changes values of its attributes, or in other words its parts became unsharp. One solution is that the brain extracts as elementary parts so-called "basic features" [7].

Our eyes constantly perceive changes in light colors and intensities. From these sensations our brain extracts features related to different objects. The "basic features" were identified in psychophysical experiments as elementary features that can be extracted in parallel. Evidence of parallel extraction comes from the fact that their extraction time is independent of the number of objects. Other features need serial search, so that the time needed to extract them is proportional to the number of objects. The high-level serial process is associated with the integration, and consolidation of items and with a conscious report. Other, low-level parallel processes are rapid, global, related to high efficiency categorization of items and largely unconscious [7]. Our work is related to the constitution of decision rules extracting basic features from the visual stream.

We have suggested previously [3] that the brain may use the multi-valued logic in order to test learned predictions about object attributes by comparing them with actual stimulus-related hypotheses. Neurons in V4 integrate object's attributes from its parts in two ways: one is relate to local excitatory-inhibitory interactions described here as SIR (stimuli interaction rules), and another way by changing possible part properties using feedback connections tuning lower visual areas. Different neurons have different SIRs watching objects by multiple "unsharp windows" (Figs. 1, 2). If object's attributes fit to the unsharp window, neuron sends positive feedback [8] to lower areas filters which in end-effect sharpen the attribute-extracting window changing neuron response from class 1 to class 2 (Fig. 1 J and K).

In summary, we have shown that using rough set theory we can divide stimulus attributes in relationships to neuronal responses into different concepts. Even if most of our concepts were very rough, they determine rules on whose basis we can predict neural responses to new, natural images.

References

1. Pawlak, Z.: Rough Sets-Theoretical Aspects of Reasoning about Data. Kluwer Academic Publishers, Boston, London, Dordrecht (1991)
2. Pollen, D.A., Przybyszewski, A.W., Rubin, M.A., Foote, W.: Spatial receptive field organization of macaque V4 neurons. Cereb Cortex 12, 601–616 (2002)
3. Przybyszewski, A.W.: Checking brain expertise using rough set theory. Rough Sets and Intelligent System Paradigms, 746–755 (2007)
4. David, S.V., Hayden, B.Y., Gallant, J.L.: Spectral receptive field properties explain shape selectivity in area V4. J. Neurophysiol. 96, 3492–3505 (2006)
5. Zadah, L.A.: Toward a perception-based theory of probabilistic reasoning with imprecise probabilities. Journal of Statistical Planning and Inference 105, 233–264 (2002)
6. Przybyszewski, A.W., Linsay, P.S., Gaudiano, P., Wilson, C.: Basic Difference Between Brain and Computer: Integration of Asynchronous Processes Implemented as Hardware Model of the Retina. IEEE Trans. Neural Networks 18, 70–85 (2007)
7. Treisman, A.: Features and objects: the fourteenth Bartlett memorial lecture. Q. J. Exp. Psychol. A 40, 201–237 (1988)
8. Przybyszewski, A.W., Gaska, J.P., Foote, W., Pollen, D.A.: Striate cortex increases contrast gain of macaque LGN neurons. Vis. Neurosci. 17, 485–494 (2000)

A Multi-objective Optimal Approach for UAV Routing in Reconnaissance Mission with Stochastic Observation Time*

Xingguang Peng and Xiaoguang Gao

School of Electronics and Information, Northwestern Polytechnical University,
Xi'an, P.R. China
pxg0510@gmail.com

Abstract. The multiple Unmanned Aerial Vehicles (UAVs) reconnaissance problem with stochastic observation time (MURSOT) is modeled by modifying the typical vehicle routing problem with stochastic demand (VRPSD). The objective consists in optimizing mission duration, total time and the quantity of UAVs. This multi-objective optimization problem is solved using a steady-state multi-objective evolutionary algorithm MOEA with $\varepsilon - dominance$ conception. In this paper, we propose a heuristic evolutionary operation (HEO) using Insert-to-Nearest Method (INM). Route Simulation Method (RSM) is presented in details to estimate the expected cost of each route and this method is designed especially for our MURSOT. The developed algorithm is further validated on a series of test problems adapted from Solomon's vehicle routing problems. Experimental results show that the INM is capable of finding better solutions in contrast and its advantage is more remarkable as the size of the problem become larger.

1 Introduction

From a practical point of view, the reconnaissance mission consists in allocating a fleet of UAVs to locations or targets with non-deterministic observation demand. Routes are assumed to start and end at the base. As mentioned in [1], a location or target is scouted by an UAV during a single visit and each visit involves a task composed of a variety of subtasks such as detection and identification/recognition. Given the nature of the locations or targets and the subtasks involved in a visit, required observation time may be non-deterministic. Objectives consist in optimizing mission duration, total travel time of the fleet and the fleet size. Taking regard of the nature of the reconnaissance mission, the maximal patrol time of each UAV is limited and a refueling is permitted. We assume that a homogeneous fleet of UAVs fly at a constant speed during the entire mission and the existence of a real-time collision avoidance procedure. As a result, this MURSOT description can be naturally linked to the VRPSD.

* This work is supported by NSFC Grant #60774064 to Xiaoguang Gao.

A. An et al. (Eds.): ISMIS 2008, LNAI 4994, pp. 246–255, 2008.

VRPSD is one issue of Stochastic Vehicle Routing Problems (SVRPs), in which elements of the problems such as the set of customers visited, the customers demands, and the travel costs, are modeled as stochastic variables with known probability distribution. As the non-deterministic character, in VRPSD, one important issue is to evaluate the routing solution with stochastic demands. In this paper, we introduce RSM to access the expected costs of solutions. In RSM, route simulation is executed once to estimate one routing solution with a set of demands randomly generated based on the demand probability distribution, then, the average expected cost of such solution can be obtained by repeating this procedure several times.

VRPSD is a NP-hard problem like most Vehicle Routing Problems (VRPs). Evolutionary computation has shown powerful ability to solve the NP-hard problems in both theory and practice. The use of multi-objective evolutionary algorithms (MOEAs) has been gaining significant attention in recent years. Many researchers have done lots of significant contributions and proposed some algorithms in [7,8,9,10], such as NAGS-II, SPEA2, PAES, PESA, etc. In this paper, we use a steady-state MOEA with $\varepsilon - dominance$ conception to solve our problem at hand.

A variable-length chromosome representation proposed in [5] is used and a HEO is proposed based on the original non-heuristic evolutionary operation (NHEO) proposed in [2]. In our HEO, the proposed INM is combined to guide the evolutionary operation process.

This paper is organized as follows: Section 2 presents the mathematical model of MURSOT. Section 3 presents the approach for solving the MURSOT. Section 4 presents the simulation results and analysis. Conclusions are drawn in Section 5.

2 Problem Formation

MURSOT is defined on a complete graph $G = (T, A, D)$, where $T = 1, 2, ..., N_T$ is a set of nodes (targets or locations) with node 0 denoting the base, $A = \{(i, j) : i, j \subset T, i \neq j\}$ is a set of arcs joining the nodes, and $D = \{d_{ij} : i, j \in T, i \neq j\}$ is the travel cost (flight time) between nodes. The travel cost d_{ij} is determined by the patrol velocity of the UAV and the Euclidean distance between node i and j. A homogeneous fleet of UAVs denoted by $V = \{1, 2, ..., N_V\}$ with limited patrol time Q have to scout a set of targets with stochastic observation demands. The mission can be assigned to at most V_{max} UAVs, that is, $N_V < V_{max}$. Each UAV is allocated to a set of locations or targets to carry out reconnaissance task composed by some subtasks, the time of finishing each subtask stochastically depends on the airborne detection system and the actual situation. For example, operators in the control station may prolong observation time if they feel the real-time environment of a location is more complex than they have forecasted or the airborne sensors do not work well, and terminate reconnaissance task for a location or target if they consider there is no need to continue. The observation demands of locations are stochastic variables $\xi_i (i = 1, 2, ..., n)$, independently distributed with known distributions. The actual demand of each location is only known when the corresponding vehicle arrives. It is also assumed that $\xi_i + 2d_{0i} \leq Q$ and ξ_i follows a discrete probability distribution $p_{ik} = Prob(\xi_i = k), k + 2d_{0i} \leq Q$. A feasible solution for each UAV is a permutation of the locations or targets $r_i = (r_i(1), r_i(2), ..., r_i(n_i)), i \in V$, starting at the base(that is, $r_i(1) = 0$), where n_i denotes the number of targets allocated to vehicle i, and it is called a priori tour. The vehicle visits the locations in the order given by a priori tour, then it has to decide, according to the actual observation demand of a location, whether to proceed to the next location or to return base to refuel in a deterministic period of d_{fuel} time because of route failure i.e. the duration starts from the vehicle leaving base to current state has exceeded the vehicle's patrol time limitation. The goal of MURSOT is to find an a priori tour that minimizes the expected mission duration, total patrol time of the fleet of UAVs and the fleet size simultaneously. The first objective is to finish the whole reconnaissance mission as soon as possible, the second one is to minimize the mission cost and the last one is to minimize the resource spent on the mission.The MURSOT model can be mathematically formulated as follows:

$$(\text{MURSOT}) \ \min f = (f_1, f_2, f_3) \tag{1}$$

$$f_1 = max\{CostV(i)\}, i \in V \tag{2}$$

$$f_2 = \sum CostV(i), i \in V \tag{3}$$

$$f_3 = N_V \tag{4}$$

Where $CostV(i)$ is obtained by RSM which is described in the next section.

3 Problem-Solving Approach

3.1 Route Simulation Method

The RSM is introduced to calculate the cost of a particular route, and its main idea is that a routing result is simulated step by step to estimate the cost. As mentioned above, due to the limited patrol time of UAV, the refueling operation is permitted when a routing failure takes place. In other words, the targets in a route are scouted in sequence, and at each step, a decision should be made whether the UAV should scout the next target or return to base to refuel. In the case where refueling operation is executed, the cost spent on going back to base and returning corresponding target should be taken into account. In addition, the time spent on refueling will be considered and the duration is a deterministic one i.e. d_{fuel} . The algorithm of RSM is given in Fig. 1.

In RSM the observation time of each target is obtained with the known distribution. Due to the stochastic nature of the cost, there is a need to repeat the RSM N_{RSM} times for every route of a particular solution, using a different set of observation time randomly obtained based on the known distribution each time and then taking the average to obtain the expected mission duration and the total patrol time of the fleet of UAVs. The algorithm of solution evaluation is shown in Fig. 2.

Algorithm 1. Route Simulation Method (RSM)

(1) Current cost $Cost_{current} = 0$, total cost $Cost_{total} = 0$, iterative times $i = 0$ and equal flag $f_{eaqual} = false$
(2) Regarding route r_k, $k = 1, ..., n_k$,
  If $f_{eaqual} = false$ is $false$
    If $Cost_{current} + d_{r_k(i),r_k(i+1)} + d_{r_k(i+1),0} \geq 0$
      $f_{equal} = ture$
      $Cost_{total} = Cost_{total} + d_{0,r_k(i)} + d_{0,r_{k(i+1)}} + d_{fuel}$
    Else
      $Cost_{total} = Cost_{total} + d_{r_k(i),r_k(i+1)}$
      $Cost_{current} = Cost_{current} + d_{r_k(i),r_k(i+1)}$
  Else $f_{equal} = false$
(3) If $Cost_{current} + \xi_{r_k(i+1)} > Q$
    $Cost_{total} = Cost_{total} + \xi_{r_k(i+1)} + 2 \times d_{0,r_k(i+1)} + d_{fuel}$
    $Cost_{current} = Cost_{current} + d_{0,r_k(i+1)} - Q$, go to (7)
(5) $Cost_{total} = Cost_{total} + \xi_{r_k(i+1)}$
(6) $Cost_{current} = Cost_{current} + \xi_{r_k(i+1)}$
(7) If $i < n_k - 1$ then $i = i + 1$, go to (2)
(8) Exit, output $Cost_{total}$

Fig. 1. Algorithm of Routing simulation method

Algorithm 2. Solution evaluation

(1) i=1
(2) If $i >$ routing number N_V, go to (11)
(3) $j = 1$, average mission duration $t_{mission}(i) = 0$, average total cost $t_{total}(i) = 0$
(4) If $j > N_{RSM}$, go to (9)
(5) Regarding route r_k, generate a set of stochastic observation time ξ_{r_i}, $r_i = r_i(1), ..., r_i(n_i)$
(6) Estimate r_i using RSM to expect mission duration t_1 and total cost t_2
(7) $t_{mission}(i) = t_{mission}(i) + t_1$, $t_{total}(i) = t_{total}(i) + t_2$
(8) $j = j + 1$, go to (4)
(9) $t_{mission}(i) = t_{mission}(i) \setminus N_{RSM}$, $t_{total}(i) = t_{total}(i) \setminus N_{RSM}$
(10) $i = i + 1$, go to (2)
(11) The solutions expected mission duration $t_{mission} = max(t_{mission}(i))$ and total cost $t_{total} = \sum_{i=1}^{N_V} t_{total}(i)$
(12) Exit, output $t_{mission}$, t_{total} and N_V

Fig. 2. Algorithm of solution evaluation

3.2 A Steady-State Multi-objective Optimal Algorithm

In this section, we introduce a steady-state MOEA based on the $\varepsilon - dominance$ conception proposed in [3], epsilon MOEA. Firstly, the $\varepsilon - dominance$ conception is necessary to be introduced. Assumed that there are n objective functions $f_i, (i = 1, 2, ..., n)$ for maximizing, the objective space along the $i - th$ objective can be divided into many small fragments by allowable tolerance $\varepsilon_i, (i = 1, 2, ..., n)$. So the whole objective space will be divided into lots of hyper-boxes according to each i and each hyper-box only allows one solution to occupy. As shown in Fig. 3, area $ABOC$ is dominated by solution A according to traditional dominance conception whereas the whole region $A^{'}B^{'}OC^{'}$ is dominated by A in the sense of the $\varepsilon - dominance$.

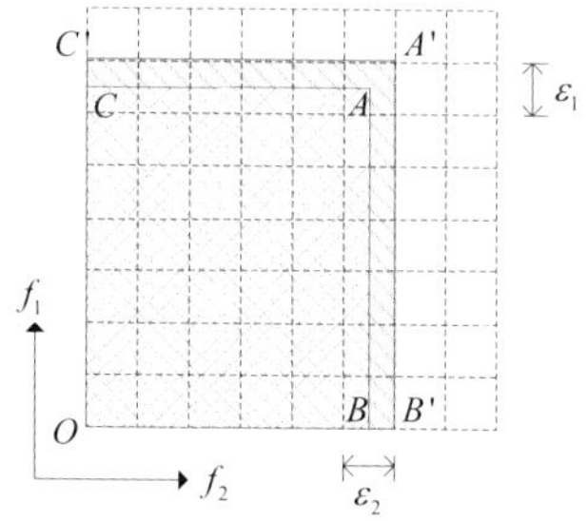

Fig. 3. The $\varepsilon - dominance$ conception for maximizing f_1 and f_2

As for the epsilon MOEA, it is a steady-state MOEA with elitist approach. At each iterative step, a solution S_p selected from population will recombine with a solution S_e randomly chosen from the elitist archive to generate an offspring S_o. Then, a decision whether to include S_o in population and archive is made according to $\varepsilon - dominance$ conception, which is described in details in [4].

3.3 Genetic Representation

As using evolutionary algorithm, we must transfer the real-world problem by genetic representation. As to MURSOT, we introduce the variable-length chromosome proposed in [5]. Such chromosome encodes a complete solution, including the quantity of vehicles and the customers served by these vehicles as shown in Fig. 4.

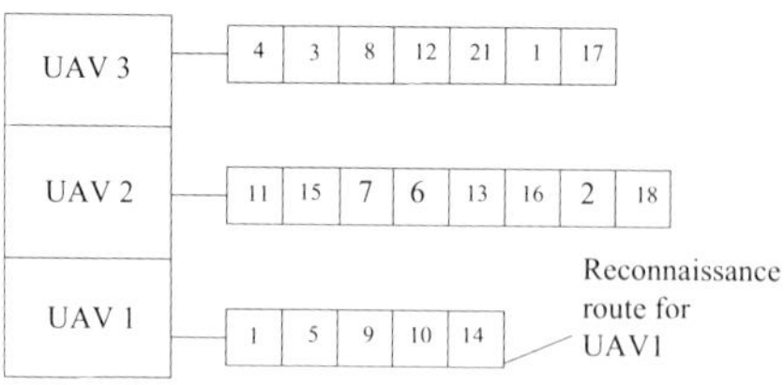

Fig. 4. Chromosome representation

3.4 Heuristic Evolutionary Operation

In this paper, we propose a HEO based on the route-exchange crossover and multi-mode mutation proposed in [5] and extended in [2]. Indeed, our crossover operation is the same as the original one, but we make the original mutation operation heuristic using the INM.

In the route-exchange crossover procedure, two routes are selected from two chromosomes which are chosen randomly from the population, and each route has the minimal cost in the corresponding chromosome. Then, each route is inserted into the other chromosome as the first route. In the case where one of the selected chromosome has only one route, a segment of the route is randomly selected to exchange with the best route of the other chromosome. After the route exchange, the duplicated customers in a chromosome should be deleted. It should be noted that these customers are deleted from the original routes while the newly inserted route is left intact. In order to increase the diversity of chromosomes to explore a lager search space, a random shuffling operator will be executed to shuffle the remanding routes with a probability $p_{shuffle}$. More details about the route-exchange crossover can be found in [2].

INM means that one customer should be inserted into a route in front of or behind the nearest customer, that is, in a route, the sequence of the newly inserted customer is identified by the nearest customer around it. As illustrated in Fig. 5, a new customer D is assigned into the original route $A-B-C-A$, and customer B is the nearest one to D. So, there will be two possible new routes after insertion of D, $A-D-B-C-A$ and $A-B-D-C-A$, and which route should be chosen depends on their travel cost. Let $Dis(C_1, C_2)$ denote the Euclidean distance between customer C_1 and C_2. If the value of $Dis(A, D) + Dis(B, C)$ is lager than $Dis(A, B) + Dis(D, C)$, the first route is worse.

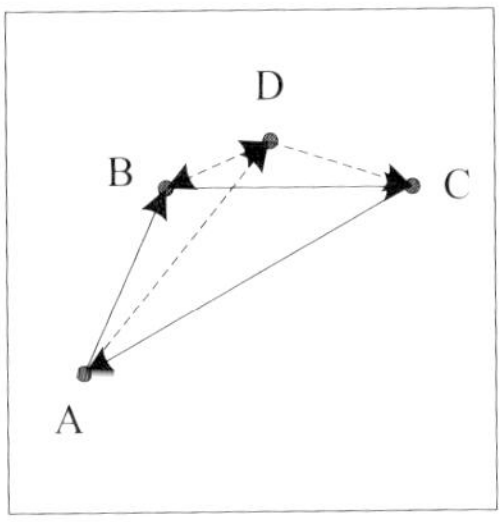

Fig. 5. INM is illustrated

Our heuristic mutation operation combines the multi-mode mutation operation proposed in [2] with INM. As a result, there are three heuristic mutation operations as follows:

- Heuristic partial swap: The operation involves a number of swap moves and for each move, two routes will be randomly chosen. A segment is then randomly selected from each route and swapped to the other route. Each new

segment inserts its elements one by one with INM to the route where the other segment is chosen. In addition, a mechanism is applied to guarantee that the same two routes will not be selected twice in a particular swap operation.
- Heuristic merge shortest route: This operation searches for two routes of the chromosome with the minimal travel cost and inserts one route into the other with INM.
- Heuristic split longest route: This operation searches for the route with the maximal travel cost and breaks the route into two segments at a random point. The new segment after the random point will be inserted into an empty route with INM.

There is no bias among these operations, and in a particular mutation operation, one of these heuristic operations is selected randomly. At the end of the heuristic multi-mode mutation operation, the random shuffling operation is applied to every route of the mutated chromosome with a probability equal to the shuffle rate $p_{shuffle}$.

4 Computational Experiment

The parameter settings for our simulation are follows: Population size = 800; Epsilon for objective 1 = 10; Epsilon for objective 2 = 10; Epsilon for objective 3 = 1; Maximal fleet size = 10; Crossover rate = 0.9; Mutation rate = 0.4; Shuffle rate = 0.1; Route simulation times = 10; Deterministic refueling time = 5; Algorithm terminates when the epsilon MOEA has iterated 10000 steps. The test problem is constructed based on the Solomon's benchmark problems which can be found in [6] by adding an observation demand standard deviation to each target. Regarding the space, we just consider the "101" series problems, that is 25_R_101, 50_RC_101, 100_C_101 etc., and name them as Solomon_25(50 or 100)_R(C or RC). The original demand quantity is treated as the mean observation time of each target and the standard deviation is uniformly generated

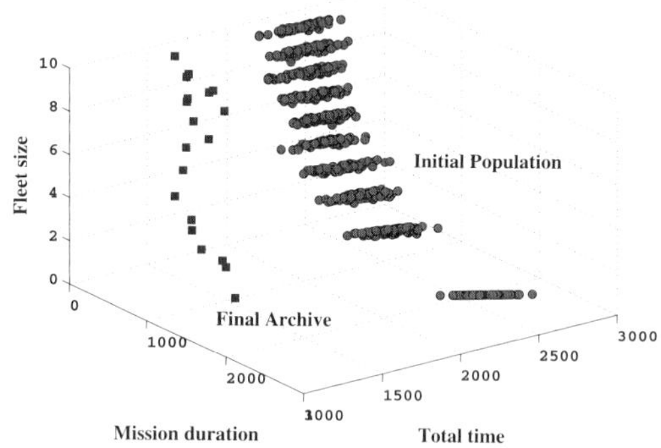

Fig. 6. Comparison between Initial population and final archive

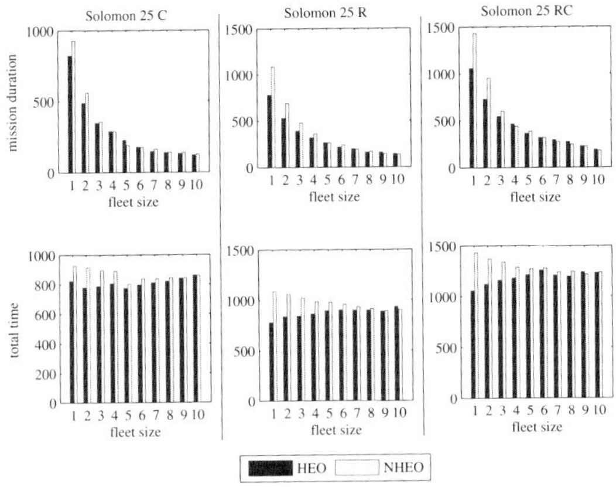

Fig. 7. Results of Solomon 25 problem

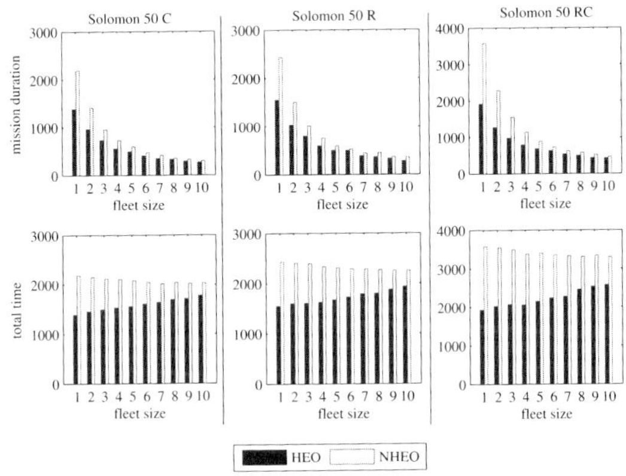

Fig. 8. Results of Solomon 50 problem

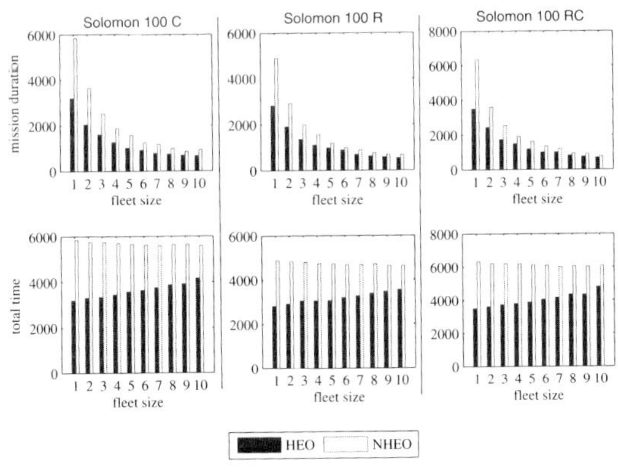

Fig. 9. Results of Solomon 100 problem

between zero and one third of the mean value. The original maximal vehicle capacity is treated as the maximal patrol time of each UAV. The velocity of the UAVs is assumed to be a constant value and the distance over which each UAV flies in one time unit is equal to one distance unit according to the maps of the Solomon series test problems.

4.1 Performance of the Steady-State MOEA with INM

To verify the ability of our algorithm for finding good non-dominated solutions, we compare the initial population and the solutions included in the final archive. Taking the Solomon_50_RC for instance, the space distribution of the solutions in initial population and the final archive are compared, which is illustrated in Fig. 6.

The result shows that the initial population is improved greatly by our algorithm and the number of the final solutions is bounded due to the $\varepsilon - dominance$ conception.

4.2 Comparison between HEO and NHEO

As for the first objective (mission duration), the more UAVs are assigned to carry out a mission the shorter the mission duration is. In the extreme case, if there are N_T UAVs to be assigned, the mission duration will be the shortest and equal to the maximal scouting time of these UAVs. In $\varepsilon - dominance$ point of view, in each case of the fleet size, there will be a non-dominated solution in the final archive after the termination of the optimization process. Consequently, as for each quantity of the UAVs, the mission duration and total time will be the criterions to evaluate the performance of the HEO and the original NHEO proposed in [2]. Because of the stochastic nature of MURSOT, each instance is run 10 times and the average of the mission duration and the total time are taken separately for comparison. Fig. 7-9 show the computational results.

It is shown that our HEO using INM could enhance the performance efficiently. Indeed, INM can guide the mutation operation to mutate a chromosome especially on the sequence in which the targets are scouted. In addition, the more complex the test problem is, the greater the INM enhances the performance.

5 Conclusion

The VRPSD has been modified to model the MURSOT which is inherently a multi-objective optimization problem that involves the optimization of routes for multiple UAVs to minimize the mission duration, total time and vehicles quantity simultaneously. Route simulation method has been presented in details to estimate the expected cost of each route. Given the nature of the reconnaissance mission and the maximal patrol time constraint of each UAV, the refueling has been taken into account. A steady-state MOEA with $\varepsilon - dominance$ conception has been equipped to optimize these multiple objectives and a list of non-dominated solutions can be archived after the termination of the optimization procedure. A multiple integer string genotype has been presented to encode a solution for MURSOT, and an efficient evolutionary operation using INM has been proposed. The computational results have shown that the proposed heuristic evolutionary operation could enhance the performance outstandingly as the problem size growing.

References

1. Berger, J., Barkaoui, M., Boukhtouta, A.: A hybrid genetic approach for airborne sensor vehicle routing in real-time reconnaissance missions. Aerospace Science and Technology 11(4), 317–326 (2007)
2. Tan, K.C., Cheong, C.Y., Goh, C.K.: Solving multiobjective vehicle routing problem with stochastic demand via evolutionary computation. European Journal of Operational Research 117(2), 813–839 (2006)
3. Laumanns, M., Thiele, L., Deb, K., Zitzler, E.: Combining convergence and diversity in evolutionary multiobjective optimization. Evolutionary Computation 10(2), 263–282 (2002)
4. Deb, K., Mohan, M., Mishra, S.: A fast multi-objective evolutionary algorithm for finding well-spread Pareto-optimal solutions. Technical Report KanGAL-TR-200302, Department of Mechanical Engineering Indian, Institute of Technology Kanpur (2003)
5. Tan, K.C., Lee, T.H., Chew, Y.H., Lee, L.H.: A hybrid multiobjective evolutionary algorithm for solving truck and trailer vehicle routing problems. In: Proceedings of the 2003 Congress on Evolutionary Computation, pp. 2134–2141 (2003)
6. The VRP Web, Available: `http://neo.lcc.uma.es/radi-aeb/WebVRP`
7. Deb, K., Pratap, A., Agarwal, S., Meyarivan, T.: A fast and elitist multiobjective genetic algorithm: NAGS-II. Evolutionary Computation 6(2), 182–197 (2002)
8. Zitzler, E., Laumanns, M., Thiele, L.: SPEA2: Improving the strength Pareto evolutionary algorithm. Technical Report TIK-TR-103, Department Information Technology and Electrical Engineering, Computer Engineering and Networks Lab (2001)
9. Knowles, J.D., Corne, D.W.: The Pareto archived solution strategy: A new baseline algorithm for multiobjective optimization. In: Proceedings of the 1999 IEEE Congress on Evolutionary Computation, pp. 98–105 (1999)
10. Corne, D.W., Knowles, J.D., Oates, M.J.: The Pareto envelope-based selection algorithm for multiobjective optimization. In: Proceedings of the Parallel Problem Solving from Nature VI Conference, pp. 839–848 (2000)

Hybrid Unsupervised/Supervised Virtual Reality Spaces for Visualizing Gastric and Liver Cancer Databases: An Evolutionary Computation Approach

Alan J. Barton and Julio J. Valdés

National Research Council Canada, M50, 1200 Montreal Rd., Ottawa, ON K1A 0R6
alan.barton@nrc-cnrc.gc.ca,
julio.valdes@nrc-cnrc.gc.ca
http://iit-iti.nrc-cnrc.gc.ca

Abstract. This paper expands a multi-objective optimization approach to the problem of computing virtual reality spaces for the visual representation of relational structures (e.g. databases), symbolic knowledge and others, in the context of visual data mining and knowledge discovery. Procedures based on evolutionary computation are discussed. In particular, the NSGA-II algorithm is used as a framework for an instance of this methodology; simultaneously minimizing Sammon's error for dissimilarity measures, and mean cross-validation error on a k-nn pattern classifier. The proposed approach is illustrated with two examples from cancer genomics data (e.g. gastric and liver cancer) by constructing virtual reality spaces resulting from multi-objective optimization. Selected solutions along the Pareto front approximation are used as nonlinearly transformed features for new spaces that compromise similarity structure preservation (from an unsupervised perspective) and class separability (from a supervised pattern recognition perspective), simultaneously. The possibility of spanning a range of solutions between these two important goals, is a benefit for the knowledge discovery and data understanding process. The quality of the set of discovered solutions is superior to the ones obtained separately, from the point of view of visual data mining.

1 Introduction

The World Health Organization (WHO) states that cancer is one of the leading causes of death in the world (`http://www.who.int/cancer/en/`) and that there are more than 100 types of cancers in which any part of the body may be affected. In particular, among men, the 5 most common types of cancer that kill are (in order of frequency): lung, stomach, liver, colorectal and oesophagus. As such, a previous study investigated lung cancer [14] and this new study investigates stomach and liver cancers. The presented approach provides the possibility of obtaining a set of spaces in which the different objectives are expressed in different degrees, with the proviso that no other spaces could improve any of the considered criteria individually (if spaces are constructed using the solutions along the Pareto front). This strategy clearly represents a conceptual improvement in comparison with spaces computed from the solutions obtained by single-objective optimization algorithms. A VR technique for visual data mining

A. An et al. (Eds.): ISMIS 2008, LNAI 4994, pp. 256–261, 2008.

on heterogeneous, imprecise and incomplete information systems was introduced in [12,13] (see also `http://www.hybridstrategies.com`).

2 The Multi-objective Approach: A Hybrid Perspective

An enhancement to the traditional evolutionary algorithm [1], is to allow an individual to have more than one measure of fitness within a population (e.g. a weighted sum [2]). Multi-objective optimization, however, relies on the concept of a Pareto Front [10] of best current solutions, rather than a single best solution. One particular algorithm for multi-objective optimization is the elitist non-dominated sorting genetic algorithm (NSGA-II) [5], [4], [3], [2]. Following a principle of parsimony this paper will consider the use of only two criteria, namely, Sammon's error (Eq-3) for the unsupervised case and mean cross-validated classification error with a k-nearest neighbour pattern recognizer for the supervised case. Clearly, more requirements can be imposed on the solution by adding the corresponding objective functions. The proximity (or similarity) of an object to another object may be defined by a distance (or similarity) calculated over the independent variables and can be defined by using a variety of measures. In the present case a normalized Euclidean distance is chosen:

$$d_{\overleftarrow{x}\overleftarrow{t}} = \sqrt{(1/p)\sum_{j=1}^{p}(x_{ij} - t_{kj})^2} \tag{1}$$

Examples of error measures frequently used for structure preservation are:

$$\text{S stress} = \sqrt{\frac{\sum_{i<j}(\delta_{ij}^2 - \zeta_{ij}^2)^2}{\sum_{i<j}\delta_{ij}^4}}, \tag{2}$$

$$\text{Sammon error [11]} = \frac{1}{\sum_{i<j}\delta_{ij}} \frac{\sum_{i<j}(\delta_{ij} - \zeta_{ij})^2}{\delta_{ij}} \tag{3}$$

$$\text{Quadratic Loss} = \sum_{i<j}(\delta_{ij} - \zeta_{ij})^2 \tag{4}$$

For heterogeneous data involving mixtures of nominal and ratio variables, the Gower similarity measure [6] has proven to be suitable. This measure can be easily extended for ordinal, interval, and other kind of variables.

2.1 Public Data

Each sample in this study is a vector in a high dimensional space. Direct inspection of the data structure and of the relationships between the descriptor variables (the genes) and the type of sample (normal/gastric cancer or control/liver tumor), is impossible. Moreover, within the collection of genes there is a mixture of potentially relevant genes with others which are irrelevant, noisy, etc.

Gastric Cancer: Gene expressions were compared in [7] to gain molecular understanding of gastric cancer. The public data contains 30 patient samples with 2 classes (8 samples of noncancerous gastric tissues and, 22 samples of primary human advanced gastric cancer tissues), with 7,129 attributes (of which 34 values were missing) and was obtained from `http://www.ncbi.nlm.nih.gov/projects/geo/gds/gds_browse.cgi?gds=1210`.

Liver Cancer: Gene expressions were compared in [8] to gain molecular understanding of similarities between livers from zebrafish (Danio rerio) and 4 human tumor types (liver, gastric, prostate and lung). The public data contains 20 zebrafish samples with 2 classes (10 control samples and, 10 samples of zebrafish liver cancer), with 16,512 attributes and was obtained from `http://www.ncbi.nlm.nih.gov/projects/geo/gds/gds_browse.cgi?gds=2220`.

2.2 Results

Two sets of 100 non-dominated solutions were found (Fig-1(a) and Fig-2(a)) using the experimental settings in Table-1 for which the true location of the Pareto front is unknown. From these, three solutions were selected to investigate the *i)* best supervised solutions (Fig-1(b) and Fig-2(b)), resolving the respective classes at the cost of possible space distortions, *ii)* best unsupervised solutions (Fig-1(d) and Fig-2(d)), and *iii)* compromised solutions (Fig-1(c) and Fig-2(c)), of both class separation and internal data structure preservation.

Table 1. Experimental settings for computing the pareto-optimal solution approximations by the multi-objective genetic algorithm (PGAPack [9] extended by embedding NSGA-II)

NumObjects	30 for Gastric Data	20 for Liver	
population size	100	number of generations	500
chromosome length	= 3· NumObjects	ga seed	101
No. new inds. in $(i+1\text{st})$ pop.	20	objective functions should be	minimized
chromosome data representation	real	crossover probability	0.8
crossover type	uniform (prob. 0.6)	mutation probability	0.4
mutation type	gaussian	selection type	tournament
tournament probability	0.6	mutation and crossover	yes
population initialization	random, bounded	lower bound for initialization	-2
upper bound for initialization	2	fitness values	raw
stopping criteria	maximum iterations	restart ga during execution	no
parallel populations	no		
number of objectives	2	number of constraints	0
pre-computed diss. matrix	Gower dissimilarity		
evaluation functions	mean cross-validated k-nn error and Sammon error		
cross-validation (c.v.)	5 folds	randomize before c.v.	yes
knn seed	101	k nearest neighbors	3
non-linear mapping measure	Sammon	dimension of the new space	3

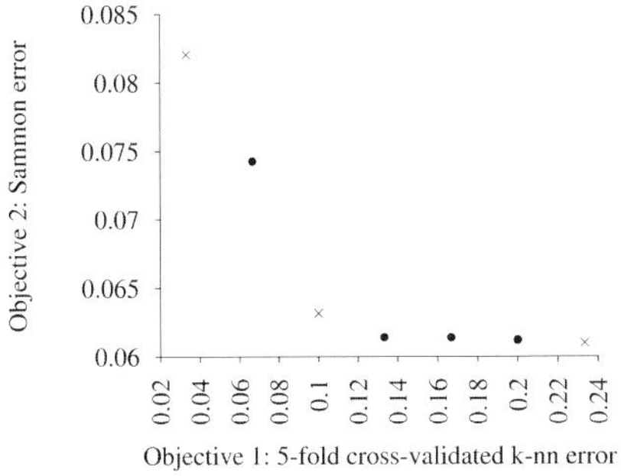

(a) Front obtained by multi-objective optimization (NSGA-II) that approximates the true Pareto Front for Gastric data.

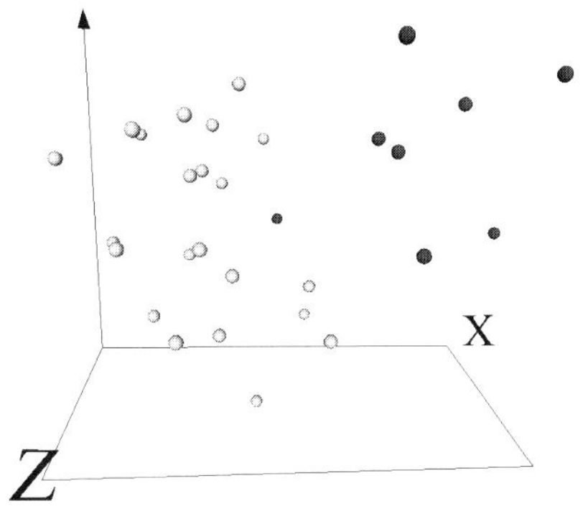

(b) Computed minimum solution for objective 1 is chromosome 0 (5-fold CV k-nn Error: 0.0333333, Sammon Error: 0.0820461)

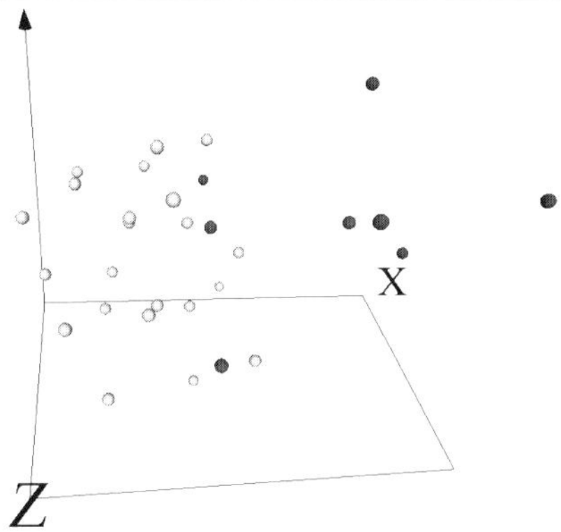

(c) Computed tradeoff solution for both objectives is chromosome 2 (5-fold CV k-nn Error: 0.100000, Sammon Error: 0.0631712)

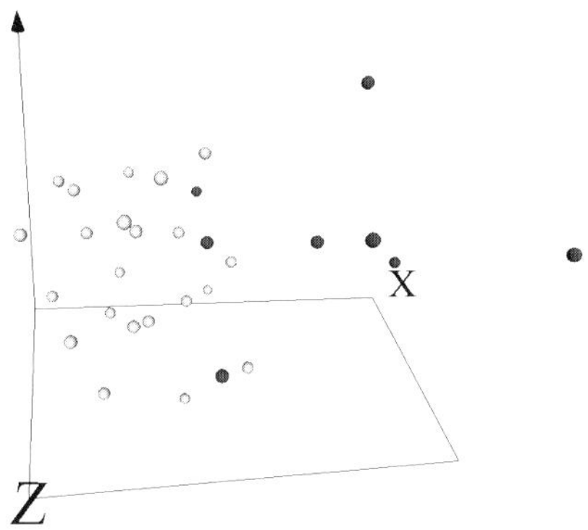

(d) Computed minimum solution for objective 2 is chromosome 1 (5-fold CV k-nn Error: 0.233333, Sammon Error: 0.0609954)

Fig. 1. Set of computed 100 multi-objective solutions for gastric cancer dataset. Those along the Pareto front approximation progressively span the extremes between minimum classification error and minimum dissimilarity loss. 3 solutions were selected and snapshots of computed VR spaces taken. Geometries: "light grey spheres" = normal samples, "dark grey spheres" = cancer samples. Behavior = static. The axis in the 3D views are highly non-linear maps from the original space (7, 129 dimensions) to the respective 3D spaces.

In general, different mappings lead to similar 3D visual representations; indicating good solution reproducibility even under the condition of potentially large amounts of attribute noise, redundancy, and irrelevancy within the sets of 7, 129 and 16, 512 original attributes. The major differences lie with local discrepancies with respect to the placement of some objects, which would need to be investigated further. For example, the object near the origin of Fig-2(b-d) is located differently in the three spaces.

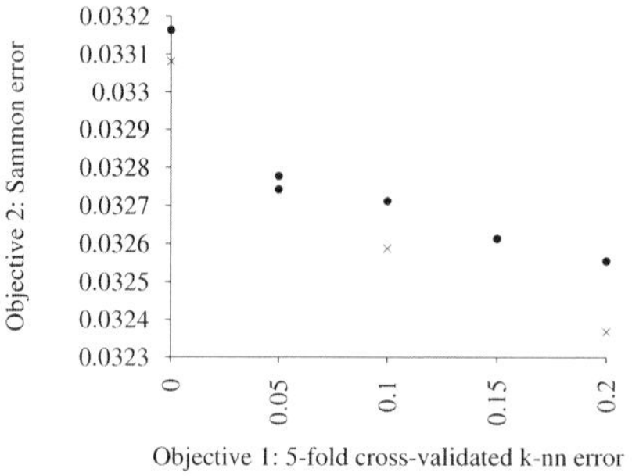

(a) Front obtained by multi-objective optimization (NSGA-II) that approximates the true Pareto Front for Liver data.

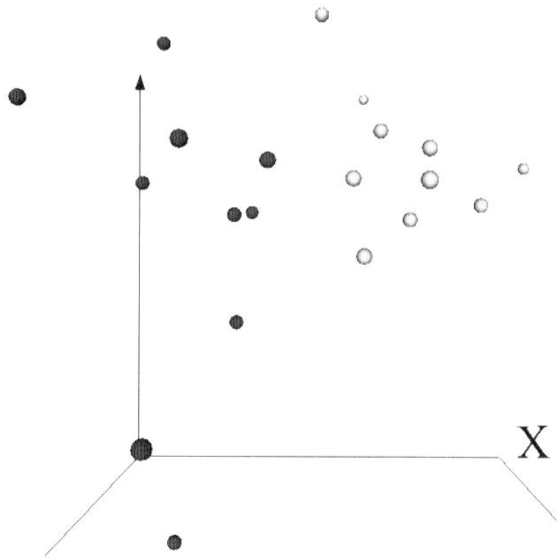

(b) Computed minimum solution for objective 1 is chromosome 9 (5-fold CV k-nn Error: 0.00000, Sammon Error: 0.0330806)

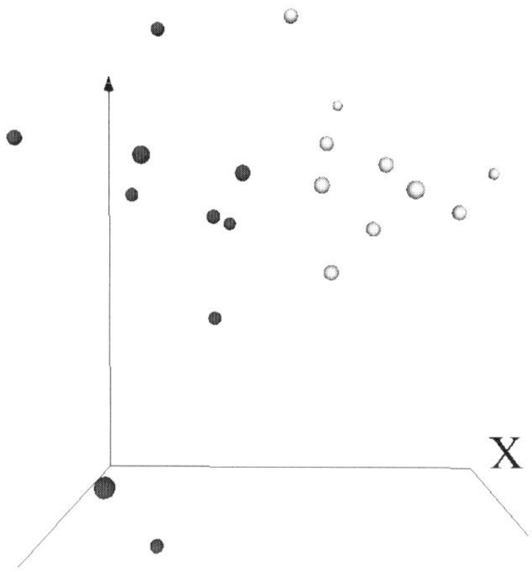

(c) Computed tradeoff solution for both objectives is chromosome 8 (5-fold CV k-nn Error: 0.100000, Sammon Error: 0.0325873)

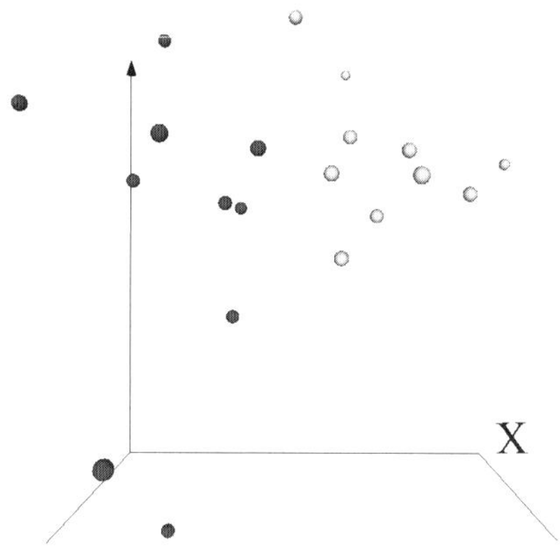

(d) Computed minimum solution for objective 2 is chromosome 0 (5-fold CV k-nn Error: 0.200000, Sammon Error: 0.0323679)

Fig. 2. Set of computed 100 multi-objective solutions for liver cancer dataset. Those along the Pareto front approximation progressively span the extremes between minimum classification error and minimum dissimilarity loss. 3 solutions were selected and snapshots of computed VR spaces taken. Geometries: "light grey spheres" = control samples, "dark grey spheres" = liver tumor samples. Behavior = static. The axis in the 3D views are highly non-linear maps from the original space (16, 512 dimensions) to the respective 3D spaces.

3 Conclusions

Analysis of high dimensional genomic data collected in the framework of Gastric and Liver cancer research was performed within the context of visual data mining and knowledge discovery research. Sequences of visualizations showing progression from spaces with minimum class separation and poor similarity preservation to spaces with reversed characteristics were reported. Solutions with reasonable compromises between the two criteria were identified. These preliminary research results expand the set of previously investigated real world cancer data sets. They also show the large potential for such an approach. Further investigations are required.

References

1. Bäck, T., Fogel, D.B., Michalewicz, Z.: Handbook of Evolutionary Computation. Institute of Physics Publishing and Oxford Univ. Press, New York, Oxford (1997)
2. Burke, E.K., Kendall, G.: Search Methodologies: Introductory Tutorials in Optimization and Decision Support Techniques. Number 0-387-23460-8. Springer Science and Business Media, Inc., 233 Spring Street, New York, NY 10013, USA (2005)
3. Deb, K., Agarwal, S., Meyarivan, T.: A fast and elitist multi-objective genetic algorithm: Nsga-ii. IEEE Transaction on Evolutionary Computation 6(2), 181–197 (2002)
4. Deb, K., Agarwal, S., Pratap, A., Meyarivan, T.: A fast elitist non-dominated sorting genetic algorithm for multi-objective optimization: Nsga-ii. In: Proceedings of the Parallel Problem Solving from Nature VI Conference, Paris, France, September 16-20, pp. 849–858 (2000)
5. Deb, K., Pratap, A., Agarwal, S., Meyarivan, T.: A fast and elitist multi-objective genetic algorithm: Nsga-ii. Technical Report 2000001, Kanpur Genetic Algorithms Laboratory (KanGAL), Indian Institute of Technology Kanpur (2000)
6. Gower, J.C.: A general coefficient of similarity and some of its properties. Biometrics 1(27), 857–871 (1973)
7. Hippo, Y., Taniguchi, H., Tsutsumi, S., Machida, N., Chong, J., Fukayama, M., Kidama, T., Aburatani, H.: Global Gene Expression Analysis of Gastric Cancer by Oligonucleotide Microarrays. Cancer Research 62, 233–240 (2002)
8. Lam, S.H., Wu, Y.L., Vega, V.B., Miller, L.D., Spitsbergen, J., Tong, Y., Zhan, H., Govindarajan, K.R., Lee, S., Mathavan, S., Krishna Murthy, K.R., Buhler, D.R., Liu, E.T., Gong, Z.: Conservation of gene expression signatures between zebrafish and human liver tumors and tumor progression. Nature Biotechnology 24, 73–75 (2006)
9. Levine, D.: Users Guide to the PGAPack Parallel Genetic Algorithm Library. Argonne National Laboratory, 9700 South Cass Avenue, Argonne, IL 60439 (January 1996)
10. Pareto, V.: Cours D'Economie Politique, vol. I and II. F. Rouge, Lausanne (1896)
11. Sammon, J.W.: A non-linear mapping for data structure analysis. IEEE Trans. Computers C18, 401–408 (1969)
12. Valdés, J.J.: Virtual reality representation of relational systems and decision rules. In: Hajek, P. (ed.) Theory and Application of Relational Structures as Knowledge Instruments, Prague (November 2002), Meeting of the COST Action 274
13. Valdés, J.J. (ed.): VR representation of information systems and decision rules. LNCS (LNAI), vol. 2639, pp. 615–618. Springer, Heidelberg (2003)
14. Valdés, J.J., Barton, A.J.: Hybrid Unsupervised/Supervised Virtual Reality Spaces for Visualizing Cancer Databases: An Evolutionary Computation Approach. In: Sandoval, F., Prieto, A.G., Cabestany, J., Graña, M. (eds.) IWANN 2007. LNCS, vol. 4507, Springer, Heidelberg (2007)

Self-calibrating Strategies for Evolutionary Approaches that Solve Constrained Combinatorial Problems*

Elizabeth Montero and María-Cristina Riff

Departamento de Informática
Universidad Técnica Federico Santa María
Valparaíso, Chile
{Elizabeth.Montero,Maria-Cristina.Riff}@inf.utfsm.cl

Abstract. In this paper, we evaluate parameter control strategies for evolutionary approaches to solve constrained combinatorial problems. For testing, we have used two well known evolutionary algorithms that solve the Constraint Satisfaction Problems GSA and SAW. We contrast our results with REVAC, a recently proposed technique for parameter tuning.

Keywords: Parameter Control, Evolutionary Algorithms.

1 Introduction

The process of finding adequate parameter values for evolutionary algorithm is a time-consuming task and considerable effort has already gone into automating this process [5]. Taking into account that a run of an evolutionary algorithm is an intrinsically dynamic adaptive process, a dynamic adaptation of the parameters during the search could help to improve the performance of the algorithm [1], [2], [8], [9], [4]. The key idea of adapting parameters is to monitor the search to be able to trigger actions, in order to improve the performance of the evolutionary algorithms. Constrained combinatorial problems are hard problems and some evolutionary approaches have been proposed in the literature to solve them. In this paper, we evaluate our strategies using two well-known evolutionary algorithms that solve Constraint Satisfaction Problems: GSA [12] and SAW [11]. From the control point of view they have interesting characteristics:

– GSA works with a micro-population of 6 individuals. At each generation just one operator is applied, furthermore all operators generate just one child. Thus, two successive populations only differ by one individual.
– SAW[1] uses an order-based representation. It has two operators. It was reported as the best evolutionary algorithm to solve CSPs.

* The authors were supported by the Fondecyt Project 1080110.
[1] We have obtained the problems and the SAW code from the web page http://www.xs4all.nl/bcraenen/resources.html

A. An et al. (Eds.): ISMIS 2008, LNAI 4994, pp. 262–267, 2008.

This paper is organized as follows: In the next section we briefly present the related work. In section 3 we introduce the adaptive techniques. In section 4 we present the performance evaluation. Finally, section 5 presents the conclusions and future work.

2 Related Work

The parameter selection process is classified into two different methods: *tuning* and *parameter control* [3]. Tuning searches the "best" parameter values for an evolutionary algorithm. Recently, a new technique called REVAC [10] has been proposed to do tuning in an efficient way. Many approaches have been proposed in the research community to do parameter control. They allow the algorithm to change its parameters values during its search [7], [1]. Another interesting method has been proposed in [6]. In this approach an external agent receives the search information from the genetic algorithm, and does a reinforcement learning task giving new parameter values to the algorithm. In this approach the Genetic Algorithm is not itself adaptive. A recent research [4] allow to change the population size of a genetic algorithm during the search. Gómez introduced the Hybrid Adaptive Evolutionary Algorithm HAEA,[13]. His algorithm randomly modifies the operator probabilities and it has been applied to solve continuous optimization problems.

3 Parameter Control Strategies

Our aim is to evaluate two kinds of strategies that allow parameter adaptation in an evolutionary approach according to the problem in hand. We propose two types of reinforcement control: A Self-Adaptive and An Adaptive one. The key idea for both control strategies is that an operator receives a reward when its application improves the search of the algorithm. Analogously, it receives a penalty when its application generates individual with worse fitness value than its parents. Both the rewards and the penalties strongly depend on the evaluation function value.

Definition 1. *Given a set of parents, a fitness function F to be maximized and a genetic operator O_k selected to be applied using the probability P_h to generate an offspring C_j. We define a success measure for the operator O_k in its $a-th$ application, $S_a(O_k)$ as:*

$$S_a(O_k) = F(C_j) - F(P) \tag{1}$$

where $F(C_j)$ and $F(P)$ are the respective fitness of the child and the average fitness of its parents. The value of $S_a(O_k)$ is positive when the child generated by the operator O_k in its a_{th} application has a better evaluation than the average of its parents.

Definition 2. *Given the success measure $S_a(O_k)$ for each operator O_k we define a quality measure $Q_a(O_k)$ as follows:*

$$Q_a(O_k) = \begin{cases} S_a(O_k) \text{ for Self-Adaptive Control} \\ \overline{S_a(O_k)} \text{ for Adaptive Control} \end{cases} \tag{2}$$

where $\overline{S_a(O_k)}$ is the average success measure of the operator O_k.

Remark 1. The self-adaptive control is looking for individual improvement instead of adaptive control is searching for a population improvement.

We need to distinguish between a positive and a negative behaviour. The following two definitions explicitly express that:

Definition 3. *Given a set of operators O_k, $k = o_1, .., o_M$ with a quality measure $Q_a(O_k) \geq 0$, $M \leq p$, we define $Max - i_l$ as the maximum improvement done by the operators during the last l generations as:*

$$Max - i_l = Argmax_{a=1,..,A_k, k=o_1,..,o_M}(Q_a(O_k)) \tag{3}$$

where A_k is the number of applications of the operator O_k in the last l generations and p is the number of operators.

Definition 4. *Given a set of operators O_k, $k = o_1, .., o_M$ with a quality measure $Q_a(O_k) < 0$, $M \leq p$, we define $Max - d_l$ as the maximum degradation done by the operators during the last l generations as:*

$$Max - d_l = Argmax_{a=1,..,A_k, k=o_1,..,o_M}(|Q_a(O_k)|) \tag{4}$$

where A_k is the number of applications of the operator O_k in the last l generations, and p is the number of operators.

We use both maximum improvement and degradation values to reward or to penalize each operator using the following definition:

Definition 5. *Given the current probability value of the operator O_k $Pr_c(O_k)$, and its quality measure $Q_a(O_k)$. We change the O_k probability according to:*

$$Pr(O_k) = Pr_c(O_k) + \rho * \frac{Q_a(O_k)}{Max_l} \tag{5}$$

where,

$$Max_l = \begin{cases} Max - i_l \text{ if } Q_a(O_k) \geq 0 \\ Max - d_l \text{ otherwise} \end{cases} \tag{6}$$

and ρ is included for providing smooth parameter adjustments, as it is required for any effective dynamic parameter control strategy. Its value is fixed at 0.1.

Remark 2. For the self-adaptive control the childs generated by the operator O_k inherit this new value of the probability of the operator O_k. Each of them include it in its representation. For adaptive control the following generation will use this value as the probability of the operator O_k for the whole population.

4 Experimental Results

The goal of the following experiments is to evaluate the performance of both kinds of dynamic control strategy. We have done two set of tests. One to evaluate the adaptive strategy and the second one to evaluate the self-adaptive strategy.

In order to present a more complete comparison we have implemented REVAC. REVAC is a good method for tuning and we applied it to the algorithms to obtain a finer tuning than the hand-made one. The hardware platform for the experiments was a PC AMD Athlon XP, 3.2 Ghz with 512 MB RAM under the Linux Debian 3.1 operating system. The algorithm has been implemented in C++.

4.1 Evaluation for GSA

We test GSA using the same benchmarks of the paper that introduced this algorithm. These benchmarks are reported in the literature as the hardest random generated CSP.

Benchmarks for GSA. We have used the CSP random generator from [11]. We use the model B with 15 variables and domain size of 15. The values for p_1 and p_2 belong to the set $\{0.1, 0.3, 0.5, 0.7, 0.9\}$. We consider 25 different configurations and we select 7 of them which are in the hardest zone. For each of these 7 configurations we have generated 30 different problem instances. For each problem the algorithm runs 10 times with different initial seed. The maximum number of iterations is fixed in 50.000.

The original reported GSA has been defined using a population size of 6 individuals. However, REVAC obtained a population size value of 3. The success rates found by using the reported values, using REVAC values and with our strategy are shown in table 1. The number of problems solved by the algorithm has been increased using both techniques. The adaptive strategy and REVAC has similar success rates, however using our strategy the algorithm itself adjusts its parameters values during its search according to the problem at hand.

Table 1. % Success Rates for GSA

Category	Reported	REVAC	Adaptive Control
d0.3_t0.7	60.7	89.36	92.69
d0.5_t0.5	84.4	91.69	92.35
d0.5_t0.7	64.3	99	99.33
d0.7_t0.5	16.7	86.37	88.03
d0.7_t0.7	70.1	99.66	99.66
d0.9_t0.5	3.3	98	98.33
d0.9_t0.7	73.2	99.66	99.66

4.2 Evaluation of the Self-adaptive Strategy

In this section we present the evaluation of our self-adaptive strategy. We compare the success rate of the algorithms using REVAC for tuning and two versions of our self-adaptive control: a light-version with random penalties/rewards and the version with penalties/rewards according to the fitness improvement.

Figure 1 shows the results for GSA. Among these methods the random penalties/rewards version shown a lower performance. REVAC and the self-adaptive technique SA_c obtain similar success rates Table 2 shows the results for SAW. The reported results of SAW are improved for all the parameter control strategies compared in this work. REVAC and our self-adaptive techniques obtain similar results.

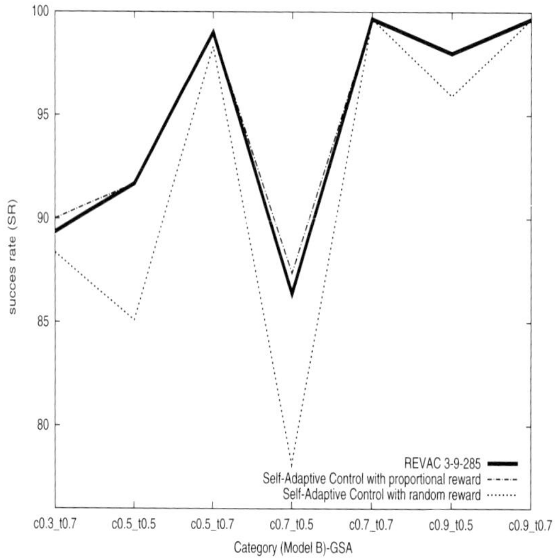

Fig. 1. Tuning v/s Self-Adaptive Strategies comparison of GSA

Table 2. Success rates for SAW

t	Reported 100-50-25	REVAC 21-2-46	Self-Adaptive fitness	Self-Adaptive random
0.24	100	100	100	100
0.25	100	100	100	100
0.26	100	100	100	100
0.27	100	100	100	100
0.28	100	100	100	100
0.29	100	100	100	100
0.30	98.8	100	100	100
0.31	68.88	75.29	72.51	73.08
0.32	29.6	30.67	31.87	28.68
0.33	2	9.96	9.87	9.16

5 Conclusions

We use in this paper two strategies for dynamic parameter control, an adaptive and a self-adaptive one. Both strategies help GSA and SAW to improve their performance. Both algorithms improve their previously reported performance using REVAC for tuning. However, despite REVAC being a good technique for tuning, it is a time consuming computational task. The time invested by REVAC is larger than the overhead included on the algorithm using our strategies. For a future work a promising research area is the collaboration between various parameter control strategies.

References

[1] Davis, L.: Adapting Operator Probabilities in Genetic Algorithms. In: Proceedings of 3rd. International Conf. on Genetic Algorithms and their Applications (1989)
[2] Deb, K., Agrawal, S.: Understanding Interactions among Genetic Algorithms Parameters. Foundations of Genetic Algorithms 5, 265–286 (1998)
[3] Eiben, A., Hinterding, R., Michalewicz, Z.: Parameter Control in Evolutionary Algorithms. IEEE Transactions on evolutionary computation 3(2), 124–141 (1999)
[4] Eiben, A.E., Marchiori, E., Valko, V.A.: Evolutionary Algorithms with on-the-fly Population Size Adjustment. In: Yao, X., Burke, E.K., Lozano, J.A., Smith, J., Merelo-Guervós, J.J., Bullinaria, J.A., Rowe, J.E., Tiňo, P., Kabán, A., Schwefel, H.-P. (eds.) PPSN 2004. LNCS, vol. 3242, pp. 41–50. Springer, Heidelberg (2004)
[5] Hinterding, R., Michalewicz, Z., Eiben, A.: Adaptation in Evolutionary Computation: A Survey. In: Proceedings of 4th. IEEE International Conf. on Evolutionary Computation (1997)
[6] Pettinger, J., Everson, R.: Controlling Genetic Algorithms with Reinforcement Learning. In: Proceedings of the GECCO 2002 (2002)
[7] Riff, M.-C., Bonnaire, X.: Inheriting Parents Operators: A New Dynamic Strategy to improve Evolutionary Algorithms. In: Hacid, M.-S., Raś, Z.W., Zighed, A.D.A., Kodratoff, Y. (eds.) ISMIS 2002. LNCS (LNAI), vol. 2366, pp. 333–341. Springer, Heidelberg (2002)
[8] Smith, J., Fogarty, T.C.: Operator and parameter adaptation in genetic algorithms. Soft Computing 1(2), 81–87 (1997)
[9] Tuson, A., Ross, P.: Adapting Operator Settings in Genetic Algorithms. Evolutionary Computation 2(6), 161–184 (1998)
[10] Nannen, V., Eiben, A.E.: Relevance Estimation and Value Calibration of Evolutionary Algorithm Parameters. In: Proceedings of the Joint International Conference for Artificial Intelligence (IJCAI) (2006)
[11] Craenen, B.G.W., Eiben, A.E., van Hemert, J.I.: Comparing evolutionary algorithms on binary constraint satisfaction problems. IEEE Transactions on Evolutionary Computation 7(5), 424–444 (2003)
[12] Dozier, G., Bowen, J., Homaifar, A.: Solving Constraint Satisfaction Problems Using Hybrid Evolutionary Search. IEEE Transactions on Evolutionary Computation 2(1), 23–33 (1998)
[13] Gomez, J.: Self Adaptation of Operator Rates in Evolutionary Algorithms. In: Deb, K., et al. (eds.) GECCO 2004. LNCS, vol. 3102, pp. 1162–1173. Springer, Heidelberg (2004)

Towards Fuzzy Query Answering Using Fuzzy Views – A Graded-Subsumption-Based Approach*

Allel Hadjali and Olivier Pivert

IRISA/Enssat, Technopole Anticipa BP 80518,
22300 Lannion, France
{hadjali,pivert}@enssat.fr

Abstract. This paper deals with fuzzy query processing in a distributed database context. In the framework considered, user queries and view descriptions may include fuzzy predicates. A satisfaction threshold is also associated with any user query. The idea is to match a fuzzy query against the fuzzy views in order to select only the views that will provide answers whose satisfaction degree is over the threshold specified by the user. The main tool for achieving this task is a graded subsumption mechanism, based on a fuzzy implication, which is discussed in a detailed way.

1 Introduction

A recent phenomenon in the world of information systems is the emergence of decentralized approaches to data sharing, illustrated in particular by peer-to-peer systems (P2P) [11]. In such a context, due to the bandwith costs related to query propagation, it becomes crucial to forward the query only to the (more or less) relevant sources. Many research works have been performed in the last decade about the problem of answering (regular) queries using (regular) views [10].

On the other hand, some recent works have also acknowledged the need for flexible ways to access information. The overall objective is to take into account user preferences in order to produce discriminated answers to a query, making the assumption that the user wants to get the best answers first. An example is the "top-k query" approach [7] where the system interprets the selection conditions in a flexible way so as to retrieve the best k answers. A more general approach is that based on the fuzzy set theory, which allows for a larger variety of flexible terms and connectors [6].

In this paper, we tackle the problem of fuzzy query processing in a context of large-scale distributed relational databases. It is assumed that data sources are accessible through fuzzy views (a fuzzy view contains a set of tuples μ/t where μ is a satisfaction degree expressing the extent to which the tuples satisfy the fuzzy constraint that defines the view). Indeed, we believe than in certain cases, it is more convenient for a data source administrator to describe his/her data by means of a fuzzy description rather than a Boolean one. It is the case in particular when one wants to avoid introducing artificial boundaries for characterizing a given set of

* Work partially funded by the ANR Research Grant ANR-05-MMSA-0007.

A. An et al. (Eds.): ISMIS 2008, LNAI 4994, pp. 268–277, 2008.

objects. For instance, it may be desirable to describe a set of hotels as "medium-priced hotels" (with fuzzy boundaries) rather than with a crisp interval such as price in [75, 120] which would discard some potentially interesting hotels. Another rationale for using fuzzy views is that in some cases the data source administrator may want to hide the precise values of some attributes for confidentiality reasons.

The main objective here is to determine the set of views which only contain satisfactory answers to a given fuzzy user query (an answer is considered satisfactory if its satisfaction degree with respect to the fuzzy query is over a user specified threshold associated with the query). In order to determine the satisfactory combinations of views, a matching mechanism is used, which measures the subsumption degree between the description of a fuzzy view and the description of the fuzzy query submitted by the user. This matching mechanism is based on a graded inclusion, itself relying on a fuzzy implication. We show that for certain fuzzy implications, it is possible, from the subsumption degree obtained and the minimal satisfaction degree associated with the tuples in a given fuzzy view, to infer whether the view qualifies or not (i.e., contains only tuples whose satisfaction degree w.r.t. the query is over the specified threshold or not). This result makes it possible to envisage an approach where a fuzzy query can be rewritten in terms of a set of conjunctions of fuzzy views, each view being associated with a minimal satisfaction threshold (in that sense, the fuzzy views can be called parameterized).

The rest of the paper is structured as follows. In section 2, different basic notions related to fuzzy queries, fuzzy implications and graded inclusion are presented. In section 3, we show how, for certain fuzzy implications, it is possible to deduce a condition about the satisfaction of a given view from the subsumption degree between the view and the query considered, as well as the minimal degree of membership of the tuples to the view. Section 4 is devoted to related works. Section 5 concludes the paper and outlines some directions for future work.

2 Background

2.1 Some Basic Fuzzy Notions

Let us recall that fuzzy set theory [16] aims at representing sets whose boundaries are not sharp. A fuzzy set F defined on a domain X is associated with a membership function μ_F from X into the unit interval [0, 1]. The closer to 1 the membership degree $\mu_F(x)$, the more x belongs to F. The support $S(F)$ and the core $C(F)$ of a fuzzy set F are defined respectively as the following two crisp sets:

$$S(F) = \{x \in X \mid \mu_F(x) > 0\}$$
$$C(F) = \{x \in X \mid \mu_F(x) = 1\}$$

In the database domain, fuzzy set theory can serve as a basis for defining a flexible querying approach [6]. The key concept is that of a fuzzy relation, i.e., a relation designed as a fuzzy subset of Cartesian products of domains. Thus, any such fuzzy relation r can be seen as made of weighted tuples, denoted by μ/t, where μ expresses the extent to which tuple t belongs to the relation, i.e., is compatible with the concept conveyed by r. Of course, since regular databases are assumed to be queried, initial relations (i.e., those stored in the database) are special cases of fuzzy relations where all the tuple weights are equal to 1.

Example 1. Let us consider a database with the relation employee(num, name, salary, age, living-city). From a given initial extension of this regular relation, it is possible to get the intermediate fuzzy relation fy-emp shown in Table 1 containing those employees who are "fairly young". It is assumed that the memberhsip function associated with the flexible predicate "fairly young" is defined as follows: $\mu_{fy}(x) = 0$ if age $\geq$ 45, $\mu_{fy}(x) = 1$ if age $\leq$ 30, linear in between. It can be noticed that no element is a full member of the fuzzy relation fy-emp since no employee reaches the maximal degree 1. In the fuzzy relation obtained, only the tuples t such that $\mu_{fy}(t) > 0$ appear. ♦

Table 1. The extension of the relation fy-emp

num	name	salary	age	living-city	degree
76	martin	12500	40	New-York	0.3
26	tanaka	12000	37	Chiba	0.4
12	smith	12000	39	London	0.4
55	lucas	13000	35	Miami	0.8

The regular relational operations can be straightforwardly extended to fuzzy relations by considering fuzzy relations as fuzzy sets on the one hand and by introducing gradual predicates in the appropriate operations (selections and joins especially) on the other hand. Hereafter, we just recall the definition of the selection. The definitions of the other fuzzy relational operators can be found in [6].

$$\mu_{\text{select}(r,\ \varphi)}(t) = \top(\mu_r(t), \mu_\varphi(t)) \text{ where } \varphi \text{ is a fuzzy predicate.}$$

For more details about query language aspects, the reader may refer to [4] where a fuzzy SQL-like language is described.

2.2 Graded Inclusions and Fuzzy Implications

If A and B denote two crisp sets built on X, the usual way for defining the inclusion of A in B is:

$$(A \subseteq B) \Leftrightarrow (\forall\, x \in X, x \in A \Rightarrow x \in B) \quad (1)$$

This definition can be extended canonically to two fuzzy sets E and F as:

$$(E \subseteq F) \Leftrightarrow (\forall\, x \in X, \mu_E(x) \rightarrow_{\text{R-G}} \mu_F(x)) \quad (2)$$

where $\rightarrow_{\text{R-G}}$ stands for Rescher-Gaines implication: $p \rightarrow_{\text{R-G}} q = 1$ if $p \leq q$, 0 otherwise, but this view does not enable to distinguish between quite different situations, as illustrated hereafter.

Example 2. Let E_1 and E_2 be the two fuzzy sets:

$E_1 = \{1/a + 1/b\}$, $E_2 = \{0.5/a + 0.2/b\}$.

None of them is included in $F = \{0.4/a + 0.7/b\}$. However, it is more obvious for E_1 (which indeed contains F) than for E_2 where the grade of a is just a bit too high. ♦

An inclusion whose result values are taken in the unit interval allows to account for set inclusion in a finer way. To do so, a logical approach in the spirit of formulas 1 and 2 can be adopted [9]:

$$\deg(E \subseteq F) = \min_{x \in X} (\mu_E(x) \rightarrow \mu_F(x)) \tag{3}$$

where $\rightarrow$ denotes a *fuzzy implication operator*, i.e., a mapping from $[0, 1]^2$ into $[0, 1]$. There are several families of fuzzy implications, notably R-implications [9]:

$$p \rightarrow_{\text{R-i}} q = \sup_{u \in [0, 1]} \{u \mid \top(u, p) \leq q\} \tag{4}$$

It is possible to rewrite these implications as:

$$p \rightarrow_{\text{R-i}} q = 1 \text{ if } p \leq q, f(p, q) \text{ otherwise}$$

where $f(p, q)$ expresses a degree of satisfaction of the implication when the antecedent (p) exceeds the conclusion (q). The implications of Gödel ($p \rightarrow_{\text{Gö}} q = 1$ if $p \leq q$, q otherwise), Goguen ($p \rightarrow_{\text{Gg}} q = 1$ if $p \leq q$, q/p otherwise) and Lukasiewicz ($p \rightarrow_{\text{Lu}} q = 1$ if $p \leq q$, $1 - p + q$ otherwise) are the three most used R-implications and they are obtained respectively with the norms $\top(x, y) = \min(x, y)$, $\top(x, y) = xy$ and $\top(x, y) = \max(x + y - 1, 0)$.

As to S-implications [9], they generalize the (usual) material implication $p \Rightarrow q =$ ((**not** p) **or** q) by:

$$p \rightarrow_{\text{S-i}} q = \perp(1 - \text{p}, q) \tag{5}$$

The minimal element of this class, namely Kleene-Dienes implication obtained with $\perp = \max$ expresses the inclusion of the support of E in the core of F (1 is reached then). This is also the case for Reichenbach implication (obtained with the norm product).

3 From Graded Subsumption to α-Satisfactory Answers

3.1 Position of the Problem

We consider two fuzzy concepts V (representing a view) and Q (representing a query) such that $\deg(V \subseteq Q) = \delta$, which will be denoted in the following by $V \subseteq_\delta Q$. One assumes that the user, along with query Q, specified a minimal satisfaction degree $\alpha > 0$. The ultimate objective is to determine a threshold β (if it exists) such that V_β – the β-level cut of V defined as: $V_\beta = \{x \in V \mid \mu_V(x) \geq \beta\}$) – contains only elements which are α-satisfactory relatively to Q (i.e., elements x such that $\mu_Q(\text{x}) \geq \alpha$). If such a β exists, then V_β can be returned to the user. In other terms, the problem is, for each graded inclusion considered, to identify a function f such that $\beta = f(\alpha, \delta)$.

To do so, we will use as a starting point the two following well-known results in fuzzy set theory. Let $x, y, n, m \in [0, 1]$. Let $x \geq n$ and $x \rightarrow y \geq m$. Then i) under Kleene-Dienes implication, we have: if $n > 1 - m$ then $y \geq m$ and ii) under an

R-implication based on a t-norm ⊤, we have: $y \geq \top(n, m)$ [15]. In this paper, we go beyond these results by extending them to the case where sets of items are considered (and not only individual items) and to other fuzzy implications.

In the following two subsections, we study graded inclusions based on the most commonly used R-implications (namely Gödel's, Goguen's and Lukasiewicz') and S-implications (namely Kleene-Dienes' and Reichenbach's).

3.2 R-Implications

Let us first consider graded subsumption based on an R-implication. Let us start from the result recalled in 3.1 about R-implication. It is straightforward to show that it extends to a set of *k* elements the following way:

Proposition 1. If $\min_{i\in[1..k]} p_i \geq n$ and $\min_{i\in[1..k]} (p_i \rightarrow q_i) \geq m$ (where → denotes an R-implication whose generator is the t-norm ⊤) then $\min_{i\in[1..k]} q_i \geq \top(n, m)$.

From Proposition 1, one can infer the following theorem.

Theorem 1. Let *V* and *Q* be two fuzzy concepts such that $V \subseteq_\delta Q$ where the inclusion is based on an R-implication. Then, $\forall x \in V_\beta$ we have $\mu_Q(x) \geq \top(\beta, \delta)$ where ⊤ is the triangular norm underlying the considered R-implication.

Remark 1. Let us denote by V_{ext} the extension of view *V*. It is worthy to notice that $\min_{x\in V_{ext}} (\mu_{V_{ext}}(x) \rightarrow \mu_Q(x)) \geq \delta$. Indeed, $\delta = \deg(V \subseteq Q)$ is computed over all the values of the domain (using the membership functions of the fuzzy predicates associated with *V* and *Q*) whereas $\min_{x\in V_{ext}} (\mu_{V_{ext}}(x) \rightarrow \mu_Q(x))$ only takes into account the values which are actually in the extension of the view *V*.

Gödel's implication

We have: $p \rightarrow_{Gö} q = 1$ if $q \geq p$, q otherwise. Let us recall that Gödel's t-norm is defined as: $\top_{Gö}(x, y) = \min(x, y)$.

From Theorem 1, we get: $\forall x \in V_\beta$, $\mu_Q(x) \geq \min(\beta, \delta)$.

Now, $\min(\beta, \delta) \geq \alpha \Rightarrow \mu_Q(x) \geq \alpha$, thus $\beta \geq \alpha$ and $\delta \geq \alpha \Rightarrow \mu_Q(x) \geq \alpha$.

Let us suppose that we are interested in obtaining α-satisfactory answers to query *Q*. The first step is to look for the fuzzy views V_i which are subsumed at a degree $\delta \geq \alpha$ by *Q*. Then, for each such V_i, the second step is to compute the α-level cut of V_i, which contains only α-satisfactory answers to *Q*.

Goguen's implication

We have: $p \rightarrow_{Gg} q = 1$ if $q \geq p$, q/p otherwise. Let us recall that Goguen's t-norm is defined as: $\top_{Gg}(x, y) = x.y$.

From Theorem 1, we get: $\forall x \in V_\beta$, $\mu_Q(x) \geq \beta\delta$.

Now: $\beta\delta \geq \alpha \Rightarrow \mu_Q(x) \geq \alpha$ thus: $\beta \geq \alpha/\delta \Rightarrow \mu_Q(x) \geq \alpha$. Notice that for the condition to be useful, one needs to have $\alpha \leq \delta$ since by definition $\beta \leq 1$.

Using the subsumption based on Goguen's implication, for each fuzzy view V_i such that $\delta \geq \alpha$, the (α/δ)-level cut of V_i contains only α-satisfactory answers to *Q*.

Lukasiewicz implication
We have: $p \rightarrow_{Lu} q = \min(1, 1 - p + q)$. Let us recall that Lukasiewicz norm is defined as: $\top_{Lu}(x, y) = \max(0, x + y - 1)$.

From Theorem 1, we get: $\forall x \in V_\beta$, $\mu_Q(x) \geq \max(0, \beta + \delta - 1)$.

Now, $\max(0, \beta + \delta - 1) \geq \alpha \Rightarrow \mu_Q(x) \geq \alpha$, thus $\beta \geq \alpha - \delta + 1 \geq \alpha \Rightarrow \mu_Q(x) \geq \alpha$.

Using Lukasiewicz's implication, for each fuzzy view V_i such that $\delta \geq \alpha$, the $(\alpha - \delta + 1)$-level cut of V_i contains only α-satisfactory answers to Q.

3.3 S-Implications

Let us now consider graded subsumption based on an S-implication. The two most commonly used S-implications (along with Lukasiewicz' which is both an R- and an S-implication), namely Kleene-Dienes' and Reichenbach's are investigated.

Let us start from the result recalled in 3.1 about Kleene-Dienes implication. It is straightforward to show that it extends to a set of k elements the following way:

Proposition 2. If $\min_{i \in [1..k]} p_i \geq n$ and $\min_{i \in [1..k]} (p_i \rightarrow_{KD} q_i) \geq m$ (where $\rightarrow_{KD}$ denotes Kleene-Dienes' implication), and if $n > 1 - m$, then $\min_{i \in [1..k]} q_i \geq m$.

Kleene-Dienes' implication
We have: $p \rightarrow_{KD} q = \max(1 - p, q)$. From Proposition 2, we get the following result.

If $\min_{x \in V} \mu_V(x) \geq \beta$ and $\min_{x \in V} (\mu_V(x) \rightarrow_{KD} \mu_Q(x)) \geq \delta$, and if $\beta > 1 - \delta$, then $\min_{x \in V} \mu_Q(x) \geq \delta$.

The previous statement can be rewritten as:

If $V \subseteq_\delta Q$ and $x \in V_\beta$, and if $\beta > 1 - \delta$, then $\mu_Q(x) \geq \delta$, which leads to the following theorem.

Theorem 2. Let V and Q be two fuzzy concepts such that $V \subseteq_\delta Q$ under Kleene-Dienes' semantics. Then, $\forall x \in V_\beta$, if $\beta > 1 - \delta$, we have $\mu_Q(x) \geq \delta$.

Now: $\beta > 1 - \delta$ and $\delta \geq \alpha \Rightarrow \mu_Q(x) \geq \alpha$.

Using a subsumption mechanism based on Kleene-Dienes' implication, for each fuzzy view V_i such that $\delta \geq \alpha$, the *strict* $(1 - \delta)$-level cut of V_i contains only α-satisfactory answers to Q.

Reichenbach implication
We have: $p \rightarrow_{Rb} q = 1 - p + pq$ and consequently we get:

$$\delta = \inf_{x \in U} (1 - \mu_V(x) + \mu_V(x).\mu_Q(x))$$

Therefore: $\forall x \in U,\ 1 - \mu_V(x) + \mu_V(x).\mu_Q(x)) \geq \delta$

hence: $\forall x \in U,\ \mu_V(x)(1 - \mu_Q(x)) \leq 1 - \delta$

If $\mu_V(x) > 0$, we have:

$$\forall x \in U,\ 1 - \mu_Q(x) \leq (1 - \delta)/(\mu_V(x)$$

and: $\forall x \in U,\ \mu_Q(x) \geq 1 - ((1 - \delta)/\mu_V(x))$.

Theorem 3. Let V and Q be two fuzzy concepts such that $V \subseteq_{\delta} Q$. Under Reichenbach semantics, we have: if $\mu_V(x) \geq \beta$ then $\mu_Q(x) \geq 1 - ((1 - \delta)/\beta)$.

Using the subsumption based on Reichenbach's implication, for each fuzzy view V_i such that $\delta \geq \alpha$, the β-level cut of V such that $\beta = (1 - \delta)/(1 - \alpha)$ contains only α-satisfactory answers to Q.

Remark 2. Notice that for each fuzzy (R- or S-) implication studied, what we get is a *sufficient* condition on the fuzzy views, but not a necessary one. In other terms, one has the guarantee of obtaining *only* satisfactory views, but not necessarily *all of them.*

3.4 Example

Let us consider a user interested in finding a *medium-priced* hotel room. Let Q be the query: "rate is medium" that he/she submits to a decentralized database system, along with a minimal desired satisfaction degree $\alpha = 0.5$. Let V_1, V_2 and V_3 be three fuzzy views accessible through the system. The descriptions of these views are respectively:

V_1: "rate is *around \$75*", V_2: "rate is *reasonable*", V_3: "rate is *rather high*".

where *around \$75*, *reasonable* and *rather high* are associated with the three membership functions depicted in Fig. 1.

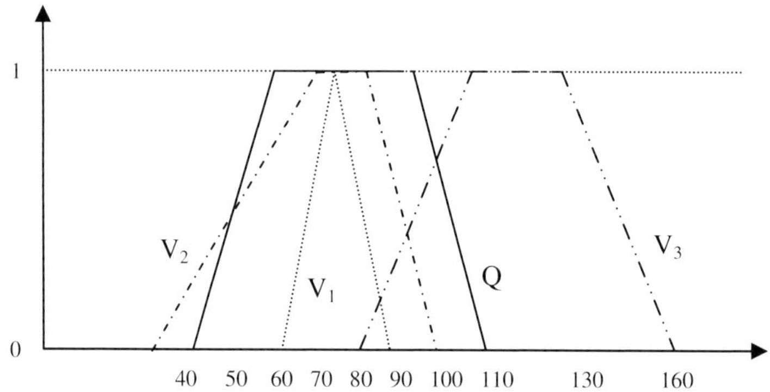

Fig. 1. A fuzzy query and three fuzzy views

Using Gödel's and Goguen's implication-based subsumption, one gets:

$\deg(V_1 \subseteq Q) = 1$, $\deg(V_2 \subseteq Q) = 0$, $\deg(V_3 \subseteq Q) = 0$.

Using Lukasiewicz' implication-based subsumption, one gets:

$\deg(V_1 \subseteq Q) = 1$, $\deg(V_2 \subseteq Q) = 0.67$, $\deg(V_3 \subseteq Q) = 0$.

Using Kleene-Dienes' implication-based subsumption:

$\deg(V_1 \subseteq Q) = 0.78$, $\deg(V_2 \subseteq Q) = 0.44$, $\deg(V_3 \subseteq Q) = 0$.

And with Reichenbach's implication-based subsumption:

$\deg(V_1 \subseteq Q) = 0.94$, $\deg(V_2 \subseteq Q) = 0.67$, $\deg(V_3 \subseteq Q) = 0$.

Thus, with Gödel's, Goguen's and Kleene-Dienes' implications, only V_1 will be accessed in order to obtain 0.5-satisfactory answers, whereas with the two other fuzzy implications, both V_1 and V_2 will be. ♦

3.5 Outline of a Query Processing Strategy

Let us now outline the way a fuzzy query can be processed in a context of large-scale distributed databases. As an example of such a system, let us consider a P2P system organized according to a super-peer architecture [1]. Let us assume that each super-peer stores the fuzzy view definitions associated with every of its descendant nodes. When a peer receives a fuzzy query Q (associated with a minimal satisfaction threshold α), it forwards it to its super-peer which computes the subsumption degree between each view associated with its descendant nodes (data sources) and the query. If the subsumption degree is above α, the super-peer computes the degree β that will be used to access the β-cut of the view, and the fuzzy query is forwarded (along with β) to the corresponding peer which processes it and returns the result to its super-peer. When the super-peer which forwarded the query has received all the results, it merges them into a rank-ordered list (according to the satisfaction degrees related to Q, cf. Remark 3 below) and delivers it to the user.

Remark 3. Let us emphasize that a post-processing of the answers is necessary if one wants to rank them according to their satisfaction degree relatively to Q. As a matter of fact, the tuples retrieved from a given view are associated with a satisfaction degree relative to the fuzzy concept underlying the view. Therefore, in order to have a satisfactory ranking w.r.t. the user query, one must compute $\mu_Q(x)$ for each tuple x retrieved.

4 Related Works

To the best of our knowledge, no previous research work has addressed the issue of fuzzy query answering using fuzzy views. Nevertheless, there exists a few works about answering top-k queries using views [8, 13, 14] and about processing skyline queries in P2P systems [12]. Even though these approaches share with the present contribution the objective of enriching decentralized database systems with flexible querying capabilities, they necessarily differ a lot on the techniques used, since i) top-k queries take into account a very limited form of flexibility in the selection conditions (only conditions of the type attribute = constant are considered, and transformed into attribute ≈ constant) and the goal is not to obtain answers which are good enough, but to obtain the best answers (even if these answers are all mediocre), ii) the skyline approach rests on a Pareto ordering of the answers and looks for the elements which are not dominated by any other; thus only a partial order is obtained since there is no global scoring function used (contrary to the fuzzy-set-based approach which assumes commensurability between the degrees coming from different predicates).

5 Conclusion

In this paper, we considered the problem of querying distributed databases by means of fuzzy queries and studied how fuzzy views could help focus on the relevant sources only (thus reducing the bandwith cost and the overall processing cost). We have shown that it was possible to determine subsets (β-level cuts) of the views such that these subsets only contain answers whose satisfaction degree relatively to the query is above a user-defined threshold α. The approach proposed is based on a graded subsumption mechanism resting on a graded implication. In this preliminary work, we have considered the case where the user query only involves one fuzzy predicate, and this work should now be extended to conjunctive fuzzy queries. Another extension should consist in considering conjunctions of fuzzy views in the rewriting process, and not only elementary views.

As mentioned in the remark from subsection 3.2, the degree of subsumption based on the membership functions associated respectively with the query and the view can be significantly smaller than the actual subsumption degree between the extension of the fuzzy view and the query. This implies that some views may be discarded whereas they are in fact subsumed by the query (in terms of their actual content). A solution to improve this state of fact could be to use some summaries of the (content) of the views [2] in order to compute a more realistic subsumption degree. This issue is left for future research. Another perspective is to study the use of a tolerant subsumption mechanism [3, 5] in the query answering process.

References

1. Bellahsene, Z., Roantree, M.: Querying Distributed Data in a Super-Peer Based Architecture. In: Galindo, F., Takizawa, M., Traunmüller, R. (eds.) DEXA 2004. LNCS, vol. 3180, pp. 296–305. Springer, Heidelberg (2004)
2. Bosc, P., Hadjali, A., Jaudoin, H., Pivert, O.: Flexible querying of multiple data sources through fuzzy summaries. In: Proc. 2nd International Workshop on Flexible Database and Information System Technology (FlexDBIST 2007), in conjunction with DEXA 2007, pp. 350–354 (2007)
3. Bosc, P., Hadjali, A., Pivert, O.: On a proximity-based tolerant inclusion. In: Proc. of the 5th Conference of the European Society for Fuzzy Logic and Technology (EUSFLAT 2007), pp. 343–350 (2007)
4. Bosc, P., Pivert, O.: SQLf: a relational database language for fuzzy querying. IEEE Transactions on Fuzzy Systems 3, 1–17 (1995)
5. Bosc, P., Pivert, O.: About approximate inclusion and its axiomatization. Fuzzy Sets and Systems 157, 1438–1454 (2006)
6. Bosc, P., Prade, H.: An Introduction to the Treatment of Flexible Queries and Uncertain or Imprecise Databases. In: Motro, A., Smets, P. (eds.) Uncertainty Management in Information Systems, pp. 285–324. Kluwer Academic Publishers, Dordrecht (1997)
7. Bruno, N., Chaudhuri, S., Gravano, L.: Top-k selection queries over relational databases: mapping strategies and performance evaluation. ACM Transactions on Database Systems 27, 153–187 (2002)
8. Das, G., Gunopoulos, D., Koudas, N.: Answering Top-k Queries Using Views. In: Proc. of the VLDB 2006 Conf., pp. 451–462 (2006)

9. Fodor, J., Yager, R.R.: Fuzzy-set theoretic operators and quantifiers. In: Dubois, D., Prade, H. (eds.) Fundamentals of Fuzzy Sets. The Handbook of Fuzzy Sets Series, pp. 125–193. Kluwer Academic Publishers, Dordrecht (1999)
10. Halevy, A.Y.: Answering queries using views: A survey. The VLDB Journal 10(4), 270–294 (2000)
11. Halevy, A.Y., Ives, Z.G., Madhavan, J., Mork, P., Suciu, D., Tatarinov, I.: The Piazza Peer Data Management System. IEEE Trans. Knowl. Data Eng. 16, 787–798 (2004)
12. Hose, K.: Processing Skyline Queries in P2P Systems. In: Proc. of the VLDB 2005 PhD Workshop, pp. 36–40 (2005)
13. Marian, A., Bruno, N., Gravano, L.: Evaluating Top-k Queries over Web-Accessible Databases. ACM Trans. on Database Syst. 29, 319–362 (2004)
14. Nejdl, W., Siberski, W., Thaden, U., Balke, W.T.: Top-k Query Evaluation for Schema-Based Peer-to-Peer Networks. In: Proc. of the 3rd International Semantic Web Conf (ISWC 2004), pp. 137–151 (2004)
15. Straccia, U.: Description Logics with Fuzzy Concrete Domains. In: Proc. 21st Conference on Uncertainty in Artificial Intelligence (UAI 2005), pp. 559–567 (2005)
16. Zadeh, L.A.: Fuzzy sets. Inf. Control. 8, 338–353 (1965)

Term Distribution-Based Initialization of Fuzzy Text Clustering

Krzysztof Ciesielski[1], Mieczysław A. Kłopotek[1,3], and Sławomir T. Wierzchoń[1,2]

[1] Institute of Computer Science, Polish Academy of Sciences, ul. Ordona 21, 01-237 Warszawa, Poland
[2] Institute of Informatics, Univ. of Gdansk, Wita Stwosza 57, 80-952 Gdansk
[3] Institute of Informatics, Univ. of Podlasie in Siedlce
{kciesiel,klopotek,stw}@ipipan.waw.pl

Abstract. We investigate the impact of an initialization strategy on the quality of fuzzy-based clustering, applied to creation of maps of text document collection. In particular, we study the effectiveness of bootstrapping as compared to traditional "randomized" initialization. We show that the idea is effective both for traditional Fuzzy K-Means algorithm and for a new one, applying histogram-based cluster description.

1 Introduction

In this research clustering is proposed as a tool allowing to reveal the internal structure of a collection, e.g., we are interested in grouping documents by topics, subtopics, etc. When using clustering we subsume so-called *cluster hypothesis*, according to which the documents in the same cluster behave similarly with respect to relevance to information needs.

A common approach to document representation and processing relies upon treating each textual document as a set of terms, i.e. words or phrases extracted automatically from the documents themselves. Then to each term in a given document we assign a numeric weight , representing an estimate of this term usefulness when distinguishing the document from other documents in the same collection. The weights assigned to the terms in a given document d can then be interpreted as the coordinates of d in the document space, i.e. $d = (w_1, \ldots, w_{|T|})$ is a point in $|T|$-dimensional document space D; here $|T|$ is the cardinality of the set of terms T . Among many existing weighting schemes the most popular is so-called *tfidf* scheme, see Sect. 3 for details. Here *tf*, i.e. term frequency, is a document-specific statistic, while the inverse document frequency *idf* is a global statistics characterizing a given term within an entire collection of documents. To get a deeper insight into the nature of different sub-collections of documents, we propose a context dependent weighting scheme; besides the two statistics so-called term specificity is additionally introduced. This allows to characterize the documents by a varying sets of weights $(w_1, \ldots, w_{|C(T)|})$ in different contexts[1];

[1] By a context one shall intuitively understand a sufficiently large (for statistics sake) set of documents with sufficiently uniform topics.

A. An et al. (Eds.): ISMIS 2008, LNAI 4994, pp. 278–287, 2008.

here $|C(T)|$ stands for the cardinality of the set of terms specific for a context C. In this paper we will examine the usefulness of this new "contextual" weighting scheme against the "global" scheme expressed by the *tfidf* weights.

With the vector representation we can adopt any clustering algorithm to reveal the internal structure of the documents collection. A popular approach to clustering relies upon minimizing an objective function expressing (weighted) distance between objects from a given cluster (group) and a characteristic point of this group (e.g. its center of gravity). Although simple and efficient, these algorithms heavily depend on the initial partition subjected further modification. In this paper we present a new, boosting-like, initialization scheme.

The paper is organized as follows: in section 2, fuzzy clustering approach and its drawbacks are briefly outlined. In section 3 the concept of clustering space is presented. Following that concept, FKH clustering algorithm is proposed. The key contribution of this paper is boosting-like initialization algorithm, described in section 4. Section 5 gives experimental results on FKH algorithm and proposed initialization method. Section 6 summarizes the paper.

2 Fuzzy Clustering Algorithm

Let $D = \{d_1, \ldots, d_n\}$ be a set of objects (documents in our case). Each object is described by a set of real-valued features, $d_i = (w_{i,1}, \ldots, w_{i,|T|})$, i.e. it can be viewed as a point in $|T|$-dimensional Euclidean space, $\Re^{|T|}$. The aim of cluster analysis is to partition the set D into a (usually predefined) number $K > 1$ of homogenous groups. The notion of homogeneity is understood in such a way that two objects assigned to the same class are much more similar one to another than any two objects assigned to different classes. Usually the similarity between pairs of objects is measured in terms of a distance metrics. Such an approach causes troubles in highly dimensional feature spaces, however, because, as shown in [1], with increasing dimensionality, the "contrast", i.e. relative difference between the closest and the farthest neighbor, is decreasing. In case of documents grouping the cosine measure is commonly used; it can be converted to a distance by the mapping $dist(d_i, d_j) = 1 - \cos(d_i, d_j)$.

In such a context, when similarity between pairs of objects is measured in terms of their mutual distance, a natural method for inducing partition of the set D is to use the objective function [2]

$$J_\alpha = \sum_{d \in D} \sum_{j=1}^{K} dist^2(d, \overline{v}_j) u^\alpha_{d,C(\overline{v}_j)} \tag{1}$$

where $u_{d,C(\overline{v}_j)}$ is a degree of membership of object d to j-th class, $\overline{v}_j$ is a characteristic of group C, usually defined as a (weighted) center of gravity called centroid, and $\alpha \geq 1$ is a parameter. Assuming that $u_{d,C(\overline{v}_j)} \in \{0, 1\}$, $\alpha = 1$ and that we are searching for the minimum of J_α, we obtain well known crisp K-means algorithm. When $u_{d,C(\overline{v}_j)} \in [0, 1]$ and $\alpha > 1$, minimization of the index J_α leads to the fuzzy K-means algorithm, called hereafter FKM for brevity [2].

Both these algorithms iteratively improve the initial partition matrix $U = [u_{d,C(\overline{v}_j)}]$ by modifying the centroids after the objects have been reassigned to the groups[2] computed and new cluster centers are determined. The quality of resulting partition is measured e.g. in terms of the index [2]

$$F_K(U) = \frac{1}{n} \cdot trace\left(U \cdot U^{\mathrm{T}}\right) \tag{2}$$

measuring the degree of fuzziness of the partition represented by the matrix U. This measure decreases when documents belong to more than one class and takes the lowest value for documents belonging to the same degree to all the classes.

An attractive feature of these algorithms is their numerical simplicity and linear time complexity with respect to the number of objects, what allows to process large collections of objects. They admit a number of drawbacks, however, like: (a) final partition heavily depends on the data order and on the initial data partition, (b) the algorithms are sensitive to the outliers, (c) the number of clusters must be known in advance, and finally, (d) these algorithm can be used only in case of numerical data representation.

To cope with the first drawback a number of approaches was proposed, like [6], [8], or [7]. In this paper a new method (inspired by the boosting algorithm used in machine learning) is proposed. It is especially useful when processing large datasets since it clusters only a sample of objects and re-clustering is performed when there are objects which do not fit to the existing partition. Before introducing this method, some remarks on contextual clustering are needed.

3 Clustering Space

Recall first, that when processing text documents the entries of the vectors d_i are so-called *tfidf* indices defined as, see e.g. [9]:

$$w_{t,d} = f_{t,d} \times \log \frac{n}{f_t^{(D)}} \tag{3}$$

where n is the number of documents in the collection D, $f_{t,d}$ is the number of occurrences of the term t in the document d, and $f_t^{(D)}$ is the number of documents in the collection D containing at least one occurrence of the term t.

The scheme (3) assumes that all the terms are weighted from the same (global) perspective. But it is obvious that the weight of a term can vary according to the context in which it is used. Hence we left the notion of cluster center, and we assume "relativistic" point of view treating the collection D as a continuous space, called hereafter clustering space, where each point p can be viewed as a cluster center. In such a space we define a proximity relationship among the documents, and $\mu_{d,C(p)}$ quantifies a normalized degree of membership of document d to the cluster $C(p)$ with center p, where normalization is over all $C(p)$

[2] More precisely, for a given partitioning group centers $\overline{v}_j, j = 1, \ldots, K$ are calculated and the objects are assigned to the groups with most similar centroids. Then the cycle repeats: centers $\overline{v}_j$ are updated and the objects are reassigned to the groups.

[5]. Thus, $\mu_{d,C(p)}$ plays role analogous to $u_{d,C(p)}$ in fuzzy clustering approach. Denoting $|C(p)| = \sum_{d \in D} \mu_{d,C(p)}$ the fuzzy cardinality of this cluster, and combining it with $f_{t,d}$ and $\mu_{d,C(p)}$, we introduce a notion of specificity $s_{t,C(p)}$ of the term t in the cluster $C(p)$ as $s_{t,C(p)} = f_{t,D}^{-1} \cdot \sum_{d \in D} f_{t,d} \cdot \mu_{d,C(p)}$, where $f_{t,D}$ is the total frequency of term t in the collection D. With these notions we introduce new contextual term weighting formula

$$w_{t,d,C(p)} = s_{t,C(p)} \times f_{t,d} \times \log \frac{|C(p)|}{f_{t,C(p)}} \tag{4}$$

where $f_{t,C(p)} = \sum_{\{d: f_{t,d} > 0\}} \mu_{d,C(p)}$ is the fuzzy count of documents in the collection $C(p)$ containing at least one occurrence of the term t; we assume that $w_{t,d,C(p)} = 0$ if $f_{t,C(p)} = 0$.

Now, the context-free weight *tfidf* given by the equation (3) is replaced by an averaged local weight

$$w_{t,d} = l \cdot \sum_{p \in HS} \left(\mu_{d,C(p)} \cdot w_{t,d,C(p)}\right) \tag{5}$$

where HS is the unit hyper-sphere and l is a normalizing constant placing the document d on the unit hyper-sphere.

3.1 Histograms of Term Weights

Furthermore, we can consider properties of each term individually (for a single document), or in a broader context of a given subset of documents, e.g. cluster $C(p)$ as a whole. In the latter case, the values of the weight w for a given term t for each document $d \in C(p)$ [3] are treated as observed values of a random variable with underlying continuous probability distribution. In practical cases, the continuous distribution will be approximated by a discrete one, so that the information about the random variable distribution for the term t can be summarized as a histogram $h_{t,C}$[4].

Single interval of a histogram $h_{t,C}$ represents the number of occurrences of a discretized value of the term weighting function w for the term t in the document collection C. The interval values can be in turn transformed to the relative frequencies via the plain normalization $h'_{t,C}(q) = f_{t,C}^{-1} \cdot h_{t,C}(q)$, where $f_{t,C} = \sum_{i=1}^{Q} h_{t,C}(q)$ is the total number of documents $d \in C$ containing term t. The normalized frequency distribution approximates the probability distribution of an unknown variable describing the weight of occurrence of the term t in randomly chosen document $d \in C$.

3.2 Histogram-Based Clustering Algorithm

Finally, we can define histogram-based document membership in a given context. A document d fits well to a given context (represented by a fuzzy cluster C) if its

[3] Where fuzzy cluster $C(p)$ is defined as $C(p) = \{d \in D : \mu_{d,C(p)} > 0\}$.

[4] A deeper discussion of the histograms of term weights is given in [5].

observed term weights $w_{t,d}$ are typical for the "majority" of documents in space C, i.e. the histogram-based probability of the weights observed in d is high. We expect such a document to follow some topic-specific term distribution. Thus, instead of fuzzy degree of membership u, defined in section 2 on the basis of distance from the single fuzzy center of gravity, we can define histogram-based (unnormalized) degree of membership $m_{d,C(p)}$:

$$m_{d,C(p)} = \frac{\sum_{t \in d} m_{t,C(p)} \cdot h_{t,C(p)}(q)}{\sum_{t \in d} m_{t,C(p)}} \tag{6}$$

where $m_{t,C(p)}$ is some measure of term significance in context $C(p)$[5], $h_{t,C(p)}$ is a histogram for term t and histogram index q is computed by discretization Δ of term weight $w_{t,d,C(p)}$ (i.e. $q = \Delta\left(w_{t,d,C(p)}\right)$). After normalization over all $C(p)$, unnormalized degrees $m_{d,C(p)}$ become $\mu_{d,C(p)}$.

Such a distribution-based degree of membership, computed for every context $C(p)$, after normalization gives us a vector of fuzzy-like probabilities $\mu_{d,C(p)}$, analogously to $u_{d,C(p)}$ from section 2. In section 4, we will use a matrix U of the normalized distribution-based contextual membership $\mu_{d,C(p)}$ to produce set of clusters covering diverse topical aspects of the whole document collection D.

Enriched fuzzy clusters description, based on sets of term histograms, are employed in an incremental clustering algorithm, see Algorithm 1, called Fuzzy K-Histograms, or FKH for brevity that we propose in this paper. It joins two known paradigms of clustering: the fuzzy clustering and the subspace clustering. The method differs essentially from fuzzy clustering in that it is designed solely for text data and is based on contextual vector representation and histogram-based description of vector subspaces.

Algorithm 1. Fuzzy K-Histograms, **FKH**

1. Fix the number of clusters, $K \geq 2$ and initialize the fuzzy partition matrix $U(\tau_0)$, $\mu_{d,C}(\tau_0) \in [0,1]$
2. Repeat steps 3-5 until the partition matrix stabilizes, i.e. $||U(\tau) - U(\tau-1)|| = max_{d,C} |\mu_{d,C}(\tau) - \mu_{d,C}(\tau-1)| \leq \varepsilon$)
3. For each term t in the document d, having fixed matrix $U(\tau)$, compute a new vector representation $w_{t,d,C}$ in current context $C = C(\tau)$.
4. For each context $C(\tau)$ construct a new histogram description $\{h_{t,C} : t = 1, \ldots, T\}$
5. Calculate new degrees of membership $m_{d,C}$ and normalize them to $\mu_{d,C}$ in order to get new partition matrix $U(\tau+1)$

Like FKM, the FKH algorithm starts with an initial split into subgroups, represented by a matrix $U(\tau_0)$, rows of which represent documents, and columns represent topical groups. Iteratively, we adapt (a) the document representation,

[5] In the simplest case, term significance could be equal to one for every term t. However, since majority of terms in a given context is expected to be irrelevant, in [5] we propose another measure, proportional to the weighted field under histogram plot.

(b) the histogram description of contextual groups, and (c) membership degree of documents and term significance in the individual groups. These modifications recursively lead to a precise description of a contextual subspace in terms of the membership degree of documents and significance of terms in a particular context; further an insight into the documents similarity is offered. So we can view the algorithm as a kind of reinforcement learning.

Both FKM and FKH algorithms are very sensitive to the choice of the initial fuzzy clusters memberships $\mu_{d,C}$. We can start without any knowledge of document similarity, via a random assignment of documents to a number of groups and global term weighing. But through the iterative process some terms specific for a group would be strengthened, so that class membership of documents would be modified, hence also their vector representation and, indirectly, similarity definition. As experimental results show, FKH initialized with the sampling-based method presented in the next section behaves much more stable (i.e. clusters are retained over different cross-validated document subsets) and it outperforms FKM algorithm in terms of supervised cluster quality (cf. section 5).

4 Sampling-Based Initialization Algorithm

Let us return to the main idea of boosting-like initialization algorithm. It starts from a random sample of documents drawn from the set D in case of global clustering, or from a contextual group C in case of contextual clustering. The sample size is $M = \min(1000, 0.1n)$ and the documents are drawn from the uniform distribution. The sample is clustered into K groups, where K is a parameter.

Surely, the sample drawn in such a way may appear to be too arbitrary. Thus, in case of the FKM algorithm the values of J_α and of F_K, defined in equations (1) and (2) respectively, are computed and a decision is made if the resulting partition can be accepted. It was assumed that the sample is rejected if $|F_K(U^{end}) - F_K(U^{start})| \leq \varepsilon$, where U^{start} is an initial partition matrix, while U^{end} is the partition matrix returned by the algorithm.

Next, for each cluster a corresponding histogram description is created and the spanning-tree of clusters is constructed.The tree plays a control role, e.g. it is used when the clusters are merged or their number is reduced.

If the distance (measured in terms of so-called Hellinger's divergence[6]) between any two clusters is less than a pre-specified threshold value ε – these clusters are merged. The resulting set of clusters S is saved. This ends a single step of the algorithm.

In the subsequent steps new samples of documents are drawn randomly, but this time the probability of including drawn document d into new sample equals:

$$P(d) = 1 - max\{m_{d,C(p)}: C(p) \in S\} \tag{7}$$

[6] Which is – contrary to Kullback-Leibler measure – a symmetric distance measure between the two distributions, cf. e.g. A.Basu, I.R.Harris, S.Basu, *Minimum distance estimation: The approach using density-based distances.* Handbook of Statistics, 15, 1997, pp. 21–48.

where S the set of clusters computed in previous step, and $m_{d,C(p)}$ is the degree of membership of the document d into the cluster $C(p)$, computed according to the contextual weighting scheme [4].

This way new sample is formed by the documents with low degree of membership to the existing set of clusters. The aim of such a procedure is recurrent improvement of the partition which should represent all the themes from the collection D.

Surely, this new sample is subjected to clustering and new clusters are added to the set S. Then both the contextual descriptions and the spanning tree of clusters are updated. The procedure terminates after a given number of steps or when the set S stabilizes, i.e. no cluster is modified and no new cluster is added.

5 Experimental Results

In this section we briefly comment our experiments concerning the quality of the contextual clustering by means of the Fuzzy K Means (FKM) algorithm and adaptive histogram-based algorithm (FKH). Further, we compare the influence of the boosting-like initialization on both algorithms.

In these experiments we use the following sets of documents: *20 Newsgroups*[7], *12 Newsgroups*[8], *WebKb*[9] and *Reuters*[10]. The *box-and-whisker* plot depicted on Figures 1 and 2 illustrates the process of global clustering of the *Reuters* data. Here a single box presents the distribution of the F_K index defined in equation (2) in subsequent iterations with respect to different divisions into training and testing sets (10-fold cross-validation). The horizontal bars represent median and the lower and upper sides of the box represent 25% and 75% quartiles; lastly the "whiskers" represent lower and upper extreme values and the dots – the outliers.

Let us comment variations of the F_K measure in consecutive iterations of particular clustering algorithms. In case of the FKM with random initialization (Fig. 1(a)) the algorithm almost immediately get stuck in a local minima. Another consequence of random initialization is large variance of this measure for particular validating data sets. Similar phenomena, although in a smaller scale is observed in case of the FHC algorithm, see Fig. 2(a), where we observe minor improvements of the F_K measure, but the results are saddled with large variance and highly depend on the content of the sets created during cross-validation.

The quality of the results improves remarkably when we use our boosting-like initialization. However even in this case we observe rather large variation of the degree of fuzziness characterizing particular clusters. Similar behavior has been observed for the remaining test sets.

Tables 1 and 2 present final values of the normalized mutual information (cf. e.g. [3]), measuring agreement between a priori given document categories and the identified clusters for respective test datasets. By *random* we denote

[7] `http://people.csail.mit.edu/jrennie/20Newsgroups/`
[8] A sample of 8094 messages from 12 groups; sizes vary from 326 to 1000 messages.
[9] `http://www.cs.cmu.edu/~TextLearning/datasets.html`
[10] `http://www.ics.uci.edu/~kdd/databases/reuters21578/reuters21578.html`

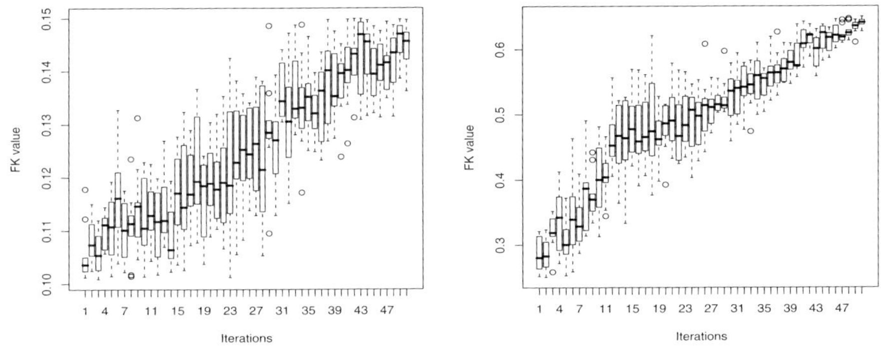

Fig. 1. Distribution of the F_K measure in consecutive iterations of FKM algorithm tested on *Reuters* data set: (a) random initialization (b) boosting-like initialization

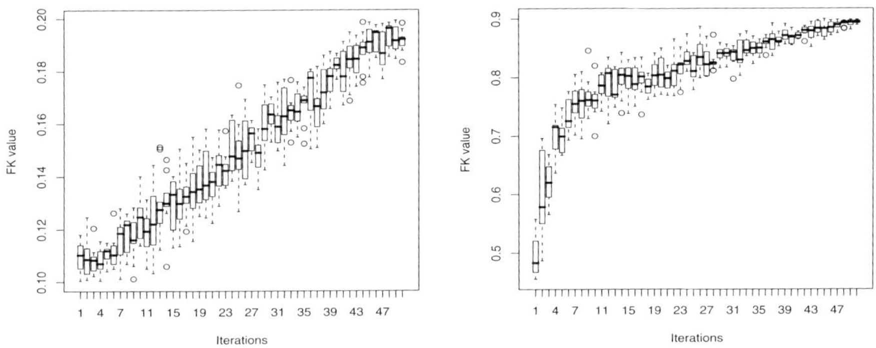

Fig. 2. Distribution of the F_K measure in consecutive iterations of FKH algorithm tested on *Reuters* data set: (a) random initialization (b) boosting-like initialization

Table 1. Normalized Mutual Information for the contextual groups generated by FKM

NMI K-Means	*12News*	*20News*	*Reuters*	*WebKb*
random/direct	0.135 ± 0.09	0.146 ± 0.08	0.102 ± 0.08	0.19 ± 0.09
initialized/direct	0.323 ± 0.06	0.364 ± 0.02	0.151 ± 0.02	0.621 ± 0.08
random/recursive	0.161 ± 0.08	0.176 ± 0.08	0.111 ± 0.1	0.189 ± 0.1
initialized/recursive	0.371 ± 0.04	0.329 ± 0.01	0.128 ± 0.03	0.674 ± 0.05

variant in which cross-validated samples are not initialized at all (i.e. initial fuzzy partition, represented by matrix U is random). On the other hand, *initialized* means initialization via the sampling algorithm described in section 4. We also considered two strategies of partitioning the data: a *direct* one, where for both FKM and FKH algorithms the number K of clusters was given as an input parameter, and *recursive* one, where the recursive splitting of large clusters into smaller ones was driven by the external criterion of thematic homogeneity. The

Table 2. Normalized Mutual Information for the contextual groups generated by FKH

NMI Histograms	*12News*	*20News*	*Reuters*	*WebKb*
random/direct	0.197 ± 0.07	0.208 ± 0.06	0.125 ± 0.08	0.587 ± 0.04
initialized/direct	0.392 ± 0.02	0.457 ± 0.01	0.326 ± 0.04	0.752 ± 0.02
random/recursive	0.256 ± 0.08	0.371 ± 0.06	0.117 ± 0.07	0.634 ± 0.04
initialized/recursive	0.453 ± 0.01	0.495 ± 0.01	0.306 ± 0.03	0.727 ± 0.02

criterion, exploiting Hellinger divergence between respective term distributions in the clusters of documents, relaxes the requirement of fixing in advance the number K of clusters. In case of four variants displayed in tables, one should note the influence of the initialization on the final results. Recurrent methods prove their advantage over the direct methods with respect to the mean value and the variance of the final result.

The robustness of the proposed approach is confirmed by in-depth analysis of the interrelationships among topical clusters identified by FKH algorithm and the document categories of examplary *20 Newsgroups* collection. The number of clusters identified by the FKH algorithm is lower than the total number of newsgroups in the collection (9 vs. 20), however, one can notice that a single cluster gathers documents from categories (i.e. newsgroups) which are closely related. For instance, one cluster is comprised mostly of documents related to religion (e.g. `alt.atheism`, `talk.religion.misc`, `talk.politics.mideast`), the other cluster contains the documents related to computer issues (`comp.graphics`, `comp.os.ms-windows.misc`, `comp.sys.ibm.pc.hardware`, `comp.sys.mac.hardware`, `comp.windows.x`), and other clusters consist of predominantly sport-related, motorization-related and science-related documents. In BEATCA search engine[11], clusters identified by FKH algorithm are used later on to construct thematical maps and visualize more subtle inter-topical proximities in graphical form. Thus, the property of grouping similar newsgroups together is an additional advantage, since all of the thematically related messages will be presented on a single map (so-called contextual map).

6 Conclusions

In this paper we explore the consequences of document cluster characterization via term (importance) distribution histograms . This idea offers a deeper insight into the role played by the terms in formation of a particular cluster. So a full profit can be taken from our earlier idea of "contextual clustering", that is of representing different document clusters in different subspaces of a global vector space. Histogram-based approach leads to many efficient algorithms for textual data processing purposes, to mention only vector space dimensionality reduction or keyword and keyphrases identification. In this paper we focus on the proposal of a fuzzy text clustering algorithm based on local ("contextual")

[11] Cf. e.g. `http://www.ipipan.eu/~klopotek/BEATCA/pdf/AIMSA2006.pdf`

document-to-cluster similarity and fuzzy clustering initialization via iterative sampling-based method in the vein of boosting.

We have observed the following properties of the presented algorithms:

- advantage of quality and stability of clustering structure identified by our FKH over the known FKM algorithm, both in direct and in recursive variant of these algorithms
- positive impact of proposed initialization method on clustering stability (i.e. reduction of variance over cross-validated data samples), in case of both FKM and FKH algorithm
- positive impact of proposed initialization method on supervised clustering quality (agreement between a priori given labels and identified clusters)
- positive impact of recursive variant of FKH algorithm on thematical homogeneity clustered together as a single contextual group

Acknowledgements. Research partly supported by the Polish state grant No. N516 01532/1906 and co-financed with the 6FP project REWERSE no. 506779.

References

1. Aggarwal, C.C., Hinneburg, A., Keim, D.A.: On the surprising behavior of distance metrics in high dimensional space. In: Van den Bussche, J., Vianu, V. (eds.) ICDT 2001. LNCS, vol. 1973, pp. 420–430. Springer, Heidelberg (2000)
2. Bezdek, J.C., Pal, S.K.: Fuzzy models for pattern recognition: Methods that search for structures in data. IEEE, New York (1992)
3. Boulis, C., Ostendorf, M.: Combining multiple clustering systems. In: Boulicaut, J.-F., Esposito, F., Giannotti, F., Pedreschi, D. (eds.) PKDD 2004. LNCS (LNAI), vol. 3202, pp. 63–74. Springer, Heidelberg (2004)
4. Ciesielski, K., Kłopotek, M.: Text data clustering by contextual graphs. In: Todorovski, L., Lavrač, N., Jantke, K.P. (eds.) DS 2006. LNCS (LNAI), vol. 4265, pp. 65–76. Springer, Heidelberg (2006)
5. Ciesielski, K., Kłopotek, M.: Towards adaptive web mining: Histograms and contexts in text data clustering. In: Berthold, M.R., Shawe-Taylor, J., Lavrač, N. (eds.) IDA 2007. LNCS, vol. 4723, pp. 284–295. Springer, Heidelberg (2007)
6. Forgy, E.: Cluster analysis of multivariate data: Efficiency versus interpretability of classification. Biometrics 21, 768–780 (1965)
7. Kaufman, L., Rousseeuw, P.J.: Finding groups in data: an introduction to cluster analysis. Wiley series in probability and mathematical statistics: Applied probability and statistics. Wiley, New York (1990)
8. MacQueen, J.B.: Some methods for classification and analysis of multivariate observations. In: Proceedings of 5-th Berkeley Symposium on Mathematical Statistics and Probability, Berkeley, pp. 281–297. University of California Press (1967)
9. Salton, G., McGill, M.J.: Introduction to Modern Information Retrieval. McGraw-Hill, New York (1983)

Cooperative Answering to Flexible Queries Via a Tolerance Relation

Patrick Bosc, Allel Hadjali, and Olivier Pivert

IRISA/Enssat, Technopole Anticipa BP 80518,
22300 Lannion, France
{bosc,hadjali,pivert}@enssat.fr

Abstract. One of the common problems that users are confronted with in their web data retrieval is overabundant answers, that is, being provided with an avalanche of responses that satisfy their queries. Most users are overwhelmed by such responses since it is difficult to examine them. In this paper, we attempt to address this issue in the context of flexible queries. The basic idea behind the solution proposed consists in modulating the fuzzy conditions involved in the user query by applying an appropriate transformation. This operation aims at intensifying the constraints of the query to make it more demanding. A transformation that relies on a convenient tolerance relation is introduced. The main features of our proposal are investigated as well.

Keywords: Cooperative answers, flexible queries, query intensification, tolerance relation, fuzzy relation.

1 Introduction

The practical need for endowing intelligent information systems with the ability to exhibit cooperative behavior has been recognized since the early' 90s. As pointed out in [8], the main intent of these systems is to provide correct, non-misleading and useful answers, rather than literal answers to user queries. Such answers allow to better serving the user's needs and expectations. It is well-known that the problem often approached in this field is the "*empty answer problem*", that is, the problem of providing the user with some alternative data when there is no data fitting his/her query. Several approaches have been proposed to deal with this issue. Some of them are based on a *relaxation* mechanism that expands the scope of the query [9][10]. See also the works done in [12][14][15][18].

Nowadays, a variety of large-scale databases, including bibliographies, scientific databases, travel reservation systems and vendors' databases, are accessible to a great number of lay users online in the web. Exploiting Web-based information sources is non-trivial because the user has no direct access to the data (one cannot for instance browse the whole target database). Thus, in their web search, users might be confronted with another problem, that is, of obtaining a *very large amount of answers* to the query asked. This is what we will call the *Overabundant Answers Problem*.

A. An et al. (Eds.): ISMIS 2008, LNAI 4994, pp. 288–297, 2008.

Facing this problem, users' desires are mainly to reduce the large set of answers and keep a manageable subset that can be easily examined and exploited. To the best of our knowledge, only little attention, however, has been paid to the overabundant answers problem in the literature. Ozawa and Yamada have addressed this issue in [16] and [17]. In [16], they suggest a method based on generating macro expressions of the queried database. Those expressions allow for providing the user with information about the *data distribution*. Then, the system identifies the appropriate attribute on which a new condition can be added to reconstruct another query. In [17], Ozawa and Yamada propose a cooperative approach that provides the user with *linguistic answers* using knowledge discovery techniques. From this information, the user can easily understand what kinds of data were retrieved and can then express a new query that shrinks the data set according to his/her interests. Let us also mention that the overabundant answers case has been pointed out in [11].

In the context of flexible queries (i.e., queries that contain gradual predicates represented by means of fuzzy sets[1] and whose satisfaction is a matter of degree), similar problems could still arise. In this context, the *empty answer problem* is defined in the same way as in the Boolean case (i.e., there is no available data in the database that *somewhat satisfies* the user query). Only few works [1-3] have addressed this problem. They mainly aim at relaxing the fuzzy requirements involved in the failing query. *Query relaxation* can be achieved by applying an appropriate transformation to gradual predicates of a failing query. Such a transformation aims at modifying a given predicate into an enlarged one by *widening its support*.

Let us now introduce the *fuzzy counterpart* of the *overabundant answer problem*. It can be stated as follows: there are too many data in the database that *fully satisfy* a flexible query Q. This means that *satisfaction degrees* of too many retrieved data are *equal to 1*. Now to cope with the issue of overabundant answers to Q, the idea is to carry on like above by transforming the fuzzy constraints contained in Q. This transformation basically aims at intensifying the query Q to make it less permissive. *Shrinking the cores* of the fuzzy sets associated to predicates in Q is the basic required property of this transformation. This property allows for reducing the width of the core and then effectively decreasing the number of answers to Q with degree *1*.

In this paper, a particular transformation to intensify the meaning of a gradual predicate P is proposed. It also relies on the notion of a *parameterized tolerance relation*. Applied to P, it aims at eroding the fuzzy set representing P by the parameter underlying the considered tolerance relation. The resulting predicate is *semantically not too far* from the original one but it is *more precise* and *more restrictive*. As will be seen, the desirable property of reducing the core is satisfied. The *intensification approach* we propose to deal with the overabundant answers problem is investigated both in single-predicate and compound queries. In the latter case, an efficient intensification strategy is discussed. It makes use of the notion of the median value of the set of answers related to each predicate involved in the query at hand.

The paper is structured as follows. Section 2 introduces a fuzzy modeling of a tolerance relation and describes a particular operation that is the key tool in our

---

[1] A fuzzy set F in the referential U is characterized by a membership function $\mu_F: U \rightarrow [0, 1]$, where $\mu_F(u)$ represents the grade of membership of u in F. Two crisp sets are of particular interest when defining a fuzzy set F: the core (i.e., $\mathcal{C}(F) = \{u \in U / \mu_F(u) = 1\}$) and the support (i.e., $\mathcal{S}(F) = \{u \in U / \mu_F(u) > 0\}$).

approach dedicated to query intensification. In section 3, we present in details the problem of overabundant answers on the one hand, and discuss how it can be solved in the case of single-predicate queries on the other hand. The intensification strategy to deal with this problem in case of flexible compound queries is investigated in section 4. Last, we briefly recall the main features of our proposal and conclude.

2 Basic Notions

Here, the notion of a *parameterized tolerance relation* is introduced. Then, we present an operation on fuzzy sets that is of interest for our problem.

2.1 Tolerance Relation

Definition 1. A tolerance relation (or a proximity relation) is a fuzzy relation E on a domain X, such that that for $x, y \in X$,

$$\mu_E(x, x) = 1 \qquad \text{(reflexivity)},$$
$$\mu_E(x, y) = \mu_E(y, x) \qquad \text{(symmetry)}.$$

The quantity $\mu_E(x, y)$ can be viewed as a grade of approximate equality of x with y. On a universe X which is a subset of the real line, an *absolute proximity* relation can be conveniently modeled by a fuzzy relation E of the form [5]:

$$\mu_E(x, y) = \mu_Z(x - y), \tag{1}$$

which only depends on the value of the difference $x - y$.

The parameter Z, called a *tolerance indicator*, is a fuzzy interval (i.e., a fuzzy set on the real line) centered in 0, such that: i) $\mu_Z(r) = \mu_Z(-r)$, this property ensures the symmetry of the proximity relation E (i.e., $\mu_E(x, y) = \mu_E(y, x)$); ii) $\mu_Z(0) = 1$ which expresses that x is approximately equal to itself to a degree 1; iii) The support $\mathcal{S}(Z)$ is bounded and is of the form $[-\omega, \omega]$ where ω is a positive real number. In terms of trapezoidal membership function (*t.m.f.*), Z is represented by the quadruplet $(-z, z, \delta, \delta)$ with $\omega = z + \delta$ and $[-z, z]$ denotes the core $\mathcal{C}(Z)$. Let us note that by this kind of proximity relation we evaluate to what extent the amount $x - y$ is close to 0. The closer x is to y, the closer $x - y$ and 0 are. Classical (or crisp) equality is recovered for $Z = 0$ defined as $\mu_0(x - y) = 1$ if $x = y$ and $\mu_0(x - y) = 0$ otherwise. Other interesting properties of the parameterized relation E are given in [5]. Furthermore we shall write $E[Z]$ to denote the proximity relation E parameterized by Z.

2.2 Dilation and Erosion Operations

Let us consider a fuzzy set F on the scalar universe X and an absolute proximity $E(Z)$, where Z is a tolerance indicator. The set F can be associated with a nested pair of fuzzy sets when using $E(Z)$ as a tolerance relation. Indeed, we can build a fuzzy set: (i) F^Z close to F, such that $F \subseteq F^Z$, this is the dilation operation; (ii) F_Z close to F, such that $F_Z \subseteq F$, this is the erosion operation.

2.2.1 Dilation Operation

Dilating the fuzzy set F by Z will provide a fuzzy set F^Z defined by

$$\mu_{F^Z}(r) = sup_s \ min\ (\mu_{E[Z]}(s,\ r),\ \mu_F(s)) = sup_s \ min\ (\mu_Z(r-s),\ \mu_F(s)), \text{ since } Z = -Z$$

$$= \mu_{F \oplus Z}(r), \text{ observing that } s + (r-s) = r. \quad (2)$$

Hence, $F^Z = F \oplus Z$, where $\oplus$ is the addition operation extended to fuzzy sets [7]. F^Z gathers the elements of F and the elements outside F which are somewhat *close to* an element in F.

Lemma 1. For $F^Z = F \oplus Z$, we have $F \subseteq F^Z$.

Thus, F^Z can be viewed as a relaxed variant of F. In terms of *t.m.f.*, if $F = (A, B, a, b)$ and $Z = (\ z, z,\ ,\ )$ then $F^Z = (A\ \ z, B + z, a +\ , b +\ )$, see Figure 1-(a). This operation can provide a basis for relaxing flexible queries involving gradual predicates as shown in [2-3]. In practice, relaxation technique is often used to support approaches for addressing the *empty answer problem*.

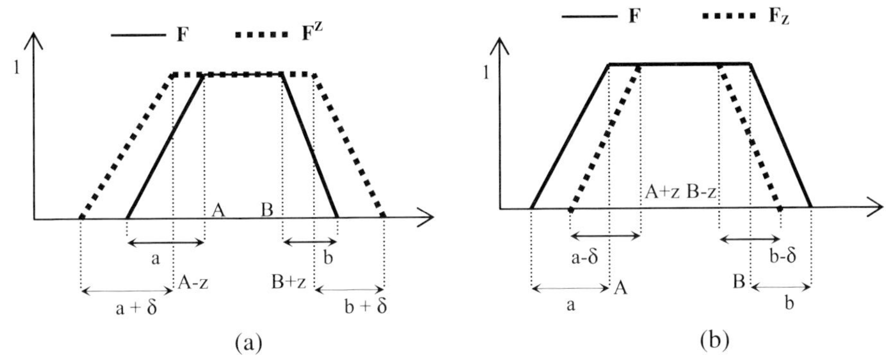

Fig. 1. (a): Dilation operation, (b): Erosion operation

2.2.2 Erosion Operation

Let $Z\ \ Y = F$ be an equation where Y is the unknown variable. It has been shown that the greatest solution of this equation is given by $\overline{Y} = F \ominus (\ Z) = F \ominus Z$ since $Z =\ Z$ and where $\ominus$ denotes the extended Minkowski subtraction defined by [6-7]:

$$\mu_{F \ominus Z}(r) = inf_s\ (\mu_Z(r\ \ s)\ \mathcal{I}_{\mathbb{T}}\ (\mu_F(s)) = inf_s\ (\mu_{E[Z]}(s, r)\ \mathcal{I}_{\mathbb{T}}\ \mu_F(s)) \quad (3)$$

where $\mathbb{T}$ is a t-norm, and $\mathcal{I}_{\mathbb{T}}$ is the implication induced by $\mathbb{T}$ defined by $\mathcal{I}_{\mathbb{T}}(u, v) = sup\{\ \ [0, 1] / \mathbb{T}(u,\ )\ \ v\}$, for $u, v\ \ [0, 1]$. We make use of the same t-norm $\mathbb{T} = min$ as in the dilation operation which implies that $\mathcal{I}_{\mathbb{T}}$ is the so-called *Gödel implication* (i.e., $p\ \ _{\text{Gödel}}\ q = 1$ if $p\ \ q$, q otherwise).

Let $(E[Z])_r = \{s,\ \mu_{E[Z]}(s, r) > 0\}$ be the set of elements that are close to r in the sense of $E[Z]$. Then, formula (3) can be regarded as the degree of inclusion of $(E[Z])_r$ in F. This means that r belongs to $F \ominus Z$ if all the elements s that are close to r are F. Now, eroding the fuzzy set F by Z results in $F_Z = F \ominus Z$. The following semantic entailment holds

Lemma 2. For $F_Z = F \ominus Z$, we have $F_Z \quad F$.

Hence, F_Z is more precise than the original fuzzy set F but it still remains not too far from F semantically speaking. If $F = (A, B, a, b)$ and $Z = (\; z, z, \;,\;)$ then $F \ominus Z = (A + z, B \quad z, a \quad , b \quad)$ provided that a and b , see Figure 1-(b). In the crisp case, $F \ominus Z = [A, B] \ominus [\; z, z\,] = [A + z, B \quad z]$ (while $F \quad Z = [A \quad z, B + z]$).

Lemma 3. The following semantic entailment holds as well: $F_Z \subseteq F \subseteq F^Z$.

In practice it may happen that one requires that the erosion operation should affect only one constituent part of a fuzzy set F (either the core or the support). Denoting by *core erosion* (resp. *support erosion*) the erosion that modifies only the core (resp. support) of F, the following proposition shows how to obtain such desirable results.

Proposition 1. Let F be a fuzzy set and $E[Z]$ a proximity relation,

- *Core erosion* is obtained using the family of tolerance indicators of the form $Z = (-z, z, 0, 0)$. In the *t.m.f.* framework if the core changes, the support will also change. Here, only the left-hand and the right-hand spreads are preserved.
- *Support erosion* is obtained using the family of tolerance indicators of the form $Z = (0, 0, \delta, \delta)$.

By this proposition, if $F = (A, B, a, b)$ the core erosion (resp. support erosion) leads to $F_Z = (A+z, B-z, a, b)$ (resp. $F_Z = (A, B, a - \delta, b - \delta)$).

3 Overabundant Answers

Let us first recall that *flexible queries* [13] are requests in which user's preferences can be expressed. Here, the fuzzy set framework is used as a tool for supporting the expression of preferences. The user does not specify crisp conditions, but fuzzy ones whose satisfaction may be regarded as a matter of *degree*. Then, the result of a query is no longer a flat set of elements but is a set of discriminated elements according to their global satisfaction w.r.t. the fuzzy constraints appearing in the query.

Let Q be a *flexible* query and let Σ_Q be the set of answers to Q when addressed to a regular database. Σ_Q contains the items of the database that *somewhat* satisfy the fuzzy requirements involved in Q. Let now Σ_Q^* denotes the set of answers that *fully* satisfy Q, i.e., each item has a satisfaction degree equal to *1*. Obviously, $\Sigma_Q^* \subseteq \Sigma_Q$ holds. Let now introduce the problem of overabundant answers and show how it can be approached by means of the parameterized proximity $E[Z]$.

3.1 Problem Definition

Definition 2. We say that Q results in *overabundant answers* if the cardinality of Σ_Q^* is too large.

It is worthy to note that definition 2 is specific to fuzzy queries and does not make sense in the Boolean queries (since in fuzzy queries, the notion of satisfaction is a matter of degree). Now, in the case of too many items that partially satisfy the query (i.e., whose degrees lie in $]0, 1[$), the solution is simple and it consists in considering just an *α-cut* (i.e. the set of elements whose degrees are greater or equal to α) of the retrieved data with an appropriate *high level*. This is why our definition only concerns retrieved data with degree *1*.

This problem often stems from the specificity of the user query that is too general. In other terms, fuzzy requirements involved in the query are not restrictive enough. To counter this problem, one can refine the query to make it more specific, so as to return a reasonable set of items. This refinement consists in intensifying the fuzzy constraints of the query in order to reduce the set Σ_Q^*. To achieve this task, a fundamental required property of the intensification mechanism is to significantly shrink the cores of the fuzzy sets associated with the conditions of the query.

A way to perform query intensification is to apply a *basic transformation* $T^{\downarrow}$ on all or some predicates involved in the query. This transformation can be applied iteratively if necessary. Some properties are required for any transformation $T^{\downarrow}$ when applied to a predicate P ($T^{\downarrow}(P)$ representing the intensified predicate):

$\mathcal{IC}_1$: $T^{\downarrow}$ does not increase the membership degree for any element of the domain, i.e., $\forall u, \mu_{T^{\downarrow}(P)}(u) \leq \mu_P(u)$;

$\mathcal{IC}_2$: $T^{\downarrow}$ must shrink the core $\mathcal{C}(P)$ of P, i.e., $\mathcal{C}(T^{\downarrow}(P)) \subset \mathcal{C}(P)$;

$\mathcal{IC}_3$: $T^{\downarrow}$ preserves the left-hand (resp. right-hand) spread of P, i.e., if $P = (A, B, a, b)$, then $T^{\downarrow}(P) = (A', B', a, b)$ with $A' - A < a$ and $B - B' < b$

The second property allows for reducing the cardinality of the core and then effectively decreasing the number of answers with degree *1*. The last property guarantees that the data excluded from the core of P remain in its support.

3.2 Intensifying Atomic Queries

Let P be a fuzzy predicate and $E[Z]$ be a proximity relation parameterized by a tolerance indicator Z of the form $(-z, z, 0, 0)$. Making use of the erosion operation, P can be transformed into a *restricted* fuzzy predicate P' defined as follows:

$$P' = T\ (P) = P_Z = P \ominus Z.$$

This transformation allows for reinforcing the meaning of the vague concept expressed by P. As previously mentioned, the resulting predicate P' contains elements r such that all elements that are close to r are in P. Hence, this transformation is not simply a technical operator acting on the membership degrees but it is endowed with a clear semantics as well. Now, if $P = (A, B, a, b)$ then $T^{\downarrow}(P) = (A + z, B - z, a, b)$, see Figure 1-(b). As can be checked, the properties $\mathcal{IC}_1$ to $\mathcal{IC}_3$ hold.

Principle of the Approach. Let $Q = P$ be an atomic query (i.e., containing a single fuzzy predicate P). Assume that $^{*}_{Q}$ is too large. In order to reduce $^{*}_{Q}$, we transform Q into $Q_1 = T\ (P) = P \ominus Z$. This intensification mechanism can be applied iteratively until the database returns a manageable set of answers to the modified query $Q_n = T^{(n)}(P) = P \ominus n\,Z$. If we take a look at the subset of $^{*}_{Q}$ resulting from the intensification process, we can consider its elements as the typical values of the concept expressed by the fuzzy set associated with Q.

intensification process, we can consider its elements as the typical values of the concept expressed by the fuzzy set associated with Q.

Controlling the intensification. We claim that semantic limits[2] of an intensification process are not as crucial as in the case of query relaxation. Indeed, the intensification process only aims at reducing the large set of answers; not at finding alternative answers. It is worthy, however, to emphasize that the query refinement must stop when the upper bound and the lower bound of the core (of the modified query Q_i) are equal to $(A + B)/2$. Indeed, the *t.m.f.* associated to Q_i is $(A + i{\cdot}z,\ B - i{\cdot}z,\ a,\ b)$. Now, since $A + i{\cdot}z \le B - i{\cdot}z$ holds we have $i \le (B - A)/2z$. This means that the maximal query refinement is obtained when the core is reduced to a *singleton*. Let us note that the risk to obtain empty answers ($\Sigma_Q = \varnothing$) during the process is excluded when $a > z$ and $b > z$ (the data that have been eliminated from $\Sigma^*_{Q_{i-1}}$ related to Q_{i-1} still belong to the support of Q_i). Now if a too specific query arises and returns $\Sigma^*_Q = \varnothing$, one can back up and try another variation (for instance, adjust the tolerance parameter Z).

Intensification Algorithm. Algorithm 1 formalizes this intensification approach (where $\Sigma^*_{Q_i}$ stands for the set of answers to Q_i).

```
Input: Q: Initial query                  /* Q := P */
       Z: tolerance Indicator            /* Z = ( z, z, 0, 0) */
  i := 0; Qi := Q;
  compute  Qi ;
  while (| Qi |is too large and i   (B - A)/2 z) do
    begin
      i := i+1;
      Qi := T (i)(P) := P ⊝ i Z;
      compute  Qi ;
    end
Output: the final result  Qi
```

Algorithm 1. Atomic query Intensification

Particular cases. For some kinds of atomic queries, the *property of symmetry* of the parameter Z is not required. Consider, for instance, the query $Q = P$ where $P = (0, 25, 0, 10)$ which expresses the concept *"young"* as illustrated in Figure 2.

Intensifying P comes down to reducing the width of its core $\mathcal{C}(P)$ and thus to come closer to the typical values of P. Then, the intensification transformation must only affect the right part of $\mathcal{C}(P)$ and preserve entirely its left part. The appropriate Z allowing this operation is $(0, z, 0, 0)$ which leads to $T^{\downarrow}(P) = (0, 25 - z, 0, 10)$.

[2] In query relaxation, semantic limits stand for the maximum number of relaxation steps that can be performed such that the final modified query Q_n is *not too far*, semantically speaking, from the original one [2].

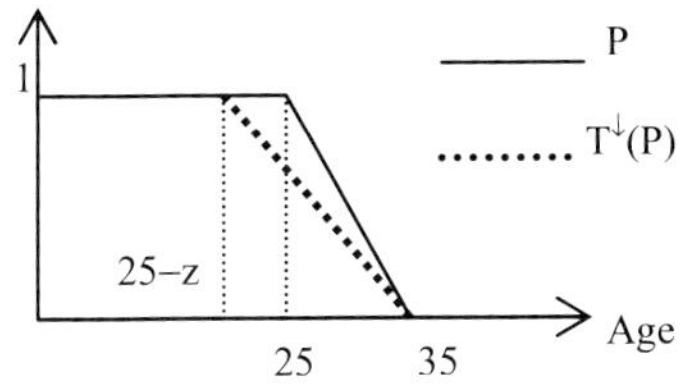

Fig. 2. Fuzzy predicate "young"

Consider now a query Q of the form $Q = \neg P$ where $\neg$ stands for the negation ($\mu_{\neg P}(u) = 1 - \mu_P(u), \forall u$) and assume that it results in overabundant answers. To solve this problem, one applies the intensification mechanism proposed to Q, i.e. shrinking the core of $\neg P$. It is easy to check that this transformation is equivalent to extending the support of P. This means that applying $T^{\downarrow}$ to Q comes down to applying a relaxation transformation $T^{\uparrow}$ to P (see [3] for more details about $T^{\uparrow}$). So, we have $T^{\downarrow}(\neg P) = T^{\uparrow}(P)$. One can easily verify that $T^{\uparrow}(\neg P) = T^{\downarrow}(P)$ holds as well.

3.3 Basic Features of the Approach

We investigate here the main features of our approach. To do this, we point out three criteria that seem to be of a major importance from a user point of view:

i) *Intensification nature*. Taking a look at the *t.m.f.* of $T^{\downarrow}(P)$, it is easy to see that the effect of the intensification applied to P in the right and the left parts is the same and amounts to z. This means that the resulting intensification is of a *symmetrical* nature.
ii) *Impact of the domain and the predicate*. Since the tolerance relation is attached to the domain attribute, the same tolerance value should be used for two predicates bearing on a same attribute[3]. So, the *relative position* of the membership function (in the domain of the attribute) has no impact on the intensifying effect. However, the *attribute domain* is identified as a major factor affecting the intensification (for instance, z will be different for the attribute "*age*" and the attribute "*salary*").
iii) *Applicability to the crisp case*. It can be easily checked that $T^{\downarrow}$ is still valid for crisp predicates. For example, if $P = (22, 30, 0, 0)$ then $T^{\downarrow}(P) = (22 + z, 30 - z, 0, 0)$.

4 Case of Conjunctive Flexible Queries

A conjunctive fuzzy query Q is of the form $P_1 \wedge \ldots \wedge P_k$, where the symbol '$\wedge$' stands for the connector *'and'* (which is interpreted by the *'min'* operator) and P_i is a fuzzy predicate. Our strategy to solve the Overabundant Answers Problem (**OAP**) in this case is still based on reinforcing the fuzzy requirements involved in Q.

Proposition 2. Let $Q = P_1 \wedge \ldots \wedge P_k$. If Q results in overabundant answers, then each of its subqueries results also in overabundant answers.

[3] It may happen that a user makes use of different tolerance parameters (z) for the same attribute. For instance, one can use a parameter z_1 for the predicate "*extremely expensive*" that is different from the parameter z_2 used for the predicate "*quite cheap*".

Lemma 4. If $Q = P_1 \wedge \ldots \wedge P_k$ results in overabundant answers, then each atomic query Q_i (*i.e.*, $Q_i = P_i$) of Q results also in overabundant answers.

As solving the OAP comes down to reducing the cardinality of the set Σ^*_Q, it suffices then to reduce the cardinality of answers (in Σ^*_Q) related to one predicate P_s (with $1 \leq s \leq k$). To do so, we apply only to P_s the intensification transformation. In practice, the question is about the predicate (i.e., attribute) to select for intensification (since a judicious choice could lead to an efficient fast intensification strategy).

Several ways can be used for selecting the predicate to be considered for intensification: one can exploit the data distribution of the queried database, or call for user intervention. In the following, we advocate another way of selecting that only exploits the set Σ^*_Q. Let $\Sigma^*_{Q(P_i)}$ be the subset of values of Σ^*_Q related to the predicate P_i, $1 \leq i \leq k$. The main idea is to take a look at the distribution of values of $\Sigma^*_{Q(P_i)}$ with respect to the core $\mathcal{C}(P_i)$. For instance, examine the location of those values regarding the bounds of the core (i.e., A_i and B_i). Let $P_i = (A_i, B_i, a_i, b_i)$, first we compute the median value, $med(P_i)$, of $\Sigma^*_{Q(P_i)}$. Then, we estimate to what extent the median $med(P_i)$ is distant from the bounds of $\mathcal{C}(P_i)$. To do this, we make use of the following index $d_i = min(\,|med(P_i) - A_i|,\ (\,|med(P_i) - B_i|)/\mathcal{L}(\mathcal{C}(P_i))$ where $\mathcal{L}(\mathcal{C}(P_i))$ denotes the width of $\mathcal{C}(P_i)$. The predicate P_s to select is such that the distance d_s is minimal. This means that the retrieved values related to P_s (i.e., $\Sigma^*_{Q(P_s)}$) are closer to the bounds of its core than the other values associated with the other predicates. It is then more efficient to apply the intensification process to P_s rather than to other predicates. This method can be formalized as follows:

- **Step1:** calculus of the median value
  for each i **in** $[1\ldots k]$
  compute $med(P_i)$
- **Step2:** distance-based index of values distribution
  for each i **in** $[1\ldots k]$
  compute $d_i = \min(|med(P_i) - A_i|,\ (|med(P_i) - B_i|)/\mathcal{L}(\mathcal{C}(P_i))$
- **Step3:** selecting the predicate P_s
  select s such that $d_s = min_{i=1,k}\ d_i$.

5 Conclusion

In this paper, we have addressed the problem of overabundant answers in the fuzzy querying setting. We have shown how it can be automatically dealt with. The key tool of the proposed approach is a tolerance relation expressed by a convenient parameterized proximity relation. Such a fuzzy relation can provide the basis to achieving query intensification of the user query. The main advantages of our proposal is the fact that it operates only on the conditions involved in the query without adding new conditions or performing any summarizing operation of the queried database. Such an approach can be useful to construct intelligent information retrieval systems that provide the user with cooperative answers. As for future work, we plan to evaluate the approach on some large practical examples.

References

1. Andreasen, T., Pivert, O.: On the weakening of fuzzy relational queries. In: 8th Int. Symp. on Meth. for Intell. Syst., Charlotte, USA, pp. 144–151 (1994)
2. Bosc, P., Hadjali, A., Pivert, O.: Towards a tolerance-based technique for cooperative answering of fuzzy queries against regular databases. In: Meersman, R., Tari, Z. (eds.) OTM 2005. LNCS, vol. 3760, pp. 256–273. Springer, Heidelberg (2005)
3. Bosc, P., Hadjali, A., Pivert, O.: Weakening of fuzzy relational queries: An absolute proximity relation-based approach. Journal of Mathware & Soft Computing 14(1), 35–55 (2007)
4. Bouchon-Meunier, B.: Stability of linguistic modifiers compatible with a fuzzy logic. In: Yager, R.R., Saitta, L., Bouchon, B. (eds.) IPMU 1988. LNCS, vol. 313, pp. 63–70. Springer, Heidelberg (1988)
5. Dubois, D., Hadjali, A., Prade, H.: Fuzzy qualitative reasoning with words. In: Wang (ed.) Computing with Words, pp. 347–366. John Wiley & Son, Chichester (2001)
6. Dubois, D., Prade, H.: Inverse operations for fuzzy numbers. In: Proc. of IFAC Symp. on Fuzzy Info. Knowledge Representation and Decision Analysis, pp. 391–395 (1983)
7. Dubois, D., Prade, H.: Possibility Theory. Plenum Press, New York (1988)
8. Gaasterland, T., Godfrey, P., Minker, J.: An overview of cooperative answering. Journal of Intelligent Information Systems 1(2), 123–157 (1992)
9. Gaasterland, T.: Cooperative answering through controlled query relaxation. IEEE Expert 12(5), 48–59 (1997)
10. Godfrey, P.: Minimization in cooperative response to failing database queries. Int. Journal of Cooperative Information Systems 6(2), 95–149 (1997)
11. Godfrey, P.: Relaxation in Web Search: A new paradigm for search by Boolean queries. Personal Communication (March 1998)
12. Huh, S.Y., Moon, K.H., Lee, H.: Cooperative query processing via knowledge abstraction and query relaxation. Advanced Topics in Databases Research 1, 211–228 (2002)
13. Larsen, H., Kacpryk, J., Zadrozny, S., Andreasen, T., Christiansen, H. (eds.): Flexible Query Answering Systems, Recent Advances. Physica (2001)
14. Motro, A.: FLEX: A tolerant and cooperative user interface databases. IEEE Transactions on Knowledge and Data Engineering 2(2), 231–246 (1990)
15. Muslea, I.: Machine learning for online query relaxation. In: 10th Int. Conf. of Knowledge Discovery and Data mining, KDD 2004, pp. 246–255 (2004)
16. Ozawa, J., Yamada, K.: Cooperative answering with macro expression of a database. In: 5th Int. Conf. IPMU, Paris, July 4-8, pp. 17–22 (1994)
17. Ozawa, J., Yamada, K.: Discovery of global knowledge in database for cooperative answering. In: 5th IEEE Int. Conf. on Fuzzy Systems, pp. 849–852 (1995)
18. Ras, Z.W., Dardzinska, D.: Failing queries in distributed autonomous information systems. In: Hacid, M.-S., Murray, N.V., Raś, Z.W., Tsumoto, S. (eds.) ISMIS 2005. LNCS (LNAI), vol. 3488, pp. 152–160. Springer, Heidelberg (2005)

Effectiveness of Fuzzy Discretization for Class Association Rule-Based Classification

Keivan Kianmehr[1], Mohammed Alshalalfa[1], and Reda Alhajj[1,2]

[1] Dept. of Computer Science, University of Calgary, Calgary, Alberta, Canada
[2] Dept. of Computer Science, Global University, Beirut, Lebanon

Abstract. This paper presents a novel classification approach that integrates fuzzy class association rules and support vector machines. A fuzzy discretization technique is applied to transform the training set, particularly quantitative attributes, to a format appropriate for association rule mining. A hill-climbing procedure is adapted for automatic thresholds adjustment and fuzzy class association rules are mined accordingly. The compatibility between the generated rules and patterns is considered to construct a set of feature vectors, which are used to generate a classifier. The reported test results show that compatible rule-based feature vectors present a highly-qualified source of discrimination knowledge that can substantially impact the prediction power of the final classifier.

1 Introduction

Classification is a technique used for prediction, which is one of the most attractive aspects of data mining. It is simply the process of building a classifier model based on some known objects and predefined classes. The task involves two major steps. First, exploring through data objects (in the training set) to find a set of classification rules which determine the class of each object according to its attributes. Second, building a classifier based on the extracted rules to predict the class or missing attribute value of unseen objects.

In our previous works [12,13], we proposed a classification framework that integrates SVM and associative classification into a novel, efficient and accurate classification technique by directly dealing with the following problems. 1) Providing more discriminative knowledge to the learning process of SVM by incorporating class association rules as rule-based feature vectors. We initially developed a set of binary rule-based feature vectors to be used as input to SVM; the result was outstanding [13]. We also used weighted rule-based feature vectors in the classifier model, and the result was promising [12]. Our framework improves the interpretability and understandability of the classification task by domain experts. 2) Developing an effective ranking method to filter out the best subset of the rules to be incorporated into the final classifier. Our algorithm improved the accuracy of the classifier by eliminating the deterioration effect of the rule ranking and selection approach in associative classification algorithms.

In this paper, we propose the use of a fuzzy discretization technique as a pre-processing step for classification, in particular appropriate for quantitative

A. An et al. (Eds.): ISMIS 2008, LNAI 4994, pp. 298–308, 2008.

attributes. Applying an association rule mining algorithm to the transformed fuzzy data set will result in a set of fuzzy class association rules which are more understandable by domain experts. We also introduce new weighted rule-based feature vectors, which are built based on the compatibility of the fuzzy class rules and data objects in the original dataset. These feature vectors represent the compatibility between fuzzy class rules and the quantitative values; we argue that using them will neutralize the impact of the discretization step in associative classifiers when applied to quantitative values. Finally, realizing the difficulty in specifying the required thresholds even by domain experts, we also adapted from [3] an effective hill-climbing method which adds to the overall robustness of the developed system by automatically suggesting appropriate threshold values. TFPC [3] is a new method which uses a different approach to select the final rule set for the classifier. It ignores the expensive coverage analysis step and obtains the classification rules directly from the association rule mining process.

The rest of the paper is organized as follows. Section 2 describes the proposed methodology. Section 3 reports the experimental results. Section 4 is conclusions.

2 The Proposed Model

Our proposed model consists of two major phases: generating fuzzy class association rules and building a classifier model based on the rules. In the first phase, the task is to extract all fuzzy class association rules (FCARs) from a discretized fuzzy training set. The extracted FCARs are then pruned to obtain the best qualified rules for building the classifier. The pruned rule set is called the discriminator FCARs set. In the second phase, all the rules in the discriminator FCARs set are first weighted based on a scoring metric strategy. Then, the compatibility of the rules from the discriminator FCARs set with every pattern in the original dataset (with quantitative values) is used to generate a set of feature vectors. Each pattern in the dataset is represented by a feature vector. A feature corresponds to an individual rule in the discriminator FCARs set. Generated feature vectors represent coverage compatibility of discriminator FCARs over the original dataset. Eventually, rule-based feature vectors are given to SVM to build a classifier model. Accuracy of the classifier is then evaluated by using the cross validation technique. In the rest of this section, components of the proposed model will be described.

2.1 Fuzzy Discretization

The problem in associative classifiers arises mainly when the dataset used for classification contains quantitative attributes. Indeed, mining association rules in such datasets is still attractive. The general approach used to address this issue is discretization, which is the process of mapping the range of possible values associated with a quantitative attribute into a number of intervals, each denoted by a unique integer label. One major problem with using discretized training set in an associative classifier is that it completely ignores the possible impact of quantitative attributes during the rule mining and classifier building stages, i.e.,

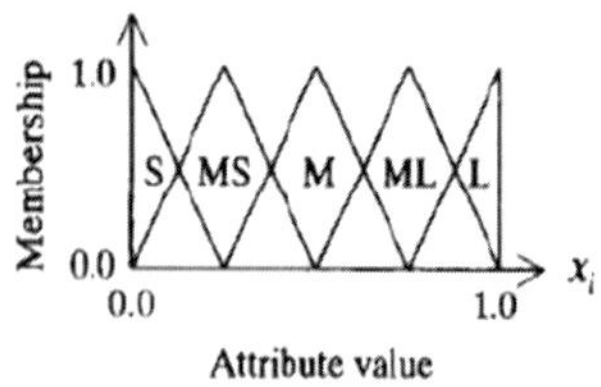

S → Small, MS → Medium-Small, M → Medium,
ML → Medium-Large, L → Large

Fig. 1. Fuzzy membership functions [10]

the extracted rules and the final classifier may not be good representatives of the original dataset due to the following reasons: 1) the sharp boundary problem where every value is represented in the antecedent of the rule by a crisp interval regardless of its degree of membership in the interval; 2) the coverage analysis for finding the best rules (training) for the final classifier is performed on the transformed dataset without considering the exact continuous values in the original training set. These effects may decrease the robustness of the classifier. To overcome this, we propose the use of a fuzzy discretization approach.

A fuzzy pattern is a sequence of linguistic values; its length is the number of attributes. For instance, the fuzzy pattern: (S, M, L, ML: Class0), means the current pattern belongs to class0 when the values of the four attributes are small, medium, large and medium large, respectively. In the proposed fuzzy discretization algorithm, we apply the five membership functions in Figure 1 to a given quantitative value, and the linguistic term that corresponds to the membership function which produces the maximum result represents the value in the generated fuzzy pattern.

2.2 Class Association Rule Mining

The method we have used for class association rule mining follows the CBA rule generator approach, called CBA-RG [9], where Apriori is adapted to find all class association rules that have support and confidence values greater than some given thresholds. A *ruleitem* has the form: $<condset, y>$, where *condset* is a set of items and y is a class label. Each *ruleitem* basically represents a rule [9]: $condset \rightarrow y$.

The algorithm generates all frequent *ruleitems* by multiple passes over the dataset. In the first pass, it counts the support of each individual *ruleitem* and determines whether it is frequent or not. In each subsequent pass, it starts with the seed set of *ruleitems* found to be frequent in the previous pass. It uses this seed set to generate new possibly frequent *ruleitems*, called candidate *ruleitems*. The actual supports for these candidate ruleitems are calculated during the pass over the dataset. At the end of each pass, it determines which of the candidate *ruleitems* are actually frequent. From this set of frequent *ruleitems*, it produces the class association rules. CBA-RG also employs the pessimistic error rate based pruning method of C4.5. Refer to [1] for further information about CBA-RG.

In our previous works [12,13], we used the unpruned rule set. However, in this study we use only rules in the pruned rule set. This makes the proposed classifier more efficient as it uses fewer rules to build the rule-based feature vectors. It is worth mentioning that the result of the class association rule mining step in our framework is a set of FCARs, which are extracted using the fuzzy patterns. A FCAR for an n-dimensional pattern classification problem is defined as follows [11]:

$$\text{Rule } R_i : \text{If } x_1 \text{ is } A_{i1} \text{ and } \ldots \text{ and } x_n \text{ is } A_{in} \text{ then Class } C \tag{1}$$

where R_i is the label of the i-th FCAR, $x = (x_1, x_2, \ldots, x_n)$ is an n-dimensional pattern vector, A_{ij} is an antecedent fuzzy set with linguistic label on the j-th axis, C is a consequent class.

2.3 Rules Weighting

In the proposed method, the significance of a FCAR is determined by using a scoring metric strategy. Our goal is to define an efficient scoring metric which is able to effectively specify a particular rule's weight that represents the rule's discrimination ability over the training set when used in the SVM classification algorithm. Most of the existing approaches weight class association rules based on the confidence value of every individual rule. Confidence is a measure of the degree of association between the consequent and the antecedent within a rule. Having a high confidence, a strong rule may not have a large representation in the training set, and consequently may not be a significant discriminative rule for the classification algorithm. In other words, in the proposed approach a FCAR is a significant discriminator if it is strong and covers more patterns in the fuzzy training set. Therefore, we believe that only considering the confidence is not a good idea to distinguish between discriminative FCARs. A better idea for weighting the rules should take into consideration their statistical importance as well, *i.e.*, the support. As a result, a good scoring metric for weighting rules should involve their support and confidence values. So, we use the scoring method introduced in [5]; however, our classification approach is different from their technique. We define the weight of a FCAR as follows.

$$W\left(R^{C_i}\right) = R^{C_i}.conf \times R^{C_i}.sup/d_{C_i} \tag{2}$$

where R^{C_i} denotes a FCAR whose antecedent is class C_i, and d_{C_i} denotes the distribution of class C_i in the fuzzy training set, *i.e.*, number of training patterns which have C_i as their class label.

First, the distribution of every individual class label in the dataset D is calculated and stored in d. Then, the weight of each rule from the FCARs is computed based on Eq. (2). A rule along with its weight is stored in a new list, called *wFCARs*.

2.4 Constructing Compatibility Feature Vectors

A feature vector has the form $f_1 = v_1 \wedge f_2 = v_2 \wedge f_3 = v_3 \wedge \ldots \wedge f_n = v_n$, where f_i is a feature (attribute) and v_i is its value. In a learning system, such as the one proposed in this study, a feature vector is used to describe the key properties of the training set to be learned by the classification algorithm. However, in order to make the learning process more understandable to domain experts, we propose to construct a new rule-based feature vector for each pattern from the original training set by utilizing the FCARs. Feature vectors constructed using our method describe the distribution compatibility of high discriminative FCARs over the original training set. That is, we check the compatibility between the rules in *wFCARs* and patterns in the original training set (with quantitative attributes). A feature in the compatibility rule-based feature vector is defined as a predicator, indicating the aforementioned compatibility. As a result, the number of features in the rule-based feature vector is equal to the number of rules within *wFCARs*. In the proposed model, a rule is compatible with a pattern if: 1) rule's consequent and pattern's class are the same, 2) they have a compatibility measure greater than zero. The compatibility of training pattern $x_p = (x_{p1}, x_{p2}, \ldots, x_{pn})$ with FCAR R_i is quantified as follows [10]:

$$\mu_i(x_p) = \mu_{i1}(x_{p1}) \times \cdots \times \mu_{in}(x_{pn}), \qquad p = 1, 2, \ldots, m, \tag{3}$$

where $\mu_{ik}(x_{pk})$ is the membership function of A_{ik}. A pair <feature, value> in a rule-based feature vector takes the following form.

$$f^{R_i} = \begin{cases} w_i \times \mu_i(x_p) & \text{if } R_i \text{ is compatible with pattern } x_p; \\ 0 & \text{otherwise.} \end{cases} \tag{4}$$

where f^{R_i} is a feature that represents rule R_i and w_i is its corresponding weight.

The process of building a set of compatibility feature vectors for a given training set D may be described as follows. For every pattern x_p from D, the algorithm checks the compatibility with all rules from *wFCARs* using the *is-Compatible* function. If rule R_i is compatible with pattern x_p, the value of the corresponding feature $f_{x_p}^{R_i}$ is set to the product of rule's weight and the compatibility measure; otherwise it is set to 0. The process incrementally adds the pair of a rule-based feature (for the recently processed rule) and its value to a vector in order to construct from the training set the rule-based feature vector FV_p that describes the compatibility of FCARs with pattern x_p. FVs is the complete set of rule-based compatibility feature vectors for the given training set.

2.5 Building Classifier Model

To deal with efficiency problems, existing association classification techniques try to reduce the number of class association rules used for building the classifier model. We take advantage of the SVM power in the context of computational complexity to use as many rules as possible in order to improve the accuracy of the classifier model. In the proposed approach, we involve in the process of building a more efficient classifier model more correlation knowledge extracted from the training set in the form of compatibility rule-based feature vectors. We argue that rule-based feature vectors provide to the SVM learner algorithm more complete and interesting information about every single pattern from the training set. The task of building a classifier model in our approach is to apply the SVM algorithm to build a classifier model. The input to the SVM algorithm is a set of compatibility rule-based feature vectors. The SVM learning algorithm uses these feature vectors to train a classifier model. We use the cross validation technique to evaluate the accuracy of our classifier model.

To predict the class label of a test instance using our classifier model, we need to generate the compatibility rule-based feature vector for the given test data; this keeps the consistency between the learning and prediction processes. However, there is a difference in the process of generating the rule-based feature vector for a test instance compared to the same process when applied to a training pattern. For the latter, we can calculate the compatibility of a discriminative fuzzy class association rule with the pattern as we already know its class label. However, for a given test instance, the class label is not known. As a result, it is not possible to apply the process with the same functionality to see whether a discriminative FCAR is compatible with a given test instance or not. To address this issue, we made the following small change to the *isCompatible* function. If the antecedent of a discriminative fuzzy rule is compatible with a particular test instance, we assume that the rule is valid for the test instance regardless of the rule's consequent and we calculate the value of the feature representing the rule in the corresponding feature vector using Eq. (4). This way, every test instance can be represented by a rule-based feature vector as well. At this stage, we can apply the classifier model to predict the class label of a given test instance by taking into consideration the consistency of the model.

2.6 Automatic Adjustment of Thresholds

Our proposed model includes two predefined thresholds: namely support and confidence that are expected to be set by the user. In our previous works [12,13], we performed experiments by varying values of these two parameters and reported the best results for the considered test data. However, such approach does not seem to be reliable for finding the best values of support and confidence because the accuracy of the final classifier depends on the arbitrary choice of thresholds. In this study, we adapted to our classifier the thresholds tuning approach TFPC-HC proposed in [3]. The approach basically performs a greedy search for finding the combination of support and confidence values that will result in the best accurate classifier for a given training set. To obtain the best combination, this basic idea has been applied using a hill-climbing procedure. The hill-climbing procedure simulates a 3-D playing area whose dimensions represent support, confidence and accuracy. The approach starts with initial support and confidence values and associated classification accuracy. Then, the procedure moves in different directions in the playing area in order to maximize the accuracy. Finally, the procedure stops when it converges. For further information about the TFPC-HC approach for threshold tuning please refer to [3].

3 Experimental Results

The "*SVM Classifier Model*" component of the system has been implemented in Java using LIBSVM [6], which is a free library for SVM classification and regression. The remaining components of the proposed model including "*Fuzzy Discretization*", "*Generating Class Association Rules*", "*Weighting Rules*" and "*Constructing Compatibility Feature Vectors*" have been developed in a Java-based software development environment. The hill-climbing procedure has been adapted for the aim of this study using the implementation provided in [4]. Finally, for the analysis, all operations are performed on a Dell Optiplex 745 with an Intel Core2 Duo 6600 @ 2.4 GHz processor and 3 GB of RAM.

For the experiments conducted to evaluate the method performance individually and in comparison with other classification methods already described in the literature, we used 6 datasets from UCI ML repository [14]. Table 1 describes these test datasets along with some related statistical information.

The main performance metric considered in this study is classification accuracy. An increase in the classification accuracy in comparison with other existing methods is a desired advantage of the proposed model. We also considered the number of generated class association rules needed for building an accurate classifier because it is a significant efficiency metric. Considering both performance and efficiency, the idle case would be when the number of rules extracted is minimized and the most accurate classifier model is built. As described earlier, to achieve this goal, our system uses an automatic threshold adjustment procedure to find the best combination of support and confidence values for building the most accurate classifier.

In term of classification accuracy, we compared the performance of our model with CBA [1], CMAR [8] and TFPC [3] from the family of associative classifiers, and with SVM as a machine learning technique. In the first set of experiments, the default values for support and confidence have been set to 1% and 50%, respectively. Table 2 lists the classification accuracy of the five methods on the 6 datasets. All accuracy rates on each dataset are obtained from 10-fold cross-validation. As can be seen from Table 2, the proposed model (named FCARSVM) dramatically outperforms all the other techniques. This shows the significance of the compatibility knowledge extracted from the training set and presented to the SVM learning algorithm in the form of rule-based feature vectors. It only gives lower accuracy for *ionosphere* dataset in comparison with SVM and

Table 1. UCI ML datasets used in the experiments [14]

Dataset	Number of Records	Number of Attributes	Number of Classes
glass	214	11	7
ionosphere	351	35	2
iris	150	5	3
pageBlocks	5473	11	5
pima	768	9	2
waveform	5000	22	3

Table 2. Comparison of classification accuracy

Dataset	Accuracy				
	CBA	CMAR	TFPC	SVM	FCARSVM
glass	68.3	75	64.5	81.3	96.72
ionosphere	31.6	90.6	85.2	91.1	89.17
iris	94	94.7	95.3	96	100
pageBlocks	90.9	90	90	92.9	95.39
pima	75.7	74.4	74.4	76.8	97.78
waveform	77.6	76.2	66.7	87	99.78

CMAR, however it is still comparable to and outperforms other methods. A deeper analysis of the obtained results may be articulated as follows.

Comparison with SVM: Our model outperforms SVM significantly in all cases except for the *ionosphere* dataset. Further, it is hard to understand the decision model made by SVM based on the properties of individual items in the dataset. This problem limits the practical use of SVM in real world applications. The proposed model takes advantages of SVM to build a fast and effective classifier, which is is more interpretable in real world classification problems. It adds a new learning layer to the classification task to make it more understandable by integrating meaningful fuzzy rules extracted from the dataset.

Comparison with single-class rule associative classifiers (CBA): Our model outperforms CBA in all cases. The CBA model ranks rules based on their confidence values; so the statistical representation of a rule in the training set is not considered. Moreover, CBA builds the classifier based on a single matching rule. As a result, it ignores some effective rules that have high distribution over the dataset and a significant discrimination ability. However, we use both support and confidence to quantify the significance of the rules to be integrated into the classifier. So, more effective rules are involved in the classifier model.

Comparison with multiple-class rules associative classifiers (CMAR): The proposed model outperforms CMAR in all cases except for the *ionosphere* dataset where our result is still comparable. Although CMAR uses some scoring metric to rank the rules and attempts to integrate high-ranked rules in the classifier model, our scoring metric weights the rules more effectively. Moreover, the CMAR classifier ignores the rule's weight while performing coverage analysis. However, we present more knowledge from the dataset to the learning process of our system by generating rule-based feature vectors by considering both rules' weights and their compatibility with all patterns.

Comparison with TFPC: Our model outperforms TFPC in all cases. Although it uses a more efficient approach to generate the class association rules, TFPC is very sensitive to the choice of support and confidence threshold. That is, the accuracy of the classifier will decrease significantly if unappropriate thresholds are selected. In comparison, as highlighted in the experiments, the proposed model can generate classifiers with acceptable accuracy as long as the choice of thresholds are reasonable by considering the statistics of the training set.

As described earlier, in case of continuous datasets, one general problem with all associative classifiers is that they use the discretized training set throughout the process and do not involve the original dataset with continuous values. However, we make use of a fuzzy discretized dataset associated with the original dataset throughout our classification process. The achieved accuracy reveals the effectiveness of the proposed fuzzy discretization approach.

In the second experiment, we conducted tests using our method after the hill-climbing approach for threshold tuning was adapted. Table 3 shows the accuracy results associated with the best support and confidence thresholds found by the

Table 3. Accuracy results when HC applied

Dataset	CBA+HC			CMAR+HC			TFPC+HC			FCARSVM+HC		
	Acc.	Best		Acc.	Best		Acc.	Best		Acc.	Best	
		S	C		S	C		S	C		S	C
glass	70.7	3.0	51.6	75.5	1.0	50.0	76.2	2.6	45.6	96.7	1.8	51.0
ionosphere	89.5	10.0	49.2	91.5	2.6	50.0	92.9	9.8	50.0	89.2	7.4	58.0
iris	94.0	1.0	50.0	94.7	2.3	50.0	95.3	1.0	50.0	100.0	7.4	58.0
pageBlocks	91.0	1.6	50.0	90.3	0.2	50.0	90.0	1.0	50.0	95.8	13.8	58.0
pima	75.7	2.8	50.0	74.5	1.6	50.0	74.9	2.3	50.0	99.1	13.8	44.0
waveform	78.2	2.6	50.0	77.2	0.6	50.0	76.6	3.2	64.3	99.9	17.4	50.5

hill-climbing procedure. Results for CBA+HC, CMAR+HC and TFPC+HC are reported from [3]. In order to keep the comparison fair, we selected the default parameter settings for the hill-climbing procedure as described in [3].

As can be seen from Table 3, in all cases except the *ionosphere* dataset, our method outperforms the rest, all combined with hill-climbing based tuning. For the *ionosphere* dataset, our result is very close and still comparable. We summarize our observations obtained from the experiments after the hill-climbing procedure was adapted to the proposed model as follows.

A large support value usually works better in our model when the input dataset is a uniformly distributed binary-class. This will have the effect of producing less rules in the system, which in turn will provide computation efficiency benefits. When the binary-class dataset is not monotonously distributed (like *pima* and *ionosphere*), medium support values perform better. For multi-class datasets with small number of classes (*waveform* has 3 uniform classes, so a high value of support like 17.4% performs well), the above guidelines still result in an efficient classifier model. However, if the number of classes in a dataset is large (say more than 5), a small support value results in a more accurate classifier (the *glass* dataset has 7 classes and a small support value of 1.8% seems appropriate). As the confidence threshold is concerned, we noticed that by increasing its value, the performance of our model does not improve. Furthermore, increasing the confidence above a certain value leads to a decrease in model accuracy. Regardless of the support value, confidence values between 45-65% have been identified to be suitable for the tested datasets, in general. Such large confidence values can significantly reduce the running time of the rule mining algorithm, while the performance of our model is still outstanding.

Table 4 displays a performance comparison in terms of the number of generated rules to construct a classifier between TFPC and our proposed method, both combined with hill-climbing based tuning. As can be seen from Table 4, in most cases, our method generates significantly fewer rules than TFPC does. This speeds up the training stage of the classification algorithm.

One might claim that the large number of generated rules will affect the effectiveness of our method. However, it should be realized that we assigned small support value for the aim of a fair comparison. Table 5 reports some statistical results obtained from the first set of experiments, where the support and confidence values were 1% and 50%, respectively. For each dataset, we also show the support and confidence values found by the hill-climbing procedure

Table 4. Number of generated rules when HC applied

Dataset	TFPC+HC			FCARSVM+HC		
	# CARs	Best		# CARs	Best	
		S	C		S	C
glass	140.0	2.6	45.6	20.0	1.8	51.0
ionosphere	91.3	9.8	50.0	2.0	7.4	58.0
iris	14.8	1.0	50.0	10.0	7.4	58.0
pageBlocks	12.8	1.0	50.0	22.0	13.8	58.0
pima	20.3	2.3	50.0	40.0	13.8	44.0
waveform	708.2	3.2	64.3	11.0	17.4	50.5

Table 5. Effectiveness of selecting large support value

Dataset	S	C	Number of Rules	Accuracy
glass	1.0	50.0	24	96.7
	1.8	51.0	20	96.7
ionosphere	1.0	50.0	2	89.2
	7.4	58.0	2	89.2
iris	1.0	50.0	13	100.0
	7.4	58.0	10	100.0
pageBlocks	1.0	50.0	42	95.4
	13.8	58.0	22	95.8
pima	1.0	50.0	121	97.8
	13.8	44.0	40	99.1
waveform	1.0	50.0	791	99.8
	17.4	50.5	11	99.9

to build the most accurate classifier. As can be seen from Table 5, the number of generated rules dramatically decreases when hill-climbing tunes the support threshold to higher value. For instance, for the *waveform* dataset, the number of generated rules decreases from 791 rules to 11 rules when the support value is varied from 1% to 17.4%, and the accuracy has slightly improved as well.

4 Summary and Conclusions

In this study, we proposed a new classification framework, which integrates association rule mining and SVM to provide users with better understandability and interpretability via an accurate class association rule-based classifier model. An existing association rule mining algorithm is adapted to extract, from the fuzzy discretized dataset, FCARs which satisfy the minimum support and confidence thresholds. Fuzzy discretization has been developed to cope with quantitative datasets. The extracted FCARs are weighted based on a scoring metric strategy in order to quantify the discrimination ability of the rules over the training set. The compatibilities of the FCARs with patterns in the training set are used to generate a set of compatibility rule-based feature vectors, which are then given to the SVM to build a classifier model. The reported results demonstrate that the proposed model dramatically improve the classification accuracy while preserving the efficiency and effectiveness of the final classifier.

References

1. Agrawal, R., Srikant, R.: Fast algorithms for mining association rule. In: Proc. of VLDB (1994)
2. Cristianini, N., Taylor, J.S.: An Introduction to Support Vector Machines and Other Kernel-based Learning Methods. Cambridge University Press, Cambridge (2000)
3. Coenen, F.P., Leng, P.: The Effect of Threshold Values on Association Rule Based Classification Accuracy. DKE 60(2), 345–360 (2007)

4. Coenen, F.P.: The LUCS-KDD TFPC Classification Association Rule Mining Algorithm, University of Liverpool (2004), www.cSc.liv.ac.uk/~frans/KDD/Software/Apriori_TFPC/aprioriTFPC.html
5. Cong, G., et al.: Mining Top-k Covering Rule Groups for Gene Expression Data. In: Proc. of ACM SIGMOD, pp. 670–681 (2005)
6. Chang, C.C., Lin, C.J.: LIBSVM: A Library for Support Vector Machines (2001), www.csie.ntu.edu.tw/~cjlin/libsvm
7. Furey, T.S., et al.: Support vector machine classification and validation of cancer tissue samples using microarray expression data. Bioinformatics 16 (2000)
8. Li, W., Han, J., Pei, J.: CMAR: Accurate and efficient classification based on multiple class-association rules. In: Proc. of IEEE ICDM (2001)
9. Liu, B., Hsu, W., Ma, Y.: Integrating Classification and Association Rule Mining. In: Proc. of ACM KDD (1998)
10. Ishibuchi, H., Nakashima, T.: Improving the performance of fuzzy classifier systems for pattern classification problems with continuous attributes. IEEE Trans. Industrial Electronics 46(6), 157–168 (1999)
11. Ishibuchi, H., Nozaki, K., Tanaka, H.: Distributed Representation of Fuzzy Rules and Its Application to Pattern Classification. Fuzzy Sets and Systems 52(1), 21–32 (1992)
12. Kianmehr, K., Alhajj, R.: Effective Classification by Integrating Support Vector Machine and Association Rule Mining. In: Corchado, E.S., Yin, H., Botti, V., Fyfe, C. (eds.) IDEAL 2006. LNCS, vol. 4224, pp. 920–927. Springer, Heidelberg (2006)
13. Kianmehr, K., Alhajj, R.: Support Vector Machine Approach for Fast Classification. In: Tjoa, A.M., Trujillo, J. (eds.) DaWaK 2006. LNCS, vol. 4081, pp. 534–543. Springer, Heidelberg (2006)
14. Merz, C.J., Murphy, P.: UCI repository of machine learning database (1996), http://www.cs.uci.edu/~mlearn/MLRepository.html

Towards a Crisp Representation of Fuzzy Description Logics under Łukasiewicz Semantics

Fernando Bobillo[1,⋆] and Umberto Straccia[2]

[1] Dpt. of Computer Science and Artificial Intelligence, University of Granada, Spain
[2] Istituto di Scienza e Tecnologie dell'Informazione (ISTI - CNR), Pisa, Italy
fbobillo@decsai.ugr.es, straccia@isti.cnr.it

Abstract. Classical ontologies are not suitable to represent imprecise nor uncertain pieces of information. Fuzzy Description Logics were born to represent the former type of knowledge, but they require an appropriate fuzzy language to be agreed and an important number of available resources to be adapted. An alternative is to use classical ontologies to represent fuzzy ontologies. To date, all of the work in this direction has restricted to the Zadeh family of fuzzy operators. In this paper, we generalize existing proposals and propose a reasoning preserving procedure to obtain a crisp representation for a fuzzy extension of the logic $\mathcal{ALCHOI}$ under Łukasiewicz semantics. This reduction makes possible to reuse a crisp representation language as well as currently available reasoners under crisp semantics.

1 Introduction

Description Logics (DLs for short) [1] are a family of logics for representing structured knowledge which have proved to be very useful as ontology languages. For instance, the standard Web Ontology Language OWL [2] can be divided in three levels, namely OWL Full, OWL DL and OWL Lite, with $\mathcal{SHOIN}(\mathbf{D})$ and $\mathcal{SHIF}(\mathbf{D})$ DLs being the subjacent formalisms of OWL DL and OWL Lite.

Nevertheless, it has been widely pointed out that classical ontologies are not appropriate to deal with imprecise and vague knowledge, which is inherent to several real-world domains. Since fuzzy logic is a suitable formalism to handle these types of knowledge, several fuzzy extensions of DLs can be found in the literature (see [3] for an overview).

Defining a fuzzy DL brings about that crisp standard languages would no longer be appropriate, new fuzzy languages should be used and hence the large number of resources available should be adapted to the new framework, requiring an important effort. An alternative is to represent fuzzy DLs using crisp DLs and to reason using their (crisp) reductions. This approach has several advantages:

- There would be no need to agree a new standard fuzzy language, but every developer could use its own language as long as he implements the reduction that we describe.

⋆ The work of F. Bobillo has been partially supported by the Spanish Ministerio de Educación y Ciencia under project TIN2006-15041-C04-01 and a FPU scholarship.

A. An et al. (Eds.): ISMIS 2008, LNAI 4994, pp. 309–318, 2008.

- We will continue using standard languages with a lot of resources available, so the need (and cost) of adapting them to the new fuzzy language is avoided.
- We will use existing crisp reasoners. We do not claim that reasoning will be more efficient, but it supposes an easy alternative to support early reasoning in future fuzzy languages. In fact, nowadays there is no reasoner fully supporting a fuzzy extension of OWL DL under Łukasiewicz semantics[1].

Under this approach an immediate practical application of fuzzy ontologies is feasible, because of its tight relation with already existing languages and tools which have proved their validity.

Although there has been a relatively significant amount of work in extending DLs with fuzzy set theory [3], the representation of them using crisp description logics has not received such attention. The first efforts in this direction are due to Straccia, who considered fuzzy $\mathcal{ALCH}$ [4] and fuzzy $\mathcal{ALC}$ with truth values taken from an uncertainty lattice [5]. Bobillo et al. widened the representation to $\mathcal{SHOIN}$ [7]. Stoilos et al. extended this work with some additional role axioms [8]. Finally, Bobillo et al. proposed a full crisp representation of $\mathcal{SROIQ}$, and optimized the process by reducing the size of the resulting ontology [9]. However, from a semantics point of view, these works restrict themselves to the Zadeh family of fuzzy operators, which has some limitations (see [6,7] for some counter-intuitive examples). This paper provides a crisp representation for a fuzzy DL under Łukasiewicz semantics, which generalizes Zadeh family.

The remainder of this work is organized as follows. Section 2 overviews some important results in fuzzy set theory. Section 3 describes a fuzzy extension of $\mathcal{ALCH}$ under Łukasiewicz semantics. Section 4 depicts a reduction into crisp $\mathcal{ALCH}$ and extends the result to $\mathcal{ALCHOI}$. Finally, Section 5 sets out some conclusions and ideas for future work.

2 Fuzzy Set Theory

Fuzzy set theory and fuzzy logic were proposed by Zadeh [10] to manage imprecise and vague knowledge. While in classical set theory elements either belong to a set or not, in fuzzy set theory elements can belong to a degree of certainty. More formally, let X be a set of elements called the reference set. A fuzzy subset A of X, is defined by a membership function $\mu_A(x)$, or simply $A(x)$, which assigns any $x \in X$ to a value in the real interval between 0 and 1. As in the classical case, 0 means no-membership and 1 full membership, but now a value between 0 and 1 represents the extent to which x can be considered as an element of X.

For every $\alpha \in [0,1]$, the *α-cut* of a fuzzy set A is defined as the set such as its elements belong to A with degree at least α, i.e. $A_\alpha = \{x \mid \mu_A(x) \geq \alpha\}$. Similarly, the *strict α-cut* is defined as $A_{\alpha+} = \{x \mid \mu_A(x) > \alpha\}$. Notice that these sets are crisp.

All crisp set operations are extended to fuzzy sets. The intersection, union, complement and implication set operations are performed by a t-norm function t, a t-conorm function u, a negation function c and an implication function i,

[1] The *fuzzyDL* reasoner (see Straccia's Web page) supports fuzzy OWL-Lite so far.

respectively. For a definition of these functions we refer the reader to [11]. We will mention throughout this paper two families of fuzzy operators, Zadeh and Łukasiewicz, which are defined as follows:

Family	t-norm $t(\alpha,\beta)$	t-conorm $u(\alpha,\beta)$	negation $c(\alpha)$	implication $i(\alpha,\beta)$
Zadeh	$\min\{\alpha,\beta\}$	$\max\{\alpha,\beta\}$	$1-\alpha$	$\max\{1-\alpha,\beta\}$
Łukasiewicz	$\max\{\alpha+\beta-1,0\}$	$\min\{\alpha+\beta,1\}$	$1-\alpha$	$\min\{1,1-\alpha+\beta\}$

Let $\otimes$, $\oplus$, $\ominus$ and $\Rightarrow$ denote the Łukasiewicz family of fuzzy operators (t-norm, t-conorm, negation and implication, respectively) and let $\wedge$, $\vee$, $\neg$ and $\rightarrow$ denote the Zadeh family. Interestingly, using the Łukasiewicz family it is possible to represent the operators of Zadeh family:

$$\neg\alpha = \ominus\alpha \qquad \alpha \wedge \beta = \alpha \otimes (\alpha \Rightarrow \beta)$$
$$\alpha \vee \beta = \neg((\neg\alpha) \wedge (\neg\beta)) \qquad \alpha \rightarrow \beta = (\neg\alpha) \vee \beta$$

3 Fuzzy $\mathcal{ALCH}$

In this section we define a fuzzy extension of the DL $\mathcal{ALCH}$ where concepts denote fuzzy sets of individuals and roles denote fuzzy binary relations. Axioms are also extended to the fuzzy case and some of them hold to a degree. Then, we will restrict ourselves to the Łukasiewicz family of fuzzy operators.

The following definition is based on the fuzzy DL presented in [12].

Syntax. Fuzzy $\mathcal{ALCH}$ assumes three alphabets of symbols, for concepts, roles and individuals. The concepts of the language (denoted C or D) can be built inductively from atomic concepts (A), atomic roles (R), top concept $\top$ and bottom concept $\bot$ according to the following syntax rule:

$$C, D \rightarrow A \mid \top \mid \bot \mid C \sqcap D \mid C \sqcup D \mid \neg C \mid \forall R.C \mid \exists R.C$$

Notice that the syntax is the same as in the crisp case.

A fuzzy Knowledge Base (KB) comprises two parts: the extensional knowledge, i.e. particular knowledge about some specific situation (a fuzzy Assertional Box or ABox $\mathcal{A}$ with statements about individuals) and the intensional knowledge, i.e. general knowledge about the application domain (a fuzzy Terminological Box or TBox $\mathcal{T}$ and a fuzzy Role Box or RBox $\mathcal{R}$).

In the rest of the paper we will assume $\bowtie \in \{\geq, \leq\}$, $\alpha \in (0,1]$, $\beta \in [0,1)$ and $\gamma \in [0,1]$. Moreover, for every operator $\bowtie$ we define its symmetric operator $\bowtie^-$ as $\geq^- = \leq$, $\leq^- = \geq$.

A *fuzzy ABox* consists of a finite set of *fuzzy assertions* of the following types:

- concept assertions $\langle a : C \geq \alpha \rangle$ or $\langle a : C \leq \beta \rangle$,
- role assertions $\langle (a,b) : R \geq \alpha \rangle$.

A *fuzzy TBox* consists of fuzzy General Concept Inclusions (fuzzy GCIs), expressions of the form $\langle C \sqsubseteq D \geq \alpha \rangle$.

A *fuzzy RBox* consists of a finite set of fuzzy Role Inclusion Axioms (fuzzy RIAs) of the form $\langle R \sqsubseteq R' \geq \alpha \rangle$.

A fuzzy axiom τ is *positive* (denoted $\langle \tau \rhd \alpha \rangle$) if it is of the form $\langle \tau \geq \alpha \rangle$, and *negative* (denoted $\langle \tau \lhd \alpha \rangle$) if it is of the form $\langle \tau \leq \beta \rangle$. $\langle \tau = \alpha \rangle$ is equivalent to the pair of axioms $\langle \tau \geq \alpha \rangle$ and $\langle \tau \leq \alpha \rangle$.

Notice that negative role assertions, GCIs or RIAs are not allowed, because they correspond to negated role assertions, GCIs and RIAs respectively, which are not part of crisp $\mathcal{ALCH}$.

Note also that for the sake of clarity we are only considering inequalities of the form $\geq$ and $\leq$, but extending the language with $>$ and $<$ is not complicated.

Semantics. A fuzzy interpretation $\mathcal{I}$ is a pair $(\Delta^{\mathcal{I}}, \cdot^{\mathcal{I}})$ consisting of a non empty set $\Delta^{\mathcal{I}}$ (the interpretation domain) and a fuzzy interpretation function $\cdot^{\mathcal{I}}$ mapping:

- every individual onto an element of $\Delta^{\mathcal{I}}$,
- every concept C onto a function $C^{\mathcal{I}} : \Delta^{\mathcal{I}} \rightarrow [0,1]$,
- every role R onto a function $R^{\mathcal{I}} : \Delta^{\mathcal{I}} \times \Delta^{\mathcal{I}} \rightarrow [0,1]$.

$C^{\mathcal{I}}$ (resp. $R^{\mathcal{I}}$) denotes the membership function of the fuzzy concept C (resp. fuzzy role R) w.r.t. $\mathcal{I}$. $C^{\mathcal{I}}(a)$ (resp. $R^{\mathcal{I}}(a,b)$) gives us the degree of being the individual a an element of the fuzzy concept C (resp. the degree of being (a,b) an element of the fuzzy role R) under the fuzzy interpretation $\mathcal{I}$.

For a t-norm $\otimes$, a t-conorm $\oplus$, a negation function $\ominus$ and an implication function $\Rightarrow$, the fuzzy interpretation function is extended to complex concepts as follows:

$$\begin{aligned}
\top^{\mathcal{I}}(a) &= 1 \\
\bot^{\mathcal{I}}(a) &= 0 \\
(C \sqcap D)^{\mathcal{I}}(a) &= C^{\mathcal{I}}(a) \otimes D^{\mathcal{I}}(a) \\
(C \sqcup D)^{\mathcal{I}}(a) &= C^{\mathcal{I}}(a) \oplus D^{\mathcal{I}}(a) \\
(\neg C)^{\mathcal{I}}(a) &= \ominus C^{\mathcal{I}}(a) \\
(\forall R.C)^{\mathcal{I}}(a) &= \inf_{b \in \Delta^{\mathcal{I}}} \{R^{\mathcal{I}}(a,b) \Rightarrow C^{\mathcal{I}}(b)\} \\
(\exists R.C)^{\mathcal{I}}(a) &= \sup_{b \in \Delta^{\mathcal{I}}} \{R^{\mathcal{I}}(a,b) \otimes C^{\mathcal{I}}(b)\}
\end{aligned}$$

A fuzzy interpretation $\mathcal{I}$ satisfies (is a model of):

- $\langle a : C \geq \alpha \rangle$ iff $C^{\mathcal{I}}(a^{\mathcal{I}}) \geq \alpha$,
- $\langle a : C \leq \beta \rangle$ iff $C^{\mathcal{I}}(a^{\mathcal{I}}) \leq \beta$,
- $\langle (a,b) : R \geq \alpha \rangle$ iff $R^{\mathcal{I}}(a^{\mathcal{I}}, b^{\mathcal{I}}) \geq \alpha$,
- $\langle C \sqsubseteq D \geq \alpha \rangle$ iff $\inf_{a \in \Delta^{\mathcal{I}}} \{C^{\mathcal{I}}(a) \Rightarrow D^{\mathcal{I}}(a)\} \geq \alpha$,
- $\langle R \sqsubseteq R' \geq \alpha \rangle$ iff $\inf_{a,b \in \Delta^{\mathcal{I}}} \{R^{\mathcal{I}}(a,b) \Rightarrow R'^{\mathcal{I}}(a,b)\} \geq \alpha$,

We say that $\mathcal{I}$ *satisfies* a fuzzy KB iff it satisfies each element in it. In the rest of the paper we will only consider fuzzy KB satisfiability, since (as in the crisp case) most inference problems can be reduced to it [13].

It can be easily shown that this fuzzy extension of $\mathcal{ALCH}$ is a sound extension of crisp $\mathcal{ALCH}$, because fuzzy interpretations coincide with crisp interpretations if we restrict the membership degrees to $\{0,1\}$.

Here in after we will concentrate on Ł-$\mathcal{ALCH}$, restricting ourselves to the fuzzy operators of the Łukasiewicz family.

4 A Crisp Representation for Fuzzy Ł-$\mathcal{ALCH}$

In this section we show how to reduce a Ł-$\mathcal{ALCH}$ fuzzy KB into a crisp knowledge base (KB). We will start by presenting our reduction procedure, then we will discuss the properties of the reduction, showing that it preserves reasoning, so existing $\mathcal{ALCH}$ reasoners could be applied to the resulting KB, and illustrate the full procedure with an example.

The basic idea is to create some new crisp concepts and roles, representing the α-cuts of the fuzzy concepts and relations, and to rely on them. Next, some new axioms are added to preserve their semantics and finally every axiom in the ABox, the TBox and the RBox is represented, independently from other axioms, using these new crisp elements.

4.1 Adding New Elements

U. Straccia showed [4] that, for a fuzzy KB fK, the set of the degrees which must be considered for any reasoning task is defined as $N^{fK} = X^{fK} \cup \{1 - \alpha \mid \alpha \in X^{fK}\}$, where $X^{fK} = \{0, 0.5, 1\} \cup \{\gamma \mid \langle \tau \bowtie \gamma \rangle \in fK\}$ [4]. This holds for fuzzy DLs under Zadeh semantics, but it is not true in general when other fuzzy operators are considered. Interestingly, in the case of Łukasiewicz logic it is true if we fix the number of allowed degrees.

In fact, let q be a natural number with $q \geq 1$. We assume a set of $q + 1$ allowed truth degrees in the fuzzy KB, i.e., $\mathcal{N} = \{0, \frac{1}{q}, \frac{2}{q}, \ldots, \frac{(q-1)}{q}, 1\}$. The following proposition shows that, using the fuzzy operators of Łukasiewicz logic to combine two truth degrees a and b, no new degrees can appear.

Proposition 1. *Let $\frac{a}{q}, \frac{b}{q} \in \mathcal{N}$. Then, under the fuzzy operators of Łukasiewicz logic, $\ominus\frac{a}{q}, \frac{a}{q} \otimes \frac{b}{q}, \frac{a}{q} \oplus \frac{b}{q}, \frac{a}{q} \Rightarrow \frac{b}{q} \in \mathcal{N}$.*

Proof. Let us consider each of the four fuzzy operators:

- $\ominus\frac{a}{q} = 1 - \frac{a}{q} = \frac{q-a}{q}$ belongs to $\mathcal{N}$: since $a \in [0, q], (q - a) \in [0, q]$.
- $\frac{a}{q} \otimes \frac{b}{q} = \max\{\frac{a}{q} + \frac{b}{q} - 1, 0\}$. If $\frac{a}{q} + \frac{b}{q} - 1 \leq 0$, then the value of the conjunction is 0, which obviously belongs to $\mathcal{N}$. Otherwise, its value is $\frac{a+b-q}{q}$ which also belongs to $\mathcal{N}$: since $a, b \in [0, q]$ and $\frac{a}{q} + \frac{b}{q} - 1 > 0$, it follows that $(a + b - q) \in [0, q]$.
- $\frac{a}{q} \oplus \frac{b}{q} = \min\{\frac{a}{q} + \frac{b}{q}, 1\}$. If $\frac{a}{q} + \frac{b}{q} > 1$, then the value of the disjunction is 1, which obviously belongs to $\mathcal{N}$. Otherwise, its value is $\frac{a+b}{q}$ which also belongs to $\mathcal{N}$: since $a, b \in [0, q]$ and $\frac{a}{q} + \frac{b}{q} \leq 1$, it follows that $(a + b) \in [0, q]$.
- $\frac{a}{q} \Rightarrow \frac{b}{q} = \min\{1 - \frac{a}{q} + \frac{b}{q}, 1\}$. If the minimum is 1, then the value of the implication obviously belongs to $\mathcal{N}$. Otherwise, the value is $\frac{q-a+b}{q}$ which also belongs to $\mathcal{N}$: since $a, b \in [0, q]$ and $1 - \frac{a}{q} + \frac{b}{q} \leq 1$, it follows that $(1 - a + b) \in [0, q]$. □

Now, we will assume that $N^{fK} = \mathcal{N}$ and proceed similarly as in [9], which creates an optimized number of new elements (concepts, roles and axioms) with respect to previous approaches.

Without loss of generality, it can be assumed that $N^{fK} = \{\gamma_1, \ldots, \gamma_{|N^{fK}|}\}$ and $\gamma_i < \gamma_{i+1}, 1 \leq i \leq |N^{fK}| - 1$. It is easy to see that $\gamma_1 = 0$ and $\gamma_{|N^{fK}|} = 1$.

Let A^{fK} and R^{fK} be the set of atomic concepts and atomic roles occurring in a fuzzy KB $fK = \langle \mathcal{A}, \mathcal{T}, \mathcal{R} \rangle$. For each $\alpha, \beta \in N^{fK}$ with $\alpha \in (0, 1]$ and $\beta \in [0, 1)$, for each $A \in A^{fK}$ and for each $R_A \in R^{fK}$, two new atomic concepts $A_{\geq \alpha}, A_{> \beta}$ and one new atomic role $R_{\geq \alpha}$ are introduced. $A_{\geq \alpha}$ represents the crisp set of individuals which are instance of A with degree higher or equal than α i.e the α-cut of A. The other new elements are defined in a similar way. The atomic elements $A_{>1}, A_{\geq 0}$ and $R_{\geq 0}$ are not considered because they are not necessary, due to the restrictions on the allowed degree of the axioms in the fuzzy KB.

A concept of the form $A_{\geq \alpha}$ may be used for instance to represent an assertion of the form $\langle a \colon A \geq \alpha \rangle$, while $A_{> \alpha}$ can be used to represent an assertion of the form $\langle a \colon \neg A \geq 1 - \alpha \rangle$. Since role negation and axioms of the form $\tau > \beta$ are not allowed, elements of the form $R_{> \alpha}$ are not needed.

The semantics of these newly introduced atomic concepts and roles is preserved by some terminological and role axioms. For each $1 \leq i \leq |N^{fK}| - 1, 2 \leq j \leq |N^{fK}| - 1$ and for each $A \in A^{fK}$, $T(N^{fK})$ is the smallest terminology containing the axioms: $A_{\geq \gamma_{i+1}} \sqsubseteq A_{> \gamma_i}, A_{> \gamma_j} \sqsubseteq A_{\geq \gamma_j}$. Similarly, for each atomic role $R \in R^{fK}$, we define $R(N^{fK})$ as the smallest terminology containing: $R_{\geq \gamma_{j+1}} \sqsubseteq R_{\geq \gamma_j}$.

4.2 Mapping Fuzzy Concepts, Roles and Axioms

Before showing how to represent the elements of the KB using these new elements, we will illustrate with an example.

Example 1. Consider a fuzzy assertion $\tau = \langle a : A_1 \sqcap A_2 \geq 0.5 \rangle$ and $\mathcal{N} = \{0, 0.25, 0.5, 0.75, 1\}$. Every model $\mathcal{I}$ of τ satisfies $\max\{A_1^{\mathcal{I}}(a) + A_2^{\mathcal{I}}(a) - 1, 0\} \geq 0.5$. Hence, it follows that $A_1^{\mathcal{I}}(a) + A_2^{\mathcal{I}}(a) - 1 \geq 0.5 \Leftrightarrow A_1^{\mathcal{I}}(a) + A_2^{\mathcal{I}}(a) \geq 1.5$. Now, we do not know exactly the degrees of truth of $A_1^{\mathcal{I}}(a)$ and $A_2^{\mathcal{I}}(a)$, but we now that they belong to $\mathcal{N}$, so there are six possibilities:

$A_1^{\mathcal{I}}(a)$	$A_2^{\mathcal{I}}(a)$	$(A_1 \sqcap A_2)^{\mathcal{I}}(a)$
0.5	1	0.5
0.75	0.75	0.5
0.75	1	0.75
1	0.5	0.5
1	0.75	0.75
1	1	1

Hence, we can think of a crisp model satisfying $a : (A_{1 \geq 0.5} \sqcap A_{2 \geq 1}) \sqcup (A_{1 \geq 0.75} \sqcap A_{2 \geq 0.75}) \sqcup (A_{1 \geq 0.75} \sqcap A_{2 \geq 1}) \sqcup (A_{1 \geq 1} \sqcap A_{2 \geq 0.5}) \sqcup (A_{1 \geq 1} \sqcap A_{2 \geq 0.75}) \sqcup (A_{1 \geq 1} \sqcap A_{2 \geq 1})$.

But this (crisp) assertion can be optimized (see below Proposition 2) since it is satisfiable if the following is: $a : (A_{1 \geq 0.5} \sqcap A_{2 \geq 1}) \sqcup (A_{1 \geq 0.75} \sqcap A_{2 \geq 0.75}) \sqcup (A_{1 \geq 1} \sqcap A_{2 \geq 0.5})$. □

Proposition 2. *Let $A \sqsubseteq B_1 \sqcup B_2$ be a crisp GCI with B_1 subsuming B_2. $A \sqsubseteq B_1 \sqcup B_2$ is satisfiable if and only if $A \sqsubseteq B_1$ is.*

Proof. If $A \sqsubseteq B_1 \sqcup B_2$ is satisfiable then there exists a model $\mathcal{I}$ such that $A^{\mathcal{I}} \subseteq B_1^{\mathcal{I}} \cup B_2^{\mathcal{I}}$. Since B_1 subsumes B_2, $B_2^{\mathcal{I}} \subseteq B_1^{\mathcal{I}}$, and hence $B_1^{\mathcal{I}} \cup B_2^{\mathcal{I}} = B_1^{\mathcal{I}}$, so $A^{\mathcal{I}} \subseteq B_1^{\mathcal{I}}$ and hence $A \sqsubseteq B_1$ holds. The other direction is similar. □

We define $\mathcal{N}^+ = \{x \in \mathcal{N} : x \neq 0\}$. Concept and role expressions are reduced using mapping ρ, as shown in Table 1. Notice that expressions of the form $\rho(A, \geq 0)$ and $\rho(A, \leq 1)$ cannot appear, because there exist some restrictions on the degree of the axioms in the fuzzy KB. The same also holds for $\top$, $\bot$ and R_A. Moreover, expressions of the form $\rho(R, \lhd\gamma)$ cannot appear either.

Axioms are reduced as in Table 2, where σ maps fuzzy axioms into crisp assertions and κ maps fuzzy TBox (resp. RBox) axioms into crisp TBox (resp. RBox) axioms. We note $\sigma(\mathcal{A})$ (resp. $\kappa(fK, \mathcal{T})$, $\kappa(fK, \mathcal{R})$) the union of the reductions of every axiom in $\mathcal{A}$ (resp. $\mathcal{T}$, $\mathcal{R}$).

Table 1. Mapping of concept and role expressions

x	y	$\rho(x,y)$
$\top$	$\geq \alpha$	$\top$
$\top$	$\leq \beta$	$\bot$
$\bot$	$\geq \alpha$	$\bot$
$\bot$	$\leq \beta$	$\top$
A	$\geq \alpha$	$A_{\geq\alpha}$
A	$\leq \beta$	$\neg A_{>\gamma}$
R_A	$\geq \alpha$	$R_{\geq\alpha}$
$\neg C$	$\bowtie \gamma$	$\rho(C, \bowtie^- 1-\gamma)$
$C \sqcap D$	$\geq \alpha$	$\bigsqcup_{\gamma_1,\gamma_2} \rho(C, \geq \gamma_1) \sqcap \rho(D, \geq \gamma_2)$ for every pair $\langle\gamma_1,\gamma_2\rangle$ such that $\gamma_1,\gamma_2 \in \mathcal{N}^+$, $\gamma_1+\gamma_2 = 1+\alpha$
$C \sqcap D$	$\leq \beta$	$\rho(\neg C, \geq 1-\beta) \sqcup \rho(\neg D, \geq 1-\beta)$
$C \sqcup D$	$\geq \alpha$	$\rho(C, \geq \alpha) \sqcup \rho(D, \geq \alpha) \bigsqcup_{\gamma_1,\gamma_2} \rho(C, \geq \gamma_1) \sqcap \rho(D, \geq \gamma_2)$ for every pair $\langle\gamma_1,\gamma_2\rangle$ such that $\gamma_1,\gamma_2 \in \mathcal{N}^+$, $\gamma_1+\gamma_2 = \alpha$
$C \sqcup D$	$\leq \beta$	$\rho(\neg C, \geq 1-\beta) \sqcap \rho(\neg D, \geq 1-\beta)$
$\exists R.C$	$\geq \alpha$	$\bigsqcup_{\gamma_1,\gamma_2} \exists\rho(R, \geq \gamma_1).\rho(C, \geq \gamma_2)$ for every pair $\langle\gamma_1,\gamma_2\rangle$ such that $\gamma_1,\gamma_2 \in \mathcal{N}^+$, $\gamma_1+\gamma_2 = 1+\alpha$
$\exists R.C$	$\leq \beta$	$\rho(\forall R.\neg C, \geq 1-\beta)$
$\forall R.C$	$\geq \alpha$	$\bigsqcap_{\gamma_1,\gamma_2} \forall\rho(R, \geq \gamma_1).\rho(C, \geq \gamma_2)$ for every pair $\langle\gamma_1,\gamma_2\rangle$ such that $\gamma_1,\gamma_2 \in \mathcal{N}^+$ and $\gamma_1 = \gamma_2 + 1 - \alpha$
$\forall R.C$	$\leq \beta$	$\rho(\exists R.\neg C, \geq 1-\beta)$

Table 2. Reduction of the axioms

$\sigma(\langle a\!:\!C \geq \alpha\rangle)$	$a\!:\!\rho(C, \geq \alpha)$
$\sigma(\langle a\!:\!C \leq \beta\rangle)$	$a\!:\!\rho(C, \leq \beta)$
$\sigma(\langle (a,b)\!:\!R \geq \alpha\rangle)$	$(a,b)\!:\!\rho(R, \geq \alpha)$
$\kappa(\langle C \sqsubseteq D \geq \alpha\rangle)$	$\bigcup_{\gamma_1,\gamma_2} \{\rho(C, \geq \gamma_1) \sqsubseteq \rho(D, \geq \gamma_2)\}$ for every pair $\langle\gamma_1,\gamma_2\rangle$ such that $\gamma_1,\gamma_2 \in \mathcal{N}^+$ and $\gamma_1 = \gamma_2 + 1 - \alpha$
$\kappa(\langle R \sqsubseteq R' \geq \alpha\rangle)$	$\bigcup_{\gamma_1,\gamma_2} \{\rho(R, \geq \gamma_1) \sqsubseteq \rho(R', \geq \gamma_2)\}$ for every pair $\langle\gamma_1,\gamma_2\rangle$ such that $\gamma_1,\gamma_2 \in \mathcal{N}^+$ and $\gamma_1 = \gamma_2 + 1 - \alpha$

4.3 Properties of the Reduction

Summing up, a fuzzy KB $fK = \langle \mathcal{A}, \mathcal{T}, \mathcal{R} \rangle$ is reduced into a KB $\mathcal{K}(fK) = \langle \sigma(\mathcal{A}), T(N^{fK}) \cup \kappa(fK, \mathcal{T}),\ R(N^{fK}) \cup \kappa(fK, \mathcal{R}) \rangle$. The reduction is reasoning preserving since the following theorem shows:

Theorem 1. *Assuming a truth space* $\{0, \frac{1}{q}, \frac{2}{q}, \ldots, \frac{(q-1)}{q}, 1\}$, *a* Ł-$\mathcal{ALCH}$ *fuzzy KB* fK *is satisfiable iff* $\mathcal{K}(fK)$ *is satisfiable.*

Example 2. Let us a consider a fuzzy KB $fK = \{\langle a : \forall R.(C \sqcap D) \geq 0.75\rangle, \langle (a,b) : R \geq 0.75\rangle, \langle b : \neg C \geq 0.75\rangle\}$ and assume a set of degrees of truth $\mathcal{N} = \{0, 0.25, 0.5, 0.75, 1\}$ ($q = 4$). Note that the TBox and the RBox are empty.

This fuzzy KB is clearly unsatisfiable. From the third assertion it follows that $C^{\mathcal{I}}(b^{\mathcal{I}}) \leq 0.25$, and it can be seen that this implies that $(C \sqcap D)^{\mathcal{I}}(b^{\mathcal{I}}) = \max\{C^{\mathcal{I}}(b^{\mathcal{I}}) + D^{\mathcal{I}}(b^{\mathcal{I}}) - 1, 0\} \leq 0.25$. But from the two former assertions it follows that every fuzzy interpretation $\mathcal{I}$ has to satisfy $(C \sqcap D)^{\mathcal{I}}(b^{\mathcal{I}}) \geq 0.5$, which is a contradiction.

Now, let us compute the crisp representation of fK. Firstly, we create some new crisp atomic concepts associated to the set of atomic fuzzy concepts $A^{fK} = \{C, D\}$ (i.e. $C_{>0}, C_{\geq 0.25}, C_{>0.25}, C_{\geq 0.5}, C_{>0.5}, C_{\geq 0.75}, C_{>0.75}, C_{\geq 1}, D_{>0}, D_{\geq 0.25}, D_{>0.25}, D_{\geq 0.5}, D_{>0.5}, D_{\geq 0.75}, D_{>0.75}, D_{\geq 1}$) and some new crisp atomic roles associated to the set of atomic fuzzy roles $R^{fK} = \{R\}$ (i.e. $R_{\geq 0.25}, R_{\geq 0.5}, R_{\geq 0.75}, R_{\geq 1}$).

Now we create some new axioms to preserve the semantics of these elements:

- $A^{fK} = \{C_{\geq 1} \sqsubseteq C_{>0.75}, C_{>0.75} \sqsubseteq C_{\geq 0.75}, C_{\geq 0.75} \sqsubseteq C_{>0.5}, C_{>0.5} \sqsubseteq C_{\geq 0.5}, C_{\geq 0.5} \sqsubseteq C_{>0.25}, C_{>0.25} \sqsubseteq C_{\geq 0.25}, C_{\geq 0.25} \sqsubseteq C_{>0}, D_{\geq 1} \sqsubseteq D_{>0.75}, D_{>0.75} \sqsubseteq D_{\geq 0.75}, D_{\geq 0.75} \sqsubseteq D_{>0.5}, D_{>0.5} \sqsubseteq D_{\geq 0.5}, D_{\geq 0.5} \sqsubseteq D_{>0.25}, D_{>0.25} \sqsubseteq D_{\geq 0.25}, D_{\geq 0.25} \sqsubseteq D_{>0}\}$,
- $R^{fK} = \{R_{\geq 1} \sqsubseteq R_{\geq 0.75}, R_{\geq 0.75} \sqsubseteq R_{\geq 0.5}, R_{\geq 0.5} \sqsubseteq R_{\geq 0.25}\}$.

Now we are ready to compute $\sigma(\mathcal{A})$, including the reduction of the three fuzzy assertions in the fuzzy KB, that is:

- $\sigma(\langle (a,b) : R \geq 0.75\rangle) = (a,b) : \rho(R, \geq 0.75) = (a,b) : R_{\geq 0.75}$.
- $\sigma(\langle b : \neg C \geq 0.75\rangle) = b : \rho(\neg C, \geq 0.75) = b : \neg C_{>0.25}$.
- $\sigma(\langle a : \forall R.(C \sqcap D) \geq 0.75\rangle) = a : \rho(\forall R.(C \sqcap D), \geq 0.75) = a : [\forall \rho(R, \geq 0.5).\rho(C \sqcap D, \geq 0.25) \sqcap \forall \rho(R, \geq 0.75).\rho(C \sqcap D, \geq 0.5) \sqcap \forall \rho(R, \geq 1).\rho(C \sqcap D, \geq 0.75)]$, where:
  - $\rho(R, \geq 0.5) = R_{\geq 0.5}$,
  - $\rho(C \sqcap D, \geq 0.25) = (C_{\geq 0.25} \sqcap D_{\geq 1}) \sqcup (C_{\geq 0.5} \sqcap D_{\geq 0.75}) \sqcup (C_{\geq 0.75} \sqcap D_{\geq 0.5}) \sqcup (C_{\geq 1} \sqcap D_{\geq 0.25})$,
  - $\rho(R, \geq 0.75) = R_{\geq 0.75}$,
  - $\rho(C \sqcap D, \geq 0.5) = (C_{\geq 0.5} \sqcap D_{\geq 1}) \sqcup (C_{\geq 0.75} \sqcap D_{\geq 0.75}) \sqcup (C_{\geq 1} \sqcap D_{\geq 0.5})$,
  - $\rho(R, \geq 1) = R_{\geq 1}$,
  - $\rho(C \sqcap D, \geq 0.75) = (C_{\geq 0.75} \sqcap D_{\geq 1}) \sqcup (C_{\geq 1} \sqcap D_{\geq 0.75})$,

It can be seen that the (crisp) KB $\mathcal{K}(fK) = \langle \sigma(\mathcal{A}), T(N^{fK}), R(N^{fK}) \rangle$ is unsatisfiable. □

Regarding the complexity, the size of the resulting KB is $\mathcal{O}(n^k)$, where k is the maximal depth of the concepts appearing in the fuzzy KB. The depth of a concept is inductively defined as follows:

- $depth(A) = depth(\neg A) = depth(\top) = depth(\bot) = 1$,
- $depth(\forall R.C) = depth(\exists R.C) = 1 + depth(C)$,
- $depth(C \sqcap D) = depth(C \sqcup D) = 1 + max\{depth(C), depth(D)\}$.

We recall that under Zadeh semantics, the size of the resulting KB is quadratic. In our case we need to generate more and more complex axioms, because we cannot infer the exact values of the elements which take part of a complex concept, so we need to build disjunctions or conjunctions over all possible degrees of truth.

Finally, our reduction procedure is modular and it could be applied to more expressive DLs. In particular, adding fuzzy nominals (indicated with the letter $\mathcal{O}$) and inverse roles (indicated with the letter $\mathcal{I}$) is straightforward because their semantics do not depend on any particular choice of fuzzy operators, so they can be dealt with in the same way as for the Zadeh family [7].

Fuzzy nominals allow to represent extensive definitions of fuzzy sets. Let o_i denote a named individual and let $\alpha_i \in (0, 1]$, for $i \in \{1, \ldots, m\}$. The syntax for fuzzy nominal concepts is $\{\alpha_1/o_1, \ldots, \alpha_m/o_m\}$ and their semantics is:

$$\{\alpha_1/o_1, \ldots, \alpha_m/o_m\}^{\mathcal{I}}(a) = \sup_{i \mid a \in \{o_i^{\mathcal{I}}\}} \alpha_i$$

The syntax for inverse roles is the same as in the crisp case, i.e. R^- denotes the inverse of the role R, and their semantics is:

$$(R^-)^{\mathcal{I}}(a, b) = R^{\mathcal{I}}(b, a)$$

As anticipated, mapping ρ can be extended in order to deal with these constructors in the following way:

x	y	$\rho(x, y)$
$\{\alpha_1/o_1, \ldots, \alpha_m/o_m\}$	$\bowtie \gamma$	$\{o_i \mid \alpha_i \bowtie \gamma, 1 \leq i \leq n\}$
R^-	$\rhd\alpha$	$\rho(R, \rhd\alpha)^-$

Hence L-$\mathcal{ALCHOI}$ can be mapped into crisp $\mathcal{ALCHOI}$.

However, we point out that how to represent transitive roles and cardinality restrictions, which are the constructs which remain to reach a fuzzy extension of $\mathcal{SHOIN}$ (and hence and eventually OWL DL), remains an open issue.

5 Conclusions and Future Work

In this paper we have shown how to reduce a fuzzy extension of $\mathcal{ALCHOI}$ under Łukasiewicz semantics, assuming a fixed set of allowed degrees of truth, into $\mathcal{ALCHOI}$. This work means an important step towards the possibility of dealing with imprecise and vague knowledge in DLs, since it relies on existing languages and tools. Our work is more general than previous approaches which provide crisp representations of fuzzy DLs under Zadeh semantics ([4,5,7,8,9]). However, from a practical point of view, the size of the resulting KB is much more complex in this case, so the practical feasibility of this approach has to be empirically verified. The idea behind our reduction is modular and could be applied to more expressive DLs. In future work we plan to extend the expressiveness of the logic and to implement the proposed reduction, studying if it can be optimized in some particular common situations.

References

1. Baader, F., Calvanese, D., McGuinness, D.L., Nardi, D., Patel-Schneider, P.F.: The description logic handbook: Theory, implementation, and applications. Cambridge University Press, Cambridge (2003)
2. McGuinness, D.L., van Harmelen, F.: OWL Web Ontology Language overview (2004), `http://www.w3.org/TR/owl-features`
3. Lukasiewicz, T., Straccia, U.: An overview of uncertainty and vagueness in description logics for the semantic web. Technical Report INFSYS RR-1843-06-07, Institut für Informationssysteme, Technische Universität Wien (2006)
4. Straccia, U.: Transforming fuzzy description logics into classical description logics. In: Alferes, J.J., Leite, J.A. (eds.) JELIA 2004. LNCS (LNAI), vol. 3229, pp. 385–399. Springer, Heidelberg (2004)
5. Straccia, U.: Description logics over lattices. International Journal of Uncertainty, Fuzziness and Knowledge-Based Systems 14(1), 1–16 (2006)
6. Hájek, P.: Making fuzzy description logics more general. Fuzzy Sets and Systems 154(1), 1–15 (2005)
7. Bobillo, F., Delgado, M., Gómez-Romero, J.: A crisp representation for fuzzy SHOIN with fuzzy nominals and general concept inclusions. In: Proceedings of the 2nd ISWC Workshop on Uncertainty Reasoning for the Semantic Web (URSW 2006), CEUR Workshop Proceedings, vol. 218 (2006)
8. Stoilos, G., Stamou, G.: Extending fuzzy description logics for the semantic web. In: Proceedings of the 3rd International Workshop on OWL: Experiences and Directions (OWLED 2007) (2007)
9. Bobillo, F., Delgado, M., Gómez-Romero, J.: Optimizing the crisp representation of the fuzzy description logic SROIQ. In: Proceedings of the 3rd ISWC Workshop on Uncertainty Reasoning for the Semantic Web (URSW 2007) (to appear, 2007)
10. Zadeh, L.A.: Fuzzy sets. Information and Control 8, 338–353 (1965)
11. Klir, G.J., Yuan, B.: Fuzzy sets and fuzzy logic: Theory and applications. Prentice-Hall, Englewood Cliffs (1995)
12. Straccia, U.: A fuzzy description logic for the semantic web. In: Fuzzy Logic and the Semantic Web. Capturing Intelligence, vol. 1, pp. 73–90. Elsevier Science, Amsterdam (2006)
13. Straccia, U.: Reasoning within fuzzy description logics. Journal of Artificial Intelligence Research 14, 137–166 (2001)

Rough Set Approximations in Formal Concept Analysis and Knowledge Spaces

Feifei Xu[1,2], Yiyu Yao[2], and Duoqian Miao[1]

[1] Department of Computer Science and Technology, Tongji University
Shanghai, 201804, P.R. China
xufeifei1983@hotmail.com, miaoduoqian@163.com
[2] Department of Computer Science, University of Regina, Regina, Saskatchewan
Canada S4S 0A2
yyao@cs.uregina.ca

Abstract. This paper proposes a generalized definition of rough set approximations, based on a subsystem of subsets of a universe. The subsystem is not assumed to be closed under set complement, union and intersection. The lower or upper approximation is no longer one set but composed of several sets. As special cases, approximations in formal concept analysis and knowledge spaces are examined. The results provide a better understanding of rough set approximations.

1 Introduction

Rough set theory [6,7] is an extension of the set theory with two additional unary set-theoretic operators known as approximation operators. One way to define approximation operators is called the subsystem-based formulation [14,17]. With respect to an equivalence relation on a finite and nonempty universe, one can construct a subsystem of the power set of the universe, which is the σ-algebra with the family of equivalence classes as a basis. The elements of the subsystem may be understood as definable or observable sets. Every subset of the universe is approximated from below and above by two sets in the subsystem.

There are two basic restrictions of the standard Pawlak model. First, a single subsystem of the power set is used. Second, the σ-algebra is closed under set complement, intersection and union. Many studies on generalized rough set approximations try to remove those restrictions. For example, in the abstract approximation space [1], topological rough set models [9,10,12,19,20], and closure rough set models [14], two subsystems are used; one for lower approximation and another for upper approximation. In the context of formal concept analysis, one considers a subsystem that is only closed under set intersection [18]. In this paper, we further generalize the rough set model by considering subsystems without these restrictions. The generalized approximations are applied to both formal concept analysis [5,18] and knowledge spaces [2,3,4].

Formal concept analysis [5,18] is developed based on a formal context, which is a binary relation between a set of objects and a set of attributes or properties. From a formal context, one can construct (objects, properties) pairs known as the

A. An et al. (Eds.): ISMIS 2008, LNAI 4994, pp. 319–328, 2008.

formal concepts [5,11]. The set of objects is referred to as the extension, and the set of properties as the intension, of a formal concept. They uniquely determine each other. The family of all formal concepts is a complete lattice. The extension of a formal concept can be viewed as a definable set of objects, although in a sense different from that of rough set theory [15,16]. The family of extensions of all formal concepts forms a subsystem of the power set of objects. This subsystem is closed under set intersection. Thus, one can immediately study approximation operators based on the subsystem introduced in formal concept analysis [18].

The theory of knowledge spaces [2,3,4] represents a new paradigm in mathematical psychology. It provides a systematic approach for knowledge assessment by considering a finite set of questions and a collection of subsets of questions called knowledge states. The family of knowledge states may be determined by the dependency of questions or the mastery of different sets of questions by a group of students. The knowledge states can be viewed as definable or observable sets. The family of knowledge states forms a subsystem of the power set of questions that is only closed under set union. Similarly, approximations can be defined based on the system of knowledge states.

The generalized subsystem-based formulation of approximation operators enables us to study approximations in two related areas of formal concept analysis and knowledge spaces. The results not only lead to more insights into rough set approximations, but also bring us closer to a common framework for studying the two related theories.

2 Subsystem-Based Formulation of Pawlak Rough Set Approximations

Suppose U is a finite and nonempty universe of objects. Let $E \subseteq U \times U$ be an equivalence relation on U. The equivalence relation divides the universe into a family of pair-wise disjoint subsets, called the partition of the universe and denoted by U/E. The pair $apr = (U, E)$ is referred to as an approximation space.

An approximation space induces a granulated view of the universe. For an object $x \in U$, the equivalence class containing x is given by:

$$[x]_E = \{y \mid xEy\}. \tag{1}$$

Intuitively speaking, objects in $[x]_E$ are indistinguishable from x. Under an equivalence relation, equivalence classes are the smallest non-empty observable, measurable, or definable subsets of U. By extending the definability of equivalence classes, we assume that the union of some equivalence classes is also definable. The family of definable subsets contains the empty set $\emptyset$ and is closed under set complement, intersection, and union. It is an σ-algebra whose basis is U/E and is denoted by $\sigma(U/E) \subseteq 2^U$, where 2^U is the power set of U.

In order to explicitly expression the role of $\sigma(U/E)$, we also denote the approximation space $apr = (U, E)$ as $apr = (U, \sigma(U/E))$. A subset of objects not in $\sigma(U/E)$ is said to be undefinable. An undefinable set must be approximated from below and above by a pair of definable sets.

Definition 1. In an approximation space $apr = (U, \sigma(U/E))$, a pair of approximation operators, $\underline{apr}, \overline{apr} : 2^U \longrightarrow 2^U$, are defined by:

$$\underline{apr}(A) = \cup\{X \in \sigma(U/E) \mid X \subseteq A\},$$
$$\overline{apr}(A) = \cap\{X \in \sigma(U/E) \mid A \subseteq X\}. \tag{2}$$

The lower approximation $\underline{apr}(A) \in \sigma(U/E)$ is the greatest definable set contained in A, and the upper approximation $\overline{apr}(A) \in \sigma(U/E)$ is the least definable set containing A. The approximation operators have the following properties: for $A, B \subseteq U$,

$$\begin{aligned}
\text{(i).}\quad & \underline{apr}(A) = (\overline{apr}(A^c))^c, \\
& \overline{apr}(A) = (\underline{apr}(A^c))^c; \\
\text{(ii).}\quad & \underline{apr}(U) = U, \\
& \overline{apr}(\emptyset) = \emptyset; \\
\text{(iii).}\quad & \underline{apr}(\emptyset) = \emptyset, \\
& \overline{apr}(U) = U; \\
\text{(iv).}\quad & \underline{apr}(A \cap B) = \underline{apr}(A) \cap \underline{apr}(B), \\
& \overline{apr}(A \cup B) = \overline{apr}(A) \cup \overline{apr}(B); \\
\text{(v).}\quad & \underline{apr}(A) \subseteq A; \\
& A \subseteq \overline{apr}(A); \\
\text{(vi).}\quad & \underline{apr}(\underline{apr}(A)) = \underline{apr}(A), \\
& \overline{apr}(\overline{apr}(A)) = \overline{apr}(A); \\
\text{(vii).}\quad & \underline{apr}(\overline{apr}(A)) = \overline{apr}(A), \\
& \overline{apr}(\underline{apr}(A)) = \underline{apr}(A).
\end{aligned}$$

Property (i) states that the approximation operators are dual operators with respect to set complement ^{c}. Properties (ii) and (iii) indicate that rough set approximations of $\emptyset$ or U equal to itself. Property (iv) states that the lower approximation operator is distributive over set intersection $\cap$, and the upper approximation operator is distributive over set union $\cup$. By property (v), a set lies within its lower and upper approximations. Properties (vi) and (vii) deal with the compositions of lower and upper approximation operators. The result of the composition of a sequence of lower and upper approximation operators is the same as the application of the approximation operator closest to A.

3 Generalized Rough Set Approximations

An approximation space $apr = (U, E)$ defines uniquely a topological space $(U, \sigma(U/E))$, in which $\sigma(U/E)$ is the family of all open and closed sets [10]. Moreover, the family of open sets is the same as the family of closed sets. The lower approximation operator defined by equation (2) is well-defined as long as the subsystem is closed under union. Similarly, the upper approximation operator is well-defined as long as the subsystem is closed under intersection. One may use two subsystems [1,13]. The subsystem for lower approximation operator must be closed under union, and the subsystem for upper approximation operator must be closed under intersection. In order to keep the duality of approximation operators, elements of two subsystems must be related to each other through set complement [13]. For further generalizations of the subsystem-based definition, we remove those restrictions.

3.1 Generalized Rough Set Approximations

The definition of generalized rough set approximations is related to the formulation of abstract approximation spaces introduced by Cattaneo [1]. We focus on set-theoretic setting and remove some axioms of an abstract approximation space.

Let $\mathcal{S}_l, \mathcal{S}_u \subseteq 2^U$ be two subsystems of 2^U. The triplet $apr = (U, \mathcal{S}_l, \mathcal{S}_u)$ is called an approximation space. We impose two conditions on $\mathcal{S}_l$ and $\mathcal{S}_u$:

$$\begin{array}{lll} \text{(a).} & \emptyset \in \mathcal{S}_l, & \emptyset \in \mathcal{S}_u; \\ \text{(b).} & U \in \mathcal{S}_l, & U \in \mathcal{S}_u. \end{array}$$

The elements of $\mathcal{S}_l$ may be understood as one family of definable or observable sets. The elements of $\mathcal{S}_u$ may be understood as another family of definable or observable sets. Our objective is to approximate an undefinable set in $2^U - \mathcal{S}_l$ from below by definable sets in $\mathcal{S}_l$ and in $2^U - \mathcal{S}_u$ from above by definable sets in $\mathcal{S}_u$.

Definition 2. In an abstract approximation space $apr = (U, \mathcal{S}_l, \mathcal{S}_u)$, the lower approximation and the upper approximation are defined by:

$$\begin{aligned} \underline{apr}(A) &= \{X \in \mathcal{S}_l \mid X \subseteq A, \forall X' \in \mathcal{S}_l (X \subset X' \Longrightarrow X' \nsubseteq A)\}, \\ \overline{apr}(A) &= \{X \in \mathcal{S}_u \mid A \subseteq X, \forall X' \in \mathcal{S}_u (X' \subset X \Longrightarrow A \nsubseteq X')\}. \end{aligned} \tag{3}$$

For simplicity, the same symbols are used for generalized approximations. The lower approximation $\underline{apr}(A)$ is the set of maximal elements of the set $\{X \in \mathcal{S}_l | X \subseteq A\}$ and the upper approximation $\overline{apr}(A)$ is the set of minimal elements of the set $\{X \in \mathcal{S}_u | A \subseteq X\}$. The definition is a generalization of Definition 1. The generalized lower and upper approximation operators have the following properties:

$$\begin{array}{ll} (1). & \underline{apr}(\emptyset) = \{\emptyset\}, \\ & \overline{apr}(\emptyset) = \{\emptyset\}; \\ (2). & \underline{apr}(U) = \{U\}, \\ & \overline{apr}(U) = \{U\}; \\ (3). & A \subseteq B \Longrightarrow (\exists X \in \underline{apr}(A), \exists Y \in \underline{apr}(B), X \subseteq Y), \\ & A \subseteq B \Longrightarrow (\exists X \in \overline{apr}(A), \exists Y \in \overline{apr}(B), X \subseteq Y); \\ (4). & X \in \underline{apr}(A) \Longrightarrow X \subseteq A, \\ & X \in \overline{apr}(A) \Longrightarrow A \subseteq X; \end{array}$$

$$\begin{aligned}(5).\quad & X \in \underline{apr}(A) \Longrightarrow \underline{apr}(X) = \{X\},\\ & X \in \overline{apr}(A) \Longrightarrow \overline{apr}(X) = \{X\};\\ (6).\quad & X \in \overline{apr}(A) \Longrightarrow \underline{apr}(X) = \{X\},\\ & X \in \underline{apr}(A) \Longrightarrow \overline{apr}(X) = \{X\}.\end{aligned}$$

They easily follow from the definition of generalized approximation operators.

3.2 Special Cases

We discuss several types of generalized rough set approximations under different conditions.

Case 1: $\mathcal{S}_l$ is closed under set union and $\mathcal{S}_u$ is closed under set intersection. If $\mathcal{S}_l$ is closed under union, the lower approximation is composed of one set defined by Definition 1. That is,

$$\underline{apr}(A) = \{\cup\{X \in \mathcal{S}_l \mid X \subseteq A\}\}\,. \tag{4}$$

Similarly, if $\mathcal{S}_u$ is closed under intersection, the upper approximation is composed of one set defined by Definition 1. That is,

$$\overline{apr}(A) = \{\cap\{X \in \mathcal{S}_u \mid A \subseteq X\}\}\,. \tag{5}$$

Case 2: $\mathcal{S}_l$ and $\mathcal{S}_u$ are dual subsystems, that is, $\mathcal{S}_u = \{X^c | X \in \mathcal{S}_l\}$ and $\mathcal{S}_l = \{X^c | X \in \mathcal{S}_u\}$. The approximations satisfy the property:

$$X \in \underline{apr}(A) \Longrightarrow X^c \in \overline{apr}(A^c). \tag{6}$$

Case 3: $\mathcal{S}_l = \mathcal{S}_u$. When $\mathcal{S}_l = \mathcal{S}_u = \mathcal{S}$, we have an approximation space $apr = (U, \mathcal{S})$. We can define the approximations as follows:

$$\begin{aligned}\underline{apr}(A) &= \{X \in \mathcal{S} \mid X \subseteq A, \forall X' \in \mathcal{S}(X \subset X' \Longrightarrow X' \nsubseteq A)\},\\ \overline{apr}(A) &= \{X \in \mathcal{S} \mid A \subseteq X, \forall X' \in \mathcal{S}(X' \subset X \Longrightarrow A \nsubseteq X')\}.\end{aligned} \tag{7}$$

Case 4: $\mathcal{S}_l = \mathcal{S}_u = \mathcal{S}$ and is closed under set complement. The approximations are the same as defined by equation (7). It also satisfies the property of equation (6).

Case 5: $\mathcal{S}_u = \mathcal{S}$ is closed under set intersection and $\mathcal{S}_l = \mathcal{S}^c$ is closed under set union. We define:

$$\begin{aligned}\underline{apr}(A) &= \{\cup\{X \in \mathcal{S}^c \mid X \subseteq A\}\}\,,\\ \overline{apr}(A) &= \{\cap\{X \in \mathcal{S} \mid A \subseteq X\}\}\,.\end{aligned} \tag{8}$$

They correspond to rough set approximations in closure systems [14]. Since a closure system is only closed under set intersection, the lower and upper approximation operators satisfy less properties, as characterized by properties (iii), (v), (vi).

Case 6: $\mathcal{S}_l = \mathcal{S}$ is closed under set union and intersection and $\mathcal{S}_u = \mathcal{S}^c$ is closed under set union and intersection. We define:

$$\begin{aligned}\underline{apr}(A) &= \{\cup\{X \in \mathcal{S} \mid X \subseteq A\}\}\,,\\ \overline{apr}(A) &= \{\cap\{X \in \mathcal{S}^c \mid A \subseteq X\}\}\,.\end{aligned} \tag{9}$$

They correspond to rough set approximations in topological spaces [10]. They are in fact the topological interior and closure operators satisfy properties (i) - (vi).

Case 7: $\mathcal{S}_l = \mathcal{S}_u = \mathcal{S}$ and is closed under set complement, intersection and union. This is the standard Pawlak rough set model.

4 Approximations in Formal Concept Analysis

Let U and V be any two finite sets. Elements of U are called objects, and elements of V are called properties. The relationships between objects and properties are described by a binary relation R between U and V, which is a subset of the Cartesian product $U \times V$. For a pair of elements $x \in U$ and $y \in V$, if $(x, y) \in R$, written as xRy, x has the property y, or the property y is possessed by object x. The triplet (U, V, R) is called a formal context.

Based on the binary relation, we associate a set of properties to an object. An object $x \in U$ has the set of properties:

$$xR = \{y \in V \mid xRy\} \subseteq V. \tag{10}$$

Similarly, a property y is possessed by the set of objects:

$$Ry = \{x \in U \mid xRy\} \subseteq U. \tag{11}$$

By extending these notations, we can establish relationships between subsets of objects and subsets of properties. This leads to two operators, one from 2^U to 2^V and the other from 2^V to 2^U.

Definition 3. Suppose (U, V, R) is a formal context. For a subset of objects, we associate it with a set of properties:

$$\begin{aligned} X^* &= \{y \in V \mid \forall x \in U (x \in X \Longrightarrow xRy)\} \\ &= \{y \in V \mid X \subseteq Ry\} \\ &= \bigcap_{x \in X} xR. \end{aligned} \tag{12}$$

For a subset of properties, we associate it with a set of objects:

$$\begin{aligned} Y^* &= \{x \in U \mid \forall y \in V (y \in Y \Longrightarrow xRy)\} \\ &= \{x \in U \mid Y \subseteq xR\} \\ &= \bigcap_{y \in Y} Ry. \end{aligned} \tag{13}$$

A pair (X, Y), with $X \subseteq U$ and $Y \subseteq V$, is called a formal concept of the context (U, V, R), if $X = Y^*$ and $Y = X^*$. Furthermore, $X = ex(X, Y)$ is called the extension of the concept, and $Y = in(X, Y)$ is the intension of the concept. The set of all formal concepts forms a complete lattice called a concept lattice, denoted by $L(U, V, R)$ or simply L.

A formal concept consists of a definable set of objects and a definable set of properties. The concept lattice is the family of all such definable concepts. Given an arbitrary set of objects, it may not be the extension of a formal concept. The set can therefore be viewed as an undefinable set of objects. Such a set of objects can be approximated by definable sets of objects, namely, the extensions of formal concepts. Approximation operators can be introduced by using the subsystem-based formulation of rough set theory, based on the combination of case 1 and case 3.

Definition 4. For a formal concept lattice L, the family of all extensions is given by:

$$EX(L) = \{ex(X,Y) \mid (X,Y) \in L\}. \tag{14}$$

The system $EX(L)$ contains the empty set $\emptyset$, the entire set U, and is closed under intersection. It defines an approximation space $apr = (U, EX(L))$. One can keep the intuitive interpretations of lower and upper approximations. That is, the lower approximation is a largest set in $EX(L)$ that is contained in A, and the upper approximation is a smallest set in $EX(L)$ that contains A. In this case, since the system is not closed under union, the smallest set containing A is unique, while the largest set contained in A is no longer unique.

Definition 5. In the approximation space $apr = (U, EX(L))$, for a subset of objects $A \subseteq U$, its upper approximation is defined by:

$$\overline{apr}(A) = \{\cap\{X \in EX(L) \mid A \subseteq X\}\}, \tag{15}$$

and its lower approximation is a family of sets:

$$\underline{apr}(A) = \{X \in EX(L) \mid X \subseteq A, \forall X' \in EX(L)(X \subset X' \Longrightarrow X \not\subseteq A)\}. \tag{16}$$

Thus, in formal concept analysis, a set can be approximated from below by several definable sets of objects.

5 Approximations in Knowledge Spaces

In knowledge spaces, one uses a finite set of universe (i.e., questions denoted by Q) and a collection of subsets of the universe (i.e., a knowledge structure denoted by $\mathcal{K}$), where $\mathcal{K}$ contains at least the empty set $\emptyset$ and the whole set Q. The members of $\mathcal{K}$ are called the knowledge states which are the subsets of questions given by experts or correctly answered by students. In knowledge spaces, there are two types of knowledge structures. One is the knowledge structure associated to a surmise relation, closed under set union and intersection. Another is the knowledge structure associated to a surmise system called a knowledge space, closed under set union. The knowledge states can be viewed as a family of definable sets of objects. An arbitrary subset of questions can be approximated by knowledge states in each of the two structures. Approximation operators are introduced by using the subsystem-based formulation of rough set theory.

In knowledge spaces, a surmise relation on the set Q of questions is a transitive and reflexive relation S on Q. By aSb, we can surmise that the mastery of a if a student can answer correctly question b. This relation imposes conditions on the corresponding knowledge structure. For example, mastery question a from mastery of question b means that if a knowledge state contains b, it must also contain a.

Definition 6. For a surmise relation S on the (finite) set Q of questions, the associated knowledge structure $\mathcal{K}$ is defined by:

$$\mathcal{K} = \{K \mid (\forall q, q' \in Q, qSq', q' \in K) \Longrightarrow q \in K\}. \tag{17}$$

The knowledge structure associated to a surmise relation contains the empty set $\emptyset$, the entire set Q, and is closed under set intersection and union. It defines an approximation space $apr = (Q, \mathcal{K})$.

Definition 7. In the approximation space $apr = (Q, \mathcal{K})$, for a subset of objects $A \subseteq Q$, we define:

$$\begin{aligned} \underline{apr}(A) &= \{\cup\{K \in \mathcal{K} \mid K \subseteq A\}\}, \\ \overline{apr}(A) &= \{\cap\{K \in \mathcal{K} \mid A \subseteq K\}\}. \end{aligned} \tag{18}$$

The definition is based on the case 1. The knowledge structure associated to a surmise relation is not closed under complement, namely, it does not satisfy the duality property.

With surmise relations, a question can only have one prerequisite. This is sometimes not appropriate. In practice, we may assume that a knowledge structure is closed only under union, called a knowledge space. A knowledge space is a weakened knowledge structure associated to a surmise relation. It is a knowledge structure associated to a surmise system.

A surmise system on a (finite) set Q is a mapping σ that associates to any element q in Q a nonempty collection $\sigma(q)$ of subsets of Q satisfying the following three conditions: 1) $C \in \sigma(q) \Longrightarrow q \in C$; 2) $(C \in \sigma(q), q' \in C) \Longrightarrow (\exists C' \in \sigma(q'), C' \subseteq C)$; 3) $C \in \sigma(q) \Longrightarrow (\forall C' \in \sigma(q), C' \nsubseteq C)$. The subsets in $\sigma(q)$ are the clauses for question q.

Definition 8. For a surmise system (Q, σ), the knowledge states of the associated knowledge structure are all the subsets K of Q that satisfy:

$$\mathcal{K} = \{K \mid (\forall q \in Q, q \in K) \Longrightarrow (\exists C \in \sigma(q), C \subseteq K)\} \tag{19}$$

They constitute the knowledge structure associated to (Q, σ). It defines an approximation space $apr = (Q, \mathcal{K})$. Any knowledge structure which is closed under union is called a knowledge space. In fact, there is a one-to-one correspondence between surmise systems on Q and knowledge spaces on Q.

Compared with the system in formal concept analysis that is closed under set intersection, knowledge spaces are opposite. Being closed under set union, the lower approximation in knowledge spaces is unique while the upper approximation is a family of sets.

Definition 9. Suppose (Q, σ) is a surmise system and $\mathcal{K} \subseteq 2^Q$ is the associated knowledge structure, closed under union. In the approximation space $apr = (Q, \mathcal{K})$, for a subset of objects $A \subseteq Q$, its lower approximation is defined by:

$$\underline{apr}(A) = \{\cup\{K \in \mathcal{K} \mid K \subseteq A\}\}, \tag{20}$$

and its upper approximation is a family of sets:

$$\overline{apr}(A) = \{K \in \mathcal{K} \mid A \subseteq K, \forall K' \in \mathcal{K}(K' \subset K \Longrightarrow A \not\subseteq K'\}. \tag{21}$$

The definition is based on a combination of case 1 and case 3. The lower approximation is the largest set in $\mathcal{K}$ that contained in A, and the upper approximation is the smallest sets in $\mathcal{K}$ that contains A. While the largest set contained in A is unique, the smallest set containing A is not unique.

6 Conclusion

We propose a subsystem-based generalization of rough set approximations by using subsystems that are not closed under set complement, intersection and union. The generalized rough set approximations are not necessarily unique, but consist of a family of sets. We investigate special cases under different conditions, including subsystems that are closed under set complement, intersection and union, as well as their combinations. To show that usefulness of the proposed generalizations, approximations in formal concept analysis and knowledge spaces are examined.

The subsystem of definable sets of objects in formal concept analysis is only closed under set intersection. There are two types of subsystems in knowledge spaces. The knowledge states of a surmise relation produce a subsystem that is closed under both intersection and union. The knowledge states of a surmise system produce a subsystem that is only closed under union. The introduction of rough set approximations to the two theories demonstrates the potential value of generalized rough set approximation operators.

Acknowledgments

The research is partially supported by the National Natural Science Foundation of China under grant No: 60775036, 60475019, the Research Fund for the Doctoral Program of Higher Education of China under grant No: 20060247039, and a Discovery grant from NSERC Canada.

References

1. Cattaneo, G.: Abstract Approximation Spaces for Rough Theories. In: Polkowski, L., Skowron, A. (eds.) Rough Sets in Data Mining and Knowledge Discovery, pp. 59–98. Physica, Heidelberg (1998)
2. Doignon, J.P., Falmagne, J.C.: Knowledge Spaces. Springer, Heidelberg (1999)
3. Duntsch, I., Gediga, G.: A Note on the Correspondences among Entail Relations, Rough Set Dependencies, and Logical Consequence. Mathematical Psychology 43, 393–401 (2001)

4. Falmagne, J.C., Koppen, M., Villano, M., Doignon, J.P., Johanessen, L.: Introduction to Knowledge Spaces: How to Test and Search Them. Psychological Review 97, 201–224 (1990)
5. Ganter, B., Wille, R.: Formal Concept Analysis: Mathematical Foundations. Springer, Heidelberg (1999)
6. Pawlak, Z.: Rough Sets. International Journal of Computer and Information Sciences 11, 341–356 (1982)
7. Pawlak, Z.: Rough Classification. International Journal of Man-machine Studies 20, 469–483 (1984)
8. Pawlak, Z., Skowron, A.: Rough Sets: Some Extensions. Information Sciences 177, 28–40 (2007)
9. Polkowski, L.: Rough Sets: Mathematical Foundations. In: Advances in Soft Computing, Physica, Heidelberg (2002)
10. Skowron, A.: On Topology in Information System. Bulletin of the Polish Academy of Sciences, Mathematics 36, 477–479 (1988)
11. Wille, R.: Restructuring Lattice Theory: an Approach Based on Hierarchies of Concepts. In: Rival, I. (ed.) Ordered Sets, pp. 445–470. Reidel, Dordecht-Boston (1982)
12. Wu, W.Z., Zhang, W.X.: Constructive and Axiomatic Approaches of Fuzzy Approximation Operators. Information Science 159, 233–254 (2004)
13. Yao, Y.Y.: On Generalizing Pawlak Approximation Operators. In: Polkowski, L., Skowron, A. (eds.) RSCTC 1998. LNCS (LNAI), vol. 1424, pp. 298–307. Springer, Heidelberg (1998)
14. Yao, Y.Y.: On Generalizing Rough Set Theory. In: Wang, G., Liu, Q., Yao, Y., Skowron, A. (eds.) RSFDGrC 2003. LNCS (LNAI), vol. 2639, pp. 44–51. Springer, Heidelberg (2003)
15. Yao, Y.Y.: A Comparative Study of Formal Concept Analysis and Rough Set Theory in Data Analysis. In: Tsumoto, S., Słowiński, R., Komorowski, J., Grzymała-Busse, J.W. (eds.) RSCTC 2004. LNCS (LNAI), vol. 3066, pp. 59–68. Springer, Heidelberg (2004)
16. Yao, Y.Y.: Concept Lattices in Rough Set Theory. In: Proceedings of 23rd International Meeting of the North American Fuzzy Information Processing Society, pp. 73–78 (2004)
17. Yao, Y.Y., Chen, Y.H.: Subsystem Based Generalizations of Rough Set Approximations. In: Hacid, M.-S., Murray, N.V., Raś, Z.W., Tsumoto, S. (eds.) ISMIS 2005. LNCS (LNAI), vol. 3488, pp. 210–218. Springer, Heidelberg (2005)
18. Yao, Y.Y., Chen, Y.H.: Rough Set Approximations in Formal Concept Analysis. In: Peters, J.F., Skowron, A. (eds.) Transactions on Rough Sets V. LNCS, vol. 4100, pp. 285–305. Springer, Heidelberg (2006)
19. Zhu, W.: Topological Approaches to Covering Rough Sets. Information Science 177, 1499–1508 (2007)
20. Zhu, W., Wang, F.Y.: On Three Types of Covering-based Rough Sets. IEEE Transactions on Knowledge and Data Engineering 19, 1131–1144 (2007)

An Ant Colony System Algorithm to Solve Routing Problems Applied to the Delivery of Bottled Products*

Laura Cruz Reyes, José F. Delgado Orta, Juan J. González Barbosa, José Torres Jimenez, Héctor J. Fraire Huacuja, and Bárbara A. Arrañaga Cruz

Instituto Tecnológico de Ciudad Madero, México
lauracruzreyes@yahoo.com,
{francisco.delgado.orta,jjgonzalezbarbosa}@gmail.com, jtj@cinvestav.mx,
hfraire@prodigy.net.mx, aralia38@hotmail.com

Abstract. This work presents a methodology of solution for the well-known vehicle routing problem (VRP) based on an ant colony system heuristic algorithm (ACS), which is applied to optimize the delivery process of RoSLoP (Routing-Scheduling-Loading Problem) identified in the company case of study. A first version of this algorithm models six variants of VRP and its solution satisfies the 100% of demands of the customers. The new version of the algorithm can solve 11 variants of VRP as a rich VRP. Experiments were carried out with real instances. The new algorithm shows a saving of two vehicles with regard to the first version, reducing the operation costs of the company. These results prove the viability of using heuristic methods and optimization techniques to develop new software applications.

Keywords: Optimization, Routing-Scheduling-Loading Problem (RoSLoP), Vehicle Routing Problem (VRP), Ant Colony System (ACS).

1 Introduction

The distribution and delivery processes are inherent to many companies producers of goods, in other cases; it is the main function of several services lenders companies. This could be intrascendent, however, the merchandises delivery on appropriate time with the minimum quantity of resources, reduces the operation costs, bringing savings between 5 to 20 % in total costs of products [1].

The Routing-Scheduling and Loading Problem immersed on the distribution and the activity of delivery of products is a combinatorial problem of high complexity, as result of the different variables that are included in it and the interdependence between them. Due to, the most of planning and logistic groups are focused on finding only a feasible solution, leaving out the process of optimization and the possibility of evaluating certain alternatives, with the multiple benefits of this way of solving problems.

* This research was supported by CONACYT and DGEST.

A. An et al. (Eds.): ISMIS 2008, LNAI 4994, pp. 329–338, 2008.

This work focuses mainly in solving routing problems applied to the distribution of bottled products in a company located in north eastern Mexico. These problems were initially solved in [2] with a basic ant colony system algorithm, which includes a method of solution of up to six variants of VRP: CVRP, VRPM, HVRP, VRPTW, VRPSD, sdVRP, which are described in section 2. Due to the complexity of the processes of the company, it was necessary to develop another implementation of this algorithm to solve simultaneously 11 variants of the pro-blem like a rich VRP variant in its method of solution.

The next sections describe the vehicle routing problem and its variants. Section 3 defines RoSLoP, section 4 presents the methodology of the solution, section 5 shows the experimentation with real instances, and sections 6 and 7 present the analysis of the results and the conclusions respectively.

2 Vehicle Routing Problem (VRP)

VRP is a classic problem of combinatorial optimization, which, consists in one or various depots, a fleet of m available vehicles and a set of n customers to be visited, which are joined through a graph $G(V,E)$, where:

$V=\{v_0,v_1,v_2,...,v_n\}$ is the set of vertex v_i, such that V-$\{v_0\}$ represents the customers and v_0 the depot. Each customer has a demand q_i to be satisfied by the depot.

$E=\{(v_i,v_j) \mid v_i,v_j \epsilon\ V\}$ is the set of edges. Each edge has an associated value c_{ij} that represents the transportation cost from v_i to v_j.

The task to solve is to get a set R of routes with a total minimum cost that starts and finalizes in the depot, where each vertex V-$\{v_0\}$ is visited only once and the length of each route must be less or equal to L. The main objective is to obtain a configuration with the minimum quantity of vehicles satisfying all the demands of the customers and visiting each facility in their established schedule.

2.1 Variants of VRP and Related Works

The most known variants of VRP add several constraints to the basic VRP such as capacity of the vehicles (CVRP) [3], independent service schedules in the facilities of the customers (VRPTW-VRPMTW) [4], multiple depots to satisfy the demands (MDVRP) [5], customers to be satisfied by different vehicles (SDVRP) [6], a set of available vehicles to satisfy the orders (sdVRP) [7], customers that can ask and return goods to the depot (VRPPD) [8], dynamic facilities (DVRP) [9], linehaul and backhaul orders (VRPB) [10], demands and stochastic schedules (SVRP) [8], multiple use of the vehicles (VRPM) [11], a heterogeneous fleet to delivery the orders (HVRP) [12], orders to be satisfied in several days (PVRP) [5], constrained capacities of the customers to dock and charge the vehicles (CCVRP) [2], thresholds of transit over the roads (rdVRP) [13] and depots that can ask for goods to another depots (DDVRP) [2].

Recent works have approached real situations of transportation with a complexity of until five simultaneous variants of VRP in real applications [14][15].

They are called rich VRP variants; and commercial applications have been developed that involve eight variants of VRP [16]. However, until now, it has not been created an efficient method of solution that approaches a considerable number of variants. Due to VRP is known by its NP-hard complexity, and its constraints are related with real-life situations, we have designed a methodology to build solutions for the assignment of routes, schedules and loads, which is detailed in section 4.

3 Scheduling-Routing-Loading Problem (RoSLoP)

RoSLoP, immersed in the logistic activity of the company case of study, involves a sub-set of three tasks. The mathematical model of RoSLoP was formulated with two classical problems: routing and scheduling through VRP and the loading through the Bin Packing Problem (BPP). Figure 1 shows RoSLoP and its relation with VRP-BPP. The case of study contains the next elements:

- A set of $ORDERS$ to be satisfied for the facilities of the customers, which are formed by boxes of products with different attributes.
- A set of n customers with independent service schedules and a finite capacity of attention of the vehicles.
- A set of depots with independent schedules of service and possibilities to request goods to other depots.
- A fleet of vehicles with heterogeneous capacity to transport goods, with a service time and a time for attention in the facilities of the customers. The attention time depends of the capacity of the vehicle and the available people to dock and charge the vehicles.
- A set of roads or edges that connect the depots with the facilities of the customers. Each road has an assigned cost C_{ij}, each one with a threshold of supported weight for the vehicles that travel through the roads.

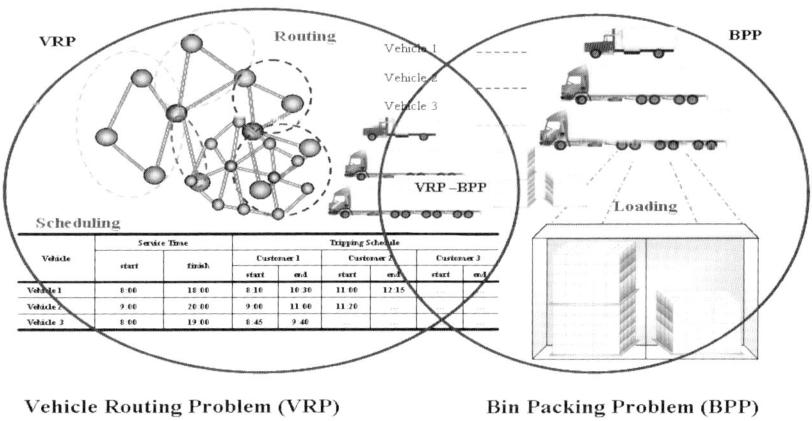

Fig. 1. RoSLoP subproblems

The objective is to get a configuration that allows to satisfy the set of $ORDERS$ in the set of facilities of the customers, minimizing the number of vehicles used and the distance of travel. This scheme of solution includes a modeling of 11 variants of VRP: CVRP, VRPTW, VRPMTW, MDVRP, SDVRP, sdVRP, VRPM, HVRP, CCVRP, DDVRP, rdVRP. They are described in section 2.1.

4 Methodology of Solution

To build feasible solutions to RoSLoP, there was created a methodology of solution based on an ant colony system algorithm (ACS), which is shown in figure 2. The assignment of routes and scheduling is solved by a basic ACS and three more components that extend the skills of this algorithm: an autoadaptative constrained list and initial and local search strategies. The DiPro algorithm, which assigns the loads, contains three main modules (construction module, assignment module and balancing module) and an optative module (filling module).

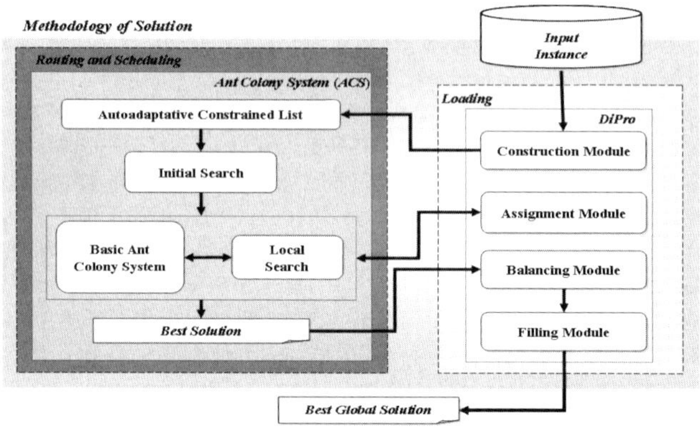

Fig. 2. RoSLoP subproblems

The construction module creates the units of load used by the algorithm; the assignment module is invoked during the process of construction of a route, each time that a customer is visited, this module determines the distribution of the load into the vehicle assigned to the customer. When the search ends and the best solution is found, this solution is improved through the balancing module and the filling module. They are executed out of line, using efficiently the time of compute, and solving up to five variants of BPP; a detailed review of DiPro is approached in [13]. Next section details the ACS that solves the transportation problem related with the solution of the associated variant rich VRP.

4.1 The Basic Ant Colony System (ACS)

Ant Colony System (ACS) is a method inspired by the behavior of the ants to find the shortest path toward their anthill. This algorithm uses two main features:

the heuristic information η_{rs}, which measures the predilection to travel through the road between two nodes v_r and v_s; and the trail information of artificial pheromone τ_{rs} (or visibility function) that computes the learned reference to travel around the road between v_r and v_s. Ants build the solution sailing through the adjacent states of the problem.

The election of the nodes in each iteration is done through a pseudo-random rule, and on-line updates executed over the information of the artificial pheromone generated by the ants. The pseudo-random selection rule is defined as follows: given an ant k located in the node v_r, $q_0 \epsilon [0,1]$ a balancing parameter between constructive focuses and q a random value in $[0,1]$, the next node s is chosen randomly with the probability distribution of the expressions 1 and 2

If $q \leqslant q_0$ then

$$s = argmax\{\tau_{rs}\eta_{rs}^{\beta}\} \qquad s \epsilon N_k(r) \tag{1}$$

Else

$$p_{rs}^{x} = \begin{cases} \dfrac{\tau_{rs}\eta_{rs}^{\beta}}{\sum\limits_{s \epsilon N_k(r)} \tau_{rs}\eta_{rs}^{\beta}} & s \epsilon N_k(r) \\ 0 & \text{Otherwise} \end{cases} \tag{2}$$

Where β is the relative importance of the heuristic information and $N_k(r)$ the set of available nodes. The rule has a double intention: $q \leqslant q_0$ then exploits the available knowledge, choosing the best choice with regard to the heuristic information and the trails of artificial pheromone. Otherwise, a controlled exploration is applied. Local update of the pheromone is done over the ant that has obtained the best solution; the trails of artificial pheromone are evaporated in all the edges used for the best ant, adding a certain evaporation rate $\rho \epsilon [0,1]$ and finally adding a determined value by the effectiveness of the found solution. Local update of artificial pheromone is shown in expression 3.

$$\tau \leftarrow (1-\rho)\tau_{rs} + \rho\Delta\tau_{rs} \tag{3}$$

Another process of on-line update is done to build different solutions for the ones already obtained. Every time that one ant travels from v_r to v_s, expression 4 is applied

$$\tau \leftarrow (1-\rho)\tau_{rs} + \rho\Delta\tau_0 \tag{4}$$

The global update $\Delta\tau_{rs}$ is computed like the inverse of the length of the shortest global solution generated by the ants; the trail of pheromone τ_0 used in the local update, is the inverse of the product of the length of the shortest global solution generated and the number of visited nodes, establishing a dependence with the size of the instance. The global update, combined with the pseudo-random rule, guides toward a more direct search. In each iteration of the algorithm, ants examine in the neighborhood of the best found solution. While the local update changes dynamically the desirable use of the roads.

4.2 Compute of the Heuristic Information

The heuristic information is used by the ACS to choose the next node to visit. The expression 5 defines the compute of the heuristic information used in the election of the customers.

$$\eta_{rs} = (\Delta t_{rs} * (ws_s + st_s) * tc_{rs})^{-1} \tag{5}$$

The factor is the difference between the current time and the arrival time to node s, ws_s represents the remaining size of the time window in s, st_s is the time of service in s and tc_{rs} is the travel cost from node r to node s, solving the variants VRPTW, VRPMTW and CCVRP. This calculation gives preference to those customers where: *a*) the needed time to arrive to the facilities starting from the actual position is the smallest, *b*) the time of attention since the current time plus the time of service is the smallest, and *c*) the traveling time is minimum. Due to the existence of the variant HVRP it is necessary to choose the next more appropriate vehicle to be used. Expression 6 defines the computation of the heuristic information implied into the selection of the vehicles; this expression satisfies the variants CVRP, VRPM, sdVRP and rdVRP.

$$\eta_v = \left[nv_v * (\overline{TM_v} + \overline{TR_v}) * \frac{tr_v}{tt_v} * idpref_v \right]^{-1} \tag{6}$$

Where η_v is the value of the heuristic information for the mobile unit v, nv_v is a bound of the quantity of travels required for the vehicle v to satisfy all the demands of $N_k(r)$, $\overline{TM_v}$ is the average of the service time in $N_k(r)$, $\overline{TR_v}$ is the time trip average of the vehicle to $N_k(r)$, tr_v is the available time for the vehicle v, tt_v is the time of attention for the vehicle v; tr_v/tt_v is a factor time of use/availability; and $idpref_v$ is the grade of predilection of use of the vehicle v. Expression 6 implies an inclination for those vehicles whose times of trip, times of service, remaining period of service and preference level are the smallest, making it possible to develop the variant SDVRP.

4.3 Autoadaptative Constrained List (ACL)

In agreement with [17], in constructive processes it is very advantageous that ants use a Constrained List of Candidates. ACL is elaborated through feasibility conditions and the distribution of the customers in the graph. The goal of ACL is to limit the global population into subsets that fulfill certain conditions, and its use allows to solve the variant DDVRP. The ACL is created following the next five steps.

Step 1. A Minimum Spanning Tree (MST) is generated including all the customers and the depot of the instance.

Step 2. The mean μ and standard deviation σ are obtained, the minimum and maximum costs associated to the roads included in the MST.

Step 3. The percentage of visibility of the associated costs to each road belonging to the MST is computed through the expression 7.

$$\theta = \left[\frac{\sigma}{2(argmax(tc_{rs}) - argmin(tc_{rs}))}\right]^{-1} \qquad (r,s)\epsilon MST \qquad (7)$$

If $\theta < 0.1$, the percentage of variability around the mean oscillates in 10%, that is because the location of the customers in the instance follows an uniform distribution. Therefore, it is possible the existence of regions in the space with more density as for the population of customers.

Step 4. The definition of the regions is carried out through the next grouping rule: if $\theta > 0.1$, then it continues with the formation of conglomerates through a hierarchical grouping, otherwise, all the customers form a single conglomerate. The threshold of acceptance ω is calculated through the expression 8. This point characterizes the autoadaptative attribute of the method.

$$\omega = 2 * argmax\,(tc_{rs}) \qquad (r,s)\epsilon MST \qquad (8)$$

Step 5. Once defined the ownership of each customer to a conglomerate, the value of the heuristic information is modified with the rule of ownership, only for the customers c_r and c_s, which belong to different groups h_i and h_j.
If $h_i \neq h_j \mid c_r \;\epsilon\; h_i \wedge c_s \;\epsilon\; h_j$ then

$$\eta_{rs} = \eta_{rs} * \frac{|H|}{|C|} \qquad c_r, c_s \epsilon C, h_i, h_j \epsilon H \qquad (9)$$

Expression 9 allows inhibiting in proportional form to the preference of a customer with regard to others, with base in the ownership to the different conglomerates and the number of these. Each iteration obtains the ACL for each depot, satisfying the variant MDVRP, this solution is optimized through the initial and local search procedures.

4.4 Initial Search

Because the exploitative focus is extremely similar to a greedy search, the possibility to use it like an initial search results in solutions of good quality over other methods with simple guideline, as the heuristic of the nearest neighbor. The expression 10 defines the preference guide in the initial search of the solution method.

$$\eta = (\Delta t_{rs} * (ws_s + st_s))^{-1} \qquad (10)$$

4.5 Local Search

For the incorporation of the Local Search to the ACS, schemes of exchange of axes were chosen: 3-opt [18] and Cross-Exchange [19], operating respectively on two routes, both include implicitly other simple operators, the first one contains by nature at the 2-opt [20], and the second, allows the use of empty segments using movements type 2-opt * [21], Relocation, and Exchange [22], making it extremely versatile.

5 Experimentation

Real instances were provided by the bottling company, which were tested using the two versions of the algorithm, the first version developed in [2], named Heuristics-Based Transportation System (HBST) and the new algorithm named the Heuristics-Based System for Assignment of Routes, Schedules and Loads (HBS-ARSL), both were developed assisting the needs of the company in two different instants of time. Both algorithms were coded in C#. A set of 12 instances of test were selected from the database of the company, which contains 312 instances classified by the date of the orders; the database contains also 1257 orders and 356 products in its catalogues. Both algorithms used a configuration of 10 ants, 5 colonies, 40 generations with the parameters of the ACS: $q_0 = 0.9$; $\beta = 1$; $\rho = 0.1$, and they were executed during two minutes. There were disposed eight available vehicles in a graph with ten edges. Results are shown in table 1.

Table 1. Solution of real instances of RoSLoP, provided by the bottled company

Instance	n	*ORDERS*	HSBT		HBS-ARSL	
			Satisfied demand(%)	Used vehicles	Satisfied demand(%)	Used vehicles
06/12/2005	4	6928	100	6	97.66	4
09/12/2005	5	7600	100	7	100	5
12/12/2005	7	11541	100	8	99.06	5
01/01/2006	6	9634	100	6	97.71	4
03/01/2006	4	5454	100	4	100	3
07/02/2006	6	12842	100	8	99.21	5
13/02/2006	5	9403	100	7	99.52	5
06/03/2006	6	9687	100	5	97.63	3
09/03/2006	6	12319	100	8	98.65	6
18/03/2006	7	11662	100	7	96.59	4
22/04/2006	8	16903	100	8	98.28	5
17/05/2006	6	11410	100	7	97.00	5
Average	6	10873.53	100	6.75	98.45	4.5

6 Analysis of Results

Table 1 shows that the algorithm HBS-ARSL obtains a saving of two vehicles in average with regard to HBST. This saving is achieved with the addition of the solution of the Autoadaptative Constrained List defined in section 6.3; however, it is observed in the column of satisfied demand that it does not solve 100% of the demands, reaching 98.45 percent due to the use of restrictions of the problem BPP, which are carried by the solution of five simultaneous variants of the problem BPP to solve the Loading Problem. As an alternative solution it was created the filling module included in DiPro, which covers the available space in the vehicles with product of more rotation, because it is not feasible for the company to send a vehicle with less than 50% of their capacity.

This demonstrates the complexity of the problem in real situations, and how the addition of restrictions in the formulation of the problem reduces the quality of the solutions in terms of satisfaction of the demand, in contrast with the saving obtained by the algorithm with regard to the number of used vehicles, achieved with the use of the Autoadaptative Constrained List, which allows the company to minimize the costs of vehicle operation related to gasoline and maintenance.

7 Conclusions and Future Work

This work showed the solution of a real problem using an Ant Colony System algorithm, which builds feasible solutions of good quality in a reasonable period of time. This shows the viability of the development of commercial applications based on heuristic methods, that applied to the planning and logistics, will allow companies the distribution and delivery of their products by obtaining significant savings for concept of transportation of them. Another future contribution can be the application of different neighborhood techniques and the addition of explorative focus to the Autoadaptative Constrained List.

References

1. Toth, P., Vigo, D. (eds.): The vehicle routing problem, SIAM Monographs on Discrete Mathematics and Applications. Philadelphia: Society for Industrial and Applied Mathematics (2001)
2. Cruz, L., et al.: A Distributed Metaheuristic for Solving a Real-World Scheduling-Routing-Loading Problem. In: Stojmenovic, I., Thulasiram, R.K., Yang, L.T., Jia, W., Guo, M., de Mello, R.F. (eds.) ISPA 2007. LNCS, vol. 4742, pp. 68–77. Springer, Heidelberg (2007)
3. Shaw, P.: Using Constraint Programming and Local Search Methods to Solve Vehicle Routing Problems. In: Maher, M.J., Puget, J.-F. (eds.) CP 1998. LNCS, vol. 1520, pp. 417–431. Springer, Heidelberg (1998)
4. Jong, C., Kant, G., Vliet, A.V.: On Finding Minimal Route Duration in the Vehicle Routing Problem with Multiple Time Windows, tech. report, Dept. of Computer Science, Utrecht Univ. (1996)
5. Mingozzi, A.: An exact Algorithm for Period and Multi-Depot Vehicle Routing Problems. Department of Mathematics, University of Bologna, Bologna, Italy (2003)
6. Archetti, C., Mansini, R., Speranza, M.G.: The Vehicle Routing Problem with capacity 2 and 3, General Distances and Multiple Customer Visits. Operational Research in Land and Resources Manangement, p. 102 (2001)
7. Thangiah, S.: A Site Dependent Vehicle Routing Problem with Complex Road Constraints. Artificial Intelligence and Robotics Laboratory, Slippery Rock University, U.S.A (2003)
8. Dorronsoro, B.: The VRP Web. AUREN. Language and Computation Sciences of the University of Mlaga (2005), `http://neo.lcc.uma.es/radi-aeb/WebVRP`
9. Leonora, B.: Notes on Dynamic Vehicle Routing. Technical Report IDSIA-05-01. IDSIA - Istituto Dalle Molle di Studi sull'Intelligenza Artificiale, Switzerland (2000)

10. Blescha, J., Goetshalckx, M.: The Vehicle Routing Problem with Backhauls: Properties and Solution Algorithms. Techincal report MHRC-TR-88-13, Georgia Institute of Technology (1993)
11. Fleischmann, B.: The Vehicle routing problem with multiple use of vehicles. Working paper, Fachbereigh Wirtschaftswissenschaften, Universitt Hamburg (1990)
12. Taillard, E.: A Heuristic Column Generation Method For the Heterogeneous Fleet VRP. Istituto Dalle Moli di Studi sull Inteligenza Artificiale, Switzerland. CRI-96-03 (1996)
13. Cruz, R., et al.: DiPro: An Algorithm for the Packing in Product Transportation Problems with Multiple Loading and Routing Variants. In: Gelbukh, A., Kuri Morales, Á.F. (eds.) MICAI 2007. LNCS (LNAI), vol. 4827, pp. 1078–1088. Springer, Heidelberg (2007)
14. Pisinger, D., Ropke, S.: A General Heuristic for Vehicle Routing Problems, tech. report, Dept. of Computer Science, Univ. Copenhagen (2005)
15. Reimann, M., Doerner, K., Hartl, R.: Analyzing a Unified Ant System for the VRP and some of Its Variants. In: Raidl, G.R., Cagnoni, S., Cardalda, J.J.R., Corne, D.W., Gottlieb, J., Guillot, A., Hart, E., Johnson, C.G., Marchiori, E., Meyer, J.-A., Middendorf, M. (eds.) EvoIASP 2003, EvoWorkshops 2003, EvoSTIM 2003, EvoROB/EvoRobot 2003, EvoCOP 2003, EvoBIO 2003, and EvoMUSART 2003. LNCS, vol. 2611, pp. 300–310. Springer, Heidelberg (2003)
16. OR/MS Today: Vehicle Routing Software Survey. United States. Institute for Operations Research and the Management Sciences (2006)
17. Dorigo, L., Gambardella, M.: Ant Colony System: A Cooperative Learning Approach to the Traveling Salesman Problem. In: Proc. IEEE Transactions on Evolutionary Computation, Belgica, vol. 1(1) (1997)
18. Bock, F.: An algorithm for solving traveling salesman and related network optimization problems. In: Fourteenth National Meeting of the Operational Research Society of America, St. Louis, MO, USA (1958)
19. Taillard, E., Badeau, P., Gendreu, M., Guertin, F., Potvin, J.Y.: A Tabu Search Heuristic for the Vehicle Routing Problem with Soft Time Windows. Transportation Science 31, 170–186 (1997)
20. Croes, G.: A method for solving traveling salesman problems. Proc. Operations Research 5, 791–812 (1958)
21. Potvin, J., Rousseau, J.M.: An Exchange Heuristic for Routing Problems with Time Windows. Proc. Journal of the Operational Research Society 46, 1433–1446 (1995)
22. Prosser, P., Shaw, P.: Study of Greedy Search with Multiple Improvement Heuristics for Vehicle Routing Problems, tech. report, University of Strathclyde, Glasgow, Scotland (1996)

Predicate Indexing for Incremental Multi-Query Optimization

Chun Jin and Jaime Carbonell

Language Technologies Institute
School of Computer Science
Carnegie Mellon University
Pittsburgh, PA 15213 USA
{cjin,jgc}@cs.cmu.edu

Abstract. We present a relational schema that stores the computations of a shared query evaluation plan, and tools that search the common computations between new queries and the schema, which are the two essential parts of the Incremental Multiple Query Optimization (IMQO) framework we proposed to allow the efficient construction of the optimal evaluation plan for multiple continuous queries.

1 Introduction

Multi-query optimization (MQO) [15], namely, finding the shared optimal query evaluation plan for multiple queries, has been widely studied, because sharing intermediate results among multiple queries can lead to significant computation reduction. But as NP-hard as it is, MQO usually has to employ heuristics to trade off between the optimality and the optimization time. MQO is particularly essential to stream databases since they tend to run multiple long-lived continuous queries concurrently, and the cost of finding the optimal plan will be amortized by continuous query evaluation. However, in reality, stream DB queries usually arrive at different times, as opposed to the assumption of synchronous arrival based on which the traditional MQO conducts the optimization as a one-shot operation. To cope with the query asynchrony and mitigate the NP-hardness, we propose a new approach, Incremental Multi-Query Optimization, by adding new query computation *incrementally* to the existing query plan with heuristic local optimization.

IMQO takes 4 steps to generate the new query evaluation plan R^*, given the existing plan R and a new query Q: 1. index R's computations and store them in persistent data structures $\mathcal{R}$; 2. identify common computations $\mathcal{C}$ between Q and R by searching $\mathcal{R}$; 3. select the optimal sharing path $\mathcal{P}$ in R that computes $\mathcal{C}$; and 4. expand R to R^* with the new computations $Q - \mathcal{C}$ that compute the final results for Q.

In this paper, we focus on the *Index* and *Search* (Steps 1 & 2) for selection-join-projection (SJP) queries; this presents the most thoroughly investigated effort so far on the most common query types (SJP). Previous work [11,10,9] discussed other query types, the sharing strategies (Step 3), and the actual continuous query plan construction (Step 4). The constructed plan will match the stream data on the fly to produce continuous query results.

A. An et al. (Eds.): ISMIS 2008, LNAI 4994, pp. 339–350, 2008.

Algorithm 1. IMQO
Input: R, Q
Output: R^*
Procedure: 1. $\mathcal{R} \Leftarrow Index(R)$);
2. $\mathcal{C} \Leftarrow Search(Q, \mathcal{R})$;
3. $\mathcal{P} \Leftarrow SelectSharing(\mathcal{R}, \mathcal{C})$;
4. $R^* \Leftarrow Expand(R, \mathcal{P}, Q - \mathcal{C})$.

Fig. 1. IMQO Algorithm

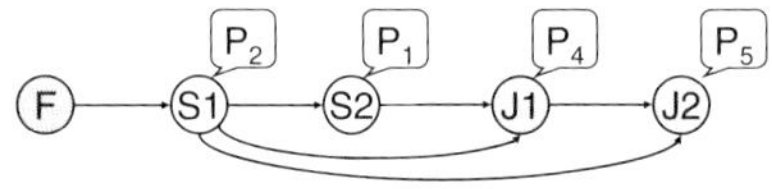

(a) The optimal query evaluation plan for Q_1

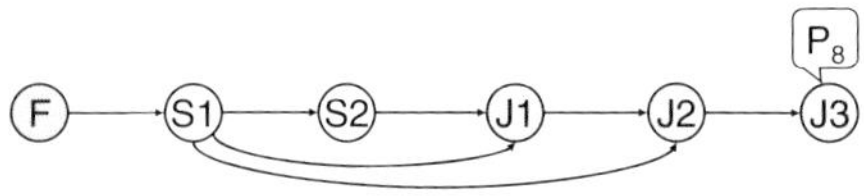

(b) Extend the plan to compute Q_2

Fig. 2. Query Evaluation Plans

In our approach, the index and search are conducted on the relational schema that stores the query plan information and is updated incrementally when new computation is added into the plan. To design the schema, we need to consider: what types of plan information should be indexed; and how to index them to permit efficient search?

For the first question, the answer depends on the plan structure. We adopt the one widely used in traditional query optimization [14]; in particular, a plan is a directed acyclic graph, constructed from the where-clause which is in conjunctive normal form (CNF). As the results, the schema indexes literal predicates, disjunctions of literals (*OR predicate*, or *ORPred*), the conjunctions of the disjunctions (*Predicate Set*, or *PredSet*), and the plan topology. The tools search the equivalence and subsumption relationships at the literal, ORPred, and PredSet layers, and the sharable nodes in the plan topology. This covers the common SJP queries.

For the second question, our solution is modeling the computations using the ER (Entity Relationship) approach and then transforming the ER model to the relational model. This allows us to utilize the ER modeling power to analyze the query plan structure, and separate the scalability solution from information access logic. In particular, the indexing and searching algorithms are realized with database updates and queries to the relational system catalogs; and the efficiency of the catalog access is gained from intrinsic database functionalities, such as primary key indexing on catalog tables.

We integrated the indexing schema and tools into ARGUS [9], a stream processing system that was built upon the commercial DBMS Oracle to process multiple complex continuous queries, and is under planning for insertion into government agency systems RDEC (*www.globalsecurity.org/military/agency/army/cecom-rdec.htm*). Empirical evaluation on ARGUS shows that with moderate acceptable cost, the schema leads to constructing shared plans that achieve up to hundreds of times speed-up in query evaluation. Due to space limit, see [9].

The proposed schema is general, being usable in stream DB systems, such as STREAM [12], Aurora [1], and NiagaraCQ [5], with minor code change. Part of the schema, in particular, the canonicalization, and the indexing and searching at the literal and ORPred layers, can also be used to enhance the common computation identification on flow-based stream processing architectures, such as TelegraphCQ.

In the remaining of the paper, we discuss the related work in Section 2, present two query examples in Section 3 to illustrate the types of computations to be indexed, present the schema design in Section 4, and conclude in Section 5.

2 Related Work

In this section, we discuss the related work on query computation indexing and searching that has been done in MQO [15], view-based query optimization (VQO)[13,2], and stream databases [12]. The major difference is that our work employs the systematic approach to analyze and model the computation and extensively expands the scope of the previous work.

Two common approaches to query indexing in MQO and VQO are bottom-up query graph search [6,3,16]. and top-down rule-based filtering[8]. The first approach performs the common predicate search with one-by-one string match through one query graph after another. The second approach identifies the sharable views by filtering out irrelevant ones with different fine-tuned tests, such as excluding the ones not containing all the required columns and the ones with more restrict range predicates.

IMQO is different from MQO and VQO. MQO focuses on one-shot optimization where queries are assumed to be available all at a time and usually uses query graph. VQO identifies the optimal materialized views to speed up the query evaluation and uses both approaches mentioned above. Therefore, MQO and VQO do not index plan structure. But IMQO indexes all materialized intermediate results across the entire shared query plan, which allows full sharability search in an efficient way.

All the known stream database systems endorse computation sharing in certain ways. But so far, most focus on the sharing strategy (Step 3), plan expansion (Step 4), and engine support. NiagaraCQ [5] focused on the strategies of selecting optimal sharing paths; Aurora [1] supported shared-plan construction with heuristic local optimization with a procedural query language; TelegraphCQ [4] realized the sharing and computation expansion on a flow-based stream processing architecture; and STREAM [12] implemented the stream processing architecture that supports shared plans.

To our knowledge, only NiagaraCQ and TelegraphCQ realized the computation indexing and searching to support IMQO on declarative queries. They applied a simple approach that identifies identical predicates and subsumptions on selection predicates which must be in the form of $Attribute\ op\ Constant$.[1] In contrast, our work supports full range of query predicates and allows equivalent ones in different format to be identified.

3 Query Examples

In this section, we present two query examples to illustrate the types of computations. The queries are formulated on the FedWire database (FED). FED contains one single data stream, comprised of FedWire money transfer transaction records. A transaction, identified by *tranid*, contains the transaction type *type_code*, date *tran_date*, amount *amount*, originating bank *sbank_aba* and account *orig_account*, and receiving bank *rbank_aba* and account *benef_account*. Consider a query Q_1 on big money transfers for financial fraud detections.

[1] Subsumption is a containment relationship between predicates. The predicate p_1 subsumes the predicate p_2, if p_2 implies p_1, denoted as $p_2 \rightarrow p_1$; then p_2 can be evaluated from the results of p_1, which reduces the amount of data to be processed.

Example 1. The query links big suspicious money transactions of type 1000 or 2000, and generates an alarm whenever the receiver of a large transaction (over $1,000,000) transfers at least half of the money further using an intermediate bank within 20 days. The query can be formulated as a 3-way self-join on F, the transaction stream table:

```
SELECT r1.tranid, r2.tranid, r3.tranid
FROM   F r1, F r2, F r3
WHERE  (r1.type_code = 1000 OR
        r1.type_code = 2000)                        -p1
AND    r1.amount > 1000000                          -p2
AND    (r2.type_code = 1000 OR
        r2.type_code = 2000)                        -p3
AND    r2.amount > 500000                           -p4
AND    (r3.type_code = 1000 OR
        r3.type_code = 2000)                        -p5
AND    r3.amount > 500000                           -p6
AND    r1.rbank_aba = r2.sbank_aba                  -p7
AND    r1.benef_account = r2.orig_account           -p8
AND    r2.amount > 0.5 * r1.amount                  -p9
AND    r1.tran_date <= r2.tran_date                 -p10
AND    r2.tran_date <= r1.tran_date + 20            -p11
AND    r2.rbank_aba = r3.sbank_aba                  -p12
AND    r2.benef_account = r3.orig_account           -p13
AND    r2.amount = r3.amount                        -p14
AND    r2.tran_date <= r3.tran_date                 -p15
AND    r3.tran_date <= r2.tran_date + 20;           -p16
```

We added two predicates $p4$ and $p6$ into the query. They can be inferred automatically [11] from $p2$, $p9$, and $p14$, and their data filtering improves the performance.

A continuous query evaluation plan should materialize some intermediate results on historical data, so they can be used to compute new results without repetitive computations over them (Rete-based query evaluation [11,5]). The materialized results can also be used for sharing among multiple queries. An effective materialization strategy is pushing down highly-selective selection predicates and materializing their results, so joins can be efficiently evaluated from much less intermediate results upon new data arrivals [7,5,11].

On the other hand, materialization should be used with caution because of the entailed disk access cost. An effective heuristic to avoid unnecessary materialization is grouping predicates based on the tables they reference and materializing the predicate groups (PredSet) [11], instead of each single predicate. So we get the PredSets: $P_1 = \{p_1, p_2\}$, $P_2 = \{p_3, p_4\}$, $P_3 = \{p_5, p_6\}$, $P_4 = \{p_7, p_8, p_9, p_{10}, p_{11}\}$, and $P_5 = \{p_{12}, p_{13}, p_{14}, p_{15}, p_{16}\}$.

Figure 2(a) shows the optimal plan to evaluate the query. We assume the selection PredSets, P_1, P_2, and P_3, are highly-selective, thus they are pushed down in the plan. Since PredSets P_2 and P_3 are equivalent, they share the same node $S1$. P_1 is subsumed by P_2 or P_3, thus P_1 can be evaluated from $S1$, instead of being evaluated from the source node F, shown as node $S2$. The subsumption sharing is useful since it reduces the amount of data to be processed to obtain $S2$. The results of P_4 and P_5 are also materialized to facilitate the efficient joins. If we assume that P_4 and P_5 are equally selective, then P_4 is evaluated first, since the size of the input to P_4 is less than that of P_5.

Consider the second query Q_2. Q_2 is similar to Q_1 except that the time span is 10 days instead of 20 days. Thus predicates p_{11} and p_{16} are substituted by $p_{17} = \{r2.tran\_date <= r1.tran\_date + 10\}$ and $p_{18} = \{r3.tran\_date <= r2.tran\_date + 10\}$, respectively; and PredSets P_4 and P_5 are by $P_6 = \{p_7, p_8, p_9, p_{10}, p_{17}\}$ and $P_7 = \{p_{12}, p_{13}, p_{14}, p_{15}, p_{18}\}$.

Since P_6 and P_7 are subsumed by P_4 and P_5 respectively, the final results of Q_2 can be evaluated from $J2$ with a selection PredSet $P_8 = \{p_{17}, p_{18}\}$, shown in Figure 2(b) as node $J3$.

From the examples, we see that the indexing schema should store several types of plan information, including literals, ORPreds, PredSets, and the plan topologies, and the searching tools should recognize the equivalence and subsumptions at the three predicate layers, and their associations with the plan topologies.

4 Predicate Indexing Schema

We describe the computation indexing schema and searching algorithms in this section. Firstly, we model the computations using the ER model methodology; then we present the relevant problems and their solutions; and finally, we derive the relational model from the first two steps.

4.1 ER Model for Plan Computations

We model the computations of a query evaluation plan as a 4-layer hierarchy. From top to bottom, the layers are topology layer, PredSet layer, ORPred layer, and literal layer.

Figure 3 shows the hierarchy for the two nodes $S1$ and $S2$ in Figure 2. For the equivalent PredSets P_2 and P_3, only P_2 is shown. For the equivalent ORPreds p_1 and p_3, only p_3 is shown, while p_1 is crossed out and dropped from the hierarchy. The dashed arrows between PredSets and literal predicates indicate subsumptions at these two layers. And the dashed arrow between nodes $S1$ and $S2$ indicates the direct topology connection and sharability between them.

The hierarchy can be presented in an ER model, as shown in Figure 4. Before transforming the ER model to the relational model, we address several issues in Sections 4.2-4.6, including rich syntax (equivalent predicates expressed differently), self-join canonicalization, subsumption identification, and topology indexing. Then the solutions are implemented in the final relational model, see Section 4.7.

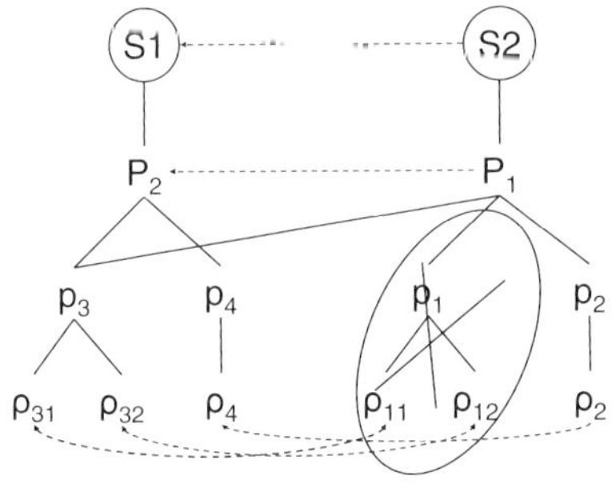

Fig. 3. Computation hierarchy. $\rho_{11} = \rho_{31}$: $type\_code = 1000$, $\rho_{12} = \rho_{32}$: $type\_code = 2000$, $\{\rho_2\} = p_2$, $\{\rho_4\} = p_4$.

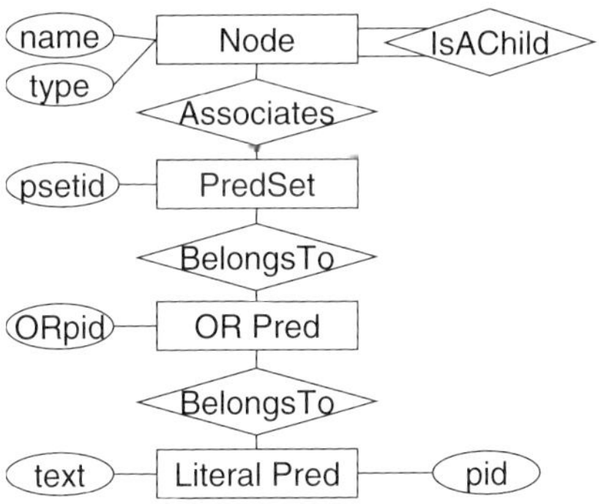

Fig. 4. Hierarchy ER model

4.2 Rich Syntax and Canonicalization

The first obstacle in the common computation identification is that semantically-equivalent literals can be expressed in different syntactic forms. For example, $t1.a < t2.b$ can also be expressed as $t2.b > t1.a$. A simple string match can not identify such equivalence. To solve the problem, we introduce a canonicalization procedure. It transforms syntactically-different yet semantically-equivalent literal predicates into the same pre-defined canonical form. Then the equivalence can be detected by exact string match.

But subsumption can still not be identified by the exact string match. For example, $t1.a > 5$ subsumes $t1.a \geq 10$. To address the problem, we apply a triple-string canonical form. For a literal ρ, we use $\gamma(\rho)$ to denote its operator, and use $L(\rho)$ and $R(\rho)$ to denote its left-hand-side and right-hand-side expressions, respectively. So ρ can be written as a triplet $L(\rho)\gamma(\rho)R(\rho)$. By making $L(\rho)$ a canonical expression of column references without constant terms, and $R(\rho)$ a canonical constant without any column references, such as $\rho : t1.a + 2 * t2.b > 5$, the subsumption can be identified by exact string match on $L(\rho)$, and comparisons on $\gamma(\rho)$ and $R(\rho)$. We apply recursive rules to transform expressions to predefined canonical format; please see [9] for the details. .

The time complexity of canonicalization is up to quadratic to the length of the predicates because of sorting. But this is not considered a problem, since the canonicalization is an one-time operation for just new queries, and the average predicate length is far less than the extent that can slow down the process noticeably.

4.3 Self-join

To allow exact-string match for finding equivalence and subsumption, table references in canonical forms should use true table name, not table alias. For example, $p4$ and $p6$ in Q_1 should be canonicalized as $F.amount > 500000$, so the equivalence can be identified. This is all right for a selection predicate or join predicate on different tables, but problematic for a self-join predicate.

For example, the self-join predicate $r1.benef\_account = r2.orig\_ account$ joins two records. The specification of joining two records is clarified by different table aliases $r1$ and $r2$. To retain the semantics of the self-join, we can not replace the table aliases with their true table names. To avoid the ambiguity or information loss, we introduce Standard Table Aliases (STA) to reference the tables. We assign $T1$ to one table alias and $T2$ to the other. To support multi-way join predicates, we can use $T3$, $T4$, and so on. In the search, we enumerate the STA assignments in the canonical form to find the predicate match.

Self-joins also present problems in the PredSet and ORPred layers. For example, an ORPred p contains two literal predicates, one is a selection predicate ρ_1: $F.c = 1000$, and the other a self-join predicate ρ_2: $T1.a = T2.b$. The canonicalized ρ_1 references the table directly, and is not aware of the STA assignment. But when it appears in p, we must identify its STA with respect to the self-join predicate ρ_2. Therefore, ρ_1's STA, $T1$ or $T2$, must be recorded when indexing p. Similar situation exists in PredSets where some ORPreds are single-table selections while others are self-joins. Thus an ORPred STA should be indexed in the PredSet in which it appears. The STA assignment must be consistent in the three-layer hierarchy. In particular, a PredSet chooses one STA assignment, and propagates it down to the ORPred layer and then the literal layer.

A 2-way self-join condition, being a literal, ORPred, or PredSet, has two possible STA assignments. And a k-way self-join has $k!$ assignments. This means that a search algorithm may try up to $k!$ times to find a match. The factorial issue is intrinsic to self-join matching, but may be addressed heuristically. In our implementation, supporting 2-way joins, the search on a self-join PredSet stops when it identifies an equivalent one from the system catalogs. If both assignments lead to the identification of subsuming PredSets, the one that has less results (indicating a stronger condition) is chosen.

4.4 Subsumption on Literals

As discussed in previous sections, subsumption is important in computation sharing. It presents in all the three predicate layers. Given a new condition p, we want to identify all conditions in the existing plan that either subsume or are subsumed by p. The former directly leads to sharing, while the later can be used to re-optimize the plan.

In this subsection, we describe how subsumptions of comparison literal predicates are detected from the triple-string canonical forms. When $L(\rho_1) = L(\rho_2)$, the subsumption between the two literals, ρ_1 and ρ_2, may exist. It is determined by the relationships between $\gamma(\rho_1)$ and $\gamma(\rho_2)$, and between $R(\rho_1)$ and $R(\rho_2)$. For example, $\rho_1 : t1.a < 1 \rightarrow \rho_2 : t1.a < 2$, but the reverse is not true. We define a subsumable relationship between pairs of operators based on the order of the right-hand-side expressions.

Definition 1. *For two literal operators γ_1 and γ_2 and an order O, we say (γ_1, γ_2, O) is a **subsumable triple** if following is true: for any pair of canonicalized literals ρ_1 and ρ_2, such that $\rho_1 = L(\rho_1)\gamma_1 R(\rho_1)$, and $\rho_2 = L(\rho_2)\gamma_2 R(\rho_2)$, if $L(\rho_1) = L(\rho_2)$, and $O(R(\rho_1), R(\rho_2))$ is true, then we have $\rho_1 \rightarrow \rho_2$.*

For example, $(<, <, Increasing)$ is a subsumable triple ($\rho_1 : t1.a < 1 \rightarrow \rho_2 : t1.a<2$, and $O(1, 2)$ is true). Figure 5 shows the implemented subsumable triples. With this, look-up queries can be formulated to retrieve the indexed subsumption literals in constant time.

It can be shown that literal subsumptions identified this way have the following property.

Theorem 1. *If $\rho \rightarrow \rho_1$, $\rho \rightarrow \rho_2$, and the subsumptions are identifiable through the subsumable triples, then either $\rho_1 \rightarrow \rho_2$ or $\rho_2 \rightarrow \rho_1$ is true.*

4.5 Subsumption on ORPreds and PredSets

As discussed in previous sections, we want to identify subsumptions on ORPreds and then on PredSets. This is proceeded from the subsumptions identified at the literals; and the results identified on the PredSets are then used to find the sharable topologies.

Given the existing plan, R, we use R_{ORPred} and $R_{PredSet}$ to denote the set of all ORPreds and the set of all PredSets in R, respectively. Given a new query Q, we use P to denote a PredSet in Q, and use p to denote an ORPred in P, namely, $p \in P$, and $P \in Q$.

γ_1	γ_2	O	γ_1	γ_2	O
$>$	$>=$	E	$<$	$<=$	E
$=$	$>=$	E	$=$	$<=$	E
$>$	$>=$	D	$>$	$>$	D
$>=$	$>=$	D	$>=$	$>$	D
$=$	$>$	D	$=$	$>=$	D
$<$	$<=$	I	$<$	$<$	I
$<=$	$<=$	I	$<=$	$<$	I
$=$	$<$	I	$=$	$<=$	I

Fig. 5. Subsumable Triples (γ_1, γ_2, O). E is equal, D is decreasing, and I is increasing.

Algorithm 2. Subsumed_ORPreds
Input: p, R
Output: $SubsumedSet(p)$
Procedure: for each literal $\rho_i \in p$, $1 \le i \le l$
  $S_{\rho_i} \Leftarrow \{p_{ijk} \Rightarrow \{\rho_{ij}\} \mid \rho_i \rightarrow \rho_{ij},\ \rho_{ij} \in p_{ijk},$
    $p_{ijk} \in R_{ORPred},\ 1 \le j \le s,\ 1 \le k \le m\}$;
  $I \Leftarrow \cap r_{i=1}^{l} S_{\rho_i}$;
  $SubsumedSet(p) \Leftarrow \{\}$;
  for each key $p' \in keys(I)$
    if $|elements(I, p')| == |p|$
      $SubsumedSet(p) + \Leftarrow p'$;

Fig. 6. Subsumption Algorithm

The problem of identifying subsumptions on ORPreds is as follows. Given an ORPred p, such that $p \in P$, and $P \in Q$, we want to find all ORPreds $p' \in R_{ORPred}$, such that p is subsumed by, subsumes, or is equivalent to p'. Similarly, given the PredSet $P \in Q$, we find all PredSets $P' \in R_{PredSet}$, such that P is subsumed by, subsumes, or is equivalent to P'.

In the rest of this subsection, we focus on the algorithm, *Subsumed_ORPreds*, which finds all the ORPreds that subsume p. We use the algorithm as the example to describe the computation representation, data structure and its operation, and the algorithm logic. We also cover other subsumption identification algorithms, which can be realized by small modifications to *Subsumed_ORPreds*. Finally, we discuss the algorithm time complexity.

Representation. We assume that each ORPred p has l literals, $\{\rho_1, \cdots, \rho_l\}$, each literal ρ_i is subsumed by s indexed literals, $\{\rho_{i1}, \cdots, \rho_{is}\}$, and each indexed literal ρ_{ij} appears in m non-equivalent ORPreds, $\{p_{ij1}, \cdots, p_{ijm}\}$, as shown in the left-hand-side of Figure 7. l is related to typical types of queries registered into the system, and thus can be viewed as a constant parameter. Similarly, we assume that each PredSet has k ORPreds, each ORPred is subsumed by t indexed ORPreds, and each ORPred appears in n different PredSets.

The algorithms assume non-redundant representations on ORPreds and PredSets. In particular, for the ORPred case, the assumption says that given an ORPred p, either indexed in $\mathcal{R}$ or in the new query Q, any literal $\rho \in p$ does not subsume any other literal $\rho' \in p$. For example, if $p = \{\rho_1\ OR\ \rho_2\}$ is non-redundant, then neither $\rho_1 \rightarrow \rho_2$ nor $\rho_2 \rightarrow \rho_1$ holds. Non-redundant PredSet representation is defined similarly. The assumption assures that all the subsumptions can be found with a single pass of the $l * s * m$ ORPreds or $k * t * n$ PredSets, based on the Theorem 2.

Theorem 2. *If the non-redundant assumption holds, then the $s * m$ ORPreds, $\{p_{ijh} | 1 \le j \le s, 1 \le h \le m\}$, whose literals subsume ρ_i, are different to each other.*

Proof. By contradiction, assume there are j_1, j_2, h_1, and h_2, such that $p_{ij_1h_1} \equiv p_{ij_2h_2}$. Then $\rho_{ij_1} \in p_{ij_1h_1}$, and $\rho_{ij_2} \in p_{ij_1h_1}$. According to Theorem 1, either $\rho_{ij_1} \rightarrow \rho_{ij_2}$ or $\rho_{ij_2} \rightarrow \rho_{ij_1}$ holds, which contradicts to the non-redundant assumption.

Note that the ORPreds across different literal predicates, such as $p_{i_1j_1k_1}$ and $p_{i_2j_2k_2}$, where $i_1 \neq i_2$, could be legitimately equivalent, and potentially are the targeted results the algorithms look for. The similar property can also be proven on the PredSet representations.

Corollary 1. *If the non-redundant assumption holds, then $s * m \leq |R_{ORPred}|$, where $|R_{ORPred}|$ is the number of ORPreds in R.*

The corollary follows from the theorem immediately. Generally, $s * m \ll |R_{ORPred}|$, since on average, ρ_i and its subsumed literals $\{\rho_{i1}, \cdots, \rho_{is}\}$ present a narrow set of semantics and only appear in a small portion of R_{ORPred}.

Data Structure. The algorithms use a data structure called 2-level hash set (2HSet) built up for literals or ORPreds to record the subsumption relationships. A 2HSet S is a set of sets, containing a set of hash keys, denoted as $keys(S)$, and each hash key $p \in keys(S)$ pointing to a set of elements, denoted as $elements(S, p)$. $keys(S)$ and all $elements(S, p)$ are hashed for constant-time accesses. For *Subsumed_ORPreds*, a 2HSet S_{ρ_i} records the ORPred set $\{p_{ijh} | 1 \leq j \leq s, 1 \leq h \leq m\}$, where each ORPred p_{ijh} in the set contains a literal ρ_{ij} that subsumes $\rho_i \in p$, as shown in Figure 7, where $keys(S_{\rho_i}) = \{p_{ijh} | 1 \leq j \leq s, 1 \leq h \leq m\}$, and $elements(S_{\rho_i}, p_{ijh}) = \{\rho_{ij}\}$.

We define a binary operation Υ-intersection $\cap_\Upsilon$ on 2HSets S_1 and S_2 to identify the intersection of their hash key sets, which represents the common part between the predicate conditions.

Definition 2. *Given two 2-level hash sets S_1 and S_2, we say S is the Υ-intersection of S_1 and S_2, denoted as $S = S_1 \cap_\Upsilon S_2$, if and only if following is true: S is a 2-level hash set, $keys(S) = keys(S_1) \cap keys(S_2)$, and for $\forall p \in keys(S)$, $elements(S, p) = elements(S_1, p) \cup elements(S_2, p)$.*

Υ-intersection preserves only the hash keys that appear in both S_1 and S_2. For any preserved hash key p, its $elements$ set is the union of p's $elements$ sets in S_1 and S_2. Υ-intersection can be computed in the time of $O(|keys(S)| * L)$ where S is the probing operand, either S_1 or S_2, in the implementation, and L is the average number of elements in $elements(S, p)$ for all $p \in keys(S)$. In Figure 7, the time of Υ-intersecting two 2HSets is $O(s * m)$.

Subsumption Algorithms. *Subsumed_ORPreds*, as shown in Figure 6, finds all ORPreds in R_{ORPred} that subsume p. It constructs all S_{ρ_i}, $1 \leq i \leq l$, Υ-intersects them to generate the final 2-level hash set I, and checks which remaining ORPreds in I subsume p. $|elements(I, p')|$ is the number of elements in $elements(I, p')$. And $|p|$ is the number of literals in p. The check condition $|elements(I, p')| = |p|$ means that if each literal in p is subsumed by some literal in p', then p is subsumed by p'.

A similar algorithm, $Subsume\_ORPreds$, finds all ORPreds in R_{ORPred} that p subsumes. The two differences from $Subsumed\_ORPreds$ are that the 2HSets are

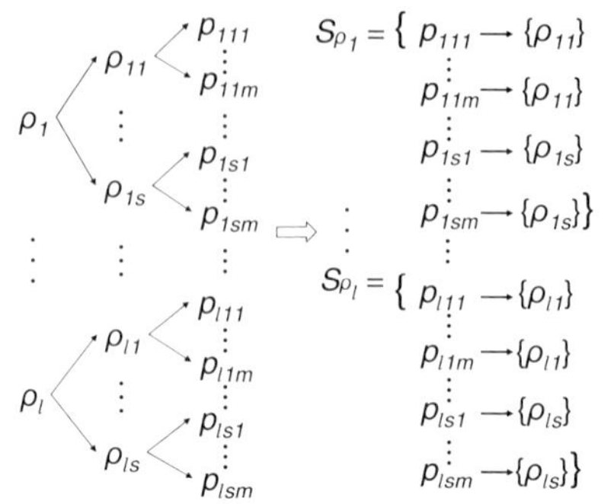

Fig. 7. ρ_i is subsumed by ρ_{ij}, $\rho_{ij} \in p_{ijk}$, $1 \leq i \leq l, 1 \leq j \leq s, 1 \leq k \leq m$. The information can be stored in 2-level hash sets.

PredIndex	**PSetIndex**	**SelectionTopology**
ORPredID	PredSetID	Node
LPredID	ORPredID	DirectParent
LeftExpr	STA	DPredSetID
Operator	**JoinTopology**	SVOA
RightExpr	Node	SVOAPredSetID
Node1	DirectParent1	JVOA1
Node2	DirectParent2	JVOA2
STA	DPredSetID	JVOAPredSetID
UseSTA	JVOA1	IsDISTINCT
	JVOA2	
	JVOAPredSetID	
	IsDISTINCT	

Fig. 8. System Catalogs

constructed from the literals that are subsumed by p's literals, and the final check condition is $|elements(I, p')| = |p'|$, meaning that if each literal in p' is subsumed by some literal in p, then p' is subsumed by p.

The algorithms can be easily extended to identify subsumptions at the PredSet layer. In that case, the hash keys are the PredSet IDs and the elements are ORPred IDs. The final check conditions dictate that a PredSet P is subsumed by another P' if P is subsumed by all ORPreds in P'. Identifying equivalence is easy given the identified subsuming and subsumed 2HSets; it is the unique ORPred or PredSet that is in the intersection of the two.

The algorithms guarantee that no redundant ORPreds or PredSets will be introduced into indexing as long as the non-redundancy assumption holds on queries.

Time Complexity. The time complexity of the ORPred-layer algorithms is $O(l*s*m)$, and that of the PredSet-layer is $O(k*l*s*m+k*t*n)$ which includes the k calls of the former. Note that $t*n \leq |R|$ given the non-redundancy assumption. $|R|$ is the number of the searchable PredSets in R and the number of nodes in R. Generally, $t*n \ll |R|$ since p usually appears only in a small portion of the indexed PredSets. Therefore, the algorithm takes only a small portion of time $O(k*l*|R_{ORPred}|+k*|R|)$ to compute.

If the sharable PredSets are searched by matching PredSets and ORPreds one by one, the searching will take the time of $O(k^2*l*|R|)$ since k new ORPreds need to match $|R|*k$ existing ORPreds and each match computes on l literal predicates. Although it is also linear to $|R|$, the factor is larger and it will be much slower on large scales.

4.6 Topology Connection

PredSets are associated with nodes. Assume that PredSet P is associated with node N, then P bears the topological connections between N and N's ancestor set $\{A\}$. In particular, the results of N are obtained by evaluating P on $\{A\}$.

The node N may be associated with multiple PredSets depending on the different types of ancestors. We define three types of ancestors for each node, direct parents ***(DParents)***, selection very original ancestor ***(SVOA)***, and join very original ancestors ***(JVOA)***.

Definition 3. *A node N's **DParents** are the set of nodes that have an edge pointing to N.*

Definition 4. *A selection node N's **SVOA** is N's closest ancestor that is either a join node or a base stream node. A join node or a base stream node N's **SVOA** is itself N.*

Definition 5. *A join node N's **JVOA**s are the closest ancestor nodes that are either join nodes (but not N) or base stream nodes. A selection node N's **JVOA**s are N's SVOA's JVOAs. And a base stream node's **JVOA** is NULL.*

We record all the three ancestor types and their associated PredSets. Each type plays an important role. DParents is necessary and sufficient to construct the plan execution code. SVOAs and JVOAs present local topological connections within and across one join depth, respectively. Their presence allows a single lookup per join depth, avoiding the chained search through DParents, to find the sharable computation.

4.7 Relational Model for Indexing

Now we convert the ER model to the relational model. A simplified version is shown in Figure 8. We made three adjustments. First, only 2-way joins are supported. Supporting multi-way joins requires a small amount of work to revise the indexing schema and the searching tools, but requires much more work in sharing strategies. In particular, multi-way joins bring back the NP-hardness, and requires more advanced heuristic optimization techniques. This will be a future work. Second, the relations that index literal predicates and ORPreds are merged into one, *PredIndex*, based on the assumption that ORPred are not frequent in queries. This allows a literal predicate to appear multiple times in *PredIndex* if it belongs to different ORPreds. But this redundancy is negligible given the assumption. The third adjustment is splitting the node topology indexing relation (The Node entity in the ER model) to two, namely, *SelectionTopology*, and *JoinTopology*, based on the observation that the topology connections on selection nodes and on join nodes are quite different.

5 Conclusion and Future Work

As part of the IMQO framework, a comprehensive computation indexing schema and a set of searching tools were presented. The schema stores shared plan computations in relational system catalogs, and the tools search the common computations between new queries and the system catalogs. We implemented the schema and tools on ARGUS to support efficient processing of a large number of complex continuous queries. The empirical evaluation on ARGUS demonstrated up to hundreds of times speed-up for multiple query evaluation via the shared plan construction [9]. The techniques would also be very useful in other IMQO-supported stream databases.

For future work, immediate extensions are supporting multi-way joins, local restructuring upon new query arrivals, and adaptive local re-optimization upon dynamic congestion detections. It is also interesting to support more advanced sharing strategies in the IMQO setting, such as identifying the minimum cover of disjoint ranges and utilizing constraints, such as foreign key constraints [8].

Acknowledgments

This work was supported in part by DTO/ARDA, NIMD program under contract NMA401-02-C-0033. The views and conclusions are those of the authors, not of the U.S. government or its agencies. We thank Christopher Olston, Phil Hayes, Jamie Callan, Minglong Shao, Santosh Ananthraman, Bob Frederking, Eugene Fink, Dwight Dietrich, Ganesh Mani, and Johny Mathew for helpful discussions.

References

1. Abadi, D.J., Carney, D., Çetintemel, U., Cherniack, M., Convey, C., Lee, S., Stonebraker, M., Tatbul, N., Zdonik, S.B.: Aurora: a new model and architecture for data stream management. VLDB J. 12(2), 120–139 (2003)
2. Blakeley, J.A., Coburn, N., Larson, P-Å.: Updating derived relations: Detecting irrelevant and autonomously computable updates. ACM Trans. Database Syst. 14(3), 369–400 (1989)
3. Chakravarthy, U.S., Minker, J.: Multiple query processing in deductive databases using query graphs. In: VLDB, pp. 384–391 (1986)
4. Chandrasekaran, S., Cooper, O., Deshpande, A., Franklin, M.J., Hellerstein, J.M., Hong, W., Krishnamurthy, S., Madden, S.R., Raman, V., Reiss, F., Shah, M.A.: TelegraphCQ: Continuous Dataflow Processing for an Uncertain World. In: CIDR (January 2003)
5. Chen, J., DeWitt, D.J., Naughton, J.F.: Design and evaluation of alternative selection placement strategies in optimizing continuous queries. In: ICDE, pp. 345–356 (2002)
6. Finkelstein, S.J.: Common subexpression analysis in database applications. In: SIGMOD Conference, pp. 235–245 (1982)
7. Forgy, C.L.: Rete: A Fast Algorithm for the Many Pattern/Many Object Pattern Match Problem. Artificial Intelligence 19(1), 17–37 (1982)
8. Goldstein, J., Larson, P.-Å.: Optimizing queries using materialized views: A practical, scalable solution. In: SIGMOD Conference (2001)
9. Jin, C.: Optimizing Multiple Continuous Queries. Ph.D. Thesis CMU-LTI-06-009, Carnegie Mellon University (2006)
10. Jin, C., Carbonell, J.G.: Incremental aggregation on multiple continuous queries. In: ISMIS, pp. 167–177 (2006)
11. Jin, C., Carbonell, J.G., Hayes, P.J.: Argus: Rete + dbms = efficient persistent profile matching on large-volume data streams. In: ISMIS, pp. 142–151 (2005)
12. Motwani, R., Widom, J., Arasu, A., Babcock, B., Babu, S., Datar, M., Manku, G.S., Olston, C., Rosenstein, J., Varma, R.: Query processing, approximation, and resource management in a data stream management system. In: CIDR (2003)
13. Roussopoulos, N.: View indexing in relational databases. ACM Trans. Database Syst. 7(2), 258–290 (1982)
14. Selinger, P.G., Astrahan, M.M., Chamberlin, D.D., Lorie, R.A., Price, T.G.: Access path selection in a relational database management system. In: SIGMOD Conference, pp. 23–34 (1979)
15. Sellis, T.K.: Multiple-query optimization. ACM Trans. Database Syst. 13(1), 23–52 (1988)
16. Zaharioudakis, M., Cochrane, R., Lapis, G., Pirahesh, H., Urata, M.: Answering complex sql queries using automatic summary tables. In: SIGMOD Conference, pp. 105–116 (2000)

SQL Queries with CASE Expressions

Jarek Gryz[1], Qiong Wang[2], Xiaoyan Qian[2], and Calisto Zuzarte[2]

[1] Department of Computer Science and Engineering
York University, Toronto, Canada
[2] IBM Laboratory, Toronto, Canada

Abstract. In recent years more and more queries are generated automatically by query managers/builders with end-users providing only specific parameters through GUIs. Queries generated automatically can be quite different from queries written by humans. In particular, they contain non-declarative features, most notorious of which is the `CASE` expression. Current query optimizers are often ill-prepared for the new types of queries as they do not deal well with procedural 'insertions'. In this paper, we discuss the inefficiencies of `CASE` expressions and present several new optimization techniques to address them. We also describe experimental evaluation of the prototype implemented in DB2 UDB V8.2.

1 Introduction

One of the great strengths of SQL is its declarative nature: it is enough to say *what* data we want without having to say *how* to get it. It is the job of the query optimizer to produce an efficient method–query evaluation plan–of retrieving the data. The declarative nature of SQL also means that it should not make a difference to the optimizer *how* a query is phrased: all semantically equivalent SQL statements are equally good (this point is often emphasized in database textbooks). Of course, this last point is not entirely true as there is only a limited number of types of query transformations that the optimizer can perform.

Current commercial query optimizers have been designed at the time when virtually all database queries were written by IT professionals. This assumption determined to some degree the types of queries that could be optimized well. In recent years, however, more and more queries are generated automatically by query managers/builders with end-users providing only specific parameters through GUIs. Query managers are both released by commercial vendors as well as custom designed in large corporations. Queries generated automatically can be quite different from queries written by humans.

1. *Size and Complexity*
   There is essentially no limit to the length of the query generated automatically. It is not unusual to see queries with over 1000 lines of SQL code. In addition, such queries can be very complex, with nesting being a common feature.
2. *Repetitiveness*
   Identical (or very similar) expressions or subquery blocks tend to appear in several different places of one query. This happens most often in different branches of the UNION query or in queries referring to the same view multiple times.

A. An et al. (Eds.): ISMIS 2008, LNAI 4994, pp. 351–360, 2008.

3. *Non-Declarative Features*
   Since query managers are written in procedural languages, their designers tend to take shortcuts and 'copy' procedures into SQL. This is most common with `CASE` expressions which encode the *IF ... THEN ... ELSE* logic of procedural languages.

None of the features listed above is new in the sense that it did not exist in queries written by humans. However, the extent and frequency with which they appear in automatically generated queries has serious ramifications for query performance. Current query optimizers are often ill-prepared for the new types of queries: they do not scale up with query size, they can discover only the most obvious similarities among query blocks [2,5], and they do not deal well with procedural 'insertions'.

There is clearly no easy way of redesigning the optimizer to handle new types of queries and we doubt that anything other than a piecemeal approach is possible. This paper is an attempt to provide just one piece of the solution. We provide several new optimization methods for queries with `CASE` expressions. In Section 2 we describe the syntax and use of `CASE` expressions. Section 3 contains a discussion of inefficiencies introduced by `CASE` expressions and our approach to alleviate them. In Section 4 we present the implementation and experimental evaluation of our techniques. We conclude in Section 5. An example of an automatically generated query is presented in the Appendix.

All techniques described in this work has been implemented as prototypes in DB2 UDB V8.2.

2 Syntax and Use of CASE Expressions

The `CASE` expression was introduced into SQL syntax a few years ago to handle conditional logic. `CASE` is just another expression so it can be placed wherever any other expression is allowed (for example, in `select`, `where`, or `order by` clauses). The syntax of `CASE` is shown in Figure 1. The *searched-when-clause* and the *simple-when-clause* are two flavors of `CASE`: the first one can use any condition for every *when*, the second compares an expression with potential values.

The utility of `CASE` is obvious: it allows you to use the *IF ... THEN ... ELSE* logic of procedural languages without having to invoke procedures. But convenience is only one advantage of `CASE` expressions: when used properly, they can improve query performance. The most prominent example is substitution of `UNION` with `CASE`. When several logical operations have to be performed against the same data, they are usually combined in a single SQL statement by means of `UNION`. This approach has the disadvantage of having to retrieve the same data multiple times. Consider the following example [4]:

```
select sum(amount) as sales, 'Dept' as key
from   salesrecords
where  Department = 'A01'
union all
select sum(amount) as sales, 'Prod' as key
from   salesrecords
```

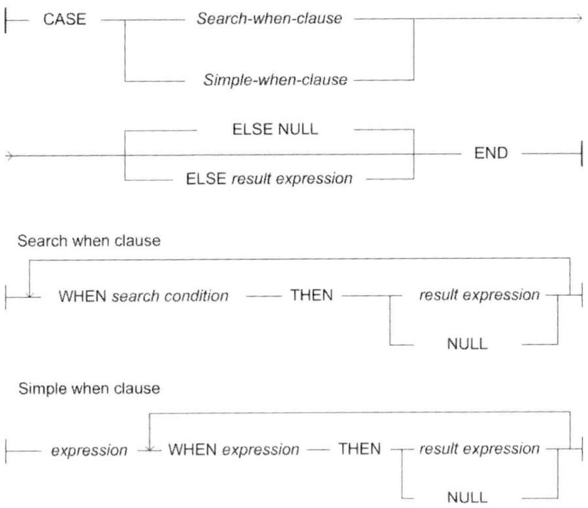

Fig. 1. Syntax of CASE expressions

```
where  Product_type = 'Garden'
union all
select sum(amount) as sales, 'Dept' as key
from   salesrecords
where  Discount = 'Y'
```

In most systems, the above query would require three scans of the `salesrecords` table. However, the query is quite simple: *What is the total sales amount occurred for products from department A01 or for a product used in the garden or a product that was on sale?*. Clearly, one scan of the table is sufficient to retrieve all the data. With `CASE`, the conditions can be included in the `select` clause of the SQL statement.

```
select
sum(CASE Department WHEN 'A01' THEN amount
                                ELSE 0
    END) as salesdept,
sum(CASE Product_type WHEN 'Garden' THEN amount
                                    ELSE 0
    END) as salesprod,
sum(CASE Discount WHEN 'Y' THEN amount
                           ELSE 0
    END) as salesdept
from salesrecords
```

Another place where `CASE` can be useful is in UPDATE statements. When the rows from a table need to be updated conditionally, the query has to be split into multiple statements or the rows have to be retrieved into the application. In both cases, performance would be worse than when a single update statement is used. Again, `CASE` expression provides a simple solution by encoding the conditions within a single update statement. For example, assume that level 2 employees will get a 5% raise and the rest 2% [4]. This can be trivially expressed as:

```
update employee
set salary = CASE level WHEN 2 THEN salary * 1.05
                                ELSE salary * 1.02
             END
```

`CASE` expression can certainly provide query performance improvements when used properly. But even the advocates [1,4] of their use, warn that adding these expressions does not come free. In particular, adding `CASE` to the `select` clause adds CPU time to processing of every row. Replacing a predicate with `CASE` expresion in the `where` clause may prevent index use as `CASE` is not indexable in most database systems unless a generated column is added with an index on it of the DBMS supports indexes on expressions. An experienced database administrator or application developer can easily avoid these pitfalls. Unfortunately, `CASE` is often introduced into SQL queries through GUIs in query managers where hand optimization is impossible. This is where they can cause dramatic performance deterioration.

3 Rewrites of Queries with CASE Expressions

3.1 CASE in the `where` Clause

We first consider a query with `CASE` expression in the `where` clause with a straigtforward rewrite. The structure of the `CASE` expression is such that WHEN conditions can be evaluated at compile time, hence can be eliminated altogether. Despite its obvious redundancy, we found such statements to be quite frequent. They are apparently generated when a static form in a GUI query manager (with the structure revealed in the `where` clause) is filled out by a user.

The typical query of this type has the following form:

Q_1:

```
select sum(c_acctbal)
from   tpcd.customer
where  (CASE WHEN 'b'= ' ' THEN 'Customer1'
             ...
             WHEN ' ' = ' ' THEN 'Customerk'
             ...
                            ELSE 'unknown'
        END) = c_name ;
```

This can be trivially simplified into:

```
select sum(c_acctbal)
from   tpcd.customer
where  'Customerk' = c_name
```

The original predicate includes a lengthy case expression. When we evaluate this predicate, we need to compute the WHEN condition for every retrieved tuple. The cost depends on the location of the first satisfied condition (in our case the k-th WHEN expression).

In our implementation, we only consider queries with a binary relational predicate: "=", "<",">", "<=", ">=", where one side is the whole case expression, the other side is an attribute (such as *c_name* in the example). Also, since the WHEN condition expression could be arbitrarily complex, we could penalize the optimization time for the general statements that will likely not have expressions simplified. Thus, in our prototype implementation, we only consider the more common scenario with the comparison of two constants. This can be extended to handle more complex expressions.

We observed that such patterns of statically computable CASE expressions can be nested as in the example below.

```
select sum(c_acctbal)
from tpcd.customer
where (CASE WHEN 'b'=' ' THEN
             (case when '13' = '14' then 'Customer1'
                              ...
                                    else 'unknown'
             end)
             ...
             WHEN ' ' = ' ' THEN
             (case when '13' = '13' then 'Customerk'
                              ...
                                    else 'unknown'
             end)
             ...
       END) = c_name ;
```

The Rewrite Rule Engine in DB2 allows to rewrite this query recursively by simply repeating the rewrite multiple times.[1]

The static rewrite described above, together with the *constant folding* rule[2] also allows removal of CASE expressions when a predicate predetermines the satisfaction of one of the WHEN conditions in the CASE expression. Consider the following example:

Q_2:

```
select *
from   T
where  (CASE WHEN T.A> 1 THEN 1
             WHEN T.A> 0 THEN 2
                         ELSE 3
       END) = T.B
       and T.A=1
```

[1] DB2 allows for multiple passes over the set of its rewrite rules [3].

[2] Constant folding is one of the current rules in DB2. It substitutes an expression in a query with a value predetermined by a predicate.

This query can be simplified to:

Q_2':
```
select *
from   T
where  2=T.B and T.A=1
```

3.2 CASE in the `select` Clause

In this section we present a popular yet very inefficient query with `CASE` expression in the `select` clause. We first describe our rewrite of a simple query Q_3 shown below and then show how it can be generalized to more complex queries.

Q_3:
```
select
SUM (case WHEN MONTH (SALE_DATE) = 1 THEN
          SALE_AMOUNT ELSE 0 end),
          ...
SUM (case WHEN MONTH (SALE_DATE) = 10 THEN
          SALE_AMOUNT ELSE 0 end)
from SALES_HISTORY
where YEAR (SALE_DATE) = 2005
```

This query has three distinct features:

- The output of the query (that is, its `select` clause) contains `CASE` expressions embedded in identical aggregate functions (SUM in Q_3).
- Each search condition of the `CASE` expression references the same column (SALE_DATE in Q_3).
- Each ELSE result expression of the `CASE` expression is a constant (0 in Q_3).

The evaluation of query Q_3 is performed as follows:

For each row of SALES_HISTORY do:

1. Apply predicate YEAR (SALE_ DATE) = 2005 If the current row does not satisfy this predicate, skip the next steps.
2. If it is the first row, prepare the output columns. In order to generate the first output column, check if the current row satisfies search condition MONTH (SALE_ DATE) = 1. If it does, store SALE_AMOUNT of the current row into a temporary buffer. Otherwise, put 0 at the temporary buffer.

   To generate the second output column, check if the current row satisfies search condition MONTH (SALE_ DATE) = 2. If it does, store SALE_AMOUNT of the current row into a temporary buffer. Otherwise, put 0 in the temporary buffer.

   Continue this effort, until reaching the completion of the process for the last output column.
3. If it is not the first row, check if the current row satisfies search condition MONTH (SALE_DATE) = 1. If it does, add SALE_AMOUNT of the current row to the number stored in the temporary buffer for the first output column. Otherwise, add 0 to the temporary buffer.

To generate the second output column, check if the current row satisfies search condition MONTH (SALE_ DATE) = 2. If it does, add SALE_AMOUNT of the current row to the number stored in the temporary buffer for the second output column. Otherwise, add 0 to the temporary buffer.

Continue this effort, until reaching the completion of the process for the last output column.

The inefficiency of query Q_3 is striking: it requires (for every row that satisfies the predicates from the `where` clause) evaluating all `CASE` expressions present in the `select` clause. This evaluation involves not only the operation specified by WHEN condition, but also a summation for the appropriate stored column. Thus, for every qualifying row of Q_3, there are 10 comparisons of strings perfomed and 10 summations computed. For some rows, that is the ones representing month 11 and 12, this computation will have no effect whatsoever on the final result.

The rewrite we propose contains three elements:

1. Add a new predicate to the `where` clause, to consider only the rows used in the summation.
2. Add a `group by` clause on the column referenced in the WHEN clause of the `CASE` expression.
3. Apply the original (modulo column renaming) `CASE` expression to the query modified through 1. and 2. above.

After the rewrite Q_3 becomes Q_3'.

Q_3':
```
select SUM (case WHEN T.id1 = 1 THEN
                T.id2 ELSE 0 end),
       ...
       SUM (case WHEN T.id1 = 10 THEN
                T.id2 ELSE 0 end)
from
(select MONTH (SALE_DATE) as id1,
        SUM(SALE_AMOUNT) as id2
 from SALES_HISTORY
 where YEAR (SALE_DATE) = 2005
        AND MONTH (SALE_DATE) in
        (1, 2, 3, 4, 5, 6, 7, 8, 9, 10)
 group by MONTH (SALE_DATE)) as T
```

Adding a new predicate to the `where` clause reduces the number of rows for which the aggregation function is computed. Grouping the rows on the column of interest is in general more efficient than doing it in a brute force way using `CASE` expression.[3] Also, performing aggregation once each group (rather than multiple times for each row in the original query) also decreases the overall computation cost. The only reason for

[3] Of course, one can contrive a case where it would not be so. We have not seen such case in our experiments.

reintroducing the CASE expression is to return the output of the query in the intended format: horizontally, rather than vertically.

The rewrite performed on Q_3 can be generalized to other queries. In our implementation we are able to handle other types of aggregation (MIN, MAX, AVG) and scenarios where the constants in ELSE clauses of the CASE expression are non-zero. For lack of space, we leave out the details here.

4 Implementation and Experimental Evaluation

The query transformations described above have been implemented as rewrite rules in DB2 Rule Engine (with a prototype in DB2 UDB V8.2). The experiments were performed on an IBM/RS6000 performance testing machine with operating system AIX 5.1, 8*752 M CPU, 16G memory, using TPCH database with size 100 MB and 1 GB.

Table 1. Experiment I

Query	100MB		1GB	
	Elapse Time (s)	CPU Time (s)	Elapse Time (s)	CPU Time (s)
T_1	6.3	2.3	58.4	23.8
T_1'	6.0	1.8	57.5	19.8
T_2	1.2	0.3	10.3	4.2
T_2'	1.2	0.2	10.1	3.1

In our first experiment we compared the performance of queries with CASE expression in the where clause. We performed the rewrites of the type described in Section 3.1 on two queries T_1 and T_2, which were simple selects on orders and parts table respectively. The two queries differred primarily on the number of WHEN clauses in the CASE expression: T_1 contained 2 and T_2 contained 10 of them. The results of the experiment are presented in Table 1.

As expected, the reduction of the CPU time grows with the number of WHEN clauses in the original queries (up to 30% for T_2'). The overall performance improvement (that is, the reduction of the elapse time) is not substantial as the query access plans and I/O costs for the original and rewritten queries are identical. The optimization is achieved solely through avoiding the evaluation of the CASE expressions for each of the qualifying rows. We point out, however, that more substantial performance improvements could be achieved if an index existed on the attribute referenced in the where clause (recall that in at least some systems CASE expressions are not indexable, so the index could not be used in the original queries).

In the second experiment, we compared the performance of queries with CASE expression in the select clause. We performed the rewrites of the type described in Section 3.2 on two queries T_3 and T_4. T_3 has the same structure as query Q_3 except that instead of a simple select it contains a join. T_4 is just like T_3, but in addition it has a group by statement. The performance results are shown in Table 2.

Table 2. Experiment II

Query	100MB		1GB	
	Elapse Time (s)	CPU Time (s)	Elapse Time (s)	CPU Time (s)
T_3	4.1	3.6	341.7	337.8
T_3'	2.8	1.9	176.5	171.2
T_4	5.3	4.9	449.6	449.4
T_4'	2.3	1.2	86.4	86.1

Two factors, discussed already in Section 3.2, contribute to the impressive (over 80% for T_4) performance improvement in this rewrite. First, a new predicate derived from the CASE expression is added to the where clause. Since not all the rows contribute to the final result of the CASE expression's embedded aggregate function output, by applying the generated predicate earlier unnecessary computation can be avoided. Furthermore, computation time for the aggregation functions can be saved by applying the aggregates on the smaller set of data vertically (using group by) in each individual group instead on the whole data set horizontally.

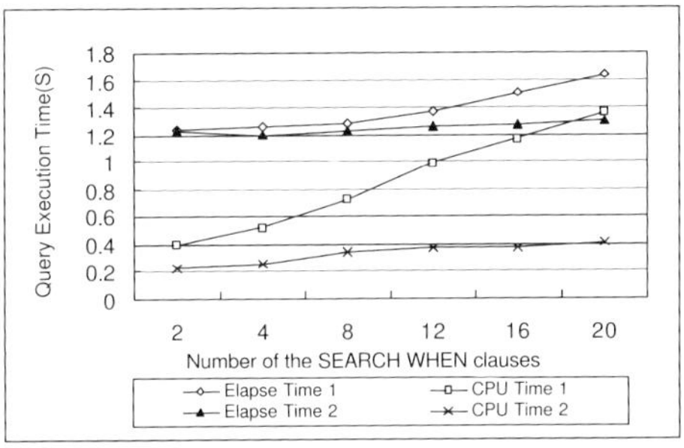

Fig. 2. Query performance as a function of the number of WHEN clauses in the CASE expression (100 MB DB)

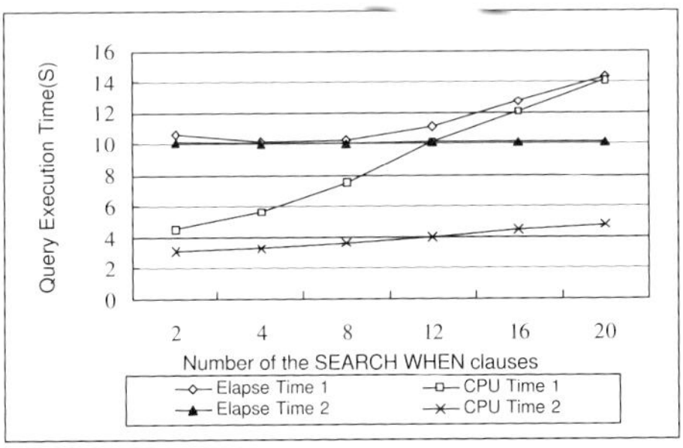

Fig. 3. Query performance as a function of the number of WHEN clauses in the CASE expression (1 GB DB)

In the last experiment we evaluated the dependence of query elapse time on the number of WHEN conditions in the `select` clause. Again, we considered a query with the structure of Q_3. We varied the number of WHEN conditions from 2 to 20 and compared the performance of the original (1) and rewritten (2) query. The graphs in Figures 2, and 3 show not only that our rewrite improves performance, but also that the optimization effect grows with the number of WHEN conditions in the query.

5 Conclusions and Future Work

In this paper we described the design and implementation of a few rewrite rules from a prototype built to optimize queries with `CASE` expressions. This work is a partial response to a new challenge to database query optimization coming from queries generated automatically. The work in this area will undoubtedly continue; our main focus is to handle multiple appearance of identical expressions within a query.

Acknowledgments

TPCH is a trademark of Transaction Processing Council.

References

1. Burns, D.: Improving sql efficiency using case, `http://oracledoug.com/case.html`
2. Dalvi, N.N., Sanghai, S.K., Roy, P., Sudarshan, S.: Pipelining in multi-query optimization. In: Proceedings of PODS, pp. 59–70 (2001)
3. Haas, L., Freytag, J., Lohman, G., Pirahesh, H.: Extensible query processing in starburst. In: SIGMOD Proceedings, pp. 377–388. ACM, New York (1989)
4. Henderyckx, J.: Version 5 case expressions: beyond sql reference. The IDUG Solutions Journal 5(3), 38–45 (1998)
5. Lehner, W., Cochrane, B., Pirahesh, H., Zaharioudakis, M.: fAST refresh using mass query optimization. In: Proceedings of ICDE, pp. 391–400 (2001)

Top-Down Compression of Data Cubes in the Presence of Simultaneous Multiple Hierarchical Range Queries

Alfredo Cuzzocrea

ICAR Institute & DEIS Department
University of Calabria
I-87036 Rende, Cosenza, Italy
cuzzocrea@si.deis.unical.it

Abstract. A novel top-down compression technique for data cubes is introduced and experimentally assessed in this paper. This technique considers the previously unrecognized case in which *multiple Hierarchical Range Queries* (HRQ), a very useful class of OLAP queries, must be evaluated against the target data cube *simultaneously*. This scenario makes traditional data cube compression techniques ineffective, as, contrarily to the aim of our work, these techniques take into consideration *one constraint* only (e.g., a given space bound). The result of our study consists in introducing an innovative *multiple-objective OLAP computational paradigm*, and a *hierarchical multidimensional histogram*, whose main benefit is meaningfully implementing an *intermediate compression* of the input data cube able to simultaneously accommodate an even large family of different-in-nature HRQ. A complementary contribution of our work is represented by a wide experimental evaluation of the query performance of our technique against both benchmark and real-life data cubes, also in comparison with state-of-the-art histogram-based compression techniques.

1 Introduction

Conventional data cube compression techniques, such as *histograms* (e.g., [1,22,6]), are devoted to drive the compression of the input data cube in dependence on *one constraint only*. Traditionally, this requirement is represented by a given space bound available to house the compressed representation of the data cube, like in conventional approaches (e.g., [25,34]). Without loss of generality, this scheme can be classified as adhering to what we call the *single-objective data cube compression paradigm*, which defines a class of methodologies wide enough to include most of the data cube compression proposals appeared in literature during the last two decades. This consolidated paradigm has been subsequently made more complex via including novel *additional requirements* to be considered *simultaneously* to the main space bound constraint, such as (*i*) compressing data cubes with the additional goal of *minimizing* the overall query error of a given query-workload (e.g., [6,16]), (*ii*) ensuring *probabilistic guarantees* over the quality of *approximate answers* evaluated against compressed data cubes (e.g., [20,21,12,16]), or (*ii*) *mediating* on the degree of approximation of the retrieved answers (e.g., [13,14]). Indeed, on a pure theoretical

A. An et al. (Eds.): ISMIS 2008, LNAI 4994, pp. 361–374, 2008.

plan, compression schemes of such a kind should be considered as still adhering to the single-objective data cube compression paradigm, due to the fact that additional goals above clearly play a secondary role with respect to the main constraint, and must be more properly intended as *application-oriented requirements* that can occur in next-generation OLAP scenarios like those drawn in [14].

More problematic issues appear when the data cube compression process must be performed in the presence of *multiple constraints*, under what we call the *multiple-objective data cube compression paradigm*, which, to the best of our knowledge, is a novel OLAP computational paradigm not considered by previous research. The main contribution of this paper consists in introducing this novel paradigm, and providing a formal framework for dealing with a significant instance of the problem underlying such a paradigm. Basically, according to this novel paradigm, the compressed representation of the input data cube is obtained as that *intermediate* (*compressed*) *representation* which accomplishes, *as more as possible*, the multiple constraints defined by the input multiple-objective computational scheme. In fact, it is worthy noticing that, in the presence of multiple constraints, it is not possible to obtain a *valid-for-all* data cube compressed representation (i.e., the compressed representation that *simultaneously* satisfies *all* the multiple constraints) so that devising *sub-optimal solutions* appears to be the most promising strategy for the so-complex computational setting dictated by the multiple-objective application scenario. To give notes on related work, the idea of introducing multiple-objective computational paradigms in order to deal with complex Database and Data Warehousing research challenges has been considered in few contexts previously, and mostly with respect to requirements defined by *multiple queries* (i.e., simultaneous queries belonging to different query classes – e.g., [37,38]). Among these initiatives, we recall: (*i*) multiple-query optimization for the view selection and materialization problem [31], (*ii*) multiple-query based data sources integration [26,17,32], (*iii*) multi-objective query processing in database systems [36,3,9] and OLAP [29], (*iv*) multi-objective query processing for specialized contexts such as data aggregation [18], complex mining tasks [27,28] and data stream processing [5,41], and, more recently, (*v*) skyline query processing [2,4,33], which aims at extracting Pareto distributions from relational data tables according to *multiple preferences*. Contrarily to the above-listed research contributions, to the best of our knowledge, there not exist in literature data cube compression techniques that take into consideration the issue of performing the compression process on the basis of multiple objectives. While this evidence clearly causes the lacking of a theoretical background for the problem investigated in this paper, at the same time it gives more relevance to the innovation carried out by our research work, and puts the basis for further research efforts in this field.

Despite theoretical issues, when a multiple-objective OLAP computational paradigm is adopted, one must first set the nature and the typology of multiple goals with respect to which the paradigm has to be implemented. Similarly to the above-listed research experiences, in our work we choose different-in-nature queries as playing the role of multiple objectives to be accommodated during the compression process. This because, in many real-life application scenarios, very often OLAP queries must be simultaneously evaluated against the target data warehouse server, as dictated by popular processes according to which advanced Data Mining and analysis tools extract useful knowledge from warehouse servers in form of more or less sophisticated statistics on multidimensional data. To give practical examples, it suffices to think of modern scenarios drawn by corporate analysis servers for *e*-commerce, *e*-government, and *e*-procurement systems. In parallel to the nature and typology of multiple goals, a *specific* query class must be set. While making this choice does not limit the range of applicability of the research presented in this paper, the general multiple-objective data cube compression paradigm we propose can be easily customized to any class of OLAP queries, from simple *range queries* [24], which are defined as the application of a SQL aggregate operator (e.g., SUM, COUNT, AVG etc) to a given range of multidimensional data, to more complex ones such as *top-k* [42] and *iceberg* [19] queries. Motivated by this assumption, in this paper we consider the class of *Hierarchical Range Queries* (HRQ), introduced by us in [15] as a meaningful extension of those defined by Koudas *et al.* in [30]. As it will be evident throughout the paper, HRQ define very useful tools for extracting *hierarchically-shaped* summarized knowledge from data warehouse servers, beyond the capabilities of conventional OLAP environments. Definition 1 introduces HRQ.

Definition 1. *Given a data cube* $\mathcal{L} = \langle \mathcal{D}, \mathcal{H}, \mathcal{M} \rangle$, *such that* (*i*) $\mathcal{D}$ *is the set of dimensions of* $\mathcal{L}$, (*ii*) $\mathcal{H}$ *is the set of hierarchies of* $\mathcal{L}$, *and* (*iii*) $\mathcal{M}$ *is the set of measures of* $\mathcal{L}$, *a Hierarchical Range Query* (HRB) Q_H *against* $\mathcal{L}$ *is a tuple* $Q_H = \langle \mathcal{T}, \mathcal{F}, \vartheta \rangle$, *such that:* (*i*) $\mathcal{T}$ *is a general tree,* (*ii*) $\mathcal{F}$ *is a domain of range queries,* (*iii*) ϑ *is a mapping function that maps queries in* $\mathcal{F}$ *onto nodes in* $\mathcal{T}$, (*iv*) *for each level* ℓ *in* $\mathcal{T}$, $\bigcap_{k=0}^{|\text{queries}(\ell)|-1} \| Q_k \| = \emptyset$, *such that* queries($\ell$) *is the set of range queries at level* ℓ *of* $\mathcal{T}$, *and* (*v*) *for each node n in* $\mathcal{T}$ *such that* depth(n) < depth($\mathcal{T}$), $\bigcup_{k=0}^{|\text{child}(n)|-1} \| Q_k \| \neq \| Q_n \|$, *where:* (*v.i*) depth(n) *denotes the depth of n,* (*v.ii*) depth($\mathcal{T}$) *denotes the depth of* $\mathcal{T}$, (*v.iii*) child(n) *denotes the set of child nodes of n,* (*v.iv*) Q_k *denotes the range query related to the node k, and* (*v.v*) $\|Q\|$ *denotes the volume of Q.*

It should be noted that, similarly to the multi-level and hierarchical nature of the target data cube $\mathcal{L}$, HRQ are multi-level and hierarchical in nature as well, meaning that, for each level ℓ of $\mathcal{T}$, the set of range queries at level ℓ, denoted by queries(ℓ), must be evaluated against the *maximal cuboid* of $\mathcal{L}$ at level ℓ, denoted by $\mathcal{L}_\ell$, i.e. the collection of (OLAP) data cells obtained by aggregating data cells of $\mathcal{L}$ at the lowest level of detail (i.e., $\mathcal{L}_0 \equiv \mathcal{L}$) with respect to *all* the aggregation levels of hierarchies in $\mathcal{H}$ at level ℓ. Due to the illustrated properties, it is a matter to note that, given a data cube $\mathcal{L}$ and a HRQ Q_H having $\mathcal{T}$ as structural tree, the answer to Q_H against $\mathcal{L}$, denoted as $A(Q_H)$, is modeled in terms of a general tree $\mathcal{Y}$ having the same topology of $\mathcal{T}$ and such that each node n stores the answer $A(Q_n)$ to the corresponding range query Q_n. Also, without loss of generality, for the sake of simplicity here we assume that (*i*) hierarchies in $\mathcal{H}$ have *all* the same depth P, and (*ii*) the depth of $\mathcal{T}$ is also equal to P.

Depending on the kind of SQL aggregate operator characterizing range queries in $\mathcal{F}$, different classes of HRQ can be obtained. In this paper, we focus on *Hierarchical Range-SUM Queries*, as SUM aggregations are very popular in OLAP, and represent the most common solution for extracting useful knowledge from massive data cubes (e.g., see 24,13,20,7]), while also encompassing the amenity of acting as baseline operators to build more complex OLAP aggregations/functions [23].

Range-SUM queries have been widely investigated in OLAP. As a consequence, in literature there exist a number of proposal dealing with the issue of answering such queries efficiently via data cube compression techniques (e.g., [24,13]). Basically, histogram-based compression approaches have been proposed, with the appreciable idea of evaluating approximate answers by means of *linear interpolation* techniques applied to buckets of the histogram (e.g., [7]) via meaningfully exploiting the *Continuous Value Assumption* (CVA) formulated by Colliat in [10], which assumes that data are uniformly distributed. Given (*i*) a data cube $\mathcal{L}$, (*ii*) the histogram Hist($\mathcal{L}$) computed on $\mathcal{L}$, and (*iii*) a range-SUM query Q against $\mathcal{L}$ such that Q overlaps the bucket set $B(Q) = \{b_0, b_1, \ldots, b_{W-1}\}$ of Hist($\mathcal{L}$), based on CVA the approximate answer to Q, denoted by $\tilde{A}(Q)$, can be obtained as follows: $\tilde{A}(Q) = \sum_{w=0}^{W-1} \frac{\| Q \cap b_w \|}{\| b_w \|} \cdot SUM(b_w)$, where SUM($b_w$) denotes the sum of (OLAP) data cells contained in b_w. To the sake of clarity, we highlight that when the CVA fails (i.e., the target data cube is characterized by *skewed*, i.e. asymmetric – e.g., see [11], distributions), *outliers* contained in $\mathcal{L}$ can seriously decrease the accuracy degree of $\tilde{A}(Q)$, as we demonstrate in [16], and also recognized in [8]. In this case, in [16] the approximate answer to Q is obtained by means of separately considering the contribution given by outliers involved by Q, and summing-up this contribution to the contribution given by classical linear interpolation. This imposes us to separately handling outliers, and originates a different formula for $\tilde{A}(Q)$, which, for the sake of simplicity, here can be modeled as: $\tilde{A}(Q) = \sum_{w=0}^{W-1} \left[\frac{\| Q \cap b_w \|}{\| b_w \|} \cdot SUM(b_w) + \mathrm{outlier}(Q \cap b_w) \right]$, where outlier($R$) is the set of outliers (of $\mathcal{L}$) involved by the region R.

The latter query evaluation scheme gives good performance when a *fixed*, *single* (range) query or, better, a workload *QWL* of queries *of the same nature and with similar geometrical characteristics* is considered, as we argue in [16], and, similarly, Bruno *et al.* in [6]. According to this experimental evidence, the compression process tries to generate buckets such that the final bucket ranges define a partition of the input data cube $\mathcal{L}$ that, as more as possible, accommodates *all* the queries of *QWL*, thus improving the accuracy of linear interpolation techniques [6,16]. To this end, the goal is two-fold: (*i*) obtaining buckets defined on uniform data (i.e., data with low variance), and (*ii*) minimizing the geometric difference between buckets and queries, in a global fashion. This provides a query performance better than the one given by computing the histogram without a fixed query-workload, as we demonstrate in [16]. The same approach could be extended in order to deal with the issue of evaluating a given *single* HRQ Q_H embedding several range queries. One can think of modeling the set of range queries embedded in Q_H as the target query-workload, and then adopting the same query evaluation scheme described above. Therefore, Q_H can be straightforwardly evaluated by iteratively evaluating range queries of its nodes, one at time. This is not conforming to the scope of this paper, which on the contrary considers the more challenging simultaneous evaluation of multiple HRQ against the target data cube.

It is worthy noticing that, in this case, linear interpolation techniques fail as ranges of queries of HRQ can be very different one from another, *at the same level as well as at different levels of the hierarchy of the target data cube* $\mathcal{L}$. This makes the previous query evaluation scheme inefficient, and requires ad-hoc solutions able to deal with the complexity and the "difficult" nature of HRQ, in a global fashion. According to the solution we propose in this paper, the final histogram-based compressed representation of $\mathcal{L}$, denoted by $\tilde{\mathcal{L}}$, is obtained as an *intermediate representation* given by *meaningfully partitioning the bucket domain defined by the different query ranges of HRQ at the same level* ℓ *of* $\mathcal{L}$, for *each* level ℓ of the $\mathcal{L}$ hierarchy. In fact, it is a matter to note that, in our specific OLAP scenario, an *arbitrary* compression of the data cube could easily origin the undesired situation in which some HRQ could take advantage from the compression process (i.e., retrieved approximate answers have a high degree of accuracy), as the final partition fits data and geometric properties of these HRQ, whereas some other HRQ could be disadvantaged (i.e., retrieved approximate answers have a low degree of accuracy), as the final partition does not fit the above-mentioned properties for such HRQ. As a consequence, our idea of generating an "intermediate" compressed representation of $\mathcal{L}$ makes sense, as, on the average, a *fair* final representation is obtained (i.e. retrieved approximate answers have acceptable/good accuracy *in most cases*). On the other hand, it should be noted that this approach has several conceptual points in common with other multiple-query data processing paradigms, such as those focusing on the view selection and materialization problem [31], where *common sub-expressions* of target queries are considered in order to obtain efficient solutions.

On a more practical plane, due to the hierarchical nature of cubes and queries, the compressed data cube $\tilde{\mathcal{L}}$ is implemented as a *hierarchical multidimensional histogram*, denoted by MQ-Hist($\mathcal{L}$), which is obtained by means of a *greedy algorithm*, namely `computeMultQHist`, that meaningfully exploits the multi-level and hierarchical nature of the input data cube $\mathcal{L}$, and defines a *top-down compression process* able to accommodate the different objectives of multiple, simultaneous HRQ against $\mathcal{L}$.

The remaining part of this paper is organized as follows. Sect. 2 describes our novel multi-query compression technique. In Sect. 3, we provide a wide experimental analysis of the proposed technique against both benchmark and real-life data cubes, and in comparison with state-of-the-art approaches. This analysis confirms the validity and the efficiency of our proposal. Finally, in Sect. 4 we derive conclusions of our work and define further research directions in this field.

2 Compressing Data Cubes under Simultaneous Multiple HRQ

We are given: (*i*) a *n*-dimensional data cube $\mathcal{L}$ having *P* hierarchical levels; (*ii*) the set of *m* HRQ that must be evaluated against $\mathcal{L}$ simultaneously, $S_{HRQ} = \{Q_{H_0}, Q_{H_1}, ..., Q_{H_{m-1}}\}$

(recall that, for the sake of simplicity, we assume that HRQ in S_{HRQ} have the same depth, P, which is also equal to the number of hierarchical levels of $\mathcal{L}$); (*iii*) the space bound $\mathcal{B}$ available for housing the compressed representation of $\mathcal{L}$, $\widetilde{\mathcal{L}}$, which is implemented by the histogram MQ-Hist($\mathcal{L}$). The multiple-query compression process we propose is accomplished according to the following multi-step approach:

- for each level ℓ of $\mathcal{L}$, such that ℓ belongs to $[0, P\text{-}1]$, starting from the bottom level 0 (i.e., according to a *bottom-up strategy*), generate, for each dimension d of $\mathcal{L}_\ell$, the *ordered union* of range bounds of queries (modeled as multidimensional points in the $\mathcal{L}_\ell$ ($\equiv \mathcal{L}$) multidimensional space) at level ℓ of HRQ in S_{HRQ} along d, thus obtaining the so-called *Multiple-Range* (MR) for the dimension d at level ℓ, denoted by $MR_{d,\ell}$ – first step finally generates, for each cuboid $\mathcal{L}_\ell$ of $\mathcal{L}$, the set of n MR, denoted by $\mathcal{MR}(\mathcal{L}_\ell) = \{MR_{0,\ell}, MR_{1,\ell}, \ldots, MR_{n\text{-}1,\ell}\}$;
- for each level ℓ, generate a *Generalized Partition* (GP) of the cuboid $\mathcal{L}_\ell$ at level ℓ, denoted by $\mathcal{G}_\ell(\mathcal{L})$, such that buckets in $\mathcal{G}_\ell(\mathcal{L})$ are obtained via (*i*) projecting, for each dimension d of $\mathcal{L}_\ell$, axes along points of MR in $\mathcal{MR}(\mathcal{L}_\ell)$, and (*ii*) storing the sum of items they contain – the collection of GP (one for each level ℓ of $\mathcal{L}$), denoted by MQ-Coll($\mathcal{L}$), constitutes the "sketch" of MQ-Hist($\mathcal{L}$), meaning that from MQ-Coll($\mathcal{L}$) we finally obtain MQ-Hist($\mathcal{L}$) according to our greedy algorithm `computeMulQHist` (described next);
- (algorithm `computeMulQHist`) for each level ℓ, obtain from $\mathcal{G}_\ell(\mathcal{L})$ of MQ-Coll($\mathcal{L}$) the so-called *Multiple-Query Partition* (MQP) of the cuboid $\mathcal{L}_\ell$, denoted by $\mathcal{P}_\ell(\mathcal{L})$, via *meaningfully merging buckets* in $\mathcal{G}_\ell(\mathcal{L})$ with the criterion that $\mathcal{P}_\ell(\mathcal{L})$ must be able to fit, *at level* ℓ, all the different query requirements of HRQ in S_{HRQ};
- return MQ-Hist($\mathcal{L}$) via hierarchically combining the P $\mathcal{P}_\ell(\mathcal{L})$.

From the described approach, it follows that the most important task of our technique is represented by algorithm `computeMulQHist`, whereas the other steps are quite obvious. For this reason, here we focus our attention on algorithm `computeMulQHist`.

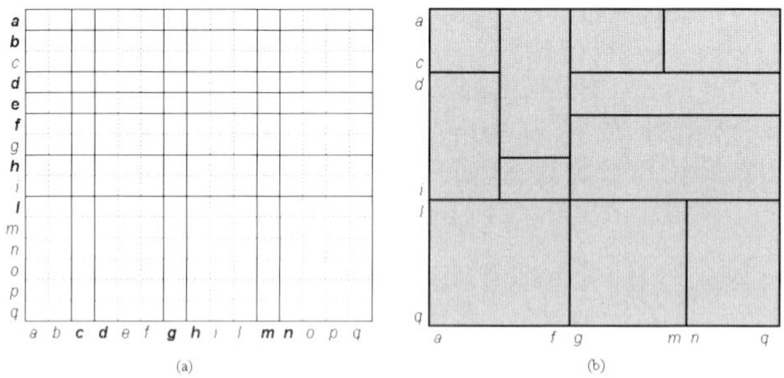

Fig. 1. An example of GP $\mathcal{G}_\ell(\mathcal{L})$ (a) and a possible corresponding $\mathcal{P}_\ell(\mathcal{L})$(b)

First, we further illustrate how MR, GP and MQP are obtained throughout a meaningful example. For the sake of simplicity, consider the following OLAP scenario (see Fig. 1 (a)): (*i*) a $|d_0| \times |d_1|$ two-dimensional cuboid $\mathcal{L}_\ell$, such that the domain of d_0 is $Dom(d_0) = \{a, b, c, d, e, f, g, h, i, l, m, n, o, p, q\}$, and the domain of d_1 is equal to that of d_0 (i.e., $Dom(d_1) = Dom(d_0)$), being the common-intended lexicographical ordering defined on both of these domains; (*ii*) three range queries, $Q_{i,\ell}$, $Q_{j,\ell}$, and $Q_{k,\ell}$ of *distinct* HRQ defined on $\mathcal{L}_\ell$, defined as follows: $Q_{i,\ell} = \langle[b, d], [c, f]\rangle$, $Q_{j,\ell} = \langle[d, e], [d, g]\rangle$, and $Q_{k,\ell} = \langle[h, i], [m, m]\rangle$. According to the third step of our multiple-query compression process, we have: (*i*) the two MR of $\mathcal{L}_\ell$ are: $\mathrm{MR}_{0,\ell} = \{b, d, e, h, i\}$ and $\mathrm{MR}_{1,\ell} = \{c, d, f, g, m\}$; (*ii*) the GP of $\mathcal{L}_\ell$, $\mathcal{G}_\ell(\mathcal{L})$, is that depicted in Fig. 1 (a); (*iii*) a possible MQP of $\mathcal{L}_\ell$, $\mathcal{P}_\ell(\mathcal{L})$, is that depicted in Fig. 1 (b).

From Fig. 1 (a), note that the so-generated buckets of $\mathcal{G}_\ell(\mathcal{L})$ are all the buckets that would allow us to provide, *at level* ℓ, approximate answers having the highest accuracy for *all* HRQ in S_{HRQ} at level ℓ. The same for the other GP of MQ-Coll($\mathcal{L}$). Obviously, it is not possible to materialize *all* the buckets of *all* the GP of MQ-Coll($\mathcal{L}$), due to the space constraint posed by the input bound $\mathcal{B}$. If this would be the case, we finally would obtain the histogram MQ-Hist($\mathcal{L}$) as corresponding to MQ-Coll($\mathcal{L}$) directly, i.e. as a *full materialization* of MQ-Coll($\mathcal{L}$). Being this impossible for reasonable configurations of the input parameters, we adopt the strategy of *obtaining the* MQP $\mathcal{P}_\ell(\mathcal{L})$ *via meaningfully merging buckets* in $\mathcal{G}_\ell(\mathcal{L})$, thus reducing the overall final size of $\mathcal{P}_\ell(\mathcal{L})$ and, as a consequence, the overall final size of MQ-Hist($\mathcal{L}$), keeping in mind the priority goal of accommodating, *as more as possible*, the multiple query constraints posed by HRQ in S_{HRQ}. As stated before, this strategy, which is implemented by algorithm `computeMulQHist`, allows us to finally compute $\mathcal{P}_\ell(\mathcal{L})$ as a *sub-optimal partition* of $\mathcal{G}_\ell(\mathcal{L})$.

To this end, `computeMulQHist` introduces a *global strategy* and a *local strategy*. The first one deals with the problem of how to explore the overall hierarchical search (bucket) space represented by MQ-Coll($\mathcal{L}$). The second one deals with the problem of how to explore the search (bucket) space represented by a given $\mathcal{G}_\ell(\mathcal{L})$.

First, consider the latter one, which properly realizes the main greedy criterion used to obtain a $\mathcal{P}_\ell(\mathcal{L})$ from a given $\mathcal{G}_\ell(\mathcal{L})$. In our approach, we adopt a strategy that is inspired by the one adopted by traditional multidimensional histograms (e.g., *MHist* [35]), i.e. *obtaining final buckets storing as more uniform data as possible via minimizing the skewness* among buckets themselves. This, in turn, has beneficial effects on the accuracy of approximate answers computed against the histogram, as widely recognized (e.g., see [11]). The difference with respect to our approach lies in the fact that traditional histograms operate on the original data cube directly, whereas MQ-Hist($\mathcal{L}$) is built starting from the bucket space defined by MQ-Coll($\mathcal{L}$), and, in turn, by each $\mathcal{G}_\ell(\mathcal{L})$ (with respect to this aspect, our proposal is reminiscent of *GenHist* histograms [22]). According to these guidelines, in the local computation of `computeMulQHist`, given the GP $\mathcal{G}_\ell(\mathcal{L})$, *we greedily select the most uniform bucket*, said b_U, among buckets of the overall bucket space of $\mathcal{G}_\ell(\mathcal{L})$, and then we explore the *neighboring buckets* of b_U in search for buckets having a homogeneity close to that of b_U, having fixed a threshold value V_U that limits the maximal difference between the homogeneity of b_U and that of its neighboring buckets. To meaningfully support this task, given a bucket b of $\mathcal{G}_\ell(\mathcal{L})$, we adopt as homogeneity definition the *greatest quadratic distance* from the average value of *outliers* in b, meaning that the less is such a distance, the more is the homogeneity of b. This approach is modeled by the function *unif*(b), defined as follows:

$$unif(b) = \frac{1}{\max\limits_{i \in Outl(b)} \{0, |b[i] - \text{AVG}(b)|^2\}} \tag{1}$$

such that (*i*) *Out*(b) is the set of outliers of b, (*ii*) $b[i]$ is the value of the i-th item of b, and (*iii*) AVG(b) is the average value of items in b. Note that we make use of the second power to ensure the convergence of (1). Outlier detection (e.g., [16]) is a non-trivial engagement. Without going into detail, in our approach this task can be implemented by means of any of the method proposed in literature (conventional or not), so that this aspect of our work is orthogonal to the main contribution. On the other hand, defining a "good" metrics of homogeneity of a given data domain (e.g., a bucket) is probing as well. Experience acquired by us in the context of outlier management for efficient query answering against compressed data cubes [16] suggests us the validity of function *unif*(•). Also, it should be noted that being the bucket merging process dependent on the threshold value V_U, different instances of this process can be easily tuned on the basis of particular application requirements.

When a set of neighboring buckets is detected and must be merged in a singleton bucket, said b_M, we impose the criterion of obtaining b_M as a *hyper-rectangular bucket* instead of an arbitrary bucket (i.e., a bucket with irregular shape). This reasonably follows the geometry of arbitrary range queries. In doing this, based on geometrical issues, we simply "throw away" protruding parts of merged neighboring buckets in such a way as to obtain a "maximal", "internal" hyper-rectangular bucket. In turn, the protruding bucket parts are then materialized as new buckets of the GP $\mathcal{G}_\ell(\mathcal{L})$, and then the same task is iterated again on the *remaining* bucket set of $\mathcal{G}_\ell(\mathcal{L})$. Fig. 2 shows some examples of the "throwing" strategy in the two-dimensional case (blocks with dashed borders represent protruding bucket parts). Comparing with *STHoles* [6], we observe that [6] proposes a "total" *hierarchical strategy* for merging buckets, whereas we propose a *planar strategy* for merging buckets, but applied to each level of our hierarchical search (bucket) space. Of course, the underlying greedy criterion is different in the two proposals.

For what regards the second aspect of `computeMulQHist` (i.e., the global strategy), we adopt an *in-depth visit* of MQ-Coll($\mathcal{L}$) starting from the aggregation ALL (i.e., from the corresponding GP at level P, $\mathcal{G}_P(\mathcal{L})$), meaning that, starting from the level P (i.e., the cuboid $\mathcal{L}_P$), when a merged bucket b_M is obtained in the GP $\mathcal{G}_\ell(\mathcal{L})$ at level ℓ, we hierarchically *move down* to the GP $\mathcal{G}_{\ell+1}(\mathcal{L})$ at level $\ell + 1$, and consider the collection of buckets of $\mathcal{G}_{\ell+1}(\mathcal{L})$ contained by b_M, and so forth. When the leaf level GP is reached (i.e., the cuboid $\mathcal{L}_0$), we re-start from the aggregation ALL. As said before, the whole process is bounded by the consumption of the storage space $\mathcal{B}$.

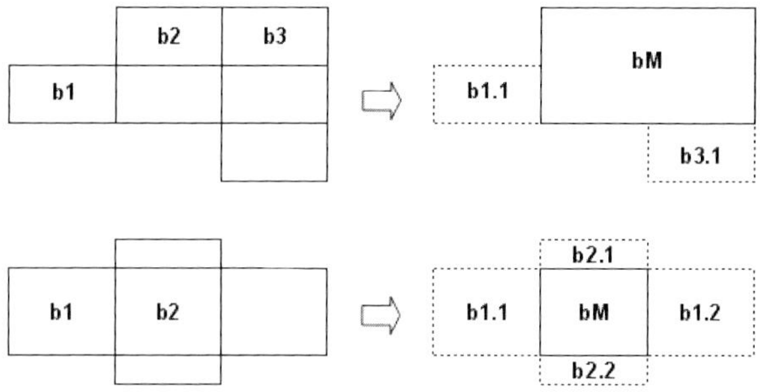

Fig. 2. Merging two-dimensional buckets (dashed blocks represent protruding bucket parts)

Basically, the above-described one defines a *top-down* approach in the computation of MQ-Hist($\mathcal{L}$). Without going into details, it should be noted that the in-depth visit approach is in favor of the idea of accommodating a large family of range queries embedded in HRQ rather than range queries referred to a *particular* level of the logical hierarchy underlying the input data cube $\mathcal{L}$. The rationale of this way to do comes from arguing that, in a typical OLAP scenario, client applications are mainly interested in querying OLAP data at granularities different from the lowest one [23]. To become convinced of this, consider a repository $\mathcal{R}$ of census data (which are particularly suitable to be processed and analyzed by means of OLAP technology) and a data cube $\mathcal{L}$ defined on top of $\mathcal{R}$ such that $\mathcal{L}$ includes, among all, the dimension *Time* with hierarchy $\mathcal{H}_{Time}$. Also, suppose that $\mathcal{H}_{Time}$ is organized as follows: *Year* $\rightarrow$ *Quarter* $\rightarrow$ *Month* $\rightarrow$ *Day*. Even if census data are available in *all* the defined temporal granularities (i.e., *Year*, *Quarter*, *Month*, and *Day*), OLAP client applications typically access and query data at the *Month* or *Quarter* or *Year* granularities mostly [23]. This evidence, combined with (*i*) the need of accommodating a large family of queries, and (*ii*) the presence of a bounded storage space (i.e., $\mathcal{B}$), gives raise to our top-down approach.

Finally, it should be noted that MQ-Hist($\mathcal{L}$) implicitly defines a *multi-level R-tree based partitioning scheme* of the input data cube $\mathcal{L}$, with the novelty that bounds of rectangular buckets at different levels are obtained by meaningfully handling and reasoning on the multiple-query scheme imposed by the OLAP scenario investigated in this paper. This data engineering model is perfectly able to capture the hierarchical nature of the compression process of $\mathcal{L}$, and, at the same time, efficiently answer SUM-based HRQ, which, as stated in Sect. 1, are the class of queries of interest for our research work. Obviously, this approach can be straightforwardly extended as to deal with different classes of HRQ, i.e. HRQ based on different kinds of SQL aggregate operators, with slight modifications of the approximate query answering engine. Also, note that the MQ-Hist($\mathcal{L}$) approach is in opposition to other proposals where the data cube partition is obtained according to a pre-fixed scheme, such as quad-tree based partitions [7], which look at the data domain of the data cube rather than to its *information content* in terms of schemas (e.g., data cube logical schemas) and models (e.g., OLAP queries and hierarchies).

3 Experimental Analysis

In order to test the effectiveness of our proposed technique, we defined experiments aiming at probing the data cube *compression performance* (or, equally, the *accuracy*) of our technique. Our experiments involve several aspects ranging from the data layer to query layer, metrics, and comparison techniques.

Data Layer. As regards the data layer of our experimental framework, we engineered two different kinds of data cubes. The usage of different classes of data cubes allowed us to submit our proposed technique to a comprehensive and "rich" experimental analysis, and, as a consequence, carefully test its performance. Data cube classes we considered are the following: (*i*) *benchmark data cubes*, which allow us to test the effectiveness of our technique under the stressing of an in-laboratory-built input, and to evaluate our technique against competitor ones on "well-referred" data sets that have been widely used in similar research experiences; (*ii*) *real data cubes*, which allow us to probe the efficiency of our technique against real-life data sets. We also conducted experiments on *synthetic data cubes*, which allow us to completely control the variation of input parameters such as the nature of OLAP data distributions. However, for space reasons, we show experimental results on benchmark and real-life data cubes, which, typically, are more probing than synthetic ones.

For what regards benchmark data sets, we considered the popular benchmark *TPC-H* [39], which is well-known in the Database and Data Warehousing research communities. By exploiting data generation routines made available at the benchmark Web site, we built a two-dimensional benchmark data cubes by means of the well-known *Microsoft Analysis Services 2000* OLAP platform. In more detail, we built a 2,000 × 2,000 two-dimensional data cube populated with 4M data cells. For what regards real-life data sets, we considered the popular data set *USCensus1990* [40], and we built a 1,000 × 1,000 two-dimensional data cube, still on top of Microsoft Analysis Services 2000.

Query Layer. As regards the input of our experimental study, we considered *random populations* of synthetic HRQ with range-SUM queries, modeled in our experimental framework by the set SQ_H. For these queries, the underlying trees have also been obtained randomly with the constraint of "covering" the different cuboids of the target data cube *as more as possible* (i.e., obtaining GP having a *large* number of buckets) at, however, a "reasonable" granularity. *Query Selectivity* ||•||, which for OLAP queries is totally equal to the query volume of the range-SUM queries, is the control parameter used to cost the complexity needed to evaluate queries in SQ_H (e.g., see [13]).

Metrics. As regards the outcomes of our experimental study, given a population of synthetic HRQ SQ_H, we introduce the *Average Relative Error* (ARE) between exact and approximate answers to queries in SQ_H, defined as follows: $\overline{\varepsilon}(SQ_H) = \frac{1}{|SQ_H|-1} \cdot \sum_{j=0}^{|SQ_H|-1} \overline{\varepsilon}(Q_{H_j})$, such that (*i*) $\overline{\varepsilon}(Q_{H_j})$ is defined as follows:

$\bar{\varepsilon}(Q_{H_j}) = \frac{1}{P} \cdot \sum_{\ell=0}^{P-1} \left[\frac{1}{|\text{queries }(\ell)|-1} \sum_{k=0}^{|\text{queries }(\ell)|-1} \varepsilon(Q_k) \right]$, and (*ii*) $\varepsilon(Q_k)$ is the *Relative Error* (RE) of the range-SUM query Q_k in Q_{H_j}, defined as follows: $\varepsilon(Q_k) = \frac{|A(Q_k) - \tilde{A}(Q_k)|}{A(Q_k)}$.

Comparison Techniques. In our experimental study, we compared the performance of our proposed technique against the following well-known histogram-based techniques for compressing data cubes: *MinSkew* by Acharya *et al.* [1], *GenHist* by Gunopulos *et al.* [22], and *STHoles* by Bruno *et al.* [6]. This because these techniques are similar to ours, and also represent the state-of-the-art for histogram-based data cube compression research. In more detail, having fixed the space budget $\mathcal{B}$ (i.e., the storage space available for housing the compressed data cube), we derived, for each comparison technique, the configuration of the input parameters that respective authors consider the best in their papers. This ensures a *fair* experimental analysis, i.e. an analysis such that each comparison technique provides its *best* performance. Furthermore, for all the comparison techniques, we set the space budget $\mathcal{B}$ as equal to the r % of *size*($\mathcal{L}$), being r the *compression ratio* and *size*($\mathcal{L}$) the total occupancy of the input data cube. As an example, r = 10 % (i.e., $\mathcal{B}$ is equal to the 10 % of *size*($\mathcal{L}$)) is widely recognized as a reasonable experimental setting (e.g., see [6]).

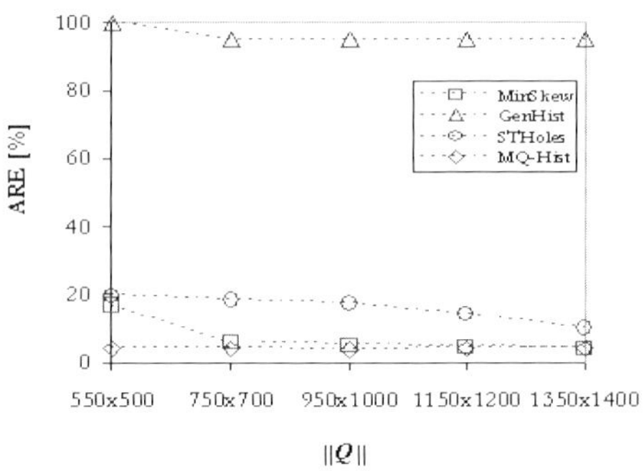

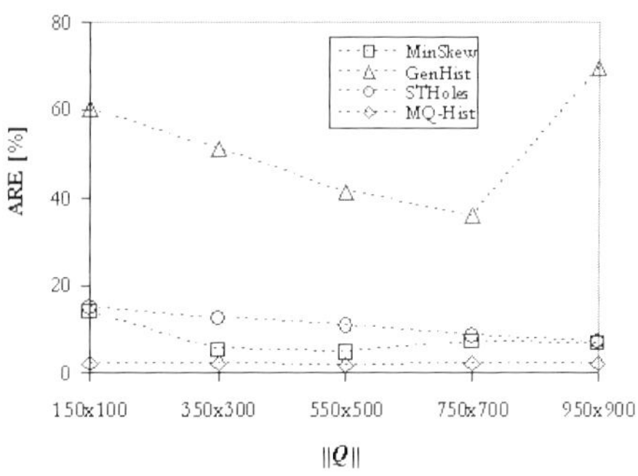

Fig. 3. ARE vs query selectivity ||Q|| on the benchmark data cube *TPC-H* (left) and on the real life data cube *USCensus1990* (right)

Experimental Results. Fig. 3 shows experimental results related to the percentage variation of ARE on both benchmark and real-life data cubes with respect to the selectivity of queries in SQ_H. This allows us to measure the *quality* of compression techniques, i.e. their capability of introducing low query approximation errors. Fig. 4 shows the results for the same experiment when ranging r on the interval [5, 20] (i.e., $\mathcal{B}$ on the interval [5, 20] % of *size*($\mathcal{L}$)), and fixing the selectivity of range-SUM queries to ||Q|| = 750 × 700 for the benchmark data cube *TPC-H*, and to ||Q|| = 350 × 300 for the real-life data cube *USCensus1990*. This allows us to measure the *scalability* of compression techniques, which is a critical aspect in approximate OLAP query answering engines (e.g., see [11]).

From the analysis of the set of experimental results on two-dimensional benchmark and real-life data cubes, it follows that compression performance of $\mathcal{MO}$-Hist($\mathcal{L}$) outperforms those of comparison techniques. Also, our proposed technique ensures a better scalability with respect to that of comparison techniques when ranging the size of the storage space $\mathcal{B}$ available for housing the compressed representation of the input data cube.

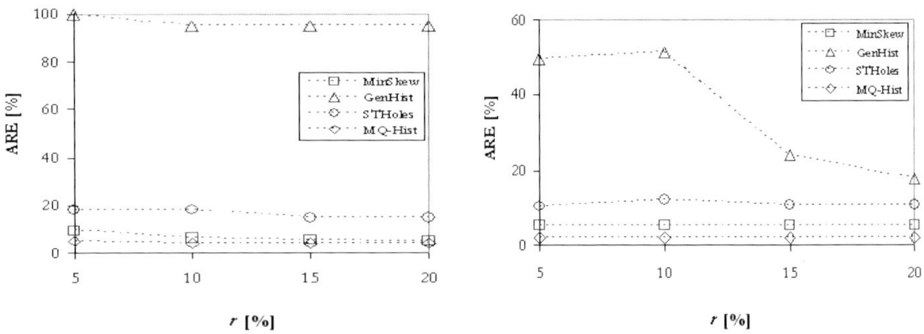

Fig. 4. ARE vs compression ratio r on the benchmark data cube *TPC-H* with $||Q|| = 750 \times 700$ (left) and on the real-life data cube *USCensus1990* with $||Q|| = 350 \times 300$ (right)

4 Conclusions and Future Work

In this paper, we introduced a novel computational paradigm for OLAP, the multi-objective data cube compression paradigm, which deals with the problem of compressing data cubes according to multiple constraints rather than only one like in traditional schemes. We demonstrated how this paradigm poses novel and previously-unrecognized challenges for next-generation OLAP scenarios. Inspired by this exciting, new line of research, we proposed an effective solution for a significant instance of the problem underlying such a paradigm, i.e. the issue of compressing data cubes in the presence of multiple, simultaneous HRB, a very useful OLAP tool for extracting hierarchically-shaped summarized knowledge from data warehouse serves. In this respect, we provided an efficient greedy algorithm for obtaining the so-called multiple-query histogram via meaningfully exploiting the hierarchical nature of cubes and queries by means of an innovative top-down approach. Finally, a wide experimental analysis on benchmark and real-life data cubes clearly demonstrated the benefits deriving from the application of our proposed technique, beyond the capabilities of state-of-the-art histogram-based data cube compression techniques.

Future work is oriented towards the following two main directions: (*i*) extending the multiple-query data cube compression framework we propose as to deal whit more complex data warehouse schemas such as *multi-measure data cubes* and *fact constellations* [23], beyond conventional data cubes like those investigated in this paper; (*ii*) extending the multiple-query data cube compression framework we propose as to deal whit OLAP aggregations different than SUM; (*iii*) studying how the approximate query answering results for multiple, simultaneous OLAP queries achieved with the proposed framework can be meaningfully exploited to handle the related problem on selectivity estimation of (multiple, simultaneous) OLAP queries; (*iv*) dealing with the relevant research challenge represented by making our framework able to handle *multi-versioning data warehouse schemas*, i.e. how to drive the multiple-query compression process in the presence of schema updates.

References

1. Acharya, S., Poosala, V., Ramaswamy, S.: Selectivity Estimation in Spatial Databases. ACM SIGMOD (1999)
2. Agrawal, R., Wimmers, E.L.: A Framework for Expressing and Combining Preferences. ACM SIGMOD (2000)
3. Balke, W.-T., Güntzer, U.: Multi-Objective Query Processing for Database Systems. VLDB (2004)
4. Börzsönyi, S., Kossmann, D., Stocker, K.: The Skyline Operator. IEEE ICDE (2001)
5. Bowman, I.T., Salem, K.: Optimization of Query Streams using Semantic Prefetching. ACM TODS 30(4) (2005)
6. Bruno, N., Chaudhuri, S., Gravano, L.: STHoles: A Multidimensional Workload-Aware Histogram. ACM SIGMOD (2001)
7. Buccafurri, F., et al.: A Quad-Tree based Multiresolution Approach for Two-Dimensional Summary Data. IEEE SSDBM (2003)
8. Chaudhuri, S., et al.: Overcoming Limitations of Sampling for Aggregation Queries. IEEE ICDE (2001)
9. Chen, Z., Narasayya, V.: Efficient Computation of Multiple Group By Queries. ACM SIGMOD (2005)
10. Colliat, G.: OLAP, Relational, and Multidimensional Database Systems. ACM SIGMOD Record 25(3) (1996)
11. Cuzzocrea, A.: Overcoming Limitations of Approximate Query Answering in OLAP. IEEE IDEAS (2005)
12. Cuzzocrea, A.: Providing Probabilistically-Bounded Approximate Answers to Non-Holistic Aggregate Range Queries in OLAP. ACM DOLAP (2005)
13. Cuzzocrea, A.: Improving Range-Sum Query Evaluation on Data Cubes via Polynomial Approximation. Data & Knowledge Engineering 56(2) (2006)
14. Cuzzocrea, A.: Accuracy Control in Compressed Multidimensional Data Cubes for Quality of Answer-based OLAP Tools. IEEE SSDBM (2006)
15. Cuzzocrea, A., et al.: Semantics-aware Advanced OLAP Visualization of Multidimensional Data Cubes. International Journal of Data Warehousing and Mining 3(4) (2007)
16. Cuzzocrea, A., Wang, W.: Approximate Range-Sum Query Answering on Data Cubes with Probabilistic Guarantees. Journal of Intelligent Information Systems 28(2) (2007)
17. Doan, A., Levy, A.Y.: Efficiently Ordering Plans for Data Integration. IEEE ICDE (2002)
18. Fan, J., Kambhampati, S.: Multi-Objective Query Processing for Data Aggregation. ASU CSE TR (2006)
19. Fang, M., et al.: Computing Iceberg Queries Efficiently. VLDB (1998)
20. Garofalakis, M.N., Gibbons, P.B.: Wavelet Synopses with Error Guarantees. ACM SIGMOD (2002)

21. Garofalakis, M.N., Kumar, A.: Deterministic Wavelet Thresholding for Maximum-Error Metrics. ACM PODS (2004)
22. Gunopulos, D., et al.: Approximating Multi-Dimensional Aggregate Range Queries over Real Attributes. ACM SIGMOD (2000)
23. Han, J., Kamber, M.: Data Mining: Concepts and Techniques. Morgan Kauffmann Publishers, San Francisco (2000)
24. Ho, C.-T., et al.: Range Queries in OLAP Data Cubes. ACM SIGMOD (1997)
25. Ioannidis, Y., Poosala, V.: Histogram-based Approximation of Set-Valued Query Answers. VLDB (1999)
26. Ives, Z.G., et al.: An Adaptive Query Execution System for Data Integration. ACM SIGMOD (1999)
27. Jin, R., et al.: A Framework to Support Multiple Query Optimization for Complex Mining Tasks. MMDM (2005)
28. Jin, R., et al.: Simultaneous Optimization of Complex Mining Tasks with a Knowledgeable Cache. ACM SIGKDD (2005)
29. Kalnis, P., Papadias, D.: Multi-Query Optimization for On-Line Analytical Processing. Information Systems 28(5) (2003)
30. Koudas, N., et al.: Optimal Histograms for Hierarchical Range Queries. ACM PODS (2000)
31. Mistry, H., et al.: Materialized View Selection and Maintenance using Multi-Query Optimization. ACM SIGMOD (2001)
32. Nie, Z., Kambhampati, S.: Joint Optimization of Cost and Coverage of Query Plans in Data Integration. ACM CIKM, 223–230 (2001)
33. Papadias, D., et al.: An Optimal and Progressive Algorithm for Skyline Queries. ACM SIGMOD (2003)
34. Poosala, V., Ganti, V.: Fast Approximate Answers to Aggregate Queries on a Data Cube. IEEE SSDBM (1999)
35. Poosala, V., Ioannidis, Y.: Selectivity Estimation Without the Attribute Value Independence Assumption. VLDB (1997)
36. Roy, P., et al.: Efficient and Extensible Algorithms for Multi-Query Optimization. ACM SIGMOD (2000)
37. Sellis, T., Ghosh, S.: On the Multiple-Query Optimization Problem. IEEE TKDE 2(2) (1990)
38. Sellis, T.: Multiple-Query Optimization. ACM TODS 13(1) (1988)
39. Transaction Processing Council: TPC Benchmark H. (2006), `http://www.tpc.org/tpch/`
40. University of California, Irvine: 1990 US Census Data (2001), `http://kdd.ics.uci.edu/databases/census1990/USCensus1990.html`
41. Wang, S., et al.: State-Slice: A New Paradigm of Multi-Query Optimization of Window-Based Stream Queries. VLDB (2006)
42. Xin, D., et al.: Answering Top-k Queries with Multi-Dimensional Selections: The Ranking Cube Approach. VLDB (2006)

Degrees of Exclusivity in Disjunctive Databases

Navin Viswanath and Rajshekhar Sunderraman

Georgia State University,Atlanta, GA 30302
nviswanath1@student.gsu.edu, raj@cs.gsu.edu

Abstract. Ever since the emergence of the relational database model, the problem of incompleteness has been studied extensively. Declarative semantics for disjunctive databases widely adopt two methodologies representing opposite ends of the spectrum - Minker's GCWA interprets disjunctions exclusively while Ross and Topor's DDR interprets them inclusively. We argue that the use of either one as the semantics is limiting. For example, in a blood group relation, saying that the blood group of John is A or B, $BG(John, A) \vee BG(John, B)$ can be interpreted exclusively while the statement supplier S1 supplies part P1 or P2 to project J1, $supplies(S1, P1, J1) \vee supplies(S1, P2, J1)$ need not necessarily be exclusive. In this paper, we present a model that allows the nature of the disjunctions to be represented explicitly and show its relation to other semantics for disjunctive databases. A notable feature of this extension is that it does not require variables in order to represent indefinite information.

1 Introduction

Consider the example of a database which maintains the time it takes (in hours) to travel between two cities by road. There might be uncertainty in the amount of time required to travel between any two cities as collected from various sources or there might be uncertainty concerning the source and destination cities itself. Consider an instance of such a relation, TRAVEL = $\{(C1, C2, 4), (C2, C3, \{3, 4\}), (\{C1, C3\}, C4, 2)\}$. Here the uncertainty is represented as a set of values, one or more of which are true, depending on whether the disjunctions are to be interpreted inclusively or exclusively. The second tuple represents the information that the time of travel between cities C2 and C3 is either 3 or 4 hours. It is obvious from the context that this disjunction is to be interpreted exclusively. Similarly, the third tuple indicates that either the time of travel between C1 and C4 or C3 and C4 is 2 hours. Such a disjunction is not necessarily exclusive. It may be the case that the time of travel is 2 hours for both pairs.

Disjunctive information has been studied in [1,2,4,6,9,10,11,12,13,14,15]. In this paper, we present an extension to the relational model that can represent various degrees of exclusivity of disjunctive data. The data structure we introduce is called an *oa-table*. An *oa-table* is a set of *oa-tuples*. An *oa-tuple* is a set of tuple sets. Each tuple set in an *oa-tuple* will be interpreted as a conjunction of tuples and an *oa-tuple* is a disjunction of tuple sets. Informally, an *oa-table* is a conjunction of disjunctions of conjuncts. For any predicate symbol R, an *oa-tuple* is of the form $\eta_1 \vee \eta_2 \vee, \ldots, \vee \eta_m$. Here η_i is a conjunction of tuples $t_{i1} \wedge t_{i2} \wedge, \ldots, \wedge t_{ik_i}$. Thus the disjunction $\eta_1 \vee \eta_2 \vee, \ldots, \vee \eta_m$ is to be viewed as the indefinite statement $(R(t_{11}) \wedge \ldots \wedge R(t_{1k_1})) \vee \ldots \vee (R(t_{m1}) \wedge \ldots \wedge$

A. An et al. (Eds.): ISMIS 2008, LNAI 4994, pp. 375–380, 2008.

$R(t_{mk_m})$). An *oa-table* reduces to an ordinary relation when for each *oa-tuple* $m = 1$ and for each $\eta_i, k_i = 1$.

The TRAVEL instance can be represented as an *oa-table* as follows:

$$\text{TRAVEL} = \{\{\{(C1, C2, 4)\}\}, \{\{(C2, C3, 3)\}, \{(C2, C3, 4)\}\}, \\ \{\{(C1, C4, 2)\}, \{(C3, C4, 2)\}, \{(C1, C4, 2), (C3, C4, 2)\}\}\}.$$

Let w_1, w_2 and w_3 be the three *oa-tuples* in the *oa-table* representation of $TRAVEL$ shown in Figure 2. Thus the instance of the relation $TRAVEL$ represents the formula $F1 \wedge F2 \wedge F3$ where,

$$\begin{aligned} F1 &= TRAVEL(C1, C2, 4) \text{ corresponding to } w_1 \\ F2 &= (TRAVEL(C2, C3, 3) \vee TRAVEL(C2, C3, 4)) \wedge \\ &\quad (\neg TRAVEL(C2, C3, 3) \vee \neg TRAVEL(C2, C3, 4)) \text{ corresponding to } w_2 \\ F3 &= TRAVEL(C1, C4, 2) \vee TRAVEL(C3, C4, 2) \text{ corresponding to } w_3. \end{aligned}$$

2 oa-Tables

In this section we formally define *oa-tables*. We discuss the notions of redundancy in an *oa-table* and present a formal definition of the information content of an *oa-table* .

A *domain* is a finite set of values. The cartesian product of domains $D_1, D_2, \ldots, D_n$ is denoted by $D_1 \times D_2 \times, \ldots, \times D_n$ and is the set of all tuples $< a_1, \ldots, a_n >$ such that for any $i \in \{1, \ldots, n\}, a_i \in D_i$. An *oa-table scheme* is an ordered list of attribute names $R =< A_1, \ldots, A_n >$. Associated with each attribute name, A_i, is a domain D_i. Then, T is an *oa-table* over the scheme R where $T \subseteq 2^{2^{D_1 \times D_2 \times, \ldots, \times D_n}}$.

An *oa-table* is a *non-empty set of oa-tuples*. An *oa-tuple* $w = \{\eta_1, \eta_2, \ldots, \eta_m\} \in 2^{2^{D_1 \times D_2 \times, \ldots, \times D_n}}, m \geq 1$ represents the formula $\eta_1 \vee \eta_2 \vee \ldots \vee \eta_m$ where $\eta_i = \{t_{i1}, t_{i2}, \ldots, t_{in}\} \in 2^{D_1 \times D_2 \times, \ldots, \times D_n}, n \geq 0$ represents the formula $R(t_{i1}) \wedge R(t_{i2}) \wedge \ldots \wedge R(t_{in})$. Note that we use the set notation and figures alternately to represent *oa-tables* due to space constraints. When the relation scheme has a single attribute, we ignore the () for each tuple for clarity. Thus, $\{(a)\}$ will be written simply as $\{a\}$.

It is possible for *oa-tables* to represent redundant information. Here, we characterize the various types of redundancies and define a $REDUCE$ operator. The mapping $REDUCE$ is used to eliminate redundant information from an *oa-table* . The following redundancies may be present in an *oa-table*:

1. $w \in T$ and $w' \in T$ and $w \subset w'$ and $\neg\exists\eta' \in w' - w$ and $\eta \subset \eta'$ and $\eta \in w$. This redundancy is eliminated by removing w' from T.
2. $w \in T$ and $w' \in T$ and $w \subset w'$ and $\exists\eta' \in w' - w$ and $\eta \subset \eta'$ and $\eta \in w$. This redundancy is eliminated by removing w from T and setting $w' = w' - \eta''$ where $(\exists\eta \in w)\eta \cap \eta'' = \varnothing$.

Given a scheme R, we define Γ_R and Σ_R as follows:

$\Gamma_R = \{T \mid T$ is an *oa-table* over $R\}$ and $\Sigma_R = \{U \mid U$ is a set of relations over $R\}$

Let T be an *oa-table* on scheme R. Then, $REDUCE(T) : \Gamma(R) \to \Gamma(R)$ is a mapping such that $REDUCE(T) = T^0$ where T^0 is defined as follows:

$$\begin{aligned} T^0 = \{w = \{\eta_1, \eta_2, \ldots, \eta_n\} \mid (w \in T) \wedge \neg(\exists w' \in T) \wedge w' \supset w \wedge \\ (\forall\eta' \in w' - w)(\neg\exists\eta \in w)\eta \cap \eta' \neq \varnothing \wedge \\ \neg(\exists w'' \in T) \wedge w'' \subset w\} \end{aligned}$$

Below is an example of $REDUCE$.

$$T = \{\{\{a\}\}, \{\{a\}, \{b\}, \{c\}\}, \{\{a\}, \{d\}, \{a, e\}\}\}$$
$$REDUCE(T) = \{\{\{a\}, \{a, e\}\}\}$$

Let $T = \{w_1, w_2, \ldots, w_n\}$ be a consistent *oa-table*. Let us denote by $atoms(w_i)$, the set of all tuples in an *oa-tuple* w_i and by $\mathcal{P}(atoms(w_i))$, its powerset. Then, $NEG(T) = \cup_{i=1}^{n} \mathcal{P}(atoms(w_i)) - w_i$. The information content of T is given by a mapping $REP : \Gamma_R \rightarrow \Sigma_R$, which is defined as follows:

$$REP(T) = M(REDUCE(T))$$

where $M(T) : \Gamma_R \rightarrow \Sigma_R$ is a mapping such that

$$M(T) = \{\{\eta_1, \eta_2, \ldots, \eta_n\} \mid (\forall i)(1 \leq i \leq n \rightarrow (\eta_i \in w_i) \wedge \neg(\exists u \in NEG(T))(u \subseteq \{\eta_1, \eta_2, \ldots, \eta_n\}))\}$$

Let $T = \{\{\{a\}, \{b\}\}, \{\{b\}, \{c\}\}\}$. Choosing a from w_1 forces us to choose c from w_2 since choosing a automatically negates b in that possible world. Hence $\{a, c\}$ is a possible world. Similarly, choosing b from w_1 forces choosing b from w_2 as well since we are assuming b to be true and c to be false in this world. Hence the second possible world is just $\{b\}$. Thus $M(T) = \{\{a, c\}, \{b\}\}$.

3 Related Work

In this section, we explore a few other disjunctive database models and compare them with *oa-tables*. The *oa-table* model is an extension of the I-tables introduced in [5,6]. I-tables fail to represent certain kinds of incomplete information. Consider the following instances of an indefinite database:

I1 : {(C1,C2,3)} and I2 : {(C1,C2,3),(C2,C3,4),(C3,C4,1)}

This set of instances does not have a corresponding I-table representation. The *oa-table* representation for the above instances is $\{\{\{(C1, C2, 3)\}\}, \{\{(C2, C3, 4), (C3, C4, 1)\}, \varnothing\}\}$.

Theorem 1. *The oa-table model is complete.*

Proof. Any indefinite database D is a set of instances $\{I_1, I_2, \ldots, I_n\}$ exactly one of which is the real world truth. Since each instance of a database is a set of tuples, the *oa-table* with the single *oa-tuple* $\{I_1, I_2, \ldots, I_n\}$ would represent D. □

Another extension of the I-tables is the E-tables discussed by Zhang and Liu in [15]. This model deals with exclusively disjunctive information. Apart from the tuples in $D_1, D_2, \ldots, D_n$, a set of dummy values $\varepsilon_1, \varepsilon_2, \ldots, \varepsilon_n$ may also be present as tuples in a relation and they are defined as $\{\varepsilon_i\} = \varnothing$. Also, $\varepsilon_i \neq \varepsilon_j$ where $i \neq j$. An E-table is a set composed of sets of tuples sets and exactly one element (tuple set) is true in a set of tuple sets making the disjunctions exclusive. This allows the representation of various forms of exclusive disjunctions in E-tables as shown in Figure 1. Here, T_1 consists of a single tuple set denoting the fact that either $T_1(a)$ or $T_1(b)$ is true (but not both). E-table T_2 contains dummy values ε_1 and ε_2 which denote dummy 'empty' tuples and

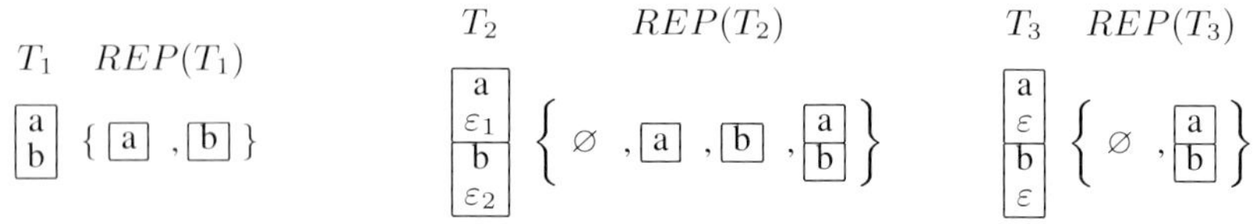

Fig. 1. E-table representations

this allows the four possible combinations shown in $REP(T_2)$. T_3 contains two tuple sets both of which contain the same dummy value ε. Choosing ε from the first tuple set forces choosing ε from the second one too since the disjunctions in tuple sets are exclusive. This allows only two possible real world scenarios as shown in $REP(T_3)$. Since the E-tables were defined to represent exclusive disjunctions, the case where the disjunction might be inclusive is not representable in E-tables. For example, purely inclusive disjunctions do not have a corresponding E-table representation. *oa-tables*, on the other hand, can be used to represent various degrees of exclusivity. For example, $REP(T_1), REP(T_2)$ and $REP(T_3)$ of Figure 1 can be captured in *oa-tables* $S1 = \{\{\{a\},\{b\}\}\}, S2 = \{\{\{a\},\varnothing\},\{\{b\},\varnothing\}\}, S3 = \{\{\{a,b\},\varnothing\}\}$ respectively and the purely inclusive case by $S4 = \{\{\{a\},\{b\},\{a,b\}\}\}$.

A third related formalism is the one discussed by Sarma et al. in [13]. That system uses a two-layer approach in which the incomplete model at the top layer has an underlying complete model. Our interest is limited to the complete model described there since the incomplete models are obtained by simply putting restrictions on the complete model. This approach, although more intuitive than the *c-tables* in [4], still introduces boolean formulas and variables. *oa-tables*, on the other hand, are variable-free and complete.

4 Relational Algebra

In this section, we define the operators of the extended relational algebra on Γ_R. We place a dot above the symbol for the operators in order to differentiate them from the standard relational operators. We also state theorems regarding the correctness of the relational operators.

Selection: Let T' be a consistent *oa-table* and F be a formula involving operands that are constants or attribute names, arithmetic comparison operators: $<, =, >, \leq, \geq, \neq$ and logical operators $\wedge, \vee, \neg$. Then $\dot{\sigma}_F(T') = REDUCE(T)$ where,

$$T = \{\{\eta_1, \eta_2, \ldots, \eta_m\} \mid w' \in T' \wedge (\forall i,j)((\eta'_i \in w' \wedge t_{ij} \in \eta'_i) \rightarrow (\eta_i = \cup_j t_{ij} \mid F(t_{ij})))\}$$

Selection in an *oa-table* is done by deleting tuples that do not satisfy the selection condition from each set of tuples.

Theorem 2. *$REP(\dot{\sigma}_F(T)) = \sigma_F(REP(T))$ for any reduced* oa-table *T.*

Projection: Projection on *oa-tables* is defined as a mapping $\dot{\pi}_A : \Gamma_R \rightarrow \Gamma_A$ as follows: Let T' be a consistent *oa-table* and $A \subseteq R$. Then, $\dot{\pi}_A(T') = REDUCE(T)$ where,

$$T = \{\{\eta_1, \eta_2, \ldots, \eta_m\} \mid w' \in T' \wedge (\forall i,j) ((\eta'_i \in w' \wedge t_{ij} \in \eta'_i) \rightarrow (\eta_i = \cup_j t_{ij}[A]))\}$$

The projection operation on an *oa-table* is simply the projection of each tuple from each tuple set in an *oa-tuple*.

Theorem 3. *$REP(\dot{\pi}_A(T)) = \pi_A(REP(T))$ for any reduced oa-table T and list of attributes A.*

Cartesian product: Let T_1 and T_2 be consistent *oa-tables* on schemes R_1 and R_2 respectively. Then, the cartesian product of T_1 and T_2, $T_1 \dot{\times} T_2 = REDUCE(T)$ where,

$$T = \{\{\eta_{11}, \eta_{12}, \ldots, \eta_{mn}\} \mid (\forall w_1 \in T_1)(\forall w_2 \in T_2) \\ ((\exists \eta_i \in w_1 \wedge \exists \eta_j \in w_2) \rightarrow (\eta_{ij} = (\cup \eta_i \times \cup \eta_j)))\}$$

The cartesian product of two *oa-tables* is obtained by taking the cartesian product of tuple sets in each *oa-tuple*.

Theorem 4. *$REP(T_1 \dot{\times} T_2) = REP(T_1) \times REP(T_2)$ for any reduced oa-tables T_1 and T_2.*

Union: Let T_1 and T_2 be consistent *oa-tables* on scheme R. Then, the union of T_1 and T_2, $T_1 \dot{\cup} T_2 = REDUCE(T)$ where,

$$T = \{w \mid w \in T_1 \vee w \in T_2\}$$

The union of two *oa-tables* is simply the set of *oa-tuples* that are present in either one of the *oa-tables*.

Theorem 5. *$REP(T_1 \dot{\cup} T_2) = REP(T_1) \cup REP(T_2)$ for any reduced oa-tables T_1 and T_2.*

Intersection: Let T_1 and T_2 be two domain compatible consistent *oa-tables*. Then, $T_1 \dot{\cap} T_2 = REDUCE(T)$ where,

$$T = \{\{\eta_{11}, \eta_{12}, \ldots, \eta_{mn}\} \mid (\forall w_1 \in T_1)(\forall w_2 \in T_2) \\ ((\exists \eta_i \in w_1 \wedge \exists \eta_j \in w_2) \rightarrow (\eta_{ij} = (\cup \eta_i \cap \cup \eta_j)))\}$$

The intersection of two *oa-tables* is the intersection of the tuple sets of the *oa-tuples* of each *oa-table*.

Theorem 6. *$REP(T_1 \dot{\cap} T_2) = REP(T_1) \cap REP(T_2)$ for any reduced oa-tables T_1 and T_2.*

For the sake of brevity, theorems 4.1-4.5 are stated without proof.

5 Conclusion and Future Work

Viewing a relational database as a model of a first order theory, although elegant, has been found to be limiting in the presence of uncertain data such as disjunctions, nulls and maybe information. In the presence of disjunctions, a set of models is considered and truth (falsity) of a sentence in first order logic is defined by its presence (absence) in all "minimal" or possible models of the database. This paper shows that these semantics do not completely capture every real world scenario. This problem arises due to the fact that the representation mechanism allows only certain types of negative information to be explicitly represented, if any at all. The paper presents a data structure, called an *oa-table* that is complete in this sense. Query processing in the model is also illustrated by extended relational operators. Other complete models introduce variables to represent real world scenarios. However, *oa-tables* are variable-free.

For future work we plan to investigate the following two issues:

1. Extending the *oa-table* model to include mixed disjunctions. i.e. disjunctions of the form $P(a) \vee Q(a)$ which were studied by Liu and Sunderraman in [7,8].
2. Extending the algebra of the *oa-table* model to compute the well founded semantics for disjunctive databases discussed in [3].

References

1. Chan, E.P.F.: A possible world semantics for disjunctive databases. Knowledge and Data Engineering 5(2), 282–292 (1993)
2. Chiu, J.-S., Chen, A.L.P.: An exploration of relationships among exclusive disjunctive data. IEEE Trans. Knowl. Data Eng. 7(6), 928–940 (1995)
3. Gire, F., Plourde, C.: A new partial semantics for disjunctive deductive databases. Fundam. Inf. 67(4), 323–342 (2005)
4. Imieliński, T., Lipski Jr., W.: Incomplete information in relational databases. J. ACM 31(4), 761–791 (1984)
5. Liu, K.-C., Sunderraman, R.: On representing indefinite and maybe information in relational databases. In: ICDE, pp. 250–257 (1988)
6. Liu, K.-C., Sunderraman, R.: Indefinite and maybe information in relational databases. ACM Trans. Database Syst. 15(1), 1–39 (1990)
7. Liu, K.-C., Sunderraman, R.: On representing indefinite and maybe information in relational databases: A generalization. In: ICDE, pp. 495–502 (1990)
8. Liu, K.-C., Sunderraman, R.: A generalized relational model for indefinite and maybe information. IEEE Trans. Knowl. Data Eng. 3(1), 65–77 (1991)
9. Lozinskii, E.L.: Plausible world assumption. In: Brachman, R.J., Levesque, H.J., Reiter, R. (eds.) KR 1989: Proc. of the First International Conference on Principles of Knowledge Representation and Reasoning, pp. 266–275. Kaufmann, San Mateo (1989)
10. Ola, A.: Relational databases with exclusive disjunctions. In: ICDE, pp. 328–336 (1992)
11. Ross, K.A., Topor, R.W.: Inferring negative information from disjunctive databases. J. Autom. Reason. 4(4), 397–424 (1988)
12. Sakama, C.: Possible model semantics for disjunctive databases II (extended abstract). Logic Programming and Non-monotonic Reasoning, 107–114 (1990)
13. Sarma, A.D., Benjelloun, O., Halevy, A., Widom, J.: Working models for uncertain data. In: ICDE 2006: Proceedings of the 22nd International Conference on Data Engineering (ICDE 2006), p. 7. IEEE Computer Society, Washington (2006)
14. Vadaparty, K.V., Naqvi, S.A.: Using constraints for efficient query processing in nondeterministic databases. IEEE Trans. Knowl. Data Eng. 7(6), 850–864 (1995)
15. Zhang, L., Liu, K.-C.: Towards a relational model for exclusively disjunctive information. In: CSC 1993: Proceedings of the 1993 ACM conference on Computer science, pp. 143–150. ACM Press, New York (1993)

The Ramification Problem in Temporal Databases: A Solution Implemented in SQL

Nikos Papadakis, Dimitris Plexousakis, Grigoris Antoniou, Manolis Daskalakis, and Yannis Christodoulou

Department of Computer Science, University of Crete, and
Institute of Computer Science, FORTH, Greece
{npapadak,dp,antoniou,mdaskal,jchrist}@ics.forth.gr

Abstract. In this paper we elaborate on the handling of the ramification problem in the setting of temporal databases. Starting with the observation that solutions from the literature on reasoning about action are inadequate for addressing the ramification problem, in our prior work we have presented a solution based on an extension of the situation calculus and the work of McCain and Turner. In this paper, we present a tool that connects the theoretical results to practical considerations, by producing the appropriate SQL commands in order to address the ramification problem.

1 Introduction

Many domains about which we wish to reason are dynamic in nature. *Reasoning about action* [10,8] is concerned with determining the nature of the world (what holds in the world state) after performing an action in a known world state, and has found application in areas such as cognitive robotics [10].

The guarantee of consistency of data that is stored in a database is a very important and difficult problem. The consistency of data is determined by the satisfaction of the integrity constraints in the different databases states (situations). A database state is considered valid (consistent) when all integrity constraints are satisfied. New situations arise as the result of action (transaction) execution. In a new situation (which includes the direct effects of the transactions) the database may be inconsistent because some integrity constraints are violated by means of indirect effects of actions. Thus, it is necessary to produce all indirect effects in order to determine the satisfaction of the integrity constraints. In a large database system with hundreds or thousands of transactions and integrity constraints, it is extremely hard for the designer to know all the effects that actions may have on the consistency of the database. The task can be greatly facilitated by a tool that systematically produces all effects (direct and indirect). The designers will be thus given the ability to realize the effects of their design, correct errors and prevent potential inconsistency problems.

We can assume that a atomic transaction is an action. In this context, the *ramification problem* [7] is concerned with the indirect effects of actions in the

A. An et al. (Eds.): ISMIS 2008, LNAI 4994, pp. 381–388, 2008.

presence of constraints. Standard solutions to this problem [2,3,4,5,6] rely on the persistence of fluents, and on the idea that actions have effects on the next situation only.

In our work we consider the ramification problem in a temporal setting. In this context, actions are allowed to have effects which may commence at a time other than the next time point, and the effects may hold only for certain time. For example, assume that if a public employee commits a misdemeanor, then for the next five months she is considered illegal, except if she receives a pardon. When a public employee is illegal, then she must be suspended and cannot take promotion for the entire time interval over which she is considered illegal. Also, when a public employee is suspended, she cannot receive a salary until the end of the suspension period. Each public employee is graded for her work. If she receives a bad grade, then she is considered a bad employee. If she receives a good grade, then she is considered a good employee and she may take a bonus if not suspended. Each public employee receives an increase and a promotion every two and five years, respectively, if not illegal.

We can identify six actions, *misdemeanor*, *take_pardon*, *good_grade*, *bad_grade*, *take_promotion* and *take_increase* and seven fluents *good_employee*, *illegal*, *take_salary*, *take_bonus*, *position*, *suspended* and *salary*. The fluent $position(p, l, t_1)$ means that the public worker is at position l for the last t_1 months while $salary(p, s, t_1)$ means that the public worker has been receiving salary s for the last t_1 months. The direct effects of the six actions are expressed in propositional form by the following rules:

$$occur(misdemeanor(p), t) \rightarrow illegal(p, t_1) \wedge\ t_1 \in [t, t+5] \quad (1)$$

$$occur(take\_pardon(p), t) \rightarrow \neg illegal(p, t_1) \wedge t_1 \in [t, \infty) \quad (2)$$

$$occur(bad\_grade(p), t) \rightarrow \neg good\_employee(p, t_1) \wedge t_1 \in [t, \infty) \quad (3)$$

$$occur(good\_grade(p), t) \rightarrow good\_employee(p, t_1) \wedge t_1 \in [t, \infty) \quad (4)$$

$$occur(take\_increase(p), t) \wedge salary(p, s, 24) \rightarrow salary(p, s+1, 0) \quad (5)$$

$$occur(take\_promotion(p), t) \wedge position(p, l, 60) \rightarrow position(p, l+1, 0) \quad (6)$$

where t is a temporal variable and the predicate $occur(misdemeanor(p), t)$ denotes that the action $misdemeanor(p)$ is executed at time t. The former four rules are dynamic and executed every time that the corresponding actions take place. The remaining two are also dynamic but are executed periodically because the corresponding actions take place periodically.

Also we have and the following integrity constraints which give rise to indirect effects of the six actions.

$$illegal(p, t_1) \rightarrow suspended(p, t_1) \quad (7)$$

$$suspended(p, t_1) \rightarrow \neg take\_salary(p, t_1) \quad (8)$$

$$\neg suspended(p, t) \wedge good\_employee(p, t) \ \rightarrow\ take\_bonus(p, t) \quad (9)$$

$$\neg good\_employee(p, t_1) \rightarrow \neg take\_bonus(p, t_1) \quad (10)$$

$$\neg suspended(p, t_1) \rightarrow take\_salary(p, t_1) \quad (11)$$

The rules (7-11) are static and executed every time. This happens because there are effects of the actions which hold for a time interval (e.g. the effect *illegal* of the action *misdemeanor* which hold 5 time point after the execution of the action). After the end of the time intervals the effects pause to hold without an action taking place. So, in a temporal setting the main assumptions of previous solutions to the ramification problem are inadequate because they based in persistent of fluent - nothing change unless an action takes place. Thus, we need new techniques. In [9] the problem has been addressed in cases where actions result in changes in the future (e.g. the promotion of an employee takes effect in two months).

In this paper, we propose a tool which produces sql commands(transcations) which in turn produce the indirect effects of other transactions. These transactions ensure that all integrity constraints will be satisfied after an update take place. The data is stored in a relational database (Oracle). The temporal data are stored in a table as triples (ID, $timestamp\_start$, $timestamp\_end$).

Our approach is based on the situation calculus [7] and the work of McCain and Turner [6]. We extended the approach of [6] by introducing duration to fluents, and by considering temporal aspects. As we have explained in [9], in a temporal database we need to describe the direct and indirect effects of an action not only in the immediately resulting next situation, but possibly for many future situations as well. This means that we need a solution that separates the current effects(dynamic rules) from the future effects (static rules). This is necessary because another action may occur between them which cancels the future effects. The approach of [6] permits this separation.

The paper is organized as follows. Section 2 describes the technical preliminaries, that are based on previous work by the authors [9]. Section 3 presents the algorithm which produces the in sql commands and a description of the system architectures.

2 Technical Preliminaries

Our solution to the ramification problem is based on the situation calculus [7]. However it is necessary to extend the situation calculus to capture the temporal phenomena, as done in the previous work [9]. For each fluent f, an argument L is added, where L is a list of time intervals $[a, b], a < b$. $[a, b]$ represents the time points $\{x|a \leq x < b\}$. The fluent f is true in the time intervals that are contained in the list L.

We define functions $start(a)$ and $end(a)$, where a is an action. The former function returns the time moment at which the action a starts while the latter returns the time moment at which it finishes. Actions are ordered as follows: $a_1 < a_2 < .... < a_n$, when $start(a_1) < start(a_2) < ..... < start(a_n)$. Actions(instantaneous) $a_1, a_2, ..., a_n$ are executed concurrently if $start(a_1) = start(a_2) = ... = start(a_n)$.

The predicate $occur(a, t)$ means that the action a is executed at time moment t. We define functions $start(S)$ and $end(S)$, where S is a situation. The former function returns the time moment at which the situation S starts while the latter returns the time moment at which it finishes.

We define as temporal situation a situation which contains all fluents with the list of time intervals in which the fluent is true. For the rest of the paper we refer to a temporal situation as situation. The function $FluentHold(S, t)$ returns the set of all fluents that are true in the time moment t. For function fluent the $FluentHold(S, t)$ returns the value that the function has at time point t. The situation S is a temporal situation. We define as non-temporal situation S in a time point t the situation $S = FluentHold(S', t)$. A transformation from a situation to another could happen when the function $FluentHold(S, t)$ returns a different set.[1]

We define two types of rules: (a) dynamic $occur(a, t) \rightarrow f([..])$ which are executed when the action a takes place; and (b) static rules $G_f([a, b]) \rightarrow f([...])$ which are executed at time point t if the fluent f is not true in the time interval $[a, b]$.[2] The execution of a dynamic or static rule has as consequence the transition to a new situation. When an action takes place we want to estimate all direct and indirects effects of the action which has as conclusion a new consistent situation.

We define as *legal* (*consistent*) a situation in which all integrity constraints are satisfied. Also, each function fluent has only one value at each time point.

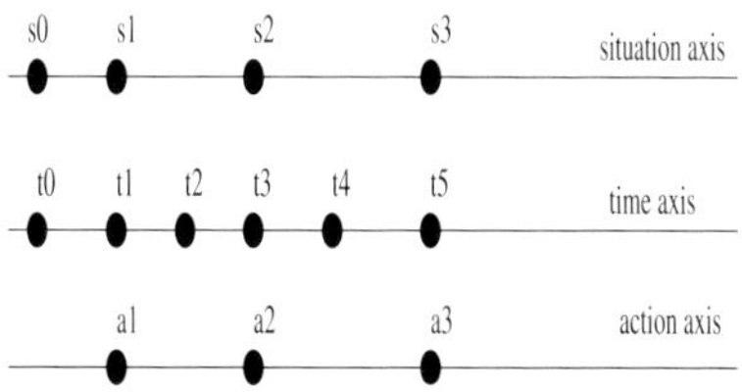

Fig. 1. The correspondence among Time-Actions-Situations

As already mentioned, the previous approaches to solve the ramification problem are inadequate in the case of temporal databases. We overcome these difficulties with the time - actions - situations correspondence that appears in figure 1. There are three parallel axes: the situations axis, the time axis and the actions axis. The database changes into a new situation when an action takes place or a static rule evaluated or cease to hold a fluent.

We base our work on the ideas of McCain and Turner [6] who propose to use *static rules* to capture the indirect effects of actions (based on integrity constraints in the particular domain), and *dynamic rules* to represent the direct effects of actions. In our approach, for each action A there is a dynamic rule of

[1] For example if $\{f_1([[7, 9]], \neg f_1([[10, \infty]]), ...\}$ means that at time point 10 we have transform from one situation to another.

[2] $G_f([a, b])$ mean that the fluent formula G_f is true at time interval $[a, b]$.

the form $A \rightarrow \bigwedge F_i(L'_i)$, where each $F_i(L'_i)$ is $f_i(L'_i)$ or $\neg f_i(L'_i)$ for a fluent f. These rules describe the direct effects of an action. Here is an example:

$occur(misdemeanor(p), t) \rightarrow illegal(p, [[t, t+5]]))$

In addition, for each fluent f we define two rules, $G(L) \rightarrow f(L)$ and $B(L) \rightarrow \neg f(L)$. $G(L)$ is a fluent formula which, when is true (at list L), causes fluent f to become true at the time intervals contained in the list L (respectively for $B(L)$). These rules encapsulate the indirect effects of an action. Here is an example: $illegal(p, [[2,7]]) \rightarrow suspended(p, [[2,7]])$.

One cornerstone of our previous work [9] is the production of the static rules from integrity constraints, according to the following ideas. We make use of a binary relation I which is produced from the integrity constraints and encodes the dependencies between fluents. The binary relation I is based on the idea that there are two kinds of integrity constraints.

$$(a) \quad G_f \rightarrow K_f \qquad (b) \quad G_f \equiv K_f,$$

where G_f and K_f are fluent propositions. For the first kind of constraints, for each $f \in G_f$ and $f' \in K_f$ the pair (f, f') is added in I. For the second kind of constraints, for each $f \in G_f, f' \in K_f$, if f can change its truth value as the direct effect of an action, then (f, f') is added in I. If f' can change its truth value as direct effect of an action then (f', f) is added in I.

The algorithm from producing static rules has been presented in [9]. The main idea is the following: when an integrity constraint is violated, find an executable static rule whose effect is to restore the violation of the constraint. The algorithm is proceedd as follows

1. At each time point t if the situation is S do
   (a) If an action $occur(a, t)$ take place at time point t then
   (b) Execute the dynamic rule which referred in action a
   (c) While there is a static rule r which is executable. Execute the r.

The following theorem has been proved.

Theorem 1. *(a) When a static rule is executable at least one integrity constraint is violated. (b) The algorithm for execution dynamic and static rules always returns a consistent situation.*

3 Transforming the Static Rules into SQL

In a relational temporal database a fluent of the form $f_1([[a,b],[c,d],[e,f]])$ is stored in a table as triples (*ID*, *timestamp_start*, *timestamp_end*). The type of *ID* is Number while the type of *timestamp_start*, *timestamp_end* is date or timestamp. In this assumption the fluent $f_1([[a,b],[c,d],[e,f]])$ will be stored in three rows (triples): $(1,a,b)$,$(1,c,d)$, $(1,e,f)$.

In order to transform and execute a rule of the form $f_1(...) \wedge f_2(...) \rightarrow f_3(...)$ in SQL we must determine the intersection of the time intervals in which the fluents f_1 and f_2 are true. In order to do that we must check the all rows which refer to f_1 with all rows which refer to f_2 and find the all the intersections of the time intervals. For example if there are two rows (1,'26-7-2007','31-7-2007') and (2,'29-7-2007','6-8-2007') then the intersection is the time interval ('29-7-2007','31-7-2007'). This means that the row (3,'29-7-2007','31-7-2007') must be inserted in the table. If we insert in the table the row (1, '4-8-2007','14-8-2007') we now have the rows: (1, '4-8-2007','14-8-2007') and (2,'29-7-2007','6-8-2007'). This means that directly after the insertion of the row (1, '4-8-2007','14-8-2007') we must insert the row (3,'4-8-2007','6-8-2007'), as conclusion of the execution of the above static rule.

In order to capture all cases of relationships between two temporal intervals we must implement all the 13 Allen temporal relations [1]. First we need to describe how we estimate the intersections between two sets of temporal intervals(e.g $f1([[..],[...],..])f2([[..],...])$, which refer to two different fluents. As we have already described above, the two sets are stored in a table as triples (ID, $timestamp\_start$, $timestamp\_end$). The following algorithm estimates the intersection between two sets of temporal intervals (the implementation in SQL).

Algorithm 1

1. Take the first pair (f_i, f_j) of fluents in the left side of the static rule.
2. For each rows $row11, row22$ s.t $row1.ID = i$ and $row2.ID = j$ do
3. s1 := row1.START; e1 := row1.FINISH; s2 := row2.START; e2 := row2.FINISH; changes := FALSE;
4. IF $s1 < s2$ AND $e1 >= s2$ AND $e1 <= e2$ THEN
   resSt := s2; resEnd := e1; changes := TRUE;
5. ELSIF $s1 >= s2$ AND $s1 <= e2$ AND $e1 >= s2$ AND $e1 <= e2$ THEN
   resSt := s1; resEnd := e1; changes := TRUE;
6. ELSIF $s1 <= s2$ AND $e2 <= e1$ THEN
   resSt := s2; resEnd := e2; changes := TRUE;
7. ELSIF $s1 >= s2$ AND $s1 <= e2$ AND $e1 > e2$ THEN
   resSt := s1; resEnd := e2; changes := TRUE;
   END IF;
8. IF changes = TRUE THEN
   INSERT INTO TABLENAME_temporary_table VALUES (currTempTableId, resSt, resEnd) ;
   END IF;

The step 4 is the case of overlaps, the step 5 is the case of during, the step 6 is the cases of equals (when hold the equal in two parts of expression) and during (when does not hold the equal in one of the two parts of the expression). The other relations are implemented in more than one steps of the algorithms. As can be seen from step 8, when a new temporal interval is estimated it is stored in a temporary table. First, we estimate the Allen relations between the time intervals of the two fluents(e.g f_1, f_2). Then, consider these two fluents as one (f_{12}) and repeat the process between this "new" fluent and the next fluent (e.g f_3). Thus we have to execute a static rule of the form $f_1(...) \wedge f_2(...) \wedge ... \rightarrow f_k$, first we execute the algorithm 1 for the first two fluents $(f_1 f_2)$. Then we consider the result as a new fluent and repeat the execution of the algorithm 1 with the next fluent. When we have to execute a static rule of the form $[f_1(...) \wedge ....] \vee [f_2(...) \wedge ...] \vee .. \rightarrow f_k$ the only thing we need to do is to execute the algorithm 1 for each part of the disjoin. The architecture of the tool which produce the SQL commands is shown in figure 2. A general description of the java logical unit follows. The input data include:

- the rules of the DB, that is the cross-correlations of fields of the DB, in form $ID1 \wedge ID2(\hat{} IDi) \rightarrow IDx$; where $ID(1, 2, i)$ are the values of the primary key field that is involved in the section for the result and IDx is the value of the primary key field with which will the result be stored in the DB.
- The name of table, as String.
- The name of column of key, any type.
- The name of the starting time column, any type.
- The name of the ending time column, any type.

The output includes:

- Code.pl: a file with SQL commands, which when it is executed in the DB in question will create the essential procedures that will safeguard its integrity.
- Drop.pl: a file with the essential SQL commands, so that the tables and procedures that were created by the program can be erased.

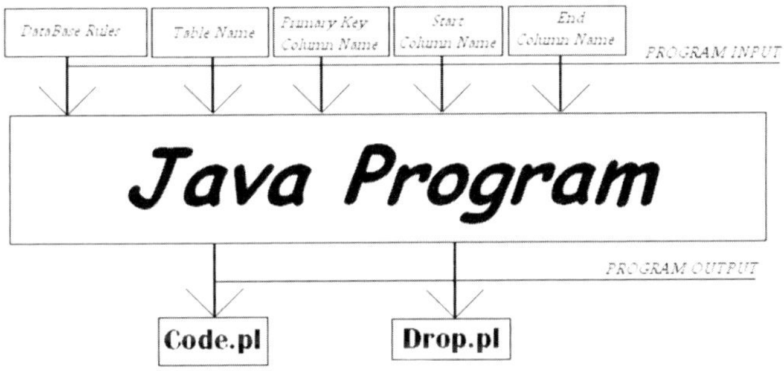

DataBase Rules	Syntax: id ^ id [^id]* -> id
Table Name	String, the name of the database table
Primary Key	Name of the column of the table that is the primary key (in any format)
Start	Name of the column of the table, containing the starting time (in any format)
End	Name of the column of the table, containing the ending time (in any format)
Code.pl	Output file, containing pl/sql code, which follows the input rules
Drop.pl	Output file, containing pl/sql code that deletes any processes and tables created by code.pl

Fig. 2. The architecture of the systems

4 Conclusions

In this paper we studied the ramification problem in the setting of relational temporal databases. More specifically, we proposed a solution and a tool which implements the solution in SQL. The key ideas of our approach are: (a)to extend the situation calculus, (b)to establish suitable correspondences between time, actions and situations, (c)to use dynamic rules to capture the direct effects of actions, and static rules to capture the indirect effects of actions, (d)to develop a system which implement the solution in sql for a relation databases.

In future work we intend to extend our system for the case that the updates refer to the past and for the case where a distinction of integrity constraints into strict and soft exists.

References

1. Allen, J.F.: Maintaining Knowledge about Temporal Intervals. Communications of the ACM 26(11), 832–843 (1983)
2. Denecker, M., Ternovska, E.: Inductive situation calculus. Artificial Intelligence archive 171(5-6), 332–360 (2007)
3. Drescher, C., Thielscher, M.: Integrating Action Calculi and Description Logics. In: Hertzberg, J., Beetz, M., Englert, R. (eds.) KI 2007. LNCS (LNAI), vol. 4667, pp. 68–83. Springer, Heidelberg (2007)
4. Giordano, L., Martelli, A., Schwind, C.: Specialization of Interaction Protocols in a Temporal Action Logic. In: LCMAS 2005 pp. 3–22 (2005)
5. Fusaoka, A.: Situation Calculus on a Dense Flow of Time. In: AAAI 1996, pp. 633–638 (1996)
6. McCain, N., Turner, H.: A causal theory of ramifications and qualifications. In: Proceedings of IJCAI 1995, pp. 1978–1984 (1995)
7. McCarthy, J., Hayes, P.J.: Some philophical problem from the standpoint of artificial intelligence. Machine Intelligence 4, 463–502 (1969)
8. Miller, R., Shanahan, M.: The Event Calculus in Classical Logic - Alternative Axiomatisations. Link. Elect. Art. in Compt. and Inform. Sc. 4(16) (1999)
9. Papadakis, N., Plexousakis, D.: Action with Duration and Constraints: The Ramification problem in Temporal Databases. International Journal of Artificial Intelligent Tools (IJTAI) 12(3), 315–353 (2003)
10. Reiter, R.: Knowledge in Action: Logical Foundations for Specifying and Implementing Dynamical Systems. MIT Press, Cambridge (2001)

Image Databases Browsing by Unsupervised Learning

Charbel Julien and Lorenza Saitta

Università del Piemonte Orientale Torino, Alessandria, Italy
{charbel,saitta}@mfn.unipmn.it

Abstract. Content-Based Image Retrieval systems provide a variety of usages. The most common one is *target* search, in which a user is trying to find a specific target image. Instead, we present, in this paper, a flexible image dataset *browsing* system. The user can browse the whole dataset looking for any "interesting" image. To this aim, images are first abstracted through a set of signatures describing their color and texture composition. Afterwards, unsupervised clustering is performed to split the image set into several clusters of "similar" images. Every cluster is represented by its centroid as an icon. The set of icons is presented to the user, who can pick one in order to see the images belonging to the cluster. Multi-dimensional scaling is used to visualize images in the same cluster by mapping the images onto a two-dimensional space. The experiments performed with a general-purpose image dataset consisting of one thousand images, categorized into ten classes, show the usefulness of the system.

Keywords: Image database Browsing, Image features, signatures, unsupervised learning, Clustering.

1 Introduction

There are two general methods for finding images in large databases: *querying* and *browsing*. Database browsing is vague since it basically means some kind of free-form navigation by the user in the image database. The query methods currently used in most Content-Based Image Retrieval (CBIR) research show problems and limitations. The search task itself can be indefinite, the user may just be looking for interesting images whatever they might be, or he/she may change the query target during a query.

An integrated browsing tool can be very useful in image retrieval applications. Designing effective interfaces for browsing is, however, not at all a trivial task. Browsing involves processing the entire database, designing some type of database visualization, and providing a tool for navigation. To assist browsing, images should be organized so that similar images are grouped together or located close to each other in the visualization.

A. An et al. (Eds.): ISMIS 2008, LNAI 4994, pp. 389–398, 2008.

Image database browsing tries to generate a visualization of an image database and to help users understanding the image database as a whole. First, low-level visual features are extracted from the images, and an *affinity* matrix is built by applying a distance measure. Using the affinity matrix as input, we use a clustering algorithm to partition the image set into subsets of similar images according to their computed distances.

Using the Multi-Dimensional-Scaling (MDS) [2], images belonging to the same cluster are visualized together. This visualization reflects the similarity among images within the same cluster.

2 Related Work

A typical CBIR system searches a database of images according to a measure of *similarity* between the query image and the images in the database; this method is known as *query by example*. Another query method, called *query-by-sketch* [3] was also proposed: the user provides the system with a sketch of the image he/she is searching for. Whatever the query method, the result is a set of images ranked by there similarity to the query. With both methods the similarity among *target* images is ignored. In order to enhance the quality of the retrieval in CBIR systems, image clustering is used with the aim to exploit the similarity not only between the query image and target images but also among the target images themselves [4] [5]. CBIR techniques assume that there exists a certain amount of mutual information between the similarity and the semantic content of the images. Clustering can improve the retrieval for any CBIR system, independently from both the similarity measure and the image features used.

Clustering systems enable the user to find images by navigating through the image database [11] [12] [13] [14] [15]. The goal is to find a mapping of the archive into clusters. The clustering provides a concise summarization and a natural way of visualization of image sets. The clustering may be performed by unsupervised or supervised methods. In unsupervised clustering the user incorporates background knowledge into the distance measure, whereas in supervised clustering a training set is needed.

Goldberger et al. [11] suggest a probabilistic framework to perform clustering: they model the color content of images by a mixture of Gaussians. Chen et al. [15] focus on the use of hierarchical tree-structures both to speed-up the search-by-query, and to effectively browse databases using similarity pyramid. The similarity pyramid groups similar images together, while allowing users to view the database at varying levels of detail. The image is represented by a feature vector of global color and texture edge histograms. In the clustering-based image retrieval system CLUE, retrieval is performed on groups of images [5]; unlike a standard CBIR system, where the images relevant to a query are presented to the user ranked by their similarity to the query image, images are grouped into a small number of classes.

3 Image Abstraction by a Set of Signature

Instead of using a fixed-size vector of features, in our approach we use a set of *signatures* to represent the image's low-level visual content. Let $E = \{\mathcal{I}_i | i = 1, ..., n\}$ be the whole set of images available in the database. The set of signatures $\mathcal{S}_i = \{S_i^{(c)}, S_i^{(cc)}, S_i^{(t)}, S_i^{(tc)}\}$ of an image $\mathcal{I}_i$ contains four single signatures: $S_i^{(c)}$ captures image *color*, $S_i^{(cc)}$ captures *color composition*, $S_i^{(t)}$ captures image *texture*, and $S_i^{(tc)}$ captures *texture composition*.

Each signature, corresponding to a given feature (color, ...), consists of a set of clusters, usually different for each individual image, grouping pixels with similar feature values. More precisely, the signature of an image $\mathcal{I}_i$, referring to a feature k, is a set of pairs $S_i^{(k)} = \{(\mathbf{s}_{i,j}^{(k)}, w_{i,j}^{(k)}) | j = 1, ...m_i\}$, with $k \in \{c, cc, t, tc\}$, $i \in \{1, 2, ..., n\}$, and $j \in \{1, ..., m_i\}$, where m_i is the number of clusters in signature $S_i^{(k)}$.

In cluster $(\mathbf{s}_{i,j}^{(k)}, w_{i,j}^{(k)})$ the first term, $\mathbf{s}_{i,j}^{(k)}$, represents the center of the cluster elements, whereas $w_{i,j}^{(k)}$ is the fraction of pixels that belong to that cluster. The center $\mathbf{s}_{i,j}^{(k)}$ is assumed as a representative of the whole cluster. The number of clusters m_i changes with the complexity of the particular image $\mathcal{I}_i$ with respect to color and texture. The representative $\mathbf{s}_{i,j}^{(k)}$ is a d-dimensional vector. In general the same vector quantization algorithms that are used to construct a palette of colors for the whole image database can be used for image signature computation, as long as they are applied to every image independently. Simple images have a short signature while complex images have long ones.

The set of signatures $\mathcal{S}_i$ $(1 \leqslant i \leqslant n)$ is then used to compute the distance between any pair of images. In particular, a distance measure is defined independently for each type of signature, obtaining thus a 4-tuple $(d_c, d_{cc}, d_t, d_{tc})$ of distances.

The distance between centers of clusters is computed using the Euclidean distance, which is also the distance measure used internally by the clustering algorithm. The global distance between two images $\mathcal{I}_1$ and $\mathcal{I}_2$ is defined as a linear combination of individual distances:

$$D(\mathcal{I}_1, \mathcal{I}_2) = \sum_{k \in \{c, cc, t, tc\}} \alpha_k \, d_k(\mathcal{I}_1, \mathcal{I}_2).$$

3.1 Signature Extraction

In this section we will go into more details about the abstraction of images through signature computation.

Composition of color and texture

To compute the signatures of *color composition* and *texture composition* of an image we use the method described in SIMPLIcity [6]. We first smooth each band of the image RGB representation using a 2D-Gaussian filter, with the aim of reducing the possible color quantization and dithering artifacts. We then transform the image representation into the LUV perceptual color space. Images are partitioned into blocks of 4 x 4 pixels. The block size is chosen as a compromise between the color/texture details and the computation time, owing to the large number of images to be processed in the image database.

Six features are extracted from each block, three concerning the color and three representing the texture of the block. The color features are the average of the *luminance* L, and of the *chrominance*, captured by the parameters U and V, which encode color information. It has been shown that the LUV color space has a good correlation with human perception.

Texture features represent energy in high frequency bands of wavelet transforms, *i.e.*, the square root of the second order moment of the wavelet coefficients in the high frequency bands. To obtain the three features describing the texture, we apply a Daubechies-4 wavelet transform [7] to the L component of the image. After a one-level wavelet transform, each 4 x 4 block is decomposed into four frequency bands, each band represented with 2 x 2 coefficients. The HL, LH, and HH bands are used to encode texture information. Let us assume the coefficients in the HL band are $\{c_{r,l}, c_{r,l+1}, c_{r+1,l}, c_{r+1,l+1}\}$. One feature is then computed as:

$$f_{HL} = (\frac{1}{4}\sum_{i=0}^{1}\sum_{j=0}^{1} c^2_{r+i,l+j})^{\frac{1}{2}} \tag{1}$$

The other two features, namely f_{LH} and f_{HH}, are computed similarly from the LH and HH bands. Afterwards, the K-Means algorithm is used to cluster the feature vectors of both color and texture separately into several classes. The number of clusters in the algorithm is determined dynamically by thresholding the average within-cluster variation.

Let $\{\mathbf{x}_1, \mathbf{x}_2, ..., \mathbf{x}_N\}$ be the observations. The role of the K-Means algorithm is to partition the observations into m groups with means $\hat{x}_1, \hat{x}_2, ..., \hat{x}_m$, in such a way that the following objective function $D(m)$ (the distortion) is minimized:

$$D(m) = \sum_{i=1}^{m} min_{1 \leqslant j \leqslant K}(x_i - \hat{x}_j)^2$$

K-Means does not specify how many clusters to choose. We adaptively choose the number of clusters by gradually increasing m and stopping when some halt condition is met. We start with $m = 2$ and stop increasing m when one of the following conditions is satisfied:

1. The distortion $D(m)$ is below a given threshold. A low $D(m)$ indicates high purity in the clusters.
2. The discrete approximation of the first derivative of the distortion with respect to m, *i.e.*, $[D(m) - D(m-1)]$, is below a threshold. A low $[D(m) - D(m-1)]$ indicates convergence in the clustering process.
3. The number m exceeds an upper bound. We force an image signature to only consists of no more than 12 elements. Usually, the segmentation process generates a much lower number of classes.

Once the clusters of color and texture have been found, the color signature $S_i^{(c)}$ and the texture signature $S_i^{(t)}$ of image $\mathcal{I}_i$ are computed; this amounts to calculate the fraction of pixels that belong to each cluster. This fraction is computed by finding, for each pixel in the image, its nearest-neighbor cluster.

Structure-Composition of color and texture

The *Structural-Composition of color*, proposed in [1], aims to capture how colors interact with each other in an image. The motivation behind this feature is that spatial relationships among color regions bring a lot of information to understand content. For instance, in a picture of a "beach" an orange region (*sun*) is completely inside a light-blue region (*sky*), while a light-blue region shares a long border with a dark-blue one (*ocean*). For a "tiger" picture, yellow and black regions sharing similar borders (*stripes*) are characterizing visual cues.

Coarse texture patterns, such as pictures of beads of different colors, could be described as "any color *bead* surrounding any other color *bead*", "some color background completely containing most colors *beads*", and so on. This idea leads to the principled formulation of the structure-composition of color.

Based on this property of color interaction, we construct a signature of *structure-composition of color*. First, the image's pixels are mapped into their nearest-neighbor cluster, built as described in the previous section. Afterwards, a *structure-composition of color* signature $S_i^{(cc)} = \{(\mathbf{s}_{i,j}^{(cc)}, w_{i,j}^{(cc)}) | 1 \leqslant j \leqslant m\}$ is calculated. Each structure-composition cluster representative $\mathbf{s}_{i,j}^{(cc)}$ is given by:

$$\mathbf{s}_{i,j}^{(cc)} = \{\{l_{ij,1}, u_{ij,1}, v_{ij,1}\}, \{l_{ij,2}, u_{ij,2}, v_{ij,2}\}\},$$

where the cluster representative is formed by a pair of colors in the LUV color space; these two colors are indeed the centers of two clusters of the *color* signature described before. The term $w_{i,j}^{(cc)}$ is the fraction of pixels that ly inside an 8-connected region in image $\mathcal{I}_i$ and that have color $\{l_{ij,1}, u_{ij,1}, v_{ij,1}\}$ and color $\{l_{ij,2}, u_{ij,2}, v_{ij,2}\}$, respectively. The clusters with a value of $w_{i,j}^{(cc)}$ below a given threshold are removed.

A signature of *structure-composition of texture* is computed using the same technique as for *structure-composition of color*, but this time the centers of clusters of *texture composition* are used instead of centers of clusters of color composition. Obviously, the *structure-composition of texture* represents how pairs of textures interact between each other.

The distance between two signatures of *structure-composition of color* can be computed using the *Earth Mover's* Distance (EMD). The distance between centers of signature clusters are taken as the minimal Euclidean distance connecting centers. More formally, let us consider two centers of two signature clusters of image $\mathcal{I}_1$ and $\mathcal{I}_2$ respectively, namely $\mathbf{s}_{1,j}^{(cc)} = \{\{l_{1j,1}, u_{1j,1}, v_{1j,1}\}, \{l_{1j,2}, u_{1j,2}, v_{1j,2}\}\}$ and $\mathbf{s}_{2,j}^{(cc)} = \{\{l_{2j,1}, u_{2j,1}, v_{2j,1}\}, \{l_{2j,2}, u_{2j,2}, v_{2j,2}\}\}$. We can compute their distance as follows:

$$d(\mathbf{s}_{i,j}^{(cc)}, \mathbf{s}_{i',j}^{(cc)}) = argmin_{\{p,q,p',q' \in\{1,2\} | p+p'=q+q'=3\}}$$

$$\{\frac{1}{2}(d(\{l_{1j,p}, u_{1j,p}, v_{1j,p}\}, \{l_{2j,q}, u_{2j,q}, v_{2j,q}\})+$$
$$d(\{l_{1j,p'}, u_{1j,p'}, v_{1j,p'}\}, \{l_{2j,q'}, u_{2j,q'}, v_{i2j,q'}\}))\}$$

The distance between two signatures of *structure-composition of texture* can be computed in exactly the same way as between the signatures of *Structure-composition of color*. The weights in the signatures are normalized to sum up to 1. We use the Euclidean distance to compute the distance between centers of clusters, so that the distance is indeed a true metric [2].

4 Image Unsupervised Clustering

Once the structure compositions of color and texture are extracted, a similarity matrix between images is computed using the EMD distance, as described above. We tested the algorithms PAM [8] (a K-Medoid clustering algorithm), K-Means, and a Spectral Graph Partitioning [9] to generate clusters. Every cluster is abstracted by his centroid in case we use K-Medoid or K-Means algorithms. In spectral graph partitioning, the representative of a cluster is the most central element, *i.e.*, the one that has the smallest sum of distances to all images within the cluster.

When a user enters a query, the set of image-centers is presented to him/her. This set of image-centers summarizes the whole database into few images, whose number is equal to the number of clusters. The user can choose one of these images and go ahead to explore the corresponding cluster. Using Multi-Dimensional-Scaling (MDS) the images in a cluster can by displayed in an intuitive way: similar images within the cluster are visualized near each other to make dense groups of similar images. A similarity matrix is build up to concisely report the distance between every pair of images; this matrix is provided as input to the clustering algorithm as well as to the MDS function. Images are known by the clustering algorithm only by their distances.

K-Means is a well known and widely used clustering algorithm. The aim of K-Means is to find the centers of natural clusters in the data. The objective is to minimize the total intra-cluster variance, or the squared error function. The number of clusters is a parameter that must be provided to the algorithm. The most common version of the algorithm uses an iterative refinement heuristic. We start by partitioning the input points into K initial sets, either at random or using some heuristic. Then the mean point, or centroid, of each set is found. A new partition, obtained by associating each point with the closest centroid, is formed. The centroids are recalculated for the new clusters, and the process is repeated by alternate application of these two steps until convergence, which is obtained when the points no longer switch among clusters, or, alternatively, centroids no longer change.

The Partitioning Around Medoids (PAM) algorithm was developed by Kaufman and Rousseeuw [8]. It finds K clusters by determining a representative object for each one. This representative object is called a *medoid*, and it is meant to be the most centrally located object within the cluster. Once the medoids have been selected, each non-selected object is associated to the most similar medoid. We first start with K random medoids, and in each subsequent iteration we swap a medoid to a new one until no more swap is possible.

A Graph Partitioning algorithm attempts to organize nodes of a graph into groups so that the within-group similarity is high, and/or the between-groups similarity is low. Given a graph $G = (V, E)$ with affinity (similarity) matrix W, a simple way to quantify the cost for partitioning nodes into two disjoint sets A and B ($A \cap B = \phi$ and $A \cup B = V$) is to compute the total weight of the edges that connect the two sets. This cost is called a *cut*:

$$cut(A, B) = \sum_{i \in A, j \in B} w_{ij}$$

Finding a bipartition of the graph that minimizes this *cut* value is known as the *minimum cut problem*. There exist efficient algorithms for solving this problem. As the *cut* does not contain any within-group information, several modified graph partition criteria can de devised, including the *Ncut* [9]:

$$Ncut(A, B) = \frac{cut(A, B)}{cut(A, V)} + \frac{cut(A, B)}{cut(B, V)}$$

Finding a bipartition with minimum *Ncut* value is an NP-complete problem. An approximated solution is proposed [9], obtained by solving a generalized eigenvalue problem:

$$(D - W)\overrightarrow{y} = \lambda D \overrightarrow{y},$$

where W is an $n \times n$ affinity matrix, and $D = diag[s_1, s_2, ..., s_n]$ is a diagonal matrix with $s_i = \sum_{j=1,\dots,n} w_{i,j}$. The eigenvector corresponding to the second smallest generalized eigenvalue is used to generate a bipartition of the graph. This process can be repeated in order to get more then two classes. The sub-graph with the maximum number of nodes is recursively partitioned. The process terminates when the bound on the number of clusters is reached or the *Ncut* value exceeds some threshold.

5 Experiments

In order to test the proposed image browsing system, we use Wang's dataset, which contains one thousand general purpose images manually selected from Corel database. The dataset contain 10 classes of 100 images each. Images are 384X256 or 256X384 pixels, compressed in JPEG format. This database has been used for classification as well as for clustering [5] tasks. The experimentation was performed on a Pentium IV CPU under Linux OS. An image takes an average 1.1 second to be converted from JPEG to raw format, and to find the set of signatures. The signatures are saved in a text file as image features. Afterwards, these features are used in the clustering algorithm to calculate the distance between images.

An example of cluster visualization using MDS is shown in Fig.1. These clusters have been found using a subset of Wang's database, namely 210 images

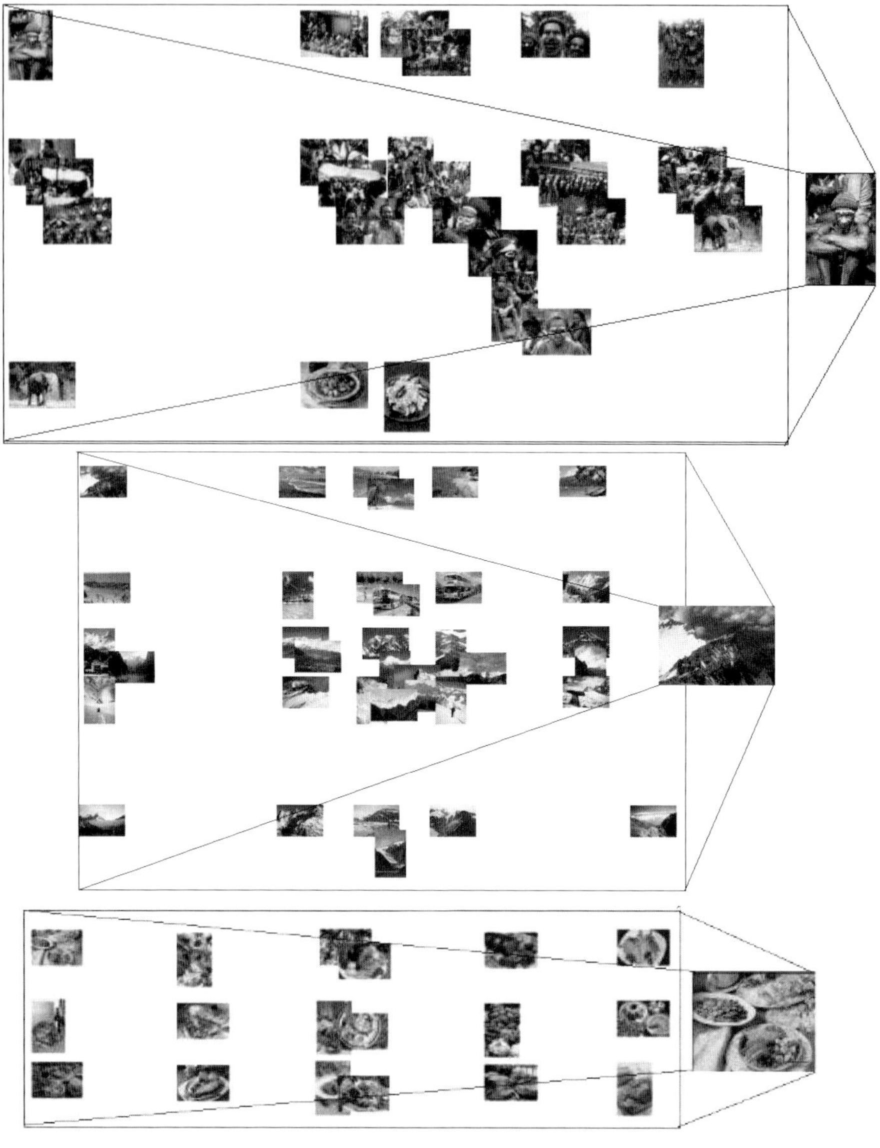

Fig. 1. Image database browsing: Clusters generated by the browsing system

from 7 classes, 30 images for class. The clustering algorithm was PAM, which achieved a 70% correct classification rate.

5.1 Cluster Quality Evaluation

The quality of clustering results is difficult to measure. In particular, one needs to find a quality measure that is not dependent on the technique used in the cluster generation process, on the representation scheme, and on the distance measure. Using a ground truth clustering database gives an independent evaluation of clustering quality. We use *Normalized Mutual Information* (NMI) [10] between true and predicted labels to measure the quality of clustering. The NMI measures the amount of information that the knowledge of one variable's value provides about the other one's. The NMI ranges between 0 and 1:

$$NMI = 2\frac{H(L) - H(L/\hat{L})}{H(L) + H(\hat{L})} \tag{2}$$

A high value of NMI indicates a strong content resemblance inside clusters. In (2) L and $\hat{L}$ are random variables corresponding to the ground-truth labels and to the labels assigned by the clustering algorithm, respectively. $H(L)$, $H(\hat{L})$ are the marginal entropies of L and $\hat{L}$, whereas $H(L/\hat{L})$ is the conditional entropy.

The results of the experiments are reported in Table 1. The PAM algorithm gives the best results in terms of clustering quality. The other two algorithms are more efficient for large image database but less reliable.

Table 1. Clustering Evaluation

Algorithm	**NMI**	**Time (m)**	**#Class**
K-Medoid	0.57	6.43	15
K-Means	0.55	1.14	15
SGP	0.43	2.03	15

6 Conclusion

In this paper we have presented an unsupervised clustering method for grouping image sets. Signatures serve as a reliable way to summarize color, texture, and structure composition of color and texture of images. The clustering enables a summarization of the image sets content. This approach can be useful for semi-automatic annotation as well as for image database indexing.

Using a fully labeled, ground truth image database we have evaluated the clustering quality. The result are very promising, and also the visual evaluation shows the good quality of clustering.

A hierarchical browsing should be envisaged to enable efficient navigation in large databases. Currently, the system does not provide the user with the possibility of browsing the image sets at varying levels of detail. However, the approach proposed in this paper can be profitably applied to small and medium size image sets.

Acknowledgments. This work was partly supported by the PRIN 2006 "Learning Hierarchical, Abstract Models from Temporal or Spatial Data" Project (N. 2006012944).

References

1. Datta, R., Ge, W., Lin, J., Wang, J.: Toward Bridging the Annotation-Retrieval gap in Image Search. In: Proceedings of ACM Multimedia Conference (October 2006)
2. Rubner, Y., Tomasi, C., Guibas, L.G.: A Metric for distribution with Applications to Image Databases. In: Proceeding of International Conference on Computer Vision, Bomby, India, pp. 59–66 (January 1998)
3. Veltkamp, R.C., Tanase, M.: Content-based Image retrieval Systems: A survey technical report UU-CS-2000-34 (October 2002)
4. Bhanu, B., Dong, A.: Concepts learning with fuzzy clustering and relevance feedback. In: Workshop on Machine learning and data Mining in Pattern recognition, July 2001, pp. 102–116 (2001)
5. Chen, Y., Wang, J.Z., Krovetz, R.: CLUE: Cluster-based Retrieval of images by Unsupervised Learning July 2004 draft (2004)
6. Wang, J., Li, J., Wiederhold, G.: SIMPLIcity: Semantic-Sensitive Integrated Matching for Picture LIbraries. IEEE Transactions on Pattern Analysis and Machine Intelligence 23(9), 947–963 (2001)
7. Daubichies, I.: Ten Lectures on Wavelets Capital City Press (1992)
8. Kaufman, L., Rousseeuw, P.J.: Finding Groups in Data: an Introduction to Cluster Analysis. Jhon Wiley & sons, Chichester (1990)
9. Shi, J., Malik, J.: Normalized cuts and image segmentation. IEEE Transaction on Pattern Analysis and machine Intelligence 22(8), 888–905 (2000)
10. Strehl, A., Ghosh, J., Mooney, R.J.: Impact of similarity measures on web-page clustering AAAI (2000)
11. Goldberger, J., Gordon, S., Greenspan, H.: Unsupervised Image-Set Clustering Using an Information Theoretic Framework. IEEE Transaction on Image Processing 15(2) (February 2006)
12. Krishnamachari, S., Abdel-Mottaleb, M.: Hierarchical clustering algorithm for fast image retrieval. In: Proceeding SPIE Conference Storage and Retrieval for image and video databases VII, San Jose (1999)
13. Krishnamachari, S., Abdel-Mottaleb, M.: Image browsing using hierarchical clustering. In: 4th IEEE Symp. Computers and Communications (July 1999)
14. Barnard, K., Duygulu, P., Forsyth, D.: Clustering art Comput. Vis. Pattern Recognit (December 2001)
15. Chen, J., Bouman, C.A., Dalton, J.C.: Hierarchical browsing and search of large image databases. IEEE Transaction Image Process 9(3) (2000)

Decision Tree Induction for Identifying Trends in Line Graphs*

Peng Wu, Sandra Carberry, Daniel Chester, and Stephanie Elzer

Dept. of Computer Science, University of Delaware, Newark, DE 19716 USA

Abstract. Information graphics (such as bar charts and line graphs) in popular media generally convey a message. This paper presents our approach to a significant problem in extending our message recognition system to line graphs — namely, the segmentation of the graph into a sequence of visually distinguishable trends. We use decision tree induction on attributes derived from statistical tests and features of the graphic. This work is part of a long-term project to summarize multimodal documents and to make them accessible to blind individuals.

1 Introduction

Information graphics (non-pictorial graphs, such as line graphs and bar charts) appear often in popular media such as *Newsweek* and *Business Week*. Such graphics generally have a message that the graphic designer intended to convey. For example, consider the information graphic shown in Figure 1, which appeared in *Business Week*. Its intended message is ostensibly that there has been a changing trend in global manufacturing utilization, falling from 2000 to 2002 and then rising until the end of 2006.

We are developing a system that reasons about a graphic and the communicative signals present in the graphic to hypothesize the graphic's intended message. The message recognition system plays an integral role in two very different projects:

1. A digital libraries project whose goal is to construct a more complete summary of a multimodal document that captures not only the article's text but also its information graphics.
2. An assistive technology project whose goal is to provide blind users with access to graphics in popular media by conveying the graphic's message via natural language.

Previous work on these projects has produced a system that can recognize the message of a simple bar chart[1], along with an interface that provides sight-impaired users with access to the message recognition system[2].

This paper provides our solution to a significant problem encountered in extending our message recognition system to line graphs. Freedman et al. [3] noted

* This material is based upon work supported by the National Science Foundation under Grant No. IIS-0534948.

A. An et al. (Eds.): ISMIS 2008, LNAI 4994, pp. 399–409, 2008.

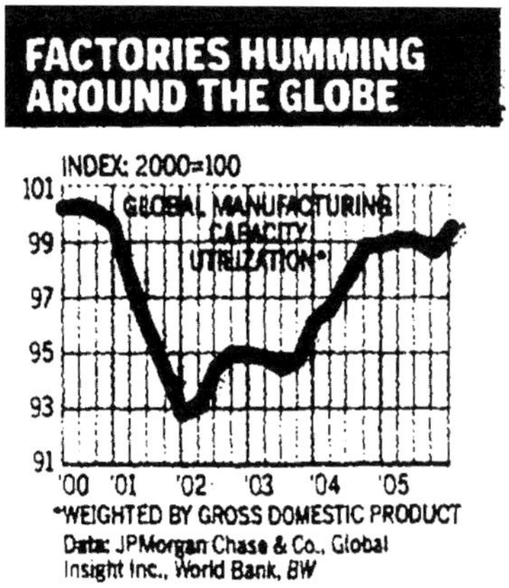

Fig. 1. Line graph with Change-trend message, from Business Week

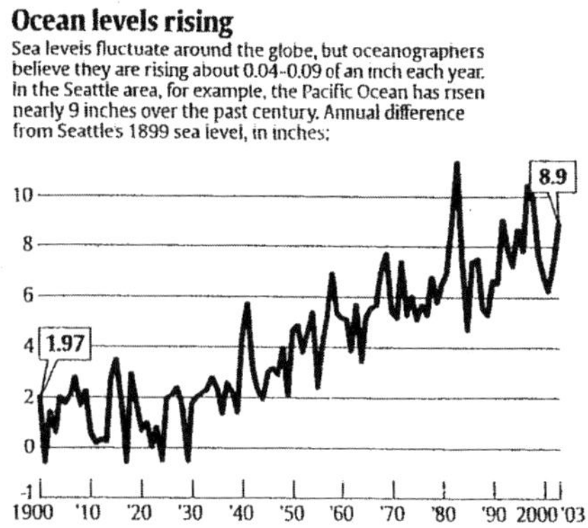

Fig. 2. Ragged line graph from USA Today

that information should be presented in a line graph if the goal is to convey a quantitative trend. It is essential that, in reasoning about a line graphs's message, we treat the line graph as capturing a sequence of visually distinguishable trends rather than as representing a set of data points connected by small line segments. For example, the line graph in Figure 2 consists of many short rises and falls, but a viewer summarizing it would be likely to regard it as consisting of a short overall stable trend from 1900 to 1930 followed by a long rising trend (both with high variance).

This paper focuses on our graph segmentation module which uses decision tree induction on a variety of attributes of the line graph to develop a model identifying how the graph should be segmented so as to capture the sequence of trends apparent in the graphic. The identified sequence of trends will then be used by the message recognition system, along with other communicative signals, to identify the graphic's intended message, such as the changing trend message for the graphic in Figure 1.

2 Related Work

Keogh et al.[4] discussed three approaches to linearly segmenting time series: sliding window, top-down, and bottom-up. In our work, we use the top-down approach. Bradley et al.[5] introduced an iterative method for smoothing data series using a Runs Test. Lin et al.[6] and Toshniwal et al.[7] discussed different ways of finding similar time series segments. Vieth[8] discussed piecewise linear regression applied to biological responses and Dasgupta et al.[9] presented an algorithm for detecting anomalies using ideas from immunology. Yu et al.[10] constructed textual summaries of time-series data sets for gas turbine engines. However, their work was concerned with identifying interesting patterns, such as spikes and oscillations, that were important for a particular problem. The above research efforts, and other related work, have mainly been concerned with detecting similar patterns or anomalies whereas the goal of our work is the identification of visually apparent trends. In addition, we use decision tree induction to investigate the contribution of a variety of different features, rather than settling on one or two features from the outset.

3 Trend Analysis in Simple Line Graphs

3.1 Sampling of Line Graphs

Our graph segmentation module works on a set of data points sampled from a representation of the original line graph. Sampling is done uniformly across the x-axis, and then additional points are added to capture change points in the graph. A Visual Extraction Module[11] is responsible for processing an electronic image and producing an xml representation of the graphic that includes all change points in the line graph, from which the sampling can be done. However to train our graph segmentation model, we scanned the 197 hard-copied line graphs and converted them to digital versions, then manually sampled each of them.

3.2 Graph Segmentation

Our graph segmentation module takes a top-down approach to identifying sequences of rising, falling, and stable segments in a graph. For example, the graph in Figure 3a should be identified as composed of three trends (short rising trend, longer falling trend, rising trend), as shown in Figure 3b.

The graph segmentation module starts with the original graph as a single segment. At each iteration, the module decides whether a segment in the current segmentation of the graph should be viewed as capturing a single trend or whether it should be split into two subsegments. If a decision is made to split the segment into two subsegments, then the segment is split at the point which is the greatest distance from the straight line connecting the two end points of

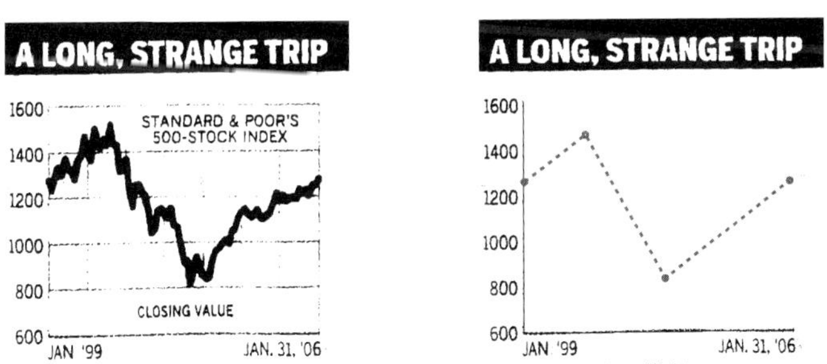

(a) Line graph with three trends (b) Trends of Figure 3a

Fig. 3. Line graph with three trends

the segment. Although this method for selecting the split point has produced good empirical results, in rare cases it can select an outlier as a split point; in future work, we will use outlier detection (see Section3.2.4) to eliminate outliers as possible split points. The graph segmentation module recursively processes each segment and stops when no segment is identified as needing further splitting. At this point, the individual segments must be represented as straight line trends. Although the least square regression line is a mathematically correct representation of a segment as a line, it does not necessarily capture the visual appearance of the trend and also results in disconnected segments representing the overall graph. Thus once the graph has been broken into subsegments, each segment is represented by a straight line connecting the segment's end points, producing a representation of the overall graph as a sequence of connected line segments, each of which is presumed to capture a visually distinguishable trend in the original graphic.

Decision tree induction is used to build a model for deciding whether to further split a segment. Thirteen attributes are considered in building the decision tree. The next four sections discuss statistical tests that are the basis for many of the attributes in our decision tree and the motivation for using them.

3.2.1 Correlation Coefficient

A trend can be viewed as a linear relation between the x and y variables. The Pearson product-moment correlation coefficient measures the tendency of the dependent variable to have a rising or falling linear relationship with the independent variable. It is obtained by dividing the covariance of two random variables x,y by the product of their standard deviation.

$$r_{xy} = \frac{n\sum x_i y_i - \sum x_i \sum y_i}{\sqrt{n\sum x_i^2 - (\sum x_i)^2}\ \sqrt{n\sum y_i^2 - (\sum y_i)^2}}$$

The correlation is 1 in the case of an increasing linear relationship, -1 in the case of a decreasing linear relationship, and some value in between for all other cases. The closer the coefficient is to either -1 or 1, the stronger the correlation between the variables; we use the absolute value of the correlation coefficient in our experiments.

Thus we hypothesize that the correlation coefficient may be useful in determining that a set of jagged short segments, such as the interval from 1930 to 2003 in Figure 2, should be captured as a single rising trend and not be split further.

3.2.2 F Test

Although the correlation coefficient is useful in detecting when a segment should be viewed as a single trend (and thus not split further), it is not sufficient by itself. For example, a long smooth rise in a line graph may overshadow a shorter stable portion of the graph (as in Figure 4), resulting in a high correlation coefficient even though the graph should be split into two segments. Similarly, a relatively flat segment, such as the line graph in Figure 5, will have a low correlation coefficient, even though it should not be split into subsegments.

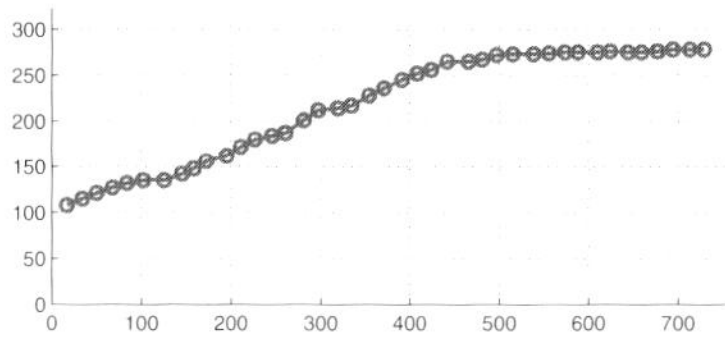

Fig. 4. Graph with high correlation coefficient but which should be treated as two trends

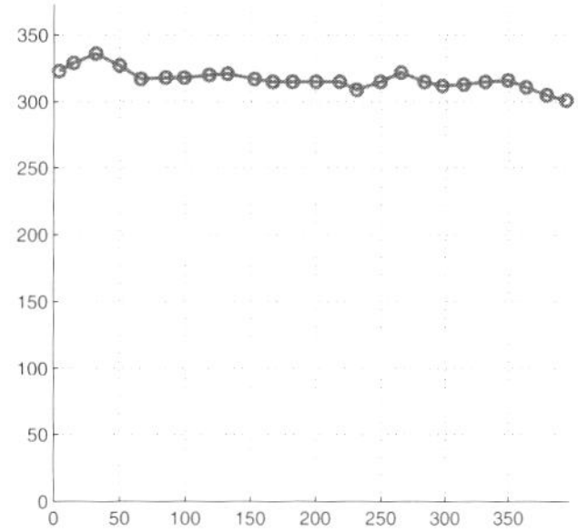

Fig. 5. Graph with low correlation coefficient, but which should be treated as a single trend

To address this, we make use of the F test[8,12] which can measure whether a two-segment regression is significantly different from a one-segment regression based on the differences in their respective standard deviations. The null hypothesis is that the two regression models are equal, suggesting that the segment need not be split further into subsegments. The F test statistic is computed as

$$\text{test statistic: } F = \frac{(RSSL - RSS)/2}{RSS/(n-4)}$$

where n is the total number of points, RSSL is the residual sum of squares of the one-phase least squares linear regression, and RSS is the residual sum of squares of the two-phase piecewise least squares linear regression. F here is distributed as an F-distribution with $(2, n-4)$ degrees of freedom as given in [12].

For each sample point x_i where $1 < i < n-1$, we test if it is appropriate to treat x_1 to x_i and x_{i+1} to x_n as two linear regressions. We use a significance level of $\alpha = 0.05$ and the critical value based on sample size given in [12]. We hypothesize that attributes based on the F test may be useful in identifying whether to split a segment into two subsegments.

3.2.3 Runs Test

In using the F-test to suggest when a segment might represent a sequence of two trends, we consider every possible way of breaking the segment into two subsegments. This is computationally impractical when considering more than two subsegments. However, we still need to recognize when a segment consists of more than two trends, such as the graph in Figure 6. For this graph, the correlation coefficient is high and the two-segment F-test fails. Thus we make recourse to the Runs Test[5]. The Runs Test detects if a regression fits the data points well. For each point, it calculates its residual from the regression line and categorizes it as +1 or −1, according to whether the residual is positive or negative. Then the number of *runs* is calculated, where a *run* is a continuous sequence of residuals which belong to the same category, such as consecutive +1 or −1. If N_+ is the number of positive residual points and N_- is the number

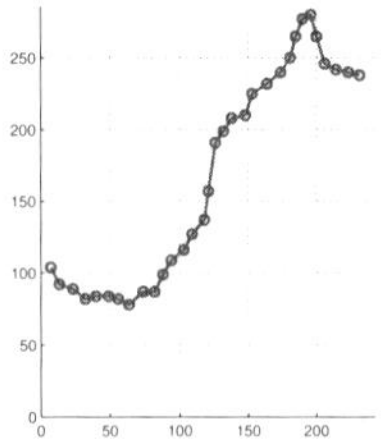

Fig. 6. Line graph with three trends in it, sampled from Business Week

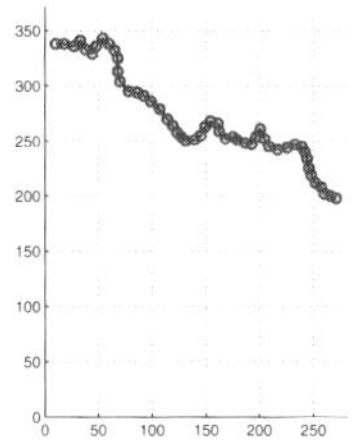

Fig. 7. Line graph of falling trend, sampled from USA Today

of negative residual points, the mean and standard deviation of the runs are approximated as

$$R_{mean} = \frac{2\ N_+\ N_-}{N_+ + N_-} + 1\ , \quad SD = \sqrt{\frac{2\ N_+\ N_-\ (2\ N_+\ N_- - N_+ - N_-)}{(N_+ + N_-)^2\ (N_+ + N_- - 1)}}$$

If the number of runs computed from the data points is sufficiently close to $R_{mean} \pm SD$, the residual is probably a reasonable approximation of the error from the regression, and this regression model may be regarded as a good fit to the data points.

In our application, we use the least square linear regression through the sampled points as a linear approximation of the segment. We use the Runs Test to check how well this linear regression can fit these data points. If the actual number of runs R is larger than $R_{mean} - SD$, then the Runs Test suggests that the segment represents a single trend. Thus we hypothesize that attributes based on the Runs Test may be helpful in inducing a decision tree for deciding when to split a segment.

Although the Runs Test appears powerful in suggesting whether a segment should be split further, it alone is insufficient. The Runs Test only uses the sign of the residual, not its value. It may suggest that the line graph in Figure 7 should be split, rather than viewing it as a single falling trend. However, other attributes, such as the correlation coefficient discussed earlier, will suggest otherwise.

3.2.4 Outlier Detection

A line graph may have one or more points that significantly diverge from the overall trend; such points perhaps should be viewed as outliers and not cause a segment to be split further. Thus we employ an outlier detection test based on residuals[13]. To detect the presence of outliers, we assume that the trend can be represented as a regression line connecting two end points; thus all the points in the segment can be represented as $y_i = b_1 + b_2\, x_i + \epsilon_i$ where ϵ_1 and ϵ_n are both 0. The residual $e_i = y_i - b_1 - b_2 x_i$ and the estimated standard deviation of e_i is

$$s_i = \hat{\sigma}\sqrt{1 - \frac{1}{n} - \frac{(x_i - \bar{x})^2}{\sum (x_i - \bar{x})^2}}$$

where $\hat{\sigma} = \sqrt{\sum e_i^2/(n-2)}$. If $\hat{\sigma}$ equals 0, there are no outliers. Otherwise, the standardized residuals $r_i = e_i/s_i$ are computed and $R_m = max|e_i/s_i|$ is used as a test statistic for outlier detection. We use a significance level of $\alpha = 0.01$ and the critical value given in [13] (based on the sample size). If R_m is greater than the critical value, outlier detection suggests the presence of an outlier in the sampled data points. If there are several r_i that exceed the critical value, then several outliers are suggested. Thus our decision tree induction includes attributes based on outlier detection.

3.3 Inducing the Model for Splitting

Table 1 presents the features of a segment in a line graph that are used to train a model for deciding whether to split a segment into subsegments. (Recall that initially the entire line graph is a single segment, which the model may recursively split into subsegments until each subsegment may be viewed as a single trend.) The first two features capture the absolute number of points both in the overall graph and in the segment under consideration, and the third feature captures the proportion of points in the segment; these features were included

Table 1. All attributes used in decision tree

ATT #	ATTRIBUTE NAME	TYPE	ATTRIBUTE DESCRIPTION
1	total number of points	numeric	number of sampling points in the whole graph
2	number of points in current segment	numeric	the number of sampling points in current segment
3	percentage of the total points	numeric	ratio of attribute2 to attribute1: indicates length of current segment as a percentage of the whole line graph
4	correlation coefficient	numeric	correlation coefficient calculated from the data points in current segment
5	F test	0 or 1	result of F test: 1, if there exists a F value greater than the critical value in F test; otherwise, 0
6	changing points in F test	numeric	the number of x_i where the F value exceeds the critical value
7	Runs Test	0 or 1	result from Runs Test: 1, when actual number of runs is less than $R_{mean} - SD$; otherwise, 0
8	actual runs	numeric	number of runs detected by runs test for current segment
9	mean runs	numeric	R_{mean} calculated in the Runs Test
10	standard deviation of runs	numeric	the SD calculated in the Runs Test
11	difference between actual runs and mean runs	numeric	the difference between actual runs and mean runs. Calculated as $\|R - R_{mean}\|/R_{mean}$
12	outlier detection	0 or 1	result from outlier detection: 1, when R_m is greater than the critical value; otherwise, 0
13	number of outliers	numeric	the number of standardized residuals r_i which are greater than the critical value

because it appeared that length of a segment or the size of a segment in relation to the overall graph might influence whether the segment should be split. The remainder of the features in Table 1 are derived from the statistical tests discussed in the previous section. The fourth feature is obtained from the correlation coefficient; the fifth and sixth features result from the F test; features 7-11 are obtained from the Runs Test; and features 12 and 13 are produced by outlier detection. The C5.0 decision tree algorithm is used to build a classification tree based on these 13 attributes. The target value of the decision tree is a binary decision, *split* or *no-split*; a decision of *no-split* indicates that the segment should be viewed as consisting of a single trend and not split further.

4 Evaluation and Analysis of Results

We collected 197 line graphs from various local and national newspapers and popular magazines such as *Business Week* and *Newsweek*. For each line graph in this set, the system computed all 13 features for the graphic and asked the human supervisor whether the graph should be split into subsegments. The values of the 13 features and the *split/no-split* decision were recorded as one instance in the training dataset. If the human supervisor indicated that the graph should be split, then the split point was computed as described in Section 3.2, the segment was split into two subsegments, and the process was repeated on the two subsegments, thereby producing additional training instances. Our 197 line graphs produced a training set containing 754 instances.

Training on this dataset using C5.0 produced the decision tree given in Table 3. The target value 1 means a *split* decision, and value 0 means a *no-split* decision. *Correlation coefficient* and *percentage of the total points* appear at the top levels of the decision tree, indicating that they are the attributes deemed most important in making the *split/no-split* decision. The *correlation coefficient* is the measure of a linear rising or falling relationship between the x and y variables. A high *correlation coefficient* seems to be the strongest predictor of whether a segment should be viewed as a single trend. *Percentage of the total points* indicates how large a particular segment is compared to the whole line graph. This attribute is important because under a global view, frequent changes over a small subsegment will be less noticeable than if the segment covered a large part of the graph. Thus the decision tree requires a higher *correlation coefficient* to make a *no-split* decision when the segment constitutes a large portion of the graph, as shown in the top half of Table 3. Similarly, when the *correlation coefficient* is low (<.815541), a decision to split is made when the segment constitutes most of the graph but more analysis is needed when the

Table 2. Confusion matrix for our model

	classified as 0	classified as 1
actual value as 0	379	76
actual value as 1	68	231

segment constitutes a smaller portion of the graphic, as shown in the lower half of Table 3.

The next two most important attributes in the decision tree are *F test* and *Runs Test*, which appear at the third and fourth levels. If the *correlation*

Table 3. Decision tree trained on 197 line graphs

```
correlation coefficient > 0.815541:
:...percentage of the total points <= 0.62963: 0 (265/5)
:   percentage of the total points > 0.62963:
:   :...correlation coefficient > 0.962782: 0 (28)
:       correlation coefficient <= 0.962782:
:       :...F test = 0:
:           :...difference between actual runs
:           :     and mean runs <= 0.052632: 1 (4)
:           :   difference between actual runs
:           :     and mean runs > 0.052632:
:           :   :...percentage of the total points <= 0.894737:
:           :       :...total number of points <= 10: 1 (2)
:           :       :   total number of points > 10: 0 (22/1)
:           :       percentage of the total points > 0.894737:
:           :       :...difference between actual runs
:           :             and mean runs <= 0.480712: 0 (28/8)
:           :           difference between actual runs
:           :             and mean runs > 0.480712: 1 (5)
:           F test = 1:
:           :...percentage of the total points > 0.866667: 1 (30/7)
:               percentage of the total points <= 0.866667:
:               :...total number of points <= 14: 0 (3)
:                   total number of points > 14:
:                   :...actual runs <= 4: 1 (3)
:                       actual runs > 4: 0 (9/2)
correlation coefficient <= 0.815541:
:...percentage of the total points > 0.894737: 1 (119/4)
    percentage of the total points <= 0.894737:
    :...Runs Test = 1: 1 (82/18)
        Runs Test = 0:
        :...F test = 1: 1 (7)
            F test = 0:
            :...percentage of the total points <= 0.392857:
                :...actual runs > 5: 0 (10)
                :   actual runs <= 5:
                :   :   number of points in current segment <= 5: 0 (31/4)
                :       number of points in current segment > 5:
                :       :...correlation coefficient <= 0.538139: 1 (10/2)
                :           correlation coefficient > 0.538139: 0 (13/3)
                percentage of the total points > 0.392857:
                :...outlier detection = 1: 1 (9/3)
                    outlier detection = 0:
                    :...total number of points > 18: 0 (21/6)
                        total number of points <= 18:
                        :...correlation coefficient > 0.725035: 0 (10/2)
                            correlation coefficient <= 0.725035:
                            :...difference between actual runs
                                  and mean runs <= 0.215768: 1 (30/4)
                                difference between actual runs
                                  and mean runs > 0.215768:
                                :...total number of points <= 17: 0 (11/3)
```

coefficient is high, but not high enough to reliably make a *no-split* decision given that the segment covers a large portion of the graph, then the *F test* becomes important in determining whether to keep this segment as a single trend (see top half of the decision tree in Table 3). This makes sense since a high *correlation coefficient* means a stronger linear relation, and the *F test* can confirm that a two-phase regression is not more appropriate than a single line regression. On the other hand, if the *correlation coefficient* is low (suggesting the lack of a linear relationship over the segment), the segment is split if it constitutes most of the graph; but when the segment constitutes less of the graph, a decision cannot reliably be made and further analysis in the form of a *Runs Test* (as a statistic measuring the goodness of fit between the data points and a single regression line) becomes important (see lower half of the decision tree in Table 3).

Leave-one-out cross validation, in which each instance is used once as test data and the other 753 instances are used for training, produces a success rate of 80.9% and the confusion matrix shown in Table 2. A baseline success rate of 60.3% would occur if we adopted the most common decision present in the training set, namely never to split a segment. Thus decision tree induction produces a relative improvement of 34.2% over the baseline model.

As examples of the segmentations produced by our system, consider the graphs in Figure 1, 2 and 3a. For the graph in Figure 1, our system generates two trends, a falling trend from 2000 to 2002 and a rising trend from 2002 to the end of 2005. And for the graph in Figure 3a, our system generates three trends as shown in Figure 3b. However, for the graph in Figure 2, our system generates an overall rising trend due to the stable segment from 1900 to 1930 not being long enough to significantly lower the correlation coefficient.

5 Conclusion and Future Work

We have presented a decision tree induction model, based on attributes derived from statistical tests and features of the graphic, for determining when to split a segment of a line graph in order to divide it into a sequence of segments representing visually distinguishable trends in the graphic. Our model has a high success rate and performs significantly better than a baseline model. In the future, the segmentation produced by our model, along with communicative signals in the graphic (such as annotation of points in the graphic or the presence of verbs such as "rising" in the caption, both of which occur in Figure 2) will be input to a Bayesian network that will hypothesize the graphic's message.

References

1. Elzer, S., Carberry, S., Zukerman, I., Chester, D., Green, N., Demir, S.: A probabilistic framework for recognizing intention in information graphics. In: Proc. of Int. Joint Conference on Artificial Intelligence (IJCAI 2005), pp. 1042–1047 (2005)
2. Elzer, S., Schwartz, E., Carberry, S., Chester, D., Demir, S., Wu, P.: A browser extension for providing visually impaired users access to the content of bar charts on the web. In: Proc. of Int. Conf. on Web Information Systems, pp. 59–66 (2007)

3. Freedman, E., Shah, P.: Toward a model of knowledge-based graph comprehension. In: Proc. of Int. Conf. on Diagram Representation and Inference, pp. 59–141 (2002)
4. Keogh, E., Chu, S., Hart, D., Pazzani, M.: An online algorithm for segmenting time series. In: Proc. of IEEE Int. Conference on Data Mining, pp. 289–296 (2001)
5. Bradley, D.C., Steil, G.M., Bergman, R.N.: OOPSEG: a data smoothing program for quantitation and isolation of random measurement error. Computer Methods and Programs in Biomedicine, 67–77 (1995)
6. Lin, J., Keogh, E., Lonardi, S., Patel, P.: Finding motifs in time series. In: Proc. of Workshop on Temporal Data Mining, pp. 53–68 (2002)
7. Toshniwal, D., Joshi, R.: Finding similarity in time series data by method of time weighted moments. In: Proc. of Australasian database conference, pp. 155–164 (2005)
8. Vieth, E.: Fitting piecewise linear regression functions to biological responses. Journal of Applied Physiology, 390–396 (1989)
9. Dasgupta, D., Forrest, S.: Novelty detection in time series data using ideas from immunology. In: Neural Information Processing Systems (NIPS) Conference (1996)
10. Yu, J., Reiter, E., Hunter, J., Mellish, C.: Choosing the content of textual summaries of large time-series data sets. Natual Language Engineering, 25–49 (2006)
11. Chester, D., Elzer, S.: Getting computers to see information graphics so users do not have to. In: Proc. of Int. Symp. on Method. for Int. Systems, pp. 660–668 (2005)
12. Beckman, R., Cook, R.: Testing for two-phase regressions. Technometrics, 65–69 (1979)
13. Tietjen, G.L., Moore, R.H., Beckman, R.J.: Testing for a single outlier in simple linear regression. Technometrics, 717–721 (1973)

Automatic Handling of Digital Image Repositories: A Brief Survey

Charbel Julien

Università del Piemonte Orientale, Alessandria, Italy
charbel@mfn.unipmn.it

Abstract. Repositories of digital images are being built in a variety of domains and for many different tasks, both for personal and for public use. Automated assistance tries to alleviate the access and to assist users during image manipulation tasks. An effective and reliable image search system has many applications. In spite of the many works in this area of research, to build a reliable and effective system is still a challenge. In this article we provide a brief survey of different techniques used to assist users when they deal with large digital image repositories.

Keywords: Image Digital library, Search approaches, Low-level features, Similarity measure, User interaction.

1 Introduction

A digital image comes from a continuous world. It is obtained from an analogue image by sampling and quantization. The process that maps the state of the real world to the digital 2D-image is very well understood. The inverse process, which tries to reconstruct the world using an image, does not exist for the simple reason that we cannot see around corners. This phenomenon is referred to as the "*sensory gap*" [1].

Images include both visual and semantic content. Visual content can be either very general or domain-specific. General visual content includes colors, texture, shape, spatial relationships. Domain-specific visual content, like human faces, is application-dependent and may require domain knowledge. Many schemes have been proposed to extract low-level visual features from images. The discrepancy between low-level visual features (such as color, texture, and shape) and high-level concepts (such as people, dogs, and trees, as perceived by humans) is referred to in the literature as the "*semantic gap*" [2]. Machine learning and statistical modeling techniques show very promising results to solve this problem.

In most practical applications the computers are pre-programmed to solve a particular task, but methods based on learning are now becoming increasingly common. The current revolution is in the use of Machine Learning to capture the variations in visual appearance rather than having the designer of the model accomplish this. Models learned from large datasets are likely to be more robust and more realistic than pre-designed models.

A. An et al. (Eds.): ISMIS 2008, LNAI 4994, pp. 410–416, 2008.

In this paper we focus on Image Retrieval Systems by Content [3] from digital image repositories. In the following we discuss the image search methods, the low-level visual descriptors, the similarity/dissimilarity measures, and the user-database interaction.

2 Image Search Methods

The earliest search techniques were based on manual annotation; first, images were manually annotated, and, afterwards, database management technologies were used to deal with images. Several popular World Wide Web search engines, such as *Images.Google* and *Flicker* from Yahoo, still use this method, where the user searches images by keywords. Newest research techniques try to identify images by their content. The reason to search images by content is that automatic search can be scalable to be applied to large image databases and are obviously objective, unlike manual annotation, which is often subjective. Recently, some works, such as ALIP [7], for instance, try to perform automatic annotation, by building a probabilistic model for every concept.

In general, image search systems should support a multitude of usage types. Users are likely to present very diverse search scenarios [8]. The most common type of search is *target search*, in which the user is trying to find a specific target image. A typical example is searching for a specific image in a personal photograph collection. Another search types is *category search*, when user is looking for images belonging to a specific class. Obviously, the class concept is user-dependent. On the other hand, in *database browsing*, the user has no idea about the target and just looks around in search of any "interesting" image. Image searches of this type are highly interactive, and the retrieval goal may change during the session.

Whatever the image search type, the common task is to extract low-level visual cues from images, and, then, to apply a distance or similarity measure to compare images among each other. Another task is to build a model for each of several concepts, for instance using a probabilistic framework for automatic annotation [6] exploiting visual features of images.

3 Low-Level Visual Descriptors

Images can be represented with different kinds of content descriptors from different levels. Until now no direct way has been found to extract high level semantic descriptors from images. Many low-level visual descriptor schemes have been proposed in the literature. Using these low-level visual descriptors, we can derive high level semantic information by inference. A comprehensive study of different CBIR systems [3] proves the value of low-level features to search images by their content.

Visual content features can be either global or local. A *global* descriptor uses the visual features of the whole image. A *local* descriptor uses the visual features of regions or objects to describe the image content. To obtain the local visual descriptors, an image is often divided into equal parts, first. Another approach consists in dividing the image into homogeneous regions according to some criterion, using region segmentation. An interesting method is to identify a complete object [10]. Each region of the image is represented by its visual feature vector. Currently, automatic object segmentation for broad domains of generic images is unlikely to succeed.

More recently, in order to capture local characteristics of an image, many systems extract *salient points* or *interest point* [9]. Salient points are locations in the image where there is a significant variation; they have been proved very effective to localize the search when the user is only interested in a portion of the image, or he/she wants to identify a specific object, being the rest of the image irrelevant.

On the other hand, signatures of color and texture are very interesting to describe the global composition of images. Signatures are used in a number of relevant work [4]. Signatures, unlike histograms, try to abstract the content of image, color and texture, using a variable number of categories, determined on each individual image through a clustering process. To compute the distance between signatures linear optimization techniques are needed such as the *Mallows* distance or the *Earth Mover's* distance [11].

In any case, *color* and *texture* remain the basic features to represent visual content either globally or locally. A set of color and texture descriptors, rigorously tested for inclusion in the MPEG-7 standard, is described in [20]; these descriptors are well suited both for natural images and for videos.

Color descriptors are very important. In content based image retrieval, many systems are based mainly on color features. Many schemes to extract relevant information of the image colors have been proposed. Color descriptors include *color space*, *color moments*, *color histogram*, *color coherence vector*, *color correlogram*, *dominant colors*, *scalable color* and *color global layout*. Color moments and color histogram are global features; they describe the distribution of colors in the image without any consideration of spatial relationships. The other color descriptors try to overcome this drawback, taking into account the color spatial relationships. The color space is used as a basis for the other color image features. Commonly used color spaces for image retrieval include RGB (Red, Green and Blue), HSV (Hue, Saturation and Value), and LUV (Luminance, U and V Chrominance information).

Texture is another important low-level visual feature. Texture plays an important role in the analysis of images. In many machine vision and image processing algorithms, simplifying assumptions are made about the uniformity of intensities. Humans recognize texture when they see it but texture is very difficult to define. Many definitions have been proposed for image texture [12],and they are grouped into four different classes [13]: *Statistical*, *Geometrical*, *Model-based*, and *Signal processing-based*. Statistical methods include *Co-occurrence matrices*, *Autocorrelation features*, and *Tamura's features*. Geometrical methods include *Voronoi tessellation* features and *Structural methods*. Model-based include *Markov Random Fields*, and *fractals*. Psychological research has provided evidence that the human brain does a frequency analysis of the image. Texture is especially suited for this type of analysis because of its properties. Signal Processing technique of texture analysis include *Fourier domain filtering*, *Wavelet transform*, and *Gabor filters*.

We present in Table 1 some experiments evaluation of different low-level visual descriptors, color, and texture. We use unsupervised clustering, k-medoid for clustering WANG image database composed of 10 classes. Images are represented by a signature of color (HSV color space), texture (Daubichies4 wavelet), structural composition of color and structural composition of texture [4]. We use F-measure [21] to evaluate the clustering accuracy according to the ground truth clustering. Numbers in table represents the F-measure regarding every class. We compute the F-measure for every cluster regarding every class and we consider the best one. This experiment try to investigate the performance of low level visual descriptors each alone and after a combination of them in representing images.

Table 1. Clustering Evaluation by color and texture

Class	Color	S.C. Color	Texture	S.C. Texture	Combination
Africa	0.42	0.38	0.33	0.28	0.50
Beach	0.32	0.45	0.23	0.31	0.39
Historical Building	0.31	0.23	0.29	0.26	0.39
Bus	0.50	0.32	0.57	0.38	0.59
Dinosaurs	0.93	0.87	0.63	0.89	0.97
Elephants	0.46	0.47	0.24	0.26	0.48
Flowers	0.42	0.57	0.36	0.48	0.47
Horses	0.68	0.72	0.41	0.36	0.61
Mountains	0.46	0.34	0.29	0.25	0.45
Foods	0.59	0.32	0.25	0.29	0.63

4 Similarity/Dissimilarity Measure

A system oriented to search images by content must by endowed with a *distance* or a *similarity* measure to compute similarity between images or between an image and a model. This measure is of fundamental importance for all automatic visual systems. Some systems use similarity, whereas others use dissimilarity or distance; indeed it is a trivial task to convert between similarity and dissimilarity.

Similarity or distance measures are closely related to low-level visual features. Many measures have been proposed, some of them applicable to histograms and some to feature vectors; among many others, we can mention the Euclidean distance, the Minkowski distance, the Cosine distance, and the Histogram Intersection distance [5]. All these distances compare histograms and feature vectors bin-by-bin. Actually, bin-by-bin comparison has the drawback that the resulting distance value is very sensitive to small changes in color or lighting conditions; to overcome this problem, different similarity/distance measures, such as the

Quadratic-form Similarity, have been suggested. Another drawback of bin-by-bin dissimilarity measures is their sensitivity to the bin size. A binning that is too coarse will not have sufficient discriminative power, while a binning that is too fine might place similar features in different bins, so that they will not be matched.

Distance between distributions or signatures can be calculated by Mallows [11] distance or Earth Mover's Distance [14]. The computation of these distances can be set as an optimization problem, solvable via linear optimization techniques.

In some automatic annotation methods, such as *ALIP* [7], the measurement of the similarity between an image and a model is performed using a probabilistic framework, namely 2D-Hidden Markov Models, other probabilistic frameworks using a mixture model [6] for computation convenience.

In the region-based image representation, many approaches have been proposed to compute similarity. As in this case images may contain different numbers of regions, it is not obvious how to compare them, in particular to decide which region in one image corresponds to which region in the other one, and an appropriate alignment has to be found. There are two main algorithms for computing image distance using homogeneous regions: Quantized Hungarian Region Matching and Integrated Region Matching (IRM) [10].

5 User and Image Database Interaction

Similarity judged by human perception is usually different from the one assessed via the feature space and the associated distance measure in artificial systems. Many content based image retrieval systems measure the similarity using a feature vector, but it is not easy to find relevant features and relevant similarity measures that imitate human visual perception. There are several reasons for this: first of all, there exists a gap between high-level concepts and low-level feature representation (*semantic-gap*). Second, human judgement may change over time. In order to solve this problem, interactive help from the user is required to map low-level features to high-level concepts; this help is called *feedback*. *Relevance feedback* is an automatic process of similarity refinement based on interactive human evaluation. In the literature many relevance feedback strategies were proposed to show its utility for CBIR systems; in this section we will consider only some of them.

Exploiting user's feedback is an iterative process, which consists of a series of consecutive queries. At every query the user provides feedback regarding the retrieval results, e.g., by qualifying images returned as either "relevant" or "irrelevant"; using this feedback, the system learns which visual features of the images are important and returns improved results to the user. The representation of individual images has an impact on the relevance feedback method; two representation schemes were used for the images: a fixed-length feature vector of global appearance, including color, texture and shape information, and a set of regions (or bag), where every region is described by color, texture and shape.

Many schemes was been proposed to exploit the feedback; some of them seek to learn a distance measure, whereas others try to incorporate the user-provided information into the search process itself. In relevance feedback, users can give positive (and negative) examples of images they are (are not) interested in. A machine learning algorithm can use labeled images to learn dissimilarity measure. In [17] a distance is learned from positive equivalence constraints by coding similarity using information theory and a probabilistic framework. Some of the relevance feedback techniques rely on re-weighting the features when the image is represented using a single feature vector [19]. In [18] a probabilistic model for clustering using pair-wise constraints, must-link and cannot-link constraints is presented. User constraints can be used in the indexing process by semi-supervised clustering, and afterwards in the search process.

A relevance feedback based on Self-Organizing Maps (SOMs) applied to image retrieval was proposed in [15]. The map surface is convolved with a window function to satisfy user constraints; a method for incorporating location-dependent information on the relative distances of the map units in the window function is presented as well.

For localized image search based on *salient points*, the multiple-instance learning approach is well suited. Each bag corresponds to an image and each feature vector in the bag corresponds to a portion of the image. Multiple instance learning has been used for both localized and classical image search tasks [9] [16].

Acknowledgments. This work was partly supported by the PRIN 2006 "Learning Hierarchical, Abstract Models from Temporal or Spatial Data" Project (N. 2006012944).

References

1. Smeulders, A., Worring, M., Satini, S., Gupta, A., Jain, R.: Content-based image retrieval at the end of the early years. IEEE transactions on patern analysis and machine intelligence 22(12) (December 2000)
2. Chen, Y., Li, J., Wang, J.: Machine Learning and Statistical Modeling Approaches to Image Retrieval. Kluwer Academic Publishers, Dordrecht (2004)
3. Remeco, C., Veltkamp, T.M.: Content-based Image retrieval Systems: A survey. technical report UU-CS-2000-34 (October 2002)
4. Datta, R., Ge, W., Li, J., Wang, J.: Toward Bridging the Annotation-Retrieval Gap in Image Search. In: Proceedings of ACM Multimedia Conference (Octobre 2006)
5. Rubner, Y.: Perceptual Metrics For Image Database Navigation A Ph.D. Dissertation submitted to the department of computer science of Stanford University (1999)
6. Li, J., Wang, J.: Real-Time Computerized Annotaion Of Pictures. In: Proceedings of ACM Multimedia Conference (Octobre 2006)
7. Li, J., Wang, J.: Automatic Linguistic Indexing of Pictures by a Statistical Modeling Approach. IEEE Transaction on Pattern Analysis and Machine Intelligence 25 (September 2003)

8. Cox, I.J., Minka, M.L.: The bayesian image retrieval system, PicHunter: Theory, implementation and psychophysical experiments. IEEE Transaction on Image Processing (2000)
9. Zhang, H., Rahmani, R., Cholleti, R., Goldman, S.: Local Image Representations Using Pruned Salient Points With Applications to CBIR. In: Proceedings of ACM Multimedia Conference (October 2006)
10. Wang, J., Li, J., Wiederhold, G.: SIMPLIcity: Semantic-Sensitive Integrated Matching for Picture LIbraries. IEEE Transactions on Pattern Analysis and Machine Intelligence 23(9), 947–963 (2001)
11. Levina, E., Bickel, P.: The earth mover's Distance is the Mallows distance: Some insights from statistics. In: International Conference on Computer vision. In: Vancouver (2001)
12. Coggins, J.M.: A framework for texture analysis based on spatial filtering Ph.D. thesis, computer science Department, Michigan state University, East Lansing, Michigan (1982)
13. Tuceryan, M.: Texture Analysis. In: Chen, C.H., Pau, L.F., Wang, P.S.P. (eds.) The Handbook of Pattern Recognition and Computer Vision, 2nd edn., pp. 207–248. World Scientific Publishing Co., Singapore (1998)
14. Rubner, Y., Tomasi, C., Guibas, L.J.: A Metric for distribution with Applications to Image Databases. In: Proceeding of International Conference on Computer Vision, Bombay, India, pp. 59–66 (January 1998)
15. Koskela, M., Laaksonen, S., Laakso, J., Oja, E.: Self-Organizing Maps as a Relevance Technique in Content-Based Image Retrieval Pattern Analysis & Applications (2001)
16. Chen, Y., Wang, J.: Image categorization by learning and reasoning with regions. Journal of machine learning Research, 913–939 (2004)
17. Hillel, A.B., Weinshall, D.: Learning Distance function by Learning similarity. In: 24th International Conference on Machine learning (June 2007)
18. Nelson, B., Cohen, I.: Revisiting Probabilistic Models for Clustering with Pair-wise Constraints. In: 24th International Conference on Machine learning (June 2007)
19. Huang, X., Chen, S.-C., Shu, M.-L., Zhang., C.: Learning and inferring a semantic space from user's relevance feedback for image retrieval ACM. In: Multimedia (2002)
20. Manjunath, B.S., Ohm, J.-R., Vasudevan, V.V., Yamad, A.: Color and texture descriptors. IEEE Transaction Circuits and Systems for video Technology (2001)
21. Yang, Y., Liu, X.: A re-examination of text categorization methods. In: Proceedings of the ACM SIGIR Conference on Research and Development in Information Retrieval (1999)

Development of the XML Digital Library from the Parliament of Andalucía for Intelligent Structured Retrieval

Juan M. Fernández-Luna, Juan F. Huete,
Manuel Gómez, and Carlos J. Martín-Dancausa

Departamento de Ciencias de la Computación e Inteligencia Artificial
E.T.S.I. Informática, Universidad de Granada, 18071–Granada, Spain
{jmfluna,jhg,mgomez,cmdanca}@decsai.ugr.es

Abstract. This paper describes the development of the XML digital library in Spanish from official documents published by Parliament of Andalucía. These documents include discussions about some important matters affecting citizens from the southern Spanish region of Andalucía. The original documents, which are organized around a very well defined structure, were published in PDF format, so the complete conversion process is explained in detail. The main reason for this format change is to allow the users of the regional chamber's website to make the most of the interesting advantages given by the structured Information Retrieval.

1 Introduction and Motivation

The Parliament of Andalucía, the autonomous region of the southern Spain, and as many other chambers, generates a group of e-documents called session diaries, published in the Internet to allow a easy access by the citizens, containing exact transcriptions of all the interventions of the members of the parliament. The sessions are organized around legislatures, usually 4 years of political activity. At present, the Portable Document Format (PDF) is used in most of the organizations, included our Parliament, to store and spread textual information.

Once the PDF documents are published, a search engine is provided so users can consult the collection by means of a query in natural language. The result of this process is a set of complete PDF documents, sorted according to their relevance degree with respect to the query, who must be inspected.

If we take into account that each session celebrated in the parliament presents a very well defined structure, as well as the fact that each document contains an exact replica of its corresponding session, the content of each PDF is organized according to a strict and rich structure that may be useful in terms of retrieval.

In the field of Information Retrieval (IR) [1], when the retrieval mechanism is able to use the structural information contained in the documents, we are dealing with so-called structured IR [2]. Then, the internal organization of the documents is employed to give back the user, instead of a whole relevant document, only those parts of them which are relevant. This means an important saving of user

A. An et al. (Eds.): ISMIS 2008, LNAI 4994, pp. 417–423, 2008.

time. Also, the user could specify in the query those specific parts in which she/he is interested (what to get) and, at the same time, those parts where she/he wishes to search in (where to look for).

Thanks to the internal organization of the text from session diaries, the legislative collection of the Parliament of Andalucía could be exploited from a structured IR perspective. In our case, this aim will be achieved by means of Garnata [3], a structured IR system based on Influence Diagrams (ID) [4].

As we take into account that PDF documents do not contain structural information, only text, this is not the most appropriate format for our purposes. But considering the nature of the session diaries, the structure is intrinsically in its content, and must be extracted in order to use it. But we must not only mine the text to look for the structure, but also we have to attach the text associated to each section of the document. The proposal for this aim is the XML standard.

In this paper we present the first stages of a project which has as one of its objectives the conversion of the whole collection of session diaries of the regional chamber into XML from PDF, in order to use it with a structured IR engine. Once the XML digital library has been built, the structured IR system would work as follows: Users would formulate queries to the system and this would retrieve the most appropriate parts of the session diaries where relevant information appears instead of whole documents. Since documents are relatively large, this implies a time saving as users do not have to find the information.

Therefore, this paper will be organized in the following parts: Section 2 will show a description of the internal organization of the session diaries. Section 3 will explain how, starting from that organization, the structure of the XML documents has been designed. The following, Section 4, will deal with the conversion process, describing the different steps to generate the XML digital collection. Section 5 will discuss the retrieval engine to return XML elements, and finally, the last section will offer some concluding remarks and future work.

2 Description of the Organization of Session Diaries

The parliamentary cycle moves around the concept of *parliamentary initiative*, which is an action taken by a member of the parliament or a politic group that could become discussed in a plenary session or in a commission of a specific area. These initiatives are included in the plenary or commission sessions and identified by means of an initiative code. Sessions are divided in two: plenary sessions, where all the members of the parliament are gathered in session debating initiatives; and commission sessions: the deputies are divided in different areas of interest (agriculture, economy, education, etc.), and discuss initiatives related to the type of commission where they belong to.

Before a session starts, the politic groups represented in the chamber agree on the agenda of the session, i.e. a sequence of initiatives, grouped by type. After this agreement the president of the parliament or the corresponding commission is in charge of leading the discussion of each point in the order.

Regarding the diary organization, it is composed of four different parts: *1. General Information section*: General information about the session itself(the legislature, date and presidency of the parliamentary session, and in case of the commission, its type). *2. Agenda*: A list of agreed initiatives grouped by their type, composed of an initiative code, its subject, and who proposes it. *3. Summary*: A detailed description of the agenda, created once the session has finished. For each initiative, grouped by their type, beside the description of the initiative (code, subject and who proposes), it is added the list of members of the parliament who participated in its debate, and the result of the vote.*4. Development of the session*: For each one of the points included in the summary section, the transcriptions of all the speeches are included.

3 Design of the XML Structure

A deeper analysis of the documents leads to the design of a DTD file containing an accurate representation of the structure that all the XML files must follow, declaring the name of the tags that will appear in the XML files, and their relationships. The DTD file has the following structure:

The root tag of the hierarchy is *diario_sesion* which can contain several *legislatura (legislature) tags*. This means that the documents are organized according to periods of 4 years, at maximum. Every *legislatura tag* contains an attribute indicating the legislature number (*numero attribute*) and a list of *diario_sesion_pa tags*, storing the session diary information. The tag contains an attribute to describe the kind of session (*tipo attribute*), and it is organized as follows:

1. General Information: The *tag diario_sesion_pa (session diary)* contains information about the session. It consists of the celebration year (*anio tag*), number of session diary (*numero_sesion tag*), and several tags which are stored in the pleno (plenary) tag: president of the session (*presidente tag*), number of plenary session (*numero_sesion_plenaria tag*) and celebration date (*fecha_celebracion tag*).

```
<!ELEMENT diario_sesion_pa  (anio, numero, pleno)>
<!ATTLIST diario_sesion_pa tipo CDATA #REQUIRED>
<!ELEMENT pleno (presidente, numero_sesion_plenaria, fecha_celebracion, orden_del_dia?,
          sumario, desarrollo)>
```

2. Agenda: The list of those initiatives to be discussed (*orden_del_dia tag*), is also stored in the *pleno tag*. It is formed by several *tipo_iniciativa tags*, which are the types of initiatives scheduled in the original agenda to be discussed. The following information of each initiative is recorded: its title (*nombre tag*), and a repetition of one of two kinds of tags or a combination of both of them: if it is a grouped debate (*debate_agrupado tag*), it contains its order (*orden tag*) and the initiatives of this grouped debate (*iniciativas tag*); if it is an individual initiative (*iniciativa tag*), its code (*expediente tag*), file number (*numero_expediente tag*), summary of the matter (*extracto tag*) and the members of parliament who discussed about this topic (*proponente tag*).

```
<!ELEMENT orden_del_dia (tipo_iniciativa)+ >
<!ELEMENT tipo_iniciativa (nombre, (debate_agrupado|iniciativas)+)>
<!ELEMENT nombre (#PCDATA)>
<!ELEMENT debate_agrupado (orden, iniciativas+)>
<!ELEMENT orden (#PCDATA)>
<!ELEMENT iniciativas, (expediente?, numero_expediente?, extracto?,proponente? )>
```

3. Summary: The summary (*sumario tag*), placed into the *pleno tag*, contains the date and the time when the session starts (*hora_inicio and fecha_incio tags*) and finishes (*hora_fin and fecha_fin tags*), and its content (*contenido tag*). Under this tag we can find the list of the points of the agenda discussed in the session (*punto_orden tag*). Inside this tag the description of the corresponding item in the agenda (*descripcion_punto tag*) is included. All the subjects (*tema tag*) discussed are placed into *punto_orden tags*. *Tema tags* store all the information about the summary of the initiative (*extracto_sumario tag*), the members of parliament who discussed (*intervienen tag*), and results of the vote (*votacion tag*).

```
<!ELEMENT sumario (hora_inicio, fecha_inicio, contenido, hora_fin?, fecha_fin?,
          rectificacion_errores?)>
<!ELEMENT contenido (punto_orden | tema)+>
<!ELEMENT punto_orden (descripcion_punto, tema+)>
<!ELEMENT tema (extracto_sumario,intervienen?, votacion*,excepcion?)>
```

4. Development: The last part of the DTD is the *desarrollo tag*, stored in the *pleno tag*, which contains the transcriptions of the discussions organized around the initiatives (t*tema_desarrollo tag*), and who talked, and what she/he said, i.e. the transcription of the speech (*intervienen_desarrollo* and (*intervencion_desarrollo tag*) organized in different paragraphs (*parrafo_desarrollo tags*).

```
<!ELEMENT desarrollo ((tema_desarrollo?,intervienen_desarrollo, intervencion_desarrollo)+)>
<!ELEMENT intervencion_desarrollo (parrafo_desarrollo+)>
```

4 The Conversion Process

The content-internal structure of the PDF files is not enough to use it in a structured IR system. The reason is that we are only able to extract plain text without any additional knowledge that tells us if a sentence extracted from the text is a common paragraph or the title of a section, for example, and this is the knowledge that we want to get. So the problem is to generate an alternative representation of the session diaries that allow feed the search engine. Then we need software that is able to extract the text contained in all the PDF documents composing the collection and organize it according to its logical structure.

The reason why we took the option of extracting the text instead of using a direct PDF-to-XML converter is that any of existing converters are not able to directly obtain an XML file marked up with the logical document organization. We have used an external PDF as text converter, specifically, *pdftotext*[1], whose output will be the input of our XML converter. The reason is that in case of changing the pdf-to-text converter the application could work without any problem and any modification of the software, provided a text file is given as

[1] Included in the Xpdf package – http://www.foolabs.com/xpdf/

input. Then, once a text file containing the textual contents of the PDF file is generated, our application creates the final XML file starting from this file as input. The software, developed in Java, performs the following steps (general steps which can be applied to any PDF collection):

By means of a *lexical analyzer*, it processes the text of these files and produces, as output, a sequence of symbols called tokens. The system introduces in the text a group of structural components or labels which correspond with the tokens, indicating the bounds of the different sections found in the document. Another task carried out by the lexical analyzer is the noise elimination, i.e. the deletion of those parts of the document which are not important from a semantic point of view. When the lexical analyzer has finished, the converter runs a *syntax analyzer* developed in javacc, in charge of generating a grammar to detect the tokens mentioned in the previous step. In order to create the XML files, we use the DOM API based on the building of a tree in memory whose nodes are XML tags. Therefore when a token is detected a new node or a group of nodes are created in the DOM tree, generating all the hierarchical structure of the XML format. Finally, the tree is converted into the corresponding XML file and validated. In the specialized literature, there are other approaches for converting PDF to XML. This is the case of [5,6,7,8].

The original collection is composed of the legislatures IV, V, VI and VII, containing the plenary and commission sessions, with a total of 1588 PDF files, with a size of 1039 MBytes. Once converted into XML its size is 392 MBytes. The reduction factor is high and the size of the XML collection is low, but growing by 100 documents per year. Information about the three first legislatures is not included as the PDF are composed of scanned images of the printed versions of the documents, and our converter is not able to work with them at this stage.

5 Use of the Digital Library: Structured IR

Once the digital library has been constructed, it will be used to feed the Garnata IR system [3], whose underlying retrieval model is based on Influence Diagrams [4,9]. The model is used to determine the optimal decision policy. More formally, an influence diagram is an acyclic directed graph containing decision nodes that represent variables that the decision maker controls directly; chance nodes, which represent random variables, i.e. uncertain quantities, which are relevant to the decision problem and cannot be controlled directly; and utility nodes that represent utility, i.e. the cost or the preference degree of the consequences derived from the decision process. They are quantified by the utility of each of the possible combinations of outcomes of their parent nodes. With respect to the quantitative part, chance nodes will store a set of probability distribution, measuring the strength of the relationships of each node with the set of parents. With respect to the utility nodes, a set of utility values is associated, specifying for each combination of values for the parents, a number expressing the desirability of this combination for the decision maker. The goal is to choose the decision that will lead to the highest expected utility (optimal policy) [9].

Garnata software implements the Context-based Influence Diagram model. Starting from a document collection containing a set of documents and the set of *terms* used to index these documents, then we assume that each document is organized hierarchically, representing structural associations of its elements, which will be called *structural units*, composed of other units or terms.

The chance nodes of the ID are the terms, T_j, and the structural units, U_i. They have associated a binary random variable, whose values could be term/unit is not relevant or is relevant, respectively. Regarding the arcs, there is an arc from a given node (either term or structural unit) to the particular structural unit node it belongs to. Decision nodes, R_i, model the decision variables. There will be one node for each structural unit. It represents the decision variable related to whether or not to return the corresponding structural unit to the user, taking the values 'retrieve the unit' or 'do not retrieve the unit'. Finally, utility nodes, V_i. We shall also consider one utility node for each structural unit, and it will measure the value of utility of the corresponding decision. In order to represent that the utility function of a decision node obviously depends on the decision made and the relevance value of the structural unit considered, we use arcs from each chance node U_i and decision node R_i to the utility node V_i. Another important set of arcs are those going from the unit where U_i is contained to V_i, which represent that the utility of the decision about retrieving the unit U_i also depends on the relevance of the unit which contains it. Finally, for each node V_i, the associated utility functions must be defined.

In the context of the Parliament of Andalucía digital library, once the user formulates a query in natural language, the system computes the probability of relevance of each structural unit. With this information, the search engine makes the decision of retrieving a unit or not considering not only its utility and relevance, but also both measures from the unit where it is contained. This process is repeated until the right unit that would contain the information need is found. Then, the system offers an ordered list of units by expected utility.

6 Conclusions

In this paper we have presented the XML digital library of session diaries from the Parliament of Andalucía. The main motivation for this conversion has been the possibility of using a structured information retrieval system to allow the access of the relevant elements with a more appropriate granularity of the retrieved items. The internal organization of the documents has been described as well as the conversion process by which, starting from the session diaries in PDF, they are converted into XML. Finally, the retrieval model is described in order to show how IDs are able to select the most interesting elements for the users.

Acknowledgment. This work has been supported by the Spanish 'Ministerio de Educación y Ciencia' and 'Consejería de Innovación, Ciencia y Empresa de la Junta de Andalucía' under Projects TIN2005-02516 and TIC-276, respectively.

References

1. Baeza-Yates, R.A., Ribeiro-Neto, B.A.: Modern Information Retrieval. ACM Press/Addison-Wesley (1999)
2. Chiaramella, Y.: Information retrieval and structured documents. In: Lectures on information retrieval, pp. 286–309 (2001)
3. de Campos, L., Fernández-Luna, J., Huete, J., Romero, A.: Garnata: An information retrieval system for structured documents based on probabilistic graphical models. In: Proc. of the 11th Int. Conf. of Information Processing and Management of Uncertainty in Knowledge-Based Systems (IPMU), pp. 1024–1031 (2006)
4. Jensen, F.V.: Bayesian Networks and Decision Graphs. Springer, Heidelberg (2001)
5. Déjean, H., Meunier, J.L.: A system for converting pdf documents into structured xml format. In: Bunke, H., Spitz, A.L. (eds.) DAS 2006. LNCS, vol. 3872, pp. 129–140. Springer, Heidelberg (2006)
6. Gurcan, A., Khramov, Y., Kroogman, A., Mansfield, P.: Converting pdf to xml with publication-specific profiles. In: Proc. of the XML Conf. (2003)
7. Hardy, M.R.B., Brailsford, D.F.: Mapping and displaying structural transformations between xml and pdf. In: DocEng., pp. 95–102 (2002)
8. Yildiz, B., Kaiser, K., Miksch, S.: pdf2table: A method to extract table information from PDF files. In: 2nd Indian Int. Conf. on AI, Pune (2005)
9. Shachter, R.D.: Probabilistic inference and influence diagrams. Oper. Res. 36(4), 589–604 (1988)

Evaluating Information Retrieval System Performance Based on Multi-grade Relevance

Bing Zhou and Yiyu Yao

Department of Computer Science, University of Regina
Regina, Saskatchewan, Canada S4S 0A2
{zhou200b,yyao}@cs.uregina.ca

Abstract. One of the challenges of modern information retrieval is to rank the most relevant documents at the top of the large system output. This brings a call for choosing the proper methods to evaluate the system performance. The traditional performance measures, such as precision and recall, are not able to distinguish different levels of relevance because they are only based on binary relevance. The main objective of this paper is to review 10 existing evaluation methods based on multi-grade relevance and compare their similarities and differences through theoretical and numerical examinations. We find that the normalized distance performance measure is the best choice in terms of the sensitivity to document rank order and giving higher credits to systems for their ability of retrieving highly relevant documents. The cumulated gain-based methods rely on the total relevance score and are not sensitive enough to document rank order.

1 Introduction

Relevance is a fundamental concept in information retrieval (IR) which has been extensively discussed from different perspectives in the past. There are two broad views of relevance: the system-oriented view regards relevance as relationships between documents and queries to a certain topic; while the user-oriented view thinks of relevance as users judgments to decide if the documents are useful to them [21]. Over the years, most IR system evaluations only consider relevance as a two-leveled judgments, that is, documents are either relevant or non-relevant to a given query. The traditional evaluation methods, such as precision and recall, are based on this binary assumption. However, documents are not equally relevant to users, some documents are more relevant, and some documents are less relevant. Relevance has multi-degrees. In modern IR systems, users can easily get a large number of relevant documents to a query which exceed the number they want to examine. Therefore, it is important for a system that can rank the most relevant documents at the top of the system output. This brings a call for evaluation methods that are able to distinguish multiple levels of relevance and give higher credits to those systems for their ability of retrieving the most highly relevant documents and ranking them at the top of the output list.

Comparing to the widely used precision and recall, the IR system evaluations based on multi-grade relevance have not received much attention until the beginning of 2000. Various of evaluation methods have been proposed [8,10,11,12,18,19,23]. Some IR

A. An et al. (Eds.): ISMIS 2008, LNAI 4994, pp. 424–433, 2008.

workshops, such as the Text REtrieval Conference (TREC) and the NII Test Collection for IR Systems (NTCIR), employed multi-grade relevance on their test collections. The main objective of this paper is to reveal the evaluation criteria based on multi-grade relevance, and compare the 10 existing evaluation methods based on multi-grade relevance through some theoretical and numerical examinations. We hope that our work can bridge the gap between the IR evaluations based on binary relevance and multi-grade relevance, and provide the references for choosing the superior evaluation methods among the others.

2 Multi-grade Relevance Scales

Over the years, researchers have argued what types of relevance scales should be used to categorize multiple levels of relevance. Eisenberg [6] introduced magnitude estimation which can use numeric, line-length or force hand grip to estimate the different relevance degrees of documents. In the experimental study of Katter [7], relevance scales have been classified into category, ranking, and ratio scales, and the category scales have been used by most of the existing evaluation methods based on multi-grade relevance.

The category scale is an ordinal scale which does not imply one document is how many times better than another document. The traditional binary relevance can be seen as a two category scale consisting of relevant or non-relevant. The TREC conference used a 3-point scale (relevant, partial relevant and non-relevant), and the NTCIR project used a 4-point scale on their test collection (highly relevant, relevant, partial relevant and non-relevant). There is an ongoing debate over the optimal number of categories of multi-grade relevance [1,9,14]. The various numbers of categories range from 2 to 11 points [4,5,15,20]. A general agreement is that a universal interpretation for the various numbers of categories does not appear to exist because people use different standards of what makes a specific number of categories optimal [3].

The difficulty in using the category scales based on multi-grade relevance suggests that a predefined relevance scale may not be suitable for measuring and reflecting user judgments. To resolve this problem, Yao [23] adopted the concept of user preference from decision and measurement theory. Using a preference relation, a user only provides the relative relevance judgments on documents without referring to any predefined relevance scale. The user preference of documents may be formally defined as a binary relation $\succ$ between any document pairs. For two documents d and d',

$$d \succ d' \quad \text{iff} \quad \text{the user prefers } d \text{ to } d'.$$

The user preference relation is rich enough to represent any category scales based on multi-grade relevance. It is not restricted to a limited number of relevance levels, and is easier for users to understand and make their judgments.

3 Evaluation Methods Based on Multi-grade Relevance

In 1966, the six evaluation criteria suggested by Cleverdon provide a foundation for designing evaluation methods [2]. These six criteria are: (1) the coverage of the collection,

(2) system response time, (3) the form of the presentation of the output, (4) user efforts involved in obtaining answers to a query, (5) recall and (6) precision. Of these criteria, precision and recall were most frequently used and still are the dominant approach to evaluate the performance of information retrieval systems. The precision is the proportion of retrieved documents that are actually relevant. The recall is the proportion of relevant documents that are actually retrieved. In spite of their successes, precision and recall are only based on binary relevance assumption and not sufficient to reflect the degrees of relevance. This disadvantage has resulted in the call for reconsidering the evaluation criteria based on multi-grade relevance.

In an ideal IR system, the retrieved documents should be ranked in a decreasing order based on their relevance degree to a certain query. That is, the more relevant documents should always be ranked ahead of the less relevant documents. Moreover, this system should be able to catch the few most relevant documents instead of retrieval a bunch of partially relevant documents. Therefore, A evaluation method based on multi-grade relevance should have the following properties: i) It should be able to compare the similarities and differences between ideal ranking given by the user judgments and actual ranking given by the IR system. The closer between the two rankings the better, ii) It should be able to give higher credit to systems that can retrieve the highly relevant documents, iii) It must be flexible and adaptable to different relevance scales. In the following subsections, we review the 10 existing evaluation methods based on multi-grade relevance, and examine how they qualify the above properties.

3.1 Sliding Ratio and Modified Sliding Ratio

The sliding ratio was proposed by Pollack [13] in 1968. In this method, the ranked system output is compared against the ideal ranking. Relevance score d_i is assigned to each ith ranked document in the output list. The scores correspond to the category rating scales and are flexible to express any categories. For example, if a 3-point scale is chosen, the possible scores of d_i are 2, 1, and 0 indicating the relevance degree as highly relevant, relevant, and non-relevant respectively. The overall ranking is quantified by the sum of d_i. The sliding ratio of the system ranking and ideal ranking is defined as:

$$sr = \frac{\sum_{i=1}^{n} d_i}{\sum_{i=1}^{n} d_{I(i)}},$$

where $d_{I(i)}$ is the relevance score of the ith ranked document in the ideal ranking.

In a good rank order, the ith document should always be ranked ahead of the jth document if $d_i \geq d_j$. Unfortunately, the sliding ratio is not sensitive to the document rank order. Sagara [17] has proposed the modified sliding ratio which takes the document ordering into account. It is defined as:

$$msr = \frac{\sum_{i=1}^{n} \frac{1}{i} d_i}{\sum_{i=1}^{n} \frac{1}{i} d_{I(i)}}.$$

That is, when a highly relevant document ranked at the bottom of the output list, its contribution to the whole system performance drops.

3.2 Cumulated Gain and Its Generalizations

The cumulated gain proposed by Jarvelin and Kekalainen [10] is very similar to the idea of sliding ratio. It assumed that the user scans the retrieved documents list, and adds a relevance score each time he finds a relevant document. The cumulated gain is defined as:

$$cg = \sum_{i=1}^{n} d_i.$$

Jarvelin and Kekalainen [10] also take the document ordering into consideration by adding a discounting function which progressively reduces the document relevance score as its rank increases. The way they used discounting is to divide the document relevance score by the log of its rank which is similar to the modified sliding ratio, except that the latter uses the division by the rank directly which makes the reduction too steeply. The discounted cumulated gain is defined as:

$$dcg = \begin{cases} \sum_{1 \le i \le b} d_i, \\ \sum_{b < i \le n} \frac{d_i}{\log_b i}, \end{cases}$$

where b is the bth ranked document in the output list.

Suppose dcg_I is the discounted cumulated gain of an ideal ranking, the normalized discounted cumulated gain up to the rth ranked document is defined as [11]:

$$ndcg = \frac{1}{r} \sum_{n=1}^{r} \frac{dcg}{dcg_I}.$$

3.3 Generalizations of Average Precision

Weighted average precision was introduced by Kando, Kuriyams and Yoshioka [12] as an extension of average precision in order to evaluate multi-graded relevance. The average precision has been widely used in IR experiments for evaluating binary relevance, and is defined as:

$$ap = \frac{1}{R} \sum_{n=1}^{N} rel(d_i) \frac{\sum_{i=1}^{n} d_i}{n},$$

where R is the total number of relevant documents for a certain query, N is the total number of documents in the ranked output, $rel(d_i)$ is a function, such that $rel(d_i) = 1$ if the document at the ith rank is relevant and $rel(d_i) = 0$ otherwise. Since the average precision is based on binary relevance, the possible relevance scores of d_i are either 1 representing relevant, or 0 representing non-relevant, so the sum of d_i is the number of relevant documents up to the nth ranked document.

The weighted average precision extends the average precision by assigning multi-grade relevance scores to d_i. The sum of d_i is the cumulated gain cg, and the cumulated gain of an ideal ranking up to the rth ranked document denoted by cg_I. The weighted average precision is defined as:

$$wap = \frac{1}{R} \sum_{n=1}^{N} rel(d_i) \frac{cg}{cg_I}.$$

However, in the case of $n > R$, the cumulated gain of the ideal ranking cg_I becomes a constant after the Rth ranked document, so it can not distinguish between two systems when one of the systems has some relevant documents ranked at the bottom of n. Sakai proposed Q-measure [19] in order to address this problem. In Q-measure, the relevance score or gain value d_i is replaced by the bonused gain $bg_i = d_i + 1$ if $d_i > 0$ and $bg_i = 0$ otherwise. So the cumulated bonused gain is defined as:

$$cbg = \sum_{i=1}^{n} bg_i.$$

Q-measure is defined as:

$$Q = \frac{1}{R}\sum_{n=1}^{N} rel(d_i)\frac{cbg}{cg_I + n}.$$

So the denominator $(cg_I + i)$ always increases after the Rth ranked document instead of remain constant.

3.4 Average Gain Ratio

In information retrieval experiments, one of the important properties that needs to be evaluated is how well the few most relevant documents are retrieved, but most evaluation methods treat them as same as the partially relevant documents. Since the amount of the partially relevant documents are usually much larger than the most relevant ones, so most evaluation methods are affected by how well the partially relevant documents are retrieved. The average gain ratio [18] is designed for giving more credits to systems for their ability of retrieving the most relevant documents. The relevant score or gain value is adjusted as $d'_{l(i)} = d_{l(i)} - \frac{R_l}{R}(d_{l(i)} - d_{(l-1)(i)})$, where l denotes the relevant level, $d_{l(i)}$ denotes the relevance score for finding an l-relevant document at rank i, and R_l denotes the number of l-relevant document. By employing this adjusted relevant score to weighted average precision, the average gain ratio is defined as:

$$agr = \frac{1}{R}\sum_{n=1}^{N} rel(d_{l(i)})\frac{cg'}{cg'_I}.$$

3.5 Normalized Distance Performance Measure

The normalized distance performance measure introduced by Yao [23] uses the user preference relation to generate a relative order of documents considered as the user ranking or ideal ranking. It measures the distance between user ranking and system ranking by examining the agreement and disagreement between these two rankings.

The distance between two rankings is defined with respect to the relationships between document pairs of these two rankings. There are three types of relationships: let the distance count 0 if two rankings agree on a document pair, count 1 if they are compatible on a document pair, and count 2 if they are contradict on a document pair. The distance function between the user ranking $\succ_u$ and system ranking $\succ_s$ is defined as:

$$\beta(\succ_u, \succ_s) = 2 * C^- + 1 * C^0 + 0 * C^+ = 2C^- + C^0,$$

where C^- is the numbers of contradictory pairs, C^0 is the numbers of compatible pairs, and C^+ is the numbers of agreeing pairs.

The notion of accelerable ranking [16,22] was suggested to be more suitable for information retrieval. Instead of measuring the distance between user ranking and system ranking, the distance performance measure measures distance between the system ranking and the acceptable ranking $\succ_a$ closest to the user ranking, and $\succ_a$ is defined as:

$$\succ_a \;=\; \succ_u \cup (\sim_u \cap \succ_s),$$

where $\sim$ is the indifferent relationship between two documents. The number of contradicting and compatible pairs between $\succ_a$ and $\succ_s$ are defined as:

$$|\succ_a \cap \succ_s^c| = |\succ_u \cap \succ_s^c| = C^-,$$
$$|\succ_a \cap \sim_s| = |\succ_u \cap \sim_s| = C^u.$$

Therefore, the distance performance measure is defined as:

$$dpm(\succ_u, \succ_s) = \beta(\succ_a, \succ_s) = 2C^- + C^0 = 2C^- + C^u.$$

The normalized distance performance measure was also proposed to measure the performance of every query equally. It is defined as:

$$ndpm(\succ_u, \succ_s) = \frac{dpm(\succ_u, \succ_s)}{\max_{\succ \in \Gamma(D)} dpm(\succ_u, \succ)} = \frac{dpm(\succ_u, \succ_s)}{dpm(\succ_u, \succ_u^c)} = \frac{2C^- + C^u}{2C},$$

where $\Gamma(D)$ denotes the set of all acceptable rankings of $\succ_u$, and $\succ_u^c$ denotes converse ranking of $\succ_u$. For example, let

$$\begin{matrix} d_1 \\ d_2 \end{matrix} \succ_u d_3 \succ_u \begin{matrix} d_4 \\ d_5 \end{matrix} \succ_u d_6$$

be a user ranking on a set of documents $D = (d_1, d_2, d_3, d_4, d_5, d_6)$, and

$$\begin{matrix} d_1 \\ d_3 \end{matrix} \succ_s \begin{matrix} d_2 \\ d_5 \end{matrix} \succ_s d_6 \succ_s d_4$$

be a system ranking. By the definition of the closest ranking to $\succ_u$, $\succ_a$ is given by:

$$d_1 \succ_a d_2 \succ_a d_3 \succ_a d_5 \succ_a d_4 \succ_a d_6.$$

The contradict pairs between $\succ_a$ and $\succ_s$ are (d_2, d_3) and (d_4, d_6), $C^- = 2$. The compatible pairs are (d_1, d_3) and (d_2, d_5), $C^u = 2$. The normalized distance performance measure is:

$$ndpm(\succ_u, \succ_s) = \frac{2C^- + C^u}{2C} = \frac{2 * 2 + 2}{2 * 13} = 3/13.$$

3.6 Average Distance Measure

The average distance measure [8] is measuring the distance between user ranking and system ranking by examining the absolute differences between system relevance estimation and user relevance estimation. Suppose D is a document set, for any document $d \in D$, let d_i denotes the relevant score of the ith document estimated by the IR system, and $d_{I(i)}$ denotes the relevance score of the ith document estimated by the user. The average distance measure is defined as:

$$adm = 1 - \frac{\sum_{d \in D} |d_i - d_{I(i)}|}{|D|}.$$

3.7 Result Analysis

After reviewing and analyzing the 10 existing evaluation methods based on multi-grade relevance, the properties of these methods and the connections between them became clear.

We find that most of these methods are based on cumulated gain, which has the similar idea with the sliding ratio method proposed back in 1968. That is, each retrieved document has been assigned a relevance score with corresponding to a predefined relevance scale, the overall ranking of the retrieved document set is quantified by the sum of relevance scores of each document, and the proportion of total relevance score of system ranking to ideal ranking indicates the performance of an IR system. The problem with cumulated gain is that it is not sensitive to the rank order of the retrieved document set. For example, if two systems retrieved the same document set with exactly opposite rank orders, their performances evaluated by cumulated gain will be the same because their total relevant scores are the same. In order to solve this problem, discounted cumulated gain, normalized discounted cumulated gain are proposed by taking the document ordering into consideration.

Some methods are generalized directly from the methods based on binary relevance in order to evaluate multi-grade relevance. The weighted average precision extends the widely used average precision by assigning multi-grade relevance scores to the retrieved documents. The problem of weighted average precision is that it is incapable to evaluated documents retrieved after rank R (i.e., the total number of relevant documents). Q-measured was proposed to address this problem by replacing the cumulated gain with bonused gain. The average gain ratio is also generalized from weighted average precision for the purpose of giving more credit to the systems that can retrieve the few most relevant documents, but it still has the same problem as weighted average precision.

Not like cumulated gain, the normalized distance performance measure and average distance measure focused on measuring the distance between system ranking and user ranking directly, so they are more sensitive to the document rank order. The normalized distance performance measure is based on a relative order of documents instead of a predefined relevance scale. The distance between system ranking and user ranking are defined by considering the relationships of document pairs of these two rankings. The average distance measure calculates the absolute differences between system relevance estimation and user relevance estimation of each document. It gives wrong evaluation results in certain cases.

4 Methods Comparison by Examples

In this section, we further compare the reviewed methods by employing them in some featured examples from two different perspectives.

First, we compare these methods in terms of their sensitivities to the document rank order. suppose that we are using a 7-point scale, and there are only five relevant documents to a given query. Let UR indicates the ideal ranking or user ranking, IRS1, IRS2, IRS3, and IRS4 represent four different IR systems. Their performance in terms of document rank order is that IRS1 is the best, IRS2 is better than IRS3, and IRS4 is the worst. Table 1 shows the actual evaluation results by the methods we discussed in the previous section. Let's briefly analyze the evaluation results. All methods are able to determine that IRS1 provides the best ranking and IRS4 provides the worst. But the methods based on cumulated gain give wrong evaluation results to the performance of IRS2 and IRS3, that is because although IRS2 provides a better ranking, but the sum of the relevance score of IRS3 is larger than IRS2. If we change the relevance score of each document in IRS2 and IRS3 so that their sum can be the same, Table 2 shows the results. All the cumulated gain based methods except the discounted cumulated gain (dcg) are able to give the right evaluation results. But unfortunately one can not adjust the relevance scores given by the system which usually decided by retrieval algorithms automatically. Therefore, the best methods in terms of document rank order is the normalized distance performance measure (ndpm). The cumulated gain based methods rely on the values of relevance and not sensitive enough to document ranking in general. The average distance measure (adm) relies on the absolute differences of the relevance score between the system ranking and user ranking, it can not provide stable evaluation results in all cases.

Second, we compare these methods in terms of giving higher credits to the IR systems for their abilities of retrieving highly relevant documents. This time we are using a 4-point scale, and there are only five relevant documents to a give query. Let IRS1, IRS2, IRS3, and IRS4 represent four different IR systems. Their performance for giv-

Table 1. Example 1: evaluation results of document rank order

Docs	d1	d2	d3	d4	d5	msr	dcg	ndcg	wap	Q	agr	ndpm	adm
UR	0.6	0.5	0.4	0.3	0.1								
IRS1	0.6	0.5	0.3	0.2	0.1	0.95	0.93	0.96	0.94	0.98	0.94	1.00	0.96
IRS2	0.5	0.3	0.4	0.2	0.1	0.79	0.77	0.78	0.79	0.93	0.78	0.90	0.92
IRS3	0.4	0.6	0.2	0.3	0.1	0.80	0.85	0.82	0.81	0.94	0.80	0.80	0.90
IRS4	0.1	0.2	0.2	0.4	0.5	0.43	0.54	0.34	0.40	0.80	0.36	0.05	0.70

Table 2. Changing the relevance score of example 1

Docs	d1	d2	d3	d4	d5	msr	dcg	ndcg	wap	Q	agr
UR	0.6	0.5	0.4	0.3	0.1						
IRS2	0.6	0.4	0.5	0.3	0.1	0.98	0.98	0.97	0.97	0.99	0.98
IRS3	0.5	0.6	0.3	0.4	0.1	0.95	0.99	0.95	0.95	0.98	0.95

Table 3. Example 2: evaluation results of retrieving highly relevant documents

Docs	d1	d2	d3	d4	d5	msr	dcg	ndcg	wap	Q	agr	ndpm	adm
UR	0.3	0.3	0.2	0.1	0.1								
IRS1	0.2	0.2	0.0	0.0	0.0	0.53	0.49	0.58	0.54	0.91	0.24	0.89	0.88
IRS2	0.0	0.3	0.2	0.1	0.1	0.47	0.63	0.46	0.50	0.89	0.55	0.63	0.94
IRS3	0.0	0.0	0.2	0.1	0.1	0.24	0.33	0.16	0.24	0.84	0.25	0.19	0.86
IRS4	0.0	0.0	0.0	0.1	0.1	0.08	0.11	0.04	0.06	0.81	0.05	0.13	0.84

ing high credits to systems which can retrieve more highly relevant documents is in a decreasing order as IRS1, IRS2, IRS3, and IRS4. Table 3 shows the actual evaluation results. Let's briefly analyze the evaluation results. The normalized distance performance measure (ndmp) provides the correct results again. All the cumulated gain-based methods except discounted cumulated gain (dcg) and average gain ratio (agr) are able to give the correct evaluation results. The average distance measure (adm) gives higher credit to IRS2 instead of IRS1 because the absolute difference between IRS2 and UR is higher.

5 Conclusions

The new features of modern information retrieval bring the call for more suitable evaluation methods. Throughput this paper, we reviewed different types of relevance scales and 10 existing evaluation methods based on multi-grade relevance. The evaluation criteria of multi-grade relevance changed compare to the traditional precision and recall. The evaluation methods based on multi-grade relevance should be able to credit the IR systems which can retrieve more highly relevant documents, provide better document rank order, and adaptable to different types of relevance scales.

Some interesting findings are revealed through the theoretical and numerical examinations. We find that that most methods are based on cumulated gain. They are able to give higher credits to IR systems for their abilities of retrieving highly relevant documents, but they are not sensitive enough to document rank order. The average distance measure is not reliable because it uses the absolute difference between system relevance estimation and user relevance estimation. Overall, The normalized distance performance measure provides the best performance in terms of the perspectives we are concerned in this paper.

References

1. Champney, H., Marshall, H.: Optimal Refinement of the Rating Scale. Journal of Applied Psychology 23, 323–331 (1939)
2. Cleverdon, C., Mills, J., Keen, M.: Factors Dermnining the Performance of Indexing Systems. Aslib Cranfield Research Project, Cranfield, UK (1966)
3. Cox, E.P.: The Optimal Number of Response Alternatives for A Scale: A Review. Journal of Marketing Research XVII, 407–422 (1980)
4. Cuadra, C.A., Katter, R.V.: Experimental Studies of Relevance Judgments: Final Report. System Development Corp., Santa Monica, CA (1967)

5. Eisenberg, M., Hu, X.: Dichotomous Relevance Judgments and the Evaluation of Information Systems. In: Proceeding of the American Scoiety for Information Science, 50th Annual Meeting, Learning Information, Medford, NJ (1987)
6. Eisenberg, M.: Measuring Relevance Judgments. Information Processing and Management 24(4), 373–389 (1988)
7. Katter, R.V.: The Influence of Scale from on Relevance Judgments. Information Storage and Retrieval 4(1), 1–11 (1968)
8. Mizzaro, S.: A New Measure of Retrieval Effectiveness (Or: What's Wrong with Precision and Recalls). In: International Workshop on Information Retrieval, pp. 43–52 (2001)
9. Jacoby, J., Matell, M.S.: Three Point Likert Scales are Good Enough. Journal of Marketing Research VIII, 495–500 (1971)
10. Jarvelin, K., Kekalainen, J.: IR Evaluation Methods for Retrieving Highly Relevant Documents. In: Proceedings of the 23rd Annual International ACM SIGIR Conference on Research and Development in Information Retrieval, New York (2000)
11. Jarvelin, K., Kekalainen, J.: Cumulated Gain-based Evaluation of IR Techniques. ACM Transactions on Information Systems 20, 422–446 (2002)
12. Kando, N., Kuriyams, K., Yoshioka, M.: Information Retrieval System Evaluation Using Multi-grade Relevance Judgments. In: Discussion on Averageable Singel-numbered Measures, JPSJ SIG Notes, pp. 105–112 (2001)
13. Pollack, S.M.: Measures for the Comparison of Information Retrieval System. American Documentation 19(4), 387–397 (1968)
14. Rasmay, J.O.: The Effect of Number of Categories in Rating Scales on Precision of Estimation of Scale Values. Psychometrika 38(4), 513–532 (1973)
15. Rees, A.M., Schultz, D.G.: A Field Experimental Approch to the Study of Relevance Assessments in Reletion to Document Searching. Cleverland: Case Western Reserve University (1967)
16. Rocchio, J.J.: Performance Indices for Document Retrieval. In: Salton, G. (ed.) The SMART Retrieval System-experiments in Automatic Document Processing, pp. 57–67 (1971)
17. Sagara, Y.: Performance Measures for Ranked Output Retrieval Systems. Journal of Japan Society of Information and Knowledge 12(2), 22–36 (2002)
18. Sakai, T.: Average Gain Ratio: A Simple Retrieval Performance Mmeasure for Evaluation with Multiple Relevance Levels. In: Proceedings of ACM SIGIR, pp. 417–418 (2003)
19. Sakai, T.: New Performance Matrics Based on Multi-grade Relevance: Their Application to Question Answering. In: NTCIR-4 Proceedings (2004)
20. Saracevic, T.: Comparative Effects of Titles, Abstracts and Full Texts on Relevance Judgments. In: Proceedings of the American Scoiety for Information Science, Learning Information, Medford, NJ (1969)
21. Vickery, B.C.: Subject Analysis for Information Retrieval. In: Proceedings of the International Conference on Scientific Information, vol. 2, pp. 855–865 (1959)
22. Wong, S.K.M., Yao, Y.Y., Bollmann, P.: Linear Structure in Information Retrieval. In: Proceedings of the 11th Annual Interna- tional ACMSIGIR Conference on Research and Development in In- formation Retrieval, vol. 2, pp. 19–232 (1988)
23. Yao, Y.Y.: Measuring Retrieval Effectiveness Based on User Preference of Documents. Journal of the American Society for Information Science 46(2), 133–145 (1995)

A Dynamic Window Based Passage Extraction Algorithm for Genomics Information Retrieval

Qinmin Hu[1] and Xiangji Huang[2]

[1] Department of Computer Science & Engineering, York University, Toronto, Ontario, Canada
vhu@cse.yorku.ca
[2] School of Information Technology, York University, Toronto, Ontario, Canada
jhuang@yorku.ca

Abstract. Passage retrieval is important for the users of the biomedical literature. How to extract a passage from a natural paragraph presents a challenge problem. In this paper, we focus on analyzing the gold standard of the TREC 2006 Genomics Track and simulating the distributions of standard passages. Hence, we present an efficient dynamic window based algorithm with a WordSentenceParsed method to extract passages. This algorithm has two important characteristics. First, we obtain the criteria for passage extraction through learning the gold standard, then do a comprehensive study on the 2006 and 2007 Genomics datasets. Second, the algorithm we proposed is dynamic with the criteria, which can adjust to the length of passage. Finally, we find that the proposed dynamic algorithm with the WordSentenceParsed method can boost the passage-level retrieval performance significantly on the 2006 and 2007 Genomics datasets.

1 Introduction and Motivation

Our major goal of the TREC Genomics Track is to create test collections for evaluation of information retrieval (IR) and related tasks in the genomics domain. The Genomics Track differs from all other TREC tracks in that it is focused on retrieval in a specific domain as opposed to general retrieval tasks, such as web searching or question answering. And the goal of the information retrieval (IR) system is to retrieve relevant information to the users. Especially the users in the biomedical domain, really desire to be provided short, specific answers to questions and put them in context by providing supporting information and linking to original sources. This motivated the TREC Genomics Track implemented a new task in 2006 that focused on passage retrieval using full-text documents from the biomedical literature [3].

For the TREC 2006 and 2007 Genomics Tracks, systems were tasked with extracting out relevant passages of text that answer topic questions and focus on retrieval of short passages (from phrase to sentence to paragraph in length) that specifically addressed an information need, along with linkage to the location in the original source document [3,4]. Here passages were required to be contiguous and not longer than one paragraph. Not only tasked to return passages of text, systems were but also measured on how well they retrieved relevant information at the passage-level, aspect-level, and document-level. It is similar to the evaluation measures in 2006 and 2007 which calculate recall and precision in the classic IR way, using the preferred TREC statistic of Mean Average

A. An et al. (Eds.): ISMIS 2008, LNAI 4994, pp. 434–444, 2008.

Precision (MAP). MAP means average precision at each point a relevant document is retrieved. The evaluation measures are also called gold standard measures which have been developed for five years since 2003 [2].

In this paper, we propose an efficient algorithm to get a dynamic window to define the passages with a WordSentenceParsed method. The center idea in the algorithm is to get reasonable passages (contiguous and not longer than one paragraph) with the criteria of word count and sentence count. The criteria is drawn from the analysis of 2006 gold standard, estimated by the regression model, and evaluated with the method of maximum likelihood. In an extensive experimental study, with the different tuning constant values of Okapi system, experiments results using the WorSentenceParsed method and the SentenceParsed method, are compared and analyzed.

2 Related Work

Most previous work on biomedical information retrieval appeared in the TREC Genomics Track from 2003 to 2007 was to address the prevalent name variation problem in biomedical text, such as query expansion, tokenization techniques, and conceptual retrieval [5,8,10,11]. Huang et al [5,10] expanded the query by generating lexical variants of the gene names in the queries using heuristic rules. Conceptual retrieval was presented to utilize domain-specific knowledge to improve its effectiveness in retrieving biomedical literature [11]. Jiang and Zhai [8] conducted a systematic evaluation of a set of tokenization heuristics on TREC biomedical text collections for ad hoc document retrieval, using two representative retrieval methods and a pseudo-relevance feedback method.

The size of the retrieved passages is also an important issue that information retrieval systems have to deal with. In the context of text retrieval, size can be defined as the length of the retrieved passage. A very short text may not contain enough information and a long text may contain either unnecessary or redundant information. It is not trivial to decide on the size that can provide the best retrieval result.

In this paper, our work focuses on how to extract the retrieval passages from full documents. There are three methods we have tried: (1) taking a natural paragraph as a passage. (2) considering taking 3 sentences as a passage. (3) simulating the criteria of word count and sentence count to get a passage. We have tried to take the natural paragraph as a passage in 2006 [6]. For the TREC 2006 and 2007 Genomics tracks, systems extract relevant passages of text that answer topic questions and focus on retrieval of short passages (from phrase to sentence to paragraph in length) that specifically addressed an information need. Here passages are required to be contiguous and not longer than one natural paragraph. Generally, users desire to get the most important with less redundant information.

We also considered to take three sentences as a passage. Since a passage is essentially a sequence of sentences, we initially measured the length of passages as the number of sentences they contain. Hence a SentenceParsed method with taking every three sentences as a passage is proposed, because for the ith sentence, it is reasonable to consider the relationship of its previous sentence and its next sentence. So we bound the

$(i-1)^{th}$ sentence, the ith sentence and the $(i+1)^{th}$ sentence to be a passage. Therefore, for every 2 passages, the overlapping are two sentences. For example,

$$Overlapping(Passage_{(i)}, Passage_{(i+1)}) = \{Sentence_{(i)}, Sentence_{(i+1)}\}$$

The problem with the SentenceParsed method is that sentences can be too short or too long. For example, a passage of three long sentences may contain redundant information. To overcome this problem, we consider using the criteria of both sentence count and word count for the length of the passages, which is the WordSentenceParsed method. Through setting up a regression model for the 2006 gold standard, we get the values of the criteria and a dynamic window based passage extraction algorithm is presented next.

3 Dynamic Window Algorithm

In this section, we will first analyze the gold standard in the TREC 2006 Genomics Track and count the numbers for sentence, passages, words for every topic, which are shown in Table 1. Then, a linear regression model is set up to simulate the distributions of words and sentences, evaluate the criteria to parse the paragraph into passages, as shown in subsection 3.1. Finally, the pseudo codes for an dynamic window based passage extraction algorithm are presented in subsection 3.2.

3.1 Linear Regression Modelling

In this model, we define the dependent variable y is modelled as a random variable because of uncertainty as to its value, given only the value of each independent variable x. We have a general linear regression equation as:

$$y_i = a + bx_i + \varepsilon_i, \qquad \varepsilon_i \sim \text{n}(0, \sigma^2) \tag{1}$$

where a is the intercept, b is the slope, and ε is the error term, which picks up the unpredictable part of the response variable y_i. And a, b, σ^2 are independent on x_i.

For $y_1, y_2, ..., y_n$, they are independent, then the probability density function of joint distribution is presented as equation (2).

$$\text{L} = \prod_{i=1}^{n} \frac{1}{\sigma\sqrt{2\pi}} \exp[-\frac{1}{2\sigma^2}(y_i - a - bx_i)^2] = (\frac{1}{\sigma\sqrt{2\pi}})^n \exp[-\frac{1}{2\sigma^2}(y_i - a - bx_i)^2] \tag{2}$$

To estimate the parameter of a, b, we use the method of maximum likelihood. Obviously, we need L to get maximum value, what we do is $\text{Q}(a, b)$ has minimum value. Equation (3) is the expression function for L. And we compute the derivation of Q with equation (4) and (5), do reduction with equation (6) and (7).

$$\text{Q}(a, b) = \sum_{i=1}^{n} (y_i - a - bx_i)^2 \tag{3}$$

$$\frac{\partial \text{Q}}{\partial a} = -2\sum_{i=1}^{n} (y_i - a - bx_i) = 0, \tag{4}$$

Table 1. 2006 Gold Standard Data Analysis

Topic ID	Sentence Count	Passage Count	Average Sentence Count	Total Word Count Per Passage	Average Word Count Per Passage
160	1116.00	527.00	2.12	19146	36.33
161	170.00	68.00	2.50	2689	39.54
162	67.00	18.00	3.72	782	43.44
163	612.00	262.00	2.34	8723	33.29
164	21.00	7.00	3.00	326	46.57
165	44.00	17.00	2.59	522	30.71
166	101.00	34.00	2.97	1372	40.35
167	823.00	208.00	3.96	14869	71.49
168	576.00	243.00	2.37	7588	31.23
169	721.00	103.00	7.00	10324	100.23
170	74.00	36.00	2.06	1148	31.89
171	124.00	50.00	2.48	1820	36.40
172	926.00	593.00	1.56	12542	21.15
173	0	0	0	0	0
174	172.00	36.00	4.78	2089	58.03
175	114.00	33.00	3.45	1609	48.76
176	45.00	14.00	3.21	609	43.50
177	33.00	9.00	3.67	419	46.56
178	23.00	7.00	3.29	370	52.86
179	37.00	13.00	2.85	537	41.31
180	0	0	0	0	0
181	2877.00	589.00	4.88	53160	90.25
182	397.00	144.00	2.76	4646	32.26
183	42.00	19.00	2.21	433	22.79
184	17.00	5.00	3.40	190	38.00
185	45.00	25.00	1.80	650	26.00
186	953.00	388.00	2.46	12941	33.35
187	13.00	3.00	4.33	392	130.67

$$\frac{\partial Q}{\partial b} = -2\sum_{i=1}^{n}(y_i - a - bx_i)x_i = 0. \tag{5}$$

$$na + (\sum_{i=1}^{n} x_i)b = \sum_{i=1}^{n} y_i, \tag{6}$$

$$(\sum_{i=1}^{n} x_i)a + (\sum_{i=1}^{n} x_i^2)b = \sum_{i=1}^{n} x_i y_i. \tag{7}$$

There is a unique solution for above equations. With the evaluation method of maximum likelihood, $\hat{a}$ and $\hat{b}$ are calculated using the following equation (8) and (9).

$$\hat{b} = \frac{n\sum_{i=1}^{n} x_i y_i - (\sum_{i=1}^{n} x_i)(\sum_{i=1}^{n} y_i)}{n\sum_{i=1}^{n} x_i^2 - (\sum_{i=1}^{n} x_i)^2} = \frac{\sum_{i=1}^{n}(x_i - \bar{x})(y_i - \bar{y})}{\sum_{i=1}^{n}(x_i - \bar{x})^2}, \tag{8}$$

$$\hat{a} = \frac{1}{n}\sum_{i=1}^{n} y_i - \frac{\hat{b}}{n}\sum_{i=1}^{n} x_i = \bar{y} - \hat{b}\bar{x}. \quad (9)$$

where

$$\bar{x} = \frac{1}{n}\sum_{i=1}^{n} x_i, \qquad \bar{y} = \frac{1}{n}\sum_{i=1}^{n} y_i.$$

Finally, the regression equation is shown as follows, where $\hat{a}$, $\hat{b}$ are parameters as the criteria for the dynamic algorithm.

$$\hat{y} = \tilde{a} + \hat{b}x \quad (10)$$

Using this regression model, we evaluate the values of $\hat{a}$ and $\hat{b}$ with the data in Table 1. The dependent variable y_i is the number in sentence count column, and the independent variable x_i is the number in passage count column. Therefore, we compute $\hat{a_1} = 3$, $\hat{b_1} = 0$. The values of $\hat{a_1} = 3$ and $\hat{b_1} = 0$ mean, for every independent passage count, that the sentence length for a passage in gold standard in the TREC 2006 Genomics Track can be a constant number averagely.

Similarly, we set that the dependent variable y_i is the number in total word count column, and the independent variable x_i is the number in passage count column. $\hat{a_2} = 45$ and $\hat{b_2} = 0$ are computed. The values of $\hat{a_2} = 45$ and $\hat{b_2} = 0$ mean, for every independent passage count, that the word count for a passage in gold standard in the TREC 2006 Genomics Track can be a constant number averagely. Therefore, we choose $\hat{a_1} = 3$ and $\hat{a_2} = 45$ as the criteria of sentence count and word count.

Input: a paragraph containing a number of sentences, $\hat{a_1} = 3$ and $\hat{a_2} = 45$ are the criteria of words and sentences
Output: a list of passages obtained from the paragraph

Method:
(1) Break the paragraph into a list of its sentences
(2) Until the list of sentences becomes empty
  (I) create an empty passage
  (II) until the passage has 45 words or 3 sentences
    (i) If there are 3 sentences within 45 words, add to passage
    (ii) Else if there are more than 2 sentences, there but less than 3 sentences, go to the end of the third sentence, add them to passage
    (iii) Else if there are more than 1 sentence but less than 2 sentences, go to the end of the second sentence, add them to passage
    (iv) Else if there is less than 1 sentence, go to the end of this sentence, add it to passage.
  (III) remove first sentence of the list if there are 2 or 3 sentences in the passage
  (IV) remove second sentence of the list only if three sentences were added
  (V) start reading from the beginning of the list

Fig. 1. The dynamic window algorithm with a WordSentenceParsed method

3.2 Algorithm

Out dynamic window based algorithm for passage extraction is presented in Figure 1. There are two major phases in this iterative algorithm. For the first phase, a natural paragraph is broken into a list of its sentences which are as the input of the second phase. For the second phase, the criteria of 45 words and 3 sentences is used to extract a passage, which is described in the step (II). In step (III) and (IV), the overlapping between passages are defined. Suppose the length of the natural paragraph is n, the time complexity of the algorithm is $\Theta(n)$. Also, a paragraph, as a input, has the maximum word count of 53,160 for Topic 181 from Table 1. Assume the maximum length of a word is 10, therefore 53,160 words take less than 1M. Hence, a paragraph can easily fit into main memory.

4 Experimental Evaluation

A series of experiments are conducted to evaluate the performance of proposed window dynamic algorithm in the biomedical domain. In this section, topics and datasets we used are described in subsection 4.1. Then, the system for the experiments is Okapi introduced in subsection 4.2. Finally, the results have been discussed and analyzed in subsection 4.3.

4.1 Datasets and Topics

For the TREC 2006 Genomics Track, a test collection of 162,259 full-text documents and 28 topics expressed as questions was assembled. Topics were derived from the set of biologically relevant questions based on the Generic Topic Types (GTTs) developed for the 2005 track [2]. These questions each had one or more aspects that were contained in the literature corpus (i.e., one or more answers to each question). They all had the general format of containing one or more biological objects and processes and some explicit relationship between them:

Biological object (1..many) ⟨—relationship—⟩ Biological process (1..many)

The biological objects might be genes, proteins, gene mutations, etc. The biological process could be physiological processes or diseases. The relationships could be anything, but were typically verbs such as causes, contributes to, affects, associated with, or regulates. We determined that four out of the five GTTs from 2005 could be reformulated into the above structure, with the exception of the first GTT that asked about procedures or methods. The Genomics website ($http://ir.ohsu.edu/genomics/$) lists all the 28 topics in 2006.

For the TREC 2007 Genomics Track, there were 36 official topics which were in the form of questions asking for lists of specific entities. The definitions for these entity types were based on controlled terminologies from different sources, with the source of the terms depending on the entity type. We gathered new information needs from working biologists. This was done by modifying the questionnaire used in 2004 to survey biologists about recent information needs. In addition to asking about information needs, biologists were asked if their desired answer was a list of a certain type of entity, such as genes, proteins, diseases, mutations, etc., and if so, to designate that entity type. Fifty information needs statements were selected after screening them against the corpus to ensure that relevant paragraphs with named entities were present, of which 36 were used as official topics and 14 used as sample topics. The Genomics website ($http://ir.ohsu.edu/genomics/$) lists all the 36 topics in 2007. [4].

4.2 The System

We used Okapi BSS (Basic Search System) as our main search system. Okapi is an information retrieval system based on the probability model of Robertson and Sparck Jones [1]. The retrieval documents are ranked in the order of their probabilities of relevance to the query. Search term is assigned weight based on its within-document term frequency and query term frequency. The weighting function used is BM25.

$$w = \frac{(k_1+1)*tf}{K+tf} * \log \frac{(r+0.5)/(R-r+0.5)}{(n-r+0.5)/(N-n-R+r+0.5)} * \frac{(k_3+1)*qtf}{k_3+qtf} \oplus k_2 * nq * \frac{(avdl-dl)}{(avdl+dl)} \tag{11}$$

where N is the number of indexed documents in the collection, n is the number of documents containing a specific term, R is the number of documents known to be relevant to a specific topic, r is the number of relevant documents containing the term, tf is within-document term frequency, qtf is within-query term frequency, dl is the length of the document, $avdl$ is the average document length, nq is the number of query terms, the k_is are tuning constants (which depend on the database and possibly on the nature of the queries and are empirically determined), K equals to $k_1 * ((1-b) + b * dl/avdl)$, and $\oplus$ indicates that its following component is added only once per document, rather than for each term.

In our experiments, the tuning constants k_1 and b are set to be different values. k_2 and k_3 are set to be 0 and 8 respectively.

4.3 Standard Evaluation

Gold standard for the TREC 2006 Genomics track has three levels of retrieval performance: passage retrieval, aspect retrieval and document retrieval [3]. Each of these provided insight into the overall performance for a user trying to answer the given topic questions. Each was measured by some variant of mean average precision (MAP). Passage-level MAP is offered as a new one in 2006 and it is also important in 2007 Genomics track. In this paper, we focus on analyzing the passage-level MAP in 2006 and 2007 tracks.

Passage MAP: As described in [3], this is a character-based precision calculated as follows: each relevant retrieved passage, precision will be computed as the fraction of characters overlapping with the gold standard passages divided by the total number of characters included in all nominated passages from this system for the topic up until that point. Similar to regular MAP, relevant passages that were not retrieved will be added into the calculation as well, with precision set to 0 for relevant passages not retrieved.

Then the mean of these average precisions over all topics will be calculated to compute the mean average passage precision.

Aspect MAP: A question could be addressed from different aspects. For example, the question "what is the role of gene PRNP in the Mad cow disease?" could be answered from aspects like "Diagnosis", "Neurologic manifestations", or "Prions/Genetics". This measure indicates how comprehensive the question is answered.

Document MAP: This is the standard IR measure. The precision is measured at every point where a relevant document is obtained and then averaged over all relevant documents to obtain the average precision for a given query. For a set of queries, the mean of the average precision for all queries is the MAP of that IR system.

4.4 Results and Discussions

We conduct a series of experiments to evaluate the effectiveness of the proposed algorithm in the biomedical domain. The data sets and the evaluation gold standard are provided by TREC Genomics. All of results are run as automatic. For the performance, TREC Genomics mainly concern the $MeanAveragePrecision$ (MAP). We also calculate the performance improvement of word-based method over the sentence-based method values under the same parameter settings, which is shown in Table 2 and 3.

In Table 2, it shows the performance with WordSentenceParsed and SentenceParsed methods for 2006 topics. The first column is Okapi tuning constant values which can be

Table 2. Passage-Level MAP Performance 2006

Tuning Constant Values		2006 Passage-Level Performance		Improvement
k1	b	WordSentencedParsed	SentenceParsed	(%)
0.40	2.00	0.0535	0.0382	39.9
0.50	**1.30**	**0.0722**	**0.0534**	35.4
0.80	1.20	0.0632	0.0475	33.1
1.00	1.00	0.0664	0.0467	**42.2**
1.20	0.75	0.0668	0.0500	33.6
1.40	0.55	0.0687	0.0485	41.7
2.00	0.40	0.0637	0.0474	34.3

Table 3. Passage-Level MAP Performance 2007

Tuning Constant Values		2007 Passage-Level Performance		Improvement
k1	b	WordSentencedParsed	SentenceParsed	(%)
0.40	2.00	0.0881	**0.0814**	8.3
0.50	**1.30**	**0.0947**	0.0784	20.8
0.80	1.20	0.0963	0.0700	37.5
1.00	1.00	0.0893	0.0758	17.7
1.20	0.75	0.0843	0.0636	32.6
1.40	0.55	0.0709	0.0455	**55.8**
2.00	0.40	0.0844	0.0758	11.3

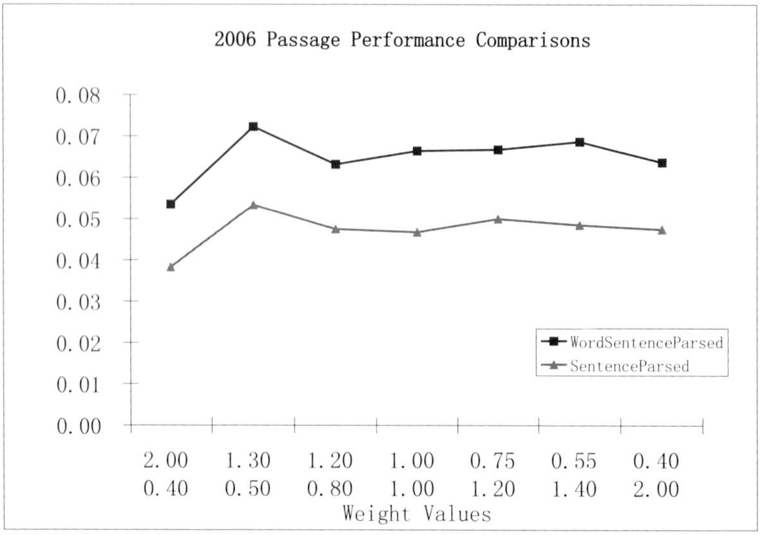

Fig. 2. Passage-Level Comparison for 2006 Topics

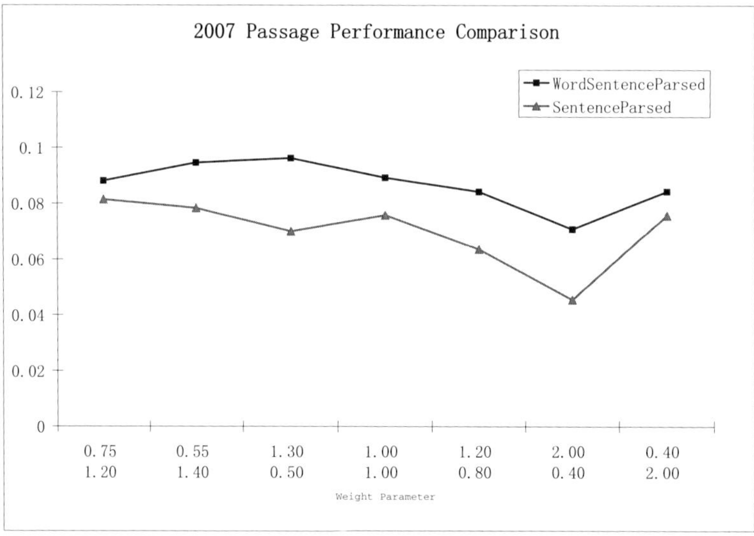

Fig. 3. Passage-Level Comparison for 2007 Topics

set up as different values given different weighting and ranking. Here we do the experiments with seven different parameter settings. The second column is passage-level MAP performance with WordSentenceParsed and SentenceParsed, we do experiments with different method on same topics automatically. The third one is the performance improvement. In Figure 2, the picture shows the performance gap between

WordSentenceParsed and SentenceParsed. In every row, with different Okapi tuning constant values, we get passage-level MAP with WordSentenceParsed and SentenceParsed. The best results are 0.0722 and 0.0534 in terms of setting the tuning constant values as $k1 = 0.50$ and $b = 1.30$. The largest improvement is 42.2% and the smallest improvement is 33.1%.

Gold Standard for the TREC 2007 Genomics Track is similar to 2006. In Table 3, same Okapi tuning constant values are set up. In Figure 3, the picture shows the performance gap between WordSentenceParsed and SentenceParsed. The best result with WordSentenceParsed is 0.0963 in terms of setting the tuning constant values as $k1 = 0.50$ and $b = 1.30$, while the best result with SentenceParsed is 0.0814 in terms of $k1 = 1.20$ and $b = 0.75$. The largest improvement is 55.8% and the smallest improvement is 8.3%. From Table 2 and 3, we see clearly that the dynamic algorithm with WordSentenceParsed method boosts the performance.

5 Conclusion and Future Work

We have proposed an efficient dynamic window based passage extraction algorithm for the purpose of biomedical information retrieval. In this paper, we learn the criteria for passage extraction through the gold standard of the TREC 2006 Genomics Track. The performance of the experimental results on the 2006 Genomics dataset, has confirmed the criteria effectively with the improvement up to 42.2%. Then a series of further experiments on the 2007 Genomics dataset have been conducted and the best improvement is up to 55.8% which shows the proposed algorithm works very well. Also, our proposed dynamic window based algorithm is independent on any domain datasets and it is easy to be implemented, because the regression model can be applied in general datasets.

In the TREC 2007 Genomics Track, the evaluation measures were a refinement of the measures used in 2006. A new character-based MAP measure (called Passage2) was added to compare the accuracy of the extracted answers, modified from the original measure in 2006 (called Passage). Passage2 treated each individually retrieved character in published order as relevant or not, in a sort of "every character is a mini relevance-judged document" approach. This was done to increase the stability of the passage MAP measure against arbitrary passage splitting techniques[4]. Therefore, in the future work, we will focus on improving passage2-level performance.

Acknowledgements

This research is supported in part by the research grant from the Natural Sciences and Engineering Research Council (NSERC) of Canada. We also would like to thank Hashmat Rohian and Damon Sotoudeh-Hosseini for their help.

References

1. Beaulieu, M., Gatford, M., Huang, X., Robertson, S.E., Walker, S., Williams, P.: (1996) Okapi at TREC-5. In: Proceedings of 5th Text REtrieval Conference. NIST Special Publication, Gaithersburg, pp. 143–166 (November 1997)

2. Hersh, W., Cohen, A., Yang, J.: TREC 2005 Genomics Track Overview. In: Proceedings of 14th Text REtrieval Conference. NIST Special Publication, Gaithersburg (November 2005)
3. Hersh, W., Cohen, A.M., Roberts, P.: TREC 2006 Genomics Track Overview. In: Proceedings of 15th Text REtrieval Conference,, November 2006, NIST Special Publication, Gaithersburg (2006)
4. Hersh, W., Cohen, A.M., Roberts, P.: TREC 2007 Genomics Track Overview. In: Proceedings of 16th Text REtrieval Conference, NIST Special Publication, Gaithersburg (November 2007)
5. Huang, X., Zhong, M., Luo, S.: York University at TREC 2005: Genomics Track. In: Proceedings of the 14th Text Retrieval Conference, NIST Special Publication, Gaithersburg (November 2005)
6. Huang, X., Hu, B., Rohian, H.: York University at TREC 2006: Genomics Track. In: Proceedings of the 15th Text Retrieval Conference, NIST Special Publication, Gaithersburg (November 2006)
7. Huang, X., Huang, Y., Wen, M., An, A., Liu, Y., Poon, J.: Applying Data Mining to Pseudo-Relevance Feedback for High Performance Text Retrieval. In: Perner, P. (ed.) ICDM 2006. LNCS (LNAI), vol. 4065, Springer, Heidelberg (2006)
8. Jiang, J., Zhai, C.: An Empirical Study of Tokenization Strategies for Biomedical Information Retrieval. In: Information Retrieval (2007)
9. Si, L., Kanungo, T., Huang, X.: Boosting Performance of Bio-Entity Recongition by Combining Results from Multiple Systems. In: Proceedings of the 5th ACM SIGKDD Workshop on Data Mining in Bioinformatics (2005)
10. Zhong, M., Huang, X.: Concept-Based Biomedical Text Retrieval. In: Proceedings of the 29th ACM SIGIR Conference, Washington, August 6-11 (2006)
11. Zhou, W., Yu, C., Neil, S., Vetle, T., Jie, H.: Knowledge-Intensive Conceptual Retrieval and Passage Extraction of Biomedical Literature. In: Proceedings of the 30th ACM SIGIR Conference, Amsterdam, July 23-27 (2007)

Mining Scalar Representations in a Non-tagged Music Database

Rory A. Lewis, Wenxin Jiang, and Zbigniew W. Raś

University of North Carolina, Dept. of Comp. Science, Charlotte, NC 28223, USA

Abstract. In the continuing investigation of the relationship between music and emotions it is recognized that MPEG-7 based MIR systems are the state-of-the-art. Also, it is known that non-temporal systems are diametrically unconducive to pitch analysis, an imperative for key and scalar analysis which determine emotions in music. Furthermore, even in a temporal MIR system one can only find the key if the scale is known or vice-versa, one can only find the scale if the key is known. We introduce a new MIRAI-based decision-support system that, given a blind database of music files, can successfully search for both the scale and the key of an unknown song in a music database and accordingly link each song to its set of scales and possible emotional states.

1 Introduction

It is known in the field of psychology and neuro-endocrinology that data from neurotransmitters in laboratories prove that certain music scales evoke measurable sensory sensations and emotions [12]. On the point of emotional analysis, the science contains a varied array of papers concerning emotions in music [10] [7]. Furthermore, it is understood in the field of Music Information Retrieval (MIR) that emotions can be mined in a closed domain [2], [6]. If a machine, given a polyphonic musical wave-form, could recognize all the instruments and the correlating notes each instrument played then, if given the key, it could calculate the scale of the music or, if given the scale, it could calculate the key and subsequently the emotions of the song thereof [11], [13]. In summation, if MIR can find the scale and key of a piece of music then it can also mine emotions. The obstacles preventing MIR methods from successfully mining emotions in music are weak Blind Source Separation (BSS) of musical instruments in a polyphonic domain, imprecise instrument identification, the inability to find a scale unless given the key or vice-versa. Putting aside the polyphonic research the authors presented in BSS [3], this paper presents a system that given a blind piece of non-polyphonic music, it first correctly determines the scale and key and then secondly, the subsequent human emotions linked to said retrieved musical scale.

The process presented builds upon the authors' processes of calculating the fundamental frequency of notes (*see* [3]) and mining music scalar theory (*see* [4]) in a music database (*see* [14]) set in a non-Hornbostel hierarchical manner (*see* [5] *and* [9]). Accordingly this paper sets forth a methodology of finding fundamental

A. An et al. (Eds.): ISMIS 2008, LNAI 4994, pp. 445–454, 2008.

frequencies in a database comprising both temporal and non-temporal attributes upon which it clusters possible scales. Next it assigns weights to each possible root and subsequently uses a heuristic-based distance algorithm to match the correct root to the correct scale.

Finding the fundamental frequency of a sound wave enables one to determine the pitch of the sound wave. The middle A above middle C has a frequency of 440 Hz in standard tuning. The frequency is doubled to make the same note an octave higher, and halved to make the same note an octave lower. The distance of all the other frequencies in contemporary tuning is calculated using semitones. The frequencies within an octave, starting from a given note and going up in the frequency scale, can be calculated using coefficients according to the following formula:

$$f_k = f_1 \cdot 2^{k/12} \tag{1}$$

where k is the number of semitones separating f_k and f_1. However, we operate in the non-temporal domain and hence we consider the transient duration as the time to reach the quasi-steady state of fundamental frequency. We calculate fundamental frequency by first computing the local cross-correlation function of the sound object, and then computing mean time to reach its maximum within each frame, and finally choosing the most frequently appearing resultant frequency in the quasi-steady status.

Let $r(i,k)$ is the normalised cross correlation of frame i with lag k. To calculate $r(i,k)$, we look at how it reaches its maximum value with ω as the maximum fundamental period expected, ours being 60ms:

$$r(i,k) = \sum_{j=1}^{m(i)+n-1} s(j)s(j-k) \Big/ \sqrt{\sum_{j=m(i)}^{m(i)+n-1} s(j-k)^2 \sum_{j=m(i)}^{m(i)+n-1} s(j)^2}, \quad k \in [1, S_r \times \omega] \tag{2}$$

where s is the audio signal, $m(i) = i * n$, where $i = 0, 1, .., M-1$ is the frame index, M is a number of frames, $n = t * sr$, where t = analysis window size, ours being 20ms, sr is a sampling rate, $k = 1, 2, ..., K$, where $K = lag = \omega * sr$.

In each frame i, the fundamental frequency is calculated in this form:

$$f(i) = \frac{S_r}{K_i/n_i} \tag{3}$$

where S_r is the sample frequency, n_i is the total number of $r(i,k)$'s local valleys across zero, where $k \in [1, K_i]$ and K_i is estimated by k as the maximum fundamental period.

Finding the fundamental frequency of a series of notes determines the relationship of the musical scales. The vast majority of scales in contemporary western music consist of 5 to 7 different notes (pitches). To calculate the number of possible scales we assert that the starting and ending notes are fixed and that there are twelve notes in an octave leaving 10 notes between the starting and ending notes. Also, we consider each note by moving from the lowest to the highest note. We cannot repeat a note and this is leaving one possible order, or scale. There are N semitones including the tonic t_1 which forms the remaining notes $t_2, ..., t_M$ in the scale which in turn are distributed over the remaining $N-1$ points. Scales can be represented using the *spiral array*, where pitch states are associated by coordinates downward along an ascending spiral [1]. Musicians represent scales in numerous forms all of which are incompatible with knowledge discovery in music. With this in mind the authors chose to represent basic score classification of music not as a music system but rather as Pawlak′*s* (see [8]) information system $S = (Y, A, V)$, called Scale Table, where Y is a set of music scales, $A = \{J^I, J^{II}, J^{III}, J^{IV}, J^V,$ *Region, Genre, Emotion, sma*$\}$ (see Table 1). Jumps between notes are represented by $J^I, J^{II}, J^{III}, J^{IV}, J^V$ which correlate to specific scales, regions and genre of music. The values $\{s, m, a\}$ of attribute *sma* should be read as *scale, mode, arpeggio.*

Table 1. Basic Score Classification Scale Table

Y	J^I	J^{II}	J^{III}	J^{IV}	J^V	Region	Genre	Emotion	sma
PentatonicMajor	2	2	3	2		Western	Blues	melancholy	s
BluesMajor	3	2	1	1	2	Western	Blues	depressive	s
PentatonicMinor	3	2	2	3		Western	Jazz	melancholy	s
BluesMinor	3	2	1	1	3	Western	Blues	dramatic	s
Augmented	3	1	3	1	3	Western	Jazz	feel-good	s
•									
•									
•									
Minor9th	2	1	4	3		neutral	neutral	not happy	a
Major11th	2	2	1	2	3	neutral	neutral	happy	a
Minor11th	2	1	2	2	3	neutral	neutral	not happy	a
Augmented	4	4				neutral	neutral	happy	a
Diminished	3	3	3			neutral	neutral	not happy	a

This table was built by our team of music experts on the basis of analyzing when composers use particular scales. Its complete version can be found in [4].

2 Experiments

To run experiments to determine both key and scale the authors focused on analyzing songs wherein a musician performed a solo. We randomly selected two songs out of a database of 422 songs, each in both *.au* and *MIDI* formats, containing solo performances, namely, Eric Clapton's "*Nobody Knows You When You're Down and Out*", and The Allman Brother's "*Blue Sky*". Upon extracting the solo sections of each song, we split the sections into phrases and bars making three sets of each song for analysis.

Using the unsegmented version of each song we submitted it to the MIRAI system (*see* [4] and [9]) for the first run of the analysis, where the pitch and

Table 2. Step 1: Computing duration for each note of each song

Blue Sky												
Note	a	$a\sharp$	b	c	$c\sharp$	d	$d\sharp$	e	f	$f\sharp$	g	$g\sharp$
Duration	5	2	15	9	128	16	0	103	2	47	11	32

Nobody Loves You												
Note	a	$a\sharp$	b	c	$c\sharp$	d	$d\sharp$	e	f	$f\sharp$	g	$g\sharp$
Duration	168	29	50	82	41	117	20	89	40	58	54	11

duration of each note is computed (see Table 2) and where each frame duration unit is 0.12 seconds.

The authors chose to focus on scalar emotions in this paper. Composers and musicians push and pull away and back towards the root and scale of a song to create tension and release. Music theory in essence dictates that these push and pulls are best suited when patterned in conjunction with two criteria of a composition: 1) Bars, which determine syncopation and rhythm of a song and 2) Phrases in the song that typically align to sentences whether aligned to verbal sentences of the singer of the piece of music, or whether aligned to musical sentences and phrases. One can typically see patterns of tension and release in the aforementioned bars and phrases. We have determined that in order to find the dominant key: First we segment each bar and phrase into notes and then categorize the music based on what scale the most notes have been played. Next, we weight this number by the likelihood value of each note when it is classified to this scale. For example, if all the notes in the music piece are grouped into k bars: $B_1; B_2; ...B_k$, with B_i corresponding to one of the scales in Table 1, then we compute a bar-score $\phi(x)$ for each $x \in Note$ (see Table 4) as

$$\phi(x) = \Big[\sum_{i=1}^{k} B_i(x)\Big] \Big/ \Big[\sum_{y \in Song} \sum_{i=1}^{k} B_i(y)\Big] \tag{4}$$

and if all the notes in the music piece are grouped into k phrases: $P_1; P_2; ...; P_k$, with each P_i, $1 \le i \le k$, corresponding to one of the scales in Table 1, then we compute a phrase-score $\psi(x)$ for each $x \in Note$ (see Table 5) as

$$\psi(x) = \Big[\sum_{i=1}^{k} P_i(x)\Big] \Big/ \Big[\sum_{y \in Song} \sum_{i=1}^{k} P_i(y)\Big] \tag{5}$$

Next, we identify the note x for which the value $[\psi(x) + \phi(x)]/2$ is maximal. In our examples *Blue Sky's* $\phi(x)$ returns note $c\sharp$ which wins with a score of 32.44% and for *Nobody Loves You*, $\psi(x)$ returns note a which wins with a score of 20.985%.

Accordingly, we set forth an algorithm and methodology to identify the scale of the song and classify push and pulls of the roots in accordance with bars and phrases.

2.1 Stage 1 of 3: Initial 100% Matches

Before we present the algorithm, we introduce the term *"Root-Matching"*.

Definition
Let $seq_1 = (j_1, j_2, ..., j_k)$ and $seq_2 = (i_1, i_2, ..., i_n)$ be any two sequences. We say that seq_1 is root-matching seq_2 if the following conditions hold:

(1) if $k \leq n$, then $(\forall m \leq k)[j_m = i_m]$,
(2) if $n \leq k$, then $(\forall m \leq n)[j_m = i_m]$.

Continuing the algorithm, as seen in Table 3, we search for 100% matches where each jump sequence calculated from Note Sequence matches with each jump corresponding to the *ith* scale (see Table 1). In other words, for each tuple in Table 3 we search for a supporting object in Table 1 such that its Jump Sequence is root-matching the Jump Sequence $(J^I, J^{II}, ..., J^V)$. The resultant was that Eric Clapton played precisely a *PentatonicDominant* in the key of f in phrase 1, a *Balinese* in the key of $f\sharp$ in phrase 2, and another *PentatonicDominant* in the key of g in phrase 8. Accordingly, *PentatonicDominant* in f, *Balinese* in $f\sharp$ and *PentatonicDominant* in g are possible candidates for the key and scale of Nobody Loves You, at this point. Similarly, The Allman Brothers played precisely a *PentatonicMajor* in the key of e in phrase 1, a *Diminished* in the key of $c\sharp$ in phrase 2, *PentatonicMinor* in the key of $c\sharp$ in phrase 4, and *BluesMajor* in the key of $c\sharp$ in phrase 5. Accordingly, *PentatonicMajor* in e, *Diminished* in $c\sharp$, *PentatonicMinor* in $c\sharp$ and *BluesMajor* in $c\sharp$ are possible candidates for the key and scale of Blue Sky, at this point.

2.2 Stage 2 of 3: Reducing the Search Space of Distance Algorithm

It is too expensive to search every possible close scale of every note according to bars and phrases. For example, in our small database of 422 songs with about 200 notes per solo, about 628 scales, 20 bars, and 10 phrases, it would require millions of calculations. To eliminate this problem, we developed a classification system that, to coin a new term, makes *musical cuts* according to weights for the purpose of invoking *Levenshtein* distance algorithm (see Section 2.3) to only search a knowledge base of relevant keys and scales that are most likely to be top candidates.

Bar Weights: We calculate the weights of fundamental frequencies in terms of bars. This is because, as mentioned above, the root of a song is often located in the first and/or last note of a bar. We store all the scores, but as a reference,

Table 3. Step 2: Find all possible scales and cut at 100% matches

Song	P (Phrase)	Note Seq.	Jump Sequence	100% Match	Scale
Blue Sky	1	*bc♯def♯g♯*	21222	*	*null*
	1	*c♯def♯g♯b*	12223	*	*null*
	1	*def♯g♯bc♯*	22232	*	*null*
	1	*ef♯g♯bc♯d*	22321	2232	*PentatonicMajor*
	1	*f♯g♯bc♯de*	23212	*	*null*
	1	*g♯bc♯def♯*	32122	*	*null*
	2	*c♯eg*	33	333	*Diminished*
	2	*egc♯*	36	*	*null*
	2	*gc♯e*	63	*	*null*
	3	*cc♯ef♯g♯*	1322	*	*null*
	3	*c♯ef♯g♯c*	3224	*	*null*
	3	*ef♯g♯cc♯*	2241	*	*null*
	3	*♯g♯cc♯e*	2413	*	*null*
	3	*g♯cc♯ef♯*	4132	*	*null*
	4	*bcc♯ef♯g♯*	11322	*	*null*
	4	*cc♯ef♯g♯b*	13223	*	*null*
	4	*c♯ef♯g♯bc*	32231	3223	*PentatonicMinor*
	4	*ef♯g♯bcc♯*	22311	*	*null*
	4	*f♯g♯bcc♯e*	23113	*	*null*
	4	*g♯bcc♯ef♯*	31132	*	*null*
	5	*c♯ef♯g*	321	32112	*BluesMajor*
	5	*ef♯gc♯*	216	*	*null*
	5	*f♯gc♯e*	163	*	*null*
	5	*gc♯ef♯*	632	*	*null*
	6	*cc♯def♯g♯*	11222	*	*null*
	6	*c♯def♯g♯c*	12224	*	*null*
	6	*cc♯def♯g♯*	22241	*	*null*
	6	*def♯g♯cc♯*	22411	*	*null*
	6	*ef♯g♯cc♯d*	24112	*	*null*
	6	*g♯cc♯def♯*	41122	*	*null*
Nobody	1	*acd♯efg*	33112	*	*null*
	1	*cd♯efga*	31122	*	*null*
	1	*d♯efgac*	11223	*	*null*
	1	*efgacd♯*	12233	*	*null*
	1	*fgacd♯e*	22331	2233	*PentatonicDominant*
	1	*gacd♯ef*	23311	*	*null*
	2	*ac♯def♯g*	41221	*	*null*
	2	*c♯def♯ga*	12212	*	*null*
	2	*def♯gac♯*	22124	*	*null*
	2	*ef♯gac♯d*	21241	*	*null*
	2	*f♯gac♯de*	12412	1241	*Balinese*
	2	*gac♯def♯*	24122	*	*null*
	3	*bcc♯def*	11121	*	*null*
	3	*cc♯defb*	11216	*	*null*
	3	*c♯defbc*	12161	*	*null*
	3	*defbcc♯*	21611	*	*null*
	3	*efbcc♯d*	16111	*	*null*
	3	*fbcc♯de*	61112	*	*null*
	4	*abcc♯dg*	21115	*	*null*
	4	*bcc♯dga*	11152	*	*null*
	4	*cc♯dgab*	11522	*	*null*
	4	*c♯dgabc*	15221	*	*null*
	4	*dgabcc♯*	52211	*	*null*
	4	*gabcc♯d*	22111	*	*null*
	5	*acc♯dd♯e*	31111	*	*null*
	5	*cc♯dd♯ea*	11115	*	*null*
	5	*c♯dd♯eac*	11153	*	*null*
	..	....	.....	..	....
	..	....	.....	..	....
	7	*ef♯aa♯cd*	23122	*	*null*
	7	*f♯aa♯cde*	31222	*	*null*
	8	*abdff♯g*	23311	*	*null*
	8	*bdff♯ga*	33112	*	*null*
	8	*dff♯gab*	31122	*	*null*
	8	*ff♯gabd*	11223	*	*null*
	8	*f♯gabdf*	12233	*	*null*
	8	*gabdff♯*	22331	2233	*PentatonicDominant*

one can see that by looking at Table 4 $c\sharp$ receives a score 33% and e receives a score 25% for Blue Sky.

Phrase Weights: We calculate the weights of fundamental frequencies in terms of duration in phrases. This is because the root of a song is often located in patterns found in context to phrases. We store all the scores, but as a reference, one can see that by looking at Table 5 $c\sharp$ receives a score 35% and e receives a score 28% for Blue Sky.

Total Weights: Summation of the weights $[\phi(x) + \psi(x)]/2$ provides us with musical cuts used to reduce the search space of a distance algorithm. Here we find that for *Blue Sky*, $c\sharp$ wins with a score of 32.44% and for *Nobody Loves You*, A wins with a score of 20.985%. This means, we can be assured that by reducing the search space to all Note Sequences (attribute *Note Seq* in Table 3) beginning with $c\sharp$, the algorithm will produce the correct *ith* scale for *Blue Sky* and similarly, by reducing the search space to all Note Sequences beginning with note a, it will also produce the correct *ith* scale for *Nobody Loves You*.

2.3 Stage 3 of 3: Calculating the Distance between Jump Sequences

We use the *Levenshtein* Distance to calculate the distance between retrieved jump sequences which are not 100% matched with our scale table. The distance between two strings is given by the minimum number of operations needed to

Table 4. Bar Weights: Calculating weights to calculate the musical cuts

Song	Note	B_1	B_2	B_3	B_4	B_5	B_6	B_7	B_8	B_9	B_{10}	B_{11}	B_{12}	ϕ
Blue Sky	a			1	1									8%
	$a\sharp$						1							4%
	b							1						4%
	c											2		8%
	$c\sharp$	1	2		1	1	1						2	33%
	d										1			4%
	$d\sharp$													0%
	e	1		1		1		1	1	1				25%
	f													0%
	$f\sharp$													0%
	g										1			4%
	$g\sharp$								1	1				8%
Nobody	a		1		1		1							19%
	$a\sharp$	1												1%
	b								1					1%
	c													
	$c\sharp$													
	d					1	1	1						19%
	$d\sharp$													
	e			1										1%
	f	1												1%
	$f\sharp$					1		1						12.5%
	g		1		1				1					19%
	$g\sharp$			1										1%

Table 5. Phrase Weights: Duration is summed for each note in each phrase in each song

Song	Note	P_1	P_2	P_3	P_4	P_5	P_6	P_7	P_8	ψ
Blue Sky	a				2		3			1%
	$a\sharp$						2			1%
	b	8			7					4%
	c	2		2	2		3			2%
	$c\sharp$	14	7	3	20	7	77			35%
	d	2					14			4%
	$d\sharp$									0%
	e	22	8	10	26	13	24			28%
	f						2			1%
	$f\sharp$	8		3	8	4	24			13%
	g		6			3	2			3%
	$g\sharp$	5		4	6		17			9%
Nobody	a	16	27		9	15	25	59	17	22%
	$a\sharp$				3	4	7	11	4	4%
	b	2		3	6	7	11	2	19	7%
	c	5	4	3	5	26	18	14	7	11%
	$c\sharp$		5	2	5	13	8	8		5%
	d		5	25	20	20	13	20	14	15%
	$d\sharp$	2				7	5	6		3%
	e	11	5	14	2	14	7	25	11	12%
	f	2		9	2			4	23	5%
	$f\sharp$		5			2	8	30	13	8%
	g	10	5		5		7	10	17	7%
	$g\sharp$		2				3	3	3	1%

transform one string into the other, where an operation is an insertion, deletion, or substitution of a single character. For example, the Levenshtein distance between "21232" and "22222" is 2, since these two edits change one into the other, and there is no way to do it with fewer than two edits.

During the process of key note searching, we use two measures, ψ and ϕ, to evaluate the importance of each note. First one is the position of the note in each bar of the song. The points are calculated by adding 1 point to one note when it occurs in the first frame or last frame in one bar. Second measure is the duration of each note in each phrase of the song. Then we get ultimate weights for each note by adding up these two measures (ratio). The key note is the one with the highest weights. And we only search the candidate key notes from the list of first notes of all the retrieved note sequences for phrases since we only consider the match among these sequences by matching the first note of each note sequence with the key note. During the matching process, we get the list of accepted candidate scales identified by key notes, among which the scales with the shortest Levenshtein distance are chosen scale patterns together with 100 percent matched patterns. Note that unlike Table 6 that shows cuts, Table 5 illustrates the process before the cuts divide each phrase.

In *Blue Sky* a $c\sharp$ Pentatonic Major has a score of 8, making it the most likely scale and key. This is correct. In *Nobody Loves You* a a Balinese has a score of 8, making it the most likely scale and key. This is correct based on the data but the input data was polluted because the input system could not correctly

Table 6. Mining All Possible scales and Cuts in "Nobody"

Song	P (Phrase)	Note Seq.	Jump Sequence	100% Match	Scale
Nobody	1	$acd\sharp efg$	33112	*	*null*
	1	$cd\sharp efga$	31122	*	*null*
	1	$d\sharp efgac$	11223	*	*null*
	1	$efgacd\sharp$	12233	*	*null*
	1	$fgacd\sharp e$	22331	2233	*PentatonicDominant*
	1	$gacd\sharp ef$	23311	*	*null*
	2	$ac\sharp def\sharp g$	41221	*	*null*
	2	$c\sharp def\sharp ga$	12212	*	*null*
	2	$def\sharp gac\sharp$	22124	*	*null*
	2	$ef\sharp gac\sharp d$	21241	*	*null*
	2	$f\sharp gac\sharp de$	12412	1241	*Balinese*
	2	$gac\sharp def\sharp$	24122	*	*null*
	3	$bcc\sharp def$	11121	*	*null*
	..	....	......	...	...
	..	....	......	...	...
	8	$dff\sharp gab$	31122	*	*null*
	8	$ff\sharp gabd$	11223	*	*null*
	8	$f\sharp gabdf$	12233	*	*null*
	8	$gabdff\sharp$	22331	2233	*PentatonicDominant*

Table 7. Final Results

Song	Jump	Matched Scale	Root	Match Count	Similarity
Blue Sky	2232	Pentatonic Major	$c\sharp$	8	100%
	32112	Blues Major	$c\sharp$	2	60%
	3223	Pentatonic Minor	$c\sharp$	4	100%
Nobody	1241	Balinese	a	8	100%
	2233	Pentatonic Dominant	a	2	60%
	43	Major	a	4	100%

assimilate polyphonic notes, which are in abundance in this piece of music. The correct scales, to humans or future MIR systems that can assimilate polyphonic sounds would be the mixture of c Spanish 8-Tone scale and c Major scale.

In *Blue Sky* a $c\sharp$ Pentatonic Major has a score of 8 making it the most likely scale and key. This is correct. In *Nobody Loves You* a a Balinese has a score of 8 making it the most likely scale and key. This is correct based on the data but the input data was polluted because the input system could not correctly assimilate polyphonic notes, which are in abundance in this piece of music. The correct scales, to humans, or future machines that can assimilate polyphonic sounds would be mixture c Spanish 8-Tone scale and c Major scale.

3 Conclusion

The algorithm worked 100% correctly on data it received by finding the correct musical cuts and then correctly reducing the search space of the distance algorithm and focusing it on the cut roots to find the scales. We randomly selected these two songs. *Blue Sky* was completely correct because in the sol there were no polyphonic notes. However, because *Nobody Loves You* had polyphonic notes, the data was skewed and thus the scale was off. But, according to the input data, albeit wrong, it received, it did correctly calculate the correct key and scale. Our future work, clearly is to start figuring out how to assimilate polyphonic notes.

Acknowledgements

This research is supported by NSF under grant IIS-0414815.

References

1. Chew, E.: Music information processing: a new application for operations researchers. Bulletin of AIROnews 7(3), 9–14 (2002)
2. Hevner, K.: Experimental studies of the elements of expression in music. American Journal of Psychology 48, 246–268 (1936)
3. Lewis, R., Zhang, X., Raś, Z.: Knowledge discovery based identification of musical pitches and instruments in polyphonic sounds. International Journal of Engineering Applications of Artificial Intelligence 20(5), 637–645 (2007)
4. Lewis, R., Raś, Z.: Rules for processing and manipulating scalar music theory. In: Proceedings of MUE 2007, IEEE Conference, Seoul, Korea, pp. 26–28 (2007)
5. Lewis, R., Wieczorkowska, A.: A Categorization of musical instrument sounds based on numerical parameters. In: Kryszkiewicz, M., Peters, J.F., Rybinski, H., Skowron, A. (eds.) RSEISP 2007. LNCS (LNAI), vol. 4585, pp. 784–792. Springer, Heidelberg (2007)
6. Li, T., Ogihara, M.: Detecting emotion in music, in ISMIR 2003 Proceed (2003), `http://ismir2003.ismir.net/papers/Li.PDF`
7. McClellan, R.: The healing forces of music. In: Element Inc., Rockport, MA (1966)
8. Pawlak, Z.: Information systems - theoretical foundations. Information Systems Journal 6, 205–218 (1991)
9. Raś, Z., Zhang, X., Lewis, R.: MIRAI: Multi-hierarchical, FS-tree based music information retrieval system. In: Kryszkiewicz, M., Peters, J.F., Rybinski, H., Skowron, A. (eds.) RSEISP 2007. LNCS (LNAI), vol. 4585, pp. 28–30. Springer, Heidelberg (2007)
10. Sevgen, A.: The science of musical sound. Scientific American Books Inc., New York (1983)
11. Sloboda, J.A., ONeill, S.A.: Emotions in everyday listening to music. In: Juslin, P.N., Sloboda, J.A. (eds.) Music and Emotion: Theory and Research, pp. 415–430. Oxford Univ. Press, Oxford (2001)
12. Valentinuzzi, M.E., Arias, N.E.: Human psychophysiological perception of musical scales and nontraditional music. IEEE/Eng Medicine and Biol. Mag. 18(2), 54–60 (1999)
13. Vink, A.: Music and Emotion, living apart together: a relationship between music psychology and music therapy. Nordic Journal of Music Therapy 10(2), 144–158 (2001)
14. Wieczorkowska, A., Synak, P., Lewis, R., Raś, Z.: Creating reliable database for experiments on extracting emotions from music. In: IIPWM 2005 Proceedings. Advances in Soft Computing, pp. 395–402. Springer, Heidelberg (2005)

Identification of Dominating Instrument in Mixes of Sounds of the Same Pitch

Alicja Wieczorkowska[1] and Elżbieta Kolczyńska[2]

[1] Polish-Japanese Institute of Information Technology, Koszykowa 86, 02-008 Warsaw, Poland
alicja@pjwstk.edu.pl

[2] Agricultural University in Lublin, Akademicka 13, 20-950 Lublin, Poland
elzbieta.kolczynska@ar.lublin.pl

Abstract. In this paper we deal with the problem of identification of the dominating instrument in the recording containing simultaneous sounds of the same pitch. Sustained harmonic sounds from one octave of eight instruments were considered. The training data set contains sounds of singular instruments, as well as the same sounds with added artificial harmonic and noise sounds of lower amplitude. The test data set contains mixes of musical instrument sounds. SVM classifier from WEKA was used for training and testing experiments. Results of these experiments are presented and discussed in the paper.

Keywords: Music information retrieval, sound recognition.

1 Introduction

Automatic recognition of the dominating musical instrument in sound mixes, when spectra overlap, is quite a difficult issue, most difficult when the interfering sounds are of the same pitch. The number of possible combinations is very high because of the number of existing instruments and sounds within their scale ranges. Therefore, it would be desirable to obtain a classifier performing such recognition, and also to train the classifier on a limited data set. The motivation for this paper was to perform experiments on selected instrument sounds, and use added artificial sounds with broadband spectrum, overlapping with the sounds under consideration, in order to check if classifiers trained this way would work for sounds mixes of real instruments. In other words, our goal was to check if using a limited number of added artificial sounds can be sufficient to train a classifier to recognize dominating musical instrument in polytimbral mix of one pitch. The main focus was on construction of the training and testing data, because if this set up is successful, we have a starting point for further experiments with other musical instrument sound mixes.

In this research, we decided to choose 8 instruments producing sustained harmonic sounds (of definite pitch), and limit the range to the octave no. 4 in MIDI notation. The sounds added to the original sounds in the training set

A. An et al. (Eds.): ISMIS 2008, LNAI 4994, pp. 455–464, 2008.

include noises and artificial sound waves of harmonic spectrum. The test set contains the original sounds mixed with sounds of other instruments, always of the same pitch. The level of the added sounds was processed to make sure that the sound of the main instrument is louder than the other sound all the time.

We have already performed experiments on training classifiers recognizing dominating musical instrument in polytimbral environment, i.e. when the sound of other instrument is accompanying the main sound. However, the pitch of the accompanying sound or sounds was usually different than the pitch of the main sound. Additionally, the added sounds were diminished in amplitude in a very simple way, i.e. by re-scaling their amplitude. Since the main and added sounds were not edited in any other way, in many cases the added sounds were starting earlier or ending later that the main ones, thus being actually louder in some parts, and obscuring the results [14]. Therefore, we decided to change the experiment set up to make sure that the main sound is actually louder all the time, and thus that the classifiers are trained properly.

2 Data for Training and Testing

The data for training and testing consist of musical instrument sounds of sustained harmonic type, also with addition of other sounds. The added sounds include artificially generated noises, white and pink, and also artificially generated harmonic waves, of triangular and saw-tooth shape, always of the same pitch as the main sound.

We have already worked on sounds mixes in previous research. However, the choice of the accompanying instrument sounds in the training and testing sets was arbitral, and the spectra were not overlapping in most cases. Also, the length and the level of accompanying sound was not normalized. As a result, a clear dependency between the quality of the classifiers and the level of accompanying sound was not observed. This is why we decided to normalize the sound length and level of the added sounds with respect to the main sounds, and make sure that the level of added sounds does not exceed the level of the main sounds at any time. The sounds added in mixes for training purposes have rich spectra, overlapping with the main sound. Artificial sounds were added for training, and musical instrument sounds were added for testing purposes. This way only a few combinations can be used for training, rather than using all possible instrument pairs. It also assures using different data for training and testing, which is important in case of sound data [8]. Since our final goal is to recognize instruments in musical recording, natural instrument mixes were used for testing.

2.1 Parameterization

The audio data we deal with represent digital sounds in .snd format, recorded stereo with 44,1 kHz sampling rate and with 16-bit resolution. Such a representation of sound wave requires parameterization for successful classification.

Features used for parameterization of musical audio data may describe temporal, spectral, and spectral-temporal properties of sounds. The research on musical instrument sound recognition conducted worldwide is based on various parameters, including features describing properties of DFT spectrum, wavelet analysis coefficients, MFCC (Mel-Frequency Cepstral Coefficients), MSA (Multidimensional Analysis Scaling) trajectories, and so on [2], [3], [4], [6], [7], [9], [13]. MPEG-7 sound descriptors can also be applied for musical sound parameterization [5], but these parameters are not dedicated to recognition of particular instruments in recordings.

The construction of feature set is an important part of creating the database for classification purposes, and the results may vary depending on the feature vector applied for the training and then testing of a classifier. In our research, we decided to use the feature vector already applied for the recognition of musical instruments in polyphonic (polytimbral) environment [15]. The feature set we have chosen consists of 219 parameters, based mainly on MPEG-7 audio descriptors, and on other parameters used in similar research. Most of the parameters represent average value of frame-based attributes, calculated for consecutive frames of a singular sound using sliding analysis window, moved through the entire sound. The calculations were performed for the left channel of stereo data, using 120 ms analyzing frame with Hamming window and hop size 40 ms; such a long analyzing frame allows analysis even of the lowest sounds. The experiments described here were performed on a limited data set (8 instruments), but we plan extend our research to much bigger set of data, and the feature set presented below can be used for this purpose [15]:

- MPEG-7 audio descriptors [5], [11], [14]:
  - *AudioSpectrumSpread* - a RMS value of the deviation of the Log frequency power spectrum with respect to the gravity center in a frame; the value was averaged through frames for the entire sound;
  - *AudioSpectrumFlatness*, $flat_1, \ldots, flat_{25}$ - parameter describing the flatness property of the power spectrum within a frequency bin for selected bins; 25 out of 32 frequency bands were used for a given frame, and the value was averaged for the entire sound;
  - *AudioSpectrumCentroid* - power weighted average of the frequency bins in the power spectrum of all the frames in a sound segment, calculated with a Welch method;
  - *AudioSpectrumBasis*: $basis_1, \ldots, basis_{165}$; spectral basis parameters are calculated for the spectrum basis functions. In our case, the total number of sub-spaces in basis function is 33, and for each sub-space, minimum/maximum/mean/distance/standard deviation are extracted to flat the vector data. Distance is calculated as the summation of dissimilarity (absolute difference of values) of every pair of coordinates in the vector. Spectrum basis function is used to reduce the dimensionality by projecting the spectrum (for each frame) from high dimensional space to low dimensional space with compact salient statistical information. The calculated values were averaged over all analyzed frames of the sound;
  - *HarmonicSpectralCentroid* - the average (over the entire sound) of the instantaneous Harmonic Centroid, calculated for each analyzing frame.

The instantaneous Harmonic Spectral Centroid is the mean of the harmonic peaks of the spectrum, weighted by the amplitude in linear scale;
    - *HarmonicSpectralSpread* - the average over the entire sound of the instantaneous harmonic spectral spread, calculated for each frame. Instantaneous harmonic spectral spread represents the standard deviation of the harmonic peaks of the spectrum with respect to the instantaneous harmonic spectral centroid, weighted by the amplitude;
    - *HarmonicSpectralVariation* - mean value over the entire sound of the instantaneous harmonic spectral variation, i.e. of the normalized correlation between amplitudes of harmonic peaks of each 2 adjacent frames;
    - *HarmonicSpectralDeviation* - average over the entire sound of the instantaneous harmonic spectral deviation, calculated for each frame, where the instantaneous harmonic spectral deviation represents the spectral deviation of the log amplitude components from a global spectral envelope;
    - *LogAttackTime* - the decimal logarithm of the duration from the time when the signal starts to the time when it reaches its maximum value, or when it reaches its sustained part, whichever comes first;
    - *TemporalCentroid* - energy weighted mean of the sound duration; this parameter shows where in time the energy of the sound is focused;
- other audio descriptors:
    - *Energy* - average energy of spectrum in the parameterized sound;
    - MFCC - min, max, mean, distance, and standard deviation of the MFCC vector, through the entire sound;
    - *ZeroCrossingDensity*, averaged through all frames for a given sound;
    - *RollOff* - the frequency below which an experimentally chosen percentage of the accumulated magnitudes of the spectrum is concentrated (averaged over all frames). It is a measure of spectral shape, used in the speech recognition to distinguish between voiced and unvoiced speech;
    - *Flux* - the difference between the magnitude of the DFT points in a given frame and its successive frame, averaged through the entire sound. This value multiplied by 10^7 to comply with the requirements of the classifier applied in our research;
    - *AverageFundamentalFrequency* (maximum likelihood algorithm applied for pitch estimation);
    - *Ratio* $r_1, \ldots, r_{11}$ - parameters describing the ratio of the amplitude of a harmonic partial to the total harmonic partials.

These parameters describe basic spectral, timbral spectral and temporal audio properties, and also spectral basis descriptor, as described in the MPEG-7 standard. The spectral basis descriptor is a series of basis functions derived from the Singular Value Decomposition (SVD) of a normalized power spectrum. In order to avoid too high dimensionality of the feature vector, a few other features were derived from the spectral basis attribute. Other audio descriptors used in our feature vector include time-domain and spectrum-domain properties of sound, used in research on audio data classification.

2.2 Training Data

The training data contain singular sounds of musical instruments, and also the same sounds with added other sounds. The audio recordings from McGill University Master Samples CDs have been used as a source of these sounds [10]. These CDs are commonly used worldwide for research on musical instrument sounds. The following instruments have been chosen:

1. B-flat clarinet,
2. cello - bowed, played vibrato,
3. trumpet,
4. flute played vibrato,
5. oboe,
6. tenor trombone,
7. viola - bowed, played vibrato,
8. violin - bowed, played vibrato.

All these instruments produce sounds of definite pitch, and their spectra are of harmonic type. Only sustained sounds were considered. We decided to use only sounds from the octave no. 4 (in MIDI notation). We also prepared the mixes of pairs of sounds, i.e. the instrumental sounds mentioned above and the following sounds:

- white noise,
- pink noise,
- triangular wave,
- saw-tooth wave.

All added sounds have broadband spectra, continuous in case of noises and harmonic in case of triangular and saw-tooth wave. Again, harmonic sounds were prepared for the frequencies from the octave no. 4. These sounds were produced using Adobe Audition [1], where only integer values were allowed. Therefore, the frequency values of the generated harmonic waves were rounded to the nearest integers, as below (standard values for A4=440 Hz shown in parentheses):

- C4 - 262 Hz (261.6),
- C#4 - 277 Hz (277.2),
- D4 - 294 Hz (293.7),
- D#4 - 311 Hz (311.1),
- E4 - 330 Hz (329,6),
- F4 - 349 Hz (349.2),
- F#4 - 370 Hz (370.0),
- G4 - 392 Hz (392.0),
- G#4 - 415 Hz (415.3),
- A4 - 440 Hz (440),
- A#4 - 466 Hz (466.2),
- B4 - 494 Hz (493.9).

Eight-second long sounds were prepared, since the longest musical instrument sounds was below 8 seconds of length. The mixes were prepared in such a way that for each pair the length of the added sound was truncated to the length of the main sound, and 0.1 s of silence replaced the beginning and the end of the added sound. Next, from the end of the silence at the beginning till 1/3 of the sound length the fade in effect was applied; similarly, fade out was applied from 2/3 of the sound. During mixing, for each instrumental sound chosen to dominate in the mix, the level of the added sound was first re-scaled to match the RMS of the main sound. Thus we assure that the main sound is louder even during transients. Next, three versions of mixes were prepared:

1. with the level of added sounds diminished to 12.5 % of the main sounds,
2. with the level of added sounds diminished to 25 % of the main sounds,
3. with the level of added sounds diminished to 50 % of the main sounds.

In each case, the mix was prepared in such as the average of the added sounds.

Altogether, the training set consisted of 96 singular sounds of musical instruments, and also these same sounds with added noises and harmonic sounds in 3 level versions as described above, i.e. 1152 mixes.

2.3 Testing Data

The data for testing consisted of mixes of instrument sounds only. As in case of the training data, the testing data were prepared in 3 versions, i.e. for the same 3 levels of added sounds. For each subset, the added sound was of the same pitch as the main sound, and was created as the average of the 7 remaining instruments from the training set, modified in amplitude as the sounds added in the training set, and diminished to the desired level. Therefore, we had 3 subsets of the test set, each one consisting of 96 sounds.

3 Experiments and Results

The classification experiments were performed using WEKA software [12]; we chose Support Vector Machine (SMO) classifier, since we have multi-dimensional data for which SVM is suitable, as it aims at finding the hyperplane that best separates observations belonging to different classes in multi-dimensional feature space. Also, SVM classifier was already reported successful in case of musical instrument sound identification [4].

General results of all experiments described in Section 2 are shown in Table 1. Detailed confusion matrices for the subsets are shown in Tables 2, 3 and 4.

The results for training on singular musical instrument sounds are not very high, but after adding mixes to the training sets the results may improve significantly. When comparing results for various levels of added sounds, we can observe that adding sounds of levels 12.5% and 50% yields significant improvement, but worsens the recognition of violin. For 25% decrease of recognition correctness is observed. In case of both 12.5% and 50% levels, the recognition of

Table 1. Results of experiments for all training and testing data

Training Set	Test Set	Classification Correctness
Singular instrument sounds only	Instrument sounds mixes with the level of added sounds diminished to 12,5% volume of the main instrument	69.7917 %
Singular instrument sounds with added artificial sound waves of 12,5% volume		81.25 %
Singular instrument sounds only	Instrument sounds mixes with the level of added sounds diminished to 25% volume of the main instrument	91.6667 %
Singular instrument sounds with added artificial sound waves of 25% volume		87.5 %
Singular instrument sounds only	Instrument sounds mixes with the level of added sounds diminished to 50% of the main instrument	69.7917 %
Singular instrument sounds with added artificial sound waves of 50% volume		81.25 %
Singular instrument sounds only	All 3 subsets together	77.0833 %
All 3 training subsets together		82.9861 %

Table 2. Results of experiments with testing on mixes with added sounds diminished in level to 12.5% of the main sound

a) training on singular musical instrument sounds only

Classified as ->	a	b	c	d	e	f	g	h
a = clarinet	11	0	0	0	0	0	1	0
b = cello	0	12	0	0	0	0	0	0
c = trumpet	0	0	12	0	0	0	0	0
d = flute	0	4	0	8	0	0	0	0
e = oboe	0	3	0	0	4	0	1	4
f = trombone	0	1	2	0	0	8	1	0
g = viola	0	5	0	0	0	0	5	2
h = violin	0	1	0	0	0	0	4	7

b) training on both singular and mixed sounds

Classified as ->	a	b	c	d	e	f	g	h
a = clarinet	10	1	0	0	1	0	0	0
b = cello	0	12	0	0	0	0	0	0
c = trumpet	0	0	12	0	0	0	0	0
d = flute	1	1	0	8	2	0	0	0
e = oboe	0	0	0	0	12	0	0	0
f = trombone	0	0	0	0	0	12	0	0
g = viola	0	3	0	0	0	0	9	0
h = violin	1	0	0	0	1	0	7	3

viola improves, and it is not so frequently mistaken with cello, but the recognition of violin worsens - it tends to be mistaken with viola.

As we can observe, adding mixes to training sets usually improves the recognition accuracy; this can be caused by the enlargement of the training set. Flute,

Table 3. Results of experiments with testing on mixes with added sounds diminished in level to 25% of the main sound

a) training on singular musical instrument sounds only

Classified as ->	a	b	c	d	e	f	g	h
a = clarinet	12	0	0	0	0	0	0	0
b = cello	0	12	0	0	0	0	0	0
c = trumpet	0	0	12	0	0	0	0	0
d = flute	0	0	0	11	0	0	1	0
e = oboe	0	0	0	0	11	0	1	0
f = trombone	0	0	0	0	0	12	0	0
g = viola	0	3	0	0	0	0	9	0
h = violin	0	0	0	0	0	0	3	9

b) training on both singular and mixed sounds

Classified as ->	a	b	c	d	e	f	g	h
a = clarinet	12	0	0	0	0	0	0	0
b = cello	0	12	0	0	0	0	0	0
c = trumpet	0	0	12	0	0	0	0	0
d = flute	0	0	0	12	0	0	0	0
e = oboe	0	0	0	0	12	0	0	0
f = trombone	0	0	0	0	0	12	0	0
g = viola	0	3	0	0	0	0	9	0
h = violin	0	0	0	0	1	0	8	3

Table 4. Results of experiments with testing on mixes with added sounds diminished in level to 50% of the main sound

a) training on singular musical instrument sounds only

Classified as ->	a	b	c	d	e	f	g	h
a = clarinet	11	0	0	0	0	0	1	0
b = cello	0	12	0	0	0	0	0	0
c = trumpet	0	0	12	0	0	0	0	0
d = flute	0	4	0	8	0	0	0	0
e = oboe	0	3	0	0	4	0	1	4
f = trombone	0	1	2	0	0	8	1	0
g = viola	0	5	0	0	0	0	5	2
h = violin	0	1	0	0	0	0	4	7

b) training on both singular and mixed sounds

Classified as ->	a	b	c	d	e	f	g	h
a = clarinet	10	1	0	0	1	0	0	0
b = cello	0	12	0	0	0	0	0	0
c = trumpet	0	0	12	0	0	0	0	0
d = flute	1	1	0	8	2	0	0	0
e = oboe	0	0	0	0	12	0	0	0
f = trombone	0	0	0	0	0	12	0	0
g = viola	0	3	0	0	0	0	9	0
h = violin	1	0	0	0	1	0	7	3

oboe and trombone are much better recognized after adding mixes. However, violin seems to be always more difficult to recognize after adding mixed sounds to each training set, and it is usually mistaken with viola (also when all available data were used - see Table 5). On the other hand, sounds of these instruments

Table 5. Results of experiments with testing on mixes with added sounds diminished in level to 12.5, 25 and 50% of the main sound

a) training on singular musical instrument sounds only

Classified as ->	a	b	c	d	e	f	g	h
a = clarinet	34	0	0	0	0	0	2	0
b = cello	0	36	0	0	0	0	0	0
c = trumpet	0	0	36	0	0	0	0	0
d = flute	0	8	0	27	0	0	1	0
e = oboe	0	6	0	0	19	0	3	8
f = trombone	0	2	4	0	0	28	2	0
g = viola	0	13	0	0	0	0	19	4
h = violin	0	2	0	0	0	0	11	23

b) training on both singular and mixed sounds

Classified as ->	a	b	c	d	e	f	g	h
a = clarinet	34	0	0	0	2	0	0	0
b = cello	0	34	0	0	0	0	2	0
c = trumpet	0	0	36	0	0	0	0	0
d = flute	0	0	0	30	4	0	2	0
e = oboe	0	0	0	0	34	0	2	0
f = trombone	0	0	0	0	0	36	0	0
g = viola	0	8	0	0	0	0	28	0
h = violin	2	0	0	0	3	0	24	7

are very similar, so mistaking these instruments is not surprising. However, correctness of recognizing viola improves after adding mixes to the training set. Viola is also quite often mistaken with cello, but 4th octave is rather high for cello, so such mistakes could have been expected.

The obtained results show quite high capabilities of classifiers trained as shown, because the recognition of the main sound in a mix of sounds of the same pitch is one of the most difficult tasks to perform in multi-timbral environment.

4 Summary and Conclusions

The purpose of this research was to perform experiments on recognizing the dominating instrument in mixes of sustained sounds of the same pitch, assuming harmonic-type spectrum of the test data. This case is the most difficult for classification, since spectra of mixed sounds overlap to high extend. The set-up of experiments for training and testing of classifiers was designed in such a way that a relatively small training data set can be used for learning. Instead of testing all possible pairs for training, we used mixes with artificial sounds with spectra overlapping with the main sounds, i.e. of the same pitch in case of added harmonic sounds, or noises. The results show that adding mixes to the training set may yield significant improvement in classification accuracy, although stringed instruments cause difficulties and cello/viola or viola/violin is most common mistake. The recognition of other instruments in most cases improves after adding mixes to the training set. However, no clear dependency was observed between the level of added sounds and correctness of classifiers trained for this level.

Our goal is to perform experiments on bigger set of data, including more instruments, and octaves. We also plan to continue our research using percussive instruments; this is one of the reasons why we chose noises for mixes. Also, experiments with training on mixes with artificial sounds are planned for sounds of different pitch, and testing on instrument sounds from outside the training set as well.

Acknowledgments. This work was supported by the National Science Foundation under grant IIS-0414815, and also by the Research Center of PJIIT, supported by the Polish National Committee for Scientific Research (KBN).

The authors would like to express thanks to Xin Zhang from the University of North Carolina at Charlotte for her help with data parameterization. We are also grateful to Zbigniew W. Raś from UNC-Charlotte for fruitful discussions.

References

1. Adobe Systems Incorporated: Adobe Audition 1.0 (2003)
2. Aniola, P., Lukasik, E.: JAVA Library for Automatic Musical Instruments Recognition, AES 122 Convention, Vienna, Austria (2007)
3. Brown, J.C.: Computer identification of musical instruments using pattern recognition with cepstral coefficients as features. J.Acoust.Soc.Am. 105, 1933–1941 (1999)
4. Herrera, P., Amatriain, X., Batlle, E., Serra, X.: Towards instrument segmentation for music content description: a critical review of instrument classification techniques. In: International Symposium on Music Information Retrieval ISMIR (2000)
5. ISO/IEC JTC1/SC29/WG11: MPEG-7 Overview, `http://www.chiariglione.org/mpeg/standards/mpeg-7/mpeg-7.htm`
6. Kaminskyj, I.: Multi-feature Musical Instrument Sound Classifier w/user determined generalisation performance. In: Proceedings of the Australasian Computer Music Association Conference ACMC, pp. 53–62 (2002)
7. Kitahara, T., Goto, M., Okuno, H.G.: Pitch-Dependent Identification of Musical Instrument Sounds. Applied Intelligence 23, 267–275 (2005)
8. Livshin, A., Rodet, X.: The importance of cross database evaluation in musical instrument sound classification: A critical approach. In: International Symposium on Music Information Retrieval ISMIR (2003)
9. Martin, K.D., Kim, Y.E.: Musical instrument identification: A pattern-recognition approach. In: 136-th meeting of the Acoustical Society of America, Norfolk, VA (1998)
10. Opolko, F., Wapnick, J.: MUMS - McGill University Master Samples. In: CD's (1987)
11. Peeters, G., McAdams, S., Herrera, P.: Instrument Sound Description in the Context of MPEG-7. In: International Computer Music Conference ICMC 2000 (2000)
12. The University of Waikato: Weka Machine Learning Project, `http://www.cs.waikato.ac.nz/~ml/`
13. Wieczorkowska, A.: Towards Musical Data Classification via Wavelet Analysis. In: Ohsuga, S., Raś, Z.W. (eds.) ISMIS 2000. LNCS (LNAI), vol. 1932, pp. 292–300. Springer, Heidelberg (2000)
14. Wieczorkowska, A., Kolczyńska, E.: Quality of Musical Instrument Sound Identification for Various Levels of Accompanying Sounds. In: Raś, Z.W., Tsumoto, S., Zighed, D. (eds.) MCD 2007. LNCS (LNAI), vol. 4944, pp. 93–103. Springer, Heidelberg (2008)
15. Zhang, X.: Cooperative Music Retrieval Based on Automatic Indexing of Music by Instruments and Their Types. Ph.D thesis, Univ. North Carolina, Charlotte (2007)

Performance Weights for the Linear Combination Data Fusion Method in Information Retrieval

Shengli Wu[1], Qili Zhou[2], Yaxin Bi[1], and Xiaoqin Zeng[3]

[1] School of Computing and Mathematics
University of Ulster, Northern Ireland, UK, BT37 0QB
{s.wu1,y.bi}@ulster.ac.uk
[2] School of Computing
Hangzhou Dianzi University, Hangzhou, China, 310018
cotzq@abertay.ac.uk
[3] Department of Computer Science
Hohai University, Nanjing, China, 210098
xzeng@hhu.edu.cn

Abstract. In information retrieval, the linear combination method is a very flexible and effective data fusion method, since different weights can be assigned to different component systems. However, it remains an open question which weighting schema is good. Previously, a simple weighting schema was very often used: for a system, its weight is assigned as its average performance over a group of training queries. In this paper, we investigate the weighting issue by extensive experiments. We find that, a series of power functions of average performance, which can be implemented as efficiently as the simple weighting schema, is more effective than the simple weighting schema for data fusion.

1 Introduction

Information retrieval as a core technology has been widely used for the WWW search services and digital libraries. In recent years, an increasing number of researchers have been working in this area and many different techniques have been investigated to improve the effectiveness of retrieval. Quite a large number of retrieval models have been proposed and experimented with various text document collections. For example, in the book "Modern Information Retrieval" written by Baeza-Yates and Ribeiro-Neto [2], 11 different retrieval models were discussed. In such a situation, data fusion, which uses a group of information retrieval systems to search the same document collection, and then merges the results from these different systems, is an attractive option to improve retrieval effectiveness.

Quite a few data fusion methods such as CombSum [4,5], CombMNZ [4,5], the linear combination method [3,9,10], the probabilistic fusion method [6], Borda fusion [1], Condorcet fusion [7], and the correlation method [12,13], have been proposed. Among them, the linear combination data fusion method is a very flexible and effective method since different weights can be assigned to different

A. An et al. (Eds.): ISMIS 2008, LNAI 4994, pp. 465–470, 2008.

systems. However, it is unclear which weighting schema is good. In some previous researches, different search methods such as golden section search [9,10] and conjugate gradient [3] were used to search suitable weights for component systems. One major drawback of these methods is their very low efficiency. Because of this, data fusion with only two or three component systems were investigated in [3] and [9,10]. In some situations such as the WWW and digital libraries, documents are updated frequently, then each component system's performance may change considerably from time to time. The weights for the systems should be updated accordingly. In such a situation, it is very difficult or impossible to use those low efficient weighting methods.

In some data fusion experiments, (e.g., in [1,8,11,13]), a simple weighting schema was used: for a system, its weight is set as its average performance over a group of training queries. There is a straightforward relationship between performance and weight. This method can be used in very dynamic situations since weights can be calculated and modified very easily. However, it has not been investigated how good this schema is. We would like to investigate this issue with extensive experiments. We shall demonstrate that, a power function weighting schema, with a power of between 2 and 6, is more effective than the simple weighting schema (which is a special case of power function, power = 1) for data fusion, though both of them can be implemented in the same way.

2 Performance Weights

Suppose we have n information retrieval systems ir_1, ir_2,..., ir_n. For a given query q, each of them provides a result r_i. Each r_i is a ranked list of documents, with an estimated relevance score for every document included. w_i is the performance weight assigned to system ir_i. Then for any document d in one or more results, the linear combination method uses the following equation to calculate its score:

$$M(d,q) = \sum_{i=1}^{n} w_i * s_i(d,q)$$

Here $s_i(d,q)$ is the normalized score of document d in result r_i, $M(d,q)$ is the calculated score of d. All the documents can be ranked using their calculated scores $M(d,q)$.

For each system ir_i, suppose its average performance over a group of training queries is p_i, then p_i is set as ir_i's weight (w_i) in the simple weighting schema, which has been used in previous research (e.g., in [1,8,11,13]). However, it is not clear how good this simple weighting schema is or is there any other effective schemas available. The purpose of our investigation is to try to find some other schemas which are more effective than the simple weighting schema but can be implemented as efficiently as the simple weighting schema. In order to achieve this, we set pw_i as a power function of p_i. Besides p_i, we used $p_i^{0.5}$, ${p_i}^{1.5}$, ${p_i}^2$, ${p_i}^{2.5}$ and ${p_i}^3$ as ir_i's weights. Note if a larger power is used for the weighting schema, then those systems with better performance have a larger impact on fusion, and those results with poorer performance have a smaller impact on fusion.

3 Experimental Results

4 groups of TREC data (2001 Web, 2003 and 2004 Robust, and 2005 Terabyte) were used for the experiment. These 4 groups of submitted results (called runs in TREC) are different in many ways from track (Web, Robust, and Terabyte), the number of results selected (32(2001), 62(2003), 77(2004), and 41(2005)), the number of queries used (50(2001 and 2005), 100(2003), and 249(2004)), to the number of retrieved documents for each query in each submitted result (1000(2001, 2003, and 2004) and 10000(2005))[1]. They comprise a good combination for us to evaluate data fusion methods.

The Zero-one linear normalization method was used for score normalization. It maps the highest score into 1, the lowest score into 0, and any other scores into a value between 0 and 1. For all the systems involved, we evaluated their average performance measured by MAP (mean average precision) over a group of queries. Then different values (0.5, 1.0, 1.5, 2.0,...) were used as powers in the power function to calculate weights for the linear combination method. In a year group, we chose m (m=3, 4, 5, 6, 7, 8, 9, or 10) component results for fusion. For each setting of m, we randomly chose m component results 200 times and carried out fusion. Two metrics were used to evaluate the fused retrieval results. They are mean average precision (MAP) and recall-level precision(RP). Besides the linear combination method with different weighting schemas, CombSum and CombMNZ were also involved in the experiment.

Tables 1-2 show the performance of the fused result in MAP and RP, respectively. Each data point in the tables is the average of 8*200*q_num measured values. Here 8 is the different number (3, 4,..., 9, 10) of component results used, 200 is the number of runs for each setting, and q_num is the number of queries in each year group. The improvement rate over the best component result is shown as well.

From Tables 1 and 2 we can see that the two measures MAP and RP are very consistent, though the RP values are usually smaller than the corresponding MAP values. Comparing CombMNZ with CombSum, CombMNZ is not as good as CombSum in all 4 year groups. With any of the weighting schemas chosen, the linear combination method performs better than the best component result, CombSum, and CombMNZ in all 4 year groups. Comparing with all different weighting schemas used, we can find that the larger the power is used for weighting calculation, the better the linear combination method performs. Two-tailed tests were carried out to compare the differences between all the data fusion methods involved. The tests show that the differences between any pair of the data fusion methods are statistically significant at a level of .000 ($p < 0.001$, or the probability is over 99.9%). From the worst to the best, the data fusion methods are ranked as follows: CombMNZ, CombSum, LC(0.5), LC(1.0), LC(1.5), LC(2.0), LC(2.5), LC(3.0).

---

[1] Some submitted results include fewer documents. For convenience, those results were not selected.

Table 1. Performance (on MAP) of several data fusion methods (In LC(a), the number a denotes the power value used for weight calculation; for every data fusion method, the improvement rate of its MAP value over the best component result is shown)

Group/ Best	Comb-Sum	Comb-MNZ	LC(0.5)	LC(1.0)	LC(1.5)	LC(2.0)	LC(2.5)	LC(3.0)
2001	0.2614	0.2581	0.2620	0.2637	0.2651	0.2664	0.2673	0.2681
0.1861	+10.44%	+9.04%	+10.69%	+11.41%	+12.00%	+12.55%	+12.93%	+13.27%
2003	0.2796	0.2748	0.2841	0.2865	0.2879	0.2890	0.2900	0.2908
0.2256	-0.71%	-2.41%	+0.89%	+1.74%	+2.24%	+2.63%	+2.98%	+3.27%
2004	0.3465	0.3434	0.3482	0.3499	0.3512	0.3522	0.3530	0.3537
0.2824	+4.40%	+3.46%	+4.91%	+5.42%	+5.82%	+6.12%	+6.36%	+6.57%
2005	0.3789	0.3640	0.3857	0.3897	0.3928	0.3952	0.3970	0.3986
0.2991	-0.89%	-4.79%	+0.89%	+1.94%	+2.75%	+3.37%	+3.85%	+4.26%

Table 2. Performance (on RP) of several data fusion methods (In LC(a), the number a denotes the power value used for weight calculation; for every data fusion method, the improvement rate of its RP value over the best component results is shown)

Group	Comb-Sum	Comb-MNZ	LC(0.5)	LC(1.0)	LC(1.5)	LC(2.0)	LC(2.5)	LC(3.0)
2001	0.2815	0.2783	0.2821	0.2838	0.2854	0.2865	0.2874	0.2882
0.2174	+6.75%	+5.54%	+6.98%	+7.62%	+8.23%	+8.65%	+8.99%	+9.29%
2003	0.2982	0.2943	0.3009	0.3024	0.3034	0.3043	0.3051	0.3058
0.2508	+0.17%	-1.14%	+1.07%	+1.58%	+1.91%	+2.22%	+2.49%	+2.72%
2004	0.3629	0.3599	0.3643	0.3656	0.3667	0.3676	0.3682	0.3687
0.3107	+3.60%	+2.74%	+4.00%	+4.37%	+4.68%	+4.94%	+5.11%	+5.25%
2005	0.4021	0.3879	0.4077	0.4112	0.4137	0.4156	0.4171	0.4183
0.3357	-1.01%	-4.51%	+0.37%	+1.23%	+1.85%	2.31%	+2.68%	+2.98%

Table 3. Percentage of the fused results whose performances (MAP) are better than the best component result

Group	CombSum	CombMNZ	LC(0.5)	LC(1.0)	LC(1.5)	LC(2.0)	LC(2.5)	LC(3.0)
2001	83.18%	79.44%	86.75%	91.25%	94.87%	97.44%	98.69%	99.38%
2003	54.62%	28.16%	65.88%	71.06%	75.25%	78.25%	81.00%	84.19%
2004	87.56%	81.69%	90.50%	92.69%	94.62%	95.88%	96.88%	97.75%
2005	50.81%	29.62%	62.87%	69.44%	75.00%	79.88%	83.00%	85.44%

For MAP, we also calculated the percentage that the fused results is more effective than the best component result, which is shown in Table 3. The figures for RP are very similar, therefore we do not present them here. From Table 3, we can see that the linear combination methods are better than CombSum and CombMNZ in all year groups.

From the above experimental results we can see that the linear combination method increases in performance with the power used for weight calculation.

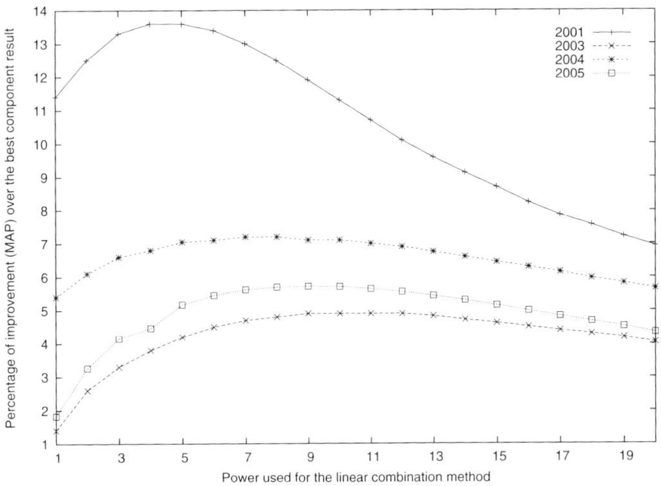

Fig. 1. Percentage of improvement (on MAP) of the linear combination method when using different powers

Since only six different values (0.5, 1, 1.5, 2, 2,5, 3) have been tested, it is interesting to find how far this trend continues. Therefore, we use more values (4, 5,... , 20) as powers for the linear combination method with the same setting as before. The experimental result is shown in Figure 1.

In Figure 1, the curve of TREC 2004 reach its maximum when a power of 4 or 5 is used. While for the three other groups, the curves are quite flat and they reach their maximum when a power of between 7 and 10 is used. It seems that, for obtaining the optimum fusion results, different powers may be needed for different sets of component results. this may seem a little strange. but one explanation for this is: data fusion is affected by many factors such as the number of component results involved, performances and performance differences of component results, dissimilarity among component results, and so on [14]. Therefore, it is likely that the optimum weight is decided by all these factors, not just by any single factor, though performances of component results is probably the most important one among all the factors. Anyway, if we only consider performance, then a power of 1, as the simple weighting schema does, is far from the optimum.

4 Conclusions

In this paper we have presented our work about assigning appropriate performance weights for the linear combination data fusion method. From the extensive experiments conducted with the TREC data, we conclude that for performance weighting, a series of power functions (e.g., a power of 2 to 6) are better than the simple weighting schema, in which the performance weight of a system is assigned as its average performance (power equals to 1). The power function schema can be implemented as efficiently as the simple weighting schema. We expect that the finding in this paper is very useful in practice. As our next stage of work, we plan to carry out some theoretical analysis to see why this is the case.

References

1. Aslam, J.A., Montague, M.: Models for metasearch. In: Proceedings of the 24th Annual International ACM SIGIR Conference, New Orleans, Louisiana, USA, September 2001, pp. 276–284 (2001)
2. Baeza-Yates, R., Ribeiro-Neto, B.: Modern Information Retrieval. ACM Press, Addison-Wesley (1999)
3. Bartell, B.T., Cottrell, G.W., Belew, R.K.: Automatic combination of multiple ranked retrieval systems. In: Proceedings of ACM SIGIR 1994, Dublin, Ireland, July 1994, pp. 173–184 (1994)
4. Fox, E.A., Koushik, M.P., Shaw, J., Modlin, R., Rao, D.: Combining evidence from multiple searches. In: The First Text REtrieval Conference (TREC-1), Gaithersburg, MD, USA, March 1993, pp. 319–328 (1993)
5. Fox, E.A., Shaw, J.: Combination of multiple searches. In: The Second Text REtrieval Conference (TREC-2), Gaitherburg, MD, USA, August 1994, pp. 243–252 (1994)
6. Lillis, D., Toolan, F., Collier, R., Dunnion, J.: Probfuse: a probabilistic approach to data fusion. In: Proceedings of the 29th Annual International ACM SIGIR Conference, Seattle, Washington, USA, August 2006, pp. 139–146 (2006)
7. Montague, M., Aslam, J.A.: Condorcet fusion for improved retrieval. In: Proceedings of ACM CIKM Conference, McLean, VA, USA, November 2002, pp. 538–548 (2002)
8. Thompson, P.: Description of the PRC CEO algorithms for TREC. In: The First Text REtrieval Conference (TREC-1), Gaitherburg, MD, USA, March 1993, pp. 337–342 (1993)
9. Vogt, C.C., Cottrell, G.W.: Predicting the performance of linearly combined IR systems. In: Proceedings of the 21st Annual ACM SIGIR Conference, Melbourne, Australia, August 1998, pp. 190–196 (1998)
10. Vogt, C.C., Cottrell, G.W.: Fusion via a linear combination of scores. Information Retrieval 1(3), 151–173 (1999)
11. Wu, S., Crestani, F.: Data fusion with estimated weights. In: Proceedings of the 2002 ACM CIKM International Conference on Information and Knowledge Management, McLean, VA, USA, November 2002, pp. 648–651 (2002)
12. Wu, S., McClean, S.: Data fusion with correlation weights. In: Proceedings of the 27th European Conference on Information Retrieval, Santiago de Composite, Spain, March 2005, pp. 275–286 (2005)
13. Wu, S., McClean, S.: Improving high accuracy retrieval by eliminating the uneven correlation effect in data fusion. Journal of American Society for Information Science and Technology 57(14), 1962–1973 (2006)
14. Wu, S., McClean, S.: Performance prediction of data fusion for information retrieval. Information Processing & Management 42(4), 899–915 (2006)

Combining Multiple Sources of Evidence in Web Information Extraction

Martin Labský and Vojtěch Svátek

Department of Information and Knowledge Engineering,
University of Economics, Prague, W. Churchill Sq. 4, 130 67 Praha 3, Czech Republic
{labsky,svatek}@vse.cz

Abstract. Extraction of meaningful content from collections of web pages with unknown structure is a challenging task, which can only be successfully accomplished by exploiting multiple heterogeneous resources. In the *Ex* information extraction tool, so-called extraction ontologies are used by human designers to specify the domain semantics, to manually provide extraction evidence, as well as to define extraction subtasks to be carried out via trainable classifiers. Elements of an extraction ontology can be endowed with probability estimates, which are used for selection and ranking of attribute and instance candidates to be extracted. At the same time, HTML formatting regularities are locally exploited.

1 Introduction

In the last decade, *web information extraction* (WIE) was dominated by two paradigms. One—*wrapper*-based—exploits regular surface-level structures found in HTML code, which can be used as anchors for the extraction. This approach is now widely adopted in industry, however, its dependence on formatting regularities limits its use for diverse categories of web pages. The other—*inductive*—paradigm assumes the presence of training data: either web pages containing pre-annotated tokens or stand-alone examples of data instances; state-of-the-art trainable IE algorithms are surveyed e.g. in [10]. Again, however, sufficient amount of appropriate training data is rarely available in practice and manual labelling is often too expensive even with the help of active learning; statistical bootstrapping alleviates this problem to some degree but at the same time it burdens the whole process with 'heavy computational machinery', whose requirements and side-effects are not transparent to a casual user of a WIE tool. In addition, both approaches usually deliver extracted information as rather weakly semantically structured; if WIE is to be used to fuel semantic web repositories, secondary mapping to *ontologies* is typically needed, which makes the process complicated and possibly error-prone.

There were recently proposals for pushing ontologies towards the actual extraction process as immediate prior knowledge. *Extraction ontologies* [2] enumerate attributes of the concepts to be extracted, their allowed values as well as higher level (e.g. cardinality or mutual dependency) constraints. Extraction ontologies are assumed to be hand-crafted based on observation of a sample of resources; however, due to their well-defined and rich conceptual structure they are superior to ad-hoc hand-crafted patterns

A. An et al. (Eds.): ISMIS 2008, LNAI 4994, pp. 471–476, 2008.

used in early times of WIE. At the same time, they allow for rapid start of the actual extraction process, as even a very simple extraction ontology may cover a sensible part of target data and generate meaningful feedback for its own redesign. It seems that for web extraction tasks where the subject of extraction evolves and does not require sophisticated NLP, extraction ontologies are the first choice. However, to achieve competitive extraction results and to prevent overfitting to a few sample resources, one must not neglect available labelled data, formatting regularities and even pre-existing domain ontologies. This was the motivation for building our WIE tool named *Ex*[1], which exploits all the mentioned resources, with central role of extraction ontologies. It has been so far tested in three domains: product catalogues of computer monitors and TVs, contact information on medical pages, and weather forecasts. First experimental results were reported in [4]. In this paper we present the principle how multiple pieces of evidence from extraction ontologies are combined during the extraction process. Section 2 uses a real-world example to explain the most important features of extraction ontologies used in *Ex*. Section 3 describes the steps of the information extraction process and, especially, the underlying pseudo-probabilistic apparatus. Finally, section 4 surveys related research, and section 5 outlines future work.

2 Ex(traction) Ontology Content — Overview and Example

Extraction ontologies in *Ex* are designed so as to extract occurrences of *attributes*, i.e. standalone named entities, and occurrences of whole *instances* of *classes*, as groups of attributes that 'belong together', from HTML pages or texts in a domain of interest. An extraction ontology defines evidence of different types which is used to identify the extractable items. Token-, character- and formatting-level patterns may address both the content and context of attributes and instances to be extracted, axioms may encode their complex constraints and relations; there are also *formatting* constraints and ranges or distributions for *numeric attribute values* and for attribute *content lengths*. The extraction ontology language of *Ex* was introduced in [4].

In *Ex*, every piece of evidence may be equipped with two probability estimates: *precision* and *recall*. The *precision* $P(A|E)$ of evidence states how probable it is for the predicted attribute or class instance A to occur given that the evidence E holds, disregarding the truth values of other evidence. For example, the precision of a left context pattern "person name: \$" (where \$ denotes the predicted attribute value) may be estimated as 0.8; i.e. in 80% of cases we expect a person name to follow in text after a match of the "person name:" string. The *recall* $P(E|A)$ of evidence states how abundant the evidence is among the predicted objects, disregarding whether other evidence holds. For example, the "person name: \$" pattern could have a low recall since there are many other contexts in which a person name could occur. Pattern precision and recall can be estimated in two ways. First, annotated documents can be used to estimate both parameters using smoothed ratios of counts observed in text. Second, when no training data are available, the user specifies one or both parameters manually. Our initial experiments indicate that extraction ontology developers are often able to specify precision and recall values with accuracy sufficient to create useful prototype IE systems.

[1] Alpha version of *Ex* is available from `http://eso.vse.cz/~labsky/ex`

```
<class id="Contact">
 <script src="contact.js" />

 <pattern recall="0.7"> ^ $title{0-3} .? $name ,? $title{0-3} </pattern>

 <axiom recall="0.8"> nameMatchesEmail($name, $email) > 0 </axiom>

 <classifier id="cls1" method="weka" classtype="attribute" features="ontology,ngram"
  name="weka.classifiers.rules.JRip" model="contacts.bin" elements="*"/>

 <attribute id="degree" type="name" card="0-4" eng="0.60">
  <content>
   <pattern recall="0.2" p="0.8"> ( Miss | Lady | Sir | MSc | MA | MPh ) .? </pattern>
   <pattern recall="1" type="format"> has_one_parent </pattern>
  </content>
 </attribute>

 <attribute id="name" type="name" card="1" eng="0.80">
  <pattern id="init"> (A|B|C|D|E|F|G|H|I|J|K|L|M|N|O|P|R|S|T|U|V|W|X|Y|Z) .? </pattern>
  <content>
   <pattern recall="0.5" p="0.8">
    <pattern src="first.txt"/> <pattern ref="init"/>? <pattern src="last.txt"/> </pattern>
   <pattern recall="0.7" p="0.4">
    <pattern src="first.txt"/> <pattern ref="init"/>? <tok type="alpha" case="CA|UC"/>
   </pattern>
   <pattern p="0.7" recall="0.5"> <phr label="cls1.name" /> </pattern>
   <length> <distribution min="1" max="6" /> </length>
   <refers> nameRefersTo($, $other) </refers>
  </content>
  <context> <pattern recall="0.1" p="0.6"> person? name :? $ </pattern> </context>
 </attribute>

 <attribute id="email" type="name" card="0-1" eng="0.60">
  <content><pattern recall="0.9" p="0.9"> <pattern ref="gen.email"/></pattern> </content>
 </attribute>
</class>
```

Fig. 1. Fragment of code of extraction ontology for contact information

Our example shows a simplified *contact information* ontology developed within the EU project MedIEQ[2], aiming to extract peoples' degrees, names and emails. Extraction results for this task were presented in [4]. The first pattern claims that 70% of *Contact* instances start with a person name or degree separated by punctuation. An axiom further claims that in 80% of cases, the person's name and email exhibit string similarity which is identified by a script function *nameMatchesEmail()*, introduced in a script "contact.js" above. Finally, a *classifier link* contracts an external trained classifier to classify all attributes of the *Contact* class. Classifications made by this classifier are used by some of the attribute content patterns defined below, e.g. for *name* we expect the classifier to have a 70% precision and a 50% recall. Other content and context patterns rely on token-level regular expressions including large named entity lists; e.g. the first content pattern for *name* claims that 80% of its matches correctly identify a person name, while it is only expected to cover about 50% of all person names since the lists are not exhaustive. In addition, a script is used by the *refers* section to match co-referring mentions of the same person *name* on a single page.

3 The Extraction Process

The inputs to the extraction process are the extraction ontology and a set of documents. First, analysed documents are tokenized and equipped with formatting structure DOM trees[3]. The actual extraction consists of four stages.

[2] `http://www.medieq.org`

[3] To parse malformed HTML we use the CyberNeko parser `http://people.apache.org/~andyc/neko/doc/html`

3.1 Attribute Candidate Generation

Attribute candidates (*ACs*) are created where at least one content or context pattern matches. Let Φ_A be the set of all evidence E_i known for an attribute A. To compute a *conditional probability estimate* $P_{AC} = P(A|E \in \Phi_A)$ of how likely the *AC* is given the presence of each evidence, we use these naive bayesian independence assumptions:

$$\forall_{E,F \in \Phi_A, F \neq E} : \; F \perp E|A, \; F \perp E|\neg A. \tag{1}$$

That is, evidence is assumed to be mutually independent within positive examples of A and outside of A. To compute P_{AC}, we use the precision $P(A|E_i)$ and recall $P(E_i|A)$ of each evidence $E_i \in \Phi_A$ and their truth values, and the prior probability $P(A)$ of encountering each attribute in text. Φ_A^+ denotes the set of evidence $E_i \in \Phi_A$ observed for that candidate, and Φ_A^- is the set of unobserved evidences $E_i \in \Phi_A$:

$$P(A|E \in \Phi_A) = \frac{P(A, E \in \Phi_A)}{P(E \in \Phi_A)} = \frac{1}{1 + \frac{P(A)}{P(\neg A)}^{|\Phi_A|-1} \Pi_{E \in \Phi_A^+} \frac{P(\neg A|E)}{P(A|E)} \Pi_{F \in \Phi_A^-} \frac{P(\neg A|\neg F)}{P(A|\neg F)}} \tag{2}$$

The $P(A|\neg F)$ member of Eq. 2 is computed according to Eq. 3; the remaining values are known from the extraction ontology. Derivation of both formulas is shown in [5].

$$P(A|\neg F) = \frac{P(\neg F|A)P(A)}{1 - \frac{P(F|A)P(A)}{P(A|F)}}. \tag{3}$$

The set of (possibly overlapping) *ACs* created during this phase is represented as an *AC lattice* spanning through the document, where each *AC* node is scored as $score(AC) = log(P_{AC})$. Apart from *ACs*, the lattice includes one 'background' node BG_w for each token w that takes part in at least one *AC*. Supposing $|AC|$ is the length of an *AC* in tokens, we define $score(BG_w) = \min_{AC, w \in AC} log(\frac{1-P(AC)}{|AC|})$ where $|AC|$ is the *AC* length in tokens. The extraction process can terminate here by extracting all *ACs* on the best path through this lattice or it may continue with instance parsing and formatting pattern induction.

3.2 Instance Candidate Generation

Initially, each *AC* is used to create a simple instance candidate *IC*={*AC*}. Then, increasingly complex *ICs* are generated bottom up from the working set of *ICs*. At each step, the highest scoring (seed) *IC* is popped from the working set and added to a *valid IC set* if it satisfies *all* ontological constraints. The seed *IC* is then extended using its neighboring *ACs* if their inclusion does not break ontological constraints. Only a *subset* of constraints is considered at this time as e.g. minimum cardinality constraints or some axioms could never get satisfied initially. An *AC* is only added as a reference if a user-defined function determines its value may corefer with an *AC* that already is part of the *IC*. The newly created ICs are added to the working set. A limited number of ACs is allowed to be skipped (AC_{skip}) between the combined *IC* and *AC*, leading to a penalization of the created IC. The IC scores are computed based on their AC content and on the observed values of evidence E known for the IC class C:

$$sc_1(IC) = exp(\frac{\sum_{AC \in IC} log(P_{AC}) + \sum_{AC_{skip} \in IC}(1 - log(P_{AC_{skip}}))}{|IC|}) \tag{4}$$

$$sc_2(IC) = P(C|E \in \Omega_C) \tag{5}$$

where $|IC|$ is the number of member ACs and Ω_C is the set of evidence known for class C; the conditional probability is estimated as in Eq. 2. By experiment we chose the Prospector [1] pseudo-bayesian method to combine the above into the final IC score:

$$score(IC) = \frac{sc_1(IC)sc_2(IC)}{sc_1(IC)sc_2(IC) + (1 - sc_1(IC))(1 - sc_2(IC))} \tag{6}$$

IC generation ends when the working set becomes empty or on a terminating condition such as after a certain number of iterations or after a time limit has elapsed. The output of this phase is the set of valid ICs.

3.3 Formatting Pattern Induction

When extracting from a single web page or web site, it often happens that a large part of valid *ICs* satisfies some unforeseen *HTML formatting pattern*. E.g. all person names and emails could reside in two dedicated columns of a table. We try to *induce* these local formatting patterns as follows. First, the best scoring path of non-overlapping *ICs* is found through the valid *IC* lattice. For each *IC* on the path, we find its nearest containing formatting *block* element (e.g. paragraph, div, table cell). We then create a *subtree of formatting elements* between the block element (inclusive) and the *ACs* comprising the *IC*. Each subtree consists of the formatting element names and their order within parent. Formatting subtrees whose relative and absolute frequencies satisfy certain thresholds are transformed into new *context patterns* indicating presence of the corresponding class, with precision and recall based on their relative frequencies. The *AC* and *IC* generation phases are then re-run for the newly created local context patterns, re-scoring and possibly yielding new *ACs* and *ICs*.

3.4 Attribute and Instance Parsing

All valid *ICs* are merged as additional nodes into the *AC* lattice created in previous steps, so that each *IC* node can be avoided by taking a path through standalone *ACs* or through background states. In the merged lattice, each *IC* node is scored by $|IC| \times score(IC)$. The merged lattice is searched for the best scoring node sequence from which all instances and standalone attributes are extracted.

4 Related Work

Most state-of-the-art WIE approaches use inductively learned models and only consider ontologies as additional structures to which extracted data are to be adapted [3], rather than directly using rich ontologies to guide the extraction process. An exception is the approach taken by Embley and colleagues [2], which even inspired our early work; the main difference is our effort to combine manually encoded extraction knowledge with HTML formatting and trained classifiers, the possibility to equip extraction evidence with probability estimates, and the pragmatic distinction we see between extraction ontologies and domain ontologies: extraction ontologies can be adapted to the way data are typically *presented* on the web while domain ontologies describe the domain semantics. A parallel stream to ontology-based IE is that relying on automated discovery of new extractable attributes from large amounts of documents using statistical and NLP methods, as in [6]. On the other hand, formatting information is heavily exploited in IE from tables [7]. Our system has a slightly different target from both these; it should allow for fast IE prototyping even in domains where there are few documents available and the content is semi-structured. Advanced automatic methods also exist for coreference resolution in attribute values [9] or for the estimate of mutual affinity among these values [8]. Our approach, again, relies on the author to supply such knowledge.

5 Conclusions and Future Work

The *Ex* information extraction system is capable of combining, in a pseudo-probabilistic manner, multiple pieces of extraction evidence, provided by the user within an extraction ontology as well as learned from annotated data and from local formatting regularities. As the next step we want to focus on bootstrapping techniques and to compare our extraction results with other approaches using standard datasets. Finally, we intend to provide support for semi-automated transformation of domain ontologies to extraction ones.

Acknowledgments

The research was partially supported by the EC under contract FP6-027026, Knowledge Space of Semantic Inference for Automatic Annotation and Retrieval of Multimedia Content - K-Space. The medical website application is carried out in the context of the EC-funded (DG-SANCO) project MedIEQ.

References

1. Duda, R.O., Gasching, J., Hart, P.E.: Model design in the Prospector consultant system for mineral exploration. Readings in Artificial Intelligence, 334–348 (1981)
2. Embley, D.W., Tao, C., Liddle, D.W.: Automatically extracting ontologically specified data from HTML tables of unknown structure. In: Spaccapietra, S., March, S.T., Kambayashi, Y. (eds.) ER 2002. LNCS, vol. 2503, pp. 322–337. Springer, Heidelberg (2002)
3. Kiryakov, A., Popov, B., Terziev, I., Manov, D., Ognyanoff, D.: Semantic annotation, indexing, and retrieval. J. Web Sem. 2, 49–79 (2004)
4. Labský, M., Nekvasil, M., Svátek, V., Rak, D.: The Ex Project: Web Information Extraction using Extraction Ontologies. In: Proc. PriCKL workshop, ECML/PKDD (2007)
5. Labský, M., Svátek, V: Information extraction with presentation ontologies. Technical report, KEG UEP, `http://eso.vse.cz/~labsky/ex/ex.pdf`
6. Popescu, A., Etzioni, O.: Extracting Product Features and Opinions from Reviews. In: Proc. EMNLP (2005)
7. Wei, X., Croft, B., McCallum, A.: Table Extraction for Answer Retrieval. Information Retrieval Journal 9(5), 589–611 (2006)
8. Wick, M., Culotta, A., McCallum, A.: Learning Field Compatibilities to Extract Database Records from Unstructured Text. In: Proc. EMNLP (2006)
9. Yates, A., Etzioni, O.: Unsupervised Resolution of Objects and Relations on the Web. In: Proc. HLT (2007)
10. Dietterich, T.G.: Machine Learning for Sequential Data: A Review. In: Caelli, T.M., Amin, A., Duin, R.P.W., Kamel, M.S., de Ridder, D. (eds.) SPR 2002 and SSPR 2002. LNCS, vol. 2396, Springer, Heidelberg (2002)

Autonomous News Clustering and Classification for an Intelligent Web Portal

Traian Rebedea[1] and Stefan Trausan-Matu[1, 2]

[1] "Politehnica" University of Bucharest, Department of Computer Science and Engineering, 313 Splaiul Independetei, Bucharest, Romania
{trebedea,trausan}@cs.pub.ro
http://www.racai.ro/~trausan
[2] Research Institute for Artificial Intelligence of the Romanian Academy, 13 Calea 13 Septembrie, Bucharest, Romania

Abstract. The paper presents an autonomous text classification module for a news web portal for the Romanian language. Statistical natural language processing techniques are combined in order to achieve a completely autonomous functionality of the portal. The news items are automatically collected from a large number of news sources using web syndication. Afterward, machine-learning techniques are used for achieving an automatic classification of the news stream. Firstly, the items are clustered using an agglomerative algorithm and the resulting groups correspond to the main news topics. Thus, more information about each of the main topics is acquired from various news sources. Secondly, text classification algorithms are applied to automatically label each cluster of news items in a predetermined number of classes. More than a thousand news items were employed for both the training and the evaluation of the classifiers. The paper presents a complete comparison of the results obtained for each method.

Keywords: News Portal, Intelligent Agent, Text Clustering, Classification, Natural Language Processing.

1 Introduction

During the last ten years, the use of the World Wide Web has dramatically increased in regard to both the number of users and the number of web domains. Recent studies show that the number of users has doubled in the last five years, exceeding today one billion [1], and the web domains have tripled to almost 100 million in the same period [2]. Thus, the volume of online information has reached an impressive quantity and there are over 20 billion web items indexed for search by Yahoo [3]. Even with the help of the PageRank algorithm [4] used by Google, the answers provided by a search engine can contain a great quantity of useless information. Another important problem on the Web is informational redundancy, because a large part of the information can be found in different sources, with slight or no variations.

The most important news sources worldwide have introduced web syndication as a new means of providing news to their readers, as well as to other web sites or applications

A. An et al. (Eds.): ISMIS 2008, LNAI 4994, pp. 477–486, 2008.

that are willing to use them. XML-based formats for web syndication were designed so that the syndicated content can be easily used by computer programs called feed readers or aggregators. These can be found in a wide range of distributions, from online aggregators to stand-alone desktop applications. Aggregators are useful as they automate the process of periodically collecting news feeds and presenting them to the user in an easy to follow manner. Still they do not solve the two essential problems: processing a large quantity of information and news redundancy. By employing natural language processing techniques, such as topic detection algorithms, the redundancy in the stream of news can be used in order to determine the most important headlines.

The main idea of this paper is the combination of web syndication with advanced text processing methods, including clustering and classification, in order to obtain a completely autonomous news portal. The paper continues with a section introducing the idea of intelligent news processing and its main concepts. The third section covers the most important techniques used for text clustering and classification. The next section contains the description of a news portal for the Romanian language that uses intelligent processing. The paper ends with conclusions and references.

2 Intelligent News Processing

When dealing with large volumes of data, it is important to find a method to determine the importance of the processed data. This is also valid for news headlines, especially when dealing with news collected from different sources. In this case, manually assigning an importance to each piece of news can be a difficult and a time consuming effort. In an initial phase of the news feeds' processing, online web portals were developed using the number of readers for each news headline as the main criteria for determining the importance of each piece of news. Although, this technique presents some important advantages over the static assignment one, like computing the importance by taking into consideration the preferences of the users, it also has a large number of disadvantages: it does not reduce the number of news headlines, it does not solve news redundancy, and it does not offer an automatic method for computing the importance of a particular news. In addition, each piece of news is attached to the source that has published it, without offering alternative sources on the same subject.

Lately, a number of researchers as well as companies worldwide have undertaken the task of intelligently processing news feeds [5, 6, 7, 8]. There are two main reasons for calling these methods intelligent. First of all, and most relevantly, this technique automatically determines the main headlines for some period of time (e.g. a day) and offers a classification of these subjects by taking into account the number of different news headlines that compose each subject. In this case, the process of determining the importance of a subject does not use the subjective opinion of the users, but the objective one of the news providers. Secondly, this methodology of processing news items uses various techniques from natural language processing and information retrieval, most notably machine learning ones.

The news taken from various feeds can be processed in two different manners. The first one is news clustering, thus determining the most important subjects, and the second one is news classification that assigns each piece of news into one of a number of predetermined news categories. In addition, some of the papers in the field of data extraction from news [5, 6] propose a more detailed processing, emphasizing specific knowledge extraction – such as persons, countries, organizations and companies' names – and using this knowledge for the grouping and classification of news. Certainly, there are multiple approaches in the field of intelligent news processing, however, most of them are still in the research stage. Assigning labels (very significant key-words or groups of key-words) to every group of news resulted from the grouping is an interesting idea. Thus, the user may follow more easily the list of subjects connected to a certain label. The central issue of this approach is represented by the quality of the labelling, as no labelling algorithms have been discovered which are able to provide satisfying results [6]. A new method is offered by the *NewsJunkie* system [7] which proposes the determination of *information novelty* for a piece of news – thus, in the case of a flow of news on a common topic, recounted over a longer period of time, the system should find the truly new information and should filter the articles recounting events that have already been presented. A similar idea is proposed by *Ranking a Stream of News* [8].

3 Text Clustering and Classification Techniques

Clustering and classification methods are widely used in natural language processing. Any piece of news can be seen as a vector, where each of its elements is associated to a keyword in the text. The values of the elements may be Boolean variables showing if the keyword appears or not in the text associated with the news, the headline and the description fetched using web syndication. A refinement to this representation is to use the frequency of the term in the text or the TF-IDF weight [9, p. 56–57].

Both clustering and classification algorithms need a measure to compute the similarity between two items, in this case text documents. The features' space used to represent documents has as many dimensions as the number of distinct terms in all the documents, regardless of the methods discussed above. Transforming metric distances into similarity measures does not offer good results for text. Therefore, other measures are used for computing similarity between text documents [11] and one of the most used is the cosine of the angle formed by the two vectors that correspond to the documents that are compared.

Data clustering [12, pp. 495–528] is a statistical analysis technique used to partition the data into subsets, called clusters, so that the data in the same group has some common characteristics. Usually, the grouping process is applied based on the proximity of the elements that need to be clustered, thus using a metric distance or a similarity measure. Used for generalization, clustering can be employed in a wide range of domains where large volumes of data are available as it exploits the redundancy found within these data.

Clustering algorithms can be divided using a wide range of criteria [11]. A first classification is possible by considering the process of forming the groups, either top-down (divisive) or bottom-up (agglomerative). Top-down processing starts with all the data considered in a single group, which is then refined into subgroups. Agglomerative algorithms consider each element as a separate group, which are then merged to form larger ones. In relation to the result of the clustering process, the algorithms can be hierarchical or flat. Hierarchical methods offer a tree structure that can be visualized as a dendrogram as a result of clustering, while flat clustering does not offer any information about the relations between the groups. In addition, the assignment of the elements into groups, divides clustering techniques into hard and soft. A hard assignment produces clusters that do not overlap, meaning that each element is part of a single group, while a soft algorithm assigns to each item a probability of being member of a certain number of groups. Hierarchical clustering almost always uses a hard assignment and is implemented using a greedy technique. These algorithms may use three distinct methods to compute the similarity between two groups. The single link clustering computes the similarity of two groups as the similarity between the most similar items in each group. The complete link method uses the similarity between the two least similar elements from each group and produces better results. Average link clustering has the performances of complete link and the complexity of single link and it defines the similarity of the groups as the average similarity between all the elements in each group.

The process of assigning a set of items to one or more categories or classes known before the process starts, is called automatic classification. A statistical classification process can be divided into two distinct stages: the training of the classifier and the effective categorization. The input data for the training phase are the distinct classes used for classification and the items provided by the environment as being part of each class. Every technique defines a different training procedure whose result is a data model, which is then used to classify new items from the universe of the problem. Text documents classification [9, pp. 124–169] is one of the applications of classifiers in natural language processing and it is used for a wide number of purposes varying from news and e-mail categorization to automatic classification of large text documents. The most relevant aspects of text classification is that text is unstructured and the number of features is very high, greater than a thousand, unlike database mining where features are usually less than a hundred.

Nearest neighbour (NN) classifiers [9, pp. 133–138] are very popular, because they are easy to train and to use, as they do not use additional mathematical constructs. The training phase is very simple and it consists of indexing the training data for each category, therefore it is very fast. For classifying a new item, the most similar k indexed documents are determined. The item is assigned to the class that has the most documents from the ones selected above. An improved variant of this classifier uses a score for each class, defined as the sum of all the documents from that class that are in the most similar k ones. The document will then be assigned to the category that has the greatest score. For improving the accuracy even more, offsets for each class can be used – these are added to the score. Training the classifier to find the best values for k and the offsets for each category can transform it into a powerful tool, that offers "an accuracy that is comparable to that of the best text classifiers" [9, p. 135]. The disadvantages are that they need more time for the classification phase and they use more memory than probabilistic based classifiers. In order to improve the performance, a greedy algorithm for features' selection can be used [9, pp. 136–143]. The classifiers that use a probability distribution for each class are very powerful for text categorization [13].

4 Intelligent News Classification in Romanian

Combining the facilities offered by web technologies like syndication with the advantages provided by text mining techniques allows the creation of a news portal that is able to function with a minimum of human intervention. It is intended to offer a viable alternative for traditional news portals based on its advantages like the autonomy towards an administrator as well as the methodology used to present the news based on the importance of the headline.

4.1 Functionality and Architecture of the Portal

The main functionalities of the web portal can be described using the following steps:

1. Periodically accessing the sites of news agencies and newspapers offering RSS and automated collecting of new syndications;
2. Introducing the news in every fresh RSS in a database, in order to offer easier access to the information;
3. Processing the text information of each fresh piece of news, by applying various computational linguistics techniques, for determining a characteristic vector associated with the news;
4. Grouping the news using a text clustering algorithm, starting from news representation in the m-dimensional space of words;
5. Classification of each group of news within a predefined category, using a regularly retrained classifier;
6. Automatically generating web pages corresponding to the most important subjects from a certain period of time, grouped in various ways, including in each category of news. These web pages constitute the final result of the operation of the portal and are visible to the users.

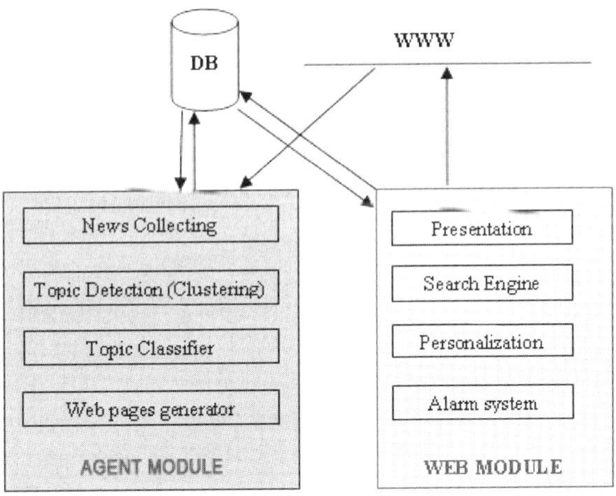

Fig. 1. The architecture of the web portal – the agent and the web modules

The primary actions described above may be run in a single stage or sequentially, following the determined order, as well as individually, at different moments in time. The method of operation is determined by the number of fresh pieces of news within a certain period and by the time interval in which the site is automatically modified.

Considering the method of operation described above, the following remarks are important: the portal is a reactive agent (it responds to changes in its environment), but it is not communicative (it can only communicate in a lesser degree, through web syndication) and non-adaptive. Apart from this behaviour as an autonomous agent, the portal must also implement a web module that will allow text search within the subjects, as well as the personalization of the portal by the user, by selecting the categories, their order and the number of news in each category.

The functionality of the portal may be broken into two different modules that are relatively independent: a module that contains the agent part of the application processing the news items and generating the web pages and a web module displaying the information and implementing search and personalization capabilities. This architecture has the advantage that it completely separates the processing of the data from the presentation and the communication with the user. The two modules communicate using a database that saves the information about the news, as presented in Figure 1. The paper is concerned only with the main functions of the agent module.

4.2 Text Processing and Clustering

The news feeds need to be processed before applying the clustering and classification techniques. Because the Romanian language uses diacritical marks (special characters) that are not used by all news sources, these need to be eliminated, regardless of the encoding scheme used by the provider. HTML tags and entities, and stop words are also eliminated from the text. The resulting text is tokenized and each term is stemmed using a special Romanian stemmer in order to reduce the number of word forms. Implementing a suffix elimination algorithm for the Romanian language is extremely difficult because the inflection rules are numerous and very complicated and they affect the inner structure of the words, not only the trailing part. Moreover, each suffix-stripping rule may bring disadvantages [14], therefore our system uses only a small set of solid rules. These rules reduce the number of terms with 20–25%, depending on the number of processed news articles. After the initial processing phase presented above, the characteristic vector for each piece of news is extracted. Because the features' space is very large and an item contains only a few terms, the vectors are very sparse and, therefore, using an ordered set representation is very useful.

The clustering algorithm uses a hierarchical bottom-up strategy, with hard assignment and average-link used for computing inter-cluster similarity. At each step, two clusters are merged if their similarity is greater than a threshold. Actually, the algorithm uses two different thresholds, a higher value for creating a first set of clusters that contain very similar items, and a lower one. The upper threshold avoids merging news items that have distinct subjects that are still similar, because it will ensure that the clusters formed in this first phase cover only one subject. The use of two distinct thresholds as well as their values was determined during the tests of the clustering process. The tests started with a minimal value for only a threshold that was increased in order to produce the results that fitted best with the results of human observers.

Table 1. Clustering results

Number of items	Frequency-based space			Boolean-based space		
	Number of groups	Avg. dim.	Max. dim.	Number of groups	Avg. dim.	Max. dim.
666	545	1.22	10	556	1.20	6
674	524	1.29	14	510	1.32	38
641	530	1.21	12	532	1.20	13
545	440	1.24	12	430	1.27	35
644	520	1.24	22	533	1.21	15
780	650	1.20	14	641	1.22	39
1024	828	1.22	18	845	1.20	10

Computing the similarity is very important for hierarchical clustering, therefore all the measures from chapter 2.1 were used in order to find the best one. For frequency-based vector spaces, cosines similarity defined as the dot product of the vectors divided by the product of their norms was chosen and for boolean-based spaces a measure was defined by the authors, which was adapted from the cosine theorem. The results obtained using these methods are provided in table 1.

4.3 The Classification of the News Topics

Determining the hottest subjects improves the quality of the information that is presented to the users, but some of them are only willing to read the news concerning a particular topic of interest. In order to achieve this demand, a classification of the news clusters has been implemented. The categorization of news clusters has the main advantage that a cluster holds more features' information and, therefore, is more probable to be correctly classified than a single piece of news. The categories that have been used are Romania, Politics, Economy, Culture, International, Sports, High-Tech and High Life.

Three variations of Nearest Neighbours were implemented and compared: the first one is simple k-NN, while the second one considers k-NN classifiers with a score for each category computed by summing the similarities of each of the k documents that are in the same class. The third method is not exactly a NN as it uses a slightly different approach. Instead of determining the similarity between the document that needs to be classified and its nearest neighbours, this algorithm computes the similarity between the document and the centroid of each category, choosing the one to which it is most similar. For this reason the method is called nearest centre (NC) or centre-based NN and it works very well when the objects from a category are evenly distributed around its centroid. Unlike, the classical NN classifiers, this method needs a simple training phase that computes the features' vector of each category by summing the vectors of each element belonging to that category from the training set.

Clearly, as the categories of the providers differed from those used within the portal, a mapping of provider categories was necessary in order for them to fit as well as possible with those of the portal. Thus, using this training set, the classifiers presented above were compared using cross-validation, in order to determine the one with the greatest accuracy. The training set was divided into two subsets: two thirds of the training set were used for the training phase and the last third was used for the classification. The number of elements in each subset, for each category, can be found in table 2.

Table 2. Dimensions of the training and validation data sets used for evaluating the classifiers

Category	Dimension of the training data set	Dimension of the validation data set
Romania	640	325
Politics	313	157
Economy	182	92
Culture	88	45
International	481	241
Sport	316	158
High-Tech	76	38
High Life	84	43
Total	2180	1099

The number of articles is not very high and it is unequally distributed, as there is a difference in size between the category with the least articles in the training set and the one with the most articles. These have affected the training process and, therefore, the performance of the classifiers. The presented parameters will improve as the training corpus will grow.

Using the data mentioned above, the following classifiers were tested: one of the nearest centre type (NC), one of the NN type and two of the k-NN type (for k = 3 and k = 5). The tested parameters were: the accuracy of the classifier (defined as the number of correctly classified articles over the total number of news), duration of training and duration of classification. In addition, for each classifier the *confusion matrix* was generated, however, it will not be presented in this paper. For each classifier, both similarity criteria were used.

Table 3, presenting the accuracy of the classifiers, indicates that frequency based models are more precise then binary based ones, regardless of the algorithm employed. The best accuracy is offered by the *nearest centre* algorithm (NC), which outclasses with a small margin the NN and 3-NN algorithms, using the sum of the similarity of the nearest neighbours for each category. Although the NC algorithm needs a more time-consuming training phase, it compensates this with its classification speed – being a few times faster than the NN algorithms.

Table 3. Accuracy obtained for each of the implemented classifiers

Classifier type	Training time (s)	Classification time (s)	Accuracy (%)
NC – cosine	279	17	64
NC – binary	280	18	63
NN – cosine	1	60	61
3-NN – cosine	1	59	57
5-NN – cosine	1	59	54
3-NN – cosine (sum)	1	59	58
5-NN – cosine (sum)	1	59	56

Table 4. The results of the validation phase for the Nearest Centre classifier, using cosines similarity. Precision and recall are shown for each category.

	Rom	Pol	Eco	Cul	Int	Spo	Teh	HL	**Tot**	Prec
Rom	177	28	32	11	37	12	10	18	**325**	0.54
Pol	15	117	10	0	12	2	1	0	**157**	0.75
Eco	16	2	57	2	10	1	2	1	**92**	0.63
Cul	11	1	0	21	3	3	2	3	**45**	0.48
Int	37	12	16	4	162	3	3	4	**241**	0.67
Spo	5	3	7	4	7	128	1	3	**158**	0.81
Teh	1	0	9	1	0	0	27	0	**38**	0.71
HL	6	0	5	10	4	0	2	16	**43**	0.37
Tot	**268**	**163**	**136**	**53**	**235**	**149**	**48**	**45**		
Rec	0.66	0.71	0.42	0.40	0.69	0.86	0.56	0.36		

Average recall = 0.59, Average precision = 0.62, Accuracy = 0.64, F1 = 0.61.

The NC type classifier, using cosine similarity, was the best one to employ both from the point of view of accuracy, as well as of duration of classification. In table 4, the confusion matrix for this classifier is presented. In addition, its descriptive parameters are also offered: precision and recall, for each category as well as for the entire classifier, as a mean of the values for each category, global accuracy and the F1 parameter.

5 Conclusions

The paper presents an alternative to classical news portals, designed to solve the problems of large amounts of news and of information redundancy, by using the latter as an advantage. Moreover, the portal uses web syndication and natural language processing techniques in order to achieve an autonomous, human independent functionality that may be called intelligent as it automatically determines the importance of the news headlines and their category.

Web feeds offer a simple solution for fetching news from a large number of sources and clustering can be used to exploit similar news and to group them into a single topic. The user is then presented with only the news topics, ordered by the number of pieces of news that cover each topic. Moreover, one has the possibility to choose their favourite news source with an article on that topic.

Automatic classification of the news topics has an important advantage over the classification of each piece of news because the features of a cluster are more consistent than the ones of a single piece of news. This is of critical importance in this application because syndication offers only a short brief of each piece of news.

The further development of the portal has two main directions: improving the clustering and classification techniques and, a more challenging one, making it language independent or, at least, multilingual.

References

1. Internet World Stats: World Internet Usage and Population Statistics (2006), `http://www.internetworldstats.com/stats.htm`
2. Netcraft: May 2006 Web Server Survey (2006)

3. Yahoo! Search Blog: Our Blog is Growing Up And So Has Our Index (2005), http://www.ysearchblog.com/archives/000172.html
4. Page, L., Brin, S., Motwani, R., Winograd, T.: The page-rank citation ranking: Bringing order to the web. Technical Report, Stanford University (1998)
5. Ueda, Y., Oka, M., Yamashita, A.: Evaluation of the Document Categorization in "Fixed-point Observatory". In: Proceedings of NTCIR-5 Workshop Meeting, Tokyo (2005)
6. Toda, H., Kataoka, R.: A Clustering Method for News Articles Retrieval System. In: Special interest tracks and posters of the 14th international conference on World Wide Web, Chiba (2005)
7. Gabrilovich, E., Dumais, S., Horvitz, E.: Newsjunkie: Providing Personalized Newsfeeds via Analysis of Information Novelty. In: Proceedings of the Thirteenth International World Wide Web Conference (2004)
8. del Corso, G., Gulli, A., Romani, F.: Ranking a Stream of News. In: Proceedings of the 14th international conference on World Wide Web, Chiba, pp. 97–106 (2005)
9. Chakrabarti, S.: Mining the Web: Discovering Knowledge from Hypertext Data. Morgan Kaufmann Publishers, San Francisco (2002)
10. Ullman, J.: Data Mining Lecture Notes (2000)
11. Strehl, A.: Relationship-based Clustering and Cluster Ensembles for High-dimensional Data Mining, Doctoral dissertation, University of Texas at Austin (2002)
12. Manning, C., Schutze, H.: Foundations of Statistical Natural Language Processing. MIT Press, Cambridge, Massachusetts (2003)
13. Dumais, S., Platt, J., Heckerman, D., Sahami, M.: Inductive Learning Algorithms and Representations for Text Categorization. In: Proceedings of the Seventh International Conference on Information and Knowledge Management, Bethesda, pp. 148–155 (1998)
14. Porter, M.F.: An algorithm for suffix stripping. Journal of the Society for Information Science 3(14), 130–137 (1980)

On Determining the Optimal Partition in Agglomerative Clustering of Documents

Ahmad El Sayed, Hakim Hacid, and Djamel Zighed

ERIC Laboratory - University of Lyon
5, avenue Pierre Mendès, France
69676 Bron cedex, France
{asayed,hhacid,dzighed}@eric.univ-lyon2.fr

Abstract. The goal of any clustering algorithm producing flat partitions of data, is to find both the optimal clustering solution and the optimal number of clusters. One natural way to reach this goal without the need for parameters, is to involve a validity index in a clustering process, which can lead to an objective selection of the optimal number of clusters. In this paper, we provide an evaluation of the major relative indices involving them in an agglomerative clustering algorithm for documents. The evaluation seeks the indices ability to identify both the optimal solution and the optimal number of clusters. Then, we propose a new context-aware method that aims at enhancing the validity indices usage as stopping criteria in agglomerative algorithms. Experimental results show that the method is a step-forward in using, with more reliability, validity indices as stopping criteria.

1 Introduction

The goal of any clustering algorithm is indeed to find the optimal clustering solution with the optimal number of clusters. Classical clustering algorithms roughly base their decisions on a similarity matrix defined between elements. Seeking the optimal flat clustering solution given by a hierarchical algorithm, which is roughly our goal in this paper, one would define a stopping criterion, such as an "arbitrary" estimated number of clusters. This is a difficult and often ill-posed problem since final performance depends on subjectively chosen parameters that do not necessarily fits the dataset. A natural way to resolve this issue is to evolve with an algorithm while optimizing a specific validity index [9] that quantifies the quality of several clustering solutions across different numbers of clusters, in order to select finally the optimal solution. This kind of algorithms are also called incremental [3]. A typical "incremental" agglomerative algorithm looks like any ordinary agglomerative one with the only difference of merging, at each iteration, the pair of clusters optimizing a validity index rather than merging the closest pair.

We can distinguish three kinds of validity indices, external, internal, and relative indices [5,6]. (1) External indices compares a clustering solution to a predefined structure reflecting the desired result (e.g. *F-Score* measure, entropy,

A. An et al. (Eds.): ISMIS 2008, LNAI 4994, pp. 487–496, 2008.

Jaccard). (2) Internal indices evaluate a solution by comparing it to a predefined structure extracted using only quantities and features inherited from the dataset itself (e.g., *CPCC, Hubert* τ *statistic*). (3) Relative indices compares a clustering solution to another one obtained with the same algorithm but with different parameters or hypotheses (e.g., *DB, Dunn*).

The work in this paper can be divided into two parts: On the first hand, we provide an experimental comparison between the major relative indices after involving them in an incremental agglomerative clustering algorithm for several documents datasets. We study their ability to identify both the optimal clustering solution and the optimal number of clusters. On the other hand, we present a method to enhance the clustering process with context-awarenes in order to enable the usage of relative indices as a stopping criteria.

The rest of this paper is organized as follows: After a quick overview on relative validity indices in the next Section, we present our method to enable the usage of these indices as stopping criteria in Section 3. Expertimental results are presented in Section 4. We conclude by summarizing and drawing some future works in Section 5.

2 Relative Validity Indices

Relative validity indices proposed in the litterature turn around two main points: maximizing the compactness between elements within the same clusters (intra-cluster), and maximizing the separation between elements within distinct clusters (inter-cluster). According to their behavior, we can distinguish two kinds of relatives indices:

Indices Following Systematically the Number of Clusters Trend: In this case, the optimal k is often chosen by inspecting the graph to take the plot having a significant local change (jump or drop) in the values of a relative validity index VI, appearing like a "knee". Among indices developed for generic clustering purposes, we can find: CH by Calinsky[15], KL, $Diff$ [8], The modified Hubert τ statistic [13]. Another set of indices are developed by Zhao [16] specifically for document clustering: $I1, I2, E1, H1, H2$.

Indices Not Following Systematically the Number of Clusters Trend: With these indices, the optimal k is more easily chosen as the point on the graph maximizing/minimizing VI. Among indices developed for generic clustering purposes, we can find: Dunn [4], modified Dunn (m-dunn) [1], Davies-Bouldin (DB) [2], $RMSSDT, SPR, RS, CD$, [14], SD, S_Dbw [6], SF [11]. Among indices developed for document clustering purposes, we can find those proposed by [10]: $C1, C2, C3, C4$.

In an attempt to overcome the high computational cost of most relative indices, we propose the $H3$ index, which is less expensive than the others since it deals with centroids to quantify the inter-cluster and the intra-cluster dissimilarities. $H3$ is inspired from the $H1$, $H2$ indices proposed in [16]. The difference is that $H3$ does not follow the trend of k after having removed its sensitivity to k in an ad-hoc manner. As a matter of fact, the intra-cluster similarity decreases as k decreases, thus the quality of clustering continuously deteriorates from an intra-cluster point of view. We consider that an optimal partition is reached, when the average of inter-cluster similarities is no more able to overwhelm the intra-cluster deterioration. Thus, we define $H3$ as follows:

$$H3 = \frac{\sum_{i=1}^{k} n_i . \sum_{j=1}^{n_i} sim(e_j, S_i)}{(\sum_{i=1}^{k} sim(S_i, S))/k}$$

where sim denotes the similarity between two objects, S_i denotes the centroid of cluster C_i containing n_i elements, e_j denotes an element (word) within C_i, and S denotes the collection centroid which is the average vector of all cluster's centroids.

3 Exploring Indices Usage as a Stopping Criteria

3.1 Problem Definition

Currently, the classical usage of VI for determining the optimal clustering solution is performed a-posteriori after evaluating the output of a clustering algorithm with all the possible k values[10,6,9]. Actually, once reaching the optimal solution, all the remaining actions are obviously an effort and time waste. A challenging goal would be to develop an incremental algorithm able to stop once reaching the "right" optimal solution by means of a specific VI. An intuitive way to resolve this issue is to go on with the clustering process until reaching a point where no improvement can be done with any (merging) action. However, such an ad-hoc method suffers from ignoring, at a specific phase, whether it has truly reached the optimal solution or a better solution will come afterward if it accepts a performance decrease at the current phase. The major problem is that indices use too much local information to take a global decision, e.g. stopping the process.

3.2 Enhancing Indices with Context-Awareness

We developed a method that aims at enhancing the clustering process with context-aware decisions along with validity indices. The notion of *context* was introduced to cluster analysis in [7]. Context was typically considered to offer a more reliable calculation of similarities between objects by taking into account their relative nearest neighbor objects. Our goal by involving context in clustering is totally different; actually, we aim to involve context in order to enhance the ability of indices to be used as stopping criteria where a First Drop (FD) in their performance can more relevantly indicates reaching the optimal solution.

The global procedure is described as follows. At a specific step of the clustering process, rather than merging the two clusters into a new cluster candidate S_p optimizing VI, the method merges the two clusters improving VI, and providing a new cluster S_p with the minimal risk of performance degredation at future iterations calculated by means of a Context Risk CR measure that we define. This will surely implies a slower optimization in VI, but has the advantage of continuously pushing, as much as possible, risky merging actions entailing possible future degradations for later processing. Thus, taking the "safest" action at each step leads an expected degradation to occur as late as possible during the process. This will allow the algorithm to consider, more relevantly, a point (b), direcly before a First performance Drop (FD), as the optimal clustering solution.

A Context Risk CR is calculated for each new candidate cluster S_p basing on its context denoted by its K-Nearest Neighbors KNN (i.e. clusters). We assume that an ideal S_p with a minimal risk CR would be a cluster having its KNN too close or/and too distant to its centroid. More precisely, we decompose the context space of a candidate cluster into three layers (See Fig. 1):

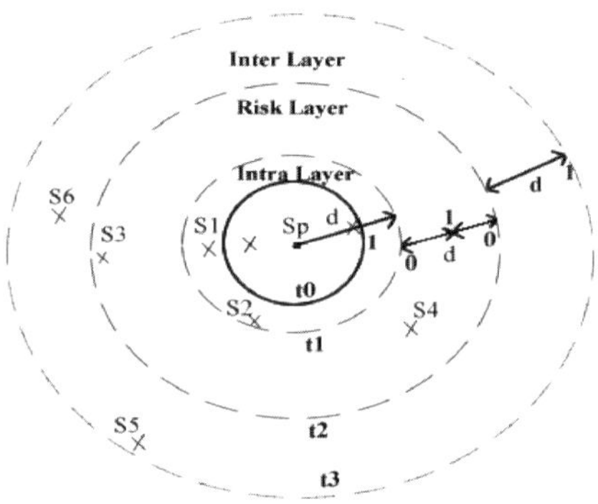

Fig. 1. The three layers context space of a candidate new cluster S_p

- Intra layer: Clusters within this layer are too close to S_p, thus they are likely to be merged into S_p in future iterations without significantly degradating the intra-cluster compactness.
- Risk layer: Clusters within this layer could degradate the clustering solution, whether on the inter or intra cluster level. Actually, clusters inside this layer, if merged into S_p, would contribute to a significant degradation in the intra-cluster compactness, and if not merged into S_p would not contribute to any significant amelioration in the inter-cluster separation.
- Inter layer: Clusters within this layer are too distant from S_p , thus they are not likely to be merged into S_p in future iterations, and keeping them outside S_p would contribute to an amelioration in terms of the inter-cluster separation.

The three layers are delimited by four thresholds $t0$, $t1$, $t2$, $t3$ obtained by means of a distance[1] d that we define as the radius of a candidate cluster S_p augmented by the standard deviation of radius values obtained following the u

[1] Indeed, the same could be applied with similarity.

previous mergings. We fixed u to 10 in our experiments. A radius is computed by the maximum distance between a cluster's centroid and an element within the cluster. Thus d is defined as follows:

$$d_k(S_p) = radius(S_p) + StDev(radius(S_{k-u}..S_{k-1}))$$

Subsequently, the four thresholds are defined as: $t_0(k) = 0$, $t_1(k) = d$, $t_2(k) = 2 * d$, $t_3(k) = 3 * d$. For calculating a Context Risk rate CR for S_p, we adopted the following formula:

$$CR(S_p) = (\sum_{i=1}^{n1} R(S_i, S_p) - \sum_{j=1}^{n2} I1(S_j, S_p) - \sum_{h=1}^{n3} I2(S_h, S_p))/K$$

$R(S_i, S_p)$, $I1(S_j, S_p)$, $I2(S_h, S_p)$ denote the score given for a cluster S situated respectively in the risk layer, intra layer, and inter layer. K refers to a predefined number of nearest neighbors, which we fix to 10. $n1$, $n2$, $n3$ denote the number of clusters situated respectively in the risk, intra and inter layers. All the scores are distributed along a [0,1] range according to their distances with the centroid of S_p(See Fig. 1). Consequently, CR varies between -1 (for a minimal risk) and 1 (for a maximal risk). For a contextual cluster S_x, having a distance d_x with the centroid of S_p, the scores are calculated with respect to the following conditions:

$$\left\{\begin{array}{l} if\ d_x < t_1 \Rightarrow I1(d_x, S_p) = \frac{t_1 - d_x}{d} \\ else\ if\ d_x > t_2 \Rightarrow I2(d_x, S_p) = \frac{d_x - t_2}{d} \\ else\ if\ d_x > t_3 \Rightarrow I2(d_x, S_p) = 1 \\ else\ if\ t_1 \leq d_x < (t1 + t2)/2 \Rightarrow R(d_x, S_p) = \frac{d_x - t_1}{d/2} \\ else\ if\ (t1 + t2)/2 \leq d_x \leq t2 \Rightarrow R(d_x, S_p) = \frac{t_2 - d_x}{d/2} \end{array}\right\}$$

Note that CR is not computationally expensive. Given p clusters at a specific iteration, we compute CR for M clusters candidates resulted from the merging of the M closest pairs of clusters. Therefore, at a given iteration, the complexity of CR is $O(M.p + M.K + M.K^2)$. Furthermore, we argue that the parameters M and K have "second order" effect on the results. In other words, they are not "critical" parameters, and their choices depend solely to which extend we are able to augment the complexity of the algorithm.

4 Experimental Study

Our experimental study was carried out on a benchmark of two different datasets (DS1, DS2) containing topic-assigned documents extracted from the Reuters corpus[2]. Since, we are dealing with a crisp algorithm, only documents assigned to a single topic are considered. Finally, DS1 and DS2 were constituted of 100 and 200 documents containing 22 and 24 distinct topics respectively.

Documents are represented by the vector-space model [12] in a multidimensional space, where each dimension represents a term expressed by its *tf.idf*

[2] Reuters corpus, volume 1, english language, release date: 2000-11-03.

weight. To calculate similarity between two documents, we use the *cosines coeficient* after normalizing each document vector d to be of unit length ($||d_{tfidf}|| = 1$). The similarity formula is then: $sim(d_i, d_j) = cos(d_i, d_j)$.

4.1 Evaluating Validity Indices

We provide an experimental evaluation of the major validity indices when applied in the context of document agglomerative clustering (mean-linkage). Having documents categorized into topics by an expert, one can define such artificial structure as the ideal "Gold Standard" output for a clustering method. Our experiments include 8 algorithms after having run the agglomerative algorithm separately along with each of the 8 VI (i.e., *C1, C2, C3, C4, H3, DB, Dunn, Dunn-like*). Then, each solution provided at each level of the clustering process is evaluated by means of the target VI (predicted quality) and the *FScore* measure (real quality).

On Evaluating a Clustering Solution. We report in Figs. 2,3 the indices results evaluated from three different angles:

- Their correlation with the *F-Score*, which represents to which extend a relative index can behave similarly to an external index.
- The optimal *F-Score* reached across all the number of clusters, which represents the optimal clustering quality that a VI can reach if it shares the same optimal solution with the *F-Score*. Obtained values express also to which extend (merging) actions based on a given index can lead to correct/incorrect classifications among clusters.
- The *F-Score* reached at the optimal value of the target VI, which depends on a VI ability for both evaluating a solution and approaching the optimal k. As we can see on the graph, F values here are considerably far from reaching their optimal ones unless for the $H3$ index. This is especially related, at out opinion, to the indices weakness in reaching the optimal k.

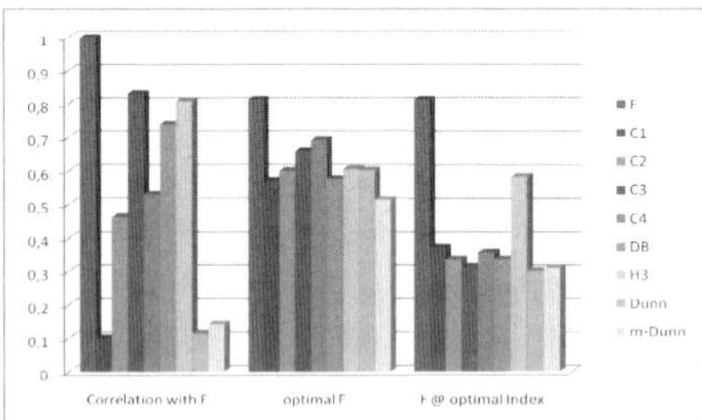

Fig. 2. Indices ability for evaluating a clustering solution and reaching the optimal F-score on DS1

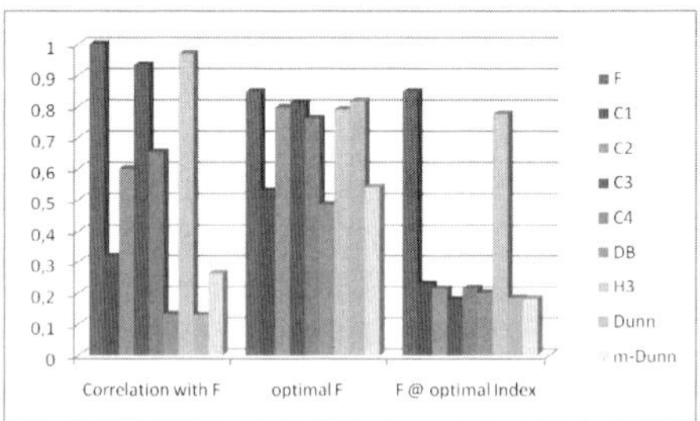

Fig. 3. Indices ability for evaluating a clustering solution and reaching the optimal F-score on DS2

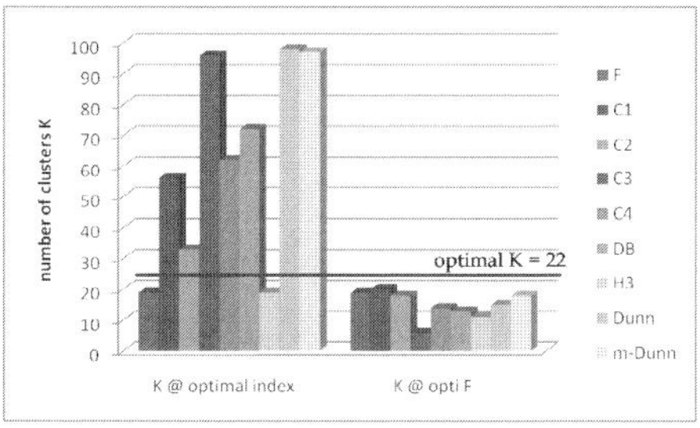

Fig. 4. Indices ability of reaching the optimal number of clusters on DS1 which is set to 22

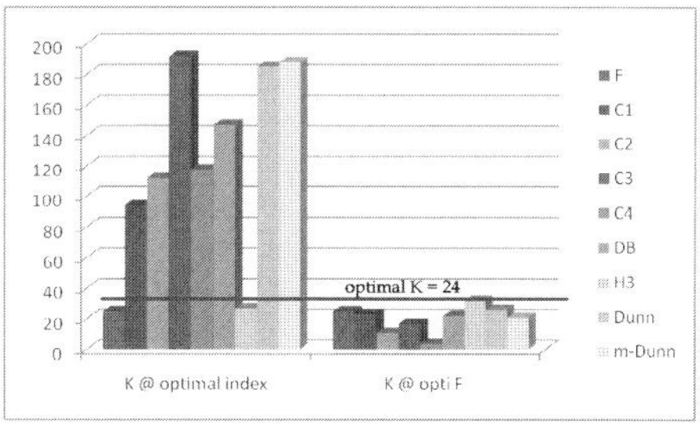

Fig. 5. Indices ability of reaching the optimal number of clusters on DS2 which is set to 24

On Determining the Optimal Number of Clusters. We define the optimal k as the number of distinct topics in a dataset. We provide in Figs. 4,5 the indices results evaluated from two different angles:

- The number of clusters at the optimal value of a target VI, which represents to which extend a VI, with its actual trend for determining the optimal k, is able to approach the real optimal k value.
- The number of clusters at the optimal value of the F-$Score$, which represents to which extend a VI, if it had the trend of F-$Score$ for determining the optimal k, is able to approach the real optimal k value.

Discussion. A first observation is the high correlations that most indices have with the $FScore$, which means that they have comparable behaviors to an external index. However, by considering the top-ranked solutions in terms of a VI, relatively poor solutions are provided in most cases, comparing to the optimal solutions that could have been reached. In our opinion, the gaps are due to the rigid trends of VI to the optimal k over different datasets. The only exception is the $H3$ index. In fact, its high ability for reaching the optimal k is surely behind

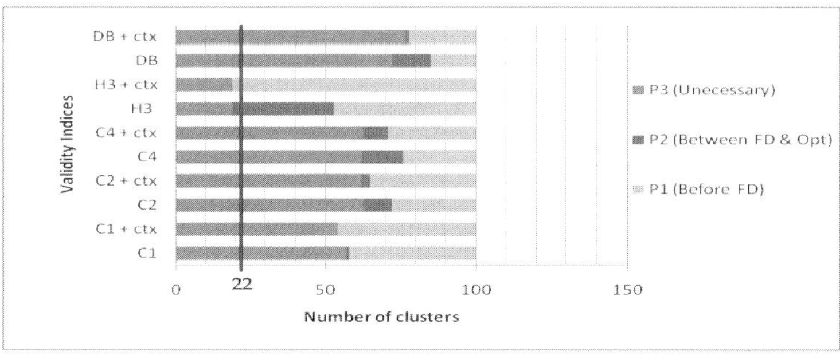

Fig. 6. The added-value of the context-aware method in approaching the optimal clustering solution

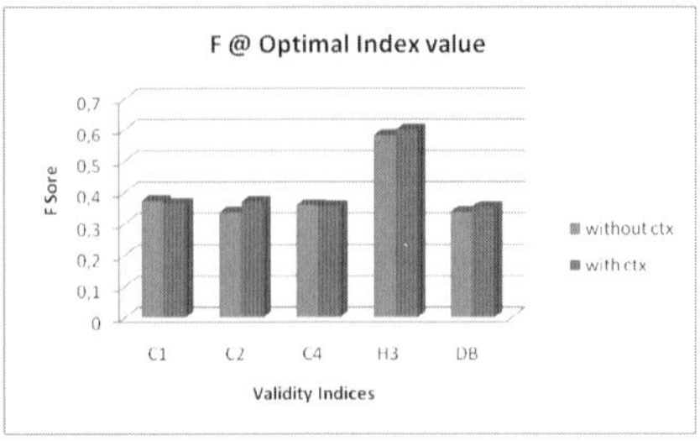

Fig. 7. F-Scores obtained at the optimal validity indices values used with and without context

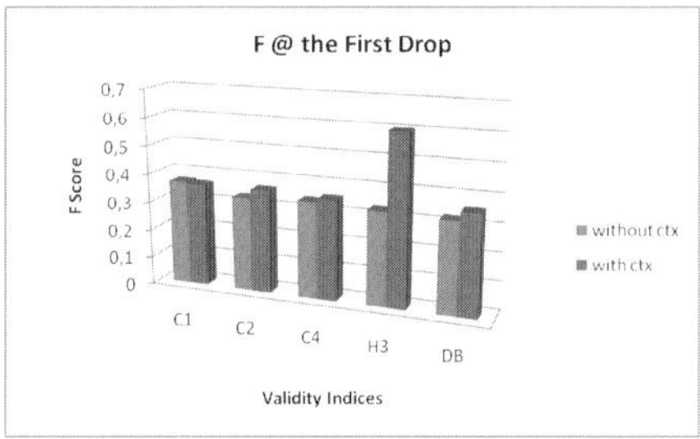

Fig. 8. F-Scores obtained before FD with and without using context

its high ability for reaching high-quality partitions. Thus, we argue that, in our specific application, the $H3$ index seems to be the most reliable index to involve in an agglomerative algorithm.

4.2 Evaluating the Context-Aware Method

Concerning our context-aware method, we excluded for our experiments some indices that showed to be unappropriate because they provide either too unstable plots (i.e. *Dunn*, *m-Dunn*) to be stabilized, or too stable results in our datasets to show clearly the effect of a context enhancement (i.e. $C3$)[3]. For each VI, we evaluate in Fig.6 its usage as a stopping criteria with and without adding context-awareness to the clustering decisions. As shown on the graph, the complete agglomerative clustering process $k = n \rightarrow 1$ is divided into three parts:

- P1: represents the part from the initial set to the last point before FD. Thus, using a VI as a stopping criteria will lead the process to the last point of P1.
- P2: represents the part from FD until the optimal clustering solution, which form the part that must have been processed but would not if a VI is used as a stopping criteria.
- P3: represents the part from the optimal solution until the root cluster, which form the unnecessary part that would be performed in vain if a VI is not used as a stopping criteria.

By observing the graph in Fig.6, we can easily notice the added-value of the context-aware method which contributes to considerably reduce P2, since FD occurs remarkably closer to the optimal solution, defined in terms of a VI. Moreover, results in Fig. 7 shows that an optimal solution obtained by involving context has a comparable and sometimes better quality to the one obtained without any context involvement. Differences between the *F-Score*

[3] Due to space constraints, we limited our context-aware evaluation to DS1.

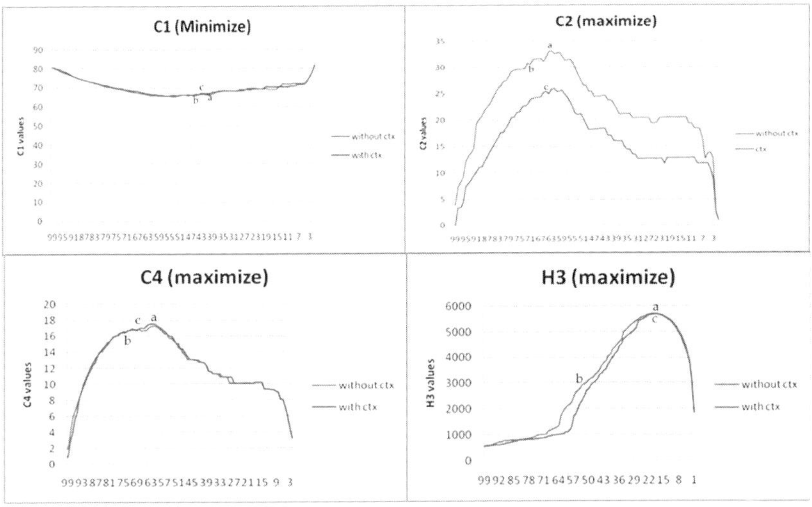

Fig. 9. The indices values along the different number of clusters, with and without using context

values at FD attained with and without context are illustrated in Fig.8. In average, using the context-aware method contributed to a 7% increase in the *F-Score* at FD.

In Fig.9, we can show the effect of adding context on each VI's plot. We set three points on each graph: (a) the optimal intended solution, (b) the point where an algorithm would stop if no context were involved, (c) the point where an algorithm would stop when involving context. Interestingly, the context-aware method appears to be very sensitive to the risk degree of a VI, taking different "levels of precautions". Thus, for relatively "safe" indices like $C1$ and $C4$, the decisions taken are very close to those taken without context, which results of two similar curves. However, for "risky" indices like $C2$ and $H3$, their values with context evolve relatively slowly, keeping a wider gap between curves.

5 Conclusion and Future Works

To wrap up, at a first stage, we studied the usage of relative validity indices in a agglomerative document clustering algorithm for evaluating a clustering solution and detecting the optimal number of clusters. At a second stage, we explored the feasibility of using these indices as a stopping criteria along with a context-aware method. Our experimental results appear very promising. However, a larger-scale evaluation is still needed before assessing the method's efficiency in real world applications. Given the fact that documents usually belong to multiple topics, a better evaluation is planned under an incremental soft clustering algorithm allowing overlaps between clusters.

References

1. Bezdek, J.C., Li, W., Attikiouzel, Y., Windham, M.P.: A geometric approach to cluster validity for normal mixtures. Soft Comput. 1(4), 166–179 (1997)
2. Davies, D.L., Bouldin, D.W.: A cluster separation measure. IEEE Transactions on Pattern Analysis and Machine Intelligence 1(2) (1979)
3. Duda, R.O., Hart, P.E., Stork, D.G.: Pattern classification. John Wiley & Sons, Chichester (2001)
4. Dunn, J.C.: Well separated clusters and optimal fuzzy paritions. Journal Cybern 4, 95–104 (1974)
5. Halkidi, M., Batistakis, Y., Vazirgiannis, M.: Cluster validity methods: Part i. SIGMOD Record 31(2), 40–45 (2002)
6. Halkidi, M., Batistakis, Y., Vazirgiannis, M.: Clustering validity checking methods: Part ii. SIGMOD Record 31(3), 19–27 (2002)
7. Jarvis, R.A., Patrick, E.A.: Clustering using a similarity measure based on shared near neighbors. IEEE Trans. Comput. 22(11), 1025–1034 (1973)
8. Krzanowski, W.J., Lai, Y.T.: A Criterion for Determining the Number of Groups in a Data Set Using Sum-of-Squares Clustering. Biometrics 44, 23–34 (1988)
9. Milligan, G.W., Cooper, M.C.: An examination of procedures for determining the number of clusters in a data set. Psychometrika V50(2), 159–179 (1985)
10. Raskutti, B., Leckie, C.: An evaluation of criteria for measuring the quality of clusters. In: IJCAI, pp. 905–910 (1999)
11. Saitta, S., Raphael, B., Smith, I.F.C.: A bounded index for cluster validity. In: MLDM, pp. 174–187 (2007)
12. Salton, G.: Automatic Text Processing: The Transformation, Analysis, and Retrieval of Information by Computer. Addison-Wesley, Reading (1989)
13. Sergios Theodoridis, K.K.: Pattern recognition. Academic Press, London (1999)
14. Sharma, S.: Applied multivariate techniques. John Wiley and Sons, Chichester (1996)
15. Calinski, T., Harabasz, J.: A dendrite method for cluster analysis. Communications in Statistics 3, 1–27 (1974)
16. Zhao, Y., Karypis, G.: Empirical and theoretical comparisons of selected criterion functions for document clustering. Machine Learning 55(3), 311–331 (2004)

Ontological Summaries through Hierarchical Clustering

Troels Andreasen, Henrik Bulskov, and Thomas Vestskov Terney

Department of Computer Science,
Roskilde University,
P.O. Box 260, DK-4000 Roskilde, Denmark
{troels,bulskov,tvt}@ruc.dk

Abstract. One approach to deal with large query answers or large collections of text documents is to impose some kind of structure to the collection for instance by a grouping into clusters of somehow related or close items. Another approach is to consider characteristics of the collection for instance by considering central and/or as a frequent keywords possibly taken from a background vocabulary or a more thorough structuring of background knowledge, like taxonomies or ontologies. In this paper we present a preliminary approach to combine these directions. More specifically we address an approach where conceptual summaries can be provided as answers to queries or survey over a document collection. The general idea is to apply a background knowledge ontology in connection with a combined clustering and generalization of keywords.

Preliminary experiments with Wordnet as background knowledge and excerpts from Semcor as data are presented and discussed.

1 Introduction

Summarization is a process of transforming sets of similar low level objects into more abstract conceptual representations [11] and more specifically a summary for a set of concepts is an easy to grasp and short description – in the form of a smaller set of concepts. For instance $\{car, house\}$ as summary for $\{convertible, van, cottage, estate\}$ or $\{dog\}$ as summary for $\{poodle, alsatian, golden\ retriever, buldog\}$.

In this paper we present two different directions to conceptual summaries as answers to queries. In both cases an ontology plays a key role as reference for the conceptualization. The general idea is from a world knowledge ontology to form a so-called "instantiated ontology" by restricting to a set of instantiated concepts.

First we consider a strictly ontology based approach where summaries are derived solely from the instantiated ontology. Second we consider conceptual clustering over the instantiated concepts based on a semantic similarity measure like e.g. shortest path [7]. The semantic grouping that results from the clustering process is then summarized using the least upper bounds of the clusters.

A. An et al. (Eds.): ISMIS 2008, LNAI 4994, pp. 497–507, 2008.

2 Ontology Representation and Modelling

Most importantly an ontology constitutes a hierarchy of concepts ordered according to inclusion (IS-A). However, in addition concepts may be related by semantic relations.

We define a generative ontology framework where a basis ontology situates a set of atomic concepts $\mathcal{A}$ in a concept inclusion lattice. A concept language (description language) defines a set of well-formed concepts, including both atomic and compound term concepts.

The concept language used here, ONTOLOG[3], is a lattice-algebraic description language. Semantic relationships r are introduced algebraically by means of a binary operator (:), known as the Peirce product $(r : \varphi)$, which combines a relation r with an expression φ. The Peirce product is used as a factor in conceptual products, as in $x \times (r\colon y)$, which can be rewritten to form the feature structure $x[r\colon y]$, where $[r\colon y]$ is an *attribution* of (an attribute/value pair for) the concept x.

Thus compound concepts can be formed by attribution. Given atomic concepts $\mathcal{A}$ and semantic relations $\mathcal{R}$, the set of well-formed terms $\mathcal{L}$ is:

$$\mathcal{L} = \{\mathcal{A}\} \cup \{x[r_1 : y_1, ..., r_n : y_n] | x \in \mathcal{A}, r_i \in \mathcal{R}, y_i \in \mathcal{L}\}$$

Thus compound concepts can have multiple as well as nested attributions. For instance with $\mathcal{R}= \{\text{WRT}, \text{CHR}, \text{CBY}, \text{TMP}, \text{LOC}, \ldots\}$[1] and $\mathcal{A}$ ={*entity, physical_entity, abstract_entity, location, town, cathedral, old*} we get:

$$\begin{aligned}\mathcal{L} = \{&entity, physical\_entity, abstract\_entity,\\ &location, town, cathedral, old,\\ &\ldots, cathedral[\text{LOC}\colon town, \text{CHR}\colon old], \ldots\\ &cathedral[\text{LOC}\colon town[\text{CHR}\colon old]], \ldots\}\end{aligned}$$

Sources for knowledge base ontologies may have various forms. Typically a taxonomy can be supplemented with, for instance, word and term lists as well as dictionaries for definition of vocabularies and for handling of morphology. The well-known and widespread resource WordNet is among the more interesting and useful resources for general ontologies.

We will not go into details on the modeling here but just assume the presence of a taxonomy $\mathcal{T}$ over the set of atomic concepts $\mathcal{A}$. $\mathcal{T}$ and $\mathcal{A}$ express the domain and world knowledge provided.

Based on $\hat{\mathcal{T}}$ the transitive closure of $\mathcal{T}$ we can generalize to an inclusion relation "$\leq$" over all well-formed terms of the language $\mathcal{L}$ by the following [6]:

$$\begin{aligned}\text{“}\leq\text{”} = &\\ &\hat{\mathcal{T}}\\ &\cup\{<x[\ldots, r\colon z], y[\ldots]> \mid <x[\ldots], y[\ldots]> \in \hat{\mathcal{T}}\}\\ &\cup\{<x[\ldots, r\colon z], y[\ldots, r\colon z]> \mid <x[\ldots], y[\ldots]> \in \hat{\mathcal{T}}\}\\ &\cup\{<z[\ldots, r\colon x], z[\ldots, r\colon y]> \mid <x, y> \in \hat{\mathcal{T}}\}\end{aligned}$$

where repeated ... denote zero or more attributes of the form $r_i\colon w_i$.

[1] For *with respect to, characterized by, caused by, temporal, location*, respectively.

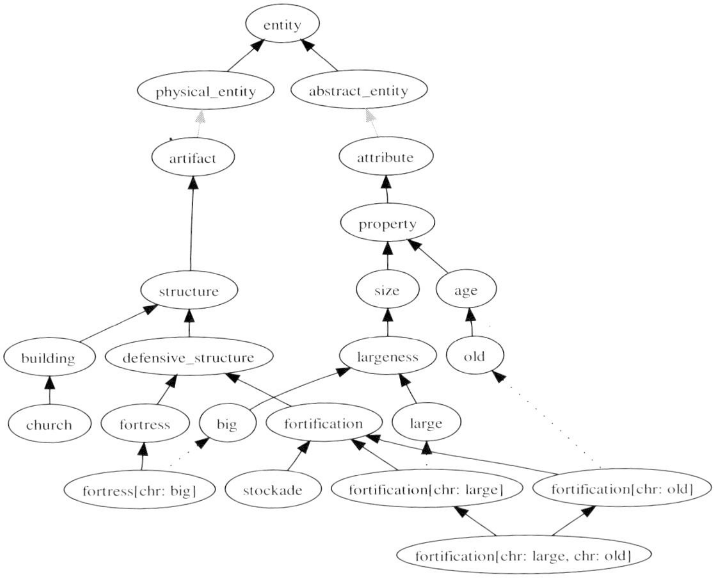

Fig. 1. An instantiated ontology based on a Wordnet ontology and the set of instantiated concepts $\{church, fortress[\text{CHR}: big], stockade, fortification[\text{CHR}: large, \text{CHR}: old]\}$

The general ontology $\mathcal{O} = (\mathcal{L}, \leq, \mathcal{R})$ thus encompasses a set of well-formed expressions $\mathcal{L}$ derived in the concept language from a set of atomic concepts $\mathcal{A}$, an inclusion relation generalized from the taxonomy relation in $\mathcal{T}$, and a supplementary set of semantic relations $\mathcal{R}$. For $r \epsilon \mathcal{R}$, we obviously have $x[r: y] \leq x$, and that $x[r: y]$ is in relation r to y. Observe that $\mathcal{O}$ is generative and that $\mathcal{L}$ therefore is potentially infinite.

Given a general ontology $\mathcal{O} = (\mathcal{L}, \leq, \mathcal{R})$ and a set of concepts C the instantiated ontology $\mathcal{O}_C = (\mathcal{L}_C, \leq_C, \mathcal{R})$ is a restriction of $\mathcal{O}$ to cover only the concepts in C and corresponds to "upper expansion" $\mathcal{L}_C$ of C in $\mathcal{O}$

$$\mathcal{L}_C = C \cup \{x | y \in C, y \leq x\}$$
$$\text{"}\leq_C\text{"} = \text{"}\leq\text{"} \cap (\mathcal{L}_C \times \mathcal{L}_C) = \{<x, y> | x, y \in \mathcal{L}_C, x \leq y\}$$

Figure 1 shows an example of an instantiated ontology. The general ontology is based on (and includes) WordNet and the ontology shown is "instantiated" with respect to the following set of concepts:

$$C = \{church, fortress[\text{CHR}: big], stockade,$$
$$fortification[\text{CHR}: large, \text{CHR}: old]\}$$

Apart from the inclusion relation "$\leq$" we will also use below a relation "$<$". The latter obviously refers to the strict variant of the former.

3 Conceptual Querying by Summaries

Conceptual querying as indicated above concerns retrieval of concepts appearing in an instantiated ontology – thus at a conceptual level to investigate the concepts appearing, or the content of the documents holding these concepts. Summaries are very useful for this purpose but these are obviously not the only possible means of conceptual querying. For a general discussion on operators for conceptual querying an instantiated ontology we refer to [1]. Here we discuss summaries only.

Summaries are intended to describe in principle any collection of text such as a single document, a set of documents, an entire database or a query result.

We shall assume the presence of some kind of filtering or extraction mechanism that for a given text collection can produce the set of concepts appearing in the text. A simple approach to this kind of extraction is to consider patterns of word classes such that for instance "A B", where A is an adjective and B a noun, maps to "A [CHR:B]". We will, however, not cover this aspect in the present paper but refer to [10] for details on the matter. So the issue here is how to produce a summary for a given set of concepts $C = \{c_1, \ldots, c_n\}$.

A summary for a set of concepts is an easy to grasp and short description – in the form of a smaller set of concepts. Good characteristics for summaries are not obvious, but loosely we can say that if C covers several distinct aspects a summarizing description should include these, thus intuitively with two distinct aspects as in $\{convertible, van, cottage, estate\}$ we should probably have two concepts ($\{car, house\}$) in the summary and with one single aspect as in $\{poodle, alsatian, golden\ retriever, buldog\}$ we should have only a single summarizer ($\{dog\}$).

We will not go into more details on characteristics on summaries here, but refer to [11] for considerations on this issue.

One approach to provide summaries is to divide the set of concepts into groups or clusters and to derive for each a representative concept – for instance the least upper bound (lub) for the group.

Summary queries are to be understood as follows. We assume an instantiated ontology corresponding to a set of documents (the target database) and a query retrieving a subset of these documents. Corresponding to this subset we have a set of concepts – those appearing in documents in the subset.

We briefly introduce two directions for deriving summaries below: one based directly on connectivity in the ontology and the other drawing on statistical clustering applying similarity measures.

3.1 Connectivity Clustering

Connectivity Clustering is clustering based solely on connectivity in the ontology O_C. More specifically the idea is to cluster a given set of concepts based on their connections to common ancestors, for instance grouping two siblings due to their common parent, and in addition to replace the group by the common ancestor. Thus rather than, when taking a bottom-up view, moving towards a smaller number of larger clusters, connectivity clusting is about moving towards a smaller number of more general concepts.

For a set of concepts $C = \{c_1, \ldots, c_n\}$ we can consider as *generalizing description* a new set of concepts $\delta(C) = \{\widehat{c}_1, \ldots, \widehat{c}_k\}$, where $\widehat{c}_i$ is either a concept generalizing concepts in C or an element from C. Each generalizer in $\delta(C)$ is a lub of a subset of C, $\widehat{c}_i = lub(C_i)$, where $\{C_1, \ldots, C_k\}$ is a division (clustering) of C. Notice that the lub of a singleton set is the single element in this.

We define *most specific generalizing description* $\delta(C)$ for a given $C = \{\widehat{c}_1, \ldots, \widehat{c}_k\}$ as a description restricted by the following properties.

(a) $\forall \widehat{c} \in \delta(C) : \widehat{c} \in C \vee \exists c', c'' \in C \wedge c' \neq c'' \wedge c' < \widehat{c} \wedge c'' < \widehat{c}$
(b) $\forall \widehat{c}', \widehat{c}'' \in \delta(C) : \widehat{c}' \not< \widehat{c}''$
(c) $\forall c', c'' \in C, \widehat{c}' \in \delta(C), \neg \exists x \in L_C : c' \leq x \wedge c'' \leq x \wedge x \leq \widehat{c}'$

Thus (a) restricts $\delta(C)$ to elements that either originate from C or generalize two or more concepts from C. Secondly (b) restricts $\delta(C)$ to be without redundance (no element of $\delta(C)$ may be subsumed by another element) and thirdly (c) reduces to the most specific in the sense that no subsumer for two elements of C may be subsumed by an element of $\delta(C)$.

Notice that $\delta(C)$ has the same form as C as a subset of $\mathcal{L}_C$, and that we therefore can refer to an m'th order summarizer $\delta^m(C)$. Obviously, to obtain an appropriate description of C we will in most cases need to consider higher orders of δ. At some point m we will in most cases have that $\delta^m(C) = Top$, where Top is the top element in the ontology. Exceptions are firstly when a more specific single summarizer is found and secondly when Top has only one successor. In the latter case we will only reach the single topmost concepts with more than one successor.

The most specific generalizing description $\delta(C)$ for a given C is obviously not unique and there are several different sequences of most specific generalizing descriptions of C from C towards TOP. However, a reasonable approach would be to go for the largest possible steps obeying the restrictions for δ, as done in the algorithm below.

ALGORITHM – Connectivity summary
INPUT: Set of concepts $C = \{c_1, \ldots, c_n\}$
OUTPUT: A most specific generalizing description $\delta(C)$ for C.
1) Let the instantiated ontology for C be $\mathcal{O}_C = (\mathcal{L}_C, \leq_C, \mathcal{R})$
2) Let $U = \{u | u \in \mathcal{L}_C \wedge \exists c_i, c_j \in C : c_i < u \wedge c_j < u\}$,
3) $U' = U - \{u | u \in U \wedge \exists v \in U : v < u\}$ and
4) $L = \cup_{u \in U'} \{c | c \in C \wedge c < u\}$
5) set $\delta(C) = C \cup U' - L$

In 2) all concepts U that generalize two or more concepts are derived. Notice that these may include concepts from C when C contains concepts subsuming other concepts. In 3) U is reduced to the most specific generalizers U'. 4) defines the set of concepts that specializes the generalizers in U' and 5) derives $\delta(C)$ from

C by adding the most specific generalizers and subtracting concepts specializing these.

With reference to figure 1 we have for instance:

- $C = \{church, fortification[\text{CHR}: old], stockade, fortress[\text{CHR}: big]\}$
- $\delta(C) = \{church, fortification, fortress[\text{CHR}: big]\}$
- $\delta^2(C) = \{church, defensive\_structure\}$
- $\delta^3(C) = \{structure\}$

The chosen approach, taking the largest possible steps where everything that can, will be grouped, is of course not the only possible. If we alternatively want to form only some of the possible clusters complying with the restrictions some kind of priority mechanism for selection is needed.

Among important properties that might contribute to priority are *deepness*, *redundancy* and *support*. The deepest concepts, those with the largest depth in the ontology, are structurally and thereby often also conceptually the most specific concepts, thus collecting these first would probably lead to a better balance with regard to how specific the participating concepts are in candidate summaries. Redundancy, where participating concepts include (subsumes) others, is avoided as regards more general concepts introduced (step 3 in the algorithm). However redundancy in the input set may still survive so priority could also be given to remove this first. In addition we could consider support for candidate summarizers. One option is simply to measure support in terms of number of subsumed concepts in the input set while more refinement could be obtained by also taking frequencies of concepts as well as their distribution in documents[2] in the original text into consideration. Support may guide the clustering in several ways. It indicates for a concept how much it covers in the input and can thus be considered as an importance weight for the concept as summarizer for the input. High importance should probably infer more reluctance as regards further generalization.

3.2 A Hierarchical Similarity-Based Approach

While we may consider connectivity clustering to ontology-based in a genuine sense it is not the only possible direction. Alternatively also clustering applying given similarity measures over the set of concepts should be considered. Obviously, if the measure is derived from an ontology, and thereby do reflect this, then so will the clustering.

Various approaches have been proposed to derive similarity or distance from ontology. A simple *Shortest Path* approach is given in [7]. Among the more refined approaches proposed are *Weighted Shared Nodes* [9] and *Information Content* [8]. The former uses the density of upper bounds shared by the concepts while the latter reflects the probability of encountering concepts in a corpus to define the similarity between concepts. It appears that approaches reflecting also frequencies of concepts, as Information Content, are especially useful to support

[2] Corresponding to term and document frequencies in Information Retrieval.

summarization – just as term and document frequencies are useful in Information Retrieval. We will assume an ontology based similarity measure *sim* below but make no further assumptions of type and characteristics of this measure.

We may expect a similar pattern in derivation of summaries in an approach based on similarity when the similarity measure closely reflects connectivity in the ontology, as a shortest path measure does.

With a given path-length dependent similarity measure derived from the ontology an *lub*-centered, agglomerative, hierarchical clustering can be performed as follows.

Initially each "cluster" corresponds to an individual element of the set to be summarized. At each particular stage the two clusters which are most similar are joined together. This is the principle of conventional hierarchical clustering. However rather than replacing the two joined clusters with their union as in the conventional approach they are replaced by their *lub*. Thus given a set of concepts $C = \{c_1, \ldots, c_n\}$ summarizers can be derived as follows.

ALGORITHM – Hierarchical clustering summary
INPUT: Set of concepts $C = \{c_1, \ldots, c_n\}$
OUTPUT: Generalizing description $\delta(C)$ for C.

1) Let the instantiated ontology for C be $\mathcal{O}_C = (\mathcal{L}_C, \leq_C, \mathcal{R})$
2) Let $T = \{\langle x, y \rangle | sim(x, y) = max_{z,w \in C}(sim(z, w))\}$,
3) Let $U = \{u | u \in \mathcal{L}_C \wedge \exists x, y \in \mathcal{L}_C : x < u \wedge y < u\}$,
4) $U' = U - \{u | u \in U \wedge \exists v \in U : v < u\}$ and
5) $L = \{x | \langle x, y \rangle \in T \vee \langle y, x \rangle \in T\}$
6) set $\delta(C) = C \cup U' - L$

As was also the case with the connectivity clustering, to obtain an appropriate description of C we might have to apply δ several times and at some point m we have that $\delta^m(C) = Top$.

4 Summarization Examples with WordNet

Preliminary experiments have been performed on texts from SEMCOR 2.0 [4] on connectivity clustering as well as hierarchical clustering.

SEMCOR is a subset of the documents in the Brown corpus which has the advantage of being semantically tagged with senses from WordNet [5]. We show below results of summarizations of the following text.

> *Greases*, *stains*, and miscellaneous *soils* are usually sorbed onto the *soiled surface*. In most *cases*, these *soils* are taken up as *liquids* through *capillary action*. In an essentially static *system*, an *oil* cannot be replaced by *water* on a *surface* unless the *interfacial tensions* of the *water phase* are reduced by a *surface-active agent*.

Words in italics indicate the initial set of concepts, in this case nouns that are mapped into WordNet. Notice that due to the use of SemCor there is no attribution in the initial set of concepts.

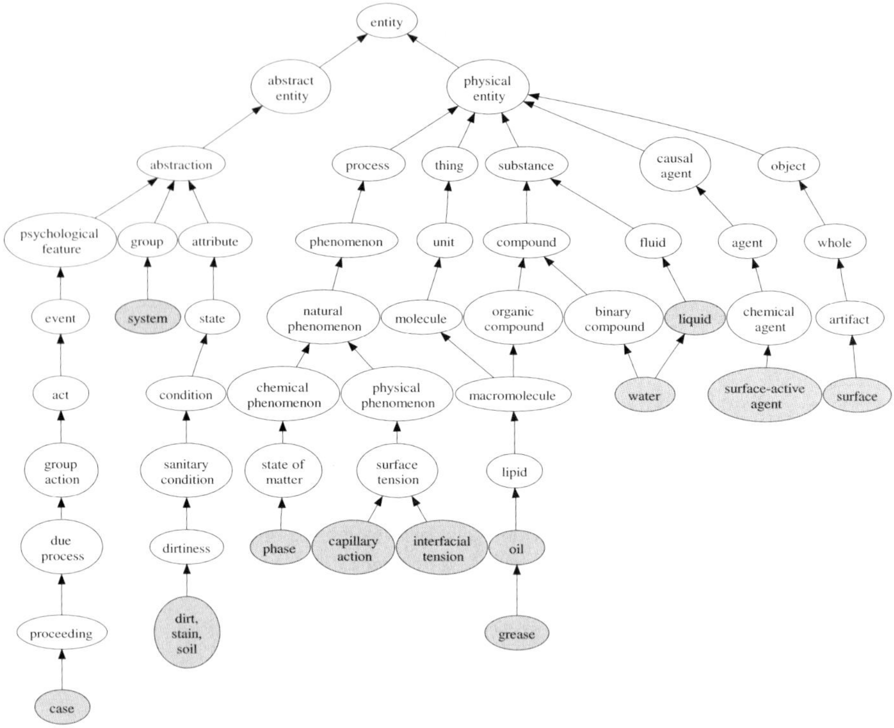

Fig. 2. An instantiated ontology based on a paragraph from SEMCOR

The 17 nouns appearing in the paragraph are mapped into 12 WordNet synsets constituting the set of elements C. Based on this mapping the instantiated ontology can be created which is depicted in figure 2 where shaded nodes indicate concepts in C. The connectivity clustering will thus include the steps:

$C = \{$*case, system, dirt, phase, capillary action, interfacial tension, grease, oil, water, liquid, surface-active agent, surface*$\}$

$\delta(C) = \{$*abstraction, phase, surface tension, oil, water, liquid, surface-active agent, surface*$\}$

$\delta^2(C) = \{$*abstraction, natural phenomenon, liquid, compound, surface-active agent, surface*$\}$

$\delta^3(C) = \{$*abstraction, natural phenomenon, substance, surface-active agent, surface*$\}$

$\delta^4(C) = \{$*abstraction,physical entity*$\}$

$\delta^5(C) = \{$*entity*$\}$

The example well illustrates some of the challenges when applying connectivity clustering. First, the set of concepts are reduced by a larger factor at the first steps of the clustering compared to the steps close to the end of the clustering process. This is due to the "greedy nature" of connectivity clustering where as many clusters as possible are merged at each step, and where clusters conceptually far apart are merged because the resulting cluster is the most specific generalizing description. The merge of *case, system* and *dirt* illustrates this aspect. Second, small summaries tend not surprisingly to be very general.

The hierarchical clustering as introduced above will follow the following steps using shortest path as similarity measure:

$$
\begin{aligned}
C &= \{\textit{case, system, dirt, phase, capillary action, interfacial tension, grease,}\\
&\quad \textit{oil, water, liquid, surface-active agent, surface}\}\\
\delta(C) &= \{\textit{case, system, dirt, phase, capillary action, interfacial tension, oil, liq-}\\
&\quad \textit{uid, surface-active agent, surface}\}\\
\delta^2(C) &= \{\textit{case, system, dirt, phase, surface tension, oil, liquid, surface-active}\\
&\quad \textit{agent, surface}\}\\
\delta^3(C) &= \{\textit{case, system, dirt, natural phenomenon, oil, liquid, surface-active}\\
&\quad \textit{agent,surface}\}\\
\delta^4(C) &= \{\textit{case, system, dirt, physical entity, oil, surface-active agent, surface}\}\\
\delta^5(C) &= \{\textit{case, system, dirt, physical entity, oil}\}\\
\delta^6(C) &= \{\textit{case, dirt, entity, oil}\}\\
\delta^7(C) &= \{\textit{case, dirt, entity}\}\\
\delta^8(C) &= \{\textit{case, entity}\}\\
\delta^9(C) &= \{\textit{entity}\}
\end{aligned}
$$

It is evident that compared to the connectivity clustering a hierarchical clustering based on shortest path will preserve concepts deep in the ontology until late in the clustering. However, this is due to shortest path as a similarity measure rather than the hierarchical clustering. Using shared nodes [9], a measure based on the cardinality of the shared upper bounds, would e.g. result in *dirt* and *case* being merged at an earlier step. Naturally, an important future issue is therefore to examine how the different similarity measures affect the clustering process.

5 Concluding Remarks

In this paper we have considered approaches to ontology-based conceptual summaries to provide means for advanced querying that retrieve concepts describing documents, rather than documents directly. The principles for conceptual summarization are presented as related to so called instantiated ontologies – a conceptual structure reflecting the content of a given document collection and therefore in particular well suited as target for conceptual querying. However the summaries introduced are not dependent on this notion.

It is obvious that such summary development should be guided by experiments within a framework that include a realistic general world knowledge resource. Preliminary studies have been done with WordNet as primary ontology source and SemCor as corpus (Information base).

5.1 Perspectives and Future Work

Replacing clusters with their least upper bounds according to the given instantiated ontology as described above is a generic principle that can be applied in connection with any clustering principle – hierarchical as well as partitive. For a clustering approach that impose a grouping $\{C_1, \ldots, C_k\}$ to a given set of concepts $C = \{c_1, \ldots, c_n\}$ we can simply provide the set of lub's $\{\widehat{c}_1, \ldots, \widehat{c}_k\} = \{lub(C_1), \ldots, lub(C_k)\}$ for the division of C as summary. So obviously partitive approaches should be investigated in more detail for summarization.

In addition approaches to softening of derived results from generalization in the ontology should be considered. First of all importance of clusters in terms of their sizes can be taken into account. The summary can be modified by the support of the generalizing concepts, $support(x, C)$, that for a given concept specifies the fraction of elements from the set C covered:

$$support(x, C) = \frac{|\{y | y \in C, y \leq x\}|}{|C|}$$

leading to a fuzzyfied (weighted) summary, based on the division (crisp clustering) of C into $\{C_1, \ldots, C_k\}$:

$$\Sigma_i support(lub(C_i), C)/lub(C_i)$$

Especially in partitive clustering we might expect noise and outliers to influence the resultof summarization and to approach this problem we may consider the very notion of least upper bound (lub).

A soft definition of lub for a set of concepts C should comprise "upper boundness" as well as "leastness" (or "least upperness"). These would naturally for a candidate u express respectively the portion of concepts in C that are generalized by u and the degree to which u is least upper with regard to one or more of the concepts in C. Preliminary definitions and discussion on these can be found in [2].

References

1. Andreasen, T., Bulskov, H.: On Browsing Domain Ontologies for Information Base Content. In: Melin, P., Castillo, O., Aguilar, L.T., Kacprzyk, J., Pedrycz, W. (eds.) IFSA 2007. LNCS (LNAI), vol. 4529, Springer, Heidelberg (2007)
2. Bulskov, H., Andreasen, T., Terney, T.V.: Conceptual Summaries as Query Answers. In: Proceedings NAFIPS 2007 (2007)
3. Nilsson, J.F.: A logico-algebraic framework for ontologies – ONTOLOG. In: Jensen, P.A., Skadhauge, P. (eds.) First International OntoQuery Workshop, University of Southern Denmark (2001)
4. Miller, G.A., Chodorow, M., Landes, S., Leacock, C., Thomas, R.G.: Using a semantic concordance for sense identification. In: Proc. of the ARPA Human Language Technology Workshop, pp. 240–243 (1994)
5. Miller, G.A.: Wordnet: a lexical database for english. Commun. ACM 38(11), 39–41 (1995)

6. Bulskov, H., Knappe, R., Andreasen, T.: On querying ontologies and databases. In: Christiansen, H., Hacid, M.-S., Andreasen, T., Larsen, H.L. (eds.) FQAS 2004. LNCS (LNAI), vol. 3055, Springer, Heidelberg (2004)
7. Rada, R., Mili, H., Bicknell, E., Blettner, M.: Development and application of a metric on semantic nets. IEEE Transactions on Systems, Man, and Cybernetics 19(1), 17–30 (1989)
8. Resnik, P.: Semantic similarity in a taxonomy: An information-based measure and its application to problems of ambiguity in natural language (1999)
9. Andreasen, T., Knappe, R., Bulskov, H.: Domain-specific similarity and retrieval. In: Proceedings IFSA 2005, pp. 496–502. Tsinghua University Press (2005)
10. Andreasen, T., Jensen, P.A., Nilsson, J.F., Paggio, P., Pedersen, B.S., Thomsen, H.E.: Content-based text querying with ontological descriptors. Data Knowledge Engineering 48(2), 199–219 (2004)
11. Yager, R.R., Petry, F.E.: A Multicriteria Approach to Data Summarization Using Concept Hierarchies. IEEE Trans. on Fuzzy Sys. 14(6) (2006)
12. Lee, D., Kim, M.: Database Summarization using fuzzy ISA hierarchies. IEEE Trans. on Sys. Man and Cyb. 27(1) (1997)
13. Kim, D.-W., Lee, K.H., Lee, D.: Fuzzy clustering of categorical data using fuzzy centroids. Elsevier Sciencedirect (2004)
14. Huang, Z., Ng, M.K.: A Fuzzy k-Modes Algorithm for Clustering Categorical Data Zhexue Ieee Trans. Ieee Trans. on Fuzzy Sys. 7, 4 (1999)

Classification of Web Services Using Tensor Space Model and Rough Ensemble Classifier

Suman Saha, C.A. Murthy, and Sankar K. Pal*

Center for Soft Computing Research, Indian Statistical Institute
{ssaha_r,murthy,sankar}@isical.ac.in

Abstract. The transition of the World Wide Web from a paradigm of static Web pages to one of dynamic Web services raises a new and challenging problem of locating desired web services. With the expected growth of the number of Web services available on the web, the need for mechanisms that enable the automatic categorization to organize this vast amount of data, becomes important. In this paper we propose Tensor space model for data representation and Rough Set based approach for the classification of Web services. The proposed tensor space model captures the information from internal structure of WSDL documents along with the corresponding text content. Rough sets are used here to combine information of the individual tensor components for providing classification results. Two step improvement on the existing classification results of web services has been shown here. In the first step we achieve better classification results over existing, by using proposed tensor space model. In the second step further improvement of the results has been obtained by using Rough set based ensemble classifier.

Keywords: WSDL, classification, rough Set, tensor space, internal structure.

1 Introduction

A major limitation of the Web services technology is that finding and composing services requires manual effort. This becomes a serious burden with the increasing number of Web services. Describing and organizing this vast amount of resources is essential for realizing the web as an effective information resource. Web Service classification has become an important tool for helping discovery and integration process to organize this vast amount of data. For instance, for categorization in the UDDI registry, one needs to divide the publicly available Web Services into a number of categories for the users to limit the search scope. Moreover, Web Services classification helps the developer to build integrated Web Services. Traditionally, Web Service classification is performed manually

* The authors would like to thank the Department of Science and Technology, Government of India, for funding the Center for Soft Computing Research: A National Facility. This paper was done when one of the authors, S. K. Pal, was a J.C. Bose Fellow of the Government of India.

A. An et al. (Eds.): ISMIS 2008, LNAI 4994, pp. 508–513, 2008.

by domain experts. However, human classification is unlikely to keep pace with the rate of growth of the number of Web Services. Hence, as the web continues to increase, the importance of automatic Web Service classification becomes necessary.

The problem of the automatic classification of Web services has been addressed in the literature with the help of two main approaches, *a*) text classification approach [4] and *b*) semantic similarity based classification approach [6]. Text classification is a long-standing problem in information retrieval (IR). Most solutions to this problem are based on term frequency analysis [3,1]. These approaches are insufficient in the web service context because text documentations for web-service operations are highly compact, and they ignore structure information that aids capturing the underlying semantics of the operations. Moreover, preprocessing techniques for HTML documents corresponding to static web pages are not adequate to preprocess WSDL documents corresponding to dynamic Web Services.

Work in the area of semantic similarity based classification approach has developed several methods that try to capture clues about the semantics similarity, and suggests classification based on them [2,5]. Such methods include linguistic analysis, structural analysis, use of domain knowledge and previous classification experience [7]. But these methods suffer from lack of annotation which is a manual process. Annotating the collection of web services is infeasible, and we rely on only the information provided in the WSDL file and the UDDI entry.

We treat the determination of a web services category as a tag based text classification problem, where the text comes from different tags of the WSDL file. Unlike standard texts, WSDL descriptions are highly structured. Our experiments demonstrate that selecting the right set of features from this structured text improves the performance of a learning classifier. By combining different classifiers it is possible to improve the performance even further, although for both the feature selection and the combination no general rule exists. The given document is split into different texts using the provided tag information.

2 Proposed Method

In this article we propose tag based tensor space model for the representation of web service documents and Rough Set based approach for its classification. Splits of the features has been performed based on tag set existing in the WSDL documents corresponding to web services. Tensor space model has been used to represent the services according to tag structure. Base level classification has been performed on individual tensor components. Finally combined classification has been obtained by using rough set based ensemble classifier.

2.1 TSM

In this paper, we propose a novel tag based Tensor Space Model (TSM) for web services representation. The proposed TSM is based on tag set existing in the

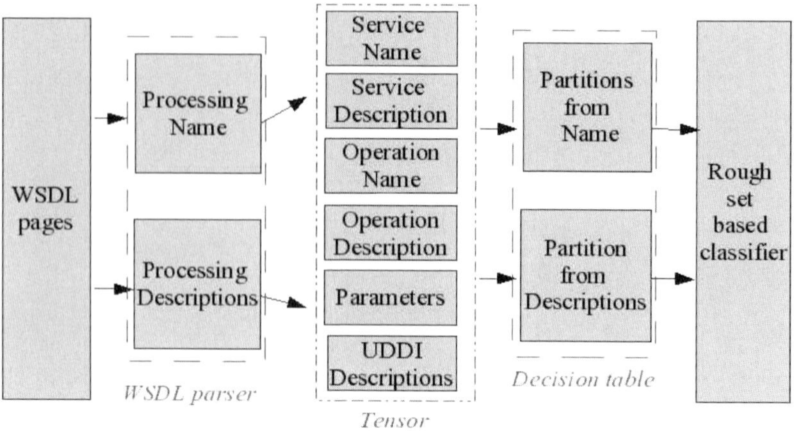

Fig. 1. Block diagram of proposed method

WSDL documents and offers a potent mathematical framework for analyzing the multifactor structure of WSDL documents. Tensor is a term in Algebraic Geometry. It is a generalization of the concepts of vectors and matrices in the area of linear algebra. A vector is called 1-order tensor and a matrix is called a 2-order tensor. Tag based TSM for web services consists of a two dimensional tensor, where one dimension represents the tags of WSDL and another dimension represents the terms extracted from WSDL. WSDL documents are tokenized with syntactic rules and canonical forms. First we select a set of relevent tags from a WSDL document. For each tag an individual tensor component is constructed. A tensor component is a vector, which represents the terms found in the text under a particular tag. The tag based tensor space model captures the structural representation of WSDL.

2.2 Base Level Classifications on TSM

We now describe how we generate partitions for each one of the components of the tensor using classifiers.

Partitions from components corresponding to names: We consider the terms in a name as a bag of words. We have constructed three different bags from service name, operation names and input/output parameter names respectively and constructed three tensor components from each of this bags. Classification algorithm is applied on these tensor components after preprocessing. We obtain three different partitions from three different tensor components corresponding to names of service, operations and parameters.

Partitions from components corresponding to description: To obtain the partitions from descriptions corresponding to services, operations and parameters, we consider the documentation as a bag of words. Word stemming and stopword removal have been performed to pre process the data. Classification algorithm is applied on the preprocesed bags to obtain partitions from the tensor components corresponding to service description, operation description and parameter description.

2.3 Rough Ensemble Classification

Our approach named RSM is designed to extract decision rules from trained classifier ensembles that perform classification tasks [9]. RSM utilizes trained ensembles to generate a number of instances consisting of prediction of individual classifiers corresponding to each tensor component as condition attribute values and actual class as decision attribute value. Then a decision table is constructed with one instance in each row. Once the decision table is constructed, rough set attribute reduction is performed to determine core and minimal reduct [8]. The classifiers corresponding to minimal reduct are then taken to form classifier ensemble for RSM classifier system. From the minimal reduct, the decision rules are computed by finding mapping between decision attribute and condition attributes. These decision rules obtained by rough set technique are then used to perform classification task.

3 Experimental Results

3.1 Data Sets

We gathered a corpuses of web services from SALCentral and webservicelist, two categorized web service indices. Details of the corpuses have been given below.

Salcentral dataset: Business-22, Communication-44, Converter-43, Country Info-62, Developers-34, Finder-44, Games-42, Mathematics-10, Money-54, News-30, Web-39.

Webservicelist dataset: Access & Security-27, Address / Locations-57, Business & Finance-97, Developer Tools-54, Content & Databases-24, Politics & Government-56, Online Validations-26, Stock Quotes-31, Search & Finders-22, Sales Automation-20, Retail Services-30.

3.2 Results

Improvement in each step over the existing classification results of web services has been shown here. In the first step we achive better classification results by using proposed tensor space model. In the second step further improvement of the results has been obtained by using Rough set based classifier.

In the Fig. 2 percentage accuracies of classifications have been compared on two different representation models. WSDL documents corresponding to above datasets have been represented in vector space model and tag based tensor space model respectively. Three well known classifiers, naive bayes (NB), support vector machines (SVM) and decision trees (C4.5) havs been considered to provide

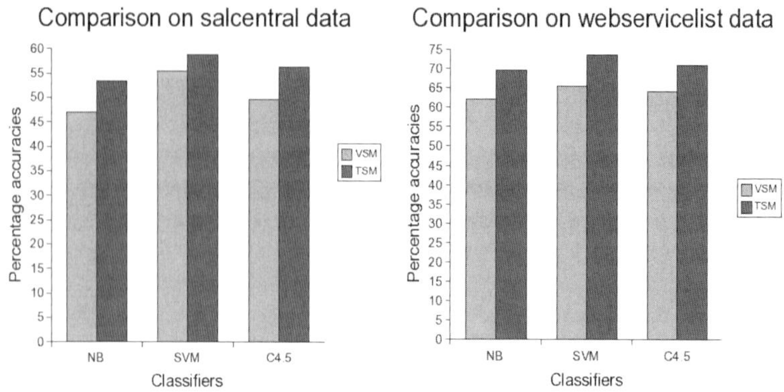

(a) salcentral (b) webservicelist

Fig. 2. VSM vs. TSM

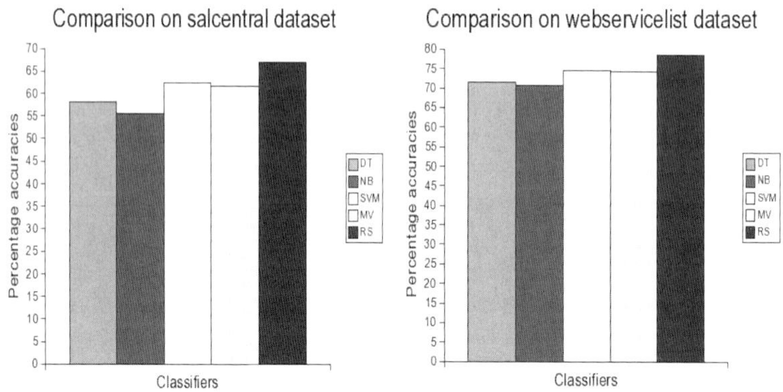

(a) salcentral (b) webservicelist

Fig. 3. Comparison of percentage accuracies of classifiers on salcentral and webservicelist datasets

classification results in two different models. Note that classification results in tensor space model have been computed on individual tensor components and combined with majority voting. The results show that classification on tensor space model provides better percentage accuracy than vector space model for both the datasets and for all classifiers considered.

In the Fig.3 bar chart of percentage accuracies of combining classifiers have been given. Here all the classifiers have been tested on meta data generated by base level classifiers from each tensor components. Naive bayes (NB), support vector machines (SVM), decision trees (DT), majority vote and rough set (RS) are applied to combine the output of base level classifiers corresponding to individual tensor components. Results show that, RSM provides better classification results than other methods on both datasets considered.

4 Conclusion

We discussed the problem of classifying a web service into a category and treated it as a split marge classification problem. Splits of the features have been performed based on tag set exists in the WSDL documents corresponding to web services. Tensor space model has been used to represent the services according to tag structure. Base level classification has been performed on individual tensor components. Finally combined classification has been obtained by using rough set based ensemble classifier. Two step improvement on the existing classification results of web services has been shown. In the first step we achieve better classification results by using proposed tensor space model. In the second step further improvement of the results has been obtained by using Rough set based ensemble classifier.

References

1. Cai, D., He, X., Han, J.: Tensor space model for document analysis. In: SIGIR 2006: Proceedings of the 29th annual international ACM SIGIR conference on Research and development in information retrieval, pp. 625–626 (2006)
2. Dong, X., Halevy, A., Madhavan, J., Nemes, E., Zhang, J.: Similarity search for web services. In: Proc. of VLDB (2004)
3. Wong, A., Salton, G., Yang, C.S.: A vector-space model for information retrieval. Journal of the American Society for Information Science 18, 13–620 (1975)
4. Heb, A., Kushmerick, N.: Learning to attach semantic metadata to web services. In: The IEEE International Conference (2003)
5. Yu, J., Guo, S., Su, H., Zhang, H., Xu, K.: A kernel based structure matching for web services search. In: WWW 2007: Proceedings of the 16th international conference on World Wide Web, pp. 1249–1250 (2007)
6. Bruno, M., Canfora, G., Di Penta, M., Scognamiglio, R.: An approach to support web service classification and annotation. In: The 2005 IEEE International Conference on e-Technology, e-Commerce and e-Service, pp. 138–143 (2005)
7. Oldham, N., Thomas, C., Sheth, A.P., Verma, K.: Meteor-s web service annotation framework with machine learning classification. In: Cardoso, J., Sheth, A.P. (eds.) SWSWPC, vol. 3387, pp. 137–146 (2004)
8. Pawlak, Z.: Rough Sets: Theoretical Aspects of Reasoning About Data, pp. 41–48. Kluwer, Boston (1991)
9. Saha, S., Murthy, C.A., Pal, S.K.: Rough set based ensemble classifier for web page classification. Fundamentae Informetica 76(1-2), 171–187 (2007)

Agent-Based Assistant for e-Negotiations

Simone A. Ludwig

Department of Computer Science, University of Saskatchewan, Saskatoon, Canada
ludwig@cs.usask.ca

Abstract. Knowledge about conflict styles and time pressure during a negotiation are important factors in a negotiation. This knowledge is used to model an agent-based assistant for e-negotiations. The idea of the proposed method is to model a utility concession function depending on the conflict style behaviour of a negotiator. Negotiators, prior to engage in e-negotiations, are asked to fill in a questionnaire designed to measure the conflict mode and specify their reservation levels. The agent-based assistant uses reservation levels and a concession-making model to propose the concession and timing of offers and attributes e.g. in a multi-attribute negotiation. The concession-making model is constructed in the utility space and it is constructed using the Thomas-Kilmann Conflict Mode Instrument and negotiation data from an experiment conducted by human negotiators.

1 Introduction

Negotiation, for decades, has been a central subject of study in disciplines such as economy, game theory and management. When discussing negotiation, it is important to distinguish between *negotiation protocol* and *negotiation strategy*. The protocol determines the flow of communication between the negotiating parties, dictating who can say what and when. It provides the rules by which the negotiating parties must abide if they are to interact. The protocol is necessarily known to all negotiation participants. The strategy, on the other hand, is the way in which a given party acting within the protocol rules makes an effort to get the best outcome from the negotiation. Strategy is used to determine, for example, when and what to concede, and when to hold firm. The strategy of each participant is therefore necessarily private.

Electronic negotiations are business negotiations conducted electronically e.g. via the Internet. The underlying IT infrastructure makes it possible for e-negotiation systems to offer features such as graph support, decision analysis and communication management. The goals of supporting negotiations through information technology are to reduce transaction costs in e-negotiations, to find and suggest an optimal deal, to conduct checks during the negotiation (e.g., that they conform to the accepted protocol), to offer decision support, and to provide argumentation support for human or software agents.

The benefit of e-negotiations is the support of negotiations via information technology offers decision support for human or software agents. To some extent, agent technology can be helpful in automating or assisting the buyer with the need identification stage. Specifically, agents can play an important role for those purchases that are repetitive (e.g. supplies) or predictable (e.g. habits) [1].

A. An et al. (Eds.): ISMIS 2008, LNAI 4994, pp. 514–524, 2008.

Interest in the automation of negotiations through the use of multi-agent systems has been stimulated to a great extent by the vision of software agents negotiating with other software agents to buy and sell goods and services on behalf of their principals in a future Internet-based global marketplace [2]. Until now, research has focused on accounting for particular interactions among agents by developing and improving specifically tailored negotiation protocols and strategies.

In any negotiation involving agents it is important that the agent is able to adequately represent the principals' interests. However, the process by which this knowledge is acquired is normally not taken into consideration [3]. In order to overcome this shortage, a possible approach is presented in this paper taking knowledge into account to model principals' objectives. The approach for the agent-based assistant for e-negotiations consists of a concession-making model which is constructed in the utility space and it is constructed using the Thomas-Kilmann Conflict Mode Instrument and negotiation data from an experiment conducted by human negotiators using the Negotiation platform Invite.

The remainder of this paper is structured as follows. Section 2 summarizes and discusses several related approaches. In Section 3 the research method is proposed including modeling and mapping of the conflict styles and the consideration of the opponent's concession making to a utility concession graph. Section 4 explains the model in more detail using a contract negotiation example. Section 5 concludes this paper by presenting the findings and discussing the shortcomings regarding further work.

2 Related Work

The construction of efficient and effective algorithms enabling software agents to be successful and obtain acceptable outcomes is one of the most active areas in agent-supported and automated negotiations. It is also important that software agents, alike human agents, represent the principal as closely as possible and are able to negotiate on behalf of their principals. For this to be effective, software agents must learn the principals' interests, strategies, preferences and prejudices in a given domain. Without this, software agents cannot execute their task appropriately. The acquisition of such knowledge is, therefore, an essential requirement for applying negotiating agents in practice, in particular:

- Exactly what knowledge an principal needs to impart to their agent in order to achieve high fidelity negotiation behaviour; and
- Ways in which this knowledge can be effectively acquired from the principal.

Guo, Mueller et al. [4] investigate how agents act on behalf of their principals in e-negotiations by eliciting information about the principal's preference structures. Using a multi-attribute utility theoretic model of user preferences, they propose an algorithm which enables the agent to learn the utility function over time. The learning method is based on an evolutionary framework with three-step learning in each generation. It combines population-based evolution with the possibility to apply external knowledge, and with individual learning through simulated annealing for further refinement of the solution.

Luo, Jennings et al. [5] analyze an automated negotiation model whereby user trade-off preferences were found to play a fundamental role in negotiation. With the method proposed user trade-off preferences were captured, modeling the main commonalities of trade-off relations and reflecting users' individualities. The basic idea behind the method is the following. First, the system queries the user about choice features in order to determine which attributes the trade-off relations exist between. Second, in order to determine the shape of the trade-off curve, the system queries the user about the relative importance degree of one attribute against another and about some features of trade-off curves. Finally, the system queries the user about his/her satisfaction degree for each trade-off alternative.

Luo, Jennings et al. [3] devised a default-then-adjust acquisition technique, whereby the system conducts a structured interview with the user to suggest the attributes of the trade-off, and then it asks the user to adjust the default preferences of the trade-off alternatives.

The goal of the described related work is the modeling of user's preferences and trade-off alternatives. The modeling and acquiring of knowledge is done using many different approaches such as learning algorithms and modeling of a range of strategies and tactics to acquire necessary domain knowledge.

The proposed approach makes use of the idea of an interview whereby Luo et al. used a default-then-adjust acquisition technique to extract negotiation knowledge. The proposed approach differs as it uses the Thomas-Kilmann conflict mode instrument designed as a questionnaire to identify the conflict mode, thereby acquiring behavioural knowledge of the negotiator. This behavioural knowledge is then used together with data from a conducted experiment, the Invite experiment, to construct a concession model for the negotiator, in particular to help inexperienced negotiators. The concession model corresponding to a particular conflict style will be modeled in the assistant and the assistant will support the negotiator by suggesting the concession to be made at a particular time during the negotiation.

3 Concession Modeling Approach

The Thomas-Kilmann Conflict Mode Instrument (TKCMI) is a commonly used psychological assessment tool and measures the five different behavioural classifications proposed by the Dual Concern Model: which was introduced by Blake and Mouton [6]. Thomas and Kilmann [7] developed and extensively tested a questionnaire in order to elicit conflict modes posited by Blake and Mouton's model. The questionnaire is a useful tool for probing bargaining styles in consulting. Shell [8] summarized his findings of the usefulness as follows:

- Ease of administration (it takes only about ten minutes to take and score);
- Relative freedom from social desirability biases in the way statements in the instrument are presented;
- Conflict styles that match up with strategy concept widely used in the negotiation literature; and
- Significant congruence between the classifications and their perceptions of their own behaviour across a set of simulations.

Thomas and Kilmann [7] did not develop the measures with bargaining or negotiation in mind. Rather, they were interested in finding a measurement device for probing the validity and independence of the five conflict modes hypothesized by Blake and Mouton. However, the Dual Concern Model had been plagued by problems as the variance in results appeared to be strongly linked to subjects' desire to exhibit socially desirable traits rather than to their actual preferences for one conflict mode. Thomas Kilmann addressed this problem by pairing simple, equally desirable or undesirable phrases representing each conflict attitude and forcing subjects to choose between the statements in each pair.

The five conflict styles are described as:

1. *Competing*: High assertiveness and low cooperativeness. The goal is to win.
2. *Avoiding*: Low assertiveness and low cooperativeness. The goal is to delay.
3. *Compromising*: Moderate assertiveness and moderate cooperativeness. The goal is to find a middle ground.
4. *Collaborating*: High assertiveness and high cooperativeness. The goal is to find a win-win solution.
5. *Accommodating*: Low assertiveness and high cooperativeness. The goal is to yield.

3.1 Invite Platform

The Invite software [9] is a negotiation support system platform mainly developed for the protocol-driven generation of systems. Their purpose is primarily educational: they are used to teach the subject of negotiation. The major features of the Invite platform are:

- Implementation of a negotiation methodology, in particular the process model and its various activities.
- Support for multiple, concurrent negotiation protocols, decision models, and interfaces.
- Provisioning of an intuitive web-based user interface.

The Invite platform allows users to negotiate a case independently of time and place restrictions. The system provides the user with general and private information about a case, allows to rate the issues and options, allows to send messages and offers, and provides a history to view exchanges in a tabular and graphical form.

This platform is under experimental use and different protocols are investigated. The different protocols are distinct from each other by the availability of analytical support and the provision of predetermined preferences. The experiment has three stages: pre-negotiation stage (questionnaire), negotiation stage and post-negotiation stage (questionnaire). The pre-questionnaire stage consists of the TKCMI, quiz, expectations and BATNA (Best Alternative To a Negotiated Agreement), case ratings of issues and options. For the negotiation stage the Invite system is used, and for the post-questionnaire stage questions about system adoption and the user's and opponent's conflict modes are asked. 88 participants successfully negotiated a case of a contract negotiation between a singer and a music agency. Out of these 88 negotiations, 48 reached an agreement. This sample data was used for the model of the proposed method which is described further below.

3.2 Concession Analysis of Negotiation Data

Data was extracted from the database of the Invite system of successful negotiations, in particular the TKCMI questions and negotiation graphs. The five conflict styles were calculated from the TKCMI questions. The negotiation graphs were taken and, where possible, the distribution of the curve categorized into convex, linear and concave distribution. A convex distribution means that large concessions are made first and then at the end only small concessions. A linear distribution means that equal concessions are made each time step, and a concave distribution characterizes small concessions at the beginning with larger concessions made at the end. Additionally, the concession of each timestamp was taken to calculate the relative concession made and the absolute value.

By analysing the data it was found that the conflict style collaborating was not strongly represented and was therefore discarded from the model. The conflict style avoiding was deliberately discarded as this conflict style describes behaviour that is unassertive and uncooperative, which means that this person would not pursue a negotiation in the first place. Hence, the avoiding conflict style was not considered for the model.

For the purpose of this study a grid was used which, for each approach, divides the TKCMI scores into three groups indicating the strength of an approach. The grid was developed by Shell [8] who collected the scores from over 1600 executives participating in negotiation training sessions. Considered are only approaches which have a strong presence as measured by the top 25% of the responses. That is, placing the limitation on the minimum score of:

- 7 or more for competing;
- 9 for compromising; and
- 6 for accommodating.

By applying the above score conditions to the questionnaire data, three types of negotiation profiles were obtained: purely compromising; a mix of competitive and compromising, and purely accommodating. For each profile type three types of functions: concave, convex and linear were fit and the accuracy was checked based on the 88 samples available. Out of the 88 samples, only 48 samples reached an agreement and therefore could be used. Out of these 48 samples 15 samples had concave concession curves, 21 had a convex concession curves and 12 showed a linear distribution. Table 1 presents the three profiles, associated function types and their accuracies based on the top 25%. It shows the summary of TKCMI measures with regard to the curve distribution.

The accuracy for each curve distribution is given in the last column and indicates how many TKCMI/curve distribution pairs match the proposed profiles, e.g., for profile 1, 10 out of 15 matched the proposed model.

Table 1. TKCMI measures and curve distribution

Profile	Compet.	Comprom.	Accommod.	Curve	Accuracy
P1		1		concave	67%
P2	1	1		convex	62%
P3			1	linear	67%

In summary, high values in compromising dictate a concave distribution curve, high values in competing and compromising result in a convex distribution and high values in accommodating result in a linear distribution. The gradient of the curve distribution is determined by the reservation level and the counterpart's first and second offer.

3.3 Utility Concession Modeling

The approach to negotiations is qualitative and indicative in nature. The assignment of shapes of the utility concession function is based on the user's conflict style and reservation values. A precise form of the utility concession graph needs to be determined depending the results given in Table 1. The equation for the utility concession graph can be constructed as follows, assuming the shape of the utility concession function to be exponential:

$$u = \delta_i \cdot u_1 + \delta_j \cdot u_2 + \delta_k \cdot u_3 \tag{1}$$

where:

$$u_1 = u_x - \alpha \cdot e^{\tau} \tag{1a}$$

is the concave distribution of the utility concession curve, and u_x is a constant value;

$$u_2 = u_x - \alpha \cdot \tau \tag{1b}$$

is the linear distribution of the utility concession curve;

$$u_3 = u_x - \alpha \cdot e^{-\tau} \tag{1c}$$

is the convex distribution of the utility concession curve;

$$\delta_{i,j,k} = \begin{cases} 1 & for\ i, j, k = 1 \\ 0 & for\ i, j, k = 0 \end{cases} \tag{1d}$$

whereby i, j, k are the factors for the concave, linear and the convex distribution respectively.

Factor τ is determined by three given points, one is the start point, the other is the end point and the third point is obtained from the analysis of the counterpart activity and normalized by *T*, the overall negotiation time. The constant value c is determined by the normalization of the utility rating, and α is a constant value.

3.4 Utility Concession Flowchart

The concession-making and offer construction of the agent-based assistant during a negotiation is done as follows. The first rule is to wait until the counterpart has made the first offer. As soon as the counterpart sends an offer the assistant checks whether the reservation level is reached. If so, the assistant suggests agreeing to this offer. If not, the assistant constructs the first offer with a utility of 100 and sends it to the counterpart. After this, the assistant waits until the counterpart makes the second offer. Once this offer is received, the assistant checks whether the reservation level is reached; if so the assistant suggests the negotiator to agree to the offer, if not then the concession made by the counterpart is evaluated and a new offer is constructed with an equal concession. Depending on the time line, as given by the constructed utility concession curve, this offer is sent to the counterpart at the calculated time. This process is repeated until an agreement is achieved or until the negotiation time has finished.

4 Agent-Based Assistant Using the Concession Model – An Example

The concession model and its use by the agent-based assistant are illustrated with an example of a contract negotiation between an artist (singer and song writer) and an entertainment promotion agency. There are four issues which the parties need to agree on, these are, number of new songs, royalties for CDs, contract signing bonus and number of promotional concerts.

Table 2. Ratings reflecting both negotiators' preferences

Issue	Option	Agency's option rating	Artist's option rating
Number of promotional concerts (per year)	5	30	0
	6	25	5
	7	5	25
	8	0	30
Number of songs	11	0	0
	12	5	15
	13	30	20
	14	25	15
	15	15	0
Royalties for the CDs (% of revenue)	1.5	20	0
	2.0	10	10
	2.5	5	15
	3.0	0	20
Contract signing bonus ($)	125,000	20	0
	150,000	10	15
	200,000	0	30

The parties negotiate using a NSS (Negotiation Support System) and, in addition, the artist uses an assistant. Prior to the negotiation, the assistant asks the artist to:

1. Fill in the TKCMI questionnaire;
2. Formulate the reservation levels for each of the four issues;
3. Set the time by which the negotiation should be completed; and
4. Engage in the preference elicitation and utility construction scheme.

The assistant uses the questionnaire to determine that the artist is a highly compromising person. This means that the artist's profile is P1 and the concession function is concave. Then the assistant normalizes the utility values to 0-100 and calculates the utility value for the alternative constructed with the four reservation levels; the obtained value is 60. At this point the assistant could engage in interaction with the artist and verify the way the reservation levels should be treated. The assistant needs to learn of the alternatives which yield a higher utility value than 60. For the sake of simplicity we assume here that every alternative which yields utility of 60 and higher is acceptable.

For this example, the issue ratings of both negotiators are assumed as shown in Table 2. Furthermore, Table 3 displays the offers exchanged by both negotiators and the corresponding utility ratings.

Table 3. Offer sequence and the artist's utility values

No	Agency's offer	Utility	Artist's offer	Utility	Utility concession
1	[5,13,1.5,125.000]	20	[8,13,3.0,200.000]	100	-
2	[6,13,1.5,125.000]	25	[7,13,3.0,200.000]	95	5
3	[6,13,2.0,125.000]	35	[7,13,2.0,200.000]	85	10
4	[6,13,2.5,125.000]	40	[7,13,3.0,150.000]	80	5

In Fig. 2, the sequence diagram of this example can be seen. The NSS helps the negotiators in their negotiation by providing a platform for conducting negotiations on the internet.

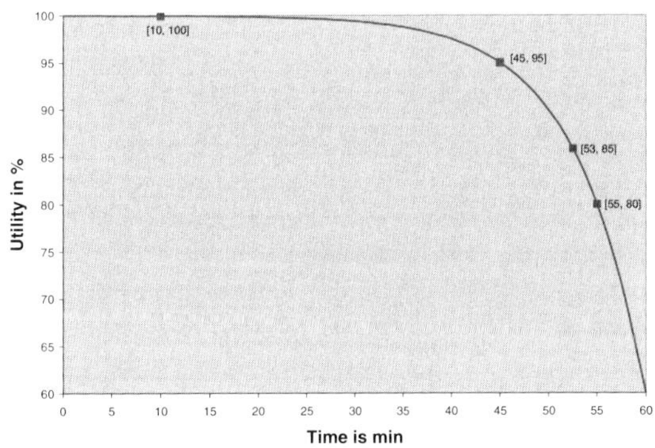

Fig. 1. Utility concession graph of the artist

The artist waits until the counterpart, the agency, makes the first offer. When the agency sends the first offer (see offer package in Table 3) after 5 min., the assistant selects the first offer with utility of 100 (see offer package in Table 3) and sends it back to the counterpart after 10 min. 13 min. later the counterpart sends the second offer (see offer package in Table 3). At this time the curve distribution is determined by three points and the utility concession curve as shown in Fig. 1 is selected and used by the agent. In particular, equation (1b) is used whereby the coefficient α is set to 0.01 (normalization for 100 range; time of negotiation 1h) and τ is defined as time t divided by a factor of 7.325. Please note, that this factor is a constant value which is determined by 3 points as mentioned previously, the starting and end point, and the point of the second offer from the counterpart (determined concession made between the first and second offer and the time the second offer was sent).

The assistant calculates the concession made (utility concession of 5) and selects the next offer with the same concession based on the modeled utility concession graph in Fig. 1 (see offer package in Table 3). Please note that the concession from the counterpart is calculated based on the negotiator's own preference setting. This implies that the negotiator should wait until t = 45 min. before this offer is sent to the counterpart.

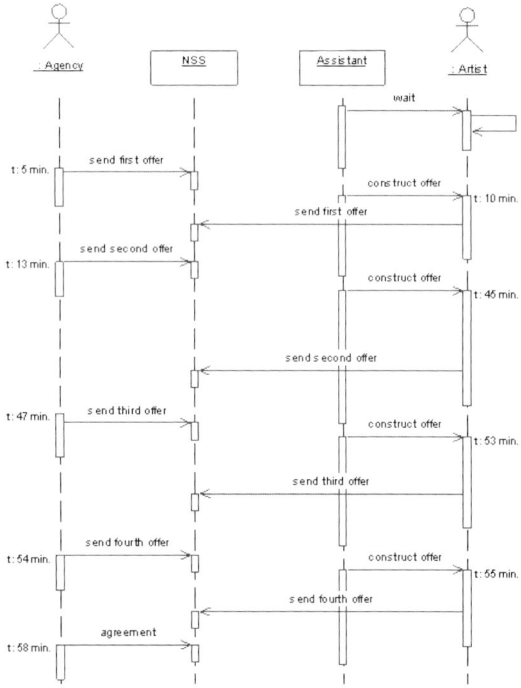

Fig. 2. Sequence diagram of negotiation

The counterpart reacts with a counter offer at 47 min. in the negotiation (see offer package in Table 3). The assistant calculates the concession made which is a utility concession of 10 and selects the next offer (see offer package in Table 3). This offer is sent at 53 min. and followed by a counter offer (see offer package in Table 3) at 54 min. After calculating the concession made by the counterpart the assistant suggests the negotiator the next offer (see offer package in Table 3) which is sent after 55 min. The counterpart agrees to the offer 2 min. before the negotiation terminates.

5 Conclusion

The proposed approach of concession modeling is based on negotiation data taken from the experiments conducted by the Invite system. The concession model is constructed using the Thomas-Kilmann Conflict Mode Instrument and the utility concession graphs. Results showed that people who had high compromising values had a concave utility concession graph, people who had high accommodating values had a linear graph and people who had a high value for competing and compromising had a convex utility concession graph. An agent-based assistant uses this concession model to help inexperienced human users during a negotiation, by suggesting possible offers to send at calculated times depending on the constructed utility concession graph.

This approach is a good first step in assisting inexperienced human negotiators based on the user's conflict behaviour and style in e-negotiations. However, more research is necessary to refine the concession model. For example, one of the problems of capturing the opponent's behaviour is that assumptions based on own preferences are made. This however is a drawback which is eminent in all negotiation scenarios where preferences are not revealed. Instead of only focusing the calculation of the opponent's concession during each offer cycle on the own preferences, a fuzzy approach could allow for a better prediction of the opponent's concession making.

Furthermore, the assistant can only deal with negotiators falling into one of the proposed categories. For example, people with a strong conflict style in collaborating are not accounted for. This clearly needs to be investigated further.

Another issue which needs to be addressed is the sample data. The sample size of 88, available for this investigation, was clearly too small for accurate results as evident from the accuracy values shown in Table 1. It is intended to expand this work as soon as more data from the negotiation experiments, which are currently underway, becomes available.

Acknowledgement

The author would like to thank Gregory Kersten for his helpful discussions and useful suggestions. This research was partially supported by grants from the Initiative for New Economy of the Social Sciences and Humanities Research Council Canada, and the Natural Science and Engineering Research Council Canada.

References

1. Boudriga, N., Obaidat, M.S.: Intelligent Agents on the Web: A Review. Computing in Science and Engineering 06, 35–42 (2004)
2. Tamma, V., Phelps, S., Dickinson, I., Wooldridge, M.: Engineering Applications of Artificial Intelligence 18, 223–236 (2005)
3. Luo, X., Jennings, N.R., Shadbolt, N.: Acquiring User Tradeoff Strategies and Preferences for Negotiating Agents: A Default-Then-Adjust Method. International Journal of Human Computer Studies (2005)

4. Guo, Y., Mueller, J.P., Weinhardt, C.: Learning User Preferences for Multi-attribute Negotiation: An Evolutionary Approach. In: Mařík, V., Müller, J.P., Pěchouček, M. (eds.) CEEMAS 2003. LNCS (LNAI), vol. 2691, pp. 303–313. Springer, Heidelberg (2003)
5. Luo, X., Jennings, N.R., Shadbolt, N.: Knowledge-based acquisition of tradeoff preferences for negotiating agents. In: Proceedings of 5th International Conference on Electronic Commerce, ACM Press, Pittsburgh, Pennsylvania, pp. 138–149. ACM Press, Pittsburgh, Pennsylvania (2003)
6. Blake, R.R., Mouton, J.S.: The managerial Grid. Gulf Publications, Houston (1964)
7. Kilmann, R.H., Thomas, K.W.: Developing a Forced-Choice Measure of Conflict-Handling Behavior: The MODE Instrument. Educational and Psychological Measurement 37, 309–325 (1977)
8. Shell, G.R.: Bargaining Styles and Negotiation: The Thomas-Kilmann Conflict Mode Instrument in Negotiation Training. Negotiation Journal 17, 155–174 (2001)
9. Strecker, S., Kersten, G., Kim, J.-B., Law, K.P.: Electronic Negotiation Systems: The Invite prototype. In: Multikonferenz Wirtschaftsinformatik, Track Collaborative Business, Gesellschaft für Informatik e.V., Passau, Germany (2006)

Local Soft Belief Updating for Relational Classification

Guoli Ding[1], R.F. Lax[1], Jianhua Chen[2], Peter Chen[2], and Brian D. Marx[3]

[1] Dept. of Mathematics, LSU, Baton Rouge, LA 70803
[2] Dept. of Computer Science, LSU, Baton Rouge, LA 70803
[3] Dept. of Experimental Statistics, LSU, Baton Rouge, LA 70803

Abstract. We introduce local soft belief updating, a new heuristic for taking into account relations that exist between entities in a database. Our idea applies Pearl's belief updating, but only in the first-order neighborhood of each node, thus avoiding any problems with loops. We apply our method to a classification problem using a subset of the Cora database of computer science articles, with Cora's citation graph giving the relations between entities.

1 Introduction

We consider the problem of classifying entities that have relational links, as well as their own attributes. Two applications we will discuss are the classification of certain computer science papers into a subspecialty and the classification of certain people as terrorist or non-terrorist. In the first of these applications, the attributes consist of certain keywords in the abstract of the paper, while we consider two papers to be linked if one of the papers cites the other. In the second application, the attributes may consist of things such as age, ethnic background, religion, and travel to certain countries, while two people will be considered to be linked if they have "significant contact" with each other.

The problem of relational classification has received much attention in recent years. Our method is a blend of two ideas - a simple relational classifier given by Macskassy and Provost (2005), and the probabilistic relational models (PRMs), as given by, for example, N. Friedman et al. (1999) and B. Taskar et al. (2002). We assume that a learning method has been applied to the "flat" database consisting of the entities and their attributes so that for each entity we have a vector whose ith component is the probability that this entity belongs to the ith class. Possible learning methods include naive Bayes, regression, support vector machine, or pseudo-Boolean (or fitness) functions. We also assume that we have a link matrix for each of the relational links. Such a matrix gives the probabilities that an entity belongs to a class c_i given that its adjacent entity belongs to a class c_j. In our classification application, we will calculate the link matrix from the data set by simple counting. For each entity, we then update the probability vector at that entity by viewing the probabilities at the adjacent entities as soft (or virtual) evidence and applying Pearl's theory of belief updating. We call this method local soft belief updating (LSBU).

A. An et al. (Eds.): ISMIS 2008, LNAI 4994, pp. 525–534, 2008.

While one may need to assume some classifications are (precisely) known in the training of the learning method used, we do not assume any classifications are known (in the sense of hard evidence) when we apply LSBU. In Macskassy and Provost (2003), the authors assume some of the classifications are known, and that the links have weights, rather than link matrices. They then define a simple relational classifier using a weighted average at each entity and iterate this process until either all entities are labeled or no new entities can be labeled (if there exist components of the graph in which no node has a known label). In a related paper, Macskassy and Provost (2005) assumed a "suspicion score" for each person in a network and then used a weighted average to update these scores.

In the case of PRMs, one considers a Bayesian network whose nodes correspond not only to the entities involved, but also to certain attributes of those entities. This can allow one to consider several different "types" of relations between entities. For example, in Taskar et al. (2001), the authors use the Cora database and consider both citations between papers and common authorship. For our method, we take only the entities as nodes and we assume only one "type" of relation between entities. For a PRM, one assumes that one has a generative probability model for the prior probability distribution. We do not assume that our network forms a graphical model for the prior probability distribution coming from the "flat" database; rather, we will treat the neighborhood of each node as a "local" graphical model. With a PRM, one proceeds to assume that some of the entities are known, and then one applies loopy belief propagation to update probabilities (beliefs) at all the nodes and obtain a posterior distribution. In contrast, our heuristic uses soft evidence everywhere, and we apply belief updating only to the neighborhood of each node in order to obtain an updated belief at the "center" of the neighborhood. Since the neighborhood of each node is a tree, we do not deal with any cycles (loops), and our update computations are very simple.

2 Local Soft Belief Updating

We assume we are given a database consisting of n entities $A_1, \ldots, A_n$, and that each entity has t attributes $X_1, \ldots, X_t$. Further, we assume there are m classes, $c_1, \ldots, c_m$, and we let Y_j denote the class random variable associated to A_j; i.e., $P(Y_j = c_k)$ is the probability that A_j belongs to class c_k.

We assume that we have prior knowledge of $P(Y_i = c_k)$ for i in a training subset $I \subset \{1, \ldots, n\}$ and for $k = 1, 2, \ldots, m$, and that this knowledge has been used to obtain a classification function that assigns a value to $P(Y_j = c_k)$ for each entity A_j in the database based solely on the attributes of that entity. Such a function could be obtained in several ways. For example, one could use naive Bayes, or perform a regression on this training subset with the attributes, and possibly some interaction terms, as the explanatory variables. Or, in the case of two classes 0 and 1 and binary attributes, one could use this training subset to construct a pseudo-Boolean function as in Chen et al. (2004) or Ding et al.

(2005), again with the attributes, and possibly some products of these attributes, as the variables, and then suitably scaling such a function so that it takes values between 0 and 1. Once we have such a classifier, we can establish a threshold value τ such that if $P(Y_i = 1) \geq \tau$, then A_i is classified in class 1.

So far, we have considered only the attributes of each entity. Now we also want to consider the links between entities in order to update the probabilities $P(A_j = c_k)$. One simple relational classifier was used by Macskassy and Provost (2005) in the application of terrorist detection. We consider two classes, 0 (not a terrorist) and 1 (a terrorist). For every k, Let $s(k) = P(Y_k = 1)$ denote the "suspicion score" of A_k. Macskassy and Provost consider a weighted social network where $w_{i,j}$ is the number of times that A_i and A_j are known to interact. They then modify $s(i)$ by considering the weighted average of the suspicion scores of A_i's neighbors. They update all estimates "pseudo-simultaneously" in the following manner. Put $s(i)^{(t+1)} =$

$$\alpha^{(t+1)} \cdot s(i)^{(t)} + (1 - \alpha^{(t+1)}) \cdot \left(\frac{1}{Z} \sum_{A_j \in N_i} w_{i,j} \cdot s(j)^{(t)} \right),$$

where t is the iteration step, $\alpha^{(t)} = (0.99)^t$, N_i is the set of neighbors of A_i, and $Z = \sum_{A_j \in N_i} w_{i,j}$. They perform 100 iterations and stop.

Now, consider the following situation. Suppose that A_1 meets with precisely $A_2, \ldots, A_r$, and that none of these r individuals meets with anyone outside this set of r people. Suppose that $s(1) = s(j)$ for $j = 2, \ldots, r$, and that $s(1)$ is slightly below the threshold score τ of being deemed "suspicious." With the Macskassy-Provost updating, $s(1)$ would remain constant and A_1 would not be considered suspicious. However, we do not consider this to be a desirable outcome. If A_1 is almost suspicious and A_1 meets with several other people who are also almost suspicious, then we believe A_1 should be considered suspicious. As we will show, this can be the case if we use Pearl's belief updating to revise the suspicion scores.

Given any entity A_i, let $N_i = \{j : A_j \text{ is linked with } A_i\}$. We consider the Bayesian network whose vertex set is $\{Y_i\} \cup \{Y_j : j \in N_i\}$, and whose arrows are $Y_i \rightarrow Y_j$, where $A_j \in N_i$. In the language of social networks, this graph corresponds to what is sometimes called the *first-order star* of A_i. For each entity A_k, we let $P(Y_k)$ denote the $m \times 1$ column vector $[P(Y_k = c_1) \cdots P(Y_k = c_m)]^T$. (This probability distribution is what we mean by "soft evidence.").

Fixing our attention on one specific entity, A_i, we consider $P(Y_i)$ to be the prior probability at the node Y_i. For each $j \in N_i$, we will view $P(Y_j)$ as soft (or virtual) evidence at the vertex Y_j and we want to update our belief at Y_i in light of this soft evidence, using the theory from J. Pearl (1988). In order to do this, we need to introduce the link matrix M that will be associated to every edge in our network. Given an edge $Y_i \rightarrow Y_j$, the link matrix is the $m \times m$ matrix $M = [P(Y_j = c_l | Y_i = c_k)]$, for $k, l = 1, 2, \ldots, m$. In the case of two classes, 0 and 1, the link matrix is

$$M = \begin{bmatrix} P(Y_j = 0|Y_i = 0) & P(Y_j = 1|Y_i = 0) \\ P(Y_j = 0|Y_i = 1) & P(Y_j = 1|Y_i = 1) \end{bmatrix}.$$

In the application of terrorist detection, the link matrix consists of the following conditional probabilities: if A_i has significant contact with A_j, what is the probability that A_j is (or is not) a terrorist given that A_i is (or is not) a terrorist? In this paper, we will assume that the link matrix is a symmetric matrix and that it is "universal" in the sense that it does not depend on i or j.

Using Pearl's theory of belief updating (section 4.2.3 of Pearl (1988)), each neighbor Y_j will send a message $\lambda_j(i)$ from Y_j to Y_i. In our case, this message is simply the matrix product

$$\lambda_j(i) = MP(Y_j).$$

If Y_j were the only neighbor of Y_i, and if $BEL(i)$ denotes the posterior probability, or belief (given the evidence), at Y_i, then we would have

$$BEL(i) = \alpha\lambda_j(i) * P(Y_i),$$

where $*$ is componentwise multiplication and α (a notation introduced by Pearl) is a normalizing scalar so that the components of the resulting column vector sum to 1. Since most treatments of belief updating deal with hard evidence, we give a brief explanation of what is happening in this soft evidence case. This procedure may be viewed as a weighted average of m applications of Bayes's Theorem, with the weights being the probabilities $P(Y_j = c_k)$, as follows. If we had the hard evidence $Y_j = c_1$, for example, then, by applying Bayes's Theorem, the posterior probabilities, or beliefs, at Y_i would be $BEL(Y_i = c_k) = \alpha P(Y_j = c_1|Y_i = c_k)P(Y_i = c_k)$ for $k = 1, 2, \ldots, m$. If we let $BEL(i)$ denote the column matrix $[BEL(Y_i = c_k)]^T$ for $k = 1, 2, \ldots, m$, then we have the matrix equation

$$BEL(i) = \alpha Me_1 * P(Y_i),$$

where e_1 is the column vector with a 1 in the first row and zeros elsewhere. Similarly, if we had the hard evidence $Y_j = c_l$, then the updated belief at Y_i would be given by $BEL(i) = \alpha Me_l * P(Y_i)$, where e_l is the column vector with a 1 in the lth row and zeros elsewhere. When we have the soft evidence $P(Y_j)$, the updated belief at Y_i is given by

$$\begin{aligned} BEL(i) &= \alpha \left(\sum_{k=1}^{m} P(Y_j = c_k) Me_k * P(Y_i) \right) \\ &= \alpha M (\sum_{k=1}^{m} P(Y_j = c_k) e_k) * P(Y_i) = \alpha MP(Y_j) * P(Y_i), \end{aligned}$$

thus showing that the updated belief may be viewed as a weighted average of m applications of Bayes's Theorem.

When Y_i receives these messages from all its neighbors (children), Y_i computes

$$\lambda(i) = \Pi^*_{j \in N_i} \lambda_j(i),$$

where the product Π^* here is componentwise multiplication. The reason the (componentwise) product is taken here is that, by the (local) graphical model given by the neighborhood of Y_i, the neighbors of Y_i are conditionally independent given Y_i (cf. section 4.2.3 of Pearl (1988)). Finally, Y_i computes the updated belief (or posterior probability), $BEL(i)$, by

$$BEL(i) = \alpha\lambda(i) * P(Y_i). \tag{1}$$

Example 1. We return to the application of terrorist detection. Assume that A_1 has significant contact with A_2 and A_3, and that these three people have no contact with anyone else. Assume that $P(Y_k = 1) = 0.7$ for $k = 1, 2, 3$, and that the threshold for considering someone to be suspicious is $\tau = 0.75$. As discussed above, using the relational classifier of Macskassy and Provost, A_1 is not considered to be suspicious, even after taking these contacts into account. To compute our updated belief, we will assume that the link matrix M is given by $M = \begin{bmatrix} 0.9\ 0.1 \\ 0.1\ 0.9 \end{bmatrix}$. This matrix reflects the idea that terrorists associate mostly (we assume 90% of the time) with other terrorists. We then have

$$\lambda_2(1) = \lambda_3(1) = M \begin{bmatrix} 0.3 \\ 0.7 \end{bmatrix} = \begin{bmatrix} 0.34 \\ 0.66 \end{bmatrix}.$$

Then

$$\lambda(1) = \begin{bmatrix} 0.34 \\ 0.66 \end{bmatrix} * \begin{bmatrix} 0.34 \\ 0.66 \end{bmatrix} = \begin{bmatrix} 0.1156 \\ 0.4356 \end{bmatrix}.$$

Hence,

$$BEL(1) = \alpha \begin{bmatrix} 0.1156 \\ 0.4356 \end{bmatrix} * \begin{bmatrix} 0.3 \\ 0.7 \end{bmatrix} = \begin{bmatrix} 0.1021 \\ 0.8979 \end{bmatrix}.$$

Therefore, the updated belief is that there is greater than an 89% chance that A_1 is a terrorist.

3 Simultaneous Updating

Notice that we do not change the priors (or soft evidence) at any node. Those priors come from the classification function discussed above and these would change only if that function, or attributes of a given entity, changes. The belief updating can be performed simultaneously at every Y_k by using the prior probability at Y_k and the soft evidence at the neighbors of Y_k.

We do not claim that the updated beliefs at each node form a probability distribution that is consistent with the network; indeed, it is very unlikely that this will be the case. Thus, our heuristic differs from loopy belief propagation, which does seek such a probability distribution (and which may not converge). Our updates are purely local at each node and do not "propagate" through the network. The complexity of our heuristic is linear in the number of nodes.

Example 2. Suppose we have five individuals, $A_1, \ldots, A_5$, who form the social network shown in Figure 1(a), where an edge denotes significant contact.

Now assume that our flat classification procedure has given us $P(Y_1 = 1) = .2$, $P(Y_2 = 1) = .3$, $P(Y_3 = 1) = .4$, $P(Y_4 = 1) = .8$, $P(Y_5 = 1) = 1$. We will use the same link matrix as in Example 1. To update the belief at Y_3, we use the Bayesian network shown in Figure 1(b). We find that

$$\lambda(3) = \begin{bmatrix} 0.74 \\ 0.26 \end{bmatrix} * \begin{bmatrix} 0.66 \\ 0.34 \end{bmatrix} * \begin{bmatrix} 0.26 \\ 0.74 \end{bmatrix} * \begin{bmatrix} 0.1 \\ 0.9 \end{bmatrix} = \begin{bmatrix} 0.0127 \\ 0.0589 \end{bmatrix}$$

$$BEL(3) = \alpha \begin{bmatrix} 0.0127 \\ 0.0589 \end{bmatrix} * \begin{bmatrix} 0.6 \\ 0.4 \end{bmatrix} = \begin{bmatrix} 0.2444 \\ 0.7556 \end{bmatrix}.$$

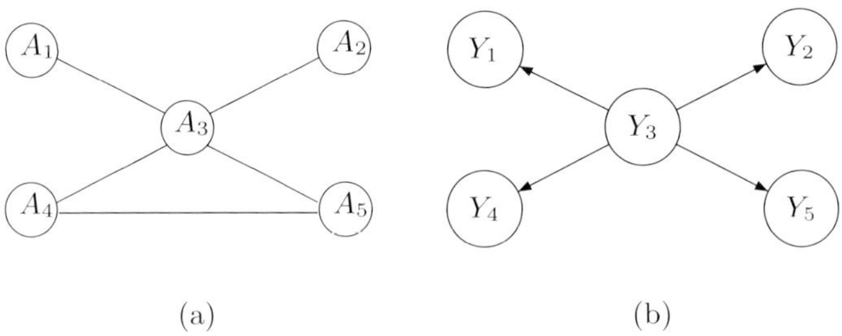

Fig. 1. (a) Social Network, (b) First-order Star of Y_3

Notice that the prior was $P(Y_3 = 1) = .4$ and A_3 had significant contact with two people with low scores and two people with high scores. But the evidence from Y_4 and Y_5 outweighed the evidence from Y_1 and Y_2 and resulted in a high (75.6%) probability that A_3 is a terrorist.

We can simultaneously update the belief at each node. For example, at Y_1 we would use the Bayesian network consisting of the two nodes Y_1 and Y_3 with the single arrow from Y_1 to Y_3. We find that $\lambda(1) = M \begin{bmatrix} 0.6 \\ 0.4 \end{bmatrix} = \begin{bmatrix} 0.58 \\ 0.40 \end{bmatrix}$ and $BEL(1) = \alpha \begin{bmatrix} 0.58 \\ 0.40 \end{bmatrix} * \begin{bmatrix} 0.8 \\ 0.2 \end{bmatrix} = \begin{bmatrix} 0.8529 \\ 0.1471 \end{bmatrix}$.

Notice that in our local soft belief updating we use the soft evidence at Y_3 that came from our flat classification, not the updated belief we calculated above. Since that flat classification gave $P(Y_3 = 1) = .4$, the probability that A_1 is a terrorist went down from .2 to .1471.

The computation of updated beliefs at each node here gives

$$BEL(1) = \begin{bmatrix} 0.8529 \\ 0.1471 \end{bmatrix}, BEL(2) = \begin{bmatrix} 0.7632 \\ 0.2268 \end{bmatrix}, BEL(3) = \begin{bmatrix} 0.2444 \\ 0.7556 \end{bmatrix},$$

$$BEL(4) = \begin{bmatrix} 0.0369 \\ 0.9631 \end{bmatrix}, \text{ and } BEL(5) = \begin{bmatrix} 0 \\ 1 \end{bmatrix}.$$

At this point, one might be tempted to repeat this process using the updated beliefs as new priors and soft evidence. But, Pearl (1988, section 4.23) makes the following observation:

> In recursive Bayesian updating, the posterior probability can be used as a new prior, relative to the next item of evidence, only when the items of evidence are conditionally independent given the updated variable X. Such recursive updating cannot be applied to networks because only variables that are separated from each other by X are conditionally independent. In general, it is not permissible to use the total posterior belief as a new multiplicative prior for the calculation.

4 The Link Matrix

In an application, if we have complete data, then the conditional probabilities in our link matrix would be determined by simple counting using the data set. In the more complicated situation of a PRM and incomplete data, Taskar et. al. (2001) used an Expectation Maximation procedure to estimate the conditional probability distributions. Alternatively, the probabilities in our link matrices could be estimated by an "expert." Clearly, the choice of link matrix is crucial with respect to the difference between the prior probabilities and the updated beliefs. For instance, if we had taken our link matrix in Example 1 to be $M' = \begin{bmatrix} 0.8\ 0.2 \\ 0.2\ 0.8 \end{bmatrix}$, then the updated belief would be $BEL(1) = [.1387 \quad .8613]^T$. Thus, the probability that A_1 is a terrorist would drop from the almost 90% in Example 1 to 86%.

We have assumed that the link matrix is a symmetric matrix. However, this is not required for our formulas, and, since we apply belief updating to the directed neighborhood of each node, one could consider link matrices that are not symmetric. But, in that case, the link matrix would not be independent of the choice of nodes.

In our applications, the diagonal entries of the link matrix will be much larger than the off-diagonal entries. A matrix $Q = [q_{ij}]$ is called (strictly) diagonally dominant if $|q_{ii}| > \sum_{j \neq i} |q_{ij}|$. So, our link matrix will likely be a diagonally dominant matrix. We note that it is well-known that a symmetric diagonally dominant matrix with positive diagonal entries is a positive definite matrix. Our link matrix is also a stochastic matrix (doubly stochastic if it is symmetric), and such a matrix has 1 as its greatest eigenvalue.

In the special case of two classes, and assuming the link matrix is symmetric, we can establish the following result that shows that the "impact" of soft evidence increases with the determinant of the link matrix. (For example, consider the change in updated belief in Example 1 when the matrix M with determinant .8 is replaced by the matrix M' above with determinant .6.) In this case, the link matrix is of the form $M = \begin{bmatrix} 1-\beta & \beta \\ \beta & 1-\beta \end{bmatrix}$, and the eigenvalues are 1 and $1 - 2\beta$ (which is also the determinant).

Proposition 1. *For* $k = 1, 2$, *let* $M_k = \begin{bmatrix} 1-\beta_k & \beta_k \\ \beta_k & 1-\beta_k \end{bmatrix}$, *where* $0 \leq \beta_k \leq 1$. *Let* $0 \leq \rho \leq 1$, *and put* $M_k \begin{bmatrix} 1-\rho \\ \rho \end{bmatrix} = \begin{bmatrix} 1-\mu_k \\ \mu_k \end{bmatrix}$. *Put* $\mathcal{O}_k = \mu_k/(1-\mu_k)$. *Assume* $\beta_1 > \beta_2$. *(This is equivalent to assuming that* $\det(M_1) < \det(M_2)$.*) Then we have: (1) if* $\rho > \frac{1}{2}$, *then* $\mathcal{O}_1 < \mathcal{O}_2$, *and (2) if* $\rho < \frac{1}{2}$, *then* $\mathcal{O}_1 > \mathcal{O}_2$.

Proof. We have

$$\frac{\mu_k}{1-\mu_k} = \frac{\rho - \beta_k(2\rho-1)}{1-\rho+\beta_k(2\rho-1)}.$$

If $\beta_1 > \beta_2$ and $\rho > \frac{1}{2}$, then $\beta_1(2\rho-1) > \beta_2(2\rho-1)$. It follows that $\frac{\mu_1}{1-\mu_1} < \frac{\mu_2}{1-\mu_2}$. If $\beta_1 > \beta_2$ and $\rho < \frac{1}{2}$, then $\beta_1(2\rho-1) < \beta_2(2\rho-1)$ and $\frac{\mu_1}{1-\mu_1} > \frac{\mu_2}{1-\mu_2}$. □

The above result no longer holds if the link matrix is not symmetric.

5 Experimental Results

We performed an experiment using Cora, a relational data set compiled by McCallum et al. (2000). This data set contains over 30,000 titles of computer science articles, as well as a citation graph containing a link when one paper cites another paper in the data set. The papers have been classified into a hierarchy containing 70 leaves. We considered a subset of this data set consisting of papers classified as being in the areas of Compression or Encryption. Since we wanted to do a "flat" classification using keywords in the abstract, we used only papers for which the field "Abstract-found" in the Cora database had the value 1. This resulted in a data set of 218 Compression papers and 247 Encryption papers.

For our (flat) learning method, we first used naive Bayes. This is a common method used for text classification and, in fact, was one of the first steps in the classification process used by McCallum et al. Using a sample consisting of 17 Compression and 17 Encryption papers, we identified 48 discriminating keywords in their abstracts by considering, for each word found in an abstract, the mutual information of the word random variable (i.e., the binary random variable Z_w that equals 1 if the word w is found in a given abstract) and the class random variable (i.e., the random variable that equals 0 for a Compression paper and 1 for an Encryption paper); cf. A. McCallum and K. Nigam (1998). These discriminating keywords are the attributes of our entities.

For our naive Bayes method, we used the Weka Multinomial Naive Bayes, as implemented in the Rapidminer program (Mierswa et. al. (2006)), although we did not count multiple occurrences of a keyword in an abstract. This means that in our experiment the probability $P(w|c)$ of observing word w given class c is given by

$$P(w|c) = \frac{1 + N_{wc}}{k + \sum_{w'} N_{w'c}},$$

where N_{wc} is the number of papers in class c whose abstract contains the word w and k is the size of our "vocabulary" (here, $k = 48$). Then, given a document d, the probability that d is in class c is

$$P(c|d) = \frac{P(c)\prod_{w\in d} P(w|c)}{P(d)}.$$

We performed a 10-times cross validation and found that this method correctly classified 417 out of the 465 papers (89.7% accuracy).

Now we proceeded to use the relational information in Cora. The citation graph in Cora contained 643 links between papers in our set of 465 Compression and Encryption papers. Of these links, only 6 of them were between a Compression paper and an Encryption paper. Therefore, for the link matrix in this application we used $M = \begin{bmatrix} .99 & .01 \\ .01 & .99 \end{bmatrix}$. From the previous section, because the determinant of M is so close to 1, soft evidence from adjacent nodes can have a large impact on the updated beliefs.

We then computed the updated beliefs using equation (1), with the soft evidence coming from the naive Bayes calculations. The result was that 432 out of 465 (92.9%) of the papers were now classified correctly. We note that it was not the case that all correct classifications were maintained when we performed our belief updating; indeed, there were three cases in which a paper was classified correctly by naive Bayes, but this classification was changed by LSBU.

We also performed a second experiment with the same data, but using logistic regression (as implemented in Rapidminer) as our (flat) learner for this binary classification problem, with the keywords as the explanatory variables. We again performed a 10-times cross validation and found that logistic regression correctly classified 401 out of the 465 papers (86.2% accuracy). When we performed LSBU using the logistic regression probabilities as our soft evidence, the result was that 416 papers (89.5%) were classified correctly.

6 Discussion

One problem we found with the Cora database is that sometimes there is no abstract in a file even though the "Abstract found" field equals 1. In fact, this was true for approximately 19% of the articles in our sample. This was the main reason our naive Bayes method did not achieve more than 89.7% accuracy, since this resulted, of course, in the "empty abstract" of these articles not having any of the keywords in it. The local belief updating was helpful in correcting some of the misclassifications due to this missing data.

One reason the increase in accuracy between the flat classification and the relational classification was not greater was that approximately 36% of the articles in our sample were "isolated vertices;" i.e., they neither cited, nor were cited by, any other paper in our sample.

We investigated the three papers that were correctly classified by naive Bayes, but whose classifications were reversed when we applied LSBU. These were encryption papers, but they dealt with encrypting codes that are frequently used in video transmission, such as turbo codes, so some of the references cited by these papers were compression papers.

7 Conclusion

We presented local soft belief updating, a new heuristic for dealing with entities that have relational links. This method uses Pearl's belief updating, but only locally at each node, instead of globally through the entire network. Thus, there are no concerns about loops, and calculations are quite simple. We showed that local soft belief updating improved classification accuracy on a subset of the Cora database of computer science articles, where our base learning method was either naive Bayes or logistic regression.

Acknowledgements

This research was partially supported by National Science Foundation grant: IIS-0326387 and AFOSR grant: FA9550-05-1-0454.

References

Chen, J., Chen, P., Ding, G., Lax, R.: A new method for learning pseudo-Boolean functions with applications in terroists profiling. In: Proceedings of the 2004 IEEE Conference on Cybernetics and Intelligent Systems, pp. 234–239 (December 2004)

Friedman, N., Getoor, L., Koller, D., Pfeffer, A.: Learning probabilistic relational models. In: Proceedings of the Sixteenth International Conference on Artificial Intelligence (IJCAI 1999), vol. 1309, pp. 1300–1309 (August 1999)

Ding, G., Chen, J., Lax, R., Chen, P.: Efficient learning of pseudo-Boolean functions from limited training data. In: Hacid, M.-S., Murray, N.V., Raś, Z.W., Tsumoto, S. (eds.) ISMIS 2005. LNCS (LNAI), vol. 3488, pp. 323–331. Springer, Heidelberg (2005)

Galstyan, A., Cohen, P.R.: Is guilt by association a bad thing? In: Proceedings of the First International Conference on Intelligence Analysis (2005)

Macskassy, S.A., Provost, F.: A simple relational classifier. In: 2nd Workshop on Multi-relational Data Mining (MRDM) at KDD 2003 (August 2003)

Macskassy, S.A., Provost, F.: Suspicion scoring based on guilt-by-association, collective inference, and focused data access. In: Proceedings of the First International Conference on Intelligence Analysis (2005)

McCallum, A., Nigam, K.: A comparison of event models for naive Bayes test classification. In: AAAI 1998 Workshop on Learning for Text Categorization (July 1998)

McCallum, A., Nigam, K., Rennie, J., Seymore, K.: Automating the construction of Internet portals with machine learning. Information Retrieval 3, 127–163 (2000)

Mierswa, I., Wurst, M., Klinkenberg, R., Scholz, M., Euler, T.: YALE: Rapid Prototyping for Complex Data Mining Tasks. In: Proceedings of the 12th ACM SIGKDD International Conference on Knowledge Discovery and Data Mining (KDD 2006) (August 2006)

Pearl, J.: Probabilistic Reasoning in Intelligent Systems. Morgan Kaufmann, San Mateo (1988)

Taskar, B., Abbeel, P., Koller, D.: Discriminative probabilistic models in relational data. In: Proceedings of the 18th conference on Uncertainty in Artificial Intelligence (UAI 2002), August 2002, pp. 485–492 (2002)

Taskar, B., Segal, E., Koller, D.: Probabilistic classification and clustering in relational data. In: Proceedings of IJCAI-01, 17th International Joint Conference on Artificial Intelligence, August 2001, pp. 870–876 (2001)

On a Probabilistic Combination of Prediction Sources

Ioannis Rousidis, George Tzagkarakis, Dimitris Plexousakis, and Yannis Tzitzikas

Institute of Computer Science, FORTH, Heraklion, Greece
{rousidis,gtzag,dp,tzitzik}@ics.forth.gr

Abstract. Recommender Systems (RS) are applications that provide personalized advice to users about products or services they might be interested in. To improve recommendation quality, many hybridization techniques have been proposed. Among all hybrids, the weighted recommenders have the main benefit that all of the system's constituents operate independently and stand in a straightforward way over the recommendation process. However, the hybrids proposed so far consist of a linear combination of the final scores resulting from all recommendation techniques available. Thus, they fail to provide explanations of predictions or further insights into the data. In this work, we propose a theoretical framework to combine information using the two basic probabilistic schemes: the sum and product rule. Extensive experiments have shown that our purely probabilistic schemes provide better quality recommendations compared to other methods that combine numerical scores derived from each prediction method individually.

Keywords: Recommender Systems, Collaborative Filtering, Personalization, Data Mining.

1 Introduction

Nowadays, most of the popular commercial systems use collaborative filtering (CF) techniques to efficiently provide recommendations to users based on opinions of other users [7][9][10]. To effectively formulate recommendations, these systems rely either upon statistics (user ratings) or upon contextual information about items.

CF, which relies on statistics, has the benefit of learning from information provided by a user and other users as well. However, RS could suffer from the sparsity problem: accurate recommendations cannot be provided unless enough information has been gathered. Other problems and challenges that RS have to tackle include: (a) the user bias from rating history (statistics), (b) the "gray sheep" problem, where a user cannot match with anyone of the other users' cliques, and (c) the cold-start problem, where a new item cannot be recommended due to the lack of any information. These problems reduce the strength of statistics-based methods. On the other hand, if RS rely merely on the content of the items, then they tend to recommend only items with content similar to those already rated by a user.

The above observations indicate that in order to provide high quality recommendations we need to combine all sources of information that may be available. To this

A. An et al. (Eds.): ISMIS 2008, LNAI 4994, pp. 535–544, 2008.

end, we propose a purely probabilistic framework introducing the concept of uncertainty with respect to the accurate knowledge of the model.

Recently, hybridization of recommendation techniques has been an interesting topic, since recommenders have different strengths over the space. In [4], two main directions on combining recommenders are presented: the first combines them in a row by giving different priorities to each one and passing the results of one as an input to the other while the second applies all techniques equally and finds a heuristic to produce the output. Each hybrid has its tradeoff. According to [4], the latter hybrids, especially the weighted recommenders, have the main benefit that all of the system's capabilities are brought to bear on the recommendation process in a straightforward way and it is easy to perform post-hoc credit assignment and adjust the hybrid accordingly. Previous works in this area [2][5][11][15][16][17] performed merging of recommendation sources as a naïve linear combination of the numerical results provided by each recommendation technique individually. In general, these approaches are not capable of providing explanations of predictions or further insights into the data. Our approach differs from the others in that the combination of distinct information sources has a pure and meaningful probabilistic interpretation, which may be leveraged to explain, justify and augment the results.

The paper is organized as follows: in Section 2, we present background theory on the predictions techniques to be used. In Section 3, we introduce the two basic probabilistic schemes for combining information sources. In Section 4, we evaluate the models resulting from our framework to conclude our work in Section 5.

2 Prediction Techniques

Many approaches for CF have been previously proposed, each of which treats the problem from a different angle, and particularly by measuring similarity between users [3][7][13] or similarity between items [6][14]. Heuristics, such as *k-nearest neighbors* (KNN), have been used when the existence of common ratings between users is required in order to calculate similarity measures. Thus, users with no common items will be excluded from the prediction procedure. This could result in a serious degradation of the coverage of the recommendation, that is, the number of items for which the system is able to generate personalized recommendations could decrease.

In a recent work [12], a hybrid method combining the strengths of both model-based and memory-based techniques outperformed any other pure memory-based as well as model-based approach. The so-called *Personality Diagnosis* (PD) method is based on a simple probabilistic model of how people rate titles. Like other model-based approaches, its assumptions are explicit, and its results have a meaningful probabilistic interpretation. Like other memory-based approaches it is fairly straightforward, operating over all data, while no compilation step is required for new data. The following section contains a description of the PD algorithm; moreover we provide an extension of this approach to an item-based and a content-based direction.

2.1 Personality Diagnosis

PD states that each user u_i, where $i = 1,2,\ldots,m$, given any rating information over the objects available, has a *personality type* which can be described as.

$$P_{u_i}^{true} = \left\{ r_{i,1}^{true}, r_{i,2}^{true}, \ldots, r_{i,n}^{true} \right\} \tag{1}$$

where P_{ui}^{true} is user's u_i vector of "*true*" ratings $r_{i,j}^{true}$ over observed objects o_j. These ratings encode users' underlying, internal preferences. Besides, we assume the existence of a critical distinction between the true and reported ratings. In particular, the true ratings $P_{u_i}^{true}$ cannot be accessed directly by the system, while the reported ratings, which are provided to the system, constitute the only accessible information. In our work, we consider that these ratings include Gaussian noise based on the fact that the same user may report different ratings depending on different occasions, such as the mood, the context of other ratings provided in the same session or on any other reason - external factor. All these factors are summarized as a Gaussian noise. Working in a statistical framework, it is assumed that a user's u_i actual rating $r_{i,j}$ over an object o_j, is drawn from an independent normal distribution with mean $r_{i,j}^{true}$, which represents the true rating of the ith user for the jth object. Specifically:

$$\Pr\left(r_{i,j} = x \mid r_{i,j}^{true} = y \right) \propto e^{-(x-y)^2 / 2\sigma^2} \tag{2}$$

where $x, y \in \left\{ 1, \ldots, |r| \right\}$ and $|r|$ denotes the number of possible rating values. It is further assumed that the distribution of rating vectors (personality types), which are contained in the rating matrix of the database, is representative of the distribution of personalities in the target population of users. Based on this, the prior probability $\Pr\left(P_{u_a}^{true} = \kappa \right)$ that the active user u_a rates items according to a vector κ, is given by the frequency that other users rate according to κ. So, instead of counting occurrences explicitly, a random variable $P_{u_a}^{true}$ is defined which takes one out of m possible values, $P_{u1}^{true}, P_{u2}^{true}, \ldots, P_{um}^{true}$ each one with equal probability $1/m$. Thus, given the ratings of a user u_a, we can apply Bayes' rule to calculate the probability that he is of the same personality type as any other user u_i, with $i \neq a$:

$$\begin{aligned} &\Pr\left(P_{u_a}^{true} = P_{u_i}^{true} \mid r_{a,1} = x_1, \ldots, r_{a,n} = x_n \right) \\ &\propto \Pr\left(r_{a,1} = x_1 \mid r_{a,1}^{true} = r_{i,1} \right) \ldots \Pr\left(r_{a,n} = x_n \mid r_{a,n}^{true} = r_{i,n} \right) \cdot \Pr\left(P_{u_a}^{true} = P_{u_i}^{true} \right) \end{aligned} \tag{3}$$

Once we have computed this quantity for each user u_i, we can find the probability distribution of user's u_a rating for an unobserved object o_j, as follows:

$$\begin{aligned} &\Pr\left(r_{a,j} = x_j \mid r_{a,1} = x_1, \ldots, r_{a,n} = x_n \right) \\ &\propto \sum_{i=1}^{m} \Pr\left(r_{a,j} - x_j \mid r_{a,j}^{true} = r_{i,j} \right) \cdot \Pr\left(P_{u_a}^{true} = P_{u_i}^{true} \mid r_{a,1} = x_1, \ldots, r_{a,n} = x_n \right) \end{aligned} \tag{4}$$

where $r_{a,j} \in \left\{ 1, \ldots, |r| \right\}$. The algorithm has a time and space complexity of the order $O(mn)$, as do the memory-based methods. According to the PD method, the observed ratings can be thought of as "*symptoms*", while each personality type, whose probability to be the cause we examine, as a "*disease*".

2.2 Feature Diagnosis

If we rotate the rating matrix by $90°$ we may consider the problem of recommendation formulation from another point of view introducing the notion of *Feature Diagnosis* (FD). Based on that, for any object o_i, where $i = 1, 2, \ldots, n$, and given any rating information from users available, a type of features can be described as:

$$F_{o_i}^{true} = \{ r_{1,i}^{true}, r_{2,i}^{true}, \ldots, r_{m,i}^{true} \} \tag{5}$$

where $F_{o_i}^{true}$ is object's o_i vector of "true" ratings $r_{j,i}^{true}$ derived from users u_j. Thus, here we assume that these ratings include Gaussian noise based on the fact that ratings of the same user on different items may be temporally related (i.e., if their popularities behave similarly over time). For example, during the period near St. Valentines' day, romance movies may be more popular than movies about war. All these factors are summarized as a Gaussian noise. These ratings encode object's underlying, internal type of features. As in PD, it is again assumed that the distribution of rating vectors (feature types) is representative of the distribution of features in the target population of objects. So, instead of counting occurrences explicitly, a random variable F_{oa}^{true} is defined that takes one out of n possible values, $F_{o1}^{true}, F_{o2}^{true}, \ldots, F_{on}^{true}$, each one with equal probability $1/n$. Finally, given the ratings of an object o_a, we can apply Bayes' rule to calculate the probability to be of the same feature type as any object o_i, with $i \neq a$:

$$\begin{aligned} &\Pr\left(F_{oa}^{true} = F_{oi}^{true} \mid r_{1,a} = x_1, \ldots, r_{m,a} = x_m \right) \\ &\propto \Pr\left(r_{1,a} = x_1 \mid r_{1,a}^{true} = r_{1,i} \right) \ldots \Pr\left(r_{m,a} = x_m \mid r_{m,a}^{true} = r_{m,i} \right) \cdot \Pr\left(F_{oa}^{true} = F_{oi}^{true} \right) \end{aligned} \tag{6}$$

Once we have computed this quantity for each object o_i, we can find the probability distribution of user's u_j rating for an unobserved object o_a using the following expression:

$$\begin{aligned} &\Pr\left(r_{j,a} = x_j \mid r_{1,a} = x_1, \ldots, r_{m,a} = x_m \right) \\ &\propto \sum_{i=1}^{n} \Pr\left(r_{j,a} = x_j \mid r_{j,a}^{true} = r_{j,i} \right) \cdot \Pr\left(F_{oa}^{true} = F_{oi}^{true} \mid r_{1,a} = x_1, \ldots, r_{m,a} = x_m \right) \end{aligned} \tag{7}$$

According to FD, the observed ratings can be thought of as "*symptoms*", while the features type as "*populations*" where symptoms may develop. The algorithm has a time and space complexity of the order $O(mn)$.

2.3 Context Diagnosis

The context of the objects, e.g. for a movie recommender any textual information on genres, can also provide useful information for recommendations. For this purpose, we define the following context vector.

$$C_{o_i}^{true} = \{ c_{i,1}^{true}, c_{i,2}^{true}, \ldots, c_{i,k}^{true} \} \tag{8}$$

where $C_{o_i}^{true}$ is the "true" context type of the object o_i according to k categories. We assume that the probability of two objects to be of the same context type, taking into account the categories in which they belong, can be derived by associating their context vectors. We calculate this probability with the following expression:

$$\Pr\left(C_{oa}^{true} = C_{oi}^{true} \mid c_{a,1}, c_{a,2}, \ldots c_{a,k} \right) \propto \frac{\left| C_{o_a} \cap C_{o_i} \right|}{\max\left(\left| C_{o_a} \right|, \left| C_{o_i} \right| \right)} \cdot \Pr\left(C_{oa}^{true} = C_{oi}^{true} \right) \tag{9}$$

where $c_{o,i}$ defines the membership of object o to category i (e.g. a 0 or 1 in an item-category bitmap matrix). The distribution of the category vectors (context types) of the objects which is available in the category matrix of the database is assumed to be representative of the distribution of context types in the target population of objects. Assuming again equal probability $1/n$, we can find the probability distribution of user's u_j rating for an unobserved object o_a based upon its context type is as:

$$\Pr\left(r_{j,a}=x_j \mid c_{a,1},c_{a,2},\ldots c_{a,k}\right) \propto \sum_{i=1}^{n}\Pr\left(r_{j,a}=x_j \mid r_{j,a}^{true}=r_{j,i}\right)\cdot\Pr\left(C_{o_a}^{true}=C_{o_i}^{true} \mid c_{a,1},c_{a,2},\ldots c_{a,k}\right) \tag{10}$$

The algorithm has a time and space complexity of the order *O(n)*, considering the number k of all categories available as a constant. According to *Context Diagnosis* (CD), the observed ratings can be thought of as "*symptoms*" which may be developed in certain "*categories*" defined by a gamut of contextual attributes.

3 Combination Strategies

In probability theory, we can find two basic combinatorial schemes for the combination of distinct information sources, namely, the product-rule and the sum–rule. In this section, we show how this theory can be applied in our case where we aim to combine the prediction techniques presented in Section 2. The described framework is purely probabilistic and we argue this is the major advantage compared to the previous works. The traditional combination used widely so far is also presented.

3.1 Product Rule

According to the product rule, we assume that the three types of information used to make predictions are independent. We apply Bayes' rule assuming that the probability of a rating value to be the predicted value is conditioned on the ratings of the user, the ratings of the object and the categories that the object belongs to, thus:

$$\Pr\left(r_{i,j} \mid P_{u_i}^{true},F_{o_j}^{true},C_{o_j}^{true}\right)=\frac{\Pr\left(P_{u_i}^{true},F_{o_j}^{true},C_{o_j}^{true} \mid r_{i,j}\right)\Pr\left(r_{i,j}\right)}{\Pr\left(P_{u_i}^{true},F_{o_j}^{true},C_{o_j}^{true}\right)} \tag{11}$$

On the equation above we neglect the denominator, which is the unconditional measurement of the joint probability density, since it is common to all rating values $r_{i,j}$ which we also consider to have equal probability. Thereby we focus only on the first term of the numerator which represents the conditional joint probability measurement distribution extracted by all the "true" vectors. We initially assumed that these vectors are conditionally independent, so:

$$\Pr\left(P_{u_i}^{true},F_{o_j}^{true},C_{o_j}^{true} \mid r_{i,j}\right)=\Pr\left(P_{u_i}^{true} \mid r_{i,j}\right)\Pr\left(F_{o_j}^{true} \mid r_{i,j}\right)\Pr\left(C_{o_j}^{true} \mid r_{i,j}\right) \tag{12}$$

Finally, by applying Bayes' rule to each one of the factors of Eq(12) we obtain the probability of a rating value as:

$$\Pr\left(r_{i,j} \mid P_{u_i}^{true},F_{o_j}^{true},C_{o_j}^{true}\right)\propto\Pr\left(r_{i,j} \mid P_{u_i}^{true}\right)\Pr\left(r_{i,j} \mid F_{o_j}^{true}\right)\Pr\left(r_{i,j} \mid C_{o_j}^{true}\right) \tag{13}$$

The argument that maximizes this expression indicates the rating that user u_i is most likely to assign to object o_j.

3.2 Sum Rule

In order to combine both PD and FD we introduce a binary variable B that refers to the relative influence of each method. When B is equal to 1, the prediction comes only from user's rating vector, while when B is equal to 0 indicates full dependency on the object's rating vector. Under these assumptions, the conditional probability can be computed by marginalization on the binary variable B. Therefore, the probability distribution of objects o_j rating by user u_i is given by

$$\begin{aligned}\Pr\left(r_{i,j} \mid P_{u_i}^{true}, F_{o_j}^{true}\right) &= \sum_B \Pr\left(r_{i,j} \mid P_{u_i}^{true}, F_{o_j}^{true}, B\right)\Pr\left(B \mid P_{u_i}^{true}, F_{o_j}^{true}\right) \\ &= \Pr\left(r_{i,j} \mid P_{u_i}^{true}, F_{o_j}^{true}, B=1\right)\Pr\left(B=1 \mid P_{u_i}^{true}, F_{o_j}^{true}\right) \\ &+ \Pr\left(r_{i,j} \mid P_{u_i}^{true}, F_{o_j}^{true}, B=0\right)\Pr\left(B=0 \mid P_{u_i}^{true}, F_{o_j}^{true}\right)\end{aligned} \quad (14)$$

By definition, $r_{i,j}$ is independent from user's ratings when $B=0$ so $\Pr(r_{i,j} \mid P_{u_i}^{true}, F_{o_j}^{true}, B=0) = \Pr(r_{i,j} \mid F_{o_j}^{true})$. The opposite holds when $B=1$, that is, $\Pr(r_{i,j} \mid P_{u_i}^{true}, F_{o_j}^{true}, B=1) = \Pr(r_{i,j} \mid P_{u_i}^{true})$. If we use a parameter ϑ to denote the probability $\Pr(B=1 \mid P_{u_i}^{true}, F_{o_j}^{true})$ we have:

$$\Pr\left(r_{i,j} \mid P_{u_i}^{true}, F_{o_j}^{true}\right) = \Pr\left(r_{i,j} \mid P_{u_i}^{true}\right)\vartheta + \Pr\left(r_{i,j} \mid F_{o_j}^{true}\right)(1-\vartheta) \quad (15)$$

To include any contextual information about the object in our conditional hypothesis we introduce another binary variable which takes the values 0 when the prediction depends solely on ratings and 1 when it relies only on the context. So, by marginalizing the binary variable as before and using a new parameter δ, we obtain:

$$\begin{aligned}&\Pr\left(r_{i,j} \mid P_{u_i}^{true}, F_{o_j}^{true}, C_{o_j}^{true}\right) \\ &= \left(\Pr\left(r_{i,j} \mid P_{u_i}^{true}\right)\vartheta + \Pr\left(r_{i,j} \mid F_{o_j}^{true}\right)(1-\vartheta)\right)(1-\delta) + \Pr\left(r_{i,j} \mid C_{o_j}^{true}\right)\delta\end{aligned} \quad (16)$$

The argument that maximizes the above expression indicates the rating that user u_i is most likely to assign to the object o_j.

3.3 Score Combination

So far, in most of the previously developed systems, the merging of information sources is carried out by a naïve linear combination of numerical scores resulting from each prediction technique individually in order to give a single prediction. In our case, where we combine scores from three different sources, the prediction is calculated as follows:

$$p_{i,j} = \begin{pmatrix} \arg\max_r \left(\Pr\left(r_{i,j} \mid P_{u_i}^{true}\right)\right)\vartheta + \\ \arg\max_r \left(\Pr\left(r_{i,j} \mid F_{o_j}^{true}\right)\right)(1-\vartheta) \end{pmatrix}(1-\delta) + \arg\max_r \left(\Pr\left(r_{i,j} \mid C_{o_j}^{true}\right)\right)\delta \quad (17)$$

4 Experimental Evaluation

We carried out our experiments using the MovieLens dataset, taken from a research recommendation site being maintained by the GroupLens project [13]. The MovieLens dataset contains 100.000 ratings, scaling from 0 to 5, derived from 943 users on 1682 movie titles (items) where each user has rated at least 20 movies. We first carried out some experiments to tune the weighting parameters ϑ and δ and then using selected values of them we tested our framework along with other algorithms. Metrics used to evaluate the quality of recommendations are the Mean Absolute Error (MAE) and the F1, which is the harmonic mean of precision and recall.

4.1 Configuration

Parameter ϑ adjusts the balance between PD and FD prediction techniques (we denote this combination with PFD), while parameter δ adjusts the balance between PFD and CD (henceforth PFCD). We vary each user's number of observed items as well as each item's number of raters to find the best possible configurations for our combination schemes.

First, we use the MAE metric to examine the sensitivity of both schemes over ϑ. For this purpose, we set the value of δ to zero. Then, we vary ϑ from zero (pure FD) to one (pure PD). We test over user-sparsity 5 and 20, and item sparsity less than 5 and less than 20. Regarding the sum-rule scheme, as Fig. 1a shows, the best results are achieved for values of ϑ between 0.3 and 0.7. For this range of values the prediction accuracy can be improved up to the 8% of the technique with the best accuracy when it is used

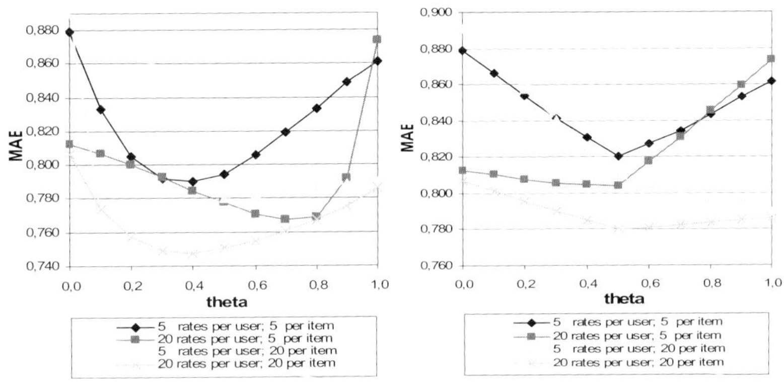

(a) sum-rule (b) score combination

Fig. 1. Impact of parameter ϑ in sum-rule and score combination schemes

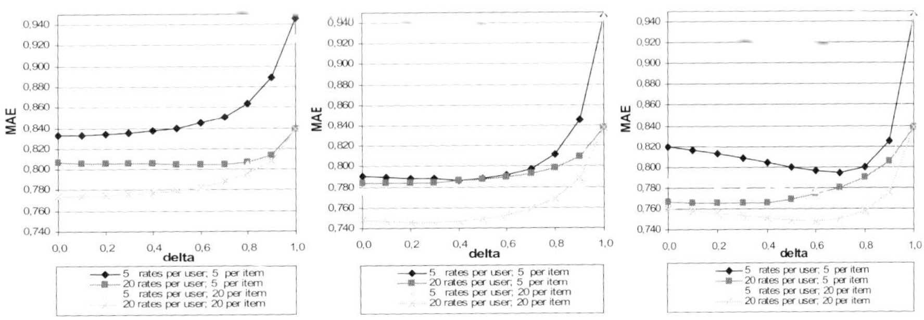

(a) $\vartheta = 0.1$ (b) $\vartheta = 0.4$ (c) $\vartheta = 0.7$

Fig. 2. Impact of parameter δ for different values of ϑ in sum-rule individually. In Fig. 1b, for score combination scheme, we obtain optimum results for values of ϑ greater than 0.5 except the third configuration since no combination of PD and FD seems to give a better MAE than PD itself.

In Fig. 2, for the sum-rule scheme, we assign to ϑ values 0.1 (denoted with PFCDs_1), 0.4 (PFCDs_2) and 0.7 (PFCDs_3) and test the sensitivity with respect to the parameter δ. Using the same configurations over user and item sparsity we vary δ from zero (pure memory-based) to one (pure content-based). Figs. 2a and 2b, for PFCDs_1 and PFCDs_2 respectively, show no clear improvement of MAE over δ. As for PFCDs_3 (Fig. 2c), we obtain the optimum results for values of δ between 0.2 and 0.8, which improve MAE almost by 4%. Based on the above observations, we tune δ to 0.1 in PFCDs_1, 0.7 in PFCDs_2 and 0.6 in PFCDs_3 to further experiment with their overall performance.

For the score combination schemes we set the value of ϑ to be equal to 0.5 (PFCDn_1) and 0.8 (PFCDn_2) and test the sensitivity regarding parameter δ. Using the same configurations over user and item sparsity we vary δ from zero (pure memory-based) to one (pure content-based). As shown in Fig. 3, using PFCDn_1 and PFCDn_2 in the recommendation process does not seem to improve the quality of prediction. Some exceptions are the sparsity configurations in which only 5 item votes are kept throughout the recommendation process (first and second configuration). After these observations we set δ to 0.3 in PFCDn_1 and 0.4 in PFCDn_2.

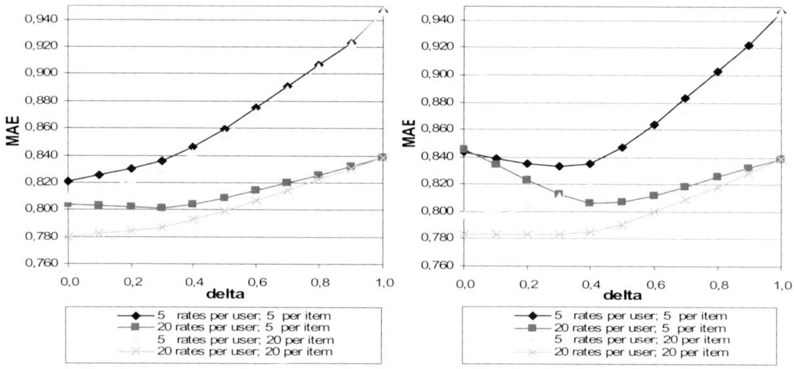

(a) $\vartheta = 0.5$ (b) $\vartheta = 0.8$

Fig. 3. Impact of parameter δ for different values of ϑ in score combination

4.2 Overall Performance

In this section, we randomly select parts of the training data and test the previous configurations along with the product-rule (we denote with PFCDp) and, moreover, with other memory-based algorithms as are the user (UBPCC) and item (IBPCC) Pearson correlation coefficient in terms of overall sparsity. The results in Table 1 indicate the superiority of the purely probabilistic schemes against the naïve score combination schemes with respect to the quality of the prediction accuracy (MAE and F1). More specifically, the purely probabilistic schemes can provide better results, up to 10%. However, this conclusion does not stand for every pair of parameters ϑ and δ - e.g., as shown in Table 1, score combination scheme PFCDn_2 outperformed the purely probabilistic scheme PFCDs_1 in terms of F1.

The results, finally, prove our initial assumption about KNN algorithms; in particular, they require the existence of common items between users. This is why the F1 metric has a decreased value for the methods UBPCC and IBPCC.

Table 1. MAE and F1 over different sparsity levels

	MAE					F1				
	20%	**40%**	**60%**	**80%**	**100%**	**20%**	**40%**	**60%**	**80%**	**100%**
PFCDs_1	0,859	0,817	0,789	0,783	0,781	0,631	0,669	0,681	0,692	0,695
PFCDs_2	0,821	0,788	0,778	0,767	0,764	0,673	0,704	0,715	0,722	0,723
PFCDs_3	0,807	0,776	0,765	0,761	0,758	0,683	0,713	0,721	0,727	0,729
PFCDn_1	0,918	0,866	0,841	0,828	0,823	0,612	0,653	0,656	0,676	0,681
PFCDn_2	0,878	0,827	0,810	0,793	0,790	0,647	0,683	0,693	0,705	0,711
PFCDp	0,812	0,780	0,771	0,760	0,758	0,677	0,709	0,719	0,725	0,727
PD	0,884	0,838	0,815	0,806	0,797	0,641	0,681	0,693	0,701	0,707
UBPCC	1,022	0,904	0,866	0,844	0,830	0,159	0,419	0,483	0,505	0,511
IBPCC	0,999	0,895	0,852	0,836	0,824	0,181	0,407	0,473	0,495	0,507

5 Discussion

In this paper, we proposed the use of the two basic combination schemes from the theory of probabilities in order to overcome accuracy issues of the RS. Results showed that purely probabilistic schemes provide better quality results than naïve linear weighting of scores derived from all techniques individually. However, the results are very sensitive to the tuning parameters - it is not clear at all how to set theta and delta in a robust way. Moreover, it is worth noticing that in most cases the product–rule which requires no tuning was outperformed slightly by the sum–rule in its best configuration (i.e. PFCDs_3). The main reason is the sensitivity in errors, which is intense in the latter case due to the factorization of prediction techniques; i.e., independence of the techniques does not always hold. For more details we refer to [8]. It is also important to notice that the combination of more than two prediction techniques does not always improve the output. Since a RS consists of a voting system we believe that this observation is related to the Arrow's Paradox [1]. Our future study will also take into account this issue.

References

1. Arrow, K.J.: Social Choice and Individual Values. Ph.D. Thesis, J. Wiley, NY (1963)
2. Billsus, D., Pazzani, M.: User Modeling for Adaptive News Access. User-Modeling and User-Adapted Interaction 10(2-3), 147–180 (2000)
3. Breese, J.S., Heckerman, D., Kadie, C.: Empirical Analysis of Predictive Algorithms for Collaborative Filtering. In: Proceedings of the 14th Conference on Uncertainty in Artificial Intelligence (UAI 1998), July 1998, pp. 43–52 (1998)

4. Burke, R.: Hybrid Recommender Systems: Survey and Experiment. User Modeling and User-Adapted Interaction 12(4), 331–370 (2002)
5. Claypool, M., Gokhale, A., Miranda, T., Murnikov, P., Netes, D., Sartin, M.: Combining Content-Based and Collaborative Filters in an Online Newspaper. In: SIGIR 1999 Workshop on Recommender Systems: Algorithms and Evaluation, Berkeley, CA (1999)
6. Deshpande, M., Karypis, G.: Item-based Top-N Recommendation Algorithms. ACM Trans. Inf. Syst. 22(1), 143–177 (2004)
7. Herlocker, J.L., Konstan, J.A., Borchers, A., Riedl, J.: An Algorithmic Framework for Performing Collaborative Filtering. In: Proc. of SIGIR (1999)
8. Kittler, J., Hatef, M., Duin, R.P.W., Matas, J.: On Combining Classifiers. IEEE Trans. Pattern Anal. Mach. Intell. 20(3), 226–239 (1998)
9. Linden, G., Smith, B., Smith, J.X.: Amazon.com Recommendations: Item-to-item collaborative filtering. IEEE Internet Computing, 76–80 (January/February 2003)
10. Papagelis, M., Plexousakis, D., Kutsuras, T.: A Method for Alleviating the Sparsity Problem in Collaborative Filtering Using Trust Inferences. In: Proceedings of the 3rd International Conference on Trust Management (2005)
11. Pazzani, M.J.: A Framework for Collaborative, Content-Based and Demographic Filtering. Artificial Intelligence Review 13(5/6), 393–408 (1999)
12. Pennock, D.M., Horvitz, E.: Collaborative Filtering by Personality Diagnosis: A Hybrid Memory- and Model-based Approach. In: Proceedings of UAI (2000)
13. Resnick, P., Iacovou, N., Suchak, M., Bergstrom, P., Riedl, J.: GroupLens: An Open Architecture for Collaborative Filtering of Netnews. In: CSCW 1994: Proceedings of the 1994 ACM Conference on Computer Supported Cooperative Work, Champal Hill, North Carolina, United States, pp. 175–186. ACM Press, New York (1994)
14. Sarwar, B., Karypis, G., Konstan, J., Riedl, J.: Item-based Collaborative Filtering Recommendation Algorithms. In: WWW 2001: Proceedings of the 10th International Conference on World Wide Web, pp. 285–295. ACM Press, Hong Kong (2001)
15. Tran, T., Cohen, R.: Hybrid Recommender Systems for Electronic Commerce. In: Knowledge-Based Electronic Markets, Papers from the AAAI Workshop, AAAI Technical Report WS-00-04. pp. 78–83. AAAI Press, Menlo Park (2000)
16. Wang, J., de Vries, A.P., Reinders, M.J.: A User-Item Relevance Model for log-based Collaborative Filtering. In: Lalmas, M., MacFarlane, A., Rüger, S.M., Tombros, A., Tsikrika, T., Yavlinsky, A. (eds.) ECIR 2006. LNCS, vol. 3936, Springer, Heidelberg (2006)
17. Wang, J., de Vries, A.P., Reinders, M.J.: Unifying User-based and Item-based Collaborative Filtering Approaches by Similarity Fusion. In: Proceedings of SIGIR (2006)

Effective Document-Oriented Telemetry Data Compression

David Maluf*, Chen-jung Hsu, Peter Tran*, and David Tran

NASA Ames Research Center, Mail Stop 269-4, Moffett Field, CA- 94087
{david.a.maluf,Chen-Jung.Hsu-1,Peter.B.Tran}@nasa.gov,
davetran@stanford.edu

Abstract. Storing vast amounts of multidimensional telemetry data presents a challenge. Telemetry data being relayed from sensors to the ground station comes in the form of text, images, audio, and various other formats. Compressing this data would optimize bandwidth usage during transmission and reduce storage resources needed at the ground level. The application of a single compression technique for all data types usually yields ineffective results. We will present a telemetry data compression algorithm that utilizes Discrete Fourier Transforms (DFTs) along with different compression algorithms for different data types, including Lempel-Ziv-Welch (LZW) and Flate for textual and numerical data and JPEG coding for images. Although these algorithms do not yield the greatest compression ratios, the Portable Document Format (PDF) standard supports decoding of all of them, which allows us to write our encoded data streams directly to a PDF file. This approach alleviates the need for traditional database storage systems. It also standardizes and simplifies the data retrieval, decoding, and viewing process. This work results in packets-oriented telemetry data encapsulated with multiple compression stream algorithms, which can be decoded, rendered and viewed by any standard PDF viewer. This paper presents the aforementioned algorithms and its development status as applicable proof-of-concept prototypes.

1 Introduction

We are currently developing a high performing ground data system utilizing a telemetry data compression technique for future space exploration and satellite applications. The algorithm combines a number of lossless compression algorithms such as LZW and Huffman coding. We use single dimensional DFTs to transform single dimensional streamed data from their respective sensors (e.g. latitude, longitude, temperature, pressure, etc.), optimizing on their periodicity.

We apply the basic principle of applying different compression algorithms to different data types to try to achieve a balance between compression ratio and data precision. Similar approaches have gained popularity in document-oriented data, such as Adobe PDF format [5], where different data streams are compressed differently.

* Corresponding author.

A. An et al. (Eds.): ISMIS 2008, LNAI 4994, pp. 545–551, 2008.

Telemetry packets are routed and stored at the kilobyte and megabyte scale, alleviating the need of traditional database storage requirements. Paging through the telemetry would require at most two packets for a continuity of the data stream. The compression scheme performs well on a suite of test telemetry data acquired from spacecraft instruments. It applies to two dimensional data images. Continuous data are truncated and optimized either towards arbitrary packet size or in signal resets. Ground system implementations are currently in the development phase.

The scope of this paper applies to packet-oriented telemetry data. Packet telemetry sends measurements at particular frequencies in bursts, whereas frame-based telemetry accumulates many measurements over time [1]. Frame-based transmission follows a fixed structure to protect against transmission errors [2]. However, with improved transmission technologies, frames become antiquated because packets offer much more flexibility both in the structure of the data as well as for transmission purposes. Packet data structure should adhere to the Space Packet Protocol standard established by the Consultative Committee for Space Data Systems (CCSDS) in 2003 [9]. Our algorithm adheres to the CCSDS packet telemetry standard, which recommends lossless. Figure 1 shows the various stages of the packet telemetry data system as defined in [8] with implementation notes about our algorithm.

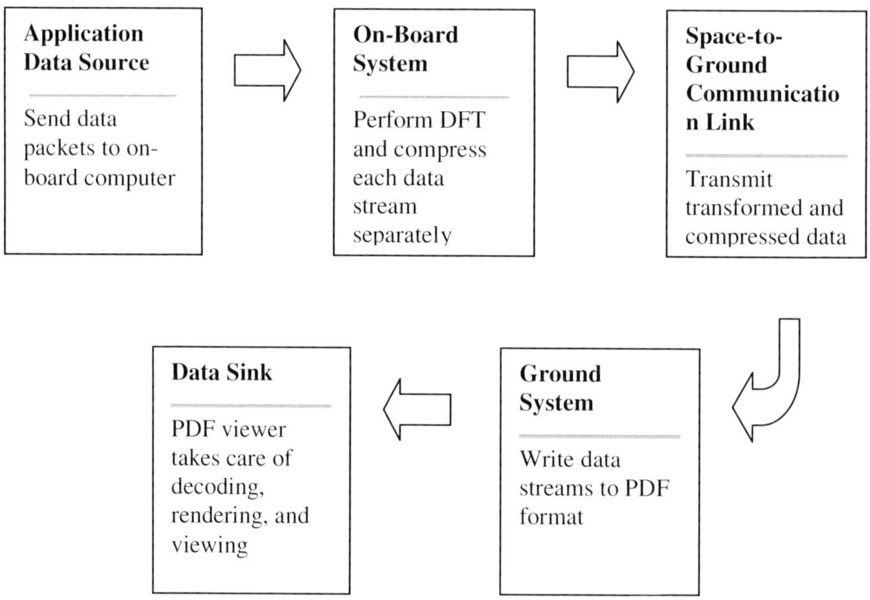

Fig. 1. Application of compression technique at different stages of the Packet Telemetry Data System *adapted from [8], pg. 2-1

We propose a document-oriented telemetry data compression technique that uses different compression algorithms for different data types. Adhering to compression algorithms supported by the Portable Document Format (PDF) convention simplifies and standardizes the presentation of the telemetry data. This process ensures that users viewing the data on different machines will see the same information because the rendering of the telemetry data in PDF document viewers is pre-specified.

2 Objectives

This paper introduces a telemetry data compression technique that combines Discrete Fourier Transforms with lossless compression algorithms such as Huffman and LZW in multiple compression streams. Adobe PDF specifications (version 1.2 or higher) [5] supports decoding of these algorithms, so any standard PDF document viewer will be able to decode and render the original data. The specific objectives of this algorithm are:

- Achieve better compression ratios while maintaining the necessary data precision by applying different encoding algorithms to different data streams.
- Further reduce the necessary storage resources required by using lossless coding techniques such as LZW and Huffman encoding and lossy techniques for images using the JPEG standard.
- Allow for efficient decoding and viewing of the original data by adhering to Adobe PDF standards, enabling the user to render and view the data in any PDF document viewer.

3 Sampling and Transform

The on-board data system processes streaming data from the application data source and truncates the continuous data into discrete samples either based on signal resets or to achieve a certain packet size. The sampling interval can be user-defined to be fit the needs of the telemetry data being measured. Previous works on telemetry data compression justifies that sampling continuous telemetry data, when done appropriately, can maintain the accuracy of the original data because it tends to be oversampled [6]. Because our algorithm focuses on the ground implementation and be paired with any sampling algorithm, the process of choosing an appropriate sampling technique, which deserves its own discussion, falls outside the scope of this paper. One such implementation is discussed in [6], which introduces an adaptive sampling technique that changes the sampling rate to try to reduce autocorrelations (redundancies) found in the data. Henceforth, we will assume that an appropriate sampling rate for the data has been chosen.

Performing one-dimensional DFTs on packets of telemetry data takes advantage of the periodicity of the stream and eliminates redundancy in the data. Multidimensional telemetry data can be decomposed into separate data streams, which can be analyzed and compressed using different algorithms. Moreover, single dimensional data streams can be transformed to the frequency domain with greater computational efficiency by eliminating the need for multi-dimensional DFTs. This technique can be applied to two-dimensional M x N images represented as a 1 x MN matrix as long as we separately save its dimensions separately. The use of the Fast Fourier Transform further optimizes the computation, allowing for $N\log_2 N$ complexity [4].

Two packets are required to be able to reconstruct the original data: the signal data containing the DFT coefficients, and secondly, the phase data, which is not stored in the frequency domain of the DFT. Before sending the packets through the space-to-ground communications link, the on-board system separates the signal data in the DFT domain and the corresponding phase data, both as single dimensional matrices of values.

4 Multiple Compression Techniques

After the DFT, depending on the data type (e.g text, numeric, image, audio, video), this technique applies different methods of compression to achieve optimal compression rates. For text, lossless encoding techniques such as LZW or FLATE are applied, and for images, JPEG encoding is used because lossiness can generally be tolerated and it is supported by the PDF specification. The compression process within JPEG encoding is analogous to our compression for Support for additional encoding techniques supported by the PDF standard (Run-length Encoding, CCITTFax, etc.) may be added later.

Although Huffman and LZW coding achieve less optimal compression rates than Golomb-Rice and other forms of lossless encoding, Adobe PDF supports decoding of both LZW and Flate, a compression technique that combines adaptive Huffman encoding and LZW [5].

Positional and velocity information in the form of textual data is followed by image data in the stream coming to the on-board system from the application data source. Fr the textual data, we apply a DFT then LZW or Flate coding. Analogously, we apply a Discrete Cosine Transform (DCT) for the JPEG and encoding using Huffman; however, we must first convert the raw image data from the RGB color space to the YCbCr subspace and perform chrominance sub-sampling; this step takes advantage of the human eye's greater relative sensitivity to luminance than chrominance. After we apply a DCT, we quantize the transform coefficients before applying Huffman coding.

By the end of all of the steps, our data stream will consist of encoded text following by a JPEG image, both ready to be passed through the space link and then written directly to a PDF-like document at the ground systems level. Along with this data stream, we must either send along another stream containing information that the PDF Decoder filters will need to decompress the data, such as which compression or transforms we used and the size of the streamed data.

5 Leveraging Off PDF Functionality

Adhering to the PDF standard allows the users to render and view the decompressed telemetry data in common PDF document viewers. Moreover, this format allows easy access, transmission, and storage of the telemetry data, eliminating the need for intermediate databases by writing the streamed telemetry data directly to PDF format. The PDF format documents containing the telemetry data can then be easily and securely stored in a local file system for intuitive user rendering using a simple PDF-compatible viewer.

PDF documents can be broken down into four principal components: the actual objects storing data, the document structure, the file structure, and the content stream [5]. The Adobe PDF standard supports eight object types: Boolean values, numbers, strings, names, arrays, dictionaries, streams, and null objects. Numerical and textual data can simply be written using numeric and string objects as specified in the PDF standard [5]. Any repeated measurement of the same data can be stored in arrays or arrays within arrays to be rendered as tables in the final PDF document.

Our algorithm uses object streams to represent the majority of the telemetry data streams because unlike strings, object streams do not require any length specification and can be encoded using any of the PDF-supported compression algorithms. For example, if telemetry data from a sensor were to relay a data stream with some unknown amount of text and numeric data followed by a series of images, our algorithm would do the following:

1. If we are just starting a PDF document, write the header, which includes any necessary metadata for later use, and write the body marker to signal the beginning of the content. Write our object marker.
2. Write the dictionary preceding the stream object, specifying its length, the appropriate decoding Filter (and the decoded length of the original data.
3. Upon receiving the text data, write the marker and commence encoding the text data using the LZW algorithm, writing to a new object stream until we encounter some kind of signal reset or a marker signaling the beginning of a new data type.
4. Write an *end stream* marker signaling the end of this encoded data (See Figure 2).

```
EXAMPLE.PDF
%Textual/Numeric Data Compressed using LZW encoding
1 0 OBJ
        <<  /Length 500
            /Filter  [/ASCII85Decode /LZW Decode]
        >>
        STREAM
              J..)6T`?p&<!J9%_[umg"B7/Z7KNXbN'S+,*Q/&"OLT'FL
            IDK#!n`$"<Atdi`¥Vn%b%)&'cA*VnK¥CJY(sF>c!Jnl@RM]W
            M;jjH6Gnc75idkL5]+cPWdR>FF(kj1_R%W_d&/jS!;iuad7h
            ?[L-F$+]]0A3Ck*$;<)CJtqi65X<W9k6Yl¥¥0McJQkDeLWdP
            N?X*al>iG1p&i;eVoK&juJHs9%;Xomop"5KatWP)lKn061la
            pKDC@¥qJ4B!!(5m+j.7F790m(Vj8818Q:_CZ(Gm1%X¥N1&u!
            FKHMB~>
        END STREAM
ENDOBJ
%Image compressed using JPEG compression
2 0 OBJ
        <<
            /Type Xobject
            /Subtype Image
            /Width 100
            /Height 100
            /Length 30000
            /Filter  [/DCTDecode]
            /DecodeParms ColorTransform 1
        >>
        STREAM
        ................................ encoded bytes for a 100x100 image, 3
      components..................................
        END STREAM
ENDOBJ
```

Fig. 2. Example of an PDF document containing two object streams, one representing encoded textual and numeric data followed by a 100x100 JPEG encoded image. *Adapted from [5], pp. 68, 84–85.

After we close this object stream, we can start writing the images in the data stream using the JPEG encoding standard. In Figure 2, the data that we write inside the stream for both the textual and image data is the compressed and transformed data we passed from the ground station through the space link. For images, in addition to simply writing the JPEG to the PDF file, we can write pertinent elements included in the Exchangeable Image File Format (EXIF) data [12] to the PDF file as well.

The content stream consists of a sequence of well-defined graphics objects (glyph, images, etc), which, along with the document structure, determine the appearance of the PDF document. PDF documents keep this data separate from our object data; sets of standard instructions about how to render different types of data can be specified at the ground level and written to the PDF document. Essentially, the content stream consists of a set of instructions that tell the PDF document viewer how to "paint" objects onto the pages [7]. This ensures a standardized rendering for multiple users viewing the same telemetry data. While the graphics objects and the document structure are related to the content stream, the latter also exists as a separate and distinct entity. Unlike the static, randomly-accessible references to text, images and other objects, the content stream must be read sequentially because it tells the viewer how to render the document. We essentially have two streams here: one stream contained our compressed data, which we write directly from our compressed data stream, and another detailing some kind of default way to render this data.

By directly writing to the PDF streams, we bypass the process of recovering the data from mass storage, decompressing it, and rendering for viewing, which can be expensive both in terms of computation time and monetary cost. Instead, compressed data will be written directly to the PDF document, which will be responsible for all decompression, rendering, and viewing.

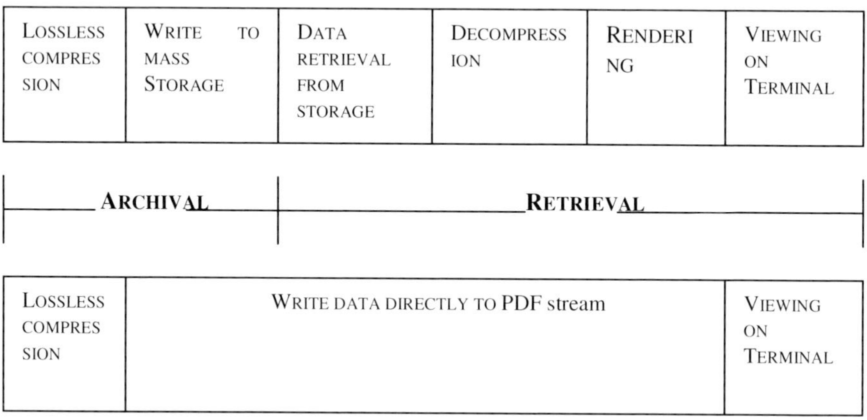

Fig. 3. The telemetry data archival and retrieval process, first with current implementations, and below, with our implementation, which reduces the complexity of the data retrieval process greatly

Because of the way a PDF file is organized, we can page through its contents to do a search much more efficiently than performing a linear keyword search as some current multimedia telemetry databases would require. Here we can make use of the document catalog to quickly narrow down the PDF file and only capture the relevant parts to search through for our desired query. Moreover, because every page is rendered independently, storing our data in PDF files will allow a user looking through the PDF files to jump to necessary pages in a non-sequential order. The data stream in the DFT frequency domain lends itself to more efficient searching than linearly searching through the uncompressed data in the time domain. Figure 3 demonstrates how much simpler our algorithm makes the telemetry data archival and retrieval process.

6 Conclusions

We have presented an algorithm that uses multiple compression streams to effectively compress multidimensional telemetry data. We adhere to the Adobe PDF standard to leverage off its rendering standards and built-in streams.

Our algorithm still needs to be tested for scalability on the level of thousands or millions of images and gigabytes and terabytes of textual and numerical data. We could build an index of PDF documents, taking advantage of the document structure, writing any metadata important for searching later on as interchange information within the PDF file in addition to the actual textual and graphical data.

Replacing existing ground-level mass storage systems with the PDF files resulting from this algorithm would reduce costs greatly. Current telemetry data storage implementations use traditional database storage; the cost of this type of storage grows quickly as the amount of data we need to store increases. Moreover, within the old framework, different rendering and viewing programs must be designed for each individual project, whereas within our algorithm, any standard PDF viewer will work.

References

1. Sayood, K.: Lossless Compression Handbook. Elsevier Science, San Diego (2003)
2. Horan, S.: Introduction to PCM Telemetering Systems, 2nd edn. CRC Press, Boca Raton (2002)
3. Gray, R.M.: Fundamentals of Data Compression. In: International Conference on Information, Communications, and Signal Processing, Singapore, September 1997, IEEE Publication, New York (1997)
4. Rao, K.R., Yip, P.C.: The Transform and Data Compression Handbook. CRC Press, Boca Raton (2001)
5. PDF Reference: Adobe Portable Document Format Version 1.7, 6th edn., Adobe Systems Incorporated (2006)
6. Staudinger, P., et al.: Lossless Compression for archiving satellite telemetry data, In: Aerospace Conference Proceedings, March 2000, vol. 2, pp. 299–304. IEEE, Los Alamitos (2000)
7. CGPDFContentStream Reference, Apple, http://developer.apple.com/documentation/GraphicsImaging/Reference/CGPDFContentStream/Reference/reference.html
8. Lossless Data Compression, Report Concerning Space Data Systems Standards, CCSDS 120.0-G-2. Green Book. Issue 2, CCSDS, Washington, D.C. (December 2006)
9. Space Packet Protocol, Recommendation for Space Data Systems Standards, CCSDS 133.0-B-1. Blue Book. Issue 1, CCSDS, Washington, D.C. (September 2003)
10. Gailly, J.-l., Adler, M.: The gzip home page, 27 July, How to compress your PDF files? Compression arithmetic for PDF files, VeryPDF.com, Inc. (2003)
11. Exchangeable image file format for digital still cameras: Exif Version 2.2, Standard of Japan Electronics and Information Technology Industries Association (April 2002)

Improving Integration with Subjective Combining of Ontology Mappings

Dennis Hooijmaijers and Markus Stumptner

ACRC, University of South Australia
Mawson Lakes Blvd, Mawson Lakes, South Australia 5095
{dennis,mst}@cis.unisa.edu.au
http://acrc.unisa.edu.au/

Abstract. Ontologies are a tool for capturing domain knowledge. All ontologies can be considered subjective in regards to the creators view of the domain. This causes difficulty for the purpose of ontology integration. RIPOSTE, a subjective ontology framework, provides the necessary mechanisms to capture the creator and their subjective opinions with each ontological resource they provide.

Mappings, between ontologies, are also subjective and can create resultant ontologies that vary depending on the integrating agent, or mapping techniques. To overcome this we extend RIPOSTE to include; mappings between ontologies, subjective opinions and the providing agents. This allows for the combining of mappings from multiple sources, creating an environment that combines the opinions of each agent. This increases the belief of agreed mappings and allows for a threshold filter to only utilise the most popular mappings when integrating the ontologies.

Keywords: Ontology, Subjective Belief, Knowledge Integration.

1 Introduction

Ontologies are a tool for capturing domain knowledge. The need to integrate ontologies is an important aspect of collaboration between enterprises. Within one domain there are three possible scenarios of ontology use: 1) The collaborators use the same ontology, 2) the collaborators use ontologies that are derived from the same ontology or 3) the ontologies are completely unrelated. In all situations the terminology and semantics of the ontology are subjective. This introduces additional problems when integrating ontologies as there is a possibility that the alignment may cause conflicts for each collaborator (i.e. including query results that are incorrect, based on that person's opinion).

Ontology mappings, like the ontologies themselves, are subjective. For each person that defines a set of mappings between two ontologies (a *map*) there is the possibility of discrepancy, which may create conflicts. Current Description Logics (DL) require the user to make a decision in the case of discrepancies and discard all but one of the possible options. By introducing probabilities to DL the ability to weight each term, relation, and mapping allows for selecting the most probable, as well as capturing them all (despite possible conflicts).

A. An et al. (Eds.): ISMIS 2008, LNAI 4994, pp. 552–562, 2008.

The RIPOSTE framework [5] provides mechanisms for capturing, and linking, authors (providers), their opinions, and the resources that they provide. This has been extended to capture ontology mappings. By using the operators *consensus* and *disjoint* [7] the ability to combine opinions about mappings allows for the ability to increase, or decrease, the probability that a mapping is correct. By integrating mappings in such a way the more probable mappings will be those that are agreed upon by multiple agents. The agreed mappings will then provide an increased *precision* and *recall* above a given *threshold*. This will improve the overall probability that the resultant ontology will provide responses to a user's queries that do not conflict with their opinions.

In this paper we provide an overview to the RIPOSTE framework and the extensions providing the ability to capture mappings. An overview of subjective logic, the two operators consensus and disjoint and how they apply to ontology integration. We evaluate our results by combining mappings and discussing threshold-ed results.

2 Subjective Ontology Framework (RIPOSTE)

This work extends the RIPOSTE framework [5] to provide additional functionality for capturing, combining and manipulating mappings between two ontologies. RIPOSTE extends OWL DL ontologies [10] by providing a meta-ontology, which captures relationships between resource providers, ontological resources and belief ratings.

An *ontology* can be described as a collection of resources that explicitly and formally conceptualise a domain model [4].

Definition 1 (Ontology). *Let R be a set of ontological resources for a domain D where*

1. $C^D \subseteq R$ *represents a set of classes* $\in D$
2. $S^D = \{\equiv, \sqsubseteq, \sqsupseteq, \perp, ., \neq\}^1 \subseteq R$ *represents a set of semantic relationships between any* C_i^D *and* $C_j^D \in D$
3. $I^D \subseteq R$ *represents a set of individuals that are instances of some* $C_i^D \in D$
4. $\chi^D \subseteq R$ *represents a set of axioms that must hold true for all* $R_i \in D$

Then an ontology is defined as $\Omega_D = (C^D, S^D, I^D, \chi^D)$.

Any ontology that a user adopts for their use is considered to be a *private ontology*. Private ontologies are a specific version of an ontology that a particular user has selected. This is important as users may use a publicly available ontology but will decide whether to evolve the ontology themselves or to adopt newer versions. This creates a fragmented environment of ontology versions that may no longer be compatible for integration.

A provider is defined as any entity that contributes an ontological resource to a private ontology.

[1] This is the sub set we use in this work out of all possible semantic relations that are defined by OWL.

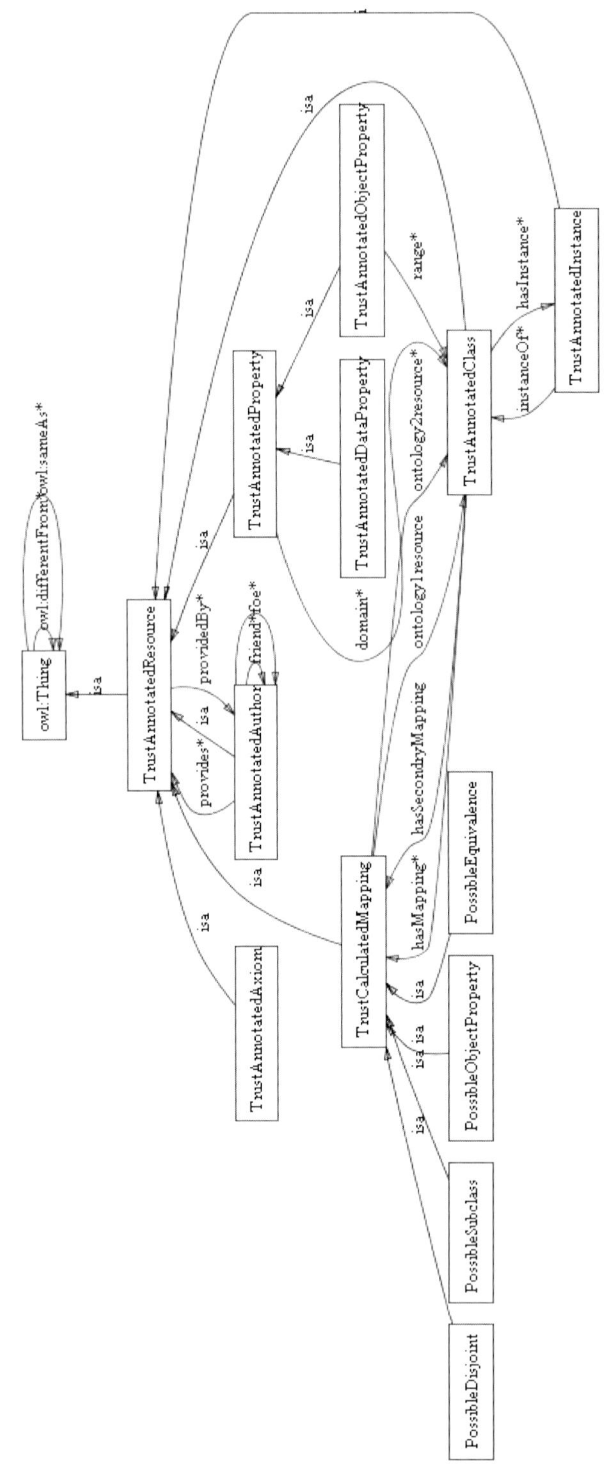

Fig. 1. Semantic Structure for Trusted Ontologies

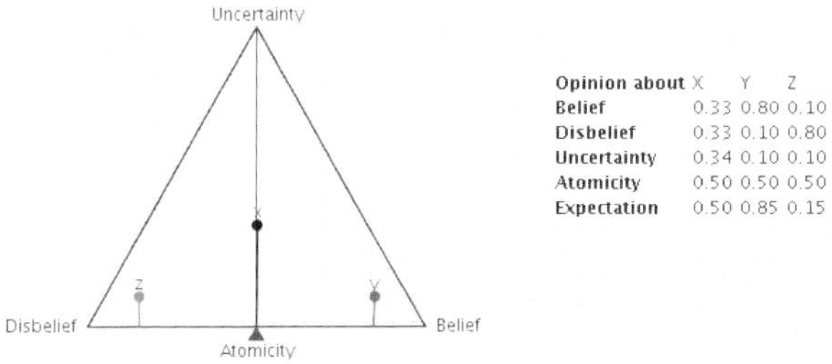

Fig. 2. Subjective Logic [7]

Definition 2 (Private Ontology). *Let Ω_S be an OWL ontology and P a provider. Then for each resource $R \in \Omega_S$ a relationship $provides(R, P)$ exists such that:*

1. $provides(R, P) \equiv providedBy(P, R)$
2. *for an agent (A) to be a provider (P) for Ω_S then:* $\exists provides(R, A) \in \Omega_S$

To capture multiple individuals and map them to their provided resources it is necessary to stipulate additional classes and semantic relationships, as shown in Fig. 1. Fig. 1 represents a *meta-ontology* that becomes the upper level of concepts of any ontology. The top level concept 'TrustAnnotatedResource' contains all properties for capturing subjective logic (*belief*, *disbelief*, *uncertainty* and a), and creates the necessary links between concepts and the providers. The subclasses of 'TrustAnnotatedResource' provide the additional properties necessary to capture each type of ontology resource. Creating the structure below 'owl:thing' allows for ontologies to be annotated, where all subclasses of 'owl:thing' in the ontology become a subclass and instance of 'trust-rated class' and all other resources become instances of the appropriate trusted version.

Ontologies annotated by RIPOSTE are able to be filtered based on authors, belief and uncertainty. The value of the property used for the filter is a *threshold* such that all resources from an ontology that fall below the value are removed.

2.1 Subjective Logic

The RIPOSTE framework uses Subjective Logic, as described by Jøsang, [8] as a basis for capturing and manipulating the probable matches between terminology. Subjective Logic is an extension to probabilistic logic that represents a provider's (P) beliefs as *opinions* (w_x^P). An opinion is the combination of belief (b_x^P), disbelief (d_x^P) and uncertainty (u_x^P) in a given ontological resource (x). Where the opinion is given by:

$$(w_x^P) = (b_x^P, d_x^P, u_x^P, a_x),\ where\ b_x + d_x + u_x = 1\ (Fig.\ 2).$$

a_x represents the size of the state space from which x is taken and in our work is, initially, kept separate from the provider's opinion as stated by Jøsang (a_x^P).

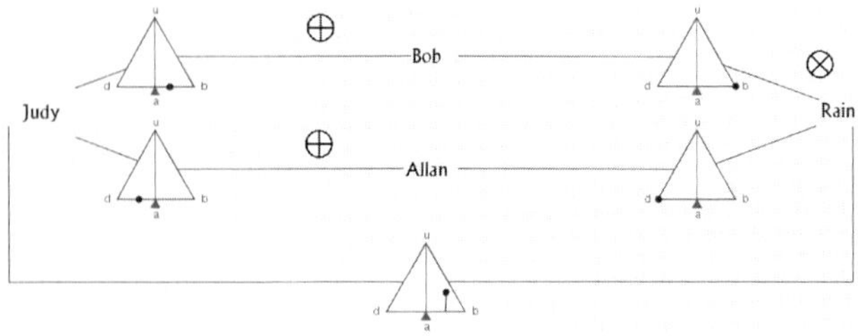

Fig. 3. Judy's belief that it will rain using Discount and Consensus, modified from [7]

For resources within the ontology we use the state space of {equivalent, disjoint, subclass, superclass, object property, unrelated} for the opinions, which represent the possible semantic relationships between two classes ($a_x = 0.17$). While for trust between authors we use the state space of {trust, distrust} $a_x = 0.5$.

The subjective logic operators, *Consensus* and *Discount*, as defined by Jøsang [7], are used to combine opinions about mappings. Consensus ($\oplus$) is used to, fairly and equally, combine two possibly conflicting opinions about a given resource and Discount ($\otimes$) is used to capture reputation (trust transitivity).

Definition 3 (Consensus [7]). *Let* $w_x^P = (b_x^P, d_x^P, u_x^P, a_x^P)$ *and* $w_x^Q = (b_x^Q, d_x^Q, u_x^Q, a_x^Q)$ *be opinions respectively held by P and Q about the same proposition x. Let* $w_x^{PQ} = (b_x^{PQ}, d_x^{PQ}, u_x^{PQ}, a_x^{PQ})$ *be the opinion such that:*

1. $b_x^{P,Q} = (b_x^P u_x^Q + b_x^Q u_x^P)/k$
2. $d_x^{P,Q} = (d_x^P u_x^Q + d_x^Q u_x^P)/k$
3. $u_x^{P,Q} = (u_x^P u_x^Q)/k$
4. $a_x^{P,Q} = \frac{(a_x^P + a_x^Q - (a_x^P + a_x^Q) u_x^P u_x^Q)}{u_x^P + u_x^Q - 2u_x^P u_x^Q}$

where $k = u_x^P + u_x^Q - u_x^P u_x^Q$ *such that* $k \neq 0$ *and* $a_x^{P,Q} = (a_x^P + a_x^Q)/2$ *when* $a_x^P, a_x^Q = 1$. *Then* w_x^{PQ} *is called the consensus between* w_x^P *and* w_x^Q, *representing an imaginary agent* $[P,Q]$*'s opinion about x, as if she represented both P and Q. By using the symbol '*$\oplus$*' to designate this operator, we define* $w_x^{PQ} \equiv w_Q^P \oplus w_x^Q$.

Definition 4 (Discount [7]). *Let P and Q be two agents where* $w_Q^P = (b_Q^P, d_Q^P, u_Q^P, a_Q^P)$ *is P's opinion about Q's advice, and let x be the proposition where* $w_x^Q = (b_x^Q, d_x^Q, u_x^Q, a_x^Q)$ *is Q's opinion about x expressed in an advice to P. Let* $w_x^{PQ} = (b_x^{PQ}, d_x^{PQ}, u_x^{PQ}, a_x^{PQ})$ *be the opinion such that:*

1. $b_x^{PQ} = b_Q^P b_x^Q$
2. $d_x^{PQ} = b_Q^P d_x^Q$
3. $u_x^{PQ} = d_Q^P + u_Q^P + b_Q^P u_x^Q$
4. $a_x^{PQ} = a_x^Q$

By using the symbol '$\otimes$*' to designate this operator, we define* $w_x^{PQ} \equiv w_Q^P \otimes w_x^Q$.

Assume *Judy* believes *Bob* is right 66% of the time and *Allan* 33% of the time, and Bob believes it will rain tomorrow while Allan believes it will not rain, then:

$$w_{bob}^{judy} = (0.66, 0.34, 0, 0.5) \; and \; w_{allan}^{judy} = (0.33, 0.67, 0, 0.5)$$

$$w_{rain}^{bob} = (1, 0, 0, 0.5) \; and \; w_{rain}^{allan} = (0, 1, 0, 0.5)$$

Then Judy's belief that it will rain (Fig. 3) is calculated by:

$$w_{rain}^{judy,bob,allan} = (w_{bob}^{judy} \oplus w_{rain}^{bob}) \otimes (w_{allan}^{judy} \oplus w_{rain}^{allan}) = (0.57, 0.14, 0.29, 0.5).$$

3 Ontology Mapping

Comparing, contrasting and resolving differences between multiple sources is part of the process of information fusion, often performed by the domain expert. This process can often be broken down to a mapping and merging problem. Knowledge of the semantics involved in the context of information can be used to discover and make decisions about appropriate merges and relationships; this is known as a mapping.

A mapping is a semantic relationship between concepts that exist in separate ontologies. Mappings provide the ability to relate the concepts in different ontologies to each other. An ontology map M, which is the set of mappings, is a declarative specification of the semantic overlap between two ontologies Ω_S and Ω_T as defined by a provider, P

$$M_{S,T}^{P} = \{m_1^{C_j^S, C_k^T}, \ldots, m_i^{C_{j'}^S, C_{k'}^T}\} \mid where \; C^S \in \Omega_S \; and \; C^T \in \Omega_T$$

There are 6 basic mappings that we consider in this work, these are related to S^D and captured in the following subsets of $M_{S,T}^P$:

- ${}_{\equiv}M_{C_S,C_T}^P$ represents the set of mappings $C_S \equiv C_T$, proposed by P.
- ${}_{\sqsubseteq}M_{C_S,C_T}^P$ represents the set of mappings $C_S \sqsubseteq C_T$ proposed by P.
- ${}_{\sqsupseteq}M_{C_S,C_T}^P$ represents the set of mappings $C_S \sqsupseteq C_T$ proposed by P. This can be captured by ${}_{\sqsubseteq}M_{C_T,C_S}^P$ as $C_T \sqsubseteq C_S$.
- ${}_{.}M_{C_S,C_T}^P$ represents the set of mappings $C_S.C_T$ proposed by P.
- ${}_{\perp}M_{C_S,C_T}^P$ represents the set of mappings $C_S \perp C_T$ proposed by P.

The final mapping of interest is unrelated ($\neq$), which represents all mappings that do not exist in the basic mappings. This is necessary for completeness of the map and to ensure that all possibilities of mappings are captured.

$${}_{\neq}M_{C_S,C_T}^P = \overline{{}_{\equiv}M_{C_S,C_T}^P \cup {}_{\sqsubseteq}M_{C_S,C_T}^P \cup {}_{.}M_{C_S,C_T}^P \cup {}_{\perp}M_{C_S,C_T}^P}$$

Additionally each set of mappings is disjoint and the following holds true:

$${}_{\sigma}M_{S,T}^P \cap {}_{\tau}M_{S,T}^P = \emptyset \mid where \; \sigma, \tau \in S^D \; and \; \sigma \neq \tau$$

3.1 Capturing Mappings in RIPOSTE

The RIPOSTE [5] meta-ontology is extended to capture mappings (Fig. 1). By introducing a mapping class (TrustCalculatedMapping) as a subclass of TrustAnnotatedResource the mappings inherit the objectProperty providedBy. This provides the necessary link to capture the provider with each of their mappings. The inherited dataProperty credibility allows for each mapping to be annotated with subjective belief.

- **TrustCalculatedMapping** provides an abstract class that captures all the necessary relationships that are required (the classes that are mapped from Ω_S and Ω_T).
- **PossibleDisjoint** provides a concrete class to capture all possible disjoint relations between classes in Ω_S and Ω_T as ${}_{\perp}M^P_{C_S,C_T}$.
- **PossibleSubclass** provides a concrete class to capture all possible subclass relations between classes in Ω_S and Ω_T as ${}_{\sqsubseteq}M^P_{C_S,C_T}$ and all possible superclass relations (${}_{\sqsubseteq}M^P_{C_S,C_T}$).
- **PossibleObjectProperty** provides a concrete class to capture all possible property relations between classes in Ω_S and Ω_T as ${}_{.}M^P_{C_S,C_T}$.
- **PossibleEquivalence** provides a concrete class to capture all possible equivalence relations between classes in Ω_S and Ω_T as ${}_{\equiv}M^P_{C_S,C_T}$.

4 Subjective Ontology Mappings

Subjective logic provides mechanisms to combine opinions, while RIPOSTE provides additional mechanisms to combine resources. This allows us to combine mappings and the opinions of the agents that provide them. The goal is to combine the maps that two agents, P and Q, provide between two ontologies, Ω_S and Ω_T, such that:

$$M^{P,Q}_{S,T} = M^P_{S,T} \times M^Q_{S,T}$$

Where each mapping, $m^{C_i,C_j} \in M^P_{S,T}$ $C_i \in \Omega_S, C_j \in \Omega_T$, is linked with a provider, P and a subjective belief of that provider $w^P_{m^{C_i,C_j}}$. When a provider Q provides a mapping between the same concepts $w^Q_{m^{C_i,C_j}} \in M^Q_{S,T}$ $C_i \in \Omega_S, C_j \in \Omega_T$, and P has an opinion w^P_Q then the opinion of P about the mapping m^{C_i,C_j} is calculated as follows:

$$w^{P,Q}_{m^{C_i,C_j}} = w^P_{m^C_i,C_j} \otimes (w^P_Q \oplus w^Q_{m^{C_i,C_j}})$$

The result can then be combined iteratively with all other providers of the same mapping. This results in reinforcing the belief in that mapping (meaning it is more likely to be correct for the current purpose of integration).

Additionally if a mapping is reinforced then all contradicting mappings need to be weakened:

$$w^{P,Q}_{m_\sigma} = w^P_{m_\sigma} \oplus (w^P_Q \oplus w^Q_{m_\tau}) \forall j \in S^D \mid where\ \sigma, \tau \in S^D\ and\ \sigma \neq \tau$$

If $C^S \equiv C^T$ then all conflicting mappings ($C^S.C^T$, $C^S \perp C^T$, $C^S \sqsubseteq C^T$, $C^S \sqsupseteq C^T$, and $C^S \neq C^T$) should have their belief decreased.

5 Experimental Evaluation

In this section we report the results of running RIPOSTE over example maps created by independent agents. Two ontologies (Ω_{sw} and Ω_{hc}) for overlapping domains using data from a charity organisation involved in social welfare were constructed. These were medium sized ontologies ($\sim$ 50 classes in each) with a significant amount of overlap. The ontologies were then mapped by a knowledge engineer to create a baseline map (M_{ke}) for result comparisons as shown in table 1.

Now lets assume that a domain expert (DE) needs to integrate Ω_{sw} and Ω_{hc} and creates a map $M^{DE}_{sw,hc}$. Let us also assume that maps for the ontologies have already been created (by domain experts Judy, Allan and a non expert Bob $M^{judy}_{sw,hc}$, $M^{allan}_{sw,hc}$, $M^{bob}_{sw,hc}$ respectively) and are available to integrate with the newly created map, $M^{DE}_{sw,hc}$.

Precision, *recall* and *F-measure* were calculated using M_{ke} by:

- Recall, R = # found mappings/ # mappings in M_{ke}
- Precision, P = # correct mappings in M_{ke}/ # found mappings
- F-measure, $F = 2 \times P \times R / P + R$

The result (table 2) shows that Judy found the most mappings while Allan was more precise (had less false positive mappings). Allan's map proved to be the best based on both precision and recall (F-measure).

The mappings were then combined using the extended RIPOSTE framework and the precison, recall and F-measure were calculated. From the initial experiment Allan had the map which corresponded closest to the knowledge engineer. It can be seen from the initial results that the recall increased (table 2) when each of the mappings were combined. This is due to Allan and Bob finding mappings that Judy did not. Additionally Precision fell as the false positive mappings of all providers are included in the final resultant map. To improve precision it is necessary to utilise a threshold, that requires the integration of subjective beliefs.

Assuming DE has an initial subjective belief regarding Judy and relies on Judy's opinion of of Allan and Bob's opinions as follows:

$$w^{DE}_{judy=(0.66,0.34,0,0.5)}, w^{judy}_{bob} = (0.66, 0.34, 0, 0.5), w^{judy}_{allan} = (0.33, 0.67, 0, 0.5)$$

and Allan, Bob and Judy all have absolute belief in their results:

$$w^{P}_{M_{sw,hc}} = (1, 0.0, 0, 0.17)\ \forall P \in \{judy, allan, bob\}\ and\ \forall m \in M_{sw,hc}$$

Table 1. Ontologies and Expert mappings

	# Classes	# Props	# Insts
Ω_{sw}	53	80	100
Ω_{hc}	46	58	93
M_{ke}	32	44	37

Table 2. Map precison, recall and F-measure

Map	P	R	F-measure
M^{judy}	0.83	0.91	0.87
M^{allan}	0.88	0.88	0.88
M^{bob}	0.71	0.53	0.61
M^{judy} & M^{bob}	0.83	0.94	0.88
M^{judy} & M^{bob} & M^{allan}	0.82	0.97	0.89

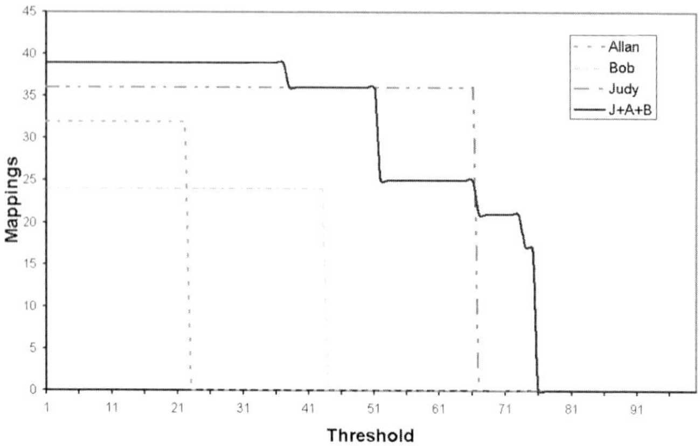

Fig. 4. A comparison of mappings at belief thresholds

Then the domain expert's opinion of mappings provided by Allan and Bob are calculated by:

$$w_m^{DE,judy,allan} = w_{judy}^{DE} \otimes w_{allan}^{judy} \otimes w_m^{allan} = (0.22, 0.45, 0.33, 0.17)$$
$$w_m^{DE,judy,bob} = w_{judy}^{DE} \otimes w_{bob}^{judy} \otimes w_m^{bob} = (0.44, 0, 0.56, 0.17)$$

Where agreement on the mappings occur the consensus operator is used combining the subjectivity i.e. if Judy and Allan have the same mapping then

$$w_m^{DE,judy,allan} = w_m^{DE,judy} \oplus w_m^{judy,allan} = (0.52, 0.27, 0.20, 0.17)$$

By using the RIPOSTE threshold mechanism it can be seen that precision is improved as the threshold increases. This is due to the fact that the mappings that are found in more than one provider's map will have their subjective belief increased and those mappings that do not have belief strengthened will be removed as shown in fig 4. In our initial experiment this increased precision due to the false positive mappings not being duplicated by all three providers, giving them a reduced belief.

6 Related Work

Jaeger also provides a mechanism for capturing probability within Terminological Logics [6] for the purpose of reasoning and this has been extended to OWL[1]. The ability to provide conditional probabilities to concepts, and uncertain expression for specific objects, provides the ability to model uncertain domains. Jaeger provides these two kinds of statements to allow probable reasoning to be applied to a terminological knowledge representation language. This provides a good framework that allows for the capturing of uncertain domains. To capture uncertain semantics it is necessary to extend these approaches so that the probabilities can be applied to the task of integration.

Automated knowledge integration has been an active research area for some time [9,13,2], but so far has mostly concentrated on knowledge assumed to be stable, certain, and (for a particular problem or domain) complete. Current approaches rely on syntactic label matching [9,13,2], semantic matching [13,11,2,12] and uncertainty [11,12] to create an ontology mapping. Syntactic mapping uses an algorithm based on pattern matching of class labels. It searches through the first ontology and then, using a set of reasoning rules, searches through the second ontology for a match. For taxonomy alignment the match is restricted to hierarchy subsumption within ontologies [9]. This is currently required by Semantic matching to locate the initial matches [13]. Additional matches are then located by searching the graph structure of the ontology and selecting matches based on criteria, such as distance, similar properties. The semantic matches can also be located using first-order logic to provide predicate axioms to map the source ontologies onto the target ontology [2].

Uncertain ontology integration [11,12] uses probabilities to improve the mappings by considering previous mappings and the likelihood of other similar mappings. OMEN [11], an extension of PROMPT [13] method, uses Bayesian Networks to assign belief values to likely matches, thereby providing more probable candidate matches. When a user accepts a match, the beliefs are updated by propagating the new evidence through the network. While DSSim [12] utilises the Dempster-Shafer framework to provide uncertainty values on possible matches by calculating the *belief mass function* to increase the precision [3] of matches based on a search query.

All of these approaches provide mechanisms for finding possible matches. The best possible match is selected and the rest are discarded. Often this can be detrimental to recall as the selected match may be incorrect and the correct match is discarded.

7 Conclusion

In this work we discuss a novel approach to integrating ontologies. By combining ontology maps from multiple sources we aimed to improve the resultant map so that the integration would contain less errors for the purpose that the ontologies were integrated.

Ontology maps are a subjective artifact that can benefit from being combined with those by other providers. By extending the RIPOSTE framework the ability to capture providers, their beliefs and the mappings can be utilised for integration. Initial experiments have shown that recall can be increased by combining ontology maps. Unfortunately this also decreases precision as more false positive mappings are introduced.

By using the RIPOSTE threshold mechanism and calculating the subjective belief in each mapping precision may be improved. This is due to the positive mappings having their subjective belief increased and the false positives having no, or minimum, consensus between the providers.

7.1 Future Work

In the future we would like to observe if more maps will improve precision and recall, or if the introduction of more false positive maps will decrease the usability of the RIPOSTE framework.

Additionally we plan to examine subjective belief transitivity. That is if B is a subclass of C and A is mapped to be equivalent to B, how will that effect the map of A is a subclass of C. By implementing this we hope to improve recall by discovering possible mappings that may be overlooked by the providers.

References

1. Ding, Z., Peng, Y.: A probabilistic extension to ontology language OWL. In: Hawaii Int. Conf. on System Sciences (2004)
2. Dou, D., McDermott, D.V., Qi, P.: Ontology translation on the semantic web. In: Meersman, R., Tari, Z., Schmidt, D.C. (eds.) CoopIS 2003, DOA 2003, and ODBASE 2003. LNCS, vol. 2888, pp. 952–969. Springer, Heidelberg (2003)
3. Ehrig, M., Euzenat, J.: Relaxed precision and recall for ontology matching. In: Integrating Ontologies. CEUR Workshop Proc., Banff, Canada, vol. 156 (2005)
4. Guarino, N.: Semantic matching: Formal ontological distinctions for information organization, extraction, and integration. In: Pazienza, M.T. (ed.) SCIE 1997. LNCS, vol. 1299, pp. 139–170. Springer, Heidelberg (1997)
5. Hooijmaijers, D., Stumptner, M.: Trust based ontology integration for the community services sector. In: Advances in Ontologies. Proc. AOW, Hobart, Australia (2006)
6. Jaeger, M.: Probabilistic reasoning in terminological logics. In: Principles of Knowledge Representation and Reasoning, pp. 305–316 (1994)
7. Jøsang, A.: A logic for uncertain probabilities. Int. Journal of Uncertainty, Fuzziness and Knowledge-Based Systems 9 (2002)
8. Jøsang, A., Ismail, R., Boyd, C.: A survey of trust and reputation systems for online service provision. In: Specker, E., Strassen, V. (eds.) Komplexität von Entscheidungsproblemen. LNCS, vol. 43, pp. 618–644. Springer, Heidelberg (1976)
9. McGuinness, D.L., Fikes, R., Rice, J., Wilder, S.: The Chimaera ontology environment. In: Proc. AAAI 2000, Austin, Texas, USA., pp. 1123–1124. AAAI Press, Menlo Park (2000)
10. McGuinness, D.L., van Harmelen, F. (eds.): OWL Web Ontology Language Overview (1994-2006)
11. Mitra, P., Noy, N.F., Jaiswal, A.R.: OMEN: A probabilistic ontology mapping tool. In: Gil, Y., Motta, E., Benjamins, V.R., Musen, M.A. (eds.) ISWC 2005. LNCS, vol. 3729, pp. 537–547. Springer, Heidelberg (2005)
12. Nagy, M., Vargas-Vera, M., Motta, E.: DSSim-ontology mapping with uncertainty. In: Proc. of the 1st Int. Workshop on Ontology Matching, Athens, Georgia, USA (2006)
13. Noy, N.F., Musen, M.A.: Prompt: Algorithm and tool for automated ontology merging and alignment. In: Proc. AAAI 2000, Austin, Texas, USA., pp. 450–455 (2000)

Text Onto Miner – A Semi Automated Ontology Building System

Piotr Gawrysiak[1], Grzegorz Protaziuk[1], Henryk Rybinski[1], and Alexandre Delteil[2]

[1] ICS, Warsaw University of Technology
[2] France Telecome R & D
{P.Gawrysiak,G.Protaziuk,H.Rybinski}@ii.pw.edu.pl,
alexandre.delteil@orange-ft.com

Abstract. This paper presents an overview of the results of the project undertaken by the Warsaw University of Technology Institute of Computer Science as a part of research agreement with France Telecom. The project goal was to create a set of tools – both software and methods, that could be used to speed up and improve a process of creating ontologies. In the course of the project a new ontology building methodology has been devised, new text mining algorithms optimized for extracting information useful for building an ontology from text corpora have been proposed and an universal text mining toolkit – TOM Platform – have been implemented.

Keywords: Natural language processing, ontologies, text mining.

1 Introduction

Ontologies have shown their importance in many application areas, such as intelligent knowledge base integration, information brokering, and natural-language processing, just to indicate few of them. Their importance is growing, as is growing on the Web the number of information repositories that need metadata enrichment and analysis. On the other hand however, their usage is still very limited by ontology engineering, which is very time-consuming and expensive. Therefore there is a growing need for automated – or at least semi-automated methods, that will be able to leverage the amount of information present in ever growing repositories of text data (e.g. obtainable via the Internet) in order to build useful ontology systems.

One can distinguish two main approaches in discovering semantic information from text corpora – knowledge-rich and knowledge-poor ones, according to the amount of knowledge they presuppose [7]. Knowledge-rich approaches require some sort of previously built semantic information, domain-dependent knowledge structures, semantic tagged training corpora, or semantic dictionaries, thesauri, ontologies, etc. (see e.g. [9,24]). In most cases, the reported methods refer to knowledge-rich methods, which require deep and specific knowledge "coded" into the algorithms, auxiliary dictionaries and/or thesauri, very much language and domain dependent. Although the knowledge rich methods may

A. An et al. (Eds.): ISMIS 2008, LNAI 4994, pp. 563–573, 2008.

bring better results than the knowledge poor ones, the requirement for deep and specific knowledge is the main limitation in using them. There is therefore a high demand for finding knowledge-poor algorithms that would give satisfactory results, especially for the cases of limited lexical resources. In this context, the most promising approach seems to be utilization of text mining techniques for discovering semantic information from text corpora in order to build or maintain ontologies (see e.g. [14]). To this date most approaches of this type have only looked at a very specific problem, e.g. how to learn the taxonomic part of ontologies, or how to find proper names. The most complete approach has been reported in the work performed by the group of AIFB (Karlsruhe University), e.g. [14,15]. Our approach follows this direction. So, as in [14] we attempt to cover the entire process of ontology building, and provide an advanced platform (named TOM), supporting the whole process. Additionally, we have incorporated to TOM a number of novel algorithms, focused mainly on enriching the ontologies lexical layer. TOM integrates text preprocessing algorithms with novel TM based algorithms [19,21,11,22], and the tools for merging partial results into the existing ontology.

In this paper we present an overview of this approach, and the set of tools. It is a semi-automated method that could help building ontologies thanks to the analysis and extraction of semantic information from large text corpora. The structure of this paper is as follows: Section 2 presents an overview of the proposed ontology building process, and Section 3 describes briefly new algorithms and methods that have been developed in order to support this process. Section 4 presents the structure of the TOM platform that has been developed specifically in order to verify experimentally the proposed algorithms and the methodology. Finally, Section 5 contains information concerning experimental results and concluding remarks.

2 Semi-automatic Ontology Building Process

In the literature many approaches to building ontologies have been introduced and discussed (see for example) [2,1,4,5,6,12,16,23]. In [8] and [17] authors present an opinion that the process of ontology building is not a rigorous engineering discipline. Nevertheless, tasks that are required in order to create ontology are quite well defined. According to e.g. [17] these include: (a) defining a domain and scope of an ontology, (b) creating a comprehensive list of concepts (classes) and their hierarchy, (c) defining relations between classes and (d) populating an ontology with instances of classes. Additionally, some auxiliary tasks, such as defining the properties of classes or preserving transitivity of some relations (e.g. taxonomy, part_of), and avoiding cycles, are usually required.

As shown in [20], there are already many publications referring to the research on automatic tools that support ontology building process in various phases. In many of above tasks (e.g. while determining relations between classes) some automatic tools can be used, to a higher or lesser extent. Some of these tasks cannot be even performed manually in a reasonable amount of time. This statement refers specifically to situations, where a huge amount of data should be processed and/or analyzed. With the text mining methods we can provide a support in such cases, however the discovered knowledge always requires human decision and intervention, thus provided tools will be always semi-automatic.

Below we sketch a method of building a domain ontology from unstructured text documents with the text mining support provided by the TOM text mining platform. We propose an approach to building an ontology from a domain specific repository. We assume that neither an a priori taxonomy nor ontology is available. The approach consists of the following steps:

1. Text extraction and preprocessing
TOM provides a variety of tools for text preprocessing. In various text mining experiments there may be different needs for defining a text unit. We have therefore introduced an option for defining granularity of the text mining process. In particular, TOM allows viewing the whole corpus as a set of documents, paragraphs or sentences. For example, for experiments aiming at discovering compound terms [19], or synonyms [21] the granularity was set to the sentence level. For discovering homonyms and homograms [22] we have performed experiments with the granularity set at the paragraph level. The process can be defined as a pipeline (Fig. 1), and the results can be used in all the other phases.

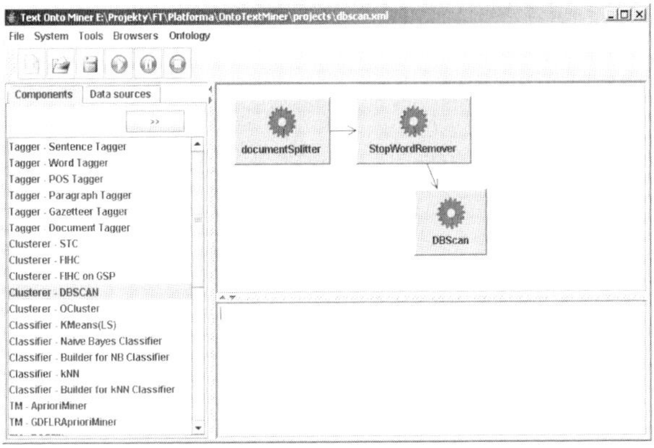

Fig. 1. Designing a preprocessing pipeline

2. Determining list of terms & purifying document representations
In this step two kinds of terms are determined: (a) words specific for a domain of interest, and (b) compound terms, including compound proper nouns (see [19] for details). For building the list (a) usually one should extract a dictionary from the given repository and subtract from it the dictionary received from a reference repository (containing a common sense texts). Next, sequences of words representing multiword terms are replaced by compound terms in the text.

3. Building a hierarchy (taxonomy)
The candidates for the hierarchy building are discovered with FIHC, and *apriori* (see [11] and [21] for details respectively). Based on it, the user selects proper candidates. It is difficult to automatically detect the hierarchy for compound proper nouns, because such terms occur in texts rather infrequently, and they are disseminated among many clusters. It is therefore reasonable in FIHC to ignore compound (proper) nouns. On the other hand, synonymy discovering

procedure [21] finds *inter alia* categories of terms. In such cases a taxonomy level can be created or adjusted manually, when needed. Then hierarchies are merged. The step ends with an ontology skeleton.

4. Discovering terms meanings
By discovering terms meanings we (1) identify the pairs of terms with close meanings, and homonyms (or in general, various meanings of some terms). To this end, we use the methods sketched in Section 3. As the method described discovers also pairs of the type (broader, narrower), part_of, or belonging to the same category, some results can be used to enrich the taxonomy obtained in the steps above. Discovered synonyms and homonyms can be used for enriching the lexical part of the ontology by relating the literals to the appropriate concepts. In the case of homonyms we also keep within the ontology various meanings of the words. The information can be then used by the user for enriching the dictionary layer and its connections to the concept layer of the ontology.

5. Discovering association relations & enriching the ontology skeleton
The process of discovering association relations is done by applying the method described in Section 3. Discovered relations can be then used by the user for building the concept layer of the ontology. In this case each term included in the discovered relations is treated as an instance of a certain class unless it is directly stated that the word is a name of a class. The OWL standard requires that each instance must have assigned a class but it may happen that for some instances appropriate classes are unknown.

It is worth noting that steps 2 and 5 are performed with the same T-GSP algorithm (though with different parameters), whereas step 4 (both homonymy and close meaningterm discovery) is performed with *apriori* [1].

3 New Text Analysis Methods

During the course of the TOM project several new text analysis methods have been developed. The methods mainly refer to discovering various relations between words and their properties relevant in the ontology development and maintenance process. In particular, the following algorithms have been elaborated and tested, and then incorporated into TOM:

1. discovering hierarchies with FIHC algorithm [6]; the algorithm is used for finding taxonomies from the set of documents; a number of improvements have been introduced to the algorithm in comparison to the original, making the algorithm much faster;
2. *apriori* based algorithm for discovering close meaning terms (synonyms, antonyms, BT/NT related terms, category/instances related terms, etc.);
3. *apriori* based algorithm for discovering homonyms and homograms;
4. T-GSP algorithm for discovering grammatically filtered frequent patterns; the algorithm can be used for discovering (proper) composed nouns, as well as association relations, depending on the grammar rules defined;

5. new (knowledge rich) methods for discovering association relations;
6. a number of rules have been elaborated in order to support the process of merging ontologies.

Some methods have been described in detail in [21,19,11,22]. A complete description is provided in [18]. Below we briefly present only short descriptions of the selected algorithms.

Discovering hierarchies based on the FIHC

Processing text corpora with the FIHC algorithm [6] generates hierarchical clustering tree. As described in [6], labels of this tree tend to form a potentially interesting taxonomy. Resulted tree comes from the frequent itemsets tree, which has taxonomic character by itself, e.g. *Warsaw University of Technology* is obviously in is a relation with University and the former itemset is a superset of the latter one. In addition, the obtained FIHC tree has the following advantages over normal frequent sets trees:

- it is not full. In the case of frequent sets, if an itemset is frequent, then all its subsets are also frequent, and all of them exist in the tree. FIHC takes only the most significant subsets.
- FIHC can merge similar siblings, so it might happen that a phrase has a parent which is not its subset.

In [11] our group have proved that the original algorithm can easily be modified to operate on closed sets and can produce exactly the same results as by means of frequent itemsets. The number of frequent closed sets is usually much lower than the number of frequent itemsets giving significant gain in time of execution and memory requirements. Moreover, we made some modifications to the algorithm which strengthen the ability of discovering taxonomies. We experimented with many modifications. The ones listed below significantly improve the taxonomy discovery:

- use of POS tagging for pattern matching & pre-filtering. It allows restricting search only for relations of a given form, e.g. only noun-noun relations. We can define any pattern which can be expressed by means of regular expressions.
- allowing obvious pruning only. Normally, the resultant tree is vastly pruned in the last phases of the FIHC algorithm. This option allows performing only obvious pruning. It means that parent-child merging is allowed only in case of one-child nodes. This option was introduced to retrieve more semantic relations. Originally FIHC was meant to produce labeled clustering tree and it was reasonable to have a small number of meaningful groups. We are not interested in clusters, instead, we want to have a rich description of semantic relations.
- pulling up labels. While performing the pruning, original FIHC leaves a parent label that is the shorter one. In our case, we do not look for cluster labels, but we need semantic relations, which are more likely hidden in longer labels. When the option of pulling labels is on, the label of the child is kept instead of the parent's one during the pruning phase.

- operating on sequential patterns instead of itemsets, which keeps words ordering, and makes results more readable and potentially more accurate. We replaced the phase of itemsets discovery with TGSP.

Discovering synonyms [21]
In our method we assume that: *The synonyms do not appear together in a sentence, but they appear quite frequently in similar contexts.* Hence, synonyms are often used with the same words. Having the pairs of terms that do not co-occur, we apply the similarity measures CSIM and ASIM [21]. The approach for generating pairs of terms that are likely to be synonyms consists of the following steps:

1. Text corpus is tagged with parts of speech, and some cleaning is done;
2. Text data are converted into transactional database, where sentences are treated as transactions (by means of [1];
3. The *apriori* based algorithm for finding frequent itemsets is executed;
4. For every frequent term being in the field of interest, frequent itemsets containing that word are found (we call the set of itemsets context of a word);
5. Finally synonymy measure is computed for every pair of words with use of their contexts. Based on this, a decision is taken whether the pair can candidate for synonymy or not.

Discovering homonyms
Distinct meanings of homonyms are indicated by various distinct contexts in which they appear frequently. This assumption is based on the distributional hypothesis [10], where the underlying idea is that "*a word is characterized by the company it keeps*". The rule is very intuitive, and therefore is applied to the proposed approaches. The problem is, however, how the notion of a context is defined. For example, it can be understood as a set of words surrounding a target word frequently enough in documents, paragraphs, or sentences. In our approach, context is evaluated as below.

Let dictionary $D = \{t_1, t_2, \ldots, t_m\}$ be a set of distinct words, called *terms*. In general, any set of terms is called a termset. The set $\mathcal{P}$ is a set of *paragraphs*, where each paragraph P is a set of terms such that $P \subseteq \mathcal{P}$.

Statistical significance of a termset X is called *support* and is denoted by $sup(X)$. $sup(X)$ is defined as the number (or percentage) of paragraphs in $\mathcal{P}$ that contain X. Clearly, the supports of termsets that are supersets of termset X are not greater than $sup(X)$.

A termset is called *frequent* if it occurs in more than ε paragraphs in $\mathcal{P}$, where ε is a user-defined support threshold. In the sequel, we will be interested in maximal frequent termsets, which we will denote by MF and define as the set of all maximal (in the sense of inclusion) termsets that are frequent.

Let x be a term. By $MF(x)$ we denote all maximal frequent termsets containing x. $MF(x)$ will be used for determining *atomic contexts* for x. A termset $X, x \notin X$, is defined as an *atomic context* of term x if $\{x\} \cup X$ is an element of $MF(x)$. The set of all atomic contexts of x will be denoted by $AC(x)$:

$$AC(x) = \{X \backslash \{x\} \mid X \in MF(x)\}.$$

Clearly, for each two termsets Y, Z in $AC(x)$, Y differs from Z by at least one term and vice versa. In spite of this, Y and Z may indicate the same meaning of x in reality. Let y be a term in $Y \backslash Z$ and z be a term in $Z \backslash Y$ and $\{xyz\}$ be a termset the support of which is significantly less than the supports of Y and Z. This may suggest that Y and Z probably represent different meanings of x. Otherwise, Y and Z are likely to represent the same meaning of x. Please, note that $\{xyz\}$ plays a role of a *potential discriminant* for pairs of atomic contexts. The set of all potential discriminants for Y and Z in $AC(x)$, denoted by $\mathcal{D}(x, Y, Z)$ is:

$$\mathcal{D}(x, Y, Z) = \{\{xyz\} \mid y \in Y \backslash Z \wedge z \in Z \backslash Y\}.$$

Among the potential discriminants, those which are relatively infrequent are called *proper discriminants*. Formally, the set of *all proper discriminants* for Y and Z in $AC(x)$ will be denoted by $\mathcal{PD}(x, Y, Z)$, and defined as follows:

$$\mathcal{PD}(x, Y, Z) = \{X \in \mathcal{D}(x, Y, Z) \mid relSup(x, X, Y, Z) \leq \delta\}, \text{ where}$$
$$relSup(x, X, Y, Z) = sup(X)/min(sup(xY), sup(xZ)), \text{ and}$$
$$\delta \text{ is a user-defined threshold.}$$

In the sequel, $relSup(x, X, Y, Z)$ is called a *relative support of discriminant X* for term x with respect to atomic contexts Y and Z.

Our proposal of determining the groups of contexts representing separate meanings of x is based on the introduced notion of proper discriminants for pairs of atomic contexts.

Atomic contexts Y and Z in $AC(x)$ are called *discriminable* if there is at least one proper discriminant in $\mathcal{PD}(x, Y, Z)$. Otherwise, Y and Z are called *indiscriminable*.

A *sense-discriminant context* $\mathcal{SDC}(x, X)$ of x for termset X in $AC(x)$ is defined as the family of the termsets in $AC(x)$ that are indiscriminable with X:

$$\mathcal{SDC}(x, X) = \{Y \in AC(x) \mid \mathcal{PD}(x, X, Y) = \emptyset\}.$$

Clearly, $X \in \mathcal{SDC}(x, X)$. Please, note that sense-discriminant contexts of x for Y and Z, where $Y \neq Z$, may overlap, and in particular, may be equal.

The family of all distinct sense-discriminant contexts, denoted by $\mathcal{FSDC}(x)$ is:

$$\mathcal{FSDC}(x) = \{\mathcal{SDC}(x, X) \mid X \in AC(x)\}.$$

Please, note that $\mid \mathcal{FSDC}(x) \mid \leq \mid AC(x) \mid$.

A given term x is defined as a *homonym candidate* if the cardinality of $\mathcal{FSDC}(x)$ is greater than 1. Final decision on homonymy is given to the user. Let us also note that the more overlapping are distinct sense-discriminant contexts, the more difficult is reusing the contexts for the meaning recognition in the mining procedures.

Merging ontologies

Having loaded an ontology, one can merge it with detected proposals. In such a process, a number of conflicts may occur, e.g. the type of an attribute is different in both ontologies. The conflicts can be resolved either automatically or manually, though, some can be resolved only manually, e.g., if we have an attribute definition of *integer* type in the base ontology and an attribute definition of *string* type in the other one, we will have a conflict that cannot be resolved automatically. The users may set the way of resolving conflicts with conflict resolving setup window, which is presented on the picture below. The available options associated with a particular kind of conflicts are organized in the hierarchy where leafs represent single conflicts.

4 The TOM Platform

The TOM platform is thought as a universal environment that allows easy experimentations with various text mining algorithms for ontology building. The system is highly modular (based on plug-in architecture), and highly portable as Java has been used as the implementation language. At the top level it consists of the following subsystems:

Text mining subsystem. The text mining subsystem enables a user to specify both data source of an experiment and a pipeline of a text mining process (Fig. 1). Such a pipeline may consists of several steps for text processing, e.g. generation of the bag of words representation of documents, splitting the text into sentences, etc. In addition some text mining algorithms may be used as a step within a pipeline, for example clustering of documents, discovering frequent multi word terms, etc.

Analysis support subsystem. The analysis support subsystem is dedicated for working with the results obtained from text mining plan-and-experiment subsystem. For each type of results (clusters, sequences, candidate homonyms, etc.) the dedicated tools supporting basic analysis are available. Also the viewer of the *ftdoc* type documents is provided.

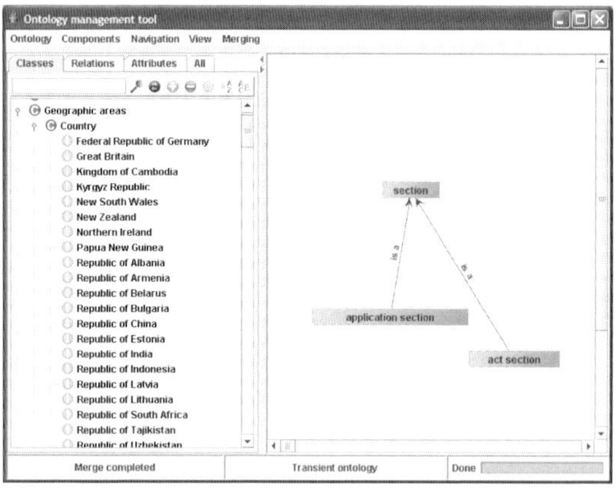

Fig. 2. Ontology managament subsystem

Ontology subsystem
The ontology subsystem is thought as a tool with a convenient graphical interface for working on ontologies, for example enriching ontologies, based on results obtained from text-mining experiments. It enables users to do various operations: such as browsing, annotating or validating ontologies. The subsystem also includes a tool by which a user may generate the owl file with proposals of new entries to ontologies. Such proposals are generated based on results obtained by using text-mining algorithm implemented in TOM.

Data storing subsystem. The data storing subsystem provides functionality concerning saving and searching for all text data used in the TOM platform. An integral part of this subsystem is an indexer (in TOM the Lucene [13] index is used). The indexer helps one to create fast-searchable database of text documents backed up with inverted index and vector document representation.

5 Project Results and Conclusions

For the experimental evaluation of the algorithms a FAO[1] document repository from the FAOLEX system was used. The full repository consists of more than 20000 national legislation documents concerning legal issues for food and agriculture. The repository is provided in various languages. We have selected 5658 documents written in English documents. Of it, we were able to extract 546.617 paragraphs and then 1.296.929 sentences. The experiments conducted on this corpus showed that especially the methods proposed by us for automatic homonymy and synonymy identification were giving very robust results (in many cases discovering information unknown previously to us, such as semantic relationship between names of various species and species groups). The hierarchy building algorithms do not eliminate the need for human intervention, they are very useful in the ontology building process, and definitely speed up the overall engineering process in ontology building – especially helpful was the ability to quickly generate a rough ontology skeleton that could be further improved by domain experts. The illustrations below present a small fragment of ontology that has been created with TOM by experiments run on the Faolex repository mentioned above.

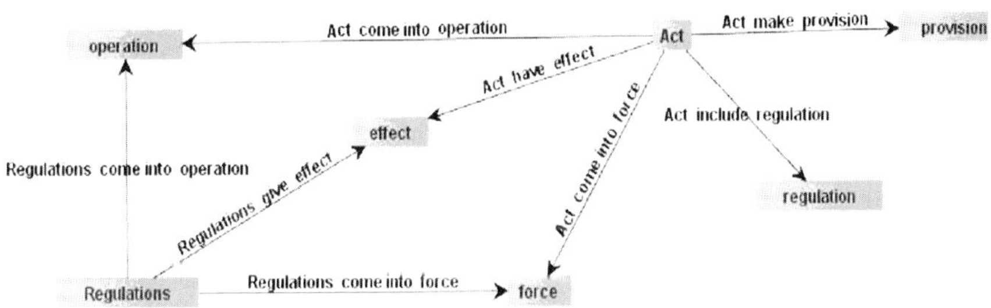

Fig. 3. A subset of ontology created from FAOEX repository with TOM

The research briefly described in this paper is far from being complete. While the initial project, that has been commission by France Telecom, has been completed successfully (i.e. the TOM platform can be readily used in order to speed up ontology building and to improve the quality of resulting ontologies) we will be extending the TOM platform system, with special emphasis on ontology merging and introduction of more knowledge-rich methods into the platform.

[1] FAO is a UN organization (Food and Agriculture Organization).

Acknowledgments. The work described in this paper results from the project funded by France Telecom.

References

1. Agrawal, R., Srikant, R.: Fast algorithms for mining association rules. In: Proc. of the 20th Int'l. Conf. on VLDB, Santiago, Chile, Morgan Kaufmann, San Francisco (1994)
2. Ahonen-Myka, H.: Finding all frequent maximal sequences in text. In: Mladenic, D., Grobelnik, M. (eds.) Proc. of the 16th Int. Con. on Machine Learning ICML 1999 Workshop on Machine Learning in Text Data Analysis, pp. 11–17 (1999)
3. Beil, F., Ester, M., Xu, X.: Frequent term-based text clustering. In: KDD 2002 (2002)
4. Byrd, R., Ravin, Y.: Identifying and extracting relations from text. In: NLDB 1999 - 4th Int. Con. on Applications of Natural Language to Information Systems (1999)
5. Faure, D., Nedellec, C.: A corpus-based conceptual clustering method for verb frames and ontology acquisition. In: LREC Workshop on Adapting Lexical and Corpus Resources to Sublanguages and Applications, Granada, Spain (1998)
6. Fung, B.C.M., Wan, K., Ester, M.: Hierarchical document clustering Using Frequent Item-sets. In: SDM 2003 (2003)
7. Grefenstette, G.: Evaluation Techniques for Automatic Semantic Extraction: Comparing Syntatic and Window Based Approaches. In: Boguraev, B., Pustejovsky, J. (eds.) Corpus processing for Lexical Acquisition, pp. 205–216. MIT Press, Cambridge (1995)
8. Guarino, N., Welty, C.: Evaluating ontological decisions with Ontoclean. Comm. of ACM 45(2) (2002)
9. Hamon, T., Nazarenko, A., Gros, C.: A step towards the detection of semantic variants of terms in technical documents. In: Proc. 36th Ann. Meeting of ACL (1998)
10. Harris, Z.: Distributional structure. Word 10(23), 146–162 (1954)
11. Skonieczny, K.M.: Hierarchical document clustering using frequent closed sets. In. Proc. IIPWM (2006)
12. Lame, G.: Using text analysis techniques to identify legal ontologie's components. In: ICAIL 2003, Workshop on Legal Ontologies & Web Based Legal Inf. Manag. (2003)
13. Lucene home page, `http://www.apache.org/lucene`
14. Maedche, A., Staab, S.: Ontology Learning, Handbook on Ontologies. Springer Series on Handbooks in Information Systems. Springer, Heidelberg (2003)
15. Maedche, A., Staab, S.: Mining Ontologies from Text. In: Dieng, R., Corby, O. (eds.) EKAW 2000. LNCS (LNAI), vol. 1937, pp. 189–202. Springer, Heidelberg (2000)
16. Morin, E.: Automatic acquisition of semantic relations between terms from technical corpora. In: Proc. 5th Int'l. Congress on TKE (1999)
17. Noy, F.N., McGuinness, D.L.: Ontology Development 101: A Guide to Creating Your First Ontology. Stanford Knowledge Systems Laboratory Technical Report KSL-01-05 and Stanford Medical Informatics Techn. Rep. SMI-2001-0880
18. Protaziuk, G., et al.: TOM Platform Reference Manual, Techn. Rep., WUT (2006)
19. Protaziuk, G., et al.: Discovering Compound and Proper Nouns. In: Kryszkiewicz, M., Peters, J.F., Rybinski, H., Skowron, A. (eds.) RSEISP 2007. LNCS (LNAI), vol. 4585, Springer, Heidelberg (2007)

20. Protaziuk, G., et al.: State of The Art on Ontology and Vocabulary Building & Maintenance Research And Applications, Techn. Rep., WUT (2006)
21. Rybinski, H., et al.: Discovering Synonyms based on Frequent Termsets. In: Kryszkiewicz, M., Peters, J.F., Rybinski, H., Skowron, A. (eds.) RSEISP 2007. LNCS (LNAI), vol. 4585, Springer, Heidelberg (2007)
22. Rybinski, H., et al.: Discovering Word Meanings Based on Frequent Termsets. In: MCD Workshop, PKDD, Warsaw (2007)
23. Velardi, P., Fabriani, P., Missikoff, M.: Using text processing techniques to automatically enrich a domain ontology. In: Proc. Int'l. Conf. on FOIS (2001)
24. Wu, H., Zhou, M.: Optimizing Synonym Extraction Using Monolingual and Bilingual Resources. In: Ann. Meeting ACL, Proc. 2nd Int'l Workshop on Paraphrasing, vol. 16, pp. 72–79 (2003)

Ontology-Driven Adaptive Medical Information Collection System

Matt-Mouley Bouamrane[1,2], Alan Rector[1], and Martin Hurrell[2]

[1] School of Computer Science
Manchester University, UK
{mBouamrane,Rector}@cs.man.ac.uk
[2] CIS Informatics, Glasgow, UK
martin.hurrell@informatics.co.uk

Abstract. Computer-based surveys and questionnaires have become ubiquitous. Yet in many cases, Information Collection Systems (ICS) offer limited support in terms of tailoring structure and content of surveys in response to user interaction. Previous techniques for content adaptation such as conditional branching do not scale well and are also hard to maintain as structural dependencies in a survey often need to be hard-coded in the system. We here propose a generic model for context-sensitive self adaptation of ICS, based on a questionnaire ontology. We illustrate the model with a description of our own medical ICS implementation and discuss the potential benefits of such system, especially in the context of tailored healthcare.

1 Introduction

Information Collection Systems (ICSs) such as online surveys and questionnaires offer many advantages. By automating the survey task, they remove the need for personal interviews and manual data entry, generating time-saving and lower cost. Although the use of ICSs is now widespread, the technology behind them generally remains surprisingly unsophisticated, typically a combination of dynamic server pages and databases. Efforts to improve ICSs are often exclusively focused on potential usability and navigation issues, typically centred on interaction with the Graphical User Interface (GUI) [1,2]. Context-sensitive adaptation to user interaction is usually either minimally addressed or simply non-existent. In a medical environment, the potential advantages offered by computer-based ICSs in comparison to traditional manual and face-to-face interviews are many fold. Automating the process can free up precious resources and let clinicians refocus their efforts on their primary mission of administrating medical care. ICSs can support the use of multiple languages and multimedia support can be provided to people with special needs (e.g. using speech for the visually impaired). In addition, the information collected by computer-based ICS is often more structured and detailed than when collected through other means [3].

However, the advantages offered by ICSs do not resolve all problems. One important issue lies in the challenges involved in designing a questionnaire which

A. An et al. (Eds.): ISMIS 2008, LNAI 4994, pp. 574–584, 2008.

remains general enough to apply to a majority of patients, while capturing at the same time critical information from individual patients. In this paper, we propose a solution to this dilemma by describing an ontology-based adaptive information collection system. By *"adaptive"* we mean a dynamic modification of the behaviour of the application (i.e. structure of the questionnaire) in response to user interaction (context-sensitive self-adaptation) [4]. The proposed method permits to iteratively capture finer-grained information with each successive step, should this information be relevant according to a questionnaire ontology. In addition, we argue that this method is robust, scalable and highly configurable. Although the method is presented in a medical context, the principles are generic.

The paper is organised as follows: we first present on overview of adaptive technology in medical ICSs, and introduce biomedical ontologies. We then present our proposal for an adaptive information collection model, using a generic questionnaire ontology. We illustrate the model with a practical adaptive ICS implementation. We conclude with a discussion on the advantages of the proposed model and its implications for tailored healthcare.

2 Background and Related Work

Adaptive Systems in Medical ICSs. The earliest implementation of adaptive behaviour in a medical questionnaire, conditional branching is probably also the most commonly used. It essentially consists in hard-coding in the ICS the dependencies between a given input and the resulting questions (e.g. through IF-THEN constructs). The method can quickly become hard to manage in complex cases. Also, systems designed on branching are hard to maintain and the sequence of potentially related questions can not easily be altered without considerable engineering work. Another approach was proposed with the OpenSDE (Open Structured Data Entry) system [5]. Originally designed as a generic method for capturing heterogeneous medical information in a structured, yet flexible way, the model was subsequently extended to create generic adaptable questionnaires. There are two elements to the adaptive survey: (i) the domain model which contains medical concepts and (ii) the actual questionnaire. Both are modelled as hierarchical trees. Adaptability is achieved by designing the questionnaires in a succession of question-answer branches: a question node is always followed by an answer node, which itself can either be followed by further questions (depending on input) or an end-node should there be no further question on this particular branch. The questionnaire tree is navigated through depth-first traversal so potential dependencies in a single question branch are explored first before moving on to the next question.

Adaptability can also be achieved by modelling a questionnaire as a Finite State Machine (FSM). This is the approach taken by the Transactional Electronic Diary, an application designed to monitor the behaviour and moods of participants during their day-to-day activities [6]. In FSM modelling, certain actions may trigger transitions from a certain state to other states depending on the fulfilment of a number of preconditions. If dependencies between the questions become more tangled, the model can also quickly become very complex.

Background. We use the term "ontology" to describe a domain knowledge formal representation, produced as a result of a knowledge engineering process [7]. This formal representation will typically be in the form of a hierarchy of concepts (called *classes*), complemented with links between various classes (through the use of *properties*). The main advantages of using ontologies in the computer domain are to be found in computational knowledge (i) *sharing* and (ii) *reuse*. The use of ontologies is particularly suited to applications in biology and medicine because these domains are knowledge-intensive and have typically witnessed an exponential growth in data [8]. OWL is a highly expressive ontology language created as part of efforts surrounding the development of the semantic web [9]. It represents a domain knowledge using formal semantics such as subsumption (hierarchical property inheritance), equivalence, disjointness, union, intersection, etc. OWL comes in several sublanguages. Using OWL-DL (Description Logic), the formal semantics expressed in an ontology can be used by a reasoner to perform certain inferences on the ontology (classification or "reasoning") and uncover relations which were not explicitly asserted in the ontology.

3 Adaptive Questionnaire Ontology Model

We will the following notation: words starting with a capital letter refers to the classes of an ontology (e.g. Question) while properties start with a lower case. We will use the "camel back" notation: CamelBack refers to a class and hasCamelBack refers to a property.

Generic Questionnaire Classes. To develop a generic questionnaire ontology, we identified and extracted structural elements from a number of medical questionnaires. The resulting subsumption is illustrated in Figure 1. We briefly describe the main classes in the ontology:

The Questionnaire class is composed of *Subquestionnaires*, which are a group of thematically related Question classes.

Subquestionnaire grouping can permit *semantic partitioning* of the questionnaire. This technique has proved to be helping patients to recall information, while reducing the number of misinterpreted questions due to an inappropriate context [2].

The StartOfQuestionnaire class points to the first question in the questionnaire and thus can permit to stop a questionnaire at any time and to later resume it from that point.

Question classes encapsulate the necessary information to determine the run-time behaviour of a questionnaire implementation. This information includes: the set of valid answers to a specific question, information on how to display the question on the UI and a set of valid actions. As an example, the UI should only allow the user to give one answer to a *MultipleChoiceQuestionWithSingleAnswer*.

The FurtherQuestion is an important class used to model the fact that adaptive questions potentially lead to further questions.

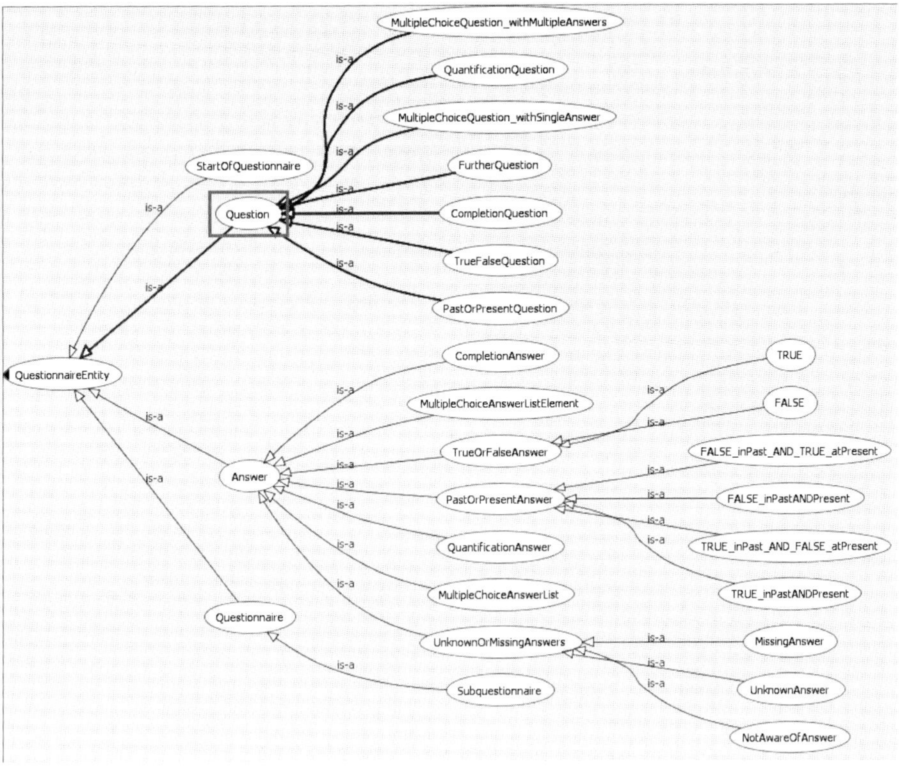

Fig. 1. The generic questionnaire ontology

Answer classes essentially mirrors the Question classes while encapsulating information subtleties which can be critical in the medical domain. Examples of such classes are *TrueInPastAndAtPresent* (e.g. "I take this medication"), *FalseInPastAndAtPresent* (e.g. "I have never taken this medication"), *TrueInPastAndFalseAtPresent* (e.g. "I use to take this medication but not anymore").

Generic Questionnaire Properties. The next step consists in defining how the questionnaire is (i) *structured* and how it (ii) *behaves*. This is done through the use of properties. Figure 2 lists the properties we found necessary to define in the ontology in order to build arbitrarily complex questionnaires. There are four main sets of properties: structural properties, composition properties, type properties and finally, adaptive properties.

The Type property *hasQAType* is simply used to define the nature of the Question or Answer classes (e.g. a Question hasQAType TrueFalseQuestion).

Structural properties are used to describe the structure of the questionnaire: *containsQuestionAbout* is used to define which are the questions contained

Fig. 2. Structural, composition and adaptive properties of the questionnaire ontology

in a Questionnaire, or Subquestionnaire. The *startsAtQuestion* property is used to link a Questionnaire to its corresponding StartOfQuestionnaire class. The *hasAssociatedSubquestionnaire* property is used to link a questionnaire, or Subquestionnaire to other Subquestionnaires. The *hasExpectedAnswers* property is used to link a specific question to its potential answers. The *hasAlwaysRelatedQuestion* can be used when one question is always followed by another.

Composition properties are used to combine classes together. An example is to combine questions (e.g. *how much* of this medication do you take *and how often*?) or combine answers from several locations in the ontology.

Adaptive properties are the most relevant to the purpose of this paper as they determine the dynamic behaviour of the questionnaire. The *hasPotentialRelatedQuestion* is the property used to flag down a question as an adaptive question. The *ifAnswerToThisQuestionIs* property is used to link the current Question to potential related questions depending on the user's answer to this question. The *ifAnswerBelongsTo* property is a generalisation of the previous property, for which the actual answer does not matter as long as it is subsumed by a certain superclass. A follow-up question to a specific answer is defined via the *thenGoToQuestion* property. The *hasFurtherQuestionIfPreviousQuestionIs* property adds context to the adaptation whereby certain questions become relevant depending on the path that led to the current question. The *addSubsequentQuestion* property is used to add more than one question at a time (e.g. first ask A *and then* ask B, where B is the subsequent question). Finally, two adaptive properties are more specifically targeted at the medical domain: they are *onlyIfAgeStateIs* and *onlyIfSexIs* as many medical questions can be age and sex specific.

We believe that one of the advantages of our proposed model is its relative simplicity. It has less than 30 classes and 30 properties, yet it permits the design of arbitrarily large and complex questionnaires, as we will demonstrate in the following sections, using a practical example of a medical questionnaire.

Adaptation Model: The Last State – Next State Model. Context sensitive adaptation methods previously described in the Related Work section of this paper, often require considerable engineering effort in order to update the system. More importantly however are issues of complexity and scalability of the adaptation method. These last two issues are closely interlinked as design complexity grows dramatically as the number of questions and the dependencies between them increase in the survey. In order to avoid this complexity and scalability pitfall, we propose an adaptation model where issues of dependencies are confined in scope within a short window. This design principle, which we call the Last State - Next State (LSNS) model, is illustrated in Figure 3. The process is somehow similar to a Markov model as it only exhibits short-term memory. At any one time, all the system needs to know in order to implement adaptability is: (i) what is the current state of the system (which question class is currently processed), (ii) what was the previous state of the system (what was the previous question and what answer was provided) and (iii) what are the potential next states (where does the system go from here given the user input). Once the next state has been selected, it becomes the new current state while the previous current state becomes the last state. This short-term memory model means that the designer of a questionnaire ontology need not worry about all possible interlinks between the various elements of the questionnaire. Instead, he only needs to focus on tackling dependencies at a single point in the survey at any one time (e.g. having got here, where do I go next?). We argue that the model is scalable and reduces the complexity of designing adaptive ICSs while providing system designers with maximum flexibility.

Modelling of an Adaptive Medical Questionnaire. The questionnaire ontology needs to models two distinct aspects: (i) the medical information and (ii) the structural and adaptive behaviour of the questionnaire. Figure 4 illustrates a questionnaire ontology generated using the Protégé-OWL development tool [10]. The left-hand side of the Protégé-OWL UI in Figure 4 shows part of the adaptive questionnaire subsumption. One specific class is highlighted: RespiratorySystemHistoryQFeature (item 1 on the figure). On the right-hand side, the UI displays the class annotation and description. A class description may include a number of equivalent classes, superclasses, disjoint

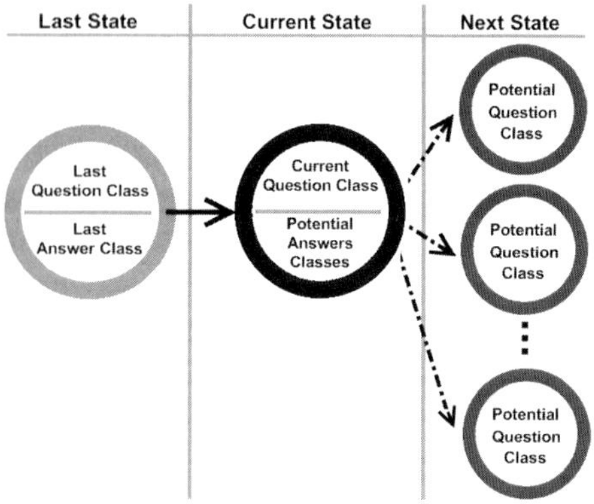

Fig. 3. The Last State – Next State (LSNS) model

classes, etc. In this case, the description corresponds to a list of superclasses of RespiratorySystemHistoryQFeature. The class description item 2 tells us that RespiratorySystemHistoryQFeature is a Question of type MultipleChoiceQuestionWithMultipleAnswers. Item 3 is a class annotation and contains the actual content of the question, which should be displayed on the UI: in this case, the string *"Which lung condition?"*. Item 4 describes where the potential answers to this question are to be found. In this case, it corresponds to the RespiratorySystemPathology class (item 5) which is defined as a MultipleChoiceAnswerList (information not visible on the screenshot). The subclasses of RespiratorySystemPathology are of the class MultipleChoiceAnswerListElement and correspond to the potential answer classes of the RespiratorySystemHistoryQFeature question (e.g. Asthma, Atelectasis, Bronchitis, etc.)

Implementation of LSNS adaptability model. Item 6 on Figure 4 defines the RespiratorySystemHistoryQFeature question as an adaptive question: it may lead to further questions depending on the user input. Let's consider RespiratorySystemHistoryQFeature to be the current state in the previously described LSNS model. Thus the system would prompt on the UI the following question: *"Which lung condition?"* and a list of possible choices, consisting of the various elements under the category respiratory system pathology (e.g. Asthma, Bronchitis, etc.) The potential next states are modelled by the items labelled 7, 8 and 9. The simplest form is the one illustrated by the items labelled 7, an example of which is: {ifAnswerToThisQuestionIs some PulmonaryEmbolism} and {thenGoToQuestion some DateOfLastClinicalEventFeature}. This means that if the user's answer to the current question is PulmonaryEmbolism, then the system will move to the DateOfLastClinicalEventFeature class (e.g. "when did this last happen?") which now becomes the new current state of the system. Items 8 are variations of adaptation whereby a number of questions are combined together and both are asked simultaneously (e.g. if you have asthma, the system needs to find out *how frequent* are the symptoms and *how severe*). Items 9 are yet

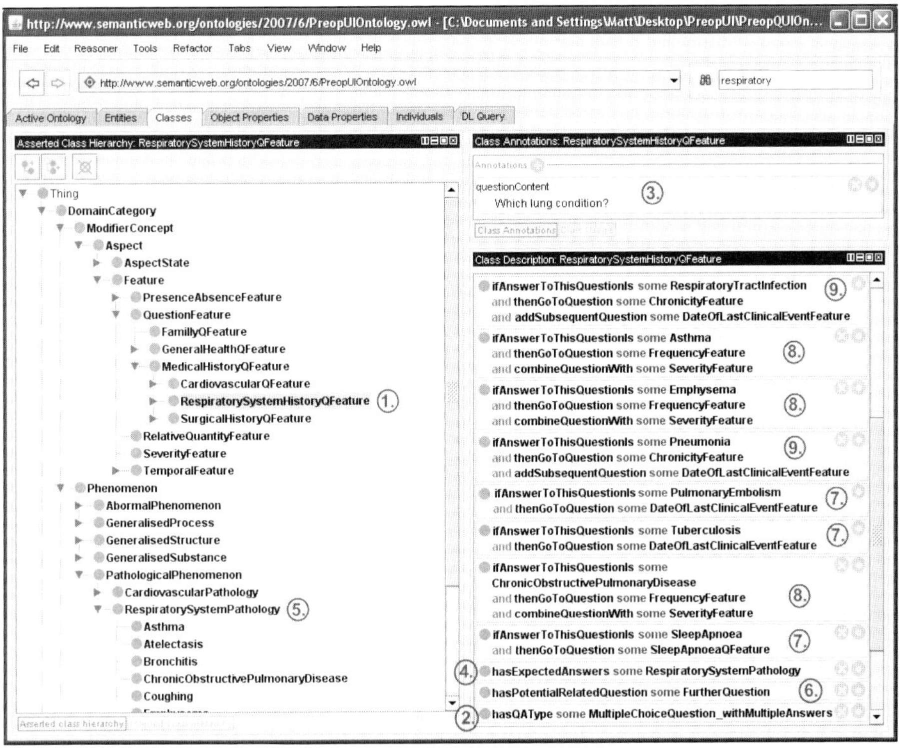

Fig. 4. Adaptive questionnaire model ontology as viewed through the Protégé-OWL User Interface

another variation whereby several questions are added but not simultaneously. Thus, a high degree of flexibility in the adaptability behaviour can be achieved through combining the various properties listed on Figure 2.

4 System Implementation

The adaptation method here described is intended to be used in a preoperative assessment software for elective (i.e. non-emergency) surgery. The aim of the software is to gather precise and reliable patient medical history so a proper and informed patient risk assessment can be performed. Initially, the software will be used in a hospital setting, under the supervision of preoperative nurses. The system could also be expanded to allow remote login for patients judged to have the necessary abilities (physical, technical and cognitive) to fill the questionnaire from home using standard internet access. Another option under consideration is to provide the software to general practitioners in primary care as a decision support tool for potential referral to specialist care.

Figure 5 illustrates the architecture of our adaptive ICS implementation. It has three main components: the user interface, the Java adaptive engine and the questionnaire ontology. Our current Java adaptive engine is implemented using the OWL 1.1 API [11]. The engine is responsible for interpreting the structural, composition and adaptive properties asserted in the ontology and to trigger appropriate responses given user input. The interaction loop is as follows: the system initially prompts the first question (corresponding to the StartOfQuestionnaire class in the ontology). Once the user has selected an answer, the adaptive engine first checks whether the current question is adaptive or not (i.e. whether it is a subclass of {*hasPotentialRelatedQuestion some FurtherQuestion*}). If it is not, the system just prompts the next question in the list. If however the current state question happens to be adaptive, the given answer is then checked against the answers which are expected to lead to potential next states. If a match is found, the system moves to the new question state. If no match is found, the next question in the list is displayed on the UI. The interaction loop is repeated until there are no more questions to be asked.

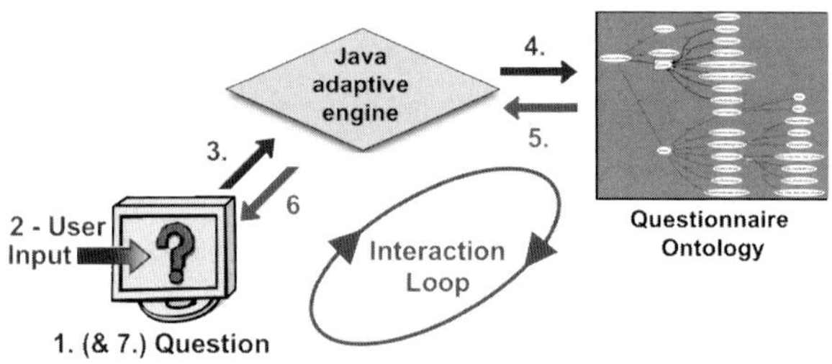

Fig. 5. The adaptive questionnaire system architecture

The separation of the adaptive engine from the underlying ontology leaves developers with considerable freedom for discretionary implementation of the run-time behaviour of the ICS, including the choice of implementation model. We chose to model our own implementation of the questionnaire as a stack, or rather, as a stack of stacks: with individual stacks corresponding to sub-questionnaires within the main questionnaire stack. The run-time behaviour of the adaptive questionnaire is illustrated by Figure 6, in which each step corresponds to a system iteration (i.e. a full Figure 5 loop). Question 1 (Q_1) does not have any adaptive properties and leads directly to Q_2 regardless of the answer (step $1 \rightarrow 2$). Question 2 does have adaptive properties, however, the user input did not trigger a call for further questions and thus also leads directly to the next question Q_3 (step $2 \rightarrow 3$). In step $3 \rightarrow 4$, unlike the previous case, the answer to Q_3 triggers the call for a further question. This additional question ($Q_{3.1}$) now seats on top of the stack (next question to appear on the UI). Finally, the answer to $Q_{3.1}$ triggers the call for three additional questions ($Q_{3.1.1}, Q_{3.1.2}$ and $Q_{3.1.3}$). These additional questions now seat on top of the question stack, in the order of priority

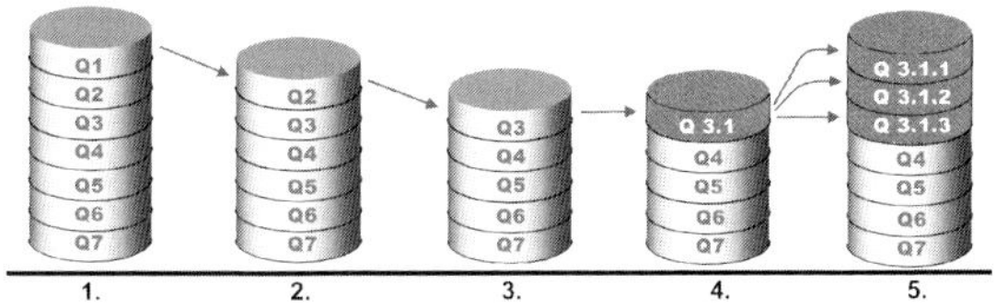

Fig. 6. Stack implementation of the adaptive questionnaire

asserted in the ontology. Depending on the adaptive properties of the remaining questions, the process of adding further questions could be iteratively repeated until the engine finally reaches the bottom of the stack (end of questionnaire).

The original questionnaire used to develop the ontology described in this paper was initially composed of 50 questions and offered very limited adaptation (conditional branching). The adaptive questionnaire reduced the minimum number of possible questions from 50 to 27: a reduction of nearly 50% for a healthy patient. On the other hand, it also increased the total number of questions to more than 90, including 22 adaptive questions and 27 multiple choice questions. This results in a far more precise patient medical history than was previously possible.

5 Discussion and Future Work

A recurring concern when implementing context sensitive self-adaptation in a medical questionnaire is: "is it safe? What are the liabilities involved and can we trust the system to make adequate judgement?" As already suggested in the introduction, ICSs generally gather better medical history than clinicians [3]. As for the safety of the system, it simply means engineering adaptation so it only applies to questions *one can afford not to ask* (e.g. such as questions $Q_{3.1.1}$ to $Q_{3.1.3}$ in Figure 6). All questions which are critical to the safety of the patient should be asked regardless (e.g. Q_1 to Q_7). Our experience with current practice of preoperative assessment in hospitals suggest that patients are systematically asked all the questions in a static questionnaire. However, certain answers may prompt clinicians to seek further information if there is any area of concern (e.g. if the patient is seeing another specialist, what are the reasons?). Our proposed adaptation method intends to replicate this investigating behaviour. While the system has the potential to reduce the number of questions and thus save time and costs for healthy patients, the emphasis is rather on collecting, *if necessary, more information* so that ultimately surgeons and anaesthetists have all relevant information to perform a proper patient risk assessment. Our original questionnaire before adaptation had a number of short-comings: a disadvantage for the patients was that the number of questions was fixed regardless of their personal circumstances. A disadvantage for clinicians was that the history taken was far from complete. A disadvantage for the system developers was that any change to the structure of the questionnaire required significant software engineering work.

Updating the system on clinical sites potentially caused delays and disruptions to the service. Because of the clean-cut separation between the questionnaire ontology and the implementation system in our proposed system, modifying the structure and behaviour of the adaptive questionnaire now only requires modifying the ontology. Future work will include a field study in clinical sites in order to measure how support for content adaptation in the ICS translates in improved personalised healthcare. Another avenue for research will be to use the reasoning power of an OWL ontology to add decision-support capabilities to the system.

References

1. Couper, M.P.: Usability evaluation of computer-assisted survey instruments. Social Science Computer Review 18(4), 384–396 (2000)
2. Norman, K.L., Friedman, Z., Norman, K., Stevenson, R.: Navigational issues in the design of online self-administered questionnaires. Behaviour & Information Technology 20(1), 37–45 (2001)
3. Bachman, J.W.: The patient-computer interview: a neglected tool that can aid the clinician. Mayo Clinic Proceedings 78, 67–78 (2003)
4. Jameson, A.: Adaptive interfaces and agents, 305–330 (2003)
5. van Ginneken, A.M., de Wilde, M., Blok, C.: Generic computer-based questionnaires: an extension to opensde. In: Proceedings of 11th World Congress on Medical Informatics, MEDINFO, pp. 688–692 (2004)
6. Vahabzadeh, M., Epstein, D., Mezghanni, M., Lin, J.L., Preston, K.: An electronic diary software for ecological momentary assessment (EMA) in clinical trials. In: Proceedings of 17th IEEE Symposium on Computer-Based Medical Systems, CBMS 2004, Bethesda, US, pp. 167–172. IEEE Computer Society, Los Alamitos (2004)
7. Gómez-Pérez, A., Fernández-López, M., Corcho, O.: Ontological Engineering. Advanced Information and Knowledge Processing series. Springer, Heidelberg (2003)
8. Yu, A.: Methods in biomedical ontology. Journal of Biomedical Informatics 39(3), 252–266 (2006)
9. Horrocks, I., Patel-Schneider, P., van Harmelen, F.: From SHIQ and RDF to OWL: The making of a web ontology language. Journal of Web Semantics 1(1), 7–26 (2003)
10. Knublauch, H., Fergerson, R.W., Noy, N.F., Musen, M.A.: The Protégé OWL Plugin: An Open Development Environment for Semantic Web Applications. In: McIlraith, S.A., Plexousakis, D., van Harmelen, F. (eds.) ISWC 2004. LNCS, vol. 3298, pp. 229–243. Springer, Heidelberg (2004)
11. Horridge, M., Bechhofer, S., Noppens, O.: Igniting the owl 1.1 touch paper: The owl api. In: Proceedings of the third International Workshop of OWL Experiences and Directions, OWLED 2007, Innsbruck, Austria (2007)

An Effective Ontology Matching Technique

Ahmed Alasoud, Volker Haarslev, and Nematollaah Shiri

Computer Science & Software Engineering, Concordia University
1455 De Maisonneuve West, Montreal, Quebec, Canada
{ahmed_a,haarslev,shiri}@cse.concordia.ca

Abstract. In this paper, we study the ontology matching problem and propose an algorithm, which uses as a backbone a multi-level matching technique and performs a neighbor search to find the correspondences between the entities in the given ontologies. A main feature of this algorithm is the high quality of the matches it finds. Besides, as the result of the initial search introduced, our algorithm converges fast, making it comparable to existing techniques.

1 Introduction

Ontology matching is a fundamental problem in sharing information and integrating ontology sources in numerous applications. We witness a continuous growth in both the number and size of available ontologies. This, on the other hand, has resulted in an increased heterogeneity in the available information. For example, the same entity could be given different names in different ontologies or it could be modeled or described in different ways. The ontology Matching Problem (OMP) is as follows: given ontologies O1 and O2, each describing a collection of discrete entities such as classes, properties, individuals, etc., we want to identify semantic correspondences between the components of these entities. This problem has been the subject of numerous studies, resulting in the development a number of useful and interesting tools and techniques.

Existing matching algorithms often focus on matching a pair of entities at a time, and hardly consider matching n entities to m entities at the same time, and correspondingly use several similarity measures to solve OMP. We view OMP as an n:m matching problem. Furthermore, to improve matching results, we believe existing methods should be used simultaneously and combined in a multi-level matching framework. This is the subject of our study in this paper. We introduce a neighbor search algorithm, with a proper initialization as an optimization for our multi-level matching algorithm proposed in [1], which improves the matching quality as well as computation time. We have developed a running program and conducted experiments. Our results indicate the proposed neighbor search is effective in improving our multi-level matching algorithm, by finding quality matches efficiently.

In Section 2 we give background definitions. Our search algorithm is introduced in Section 3. An illustrative scenario is provided in Section 4. The experiments and results are presented in Section 5. Section 6 reviews related work, and Section 7 includes concluding remarks and a discussion of future work.

A. An et al. (Eds.): ISMIS 2008, LNAI 4994, pp. 585–590, 2008.

2 Background

In this section, we provide some definitions of concepts and terms used in our work.

Definition 1 (Entity-relationships). Let S be a source ontology and T a target ontology. We use $E^S = \{s_1, s_2, \ldots, s_n\}$ and $E^T = \{t_1, t_2, \ldots, t_m\}$ to denote the set of entities in S and T, respectively. In this work, we limit ourselves to finding mappings for entities of types classes and relationships only.

Definition 2 (Similarity Matrix). This relational matrix, denoted $L(l_{ij})$, includes values in the range [0,1], called the *similarity coefficients*, denoting the degree of similarity between s_i and t_j.

Definition 3 (Matching Matrix). A matching matrix, denoted $Map_{0\text{-}1}$, is a 0-1 matrix with dimension n×m and with entries $r_{ij} \in \{0,1\}$. If $r_{ij} = 1$, it means that s_i and t_j are "matchable." They are unmatchable if $r_{ij} = 0$.

Definition 4 (Matching Space). Matching space includes all possible assignments for the matching matrix, called the *mapping space*. Every assignment is a state in the matching space and represents a solution for the ontology matching problem.

3 A Neighbor Search Algorithm

The proposed neighbor search algorithm has three phases, described in Fig. 1.

```
Algorithm Match(S, T)
begin
/* Initialization phase
     K ← 0 ;
     St_0 ←preliminary_matching_techniques(S,T);
     St_f ←St_0 ;
/* Neighbor Search phase
     St ←All_Neighbors(St_n);
     While (K++ < Max_iteration) do
/* Evaluation phase
             If score(St_n) > score(St_f) then
                     St_f ←St_n ;
             end if
             Pick the next neighbor St_n ∈ St ;
             St ← St - St_n ;
             If  St = ∅ then Return St_f ;
     end
     Return St_f ;
end
```

Fig. 1. The Search Algorithm

First, in the initialization phase, a partial set of similarity measures is applied to the input ontologies to determine a single initial sate St_0 for the search algorithm. In the second phase, we search in the neighborhood of the initial state. The neighbors of state St_0 are the mapping states that can be computed either by adding to or removing from St_0 a couple of vertices, obtained by toggling a bit in the similarity matrix L. So, the total number of the neighbor states will be $n*m$. We evaluate the neighbor states using the following score function v:

$$v = \left(Map_{0-1} \cdot L\right)/k = \sum_{i=1}^{n}\sum_{j=1}^{m} Map_{0-1}(i,j).L(i,j) \Big/ \sum_{i=1}^{n}\sum_{j=1}^{m} Map_{0-1}(i,j) \geq th \ .$$

where $K \geq \min(n,m)$ is the number of matched pairs, n is the number of entities in S, and m is the number of entities in T.
In the third phase (evaluation phase), the algorithm will apply the next level(s) similarity techniques in order to find St_f, the best possible matching state solution.

4 Illustrative Example

Consider simple examples shown in Fig. 2, which are taxonomies for computer ontologies O_1 and O_2.

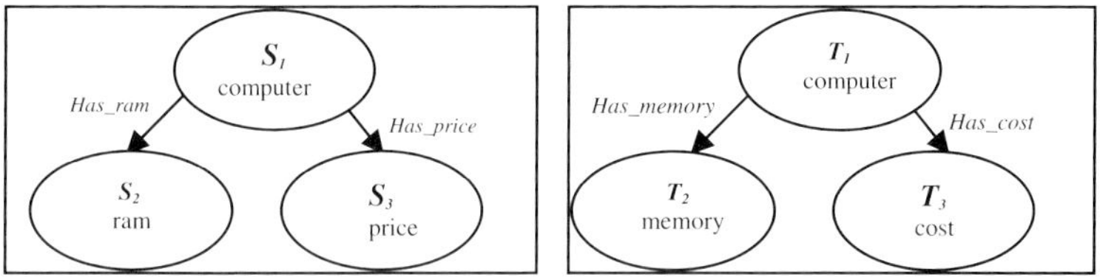

O_1 O_2

Fig. 2. Computer Ontology Examples

For ease of explanation, we only use three different similarity measures applied in two different phases. There are two similarity measures applied in the first phase to compute the initial state St_0: name similarity (Levenshtein distance) [2] and linguistic similarity (WordNet) [11]. This yields two similarity matrices for the concepts. The first matrix based on name similarity, and the second matrix based on linguistic similarity. Assuming that $th \geq 0.45$, and after normalizing the cost of the two similarity matrices, we get the matrix L. Then L is transformed into the matching matrix Map_{0-1}. Note that we are using Map_{0-1} and St_n as synonymous.

$$L = \begin{bmatrix} 1.0 & 0.4 & 0.265 \\ 0.463 & 0.534 & 0.083 \\ 0.363 & 0.158 & 0.5 \end{bmatrix} \qquad \text{Map}_{0\text{-}1} = \begin{bmatrix} 1 & 0 & 0 \\ 1 & 1 & 0 \\ 0 & 0 & 1 \end{bmatrix}$$

The binary matrix $\text{Map}_{0\text{-}1}$ above corresponds to state $St_0=\{(s_1, t_1), (s_2, t_1), (s_2, t_2), (s_3, t_3)\}$, which says entity s_1 is matched to t_1, s_2 is matched to both t_1 and t_2, and s_3 is matched to t_3. Table 1 indicates the binary matrix for other neighboring states together with their score values. In the search phase, 9 neighbors of St_0 will be evaluated to pick the best candidate(s) for the next level. To reduce the cost of the evaluation phase, we filter the neighbor states by keeping $\lceil x\% \rceil$ of the top weighted states for the next level. In phase three, we applied our structure similarity measure proposed in [1]. Finally, the search algorithm will output St_4 which has a highest overall score value, for being structurally more similar.

Table 1. Score value for each state neighbor

Neighbor number	**Matched pairs**	**Score value based on our score function V_{stn}**
St_{n1}	$\{(s_2, t_1), (s_2, t_2), (s_3, t_3)\}$	0.499
St_{n2}	$\{(s_1, t_1), (s_1, t_2), (s_2, t_1), (s_2, t_2), (s_3, t_3)\}$	0.5794
St_{n3}	$\{(s_1, t_1), (s_1, t_3), (s_2, t_1), (s_2, t_2), (s_3, t_3)\}$	0.5524
St_{n4}	$\{(s_1, t_1), (s_2, t_2), (s_3, t_3)\}$	**0.678**
St_{n5}	$\{(s_1, t_1), (s_2, t_1), (s_3, t_3)\}$	0.6543
St_{n6}	$\{(s_1, t_1), (s_2, t_1), (s_2, t_2), (s_2, t_3), (s_3, t_3)\}$	0.516
St_{n7}	$\{(s_1, t_1), (s_2, t_1), (s_2, t_2), (s_3, t_1), (s_3, t_3)\}$	0.572
St_{n8}	$\{(s_1, t_1), (s_2, t_1), (s_2, t_2), (s_3, t_2), (s_3, t_3)\}$	0.531
St_{n9}	$\{(s_1, t_1), (s_2, t_1), (s_2, t_2)\}$	0.6656

5 Experiments and Results

Case study (1): In this case study we used the OAEI 2007 benchmark test samples suite [13]. Except for case 206, which is related to French translation, in all other cases we considered, when the *precision* value was less than 1 the *recall* value was equal to 1. We noted all the systems we considered produced all the correct mappings, together with some additional unwanted mappings. The precision of our search algorithm on the other hand we observed did not fall below the *recall* value, i.e., no extra unwanted mappings returned by our framework. However, in test case 206, the reason that the matching result of our search algorithm was not fulfilled was that it did not use translating techniques as one of its underlying techniques. Fig 3 shows the comparison of matching quality of our algorithm and the other 10 systems. To measure a match quality, we have used the following indicators: *precision*, *recall*, and *F-measure*.

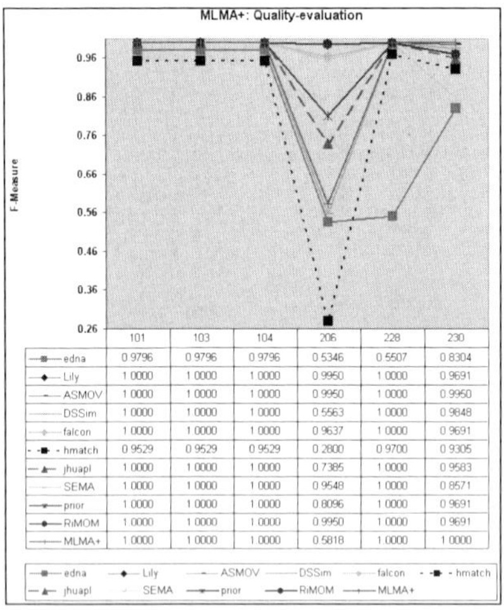

Fig. 3. Quality Comparison

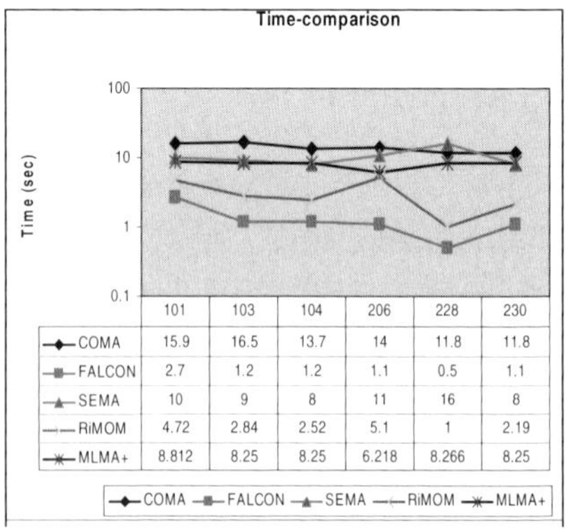

Fig. 4. Efficiency Comparison

The version computed here is the harmonic mean of precision and recall [3]. Moreover, Fig. 4 shows an approximate time comparison indicating the scalability of our search algorithm (logarithmic scale). We use $MLMA^+$ to refer to MLMA with the proposed neighbor search algorithm included.

Case study (2): In this case study we used three pairs of ontologies: (1) the MIT bibtex ontology[1] and the UMBC publication ontology[2] which are publicly available, (2) computer ontologies, and (3) ontologies about computer science departments. We have created the second and third pairs of the ontologies. The execution time in seconds for our algorithm over these test cases we measured was, 4.68, 0.547, and 1.719, respectively. A naïve implementation of MLMA would not perform as desired. The MLMA+ is polynomial with respect to the size of the search space $O((|E^S|\times|E^T|)^2)$, where $|E^S|$ is the number of entities in S. All in all, we consider the proposed algorithm as an optimization for MLMA, which we called $MLMA^+$.

6 Related Work

The RiMOM system [9] integrates multiple strategies and applies a strategy selection method to decide the strategy will rely more on it. The proposed method in [12] recommends an alignment strategy for a given alignment problem. The work in [10] has a matching engine which contains diverse libraries that support many match algorithms and strategies. Falcon-AO [8] has two Linguistics matchers and one structural matcher. The results of Falcon-AO were derived either from linguistic or structural matchers. Otherwise, the Falcon-AO results will be generated by combining both matchers with a weighting scheme. Some researchers propose a similarity metric between concepts in OWL ontologies [4] is a weighted combination of similarities of various features in OWL concept definitions. Algorithms such as the one proposed in [7] make use of derived graphs or alternative representations like pair-wise connectivity graphs.

There are three features which make our approach distinct from the aforementioned algorithms and systems. The first is that our matching results are guided by the fact that *n* entities at a time are matched to *m* entities. The second is in the way similarities are transformed into mappings and measured using our multi-match technique in order to deal with a many to many match problem. The third difference is the neighbor search method we introduced for MLMA to improve its efficiency.

7 Conclusions and Future Work

We proposed a neighbor search algorithm, which given an initial mapping state among entities in two ontologies, searches the neighboring states and returns a list of states ranked based on their evaluation scores. We incorporated this search algorithm into our multi-level match algorithm (MLMA) proposed in [1]. This results in $MLMA^+$, a framework for solving ontology match problem, which improves the efficiency of

---

[1] http://visus.mit.edu/bibtex/0.1/bibtex.owl
[2] http://ebiquity.umbc.edu/ontology/publication.owl

MLMA considerably, due to its use of the neighbor search algorithm. It proceeds by computing an initial state and then performing a search in its neighboring states. We have developed a running prototype of $MLMA^+$ and conducted experiments using some well-known benchmark ontologies. Our results indicated that the proposed search technique improved the overall performance of MLMA. A main characteristic of $MLMA^+$ is its improved efficiency over the basic MLMA obtained through the initial search. We are working on combining the search with machine learning techniques to further improve efficiency and accuracy of $MLMA^+$.

Acknowledgements: This work was supported in part by grants from Natural Sciences and Engineering Research Council (NSERC) of Canada, and by Libyan Ministry of Education.

References

1. Alasoud, A., Haarslev, V., Shiri, N.: A Multi Level Matching Algorithm for Combining Similarity Measures in Ontology Integration. In: Collard, M. (ed.) ODBIS 2005/2006. LNCS, vol. 4623, pp. 1–17. Springer, Heidelberg (2007)
2. Cohen, W., Ravikumar, P., Fienberg, S.: A Comparison of String Distance Metrics for Name-Matching Tasks. In: IJCAI 2003, pp. 3–78 (2003)
3. Do, H.H., Melnik, S., Rahm, E.: Comparison of schema matching evaluations. In: Proc. workshop on Web and Databases (2002)
4. Euzenat, J., Valtchev, P.: Similarity-based ontology alignment in OWL-Lite. In: Proc. 16th European Conference on Artificial Intelligence (ECAI 2004), Valencia, Spain (2004)
5. Euzenat, J., Mochol, M., Shvaiko, P., Stuckenschmidt, H., Svab, O., Svatek, V., Robert, W., Hage, V., Yatskevich, M.: Results of the ontology alignment evaluation initiative 2006. In: Proc. of ISWC workshop on Ontology Matching, Athens, pp. 73–95 (2006)
6. Euzenat, J., Isaac, A., Meilicke, C., Shvaiko, P., Stuckenschmidt, H., Šváb, O., Svátek, V., Robert, W., Hage, V., Yatskevich, M.: Results of the ontology alignment evaluation initiative. In: Proc. of the ISWC workshop on Ontology Matching, Busan, Korea (November 2007)
7. Hu, W., Jian, N.S., Qu, Y.Z., Wang, Y.B.: GMO: A Graph Matching for Ontologies. In: Proc. K-Cap Workshop on Integrating Ontologies, pp. 43–50 (2005)
8. Hu, W., Cheng, G., Zheng, D., Zhong, X., Qu, Y.: The results of Falcon-AO. In: Proc. Int'l. workshop on Ontology Matching (OM), Athens, Georgia, U.S.A., November 5 (2007)
9. Li, Y., Li, J., Zhang, D., Tang, J.: Results of ontology alignment with RiMOM. In: Proc. Int'l. workshop on Ontology Matching (OM), Athens, Georgia, U.S.A., November 5 (2007)
10. Massmann, S., Engmann, D., Rahm, E., Tang, J.: Results of ontology alignment with COMA++. In: Proc. Int'l. workshop on Ontology Matching (OM), U.S.A., November 5 (2006)
11. Pedersen, T., Patwardhan, S., Michelizzi, J.: WordNet:Similarity – Measuring the Relatedness of Concepts. In: Proceedings of 19th National Conf. on Airtifical Intelegence (AAAI 2004), San Jose, CA, pp. 1024–1025 (July 2004)
12. Tan, H., Lambrix, P.: A method for recommending ontology alignment strategies. In: Proceedings of the 6th International Semantic Web Conference, Busan, Korea (2007)
13. `http://oaei.ontologymatching.org/2007/benchmarks/`

A Causal Approach for Explaining Why a Heuristic Algorithm Outperforms Another in Solving an Instance Set of the Bin Packing Problem

Joaquín Pérez[1], Laura Cruz[2], Rodolfo Pazos[1], Vanesa Landero[1], Gerardo Reyes[1], Crispín Zavala[1], Héctor Fraire[2], and Verónica Pérez[2]

[1] Centro Nacional de Investigación y Desarrollo Tecnolgico (CENIDET)
Departamento de Ciencias Computacionales
AP 5-164, Cuernavaca 62490, México
[2] Instituto Tecnológico de Ciudad Madero (ITCM)
División de Estudios de Posgrado e Investigación, Cd. Madero México
{jperez,pazos,greyes}@cenidet.edu.mx

Abstract. The problem of algorithm selection for solving NP problems arises with the appearance of a variety of heuristic algorithms. The first works claimed the supremacy of some algorithm for a given problem. Subsequent works revealed that the supremacy of algorithms only applied to a subset of instances. However, it was not explained why an algorithm solved better an instances subset. In this respect, this work approaches the problem of explaining through causal modeling the interrelations between instances characteristics and the inner workings of algorithms. For validating the results of the proposed approach, a set of experiments was carried out in a study case of the Tabu Search algorithm applied to the Bin Packing problem. Finally, the proposed approach can be useful for redesigning the logic of heuristic algorithms and for justifying the use of an algorithm to solve an instance subset. This information could contribute to algorithm selection for NP-hard problems.

1 Introduction

Heuristic algorithms have been proposed as a good alternative for solving very large instances of combinatorial optimization problems[1]. Unfortunately, in real-life situations, there is usually no algorithm that outperforms all the other algorithms for all instances [2]; and therefore, the problem of selecting the best algorithm arises.

The works related to this topic have tried to analyze, in an experimental way, the behavior of heuristic algorithms in order to find the best algorithm. Cohen comments in his book [3] that the analysis of an observed behavior is made through a transition comprising three stages: description, prediction and causality. Therefore, in the case of behavior analysis of heuristic algorithms, the transition among these stages has been carried out through several related

A. An et al. (Eds.): ISMIS 2008, LNAI 4994, pp. 591–598, 2008.

works, which are presented in Table 1. Column 2 indicates for each work the analysis type: descriptive (D), predictive (P) and causal (C). Columns 3 and 4 indicate whether the algorithm analysis includes information from the description of the problem instance (ID) or from a sample of the solution space of a problem instance (PS).Columns 5, 6 and 7 indicate if the information from: algorithm behavior (AB), search trajectory (ST) and algorithm structure (AS), is considered in the analysis. Column 8 indicates if the works present formal explanations on the algorithm performance.

Table 1. Related work

Work	Analysis Type	Problem Indicators		Algorithm Indicators		AS	Formal Explanation
		ID	PS	AB	ST		
Hoos 1998 [4]	D						
Soares 2003 [5]	P	✓					
Pérez 2004 [6]	P	✓					
Hoos 2004 [7]	C	✓			✓		
Lemeire 2004 [8]	C			✓			✓
Pérez 2007 [9]	C		✓	✓			✓
This Paper	C	✓	✓	✓	✓	✓	✓

Initially, the works reported by the scientific community focused on showing the superiority of an algorithm for some specific problem [4]. Subsequent works [5], [6], [10], [11] found dominance regions for algorithms; i.e., instances subsets of a problem where each algorithm outperforms the others. Additionally, they used decision tree or k-nearest neighborhood algorithms for predicting the best algorithm for a new problem instance.The work in [7] provides informal explanations through a tabular analysis of experimental results of algorithm behavior on instances of the problem. Other works [8], [9] carry out a causal analysis using learning algorithms of structure and parameters, which provide a formal model of algorithm behavior. The work in [9] explains formally why an algorithm performs better in solving an instance set; however, its explanation is limited since it does not include other indicators of the problem and algorithm. A survey of the specialized literature, revealed the inexistence of a formal model that explains the association between indicators of problem instances and indicators of the characteristics of an algorithm that solves the instances successfully. The solution to this problem is important, since it may provide a solid foundation for the selection of algorithms for solving given instances of NP-hard problems. Therefore, the problem of explaining why an algorithm dominates in an instance region is approached in this paper. The solution approach presented permits systematically finding relations between influencing indicators (columns 3-6 of Table 1) of dominance of an algorithm and the inner workings of algorithms (column 7), in order to provide formal explanations through causal analysis.

2 A Study Case: Causal Analysis of the Performance of the Tabu Search Algorithm Applied to the Bin Packing Problem

The general approach of solution proposed is shown in Figure 1. It is described and validated through a study case that involves two variants of the Tabu Search algorithm [12] for solving 324 instances of the Bin Packing problem. These instances were randomly selected from [13], [14].

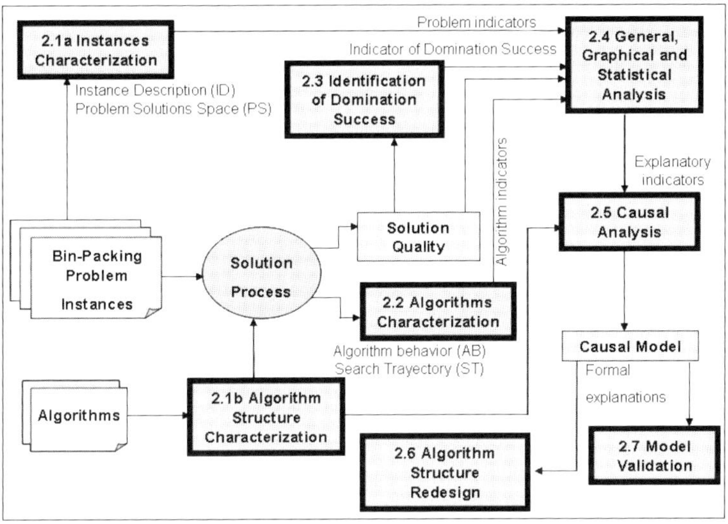

Fig. 1. General scheme of the solution approach

2.1 Characterization of Problem Instances and Algorithm Structure

a) Instances characterization. The information on the instances that constitute the study case is characterized by the following indicators:

Instance description (ID). The indicator *acc* uses problem parameters to calculate average occupied capacity of the objects, and it is explained in [6], [10], [11]. *Problem solutions space (PS).* We obtained a sample of the solutions space of each problem instance prior to algorithm experimentation. This sample was built by generating 100 random solutions, where each generated solution x is evaluated by the fitness function *f(x)*. This function is described in [15]. The variability *vss* of the fitness function values is calculated.

b) Characterization of the structure of the Tabu Search algorithm. The internal structure of variant V1 of the Tabu Search algorithm is characterized as follows: a static Tabu list of size 7, a random initial solution, a stop criterion that can be divergent DI (stop after 4000 iterations) or convergent CO (there is

Table 2. Characteristics of the variants of the Tabu Search algorithm

Variant	Tabu List		Initial Solution		Neighborhood		Stop Criterion	
	Static	Dynamic	Random	Heuristic	One	Several	DI	CO
V1	✓		✓			✓	✓	✓
V2	✓		✓		✓		✓	✓

no improvement in the solution), and several alternative methods for generating random neighboring solutions (swap(1, 0), swap(1, 1), swap(1, 2), swap(2, 2), swap(0, 1), and swap(2, 1)) [16]. Variant V2 uses only swap (1, 0) method. Table 2 shows the parts that constitute the internal structure of variants V1 and V2.

2.2 Algorithms Characterization

Variants V1 and V2 were executed 15 times for each problem instance, we observed in a pilot study a very small variance on these runs. The algorithm information is collected during the solution process and characterized using the following indicators:

Algorithm behavior (AB). The behavior of the algorithm during execution is observed and recorded using the following indicators: the variability *vfs* of the values of the fitness function of the feasible solutions and the number *nne* of neighborhoods generated by the algorithm during the search process.

Search trajectory (ST). The trajectory traced by the algorithm during the search process is characterized by the following indicators: the number *nin* of inflexion points, the number *nva* of valleys and the average size *asv* of the valleys.

2.3 Identification of Domination Success

Once the solution process ends, the solution quality is calculated by means of the performance measure *quality*, which is the ratio of the best solution found by the algorithm to the theoretical solution. This solution is the summation of the object sizes divided by the containers capacity. The domination success indicator (*success*) of each variant was established by means of two criteria: first, the variant with the best *quality* out of 15 executions of an instance is considered superior for solving this instance (*success=1*); second, if the variants have the same *quality*, the variant with the lowest value of evaluations of the fitness function is considered (*success=1*); otherwise *success=0*. The dominant cases (*success=1*) for the variants were: 296 for V1 and 28 for V2. In this study case, we analyze variant V1 and explain why it was superior with respect to V2.

2.4 General, Graphical and Statistical Analysis

The indicators derived from the measurement of the problem and algorithm parameters were prepared. Firstly, the data were normalized by min-max method, all values were set in a range of [0, 1]. Secondly, the data were discretized by the

Table 3. Intevals of indicators

Indicators	1	2	3
acc(ID)	[0, 0.4020]	[0.4020, 0.6083]	[0.6204, 1]
vss(PS)	[0, 0.1009]	[0.1010, 1]	
nne(AB)	[0, 0.1424]	[0.1425, 1]	
vfs(AB)	[0, 0.0690]	[0.0691, 1]	
nin(ST)	[0, 0.1369]	[0.1370, 1]	
nva(ST)	[0, 0.1219]	[0.1220, 1]	
asv(ST)	[0, 0.3331]	[0.3332, 1]	

MDL method [17]. Table 3 shows the intervals (1, 2 and 3) of indicators. In the following sections, the value ranges for the indicators will be denoted according to the columns of Table 3; for example, the value range [0, 0.4020] for indicator *acc*(ID) that corresponds to column 1 of the *acc*(ID) row will be denoted by *acc*=1.

After the data preparation, the indicators were analyzed to identify those that had some effect on the algorithm performance (*quality*). To this end, we performed general, graphic and statistical analyses. An example of these analyses can be found in our previous work [9]. The problem indicators that turned out to be relevant were *acc* and *vss*, and the algorithm indicators were: *vfs*, *nne*, *nin*, *nva* and *asv*.

2.5 Causal Analysis

The procedure used to build a causal model incorporates the main ideas of Cohen and Spirtes [3,18].

Specification of causal order. The construction of the causal model was carried out using the HUGIN causal inference software (www.hugin.com) and the

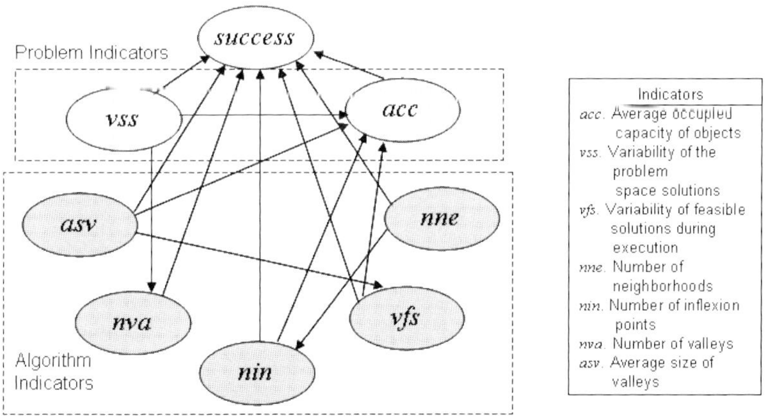

Fig. 2. Causal model of variant V1

Table 4. Causal relation functions

Causal Relation Functions	%	Exp
P(*success*=1 |*acc*=2, *vss*=2, *nne*=2, *vfs*=2, *nin*=2, *nva*=2, *asv*=2)	99	40
P(*success*=0 |*acc*=1, *vss*=1, *nne*=1, *vfs*=1, *nin*=1, *nva*=1, *asv*=1)	98	11
P(*success*=0 |*acc*=3, *vss*=1, *nne*=1, *vfs*=1, *nin*=1, *nva*=1, *asv*=1)	81	11

PC algorithm [18] with a confidence level of 95%. Figure 2 shows that problem indicators: *acc*, *vss*, and algorithm indicators *nne*, *vfs*, *nin*, *nva*, *asv* are direct causes of superiority (*success*=1) or inferiority (*success*=0) of variant V1.

Estimation of the model. Tables of conditional probability (CPT) of the indicators were calculated using the learning algorithm Expectation Maximization [19]. We focused on the most important causal relation functions of the direct causes of node *success*. Their probabilities and "experience" (Columns 2, 3) are presented in Table 4.

Model interpretation. Causal relation functions with largest values of conditional probability and "experience" were interpreted. The following explanations were inferred from these relation functions. The use of several alternate methods for constructing solutions by V1 generates a larger neighborhood (*nne*=2) when the variability of a sample of the instance solution space is large (*vss*=2) and the average capacity of the objects is medium (*acc*=2). This situation permits the algorithm to intensify its search in the solution space, generating solutions with high variability among them (*vfs*=2); i.e., the search trajectory is better suited to the problem space (large number of inflexion points (*nin=2*) and valleys (*nva*=2), large valleys (*asv*=2)), which allows V1 to enter and get out of valleys; otherwise, variant V1 is at disadvantage and V2 performs better.

2.6 Model Validation

The theoretical validation was omitted in this example because we used the PC algorithm, which guarantees the conditions of Markov, Minimality and faithfulness [18]. We used the NETICA software (Norsys Corporation, www.norsys.com) to test the model generated on some instances, for which we ignored whether variant V1 will be superior or not. We obtained a prediction percentage of 79.8%.

2.7 Algorithm Structure Redesign

The Tabu Search algorithm is described in Figure 3. Variant V1 uses several methods for generating neighboring solutions (lines 6, 7). Conclusions from previous section allow us to redesign V1 (V3) for improving its performance. Specifically, variant V3 could perform one simple method or several alternative methods to build neighboring solutions depending on the average occupied capacity of the

```
Tabu Search Algorithm
1  Begin
2    x* = a feasible solution initial; x = x*
3    tn = 7, Tenency of one solution in Tabu List
4    LCANDI = ∅; LTabu = ∅
5    Repeat
6      Building neighborhood of x using swap (0,1) or swap (1,1) or      } V1
7          swap (1,2) or swap (2,2) or swap (1,0) or swap (2,1)
8      Calculate acc, vss from parameters of problem instance
9      If (acc=2 and vss=2) Then
10         Building neighborhood of x using swap (0,1) or swap (1,1)      } V3
11             or swap (1,2) or swap (2,2) or swap (1,0) or swap (2,1)
12     Else Building neighborhood of x using swap (0,1)
13     LCANDI = list of candidate solutions, taken from neighborhood x
14     y = the best solution of LCANDI and y ∉ LTabu
15     LTabu = LTabu ∪ {(y, tn)}
16     For each e ∈ Ltabu the tenency of e is decremented
17     If the tenency e is expired Then LTabu = LTabu − {(e, tn)}
18     If f(y) < f(x) Then x = y
19   Until 4000 iterations of there is no improvement in the solution
20 End
```

Fig. 3. Tabu Search Algorithm Redesign

objects *acc* and the variability of a sample of instance solution space *vss* (lines 8-12).Preliminary experimental results show that new variant V3 attains superiority in 63.58% out of 324 instances, when contending against V1 and in 88.58% when contending against V2. The new redesign proposal V3 yields a performance improvement of 27.16% and 77.16% with respect to variants V1 and V2.

3 Conclusions

This work presents a new approach for solving the problem of explaining why an algorithm outperforms another on a set of instances using a causal analysis, which yielded encouraging results. One of the main contributions of this work is the development of indicators that characterize problem instances, algorithm structure and their performance at execution time. For validating the proposed approach, a set of experiments were carried out for generating a causal model that shows the interrelation of 2 indicators of the Bin Packing problem instances and 5 indicators of the behavior and performance of the Tabu Search algorithm. We obtained a prediction percentage of 79.8% using the model generated. The formal explanation found permitted to devise an improvement to the logic of variant V1 of the Tabu Search algorithm. Preliminary experimental results show that the new variant V3 yields a performance improvement of 27.16% and 77.16% with respect to variants V1 and V2. Finally, this approach can be contributed to understand and formalize the general problem of heuristic algorithms selection for NP-hard problems.

References

1. Garey, M.R., Jhonson, D.S.: Computers and Intractability, a Guide to the Theory of NP-completeness. W. H. Freeman and Company, New York (1979)
2. Wolpert, D.H., Macready, W.G.: No Free Lunch Theorems for Optimization. IEEE Transactions on Evolutionary Computation 1, 67–82 (1997)
3. Cohen, P.: Empirical Methods for Artificial Intelligence. The MIT Press, Cambridge (1995)
4. Hoos, H.: Stochastic Local Search Methods, Models, Applications, PhD Thesis, Department of Computer Science from Darmstadt University of Technology (1998)
5. Soares, C., Pinto, J.: Ranking Learning Algorithms: Using IBL and Meta-Learning on Accuracy and Time Results. Journal of Machine Learning 50(3), 251–277 (2003)
6. Pérez, O., Pazos, R.: A Statistical Approach for Algorithm Selection. In: Ribeiro, C.C., Martins, S.L. (eds.) WEA 2004. LNCS, vol. 3059, pp. 417–431. Springer, Heidelberg (2004)
7. Hoos, H., Smyth, K., Stutzle, T.: Search Space Features Underlying the Performance of Stochastic Local Search Algorithms for MAX-SAT. In: Yao, X., Burke, E.K., Lozano, J.A., Smith, J., Merelo-Guervós, J.J., Bullinaria, J.A., Rowe, J.E., Tiňo, P., Kabán, A., Schwefel, H.-P. (eds.) PPSN 2004. LNCS, vol. 3242, pp. 51–60. Springer, Heidelberg (2004)
8. Lemeire, J., Dirkx, E.: Causal Models for Parallel Performance Analysis. In: 4th PA3CT Symposium, Edegem, Belgium (2004)
9. Pérez, J., Cruz, L., Landero, V., Pazos, R.: Explaining Performance of the Threshold Accepting Algorithm for the Bin Packing Problem: A Causal Approach. In: Proceedings of 14th International Multi-conference, Advanced Computer Systems, Polland (2007)
10. Pérez, J., Pazos, R.: Comparison and Selection of Exact and Heuristic Algorithms. In: Laganá, A., Gavrilova, M.L., Kumar, V., Mun, Y., Tan, C.J.K., Gervasi, O. (eds.) ICCSA 2004. LNCS, vol. 3045, pp. 415–424. Springer, Heidelberg (2004)
11. Pérez, J., Pazos, R.: A Machine Learning Approach for Modeling Algorithm Performance Predictors. In: Torra, V., Narukawa, Y. (eds.) MDAI 2004. LNCS (LNAI), vol. 3131, pp. 70–80. Springer, Heidelberg (2004)
12. Glover, F.: Tabu Search - Part I, First Comprehensive Description of Tabu Search. ORSA-Journal on Computing 1(3), 190–206 (1989)
13. Beasley, J.E.: OR-Library. Brunel University (2006), `http://people.brunel.ac.uk/~mastjjb/jeb/orlib/binpackinfo.html`
14. Scholl, A., Klein, R.: `http://www.wiwi.uni-jena.de/Entscheidung/binpp/` (2003)
15. Falkenauer, E., Delchambre, A.: A Genetic Algorithm for Bin Packing and Line Balancing. In: Proceedings of the IEEE 1992 International Conference on Robotics and Automation, pp. 1186–1192. IEEE Computer Society Press, Los Alamitos (1992)
16. Fleszar, K., Hindi, K.S.: New Heuristics for One-dimensional Bin Packing. Computers and Operations Research 29, 821–839 (2002)
17. Fayyad, U.M., Irani, K.B.: Multi-Interval Discretization of Continuous-Valued Attributes for Classification Learning. In: 13th International Joint Conference of Artificial Intelligence, pp. 1022–1029 (1993)
18. Spirtes, P., Glymour, C., Scheines, R.: Causation, Prediction, and Search, 2nd edn. The MIT Press, Cambridge (2001)
19. Lauritzen, S.L.: The EM algorithm for Graphical Association Models with Missing Data. Computational Statistics Data Analysis 19, 191–201 (1995)

A Graph b-Coloring Based Method for Composition-Oriented Web Services Classification

Lyes Dekar and Hamamache Kheddouci

Université de Lyon, Lyon, F-69003, France; Laboratoire LIESP,
Université Lyon1 Batiment Nautibus (ex.710), 43 bd du 11 Novembre 1918
F-69622 Villeurbanne Cedex, France
{ldekar,hkheddou}@bat710.univ-lyon1.fr

Abstract. Web services aim to be the key of the Web development in the next years, by enabling the deployment, the sharing and the gathering of functionalities in a more simple and flexible way. The expectable growth of the number of Web services makes the need for automatic organization and discovery of Web services more important. In this paper, we propose a new classification based on Web services composition. The Web services that are often composed, are grouped together dynamically. Then, we give a graph based approach, that uses the b-coloring to partition these Web services into clusters and maintain them.

1 Introduction

With the growth of the Web services number, it is essential to organize the Web services in order to facilitate operations such as Web services invocation, discovery, maintenance, etc. In this aim, Web services must be grouped into categories. This is called the Web services classification. In this paper, we aim to regroup Web services according to the compositions made by the users. This is a new approach of the classification since all the classification methods of Web services are based on the similarity between these ones. On the other hand, we use an efficient b-coloring-based clustering method to regroup the different Web services into clusters.

Two different classification approach exist in the literature. Those based on information contained in non-semantic service descriptions and those based on semantic service descriptions. In [5], non-semantic service descriptions are used to create dynamically categories that should contain services, by using clustering techniques. For the second approach, Oldham et al. [7] give an algorithm that compares between Web service data types and domain ontology concepts, by using schemas matching techniques. The category of which the corresponding domain ontology gives the highest similarity with the service will contain this service. In [1], the authors provide a semantic-based web services classification method. It is based on the comparison of a new unclassified service with a set of already classified services. Hence, the classification heuristic is divided into three different levels of granularity, each one corresponding to the comparison

A. An et al. (Eds.): ISMIS 2008, LNAI 4994, pp. 599–604, 2008.

between two elements involved in the classification process. In this paper, we propose a new approach of Web services classification which regroups Web services that are often composed together. This new classification approach gives several advantages that motivate our study. 1- The classification can serve as a complementary aid for automatic service discovery, such that each class represents a pack of services, and generally all the services invoked in a request can be satisfied by only one pack. Then, it suffices to discover the pack that contains one service invoked in the request. 2- According to the classification, a provider can organize its Web services in order to enhance their performance and to ensure more quality of service and security. For example, a provider can regroup all the services that belong to the same class, and then that are often invoked together, in the same server. This enables to perform the composition between them more easily and more rapidly. which enhance the QoS. 3- Web services that are often used together can be reprogrammed to ensure a better interworking between them.

The method proposed in this paper uses the b-coloring of graphs for services classification. Let $G = (V, E)$ be an undirected connected and simple graph with a vertex set V and an edge set E. The b-coloring of G is a vertex coloring function c from V to the set of colors $\{1, 2, ..., k\}$ such that: 1- for each pair of adjacent vertices $(v_i, v_j) \in E$, $c(v_i) \neq c(v_j)$ (proper coloring). 2- In each color class, there exists at least one vertex having neighbors in all other color classes. Such a vertex is called a *dominating vertex*. A color that has a dominating vertex is called a *dominating color*. We call the *b-chromatic number* $\varphi(G)$ of a graph G the largest number k such that G has a *b-coloring* with k-colors. This coloring was introduced by Irving and Manlove [6] where they proved that finding the b-chromatic number of any graph is a NP-hard problem and they gave a polynomial-time algorithm for finding the b-chromatic number of trees. This parameter was also studied for other classes of graphs like power graphs of paths and cycles [2]. Effantin and Kheddouci [3] proposed a distributed algorithm for a b-coloring of graphs, while Elghazel et *al.* [4] apply the b-coloring for clustering heterogeneous data objects into groups by considering the similarity between them.

The remaining of the paper is organized as follows: in Section 2 we give our Web services clustering method. In Section 3, an evolutive Web services classification method is proposed. Finally, Section 4 concludes the paper.

2 b-Coloring-Based Approach for Composition-Oriented Web Services Classification

In this Section, a new Web services clustering method based on the b-coloring, and oriented composition is proposed. Consider the services to be clustered as an undirected edge-weighted graph $G = (V, E)$. The vertices in G represent services, the edges correspond to the relation between services, and the edge weights represent the number of times two linked services are composed. This information is presented in the Composition Weight Matrix (CWM). The clustering here consists to regroup services such that the edges between two vertices of the same cluster are large weighted, while edges between vertices of different clusters are small weighted.

2.1 The Filtered Graph

The graph G corresponding to the composition weight matrix is a complete undirected edge-weighted graph. Then, if we perform b-coloring on it for clustering, all graph vertices will have different colors, and then each cluster (color class) will contain only one service (vertex), which is not interesting. Therefore, in order to regroup the services joined by a large weighted link in the same cluster, we remove all edges with a weight larger than a threshold α. Consequently, after removing these edges, we obtain a *filtered graph* $G_{<\alpha} = (V, E_{<\alpha})$, such that $E_{<\alpha} = \{(v_i, v_j) \mid CWM(v_i, v_j) \leq \alpha\}$. Figure 2 gives the filtered graph $G_{<9}$ corresponding to the composition weight matrix given in Figure 1. After constructing the *filtered graph*, we perform b-coloring on it to obtain a Web services clustering. This is explained in the remainder of the paper.

	S_1	S_2	S_3	S_4	S_5	S_6	S_7	S_8	S_9
S_1	0								
S_2	2	0							
S_3	1	13	0						
S_4	12	3	2	0					
S_5	5	14	23	3	0				
S_6	9	14	4	5	15	0			
S_7	1	17	6	0	9	12	0		
S_8	12	19	1	18	12	4	14	0	
S_9	13	6	17	20	3	2	16	2	0

Fig. 1. A composition weight matrix (CWM)

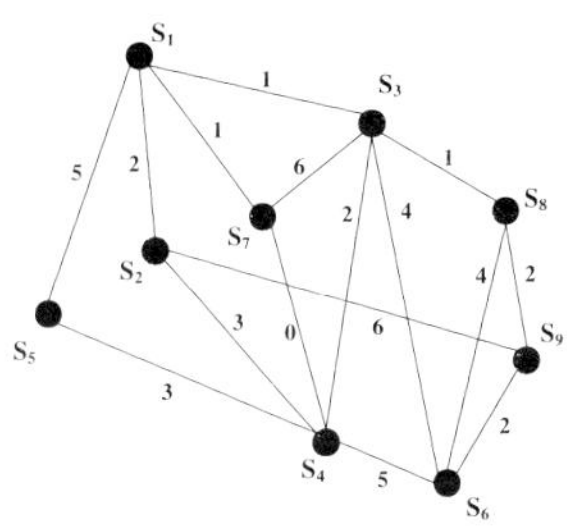

Fig. 2. The filtered graph

2.2 The Clustering Algorithm

The proposed algorithm constructs a b-coloring to obtain a partition of the filtered graph $G_{<\alpha}$ into disjoined color classes $\{C_1, C_2, ..., C_k\}$ that represent clusters. Throughout this paper, We let Δ the maximum degree of G and $c(v)$ the color of the vertex v in the graph G. For every vertex v, we define $N(v)$ its open neighborhood, as the set of vertices adjacent to v. The set of colors of $N(v)$ is denoted $N_c(v)$. We note L the color set used in the graph. For each color c used in the graph, we associate a variable $Dom[c]$ that indicates if the color c is a dominating or a non-dominating color (*true*: if c is dominating. *false*: otherwise). Finally, we define a function *weight(v,c)* that indicates the *composition weight* between the vertex v and the color c. This function is defined by: $weight(v, c) = \max\{CWM(v, v') \mid c(v') = c\}$.

The b-coloring is made in two steps: in the first, ***The coloring initialization*** (Procedure 1), the graph is initialized by coloring the graph with the maximum number of colors. In the second step, ***Find a b-coloring of*** G (Procedure 2), all non-dominating colors are removed from the graph. Let us consider the filtered graph $G_{<9}$ obtained in Figure 2. By performing the procedure 1, the filtered graph $G_{<9}$ has an initial coloring showed in Figure 3. Then, by performing the procedure 2, we obtain a b-coloring of a graph $G_{<9}$, as shown in Figure 4. The Weighted composition graph is partitioned into four color classes representing the clusters: $C_1 = \{S_1, S_4, S_9\}$, $C_2 = \{S_5, S_8\}$, $C_3 = \{S_2, S_6, S_7\}$ and $C_4 = \{S_3\}$.

Procedure 1: *Init-Coloring()*
BEGIN
$L = \{1, 2, 3, ..., \Delta + 1\}$.
Let v be the vertex with a degree Δ.
$c(v) = 1$. Add v to S.
for each vertex v_i such that
$d(v_i) = \max\{d(v) \mid v \in S\}$ **do**
Let $M = N_c(v_i) \cup \{c(v_i)\}$.
$q = 0$.
for every vertex $v_j \in N(v_i)$ such that
$c(v_j) = \emptyset$ **do**
$q = \min\{k \mid k > q, k \notin M \ and$
$k \notin N_c(v_j)\}$.
if $q \leq \Delta + 1$ **then** $c(v_j) = q$
else $c(v_j) = \min\{k \mid k \notin N_c(v_j)\}$.
endif.
Add v_j to S.
enddo.
if $N_c(v_i) = L \setminus c(v_i)$ **then**
$Dom[v_i] = 1$.
endif.
Remove v_i from S.
enddo. END

Procedure 2: *Bcoloring-Construct()*
BEGIN
for each $p \in L$ such that $p = \max\{k \mid$
$Dom[k] = false\}$ **do**
$L = L \setminus p$.
for each vertex v_i such that
$c(v_i) = p$ **do**
$K = \{k \mid k \in L \ and \ k \notin N_c(v_i)\}$.
$c(v_i) = \{c \mid weight(v_i, c) =$
$\max_{k \in K}(weight(v_i, k))\}$.
enddo.
for each vertex v_j such that
$Dom[c(v_j)] = false$ **do**
if $N_c(v_j) = L \setminus \{c(v_j)\}$ **then**
$Dom[c(v_j)] = true$.
endif
enddo.
enddo.
END

3 An Evolutive Web Services Classification Scheme

The composition weights between services can evolve over time. Then, the filtered graph can evolve too. Indeed, an edge can appear (resp.desapear) in (resp. from) the filtered graph if its weight becomes under the threshold value (resp. its weight exceeds the threshold value). Hence, in order to maintain a precise and correct classification, it is imperative to consider the filtered graph evolution (edges appearing and disappearing) to maintain the graph b-coloring and then an updated classification. Then we propose an edge-dynamic algorithm to maintain the b-coloring when edges are added or removed from the graph. As explained previously, in the b-coloring, there is at least a dominating vertex for each color class. A dominating vertex x is said *satisfied* if for any color q in the graph, there exists at least one vertex $y \in N(x)$ such that $c(y) = q$. If there exists only one such a vertex then this one is called a *Satisfaction vertex*. Any change of the

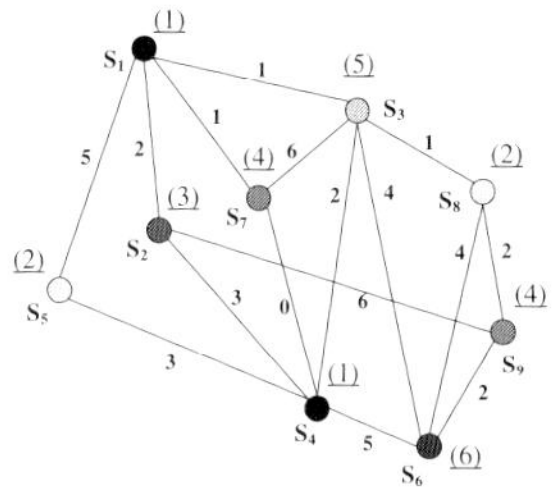

Fig. 3. A graph $G_{<9}$ Coloring initialization

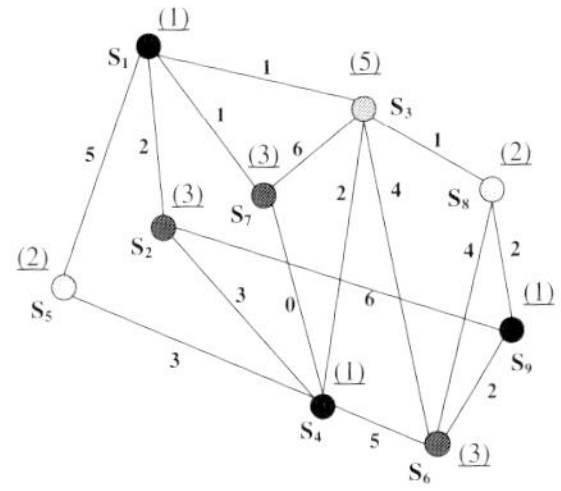

Fig. 4. The b-coloring of a graph $G_{<9}$

satisfaction vertex color can affect the b-coloring since the corresponding satisfied vertex is not anymore dominating. The vertices that are neither dominating vertices nor satisfactions vertices are called *Normal vertices*.

Finally, since all the graph is colored and each vertex belongs to a color class, then we propose to modify the definition of the function $weight(v, c)$ that indicates the composition weight between the vertex v and the color c. Then, we assume that $weight(v, c) = \frac{\sum_{c(v')=c} CWM(v,v')}{|c|}$, which means that it is equal to the average of the weights between the vertex v and the vertices colored with c.

3.1 Adding an Edge

When an edge (v, y) is added to the graph, we can distinguish three different cases, according to the endpoints of the added edge: 1- **The edge is added between a normal vertex v and a dominating vertex y such that the two vertices have different colors**: In this case, the b-coloring of the graph is not affected. However, this edge adding can enable to increase the number of colors in the graph. Then, we check if the vertex v can take a new color for which it will be dominating. 2- **The edge is added between a normal vertex v and another vertex having the same color**: In this case, the coloring is not anymore proper, and the b-coloring conditions are not verified. Then, the color of the normal vertex must be changed. Hence, we can distinguish four different cases: (A) *The vertex v is adjacent to a dominating vertex of every color in the graph*: In this case, we give the vertex v a new color for which v will be dominating. (B) *There exists at least a color c to which the vertex v is not adjacent*: In this case, we give the vertex v this color. (C) *The vertex v is adjacent to all the colors in the graph, and there exists at least one color c that appears only on a normal vertex w*: in this case, the vertex v takes the color c, which causes a not proper coloring. Then, the vertex w takes another color. Such a color always exists since the vertex w is not dominating and then is not adjacent to at least one color in the graph. (D) *The vertex v is adjacent to satisfaction and/or dominating vertices with every color in the graph and there exists at least one color c that appears on a satisfaction vertex u adjacent to v*: in this case, the vertex v takes the color c. Then, we give to the satisfaction vertex u another color not appearing in its neighborhood. This is always possible since this vertex is not dominating and then is not adjacent to all the colors. If a dominating vertex x adjacent to u is the only one for its colors then the b-coloring is not satisfied. In order to reestablish the b-coloring without systematically removing the non dominating color, we try to put the color c on another normal vertex z adjacent to x to reestablish the dominating condition. If such a vertex does not exist then we try to form another dominating vertex for the color c. Otherwise, we remove the color c from the graph. 3- **The edge is added between a satisfaction vertex v and another satisfaction or dominating vertex having the same color**: Hence, we change the color of a satisfaction vertex, which can make the dominating condition not anymore satisfied. Therefore, we perform the same actions as in the point *(D)* of the previous case.

3.2 Deleting an Edge

The removing of an edge from the b-colored graph can make the dominating condition not verified. Then, we perform the same actions as in the point *(D)* of the second case of edge adding.

4 Conclusion

In this paper, we proposed a new graph based method for composition oriented Web services classification. We first gave an algorithm that uses the b-coloring to regroups services into clusters. We also provide a second algorithm to maintain the clustering and update it. As future works, we aim to apply our method to enhance existent service discovery and composition methods.

References

1. Corella, M.A., Castells, P.: A heuristic approach to semantic web services classification. In: Gabrys, B., Howlett, R.J., Jain, L.C. (eds.) KES 2006. LNCS (LNAI), vol. 4251, Springer, Heidelberg (2006)
2. Effantin, B., Kheddouci, H.: The b-chromatic number of some power graphs. Discrete Mathematics and Theoretical Computer Science 6, 45–54 (2003)
3. Effantin, B., Kheddouci, H.: A distributed algorithm for a b-coloring of a graph. In: Guo, M., Yang, L.T., Di Martino, B., Zima, H.P., Dongarra, J., Tang, F. (eds.) ISPA 2006. LNCS, vol. 4330, pp. 430–438. Springer, Heidelberg (2006)
4. Elghazel, H., Kheddouci, H., Deslandres, V., Dussauchoy, A.: A new graph-based clustering approach: Application to pmsi data. In: IEEE ICSSSM 2006, France (2006)
5. Heb, A., Kushmerick, N.: Automatically attaching semantic metadata to web services. In: Workshop on Information Integration on the Web, Mexico (2003)
6. Irving, W., Manlove, D.F.: The b-chromatic number of graph. Discrete Mathematics 91, 127–141 (1999)
7. Oldham, N., Thomas, C., Sheth, A., Verma, K.: Meteor-s web service annotation framework with machine learning classification. In: Cardoso, J., Sheth, A.P. (eds.) SWSWPC 2004. LNCS, vol. 3387, Springer, Heidelberg (2005)

OWL-S Atomic Services Composition with SWRL Rules

Domenico Redavid[1], Luigi Iannone[2], Terry Payne[3], and Giovanni Semeraro[1]

[1] Dipartimento di Informatica
Università di Bari, Bari 70126, Italy
[2] Computer Science Dept., University of Liverpool
Ashton Building, Ashton Street L69 3BX, Liverpool UK
[3] School of Electronics and Computer Science, University of Southampton
Southampton, SO17 1BJ, United Kingdom
{redavid,semeraro}@di.uniba.it,
L.Iannone@liverpool.ac.uk,
trp@ecs.soton.ac.uk

Abstract. This paper presents a method for encoding OWL-S atomic processes by means of SWRL rules and composing them using a backward search planning algorithm. A description of the preliminary prototype implementation and a grounding in BPEL are also presented.

1 Introduction

Semantic Web (SW) aims at proposing standards, tools and languages for knowledge representation on the Web. Amongst the other issues, it deals with the provision of semantics to Web Services in order to achieve a more abstract and flexible automation. The result of this effort is the notion of Semantic Web Services (SWS) [1]. This term refers to traditional Web services that have been annotated by means of SW languages and techniques so as to make possible their automatic discovery, composition and invocation. In order to achieve that, in literature there are different approaches which produced different frameworks, among which the most widespread are OWL-S[1], and WSMO[2]. In this paper[3] our aim is the composition of OWL-S atomic processes adopting SWRL [2] as language for the representation of their IOPR (Inputs, Outputs, Preconditions and Results) models. Such SWRL descriptions are used as input to generate candidate service compositions in order to achieve a given goal. The process model denoted by services compositions can be grounded in BPEL to obtain an executable process.

2 Preliminary Considerations

In this section we report the basic notions about the OWL-S process model with some considerations on the guidelines that should be followed in order to have useful metadata for the Web services to be described.

[1] OWL-S: Semantic markup for web services, http://www.w3.org/submission/owl-s/
[2] WSMO: Web service modeling ontology d2v1.3, http://www.wsmo.org/tr/d2/v1.3/
[3] This research was partially funded by the project DIPIS (Distributed Production as Innovative System), Apulia Region Strategic Project (2006-08).

A. An et al. (Eds.): ISMIS 2008, LNAI 4994, pp. 605–611, 2008.

Each OWL-S process is based on an IOPR model. The *Inputs* represent the information that is required for the execution of the process. The *Outputs* represent the information that the process returns to the requester. *Preconditions* are conditions that are imposed over the *Inputs* of the process and that must hold for the process to be successfully invoked. Since an OWL-S process may have several results with corresponding outputs, the *Result* entity of the IOPR model provides a means to specify this situation. Each result can be associated to a result condition, called *inCondition*, that specifies when that particular result can occur. Therefore, an *inCondition* binds inputs to the corresponding outputs. It is assumed that such conditions are mutually exclusive, so that only one result can be obtained for each possible situation. When an *inCondition* is satisfied, there are properties associated to this event that specify the corresponding output (*withOutput* property) and, possibly, the *Effects* (*hasEffect* properties) produced by the execution of the process. *Effects* are changes in the state of the world. The OWL-S conditions (*Preconditions*, *inConditions* and *Effects*) are represented as logical formulas. Formally, *Input* and *Output* are subclasses of the more general class *Parameter* declared in its turn as a subclass of *Variable* in SWRL ontology. Every parameter has a type, specified using a URI. Such type is needed to refer it to an entity within the domain knowledge of the service. The type can be either a *Class* or a *Datatype* (i.e.: a concrete domain object such as a string, a number, a date and so on) in the domain knowledge. Nevertheless, we argue that providing descriptions of Web services parameters using concrete datatypes gives very little in terms of added semantics. For example, consider a service whose input has declared as *Datatype* within a knowledge domain, i.e. a string. This means that the reference knowledge model of this input parameter is a concrete XML Schema datatype instead of being an entity within a domain ontology. This mismatch becomes critical in automatic composition of services. Indeed, suppose that, during an hypothetical composition process, we need to find another service whose output will be fed into the service described above. Our composer, then, must necessarily consider those services that have as output a resource of the same type of our input parameter. In the example above, this type is string, hence every service that returns a string as an output can be composed with our service. Therefore, this would result in meaningless compositions of totally unrelated services due to the fact that parameters have been semantically poorly described. In the rest of this paper we consider only those services that have parameters declared as entities in a domain ontology (i.e. not as datatype).

3 Encoding OWL-S Atomic Processes with SWRL Rules and Composition Algorithm

In this section we explain our approach for transforming process descriptions into sets of rules expressed in an ontology-aware rule language, namely Semantic Web Rule Language (SWRL), and our composer implementation.

To our aim, it is important to underline two SWRL characteristics: every rule must respect the *safety* condition and every rule with conjunctive consequent can be transformed into multiple rules each with an atomic consequent [3]. Furthermore, we work exclusively with SWRL DL-safe rules [4] fragment. Within OWL-S, conditions (logical formulas) can be declared using languages whose standard encoding is in XML, such as SWRL. Body and head are logical formulas, whereby the OWL-S conditions can be identified with the body or with the head of a SWRL rule. Such conditions are expressed over *Input* and *Output*. Therefore, if the above requirement is met, conditions will be also expressed in terms of a domain ontology and will hence have the right level of abstraction. After these considerations, we can describe the guidelines we follow for encoding an OWL-S process into SWRL.

- For every result of the process there exists an *inCondition* that expresses the binding between inputs variables and the particular result (output or effect) variables.
- Every *inCondition* related to a particular result will appear in the antecedent of each resulting rule, whilst the *Result* will appear in the consequent. An inCondition is valid if it contains all the variables appearing in the *Result*.
- If the *Result* contains an *Effect* composed of more atoms, the rule will be split into as many rules as the atoms are. Each resulting rule will have the same inCondition as antecedent and a single atom as consequent.
- The *preconditions* are conditions that must be true in order to execute the service. Since these conditions involve only the process *Inputs*, they will appear in the antecedent of each resulting rule togheter with *inConditions*. In this work we consider always true all the *Preconditions*.

The first guideline is needed because there may be processes in which such binding is implicit in their OWL-S descriptions. Let us consider, for example, an atomic process having a single output. In this case there might be no *inCondition* binding inputs and output variables since, being the output the unique outcome, such binding is obvious. In this case, though, our encoding with SWRL rules would not be possible because the second guideline is not applicable. However, we can add a new *inCondition* that makes explicit such implicit binding. For example, suppose we have a service that returns book information whose process is declared having one input (*?process:BookName*), one output (*?process:BookInfo*), and none condition. We should write the corresponding rule as "*kb:BookTitle(?process:BookName)* → *bibtex:Book(?process:BookInfo)*", but the variable *process:BookInfo* does not appear in the antecedent of the rule, consequently this is not a valid SWRL rule. Since every service produces the output manipulating the inputs, we can suppose that there exists a predicate (*hasTransf* predicate) always true that binds every input to the output. In order to obtain valid rules, we add this predicate at antecedent of the rule obtaining the implicit *inCondition*.

The realized SWRL composer prototype implements a backward search algorithm for the composition task. It works as follows: it takes as input a knowledge base containing SWRL rules and a goal specified as a SWRL atom, and it returns every possible path built combining the available SWRL rules in order to achieve such goal. These rules comply with SWRL safety condition. In details, the algorithm performs backward chaining starting from the goal in the same fashion Prolog-like reasoners work for query answering. The difference is that this algorithm does not rely just on Horn clause but on SWRL DL-safe rules. This means that, besides the rule base, it takes into account also the Description Logic ontology to which the rules refer. The SWRL rule path found, and consequently the resulting OWL-S service composition, will be valid, in the sense that it will produce results for the selected goal, only if the SWRL rules in the path are DL safe. In other words the DL-safety means that rules are true for individuals that are

known, i.e.: they appear in the knowledge base[4]. At present, the prototype performs DL-safety check. This guarantees that the application of rules is grounded in the ABox and consequently that the services that embody those rules can be executed.

4 Example

In this section we present an example that shows the applicability of our method. The dataset of OWL-S services can be found on Mindswap Web site [5]. It is formed by four services, namely *BookFinder*, *BNPrice*, *AmazonPrice* and *CurrencyConverter*. Among them, only one service has not any inputs and outputs described as datatype in knowledge domain. All services have no declared *inConditions*, hence we assume that for each of them there is only one *Result* corresponding to the service output and there is no *Precondition* and *Effect*. To obtain SWRL rules that satisfy the requirements described in the section 3, we have modified the atomic services as follow:

- For every parameter having a datatype as type, we created a class in the domain ontology having a datatype property with the corresponding datatype as range The OWL-S descriptions have been modified assigning the newly created class to the corresponding *parameterType*.
- For each service, we create two logical formulas. The first composed of unary atoms having the *parameterType* URI as their predicate and the input as their variable, for each input. The second composed of a unary atom having the *parameterType* URI as its predicate and the output its variable. We set these two logical formulas as, respectively, the antecedent and consequent of a new SWRL rule.
- Since every service produces the output manipulating the inputs, we can suppose that there exists a predicate (*hasTransf* predicate) always true that binds every input to the output. We did this in order to guarantee the SWRL safety condition, then we

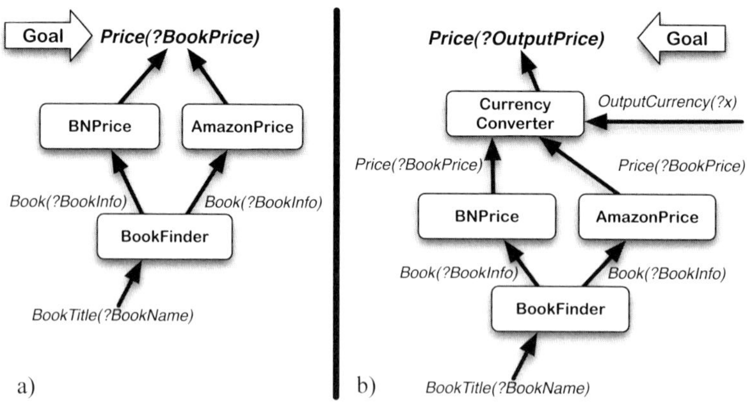

Fig. 1. Examples of composition (all concepts are defined with same namespace)

[4] It might not be the case in general, given the Open World Assumption holding in Description Logics, see [4] and chapter 2 in [5].

[5] Available at: http://www.mindswap.org/2004/owl-s/services.shtml

added *hasTransf* predicates to the antecedent of the rule built in the previous step. With this modification the antecedent can be identified with a new *inCondition*.

The obtained SWRL rule set is given as input to our composer and some resulting composition are showed in Figure 1. In a) and b) the searched goal is the same, i.e. the price of a book, but with composition b) is possible to obtain the price in a given currency. The paths having the same service as starting point (*BookFinder* service) are been joined to form a *Split-like* construct. In the next section we deal with a possible grounding of the composition plan with BPEL.

5 Grounding Service Plan with BPEL

An interesting application of our composition method is the grounding of the resulting composition plan with Business Process Execution Language (BPEL)[6]. The deficiency of BPEL is the impossibility to represent semantic information. On the other hand, OWL-S process model is designed to represent such kind of information, but, in general, such semantic information is superfluous during the execution of procesess. Since our method works at semantic abstraction level, it is possible to ground plans obtained with our SWRL composer using BPEL. BPEL makes use of the *activity* notion for process modeling. There exist two kinds of *activities* for this task i.e.: *Basic Activities* and *Structured Activities* [6]. *Basic Activities* (e.g.: "Invoke", "Receive" and "Reply") are used to model interaction between business partners. These activities can be nested in some *Structured Activities* to define BPEL workflow. Since OWL-S Atomic services can be grounded in WSDL, we have a set of services described with WSDL usable as base building blocks to create BPEL process models. The encoding between WSDL and BPEL is straightforward. The linking between building blocks is represented by the plan produced with the SWRL composer prototype. Analyzing a plan produced by SWRL composer, we note that:

- A rule path is a *Sequence* of rules (Atomic services).
- Two or more rule path can be split and then joined in one rule (Atomic service).
- No iterations are possible.

The correspondent BPEL *Activity* used to represent such plans are "Sequence" and "Flow". Moreover, it is necessary to use BPEL *Data Handling* to specify the dataflow into process model. It defines the notion of "Assignment" used between two basic activities to assign an output message or message parts (in case of complex message types) of first activity, as an input message or message part for next activity. These messages are originated from WSDL messages. Therefore it is possible to encode plans produced with SWRL composer with BPEL process model.

6 Related Work

To the best of our knowledge no approach in literature makes use of SWRL for the SWS composition. Researchers focussed either on semi-automated or fully automated

[6] Business process execution language for web services (BPEL) 1.1, http://www.ibm.com/developerworks/webservices/library/ws-bpel/ (2003).

methods for service composition, drawing inspiration especially from AI planning [7] and state machines [8]. Generally, two different approaches to perform the composition task have been adopted. One approach aims at integrating Semantic Web formalisms into classical planner methodologies. Berardi et al. [9] address the problem of automatic composition synthesis of e-Service. They developed a framework in which the exported behavior of an e-Service is described in terms of its possible executions (execution trees). Then they specialize the framework to the case in which such exported behavior (i.e., the execution tree of the e-Service) is represented by a finite state machine. In [10], the semantics underlying the DAML-S specification (the ancestor of OWL-S) has been translated into FOL, obtaining a set of axioms for describing the features of each service. By combining these axioms within a Petri Net, the authors have obtained process-based service models that enable reasoning about the interactions among the processes that form the structure of a service. Traverso and Pistore [11] propose a planning technique for the automated composition of Web services described in OWL-S process models, which can deal with nondeterminism, partial observables, and complex goals. Such technique facilitates the synthesis of plans that encode compositions of web services with the usual programming constructs, like conditionals and iterations. In [12] an approach for developing a Semantic Web service discovery and composition framework on top of the CLIPS rule-based system is presented. More specifically, it describes a methodology for using production rules over Web services semantic descriptions expressed in the OWL-S ontology.

Other approaches, in which our methodology can be framed, apply methodologies and tools developed in the field of AI planning directly on Semantic Web settings. Sirin and Parsia [13] demonstrate how an OWL reasoner can be integrated within an AI planner, called SHOP2 [14], for the SWS composition. The reasoner is used to store the world states, answer the planners queries regarding the evaluation of preconditions, and update the state when the planner simulates the effects of services.

The first type of approach foresees a translation from the Semantic Web formalisms to a dedicated formalism so that tools developed in particular research areas can be applied maintaining the same performances. On the contrary, the second type of approach foresees a porting of the algorithms and methodologies from other research fields using the Semantic Web technologies. The advantage of this approach, in which we frame our methodology, is the direct use of the Semantic Web formalisms. In this manner, we are able to use methodologies coming from more consolidated research fields exploiting the advantages that Semantic Web guarantees, i.e. a distributed knowledge base and the semantic interoperability.

7 Conclusion and Future Work

In this paper we have presented a new method that exploits SWRL for OWL-S atomic services composition. We have proved that if the OWL-S services have a meaningful semantics and valid SWRL conditions it is possible to build composer exploiting only the Semantic Web technology to achieve the composition task. Working at semantic abstraction level it is possible to map the resulting composition with XML-based languages for process management like BPEL. This work can be considered as a starting point for the solution of a broader issue like the orchestration of SWS. Future work will mainly consist of augmenting the types of services that can be encoded into SWRL rules. In other words the system should be able in the future to handle composite services as input and to produce more complex control structures (such as selection and iteration). The latter seems to be the most challenging task since it will require more powerful algorithms for the composition task. Furthermore, an interesting aspect to deal with is the management of knowledge bases when there are changes produced by the effects of a service execution. Semantic Web languages are based on Description Logics which implement monotonic reasoning. In other words, they do not provide any means for retracting or modifying the status of the knowledge base that is not adding some new facts. This is somewhat a too restrictive requirement to represent, for instance, service execution in such formalisms.

References

[1] McIlraith, S.A., Son, T.C., Zeng, H.: Semantic Web Services. IEEE Intelligent Systems 16, 46–53 (2001)
[2] Horrocks, I., Patel-Schneider, P.F., Bechhofer, S., Tsarkov, D.: OWL rules: A proposal and prototype implementation. J. of Web Semantics 3, 23–40 (2005)
[3] Lloyd, J.W.: Foundations of Logic Programming, 2nd edn. Springer series in symbolic computation. Springer, New York (1987)
[4] Motik, B., Sattler, U., Studer, R.: Query Answering for OWL-DL with Rules. Journal of Web Semantics: Science, Services and Agents on the World Wide Web 3, 41–60 (2005)
[5] Baader, F., Calvanese, D., McGuinness, D., Nardi, D., Patel-Schneider, P. (eds.): The Description Logic Handbook. Cambridge University Press, Cambridge (2003)
[6] Aslam, M.A., Auer, S., Shen, J., Herrmann, M.: Expressing Business Process Models as OWL-S Ontologies. In: Eder, J., Dustdar, S. (eds.) BPM Workshops 2006. LNCS, vol. 4103, pp. 400–415. Springer, Heidelberg (2006)
[7] Georgeff, M.P.: Planning. In: Allen, J., Hendler, J., Tate, A. (eds.) Readings in Planning, pp. 5–25. Kaufmann, San Mateo (1990)
[8] Gurevich, Y.: Evolving Algebras 1993: Lipari Guide. In: Börger, E. (ed.) Specification and Validation Methods, pp. 9–37. Oxford University Press, Oxford (1994)
[9] Berardi, D., Calvanese, D., Giacomo, G.D., Hull, R., Mecella, M.: Automatic Composition of Transition-based Semantic Web Services with Messaging. In: Böhm, K., Jensen, C.S., Haas, L.M., Kersten, M.L., Larson, P.Å., Ooi, B.C. (eds.) VLDB, pp. 613–624. ACM, New York (2005)
[10] Narayanan, S., McIlraith, S.A.: Simulation, verification and automated composition of web services. In: WWW 2002: Proceedings of the 11th international conference on World Wide Web, pp. 77–88. ACM Press, New York (2002)
[11] Traverso, P., Pistore, M.: Automated Composition of Semantic Web Services into Executable Processes. In: McIlraith, S.A., Plexousakis, D., van Harmelen, F. (eds.) ISWC 2004. LNCS, vol. 3298, pp. 380–394. Springer, Heidelberg (2004)
[12] Meditskos, G., Bassiliades, N.: A Semantic Web Service Discovery and Composition Prototype Framework Using Production Rules. In: OWL-S:Experiences and Directions Workshop at 4th European Semantic Web Conference (ESWC) (2007)
[13] Sirin, E., Parsia, B.: Planning for Semantic Web Services. In: Semantic Web Services Workshop at 3rd International Semantic Web Conference (2004)
[14] Nau, D.S., Au, T.C., Ilghami, O., Kuter, U., Murdock, J.W., Wu, D., Yaman, F.: Shop2: An HTN Planning System. J. Artif. Intell. Res (JAIR) 20, 379–404 (2003)

A Web-Based Interface for Hiding Bayesian Network Inference

C.J. Butz[1], P. Lingras[2], and K. Konkel[1]

[1] Department of Computer Science, University of Regina
Regina, Saskatchewan, Canada S4S 0A2
{butz,konkel1k}@cs.uregina.ca
[2] Department of Math and Computing Science, Saint Mary's University
Halifax, NS, Canada B3H 3C3
pawan.lingras@stmarys.ca

Abstract. *Bayesian networks* have been applied for several uncertainty management problems in the artificial intelligence and Web intelligence communities. However, one may require the use of Bayesian networks, yet lack the background knowledge to build them. Moreover, it is widely acknowledged in the Bayesian network community that understanding Bayesian network inference is an arduous task. In this paper, we solve this dilemma by proposing a Web-based interface for hiding Bayesian network inference. This approach allows a much wider audience to utilize Bayesian network inference without having to understand how the inference process is actually carried out.

1 Introduction

Bayesian networks (BNs) [11] are an established framework for uncertainty management in the artificial intelligence community. A BN consists of a *directed acyclic graph* (DAG) and a corresponding set of *conditional probability tables* (CPTs). The *probabilistic conditional independencies* (CIs) [13] encoded in the DAG indicate that the product of CPTs is a joint probability distribution. Exact inference algorithms in BNs can be broadly classified into two categories. One approach is *join tree propagation*, which systematically passes messages in a join tree constructed from the DAG of a BN (see [8] for example). The second approach to BN inference is *direct computation*, which performs inference directly in a BN. Very recently, Madsen [7] examined hybrid approaches to BN inference. Of the three hybrid approaches tested, Lazy arc-reversal (Lazy-AR) was empirically shown to be the state-of-the-art method for exact inference in discrete BNs [7]. In [3], we proposed Lazy arc-reversal with variable elimination (Lazy-ARVE) as a new approach to BN inference and illustrated its benefits over Lazy-AR.

It is a difficult challenge, however, to learn how BN inference algorithms work [5,6,12]. Jensen [6], one of the founders of BNs, explicitly states that probabilistic reasoning literature is not meant for readers looking for a way into the field. In fact, Russell and Norvig [12] state that some of the mathematics

A. An et al. (Eds.): ISMIS 2008, LNAI 4994, pp. 612–617, 2008.

and notation are unavoidably intricate even for inference in singly connected BNs [11]. Since the ideas and techniques of BNs are rather complex, they have not spread much beyond the research community responsible for them [5]. The above remarks provide clear motivation for the work now presented.

In this paper, we suggest the use of a Web-based interface which allows users to utilize the powerful Bayesian network framework without having prior experience in the area. More specifically, a user is able to use the Web for uploading a data set that will be used in reasoning, as well as for posing queries about the supplied problem domain. Our system will learn a BN from the supplied data set, and, more importantly, it will perform BN inference on behalf of the user. By serving as a buffer between the inexperienced user and the sophisticated BN inference techniques, the primary advantage of our system is that it allows a much larger audience to apply BNs in practice. The work here corresponds to the notions of *three-level architecture* in databases and *abstraction* in computer programming languages, both of which are central to their fields. While it is acknowledged that no new technical contribution is made, the advantage of this work is that it allows non-experts to take full advantage of the proven BN technology.

This paper is organized as follows. In Section 2, Bayesian networks are reviewed. In Section 3, the architecture of our prototype interface is outlined. Related works are discussed in Section 4. The conclusion is given in Section 5.

2 Bayesian Networks

Let U be a finite set of discrete random variables, each with a finite set of mutually exclusive states. It may be impractical to define a joint distribution on U directly: for example, one would have to specify 2^n entries for a distribution

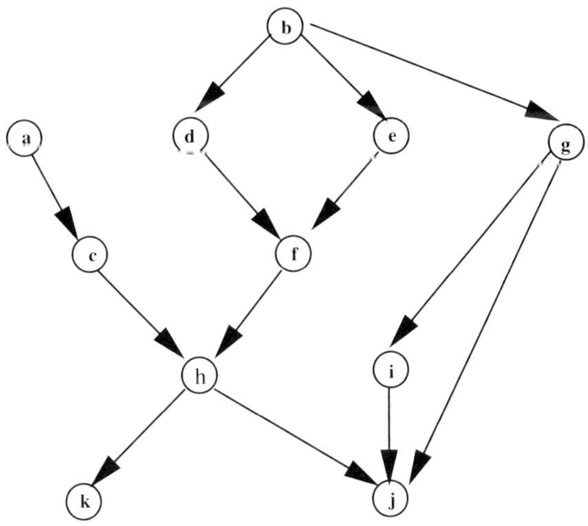

Fig. 1. A Bayesian network

over n binary variables. BNs utilize *conditional independencies* [13] to facilitate the acquisition of probabilistic knowledge.

Let X, Y and Z be disjoint subsets of variables in R. Further, let x, y, and z denote arbitrary values of X, Y and Z, respectively. We say Y and Z are *conditionally independent* given X under the joint probability distribution p, denoted $I(Y, X, Z)$, if $p(y|x, z) = p(y|x)$, whenever $p(x, z) > 0$.

A *Bayesian network* (BN) [11] is a pair $\mathcal{B} = (D, C)$. In this pair, D is a *directed acyclic graph* (DAG) on a set U of variables, and $C = \{p(a_i|P_i) \mid a_i \in D\}$ is the corresponding set of *conditional probability tables* (CPTs), where P_i denotes the *parent set* of variable a_i in the DAG D. We will use the terms BN and DAG interchangeably if no confusion arises. For example, consider the BN $\mathcal{B} = (D, C)$, where D is the DAG in Fig. 1 on $U = \{a, b, c, d, e, f, g, h, i, j, k\}$, and C is the corresponding set of CPTs. The conditional independencies encoded in the DAG D indicate that the product of the CPTs in C define a *unique* joint probability distribution $p(U)$:

$$p(U) = p(a)p(b)p(c|a)p(d|b)p(e|b)p(f|d, e)p(g|b)$$
$$p(h|c, f)p(i|g)p(j|g, h, i)p(k|h). \quad (1)$$

3 A Prototype Interface

We begin this section by outlining the main components of our Web-based interface for hiding Bayesian network inference and end it by showing some screenshots of our implementation.

The user of our system will provide two kinds of input, namely, the sample data to be reasoned with, and queries posed to the system. We assume that the former will be given once, while the latter can be posed multiple times interactively. The data is presented in tabular form and the domain of each attribute (variable) is defined as the set of values appearing for this attribute in the sample data. Our system provides a menu box where the user can upload the data onto our system. Once the sample data resides on our computer system, we apply *Netica* [10] to learn a BN from this data. At this point, we are ready for the second kind of input from the user.

Our system provides a screen in which the user can enter queries. In other words, the user can request $p(X|Y = y)$, where the collected evidence is $Y = y$ and the user is interested in the set X of target variables. For example, in a Web-based intelligent system like [4], the system may be asked for the probability of a student "passing the exam" (X), given that the student "failed the midterm examination" ($Y_1 = y_1$) and "has not completed the sample exercise questions" ($Y_2 = y_2$).

Given an input query $p(X|Y = y)$, our system computes the posterior probabilities of the variables in X using our own inference algorithm, which is based on the recent work in [3]. More specifically, our algorithm is a hybrid approach utilizing join tree propagation to guide the inference procedure. However, direct computation techniques are taken advantage of to perform the physical computation on the stored probability distributions. Moreover, our approach performs

Fig. 2. Welcoming page of our interface

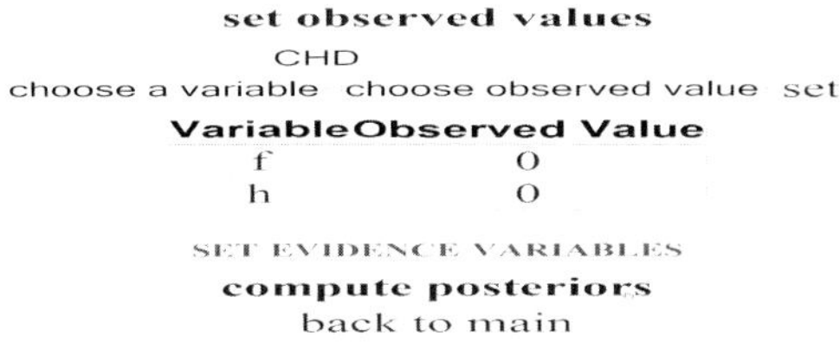

Fig. 3. Entering collected evidence $f = 0$ and $h = 0$

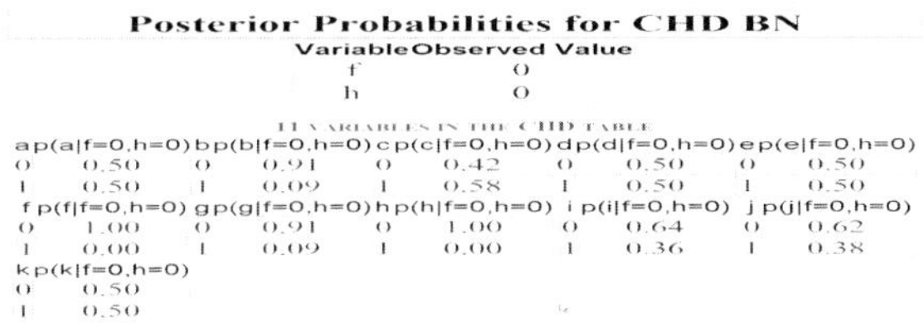

Fig. 4. Our system returns the posterior probabilities of all other variables in the BN of Fig. 1 given the collected evidence in Fig. 3

parallel computation whenever possible in order to determine the requested probabilities as soon as possible.

Once $p(X|Y = y)$ has been physically computed, our system displays the posterior probabilities for the variables in X on a Web page for the user. Based on the calculated probabilities, the user can make required decisions and take any appropriate action. This may include posing further queries to our system.

The important point is that our Web-based interface allows BNs to be utilized by the user without the user having to understand the inference process in BNs. Our system serves as a shield between the user and BN inference, thereby sparing the user the arduous task of learning how BNs work (see Section 3).

We conclude this section by demonstrating an example session of our system, which is currently only implemented as a browser-based system. The welcoming screen-shot is given in Fig. 2. As indicated, the user can either upload a data set or request posterior probabilities regarding the supplied data. Thus, the user will first upload a data set. Our system will then learn a BN from the given data. Let us assume that the BN in Fig. 1 is obtained from the supplied data. The BN CPTs obtained from the data set are not depicted as they are not pertinent to our discussion. At this point, our system is ready to perform BN inference.

The user is able to set observed values based on the evidence collected in their own application domain. Fig. 3 shows the user setting the values of variables f and h, namely, $f = 0$ and $h = 0$. Given the entered evidence, our system performs reasoning in the BN of Fig. 1 on behalf of the client and provides the posterior probabilities of the other variables, as depicted in Fig. 4. The user can make the appropriate decisions based on these returned probabilities and act accordingly.

4 Related Works

In this section, we briefly contrast our interface to Web-based BN inference with some related works.

Myllymki et al. [9] proposed a Web-based BN system to demonstrate how to build a dependency model out of a data set, and how to use it for finding interesting relations in the data. One aim of this system is to teach the user how to perform Bayesian dependency modeling and Bayesian inference. In contrast, our Web-based BN system seeks to hide these modeling and inference tasks from the user.

A Web-based BN viewer, called *BayesNet*, was suggested in [1]. Similar to the above system, BayesNet's objective is to allow viewing and working with BNs. On the contrary, we assume that the user lacks the knowledge to use BNs but still wants to take advantage of them.

It is worth mentioning that the *Uncertainty in Artificial Intelligence* (UAI) community has investigated a standard XML format for BNs, called *Bayesian networks Interchange Format* (BNIF) [2]. The stated goal of the BNIF discussions is to make it easy for UAI researchers using different BN modeling and inference tools to share models with ease. Once again, our purpose is very different - it is to allow researchers with limited BN knowledge the ability to take advantage of BNs in their Web applications.

5 Conclusion

This is the first work to suggest a Web-based interface for hiding BN inference. Our interface allows researchers and practitioners access to the established BN framework without having prior expertise in the area. It can then be seen that the work here corresponds to the notions of *three-level architecture* in databases and *abstraction* in computer programming languages, both of which are central to their fields. We have contrasted our work with other related works of similar names by emphasizing opposing objectives. Whereas these other systems use the Web to teach users the intrinsics of Bayesian network inference, our Web-based interface instead tries to shield the user from these complicated issues. While it is acknowledged that no new technical contribution is made, the work here allows any Web user to exploit BNs. We have implemented a prototype interface and have included screen shots illustrating its functionality.

References

1. http://www.webbayes.net/
2. http://research.microsoft.com/dtas/bnformat/
3. Butz, C.J., Hua, S.: An Improved Lazy-AR Approach to Bayesian network Inference. In: Nineteenth Canadian Conference on Artificial Intelligence, pp. 183–194 (2006)
4. Butz, C.J., Hua, S., Maguire, R.B.: A Web-based Intelligent Tutoring System for Computer Programming. In: IEEE/WIC/ACM International Conference on Web Intelligence, pp. 159–165 (2004)
5. Charniak, E.: Bayesian networks without tears. The AI Magazine 12(4), 50–63 (1991)
6. Jensen, F.V.: An Introduction to Bayesian Networks. UCL Press, London (1996)
7. Madsen, A.L.: An empirical evaluation of possible variations of lazy propagation. In: Proc. 20th Conference on Uncertainty in Artificial Intelligence, Banff, Canada, pp. 366–373 (2004)
8. Madsen, A.L., Jensen, F.V.: Lazy propagation: A junction tree inference algorithm based on lazy evaluation. Artif. Intell. 113(1-2), 203–245 (1999)
9. Myllymki, P., Silander, T., Tirri, H., Uronen, P.: B-course: A web-based tool for Bayesian and causal data analysis. International Journal on Artificial Intelligence Tools 11(3), 369–387 (2002)
10. Netica. Norsys: Software corp. (2000), http://www.norsys.com/netica.html
11. Pearl, J.: Probabilistic Reasoning in Intelligent Systems: Networks of Plausible Inference. Morgan Kaufmann, San Francisco (1998)
12. Russell, S.J., Norvig, P.: Artificial Intelligence: A Modern Approach. Prentice Hall, Upper Saddle River (2003)
13. Wong, S.K.M., Butz, C.J., Wu, D.: On the implication problem for probabilistic conditional independency. IEEE Transactions on Systems, Man, and Cybernetics, Part A: Systems and Humans 30(6), 785–805 (2000)

Extraction of Informative Genes from Integrated Microarray Data

Dongwan Hong[1], Jongkeun Lee[1], Sangkyoon Hong[1], Jeehee Yoon[1], and Sanghyun Park[2]

[1] Division of Information and Communication Engineering, Hallym University, Okcheon-Dong, Chuncheon, 200-702, Korea
{dwhong,jeikei,kyoons,jhyoon}@hallym.ac.kr
[2] Department of Computer Science, Yonsei University, Shinchon-Dong, Seodaemun-Gu, Seoul, 120-749, Korea
sanghyun@cs.yonsei.ac.kr

Abstract. We have recently proposed a rank-based approach as a new microarray data integration method. The rank-based approach, which converts the expression value of each sample into a rank value within the sample, enables us to directly integrate samples generated by different laboratories and microarray technologies. In this study, we show that a non-parametric scoring method can be efficiently employed for the rank-based data, and informative genes can be effectively extracted from the integrated rank-based data. To verify the statistical significance of the scoring results from the rank-based data, we compared the distribution of the score statistics to a set of distributions obtained from the randomly column-permuted data. We also validate our methods with experimental study using publicly available prostate microarray data. We compared the informative genes extracted from each individual data to the informative genes extracted from the integrated data. The results show that we can extract important prostate marker genes by directly integrating inter-study microarray data, which are missed in either single analysis.

Keywords: Informative genes selection, microarray data integration, prostate cancer, statistical significance verification.

1 Introduction

Microarray experiments enable scientists to obtain a tremendous amount of gene expression data at one time, so they are effectively used in identifying the phenotypes of diseases. In general, increasing sample size is quite desirable for more reliable and valid results. However, microarray experiments are still cost-expensive, so it is hard in reality to obtain experimental results based on a large number of samples. Thus, the experimental results from different investigations with the same research goals are somewhat different and usually contain many errors.

With the rapid accumulation of microarray data, it is of great interest and challenge to integrate inter-study microarray data to increase sample size, which leads to better experimental results. In our earlier work [1], we proposed a new

A. An et al. (Eds.): ISMIS 2008, LNAI 4994, pp. 618–627, 2008.

microarray integration method using a rank-based approach. The rank-based approach, which simply converts the expression value of each sample into a rank value within the sample, enables us directly integrate samples generated by different laboratories and microarray technologies.

In this study, we show that a non-parametric scoring method can be efficiently employed for the rank-based data, and informative genes can be effectively extracted from the integrated rank-based data. As a non-parametric scoring method, Park's method [2] is employed. However, as the scoring method compares the sample values of each gene to calculate a score, it may give slightly different score results when it is applied to the rank-based data and the actual expression value data, respectively. Here we verify the statistical significance of the scoring result from the rank-based data. We compared the distribution of the score statistics to a set of distributions which is obtained from the randomly column-permuted data. Golub's leukemia data [3] was tested, and its result was significant with the p-value of 0.0005 for the rank-based data. Then we compared the informative genes extracted from the rank-based data to the informative genes extracted from the actual expression value data. To exemplify the effectiveness of our integration method, we used three publicly available prostate microarray data. We compared the informative genes extracted from each individual data to the informative genes extracted from the integrated data. The results reveal that important marker genes are selected from the integrated data, which are missed from a single data.

2 Related Works

Experimental microarray data are organized as matrices where rows represent genes and columns represent samples. However, even when considering the microarray data with the same research goals, differences in platforms, protocols, set of genes, and scales of gene expression values lead to difficulties in integrating microarray data across experiments.

To integrate microarray data, the typical methods include *meta-analysis method* [4], *normalization and transformation method* [5,6], and *rank-based approach* [1]. Instead of comparing microarray expression values from individual experiments, *meta-analysis method* combines the results of individual experiments by using statistical technique. However, there are many cases where the individual experimental results are not reliable due to the small sample size. So the integration of these results may bring an even worse analysis. *Normalization and transformation method* transforms the gene expression values of individual experimental data into a common scale, and then integrates inter-study data [5]. A classical method is the z-score transformation [6], which normalizes the expression values with the mean and standard deviation of each sample. Statistical tests, such as fold ratio, z ratio/test [6], and t statistical test, can be applied directly to the normalized data for predicting significant changes in gene expressions. However, there is still no consensus on the best method to perform data normalization [7]. *Rank-based approach* converts the expression value of each sample into a rank value [1]. In statistical area, this method has been used as a noise reduction method [8]. Xu *et al.* [7] proposed a new classification method (top-scoring pair classifier) to select maker genes from the integrated rank-based data. However this method is only based on comparing relative expression values within each sample.

One of the difficulties in analyzing microarray data is the high dimensionality due to a large number of genes. However, only a small fraction of genes is informative for predicting significant changes in gene expressions. Currently, various methods are being presented to select informative genes precisely and effectively. Typically, informative genes are selected according to a test statistics. A *parametric method* assumes a statistical model representing the data, such as the t-statistics [9], Fisher [10], and Golub's method [3]. There are *non-parametric methods* such as TNom [11], Wilcoxon rank sum [12], and Park's method [2]. These methods define a minimum boundary and calculate the distance from the boundary as the score. On the other hand, when the gene is considered as a feature, the rank-based feature selection method [13] can be used. This method measures the significance of features and then ranks them. In this approach, the popular methods are Information Gain [13], Relief-F [14], and the method using Kendall's Correlation Coefficient [15]. However, all these methods use the gene expression values of each gene, and there is no consideration regarding the integration and normalization of the microarray data.

3 Methods

3.1 A Rank-Based Microarray Data Integration

The integration procedure of microarray data is shown as follows. First, only the experimental data of common genes are extracted from the individual microarray data, which has the same research goals. Then the expression value of each sample in each experiment is converted to a rank value within the sample. Once the expression values are changed to rank values, the integration of samples from different experiments becomes feasible. This method is simple and useful for integrating a large number of microarray samples without the need to perform any normalization. Hereafter, for simplicity, we call experimental data using the original expression values *raw data*, and experimental data using the rank values *rank data*. As the integrated data contains only the rank values rather than the actual expression values, there may be a slight loss of information. However, too big or too small expression values of each sample can be noises, which may give a negative effect on extracting informative genes. In return, we gain the robustness to external factors, such as noises.

3.2 Informative Genes Selection Method

Park's non-parametric scoring method [2] is extended and applied to the integrated microarray data. Park's method, which is proposed for a single microarray

	Normal			Cancer		
Sample no.	1	2	3	4	5	6
Sample data	95	106	20	74	69	271
Class level	0	0	0	1	1	1

↓ After Sorting

	Normal			Cancer		
Sample no.	1	2	3	4	5	6
Sample data	20	69	74	95	106	271
Class level	0	1	1	0	0	1

Score	Binary sequence						Position swapped
	0	1	1	0	0	1	3 and 4
+1	0	1	0	1	0	1	2 and 3
+1	0	0	1	1	0	1	4 and 5
+1	0	0	1	0	1	1	3 and 4
+1	0	0	0	1	1	1	

Fig. 1. An example of gene scoring

data, builds a binary sequence for a gene and calculates a score measuring how differently the genes are expressed in the two class groups, by using Kendall's Correlation Coefficient [15].

Let us explain the scoring method by using an example. Fig. 1 shows how to calculate the score of a gene with six sample data of 95, 106, 20, 74, 69, and 271. Here, we assume that each sample data represents the rank value. In this figure, samples 1, 2, and 3 represent normal class and samples 4, 5, and 6 represent cancer class. First, class label 0 is assigned the normal sample and class label 1 is assigned to the cancer sample, to obtain an initial binary sequence S = 000111, which represents the class labels of the gene data. Next, the sample data are sorted in ascending order along with the class labels. Thus, the sorted binary sequence T = 011001 is obtained, and it represents the class labels of the sorted gene data. A distance between S and T is used as the score of the gene. The distance is defined as the minimum number of swaps of neighboring 0 and 1 which is necessary to transform the sorted binary sequence into the initial binary sequence. Fig. 1 shows the process in which T = 011001 is transformed into S = 000111, and result score of 4. Suppose the number of normal samples is n_1 and the number of cancer samples is n_2, then the score ranges from 0 to $n_1 \times n_2$. Both low and high scores indicate differentially expressed genes, which are selected as informative genes.

3.3 Example

Next we illustrate data integration and informative gene selection procedures using an example. Let us consider two data, Data(A) and Data(B), which are generated independently but have the same research goals. As shown in Fig. 2, the scale of the expression values for each data is quite different and a direct integration is inappropriate. First we convert all expression values into ranks within each sample, and obtain Data(A)′ and Data(B)′ of *rank data*. As explained in Section 3.2, the score refers to the minimum number of swaps of neighboring

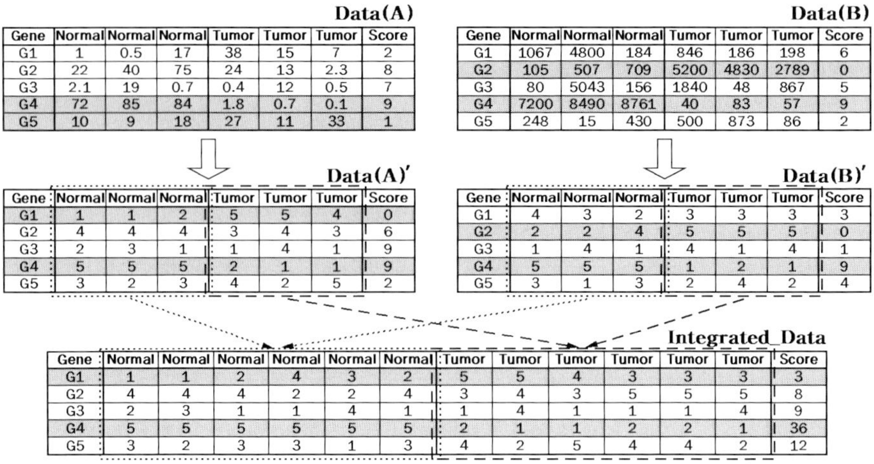

Data(A)

Gene	Normal	Normal	Normal	Tumor	Tumor	Tumor	Score
G1	1	0.5	17	38	15	7	2
G2	22	40	75	24	13	2.3	8
G3	2.1	19	0.7	0.4	12	0.5	7
G4	72	85	84	1.8	0.7	0.1	9
G5	10	9	18	27	11	33	1

Data(B)

Gene	Normal	Normal	Normal	Tumor	Tumor	Tumor	Score
G1	1067	4800	184	846	186	198	6
G2	105	507	709	5200	4830	2789	0
G3	80	5043	156	1840	48	867	5
G4	7200	8490	8761	40	83	57	9
G5	248	15	430	500	873	86	2

Data(A)′

Gene	Normal	Normal	Normal	Tumor	Tumor	Tumor	Score
G1	1	1	2	5	5	4	0
G2	4	4	4	3	4	3	6
G3	2	3	1	1	4	1	9
G4	5	5	5	2	1	1	9
G5	3	2	3	4	2	5	2

Data(B)′

Gene	Normal	Normal	Normal	Tumor	Tumor	Tumor	Score
G1	4	3	2	3	3	3	3
G2	2	2	4	5	5	5	0
G3	1	4	1	4	1	4	1
G4	5	5	5	1	2	1	9
G5	3	1	3	2	4	2	4

Integrated_Data

Gene	Normal	Normal	Normal	Normal	Normal	Normal	Tumor	Tumor	Tumor	Tumor	Tumor	Tumor	Score
G1	1	1	2	4	3	2	5	5	4	3	3	3	3
G2	4	4	4	2	2	4	3	4	3	5	5	5	8
G3	2	3	1	1	4	1	1	4	1	1	1	4	9
G4	5	5	5	5	5	5	2	1	1	2	2	1	36
G5	3	2	3	3	1	3	4	2	5	4	4	2	12

Fig. 2. An example of microarray data integration and informative gene selection

digits necessary to arrive at perfect splitting, with all the 0's on the left and all the 1's on the right. If only two genes are selected as informative genes from each data, the genes with the lowest and highest score are selected. For instance, "G5 and G4" and "G2 and G4" are selected from Data(A) and Data(B), respectively, and at the same time "G1 and G4" and "G2 and G4" are selected from Data(A)′ and Data(B)′, respectively. Notice that the extracted informative genes from *raw data* and *rank data* may be different. Next, Data(A)′ and Data(B)′ are merged and finally "G1 and G4" are selected from the Integrated_Data as informative genes.

3.4 Significance Test

A permutation test is performed to test the significance of gene scoring result for the *rank data*. We generate a random permutation of entire columns, keeping all the rank values for each samples together. A p-value is then computed by comparing the distribution obtained from the original data to the set of distributions obtained from the randomly permuted data. To calculate a p-value, a cumulative function S_i of (Eq. 1) is used. For the comparison, we use the same function which is given in [2]. S_i is the measure of how much the i-th score distribution is different from the average of all the other score distributions. Here, f_i^* represents the average of all distributions except for the score distribution of the i-th column-permuted data, and M represents the number of column-permuted data. S_0 represents the difference between the score distribution of original data and the average of the score distributions of other column-permuted data. A significance probability $P(S_i \geq S_0)$ is now calculated. Here, the requirement of $i = 1, \cdots, M$ is met. If the p-value is smaller than the significance level, we assume that the gene scoring result is significant.

$$S_i = \sum_{j=0}^{n_1 n_2} (f_i(x_j) - f_i^*(x_j))^2, \quad i = 1, \cdots, M$$
$$f_i^*(x_j) = \frac{1}{M-1} \sum_{k=1, k \neq i}^{M} f_k(x_j) \tag{1}$$

4 Results

4.1 Significance Test for Scoring Results

We applied the non-parametric scoring method described in Section 3.2 to the Golub's leukemia data [3]. Golub's data contains 38 bone marrow samples obtained from acute leukemia patients. 27 samples are from ALL class and 11 samples are from AML class. High-density oligonucleotide microarrays (produced by Affymetrix) containing 7129 probes for 6817 human genes are used. As stated in Section 3.4, a permutation test was performed for two data, *raw data* and

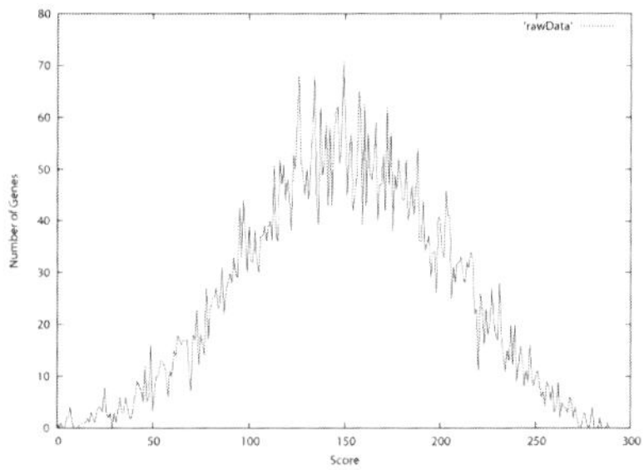

(a) The score distribution from the *raw data*

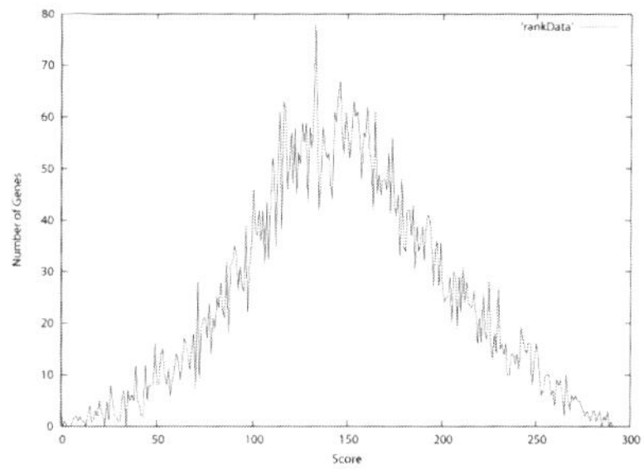

(b) The score distribution from the *rank data*

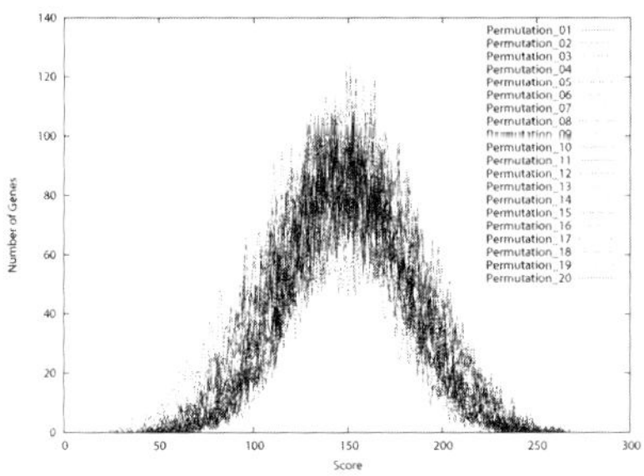

(c) The score distributions from the column-permuted *raw data*

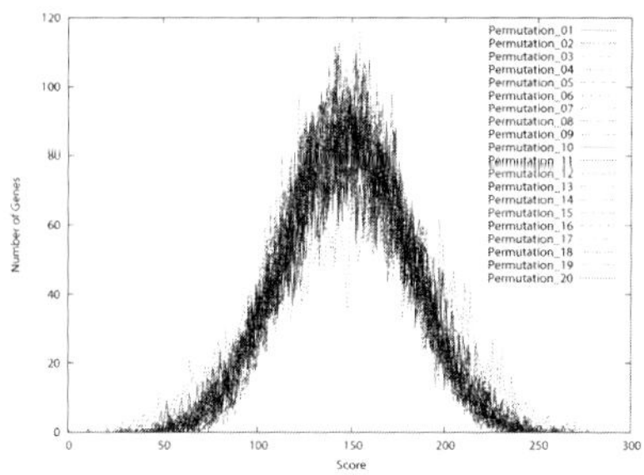

(d) The score distributions from the column-permuted *rank data*

Fig. 3. Comparison of gene score distributions using *raw data* and *rank data* (Golub's data [3] is used) *rank data*. We performed 10,000 permutations. Fig. 3 shows the score distributions from the original data and a set of randomly column-permuted data. Fig. 3-(a) shows the score distribution from the *raw data* where the score of each gene is calculated by using the gene expression values of samples. And Fig. 3-(b) shows the score distribution from the *rank data* where the score of each gene is calculated by using the rank values of samples. The results show that the two score distributions have very similar shapes and especially show heavier tails as expected, indicating many genes are differentially expressed in the two classes. Fig. 3-(c) shows a set of score distributions from the column-permuted *raw data*, and Fig. 3-(d) shows a set of score distributions from the column-permuted *rank data*. Here, only twenty score distributions are plotted for each case. The results show that the score distributions from the *raw/rank data* are more spread out with heavy tails, while the score distributions from the column-permuted *raw/rank data* are relatively concentrated with smaller variances. Based on the S_i values explained in Section 3.4, p=0.0005 was obtained from the *rank data*. Also, p=0.0053 was obtained from the *raw data*, which is consistent with the p-value reported by Park [2]. This result verifies our expectation that the scoring result from the *rank data* is statistically significant.

Next, we compared informative genes extracted from *rank data* to the informative genes from *raw data*. When top 1% of the genes are selected as informative genes, about 70% of the informative genes overlap each other. When top 5% of the genes are selected, about 76% of them overlap each other. Also, they include all 50 informative genes that were chosen by Golub's experiment [3].

4.2 Selection of Informative Genes from Integrated Data

To investigate whether more accurate informative genes can be selected from integrated data, the scoring method has been applied to the individual and integrated *rank data*. We used three prostate cancer microarray data which are publicly available. The platform of these data is Affymatrix HG_95Av2. Each data will be represented as an abbreviation of the first author of the paper, like as LaTulippe [16], Welsh [17], and Singh [18]. LaTulippe consists of 3 normal samples, 14 primary prostate cancer samples, and 9 metastatic prostate cancer samples. Welsh consists of 9 normal samples and 25 cancer samples, and Singh consists of 50 normal samples and 52 cancer samples.

As mentioned previously, we assume that larger sample size enables to extract more statistically significant genes. Also, we can expect a better statistical result when the number of test samples is almost equal to that of control samples. Singh's sample size is relatively larger than both LaTulippe's and Welsh's, and the number of its test samples is almost same as that of its control samples.

We merge LaTulippe and Welsh, using the 12600 common probe sets, to form an integrated data of increasing sample size. Here, (LaTulippe+Welsh) represents the integrated data resulting from the merging of LaTulippe and Welsh data. The scoring method is applied to the individual and integrated data, and top 1% of genes are selected as informative genes for each data. The selected informative genes are listed in Table 1.

Table 1. Comparison of informative genes extracted from LaTulippe, Welsh, Singh and (LaTulippe+Welsh)

Ranking	LaTulippe	Welsh	Singh	LaTulippe+Welsh	Ranking	LaTulippe	Welsh	Singh	LaTulippe+Welsh
1	FCGRT	MYL6	**HPN**	**ANGPT1**	64	KIAA0303	**CALM1**	AMACR	**HSPD1**
2	SOX5	CLU	**PTGDS**	**CALM1**	65	MEIS2	GJA1	**NME1**	MYL6
3	LCAT	PSIP2	**NELL2**	LPIN1	66	CDC5L	MYH11	CLDN3	TPM1
4	PNMT	**ANGPT1**	TRG@	SVIL	67	Rab11-FIP2	GSN	XBP1	SYP
5	IGF2	**DSCR1L1**	ANXA2P3	**COL4A6**	68	TRO	**GSTP1**	KIAA0977	ATP2A2
6	CYP3A5	FZD7	**HSPD1**	MEIS2	69	HSD11B1	RBPMS	SLC25A6	TACC1
7	CYP3A5	CBX7	ANXA2	CBX7	70	LDB2	TPM1	RPL12	KIAA0992
8	MDM1	KIAA0469	KLK3	LAPTM4A	71	NDN	MEIS2	GFPT1	JAM3
9	ELKS	FTO	PLA2G7	PRNP	72	ARHGEF4	CNN1	TNA	TRIP6
10	COL13A1	DMPK	PDLIM5	MYLK	73	FTO	TGFB1I1	TACSTD2	SH3GLB1
11	**ANGPT1**	RRAS	**STAC**	**GSTP1**	74	NRLN1	KANK	KIAA1157	STOM
12	CHRNA7	TRIP6	TMSNB	CLIPR-59	75	DOCK1	PMP22	AKR1B1	**HPN**
13	LDOC1	-	XBP1	GASP	76	SLK	ATP2A2	RBP1	LOC171220
14	RE2	PPP3CB	**DF**	NRLN1	77	DKFZP586A0522	CALD1	MYO6	CLU
15	GPR161	SVIL	**SPON1**	BART1	78	LAPTM4A	FLJ21174	RAP1GA1	**STAC**
16	CX3CR1	SRF	RGS10	SNX1	79	SRI	FLNA	MEG3	GNAZ
17	KIAA0888	DES	GUCY1A3	SPARCL1	80	MEIS1	CLIC4	PDIR	SLC2A5
18	LOC151584	PPP1R12B	**NME2**	**DAT1**	81	PGCP	GATM	**SC65**	FLNC
19	KIAA0534	KIAA0992	THBS4	**C7orf24**	82	TCF12	COL4A2	IL11RA	TBL1X
20	APEG1	OPTN	-	RBPMS	83	CDC42EP3	**DAT1**	**GSTP1**	ROR2
21	IGSF1	FLNA	RPL13A	GJA1	84	MAPRE1	RBPMS	ZNF146	WFS1
22	AIP1	BPAG1	SLC25A6	HOXC6	85	PPAP2B	NIFU	PHYHIP	DMPK
23	MKLN1	MYLK	**CALM1**	KIAA0725	86	STAT5B	ENO2	HOMER2	WFDC2
24	KIAA0980	**GSTP1**	SIM2	TCF8	87	SUSP1	ATP2B4	HSPA8	LDB3
25	TRO	TAZ	**DAT1**	**ANGPT1**	88	SLC2A5	LTBP1	MTHFD2	SEC23A
26	CHS1	PLS3	TSPAN-1	EDNRA	89	**SPON1**	GSTM5	**PYCR1**	SPG20
27	KPNA3	COL6A2	**C7orf24**	RBPMS	90	SYNGR1	-	ATP2C1	MXRA7
28	SNCG	FLNC	FBP1	TPM1	91	CLIPR-59	TPM2	LOC285843	ITPR1
29	D2LIC	BC008967	TACSTD1	TGFB1I1	92	VCL	TPM2	**NME1**	KRT18
30	TCF21	SDFR1	**COL4A6**	FNBP1	93	DHX38	TPM1	CYP1B1	**PYCR1**
31	ALDH1A2	CAV1	RPLP0	PPAP2B	94	BTBD3	TPM1	**PTGDS**	**NME2**
32	GSTM1	DPYSL3	ITSN1	CCND2	95	DKFZP434D1335	GASP	TRAF4	CNN1
33	ACTC	PRNP	P4HB	FEZ1	96	CETN2	ITGA8	SAT	RIMS3
34	MAP1LC3B	PLEKHC1	AGR2	DKFZP564M1416	97	MGC35048	DFNA5	ODC1	EMILIN1
35	PBX1	CCND2	-	MEIS3	98	GPRC5B	SMTN	EEF1G	FGFR1
36	SSX2IP	LMOD1	EPB41L3	CSRP1	99	**CALM1**	TEAD3	-	MYH11
37	C22orf2	RBPMS	TU3A	OPTN	100	C14orf132	BCMP1	S100A4	**NME2**
38	EFS	LPIN1	FOLH1	**GSTP1**	101	IGF1	ANXA6	TGFB3	SYNGR2
39	WFDC2	DSTN	G6PD	-	102	MADH6	CDC10	RPS10	TPM2
40	SSA2	TUBA3	MLP	MEIS1	103	RRAS	PTRF	MGC2650	DOCK1
41	FBXO7	JAM3	WSB2	RRAS	104	TGFB3	SDC4	CANX	DKFZP586A0522
42	DKFZP564M1416	FHL1	PLAB	KIAA1128	105	IGF1	CLIPR-59	ADCY3	RBPMS
43	DPT	**ANGPT1**	BUCS1	RRAS	106	RGN	SPARCL1	SH3YL1	ARMET
44	**PTGDS**	RBPMS	**GSTM4**	COX7A1	107	-	FEZ1	FASN	AKR1A1
45	**DF**	MYH11	RPL18A	ACTC	108	MADH4	MYL9	KIAA0934	RBPMS
46	SMARCD3	ITPR1	-	DMD	109	MADH4	ENIGMA	ERG	SMTN
47	**NELL2**	RBPMS	DKFZP586I2223	PLEKHC1	110	CASP9	FGF2	TM4SF2	ST5
48	BART1	**COL4A6**	RPS18	TPM2	111	PTGDS	**SC65**	U38A	DMN
49	FGFR2	SNX1	ATP6V1G1	DES	112	MLLT1	EDNRA	KIAA0746	CX3CL1
50	**ANGPT1**	COL6A1	RPS2	CDC42EP3	113	RBPMS	ACTG2	C2orf3	**NME1**
51	TGFB3	LAPTM4A	**DSCR1L1**	PTRF	114	TPS1	KIAA0476	GADD45G	PPP1R3C
52	RASA1	ACTN	DKFZP564B167	ACTG2	115	COL4A3	MEIS3	ANGPTL2	**CRYAB**
53	IGF2	TACC1	**ANGPT1**	FER1L3	116	TNFRSF4	-	CPD	EZH1
54	CX3CL1	ACTA2	CDC42BPA	SRF	117	MLLT7	KIAA0727	IMPDH2	ATP1A2
55	PRSS11	TCF8	PENK	FLNA	118	FLJ32389	RIMS3	p100	CTF1
56	**GSTM4**	MBNL1	**CRYAB**	BC008967	119	GASP	KCNMB1	KIAA0342	TPM1
57	SMARCD3	RARRES2	UAP1	VCL	120	MEIS3	RNPC1	BMP5	GPM6B
58	SPOCK3	LDB3	**NME1**	MYL9	121	TIP120B	COX7A1	CLDN8	PPP3CB
59	COL4A3	EFEMP2	**NME2**	RARRES2	122	CYLD	FNBP1	RPL14	RIL
60	CTF1	GNAI2	EEF2	SDFR1	123	ASPA	STAT5B	RCL	PRKCB1
61	RAMP2	SYP	MGC5178	LMOD1	124	DBCCR1	NID	TFPI	RGN
62	ZNF288	ST5	ATP1A1	FGFR2	125	SPINK2	KIAA0193	ALCAM	COPE
63	COL13A1	MXRA7	FLRT2	GAS1	126	CXCL13	**SC65**	RPLP2	**DSCR1L1**

Genes that were selected in common are shown in bold and underlined.

We then compared the informative genes from independently conducted data to the informative genes from the integrated data. When we compare the informative genes of Singh and (LaTulippe+Welsh), 15 genes are found in common. In contrast, when we compare the common genes of Singh and (Latulippe+Welsh+Singh) with LaTulippe and Welsh, only 9 and 10 genes are found in common, respectively.

Furthermore, among the 15 informative genes we identified several tumor marker genes such as *HPN*, *C7orf24*, *NME1*, *NME2*, *CRYAB*, and *PYCR1*. *HPN* has been identified as a marker gene of prostate cancer in recent studies [19,20]. *HPN* encodes hepsin, a cell surface transmembrane serine protease which plays an essential role in cell growth and the maintenance of cell morphology [7]. Also, *NME1* has been well known to be involved in the metastatic potential of several tumor cells, including prostate cancer cells [21]. Recently, Reference [22] reported that *C7orf24* may have an important role in cancer cell proliferation, and may be an appropriate therapeutic target molecule against cancer. However, these genes are not included in the list of LaTulippe or Welsh, either. These findings suggest that we can extract important marker genes which are missed in an individual data analysis by integrating several different microarray data.

5 Conclusion

In this paper, we showed the effectiveness of microarray integration and analysis method using *rank data*. To verify the statistical significance of the nonparametric scoring results, a random permutation test was performed for the *rank data*. With an experimental study using publicly available prostate microarray data, we also demonstrate that we can obtain more reliable and valid results from integrated data, based on a large number of samples.

Acknowledgment

This work was partially supported by the Korea Research Foundation Grant funded by the Korean Government(MOEHRD, Basic Research Promotion Fund) (KRF-2007-531-D00019) and by the Korea Science and Engineering Foundation(KOSEF) grant funded by the Korea government(MOST) (No. R01-2006-000-11106-0).

References

1. Yoon, Y.M., Lee, J.C., Park, S.H.: Building a Classifier for Integrated Microarray Datasets through Two-Stage Approach. In: Proc. IEEE Symposium on Bioinformatics & Bioengineering, vol. 6, pp. 94–102 (2006)
2. Park, P.J., Pagano, M., Bonetti, M.: A nonparametric scoring algorithm for identifying informative genes from microarray data. In: Pacific Symposium on Biocomputing, pp. 52–63 (2001)

3. Golub, T.R., et al.: Molecular Classification of Cancer: Class Discovery and Class Prediction by Gene Expression Monitoring. Science 286, 531–537 (1999)
4. Rhodes, D.R., Barrette, T.R., Rubin, M.A., Ghosh, D., Chinnaiyan, M.: Meta-Analysis of Microarrays: Interstudy Validation of Gene Expression Profiles Reveals Pathway Dysregulation in Prostate Cancer. Cancer Research 52, 4427–4433 (2002)
5. Jiang, H., Deng, Y., Chen, H.S., Tao, L., Sha, Q., Chen, J., Tsai, C.J., Zhang, S.: Joint analysis of two microarray gene-expression data sets to select lung adenocarcinoma marker genes. BMC Bioinformatics 5, 81–93 (2004)
6. Cheadle, C., Vawter, M., Freed, W., Becker, K.: Analysis of Microarray Data Using Z Score Transformation. Journal of Molecular Diagnostics 5-2, 62–73 (2003)
7. Xu, L., Tan, A.C., Naiman, D.Q., Geman, D., Winslow, R.L.: Robust prostate cancer marker genes emerge from direct integration of inter–study microarray data. Bioinformatics Advance Access 21, 3905–3911 (2005)
8. Rosner, B.: Fundamentals of Biostatistics. Thompson 6, 540–544 (2003)
9. Shamir, B.A., Yakhini, R.Z.: Clustering gene expression patterns. J. Comput. Biol., 281–297 (1999)
10. Drăghici, S., Khatri, P., Martins, R.P., Ostermeier, G.C., Krawetz, S.A.: Global functional profiling of gene expression. Genomics 81, 98–104 (2003)
11. Rogers, S., Williams, R.D., Campbell, C.: Class Prediction with Microarray Datasets. In: Bioinformatics using Computational Intelligence paradigms. Studies in Fuzziness and Soft Computing, vol. 176, pp. 119–141 (2005)
12. Deng, L., Pei, J., Ma, J., Lee, D.L.: A Rank Sum Test Method for Informative Gene Discovery. In: Proceedings of the 10th ACM SIGKDD International Conference on Knowledge Discovery and Data Mining (KDD 2004), vol. 176, pp. 410–419 (2004)
13. Witten, I.H., Frank, E.: DATA MINING Practical Machine Learning Tools and Techniques, pp. 97–112. Morgan Kaufmann, San Francisco (2005)
14. Marko, R., Igor, K.: Theoretical and Empirical Analysis of ReliefF and RReliefF. Machine Learning Journal 53, 23–69 (2003)
15. Bailey, N.: Statistical methods in biology. Cambridge University Press, Cambridge (1995)
16. LaTulippe, E., Satagopan, J., Smith, A., Scher, H., Scardino, P., Reuter, V.: Comprehensive gene expression analysis of prostate cancer reveals distinct transcriptional programs associated with metastatic disease. Cancer Res. 62, 4499–4506 (2002)
17. Welsh, J.B., Sapinoso, L.M., Su, A.I., Kern, S.G., Wang-Rodriguez, J., Moskaluk, C.A.: Analysis of gene expression identifies candidate markers and pharmacological targets in prostate cancer. Cancer Res. 61, 5974–5978 (2001)
18. Singh, D., Febbo, P., Ross, K., Jackson, D., Manola, J., Ladd, C.: Gene expression correlates of clinical prostate cancer behavior. Cancer Cell 1, 203–209 (2002)
19. Hood, B., et al.: Proteomic Analysis of Formalin Fixed Prostate Cancer Tissue. Molecular & Cellular Proteomics 4, 1741–1753 (2005)
20. Pal, P., et al.: Variants in the HEPSIN gene are associated with prostate cancer in men of European origin. Hum. Genet. 210, 187–192 (2006)
21. Bemd, G., et al.: Mass spectrometric identification of human prostate cancer-derived proteins in serum of xenograft-bearing mice. Molecular & Cellular Proteomics 5, 1830–1839 (2006)
22. Iwaki, H., et al.: A novel tumor-related protein, C7orf24, identified by proteome differential display of bladder urothelial carcinoma. PROTEOMICS - Clinical Applications 1, 192–199 (2007)

Using Data Mining for Dynamic Level Design in Games

Kitty S.Y. Chiu and Keith C.C. Chan

Department of Computing, The Hong Kong Polytechnic University,
Hung Hom, Kowloon, Hong Kong
{cssychiu,cskcchan}@comp.polyu.edu.hk

Abstract. "Fun" is the most important determinant of whether a game will be successful. Fun can come from challenges and goals, such as victory in a scenario, the accumulation of money, or the right to move to the next level. A game that provides a satisfying level of challenge is said to be balanced. Some researchers use artificial intelligence (AI) on the dynamic game balancing. They use reinforcement learning and focuses on the non-player characters. However, this is not suitable for all game genres such as a game requiring dynamic terrains. We propose to adjust the difficulty of a game level by mining and applying data about the sequential patterns of past player behavior. We compare the performance of the proposed approach on a maze game against approaches using other types of game AI. Positive feedback and these comparisons show that the proposed approach makes the game both more interesting and more balanced.

Keyword: Artificial intelligence, Game Development, Data Mining.

1 Introduction

Successful computer games are fun. Fun can come from challenges and goals, which can be many things, such as victory in a scenario, the accumulation of asset, or the right to move to the next level [1]. When a game is satisfyingly challenging at a particular level, we say that a game is balanced. Generally, game designers adjust the difficulty of the levels of a game by using pre-defined templates and random generation or rule-based system. Such approaches however provide relatively rigid responses to individual game play that can make games a poor fit for the skill levels of many players who otherwise might enjoy them. That is, they may find the game too hard or too easy. One response to this problem is to generate game levels adaptively by using game artificial intelligence (AI). This would allow the difficulty of the game level to be adjusted dynamically, a process called dynamic game balancing. [2][3][4][7] There are two approaches to dynamic game balancing: changing the behavior of the non-player characters (NPCs) [3][4] and changing parameters of the game environment. [7] These approaches, however, apply only to some specific game genres such as shooting games and role-play games.

In this paper we propose an approach to using sequential pattern mining to adjust the difficulty of the levels of games that require terrain generation. This approach is not only suitable for "shooting games" and "role-play games", other game genres, such as "puzzle games" and "racing games" can be applied. We do this by using the game log to

A. An et al. (Eds.): ISMIS 2008, LNAI 4994, pp. 628–637, 2008.

find the relationship between a player's past behaviors and the parameters of the game environment. The player's behaviors can then be predicted and the game level can be adjusted to the individual player. To replace the traditional rule-based game AI or pre-defined game level template, we use sequential pattern mining. Our proposed approach is as follows. First we detect underlying patterns in an ordered game log over a series of levels. We then use these patterns to construct sequence-generation rules that can predict the attributes of a player and those attributes are used to generate a game environment of a suitable level of difficulty.

The remainder of this paper is organized as follows. In the Section 2, we introduce some of the problems of the dynamic game balancing. In Section 3 we describe the induction learning method used in assembling the game log. In Section 4 we present our proposed approach. In Section 5, we describe the implementation and results for our approach on a test module. The final section presents our conclusion and some directions for future work.

2 Dynamic Game Balancing

A game is a series of processes that takes a player to a result; it is a series of interesting decisions. [1]. Some game genres, such as puzzle games, adventure games, role-play games and sports games challenge players by using use "levels", also called "maps", "stages" or "missions". A game level is built on a set of small challenges [2], for example an enemy, a maze, or some hybrid challenge, and the game environment [6], the number of enemies, and so on. Most of the elements of the game environment are related to the mission. A balanced game will provide a satisfying level of challenge to the player with a current level being more difficult than a previous level. Although some games allow players to adjust the basic level of difficulty, the levels of most games are pre-defined during the game design process and the overall level of the game is static. As a result, some players will feel a game is easy and others that it is hard. Adjustment of the difficulty of a game is called dynamic game balancing [2].

Dynamic game balancing uses an automatic algorithm for changing parameters, scenarios and behaviors in the game to prevent players feel frustrated during playing. There are a number of different approaches to dynamic game balancing, including modifying the behaviors of NPCs and changing the parameters of the game environment. [2] [3][6] [7]. Research into dynamic scripting focuses on the dynamic game balance [7] [8] [9] but is suitable only for games that are scripted or imply storytelling. Lee's [8] worked on the dynamic scripting for a shooter game used a Gaussian Mixture Module that modeled the player's reaction pattern. Spronck [7] [9] focused on dynamic scripting, using reinforcement learning to control the movement of the NPC. Andrade's [3] also used reinforcement learning to modify NPC behaviors. They did not, however, change the game environment or adjust the difficulty of the game level during play.

3 Learning of the Game Log

Generally, game developers use a random number approach, pre-defined rules or a template approach to build the game world and focus on the movement of NPCs (including action, instance response, and movement), user interface design (design of the characters, rendering of the graphics and layout of the world), game control, and sound effects. To reduce the development time on the game, many game level designers use classic game AI and game companies release a beta version of the game to the public or target players to evaluate the game levels. Game companies can then adjust the difficulty of the game level before official release [10]. One example of the random number approach is maze games, which almost always uses maze game AI to generate a random number as a parameter and then builds the mazes on every game level, repeating the process until the players pass all levels and reach the final goal. To ensure that the difficulty of each level is greater than the previous level, when generating a new level, some game level designers add more criteria to the maze generator while some use pre-defined templates of the game environment. After players finish Level T, the game engine immediately loads the new level T+1. The disadvantage of the random generation and pre-defined templates approaches is that they lack flexibility.

One way to make such games more flexible is to adjust the level of difficulty, that is, dynamic game balancing [2] by applying an adaptive game engine. To do this, we analyze the individual player's gameplay data, identify characteristic patterns, and then predict a player's future behavior, adjusting the game level so that it adapts to the player. The player's gameplay data will be held in a game log. Our approach analyzes this log using a multivariate pattern mining method and builds a set of prediction rules. The game log is a set of sequential patterns. It can be defined as follows: Suppose that there is an ordered sequence S of M player behavior data, Log_1, $\ldots Log_{jp},\ldots,Log_m$, where Log_{jp} is located at position p in S. Suppose also that each behavior in the sequence is described by n distinct attributes, $Attr_{1p}$, …, $Attr_{jp}$, …, $Attr_{np}$, and that in any instantiation of the description of the behavior, an $Attr_{pj}$ takes on a specific value, $val_{jp} \in$ domain($Attr_{jp}$)=$\{v_{jkl}\ k=1,\ldots,J\}$, which may be numerical or symbolic or both.

The relationships between the attributes are important and can help us to improve the accuracy of prediction. If the ith attribute of a behavior that takes on v_{il}, is always preceded at t level earlier by the behavior whose jth attributes takes on the value v_{jk}, we can conclude that the v_{j1} is dependent on v_{jk}, with a level t.

To decide if the ith attribute of a behavior in a sequence is dependent on the jth attribute $Attr_{jp}$ of the behavior at t level earlier, the chi-square test can be employed. The chi-square test uses a two-dimensional contingency table of I rows and J columns. (Table 1) (I and J respectively being the total number of values taken on by the ith and the jth attributes).Let Log_{jk} be the total number of behaviors in S whose ith attribute $Attr_{i(p+t)}$ takes on the value v_{il} and are preceded at t positions earlier by behaviors that have the characteristic v_{jk}. Let e_{lk} be the expected number of such behaviors, under the assumption that $Attr_{i\,(p+t)}$ and $Attr_{jp}$ are independent.

$$e_{lk} = \sum_{u=1}^{J} o_{lu} \sum_{u=1}^{I} o_{uk} / M' \tag{1}$$

where $M' = \sum_{l,k} o_{lk}$ is less than or equal to M (the total number of behaviors in the sequence S) due to the possibility of there being missing values in the data. The chi-square statistic can be defined as

$$X^2 = \sum_{l=1}^{I}\sum_{k=1}^{J}\frac{(o_{lk}-e_{lk})^2}{e_{lk}} = \sum_{l=1}^{I}\sum_{k=1}^{J}\frac{o_{lk}^{\;2}}{e_{lk}} - M'. \quad (2)$$

The difference between the observed and expected value could have arisen by chance. This can be determined by comparing chi-square statistic X^2 with the critical chi-square $X^2_{d,a}$, where d=$(I$-1)(J-1) is the degree of freedom and a, usually taken to be 0.05 or 0.01, is the significance level. The confidence level is (1-α)%. If X^2 is greater than the critical value, there is enough evidence to conclude that $Attr_{i(p+t)}$ is dependent on $Attr_{jp}$. One cannot conclude this if X^2 is less than $X^2_{d,\,a.}$ Finally, we can use multivariate analysis to find a set of prediction rules that describe S and can be employed to predict the characteristics of a future behavior. If the results of chi-square testing are significant, we can conclude that an attribute $Attr_{i(p+t)}$ is dependent on another attribute, $Attr_{jp}$. However, it cannot provide information as to how the observed valued of the ith attribute in a sequence is dependent on that of the jth attribute of a behavior at t positions earlier.

Table 1. A two dimensional contingency table with I rows and J columns

		Attrjp						Totals
		v_{j1}	v_{j2}	...	v_{jk}	...	v_{jJ}	
	v_{i1}	o_{11} (e_{11})	o_{12} (e_{12})	... ...	o_{1k} (e_{1k})	... ...	o_{1J} (e_{1J})	o_{1+}
	v_{i2}	o_{21} (e_{21})	o_{22} (e_{22})	... ...	o_{2k} (e_{2k})	... ...	o_{2J} (e_{2J})	o_{2+}
	. . .	. . .	. . .		. . .		. . .	. . .
Attri(p+τ)	v_{il}	o_{l1} (e_{l1})	o_{l2} (e_{l2})	... ...	o_{lk} (e_{lk})	... ...	o_{lJ} (e_{lJ})	o_{l+}
	. . .	. . .	. . .		. . .		. . .	. . .
	v_{iI}	o_{I1} (e_{I1})	o_{I2} (e_{I2})	... ...	o_{Ik} (e_{Ik})	... ...	o_{IJ} (e_{IJ})	o_{I+}
Totals		o_{+1}	o_{+2}		o_{+k}		o_{+J}	***M'***

To predict a player's future behavior, we can construct a set of prediction rules. These represent each detected dependence relation between two attributes values by using a rule in the following form:

If<*Condition*> then <*conclusion*> with certainty *W*. (3)

The condition part of the rule shows the characteristic that a behavior should possess so that the behavior at a certain level later in the player's game log history will take on the attribute value predicted in the conclusion. As such a prediction cannot usually be constructed with complete certainty; the degree of certainty has to be reflected by the weight W associated with the rule.

Suppose the attribute value v_{il}, is found and it is dependent on v_{jk} as described in the previous level. The prediction rule is constructed as follows and shows the following relationship:

If $Attr_{jp}$ of a behavior is v_{jk}, then it is with certainty W that $Attr_{i(p+t)}$ of a behavior located at t level later in the sequence has the value v_{il}, (4)

Where $W = W(Attr_{i(p+t)}=v_{il} \ / \ Attr_{i(p+t)} \neq v_{il} \mid Attr_{jp}=v_{jk})$ use to measure the amount of positive or negative evidence provided by v_{jk} supporting or refuting the behavior at t level later to have the characteristic, v_{il}.

The derivation of W is based on an information theoretic measure known as mutual information and defined between v_{jk} and v_{ik} as in [12]:

$$\begin{aligned} & I(Attr_{i(p+\tau)} = v_{il} : Attr_{jp} = v_{jk}) \\ & = \log \frac{\Pr(Attr_{i(p+\tau)} = v_{ij} \mid Attr_{jp} = v_{jk})}{\Pr(Attr_{i(p+\tau)} = v_{il})} \end{aligned} \tag{5}$$

$I(Attr_{i(p+\tau)} = v_{il} : Attr_{jp} = v_{jk})$ is positive if and only if $\Pr(Attr_{i(p+\tau)} = v_{ij} \mid Attr_{jp} = v_{jk}) > \Pr(Attr_{i(p+\tau)} = v_{il})$. Otherwise it is either negative or has a value of 0. As v_{il} of $Attr_{i(p+\tau)}$ is dependent on v_{jk} of $Attr_{jp}$, the weight of evidence can be defined as in [12]:

$$\begin{aligned} & W(Attri(p+\tau) = vil \,/\, Attri(p+\tau) \neq vil \mid Attrj[= vjk) \\ & = I(Attri(p+\tau) = vil : Attrj[= vjk) \\ & \quad - I(Attri(p+\tau) \neq vil : Attrj[= vjk) \end{aligned} \tag{6}$$

W can also be expressed as

$$\begin{aligned} & W(Attr_{i(p+\tau)} = v_{il} \,/\, Attr_{i(p+\tau)} \neq v_{il} \mid Attr_{j[} = v_{jk}) \\ & = I(Attr_{i(p+\tau)} = v_{il} : Attr_{j[} = v_{jk}) \\ & \quad - I(Attr_{i(p+\tau)} \neq v_{il} : Attr_{j[} = v_{jk}) \\ & = \log \frac{\Pr(Attr_{jp} - v_{jk} \mid Attr_{i(p+\tau)} = v_{il})}{\Pr(Attr_{jp} = v_{jk} \mid Attr_{i(p+\tau)} \neq v_{il})} \end{aligned} \tag{7}$$

The weight of evidence is a measure of the difference in the gain in information when the ith attribute of the behavior takes on the value v_{il} and when it takes on other values, given that the behavior that is t level in front has the characteristic v_{jk}. If v_{jk} provides positive evidence supporting the ith attribute of the behavior at t levels later in the sequence having the value v_{il}, the weight W is its level. W is negative if the evidence of v_{jk} is negative.

Assume that a set of behaviors is generated probabilistically in such a way that the characteristics of the behavior at a certain position depend on that of a maximum of L behaviors before it. The prediction process starts by searching though the prediction rules that determine how the characteristics of Log_M, Log_{M-1},$Log_{(M-L)+1}$ may affect the value of $Attr_{i(p+t)}$ of $Logj_{M+h}$.

We use a searching process by matching the attribute value val_{jp} (where j=1,2,…,n and p=M, M-1,(M-L)+1) of the behavior Log_M, Log_{M-1}, $Log_{(M-L)+1}$, against the subset of prediction rules whose conclusions predict what values the *i*th attribute of a behavior at *h*, *h+1*, …, *(h+L)–1* level later will take on. An attribute value that satisfies the condition part of a rule, affects the value of the *i*th attribute of the behavior at *M+h*. Then, we can retrieve the significant prediction rule of the attributes by making use of the weight of evidence.

4 Case Study

We present our implementation of our proposed solution on a testing model, a maze game (Figure 1). In this game, the player needs to control the movement of the mouse and to eat the cheese. If the mouse eats the cheese, the level is finished and moves to the next level. Other features include monsters who to block the movement of the mouse and the ability of the player to break the wall to create a new path, or to kill the monster with a limited of hammers. The parameters of the game world are the number of monsters, the number of dead end, the number of hammers, the size of a maze, the number of steps to get the cheese and the expected time to finish the game Player behaviors is saved at each level include the number of hammers used, monsters killed, walls broken, and dead ends, as well as the time taken to finish.

If the mouse eats the cheese within the expected time, the mouse is rated "Good" and a mark "B" is placed on the level performance on the game log. Otherwise, the mouse is rated "Bad" and the mark "W" is placed on the level performance on the game log. The game is implemented with Java. We also include a positive feedback approach to maintaining the balance of the game [13]. Positive Feedback is defined to

Fig. 1. Maze Game

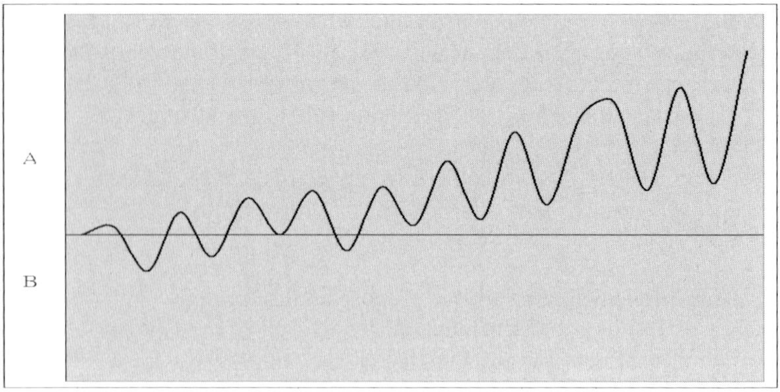

Fig. 2. The ideal balance graph

make things easier for the player when he is ahead. Figure 2 is an ideal balance graph that displays the player's progress. The horizontal axis is used for the level of difficulty and the vertical axis is used to indicate the player's performance to show who is ahead. An ideal balance graph slowly becomes unbalanced over time until the player wins the game.

We tested our solution by having ten players play three different versions of our twenty-level maze game: a version using random number generation, a version using decision trees, and a version using our adaptive game engine. With the random numbers version, all of the parameters were generated at random. With the decision tree we first go through the training phase to build a decision tree using ID3. With ID3, we discovered classification rules. We also define the "Easy" and "Hard" templates of every lever. The player who wins is the player who gets "Good" performance on a set of parameter of the game world, and that parameters correspond to the "Easy" level template of the corresponding level. If most players receive a "Bad" performance on this level, that parameters of the game level will be the "Hard" level template of the corresponding level. With the ten players we recruited for our testing, the decision tree was used to analyze their behavior on level *t*, which allowed a template of either "Hard" or "Easy" to be applied at level *t+1*.

With our proposed approach, we tested the game to define the default level template before the testing. When the tester plays, the parameter of the first 10 levels are the same, the game engine is made adaptive with the last 10 game levels. The game engine collects the game log of first 10 level and analyze and generate a list of prediction rules. Then the game engine changes the parameters of the coming game level base on those rules according to the weighting of the rules. The higher value of the weight means that parameter is more correlated with each other. Table 2 shows the game logs of one of the test players.

Table 3 shows some prediction rules generated by our proposed solution. The game engine starts to predict player's behavior on Level 10, when the parameters of the level are adjusted for the next level.

Table 2. Player's behavior with our proposed approach

NUMBER OF LEVELS	DEAD ENDS ACHIEVED	HAMMERS USED	ACTUAL TIME (ms)	MON-STERS HIT	WALLS BROKEN	DEAD ENDS	HAMMERS	MAP SIZE	MON-STERS	EXPECTED TIME (ms)
1	0	0	18046	0	0	0	0	12	0	27000
2	1	2	15091	0	0	1	2	10	0	27500
3	0	3	18326	1	2	3	4	10	3	28500
4	0	1	15112	0	1	2	3	12	2	28000
5	0	1	30183	0	1	2	3	12	2	28000
6	0	2	33558	1	1	3	4	15	3	28500
7	1	1	27209	0	1	2	3	12	3	28000
8	0	1	27229	1	0	1	2	12	4	27500
9	0	1	24225	0	1	1	2	10	3	27500
10	0	2	27490	0	0	2	3	12	0	28000
11	0	0	27389	0	0	0	0	12	0	27000
12	0	1	24455	1	0	1	2	12	2	27500
13	0	0	20690	0	0	0	0	10	4	27000
14	0	2	35496	1	1	1	2	15	2	28500
15	0	1	19337	0	1	2	2	12	2	27500
16	0	0	28714	0	0	3	0	12	1	27000
17	0	1	28500	0	1	1	2	12	1	29000
18	0	2	22471	0	2	2	2	15	0	27500
19	0	2	20500	0	2	1	3	10	4	27500
20	0	2	18500	2	0	1	2	12	2	31000

Table 3. Some prediction rules at Level 10

If two hammers are used on level p, then there is a certainly of 4.0 that the player break the two walls on level p+1.
If a monster is hit on level p, then there is a certainly of 3.9 that the player will not break the wall on the level p+2.
If the maze size is 10 on level p, then there is a certainly of 2.2 that the player's performance is B on level p+2.
If a wall didn't break on level p, then there is a certainly of 2.7 that the player will not hit the monster on level p+1.

Figure 3 shows a balance graph for our three versions of the maze game. The levels of the games are represented on the horizontal axis and the vertical axis indicates the player's performance as a comparison between the difference between the actual

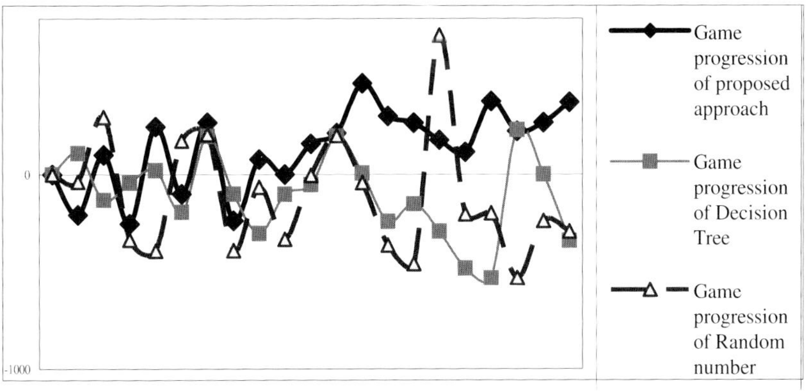

Fig. 3. Balance graph of the three versions of the maze game for one player

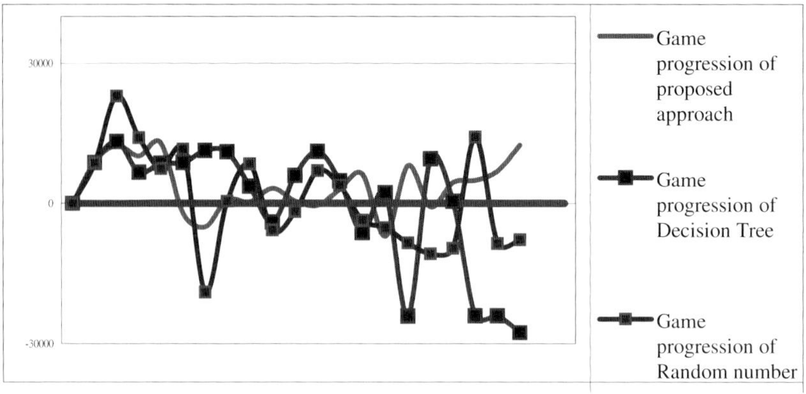

Fig. 4. Balance graph of the three versions of the maze game showing the average performance of the ten players

playing time and the expected playing time. It appears that the maze game using our proposed approach is a better fit for the ideal game progression, making it a more balanced game experience for the player. Figure 4, which compares the average performance of the ten players on the three maze games against the curve of our proposed algorithm, also shows them to be a good fit to the ideal game progression. It shows how a game becomes more balanced when the game makes use of the proposed approach.

5 Conclusions and Future Research

Our proposed algorithm differs from existing algorithms in that the game level is based on the past experience of the player, is data-driven, and the game environment is unique and is not predefined and our approach is used for building game environments. Not only can this approach build a maze, it can also help the game to build terrain, which may not be predefined. We compared the performance of the proposed approach on a maze game against approaches using other types of game AI. Positive feedback and these comparisons show that the proposed approach makes the game both more interesting and more balanced. Our approach also has applications beyond gaming. Given that a web site is also a type of maze in which users can easily get lost, our approach could be used to mine the users' visit log of a website as a set of sequential multivariate patterns. Our proposed approach would provide users with an interactive navigation menu and site owners could use the information to target users with suitable information or products when user's behaviors are predicted.

References

1. Schuytema, P.: Game Design: A Practical Approach. Charles River Media (2007)
2. Wikipedia, `http://en.wikipedia.org/`
3. Andrade, G., Ramalho, G., Santana, H., Corruble, V.: Automatic computer game balancing: a reinforcement learning approach. In: International Conference on Autonomous Agents and Multiagent Systems, pp. 1111–1112 (2005)

4. Lee, S., Jung, K.C.: Dynamic game level design using Gaussian mixture model. In: Yang, Q., Webb, G. (eds.) PRICAI 2006. LNCS (LNAI), vol. 4099, pp. 955–959. Springer, Heidelberg (2006)
5. Ryan, T.: Beginning Level Design, Part 1: Level Design Theory. Gamasutra (1999), http://www.gamasutra.com/features/19990416/level_design_01.htm
6. Hunicke, R., Chapman, V.: AI for Dynamic Difficulty Adjustment in Games. In: Challenges in Game Artificial Intelligence AAAI Workshop, San Jose, pp. 91–96 (2004)
7. Spronck, P., Sprinkhuizen-Kuyper, I., Postma, E.: Difficulty Scaling of Game AI. In: El Rhalibi, A., Van Welden, D. (eds.) GAME-ON 2004: 5th International Conference on Intelligent Games and Simulation, pp. 33–37. EUROSIS, Belgium (2004)
8. Lee, S., Jung, K.: Dynamic Game Level Design Using Gaussian Mixture Model. In: Yang, Q., Webb, G. (eds.) PRICAI 2006. LNCS (LNAI), vol. 4099, pp. 955–959. Springer, Heidelberg (2006)
9. Spronck, P., Ponsen, M., Sprinkhuizen-Kuyper, I., Postma, E.: Adaptive Game AI with Dynamic Scripting. Machine Learning 63(3), 217–248 (2006)
10. Woodcock, S.: Game AI: The State of the Industry. Gamasutra (1999), http://www.gamasutra.com/features/19990820/game_ai_01.htm
11. Osteyee, D.B., Good, I.J.: Information Weight of Evidence, the Singularity between Probability Measures and Signal Detection. Springer, Berlin (1974)
12. Adams, E.: Balancing Games with Positive Feedback. Gamasutra (2002), http://www.gamasutra.com/features/20020104/adams_01.htm

A Logic Programming Based Framework for Security Protocol Verification

Shujing Wang and Yan Zhang

Intelligent Systems Laboratory
School of Computing and Mathematics
University of Western Sydney
Penrith South DC, NSW 1797, Australia
{shwang,yan}@scm.uws.edu.au

Abstract. Security protocol analysis has been a major research topic in information security and recognised to be a notoriously hard problem. In this paper, we take the advantage of answer set programming technology to develop an effective framework to verify security protocols carrying claimed security proof under adversary models on computational complexity theory. In our approach, a security protocol, adversary actions and attacks can be formally specified within a unified logic program. Then the verification is performed in an automatic manner by computing the stable models of the underlying logic program. We use Boyd-González Nieto conference key agreement protocol as our case study protocol to demonstrate the effectiveness and efficiency of our approach.

1 Introduction

In recent years, security protocols are increasingly being used in many diverse secure electronic communications and electronic commerce applications. However, despite an enormous amount of research effort expended in design and analysis of such protocols, it is still notoriously hard. When security protocols are designed by hand, errors may creep in by combining protocols actions in ways not foreseen by the designer [3]. Some protocols have been found errors after they were published many years, even since they have been proven secure [10].

The study of cryptographic protocols has led to the dichotomization of cryptographic protocol analysis techniques between the formal methods approach and the computational complexity approach [8]. The formal methods approach is to use logic based methods including model checking and theorem proving to automatically verify a protocol. The computational complexity approach, on the other hand, adopts a reductive process which allows a proven reduction from the problem of breaking the protocol to another problem believed to be hard. These two approaches have been developed in two mostly different communities. Recently, some research works have been done to bridge the gap between them, which achieve automatic provability under classical computational models, see [2,4] for example.

In this paper, based on the answer set programming approach [5], we propose a framework to analyze security protocols that are found insecure against certain types of attacks. We use Boyd-González Nieto conference key agreement protocol as our case study protocol to demonstrate our approach.

A. An et al. (Eds.): ISMIS 2008, LNAI 4994, pp. 638–643, 2008.

2 Logic Programming Specification for Security Protocols

Modelling Security Protocols

Now we present how to model a security protocol through the case study protocol (refer to the appendix A for the complete specification program). For specification simplicity and efficiency, we simplify the case study protocol to a two-party protocol showed in Figure 2 as explained in [9]. Because in a protocol flow of Figure 1, message 1 and 2 can be sent concurrently, in the simplified protocol, we merge them into one message.

$$
\begin{array}{l}
1.\ U_1 \rightarrow U_2 : \mathcal{U} = \{U_1, U_2\}, S_{d_{U_1}}(\mathcal{U}, \{N_1\}_{e_{U_2}}), \{N_1\}_{e_{U_2}} \\
2.\ U_2 \rightarrow U_1 : U_2, N_2 \\
\qquad\qquad SK_{U_1} = H(N_1||N_2) = SK_{U_2}
\end{array}
$$

Fig. 1. Simplified Boyd-González Nieto Conference Key Agreement Protocol

Let $\mathcal{U} = \{U_1, U_2\}$. The initiator, U_1 encrypts N_1 using the public key of U_2, signs $\mathcal{U}$ and the encrypted nonce $\{N_1\}_{e_{U_2}}$, and broadcasts $\mathcal{U}$, the signature value and the encrypted nonce in message flow 1. The principal, U_2, upon receiving the initiate message, will respond with his/her identity and a random nonce in message flow 2.

The first part of protocol specification is to set up principals and their keys through predicates, $player(A)$, $agent(A)$, $ag\_id(A, N)$, and $key(K)$, where K is one of key functions. For instance, in our case study protocol, we have

$$
\begin{array}{l}
player(u_1), player(u_2), adversary(a) \\
ag\_id(u_1, 0), ag\_id(u_2, 1), ag\_id(a, 2) \\
key(pKey(A)) \leftarrow agent(A). \\
key(sKey(A)) \leftarrow agent(A). \\
key(sig\_sKey(A)) \leftarrow agent(A). \\
key(sig\_vKey(A)) \leftarrow agent(A).
\end{array}
$$

The second part is to model relationships between keys of principals. In the case study protocol, there are encryption and signature keys which are specified as follows.

$$
\begin{array}{l}
asymKeyPair(pKey(A), sKey(A)) \leftarrow agent(A). \\
asymKeyPair(sig\_sKey(A), sig\_vKey(A)) \leftarrow agent(A). \\
asymKeyPair(K_1, K_2) \leftarrow asymKeyPair(K_2, K_1).
\end{array}
$$

The third part is about message flows in a protocol. During a protocol run, we assume that if a principal A sends a message to B and the adversary does not intercept it, B will receive it at the next time. We model the assumption using the rule:

$$
\begin{array}{l}
gets(B, M, P, T+1) \leftarrow sends(A, B, M, P, T), \\
\qquad\qquad neq(A, B), not\ intercept(a, M, P, T+1).
\end{array}
$$

A protocol consists of a sequence of messages. Except the first message which is sent by the initiator of the protocol run, principals will check preconditions before they send a response message. As explained in [3], a protocol was denoted like:

$$
\begin{array}{ll}
A \rightarrow B_{i_1} : m_{i_1}, p_{i_1} & \%\ first\ message\ A\ must\ send, \ldots \\
B_{j_1} \rightarrow A : m_{j_1}, p_{j_1} & \%\ first\ message\ A\ must\ receive, \ldots \\
A \rightarrow B_{i_s} : m_{i_s}, p_{i_s} & \%\ last\ message\ A\ must\ send\ before\ m, \ldots \\
B_{i_r} \rightarrow A : m_{i_r}, p_{i_r} & \%\ last\ message\ A\ must\ receive\ before\ m, \ldots \\
A \rightarrow B : m, p &
\end{array}
$$

As showed above, principal A will sends message (m, p) to B, if we check that a sequence of messages should have been received and sent before (m, p) in a correct run. We code the following rule:

$sends(A, B, m, p, T+1) \leftarrow$
$\quad sends(A, B_{i_1}, m_{i_1}, p_{i_1}, T_{i_1}), \ldots, sends(A, B_{i_s}, m_{i_s}, p_{i_s}, T_{i_s}),$
$\quad gets(A, m_{j_1}, p_{j_1}, T_{j_1}), \ldots, gets(A, m_{i_r}, p_{i_r}, T_{i_r}), T_{j_1} > T_{i_1}, \ldots, T_{i_r} >$
$T_{i_s},$
$\quad protocol\text{-}dependant\ literals$
$contains(m, p, msg(.)).$

We consider preconditions for sending message m by principal A as actions that A has performed in previous steps according to the protocol run. Protocol dependant literals are usually to check the freshness of random nonces or timestamps and other conditions needed by particular protocols. Because we represent a message using a message id and type in predicate $sends$, we should add a fact rule, in which $contains$ is the head to denote what the message is indeed.

For instance, in our case study protocol, principal u_1 sends an initial message to start a protocol run. We model it as the following rule.

$sends(u_1, all, 0, 0, 0).$
$contains(0, 0, agset(u1, u2)).$
$contains(0, 0, sign(sig_sKey(u_1), agset(u1, u2)||enc(pKey(u_2), n(0)))).$
$contains(0, 0, enc(pKey(u_2), n(0))).$

Finally, we model the principal knowledge including the principal initial knowledge base and knowledge change during the protocol run. Each principal taking part in the protocol run has an initial knowledge base such as other principals' public keys. While sending and receiving messages, principals will hold them and derive more information by breaking or decrypting all messages for which they have a key. Their knowledge will change during the protocol run. We use predicate $holds$ to specify principals' knowledge.

For encryption and signature keys in the case study protocol, we code initial knowledge bases for principals using following rules.

$holds(A, pKey(B), 0) \leftarrow agent(A), agent(B).$
$holds(A, sig\_vKey(B), 0) \leftarrow agent(A), agent(B).$
$holds(A, sKey(A), 0) \leftarrow agent(A).$
$holds(A, sig\_sKey(A), 0) \leftarrow agent(A).$

Then we write following rules to model principals' knowledge change during the protocol run.

$$holds(A, M, T) \leftarrow gets(A, M, P, T).$$
$$holds(A, M, T) \leftarrow sends(A, B, M, P, T).$$
$$holds(A, S, T) \leftarrow holds(A, M, T), contains(M, P, S).$$
$$holds(A, S_1, T) \leftarrow holds(A, M, T), contains(M, P, S_1 || \dots || S_n).$$
$$\dots$$
$$holds(A, S_n, T) \leftarrow holds(A, M, T), contains(M, P, S_1 || \dots || S_n).$$
$$holds(A, S_1, T) \leftarrow holds(A, enc(K_1, S_1 || \dots || S_n), T), holds(A, K_2, T_1), asymKeyPair(K_1, K_2).$$
$$\dots$$
$$holds(A, S_n, T) \leftarrow holds(A, enc(K_1, S_1 || \dots || S_n), T), holds(A, K_2, T_1), asymKeyPair(K_1, K_2).$$

Modelling Attacks

In our framework, the adversary model is closely based on Bellare-Rogaway model. If protocols with claimed security under Bellare-Rogaway model are found to be violating any of the conditions in the definition of *insecurity*, they will be insecure in Bellare-Rogaway model. Moreover, the proof of the protocol will also be invalid. Based on the definition of *insecurity*, we should model SIDs and session keys of principals. The SID of a principal is the concatenation of all messages he receives and sends. We use predicate *inSidList(U,M)* to record the messages that the principal U receives and sends.

$$inSidList(U, M) \leftarrow sends(U, all, M, P, T).$$
$$inSidList(U, M) \leftarrow gets(U, M, P, T).$$

The following two rules specify that two principals have same SIDs, where the first one denotes that if a message is in the session id list of principal U_1, and not in the session id list of principal U_2, $sid\_neq\_pair(U_1, U_2)$ is true, and the second one specifies conditions which should be satisfied for two principals to have same SIDs.

$$sid\_neq\_pair(U_1, U_2) \leftarrow inSidList(U_1, M),\ not\ inSidList(U_2, M),\ neq(U_1, U_2).$$
$$same\_sid\_pair(U_1, U_2) \leftarrow not\ sid\_neq\_pair(U_1, U_2),\ not\ sid\_neq\_pair(U_2, U_1),\ neq(U_1, U_2).$$

In our case study protocol, the session key of a principal is a one-way hush function of the concatenations of random nonces of all principals taking part in the conference protocol.

$$sk(A, h(n(M_1), n(M_2))) \leftarrow holds(A, agset(B, C), T), holds(A, nonce(B, n(M_1)), T_1),\ holds(A, nonce(C, n(M_2)), T_2).$$

The following rule models that principal U_1 and U_2 have same session keys.

$$same\_sk\_pair(U_1, U_2) \leftarrow sk(U_1, h(n(M_1), n(M_2))),\ sk(U_2, h(n(M_1), n(M_2))),\ neq(U_1, U_2).$$

Consider the condition 1 in *insecurity* definition as an instance, if two non-partner oracles have the same session keys, the protocol is insecure. Here two oracles are not partners if they have different $SIDs$. The attack is modelled as follows.

$$attack \leftarrow same\_sk\_pair(U_1, U2),\ not\ same\_sid\_pair(U_1, U2).$$

Note that $same\_sk\_pair(U_1, U_2)$ denotes that principal U_1 and U_2 have same session keys and $same\_sid\_pair(U_1, U_2)$ denotes that principal U_1 and U_2 have same $SIDs$.

3 Model Checking and Verification

After specifying security protocols, adversary actions, and attacks using lanuage $\mathcal{L}_{sp}$, we merge three parts into a logic program $\mathcal{P}$ in which we add a constraint rule,

$$\leftarrow not\ attack.$$

We use *Smodels* system [11] to verify security protocols as follows: (1) through *lparse*, we obtain a finite ground logic program $\mathcal{P}^g$ from program $\mathcal{P}$; (2) Using *smodels*, we compute stable models of ground program $\mathcal{P}^g$; (3) If no stable model exists, the attack does not exist for protocol runs up to time t_{max}[1]; (4) If there is a stable model, we collect atoms representing actions, $sends$, $gets$ and $intercept$ that are true in the model, from which we can find the sequence of actions that is an attack trace.

1. At time t_0, initiator u_1 broadcasts an initial message which has three parts: the set of principals in the protocol run, $agset(u_1, u_2)$; the signature of the principal set and encrypted random nonce $n(0)$ under the public key of principal u_2, $sign(sig\_sKey(u_1), agset(u_1, u_2)||enc(pKey(u2), n(0)))$; the encryption of the encrypted random nonce $n(0)$, $enc(pKey(u2), n(0))$.
2. At time t_1, $\mathcal{A}$ receives the message and intercepts it. After modifying the principal set to $agset(a, u_2)$ and make a new signature using his own signature key, $sign(sig\_sKey(a), agset(a, u_2)||enc(pKey(u_2), n(0)))$, $\mathcal{A}$ fabricates a new message, and sends it to principal u_2. Now $\mathcal{A}$ acts as an initiator and start a different session.
3. At time t_2, principal u_2 receives the message from $\mathcal{A}$ and believes that $\mathcal{A}$ initiates a protocol run.
4. At time t_3, principal u_2 broadcasts his identifier and random number.
5. At time t_4, principal u_1 and $\mathcal{A}$ receive the random nonce of principal u_2. u_1 believes he finishes his own session with u_2, however u_2 believes he is in a different session with $\mathcal{A}$.

An attack was found in 5.460 seconds. We observe that principal u_1's SID is $(0, 8)$ and principal u_2's SID is $(11, 8)$. Then u_1 and u_2 are not partners since they do not have matching $SIDs$. u_1 believes the session key $SK_{u_1} = h(n(0)||n(8))$ is being shared with u_2, but u_2 believes the session key $SK_{u_2} = h(n(0)||n(8)) = SK_{u_1}$ is being shared with $\mathcal{A}$. Although $\mathcal{A}$ does not know the session key as $\mathcal{A}$ does not know the value of $n(0)$, he is able to send query $Reveal$ to the session with u_2 and get $SK_{u_2} = h(n(0)||n(8))$ which is same as SK_{u_1}. Our case study protocol is not secure under Bellare-Rogaway model as being claimed.

[1] t_{max} is a max time limitation set up in the logic program.

4 Conclusions

In this paper, we developed a logic programming framework in which we not only use formal verification under adversary models in the computational complexity theory, but also integrate protocol analysis into the approach. As logic programming is a declarative executable approach for knowledge representation and reasoning, in our framework, we defined a security protocol specification language $\mathcal{L}_{sp}$ under logic programming with stable model semantics which is used to specify security protocols carrying claimed security proof under adversary models. Using $Smodels$ we are able to verify the program we have modelled. As a case study, Boyd-González Nieto conference key agreement protocol has been specified, verified using our framework.

References

1. Abdalla, M., Chevassut, O., Fouque, P., Pointcheval, D.: A Simple Threshold Authenticated key Exchange from Short Secrets. In: Advances in Cryptology - Asiacrypt 2005, pp. 566–584. Springer, Heidelberg (2005)
2. Abadi, M., Rogaway, P.: Reconciling Two Views of Cryptography (The Computaiotnal Soundness of Formal Encryption). Journal of Cryptology 15(2), 103–127 (2002)
3. Aiello, L.C., Massacci, F.: Verifying Security Protocols as Planning in Logic Programming. ACM Transactions on Computational Logic 2(4), 542–580 (2001)
4. Backes, M., Jacobi, C.: Cryptographically Sound and Machine-Assisted Verification of Security Protocols. In: Alt, H., Habib, M. (eds.) STACS 2003. LNCS, vol. 2607, pp. 310–329. Springer, Heidelberg (2003)
5. Baral, C.: Knowledge Representation, Reasoning and Declarative Problem Solving. Cambridge University Press, Cambridge
6. Boyd, C., González Nieto, J.M.: Round-optimal Contributory Conference Key Agreement. In: Desmedt, Y.G. (ed.) PKC 2003. LNCS, vol. 2567, pp. 161–174. Springer, Heidelberg (2002)
7. Bellare, M., Pointcheval, D., Rogaway, P.: Authenticated key exchange secure against dictionary attacks. In: Preneel, B. (ed.) EUROCRYPT 2000. LNCS, vol. 1807, pp. 139–155. Springer, Heidelberg (2000)
8. Choo, K.R.: Refuting Security Proofs for Tipartite Key Exchange with Model Checker in Planning Problem Setting. In: The 19th IEEE Computer Security Foundations Workshop - CSFW 2006, pp. 297–308 (2006)
9. Jeong, I.R., Katz, J., Lee, D.H.: One Round Protocols for Two-Party Authenticated Key Exchange. In: Jakobsson, M., Yung, M., Zhou, J. (eds.) ACNS 2004. LNCS, vol. 3089, pp. 220–232. Springer, Heidelberg (2004)
10. Lowe, G.: Some New Attacks upon Security Protocols. In: Proceedings of the 9th IEEE Computer Security Foundatons Workshop (CSFW 1996), pp. 162–169. IEEE Computer Society Press, Los Alamitos (1996)
11. Niemela, I., Simons, P., Syrjanen, T.: Smodels: A system for answer set programming. In: Proceedings of the 8th International Workshop on Non-monotonic Reasoning (2000)
12. Paulson, L.C.: The Inductive Approach to Verifying Cryptographic Protocols. Journal of Computer Security 6, 85–128 (1998)
13. Ryan, P., Schneider, S.: An Attack on a Recursive Authentication Protocol: A Cautionary Tale. Information Processing Letters 65(15), 7–16 (1998)

Applying Cost Sensitive Feature Selection in an Electric Database

Manuel Mejía-Lavalle

Instituto de Investigaciones Eléctricas, Gerencia de Sistemas Informáticos
Reforma 113, Palmira, 62490 Cuernavaca, Morelos, México
mlavalle@iie.org.mx

Abstract. Feature selection is a crucial activity when knowledge discovery is applied to large databases, as it reduces dimensionality and therefore the complexity of the problem. Its main objective is to eliminate attributes to obtain a computationally tractable problem, without affecting the solution quality. To perform feature selection, several methods have been proposed, some of them tested over small academic datasets. In this paper we evaluate different feature selection-ranking methods over a large real world database related with a Mexican electric energy client-invoice system. Most of the research on feature selection methods only evaluates accuracy and processing time; here we also report on cost sensitive classification and the amount of discovered knowledge. Additionally, we stress the issue around the boundary that separates relevant and irrelevant features. Finally, we propose a promising feature selection heuristic based on the experiments performed, taken into account a cost sensitive classification.

1 Introduction and Motivation

The process of non-trivial extraction of relevant information that is implicit in the data is known as Knowledge Discovery in Databases (KDD), where the data mining phase plays a central role in this process [1]. It has been noted that when large databases are going to get mined, the mining algorithms get very slow, requiring too much time to process the information and sometimes making the problem intractable. One way to attack this problem is to reduce the amount of data before applying the mining process [2]. In particular, the pre-processing method of feature selection applied to the database before mining has shown to be successful, because it eliminates the irrelevant or redundant attributes that cause the mining tools to become inefficient, but preserving the classification quality of the mining algorithm. However, few works take into account, at the same time, the effect of cost sensitive classification [3].

Feature selection wrapper methods, although effective to eliminate irrelevant and redundant attributes, are very slow because they apply the mining algorithm many times, changing the number of attributes each execution time, as they follow some search and stop criteria [4]. Filter methods are more efficient, using existing techniques such as decision trees algorithms, neuronal networks, nearest neighborhood, etc., that take into account dependencies between attributes. A sub-filter technique, called ranking methods, uses some type of information gain measurement between individual attributes and

A. An et al. (Eds.): ISMIS 2008, LNAI 4994, pp. 644–649, 2008.

the class, and it is very efficient [5]; however, because it measures the relevance of each isolated attribute, they cannot detect if redundant attributes exist, or if a combination of two attributes, apparently irrelevant when analyzed independently, can be transformed into relevant [6].

In this paper we present an evaluation of different filter-ranking methods for supervised learning. The evaluation takes into account not only the classification quality and the processing time obtained after the filter application of each ranking method, but also it considers cost sensitive classification effect and discovered knowledge size, which, the smaller, the easier to interpret. We also propose a promising feature selection heuristic based on the experiments performed.

2 Application Domain Description

One the main CFE (Federal Commission of Electricity) function is to distribute to the customers the electrical energy produced in the different generating plants in Mexico. Related to distribution, CFE faces different problems that prevent it to recover certain amount of "lost income" from the 100% of the total energy for sale. 11% of the losses are due to administrative control problems, where the big problem is the illicit use of energy, that is to say, people who steal the energy and therefore they do not pay for it.

CFE has faced this problem applying different actions, one of then is using a knowledge discovery approach based on data mining to obtain patterns of behavior of the illicit customers. This alternative solution does not require a great deal of investment and it has been proven to be effective in similar cases, like credit card fraud detection [7].

The subject information to analyze is a sample of the SICOM database, a legacy system that contains around twenty tables with information about contracts, invoicing, and collection from customers across the nation. This system was not designed with the illicit users discovery in mind; nevertheless, it contains a field called *debit-type* in which a record is made if the debit is due to illicit use of energy. After joining three tables, including the one that has the *debit-type* field, a "mine" with 35,983 instances was obtained with the following attributes: *Permanent customer registry (RPU), Year, Month, debit-type, Digit, kWh, Energy, Cve-invoicing, Total, Status, Turn, Tariff, Name, Installed-load, Contract-load*, and others that altogether add up to 21 attributes. Various experiments were executed with this database to evaluate the different ranking methods as described next.

3 Evaluating Ranking Methods

3.1 Measuring the Attribute Relevance Degree

The application of filter-ranking methods to select features of a VLDB is adequate due to its low computational cost. We use *Elvira* [8] and *Weka* [9] tools, since they provide suitable and updated platforms for the easy execution of multiple experiments in a PC environment. In the presentation of the experiments the processing time has been left out because it was always very small, for example, *Elvira* obtains, in less of a second, the Mutual Information distance to measure the relevance of 21 attributes using 35,983 instances.

Although in this case the attributes appear ordered according to their relevance, we lack of a uniform criterion to decide which attributes to select. We used the Stoppiglia´s criterion [10], but modifying it as follows: instead of using a single random variable, we added three, to observe how the ranking method maintains together, or not, the random variables in the set of ranked attributes, avoiding a possible bias introduced to the result by a single random variable, that in fact is a computational pseudo-random variable. The obtained result is shown in Table 1, where variables RAND3, 2 and 1 are the boundaries of the four subsets of attributes.

We applied different ranking methods to the database (a detailed explanation of the used "distances" can be found in [6] and [11]); all the obtained results are shown in Table 1. Also, the methods: Principal Component Analysis (PCA), Information Gain, Gain Ratio and Symmetrical were explored, and they produced similar results as Chi-Square, which means that they did not obtain a significant reduction on the number of attributes. From Table 1 we observe that, although some ranking methods agree in the selection of some attributes, in general, each method produces different attribute ordering, including the position for the three random variables. (This is a very interesting result, as we will see in Table 2).

3.2 Performance Evaluation

In order to evaluate the methods, we applied the *J4.8* tree induction classifier (the Weka implementation of the last public version of *C4.5*) to the database "projected" on the attributes selected by each method. Table 2 shows the results. In all the cases, we always used the Weka's default parameters and the attributes of the first subset identified by the appearance of the first random variable (in section 3.3 we analyze this in more detail).

The feature reduction column measures the number of attributes selected against the total number of attributes. The processing time is expressed in relation to the time required to obtain a tree that includes all the attributes of the database (*complete case*). The size of the discovered knowledge is measured by the number of leaves and the number of nodes of the induced tree. The classification quality appears as the percentage of instances correctly classified using the training data (accuracy) and also using a 10-fold cross validation test. A very important column is included that considers cost sensitive or cost-benefit that it would be obtained if the discovered knowledge were applied by CFE, and assuming that each inspection has a cost of –2.5 units and that the obtained benefit of a correct prediction of an illicit is of +97.5 units. The reported cost-benefit corresponds to the application of the above mentioned 10-fold cross validation test and it is calculated considering that the *complete case* obtains a 1,000 units of benefit, and the results of the other methods are normalized with respect to the *complete case*.

In Table 2, we observe that most of the methods obtain a reduction of the number of attributes greater than 0.50 and reduce the mining algorithm processing time in an order of magnitude; a special case is Relief, that unlike the other methods whose processing time is small, Relief requires a proportion of time 9721 times greater than the time required to induce the tree by using all the attributes.

Table 1. Application of different ranking measures

Euclidean distance	Matusita distance	Kullback-Leibler 1	Kullback-Leibler 2	Shannon entropy	Bhatta-charyya	Relief	OneR	Chi-Square
fctura	fctura	fctura	fctura	kwh	kwEen	anio	factra	factra
mes	kwEen	status	mes	enrgia	fctura	mes	status	status
cIMcC	kwMen	kwEen	status	total	kwMen	factra	anio	mes
anio	RAND3	cCEto	cgInst	tarifa	RAND3	digito	tarifa	kwEen
RAND3	status	RAND3	cgCont	cgInst	toMkw	RAND3	digito	kwMcI
tarifa	cCEto	kwMen	cCEto	cgCont	toMcI	RAND2	mes	kwh
digito	toMkw	toMkw	cIMcC	kwEen	enrgia	RAND1	cIMcC	toMcI
status	enrgia	enrgia	anio	toMcI	cCEto	status	cgCont	toMcC
RAND2	toMcI	kwMcI	kwEen	kwMen	total	cgInst	cgInst	total
cIEen	total	toMcI	RAND2	toMen	RAND2	tarifa	RAND1	toMen
cgInst	kwMcI	RAND2	RAND3	toMkw	kwMcI	cgCont	toMkw	enrgia
cgCont	RAND2	kwh	toMkw	kwMcI	toMcC	cCEto	RAND2	kwMen
RAND1	kwh	total	kwh	toMcC	RAND1	cIEen	cCEto	toMkw
cCEto	toMcC	mes	kwMen	cCEto	kwh	cIMcC	kwh	cCEto
toMcC	RAND1	toMcC	cIEen	cIEen	toMen	kwEen	toMcI	cIEen
kwMcI	toMen	RAND1	enrgia	status	mes	toMen	RAND3	cgInst
toMkw	mes	cIEen	total	cIMcC	status	total	total	cgCont
toMen	cIEen	toMen	kwMcI	anio	cIEen	toMcC	toMen	anio
kwMen	cgInst	cgInst	RAND1	RAND2	cgInst	toMcI	kwEen	cIMcC
toMcI	cgCont	cgCont	tarifa	RAND1	cgCont	kwh	enrgia	tarifa
kwEen	cIMcC	cIMcC	digito	RAND3	cIMcC	kwMcI	kwMen	RAND2
total	anio	anio	toMcC	digito	anio	kwMen	cIEen	RAND3
enrgia	tarifa	tarifa	toMcI	mes	tarifa	toMkw	kwMcI	RAND1
kwh	digito	digito	toMen	fctura	digito	enrgia	toMcC	digito

Table 2. Evaluating ranking methods by inducing J4.8 trees

Method	Feature reduction	Time (100=1.8 secs)	Leaves / Nodes	Acc train / test	Cost-sensitive (test)
Complete case	0	100	21 / 41	98.41 / 97.25	1000
Mutual Information	0.80	12	5 / 9	90.86 / 90.10	444
Euclidean distance	0.80	11	3 / 5	93.89 / 93.89	520
Matusita distance	0.86	8	2 / 3	90.58 / 90.21	507
Kullback-Leibler 1	0.80	11	5 / 9	90.86 / 90.10	444
Kullback-Leibler 2	0.57	14	17 / 33	98.26 / 97.50	1001
Shannon entropy	0.14	92	23 / 45	95.52 / 93.71	876
Bhattacharyya	0.86	9	2 / 3	90.18 / 90.21	507
Relief	0.80	12 + 9721	3 / 5	93.89 / 93.89	520
OneR	0.57	15	12 / 23	96.64 / 95.95	892

With respect to the size of the discovered knowledge it is observed that almost all the methods produce trees smaller than the *complete case*. On the other hand, although apparently all the methods do not affect too much on the accuracy of the discovered knowledge, the cost-sensitive column highlights those methods that better impact on the prediction of the illicit energy use patterns.

4 Cost Sensitive with Filter-Ranking Plus Wrapper Method

Although the ranking methods are very efficient, they have a flaw in that they do not take into account the possible interdependences between attributes. Observing the obtained results mention above, we propose a heuristic that looks for to overcome such a deficiency, combining the efficiency of the ranking methods, with the effectiveness of the wrapper methods. The heuristic involves the induction of a number of decision trees considering all subsets of attributes that a method produces (the subsets appear limited by the three random variables in Table 2). Applying the previous idea, we can observe, in a computationally economic way, if some combination of attributes exists in the subsets that improves the obtained results as compared when using only the first attribute subset. For example, the application of KL-2 with three random variables produces three subsets. The induction trees produced by *J4.8* using the three subsets and a combination of these subsets are shown in Table 3. It is observed that, for this case, it does not exist a combination that significantly improves the results of the first subset, and this is why we can conclude that we have found a good solution, one that manages to reduce to the processing time and the knowledge size, without affecting the tree quality of prediction and, more important, the cost sensitive classification.

Table 3. Using feature subsets to induce *J4.8 trees*

Feature subsets	Feature reduction	Time	Leaves / Nodes	Acc train / test	Cost-sensitive (Test)
begin – RAND2	0.57	14	17 / 33	98.26 / 97.50	1001
RAND3–RAND1	0.66	12	1 / 1	79.42 / 79.45	–910
RAND1-end	0.76	11	1 / 1	79.42 / 79.42	–913
begin-RAND1	0.23	16	17 / 33	98.26 / 97.43	1001
RAND3-end	0.42	18	1 / 1	79.42 / 79.45	–910
begin-RAND2/ RAND1-end	0.33	17	21 / 41	98.41 / 97.18	992

5 Conclusions and Further Directions

Feature selection ranking methods are very efficient because they only need to calculate the relevance of each isolated attribute to predict the class attribute. The disadvantages of these methods are that no uniform criterion is provided to decide which attributes are more relevant than others, and that no mechanism is included to detect the possible interdependences between attributes.

In this article the integration of three random variables to the database is proposed to avoid a possible bias introduced to the result if a single random variable is used. We observed that, although some ranking methods agree in the selection of some attributes, in general, each method produces different attribute ordering, including the position for the three random variables. The three variables serve as subset boundaries and help to decide which attributes to select. Also, we propose to analyze the possible interdependences between attributes using the induction trees constructed on these subsets, stressing the issue around a cost sensitive classification, which is crucial when we faced real world problems. These ideas have been proven to be successful in a real world electrical energy customer-invoice database.

In the future these ideas are going to be applied to other databases and classifiers. We are considering testing over academic and real very large power system databases such as the national power generation performance database, the national transmission energy control databases, the de-regulated energy market database, and the Mexican electric energy distribution database. In particular we are going to perform more simulations using the inclusion of multiple random variables to observe its utility like criterion within the feature selection area, and using the cost sensitive in a early stage, it is to say, try to integrate cost sensitive analysis into the pre-processing feature selection phase, and not at the data mining stage.

References

1. Frawley, W., et al.: Knowledge Discovery in DBs: An Overview. In: Piatetsky-Shapiro, G. (ed.) Knowledge Discovery in Databases, pp. 1–27. AAAI/MIT, Cambridge (1991)
2. Pyle, D.: Data preparation for data mining. Morgan Kaufmann, San Francisco, California (1999)
3. Guyon, I., Elisseeff, A.: An introduction to variable and feature selection. Journal of machine learning research 3, 1157–1182 (2003)
4. Kohavi, R., John, G.: Wrappers for feature subset selection. Artificial Intelligence Journal, Special issue on relevance, 273–324 (1997)
5. Leite, R., Brazdil, P.: Decision tree-based attribute selection via sub sampling. In: Herrera, F., Riquelme, J. (eds.) Workshop de minería de datos y aprendizaje, VIII Iberamia, Sevilla, Spain, November 2002, pp. 77–83 (2002)
6. Piramuthu, S.: Evaluating feature selection methods for learning in data mining applications. In: Proc. 31st annual Hawaii Int. conf. on system sciences, pp. 294–301 (1998)
7. Stolfo, S., Fan, W., Lee, W., Prodromidis, A., Chan, P.: Credit card fraud detection using meta-learning: Issues and initial results. In: Working notes of AAAI Workshop on AI Approaches to Fraud Detection and Risk Management (1997)
8. (2003), http://www.ia.uned.es/~elvira/
9. (2003), http://www.cs.waikato.ac.nz/ml/weka
10. Stoppiglia, H., Dreyfus, G., et al.: Ranking a random feature for variable and feature selection. Journal of machine learning research 3, 1399–1414 (2003)
11. Molina, L., Belanche, L., Nebot, A.: Feature selection algorithms, a survey and experimental evaluation. In: IEEE Int. conf. on data mining, Maebashi City Japan, pp. 306–313 (2002)

Author Index

Printing: Mercedes-Druck, Berlin
Binding: Stein+Lehmann, Berlin

Lecture Notes in Artificial Intelligence (LNAI)

Vol. 5009: G. Wang, T. Li, J.W. Grzymala-Busse, D. Miao, A. Skowron, Y. Yao (Eds.), Rough Sets and Knowledge Technology. XVIII, 765 pages. 2008.

Vol. 4994: A. An, S. Matwin, Z.W. Raś, D. Ślęzak (Eds.), Foundations of Intelligent Systems. XVII, 653 pages. 2008.

Vol. 4953: N.T. Nguyen, G.S. Jo, R.J. Howlett, L.C. Jain (Eds.), Agent and Multi-Agent Systems: Technologies and Applications. XX, 909 pages. 2008.

Vol. 4946: I. Rahwan, S. Parsons, C. Reed (Eds.), Argumentation in Multi-Agent Systems. X, 235 pages. 2008.

Vol. 4938: T. Tokunaga, A. Ortega (Eds.), Large-Scale Knowledge Resources. IX, 367 pages. 2008.

Vol. 4933: R. Medina, S. Obiedkov (Eds.), Formal Concept Analysis. XII, 325 pages. 2008.

Vol. 4930: I. Wachsmuth, G. Knoblich (Eds.), Modeling Communication with Robots and Virtual Humans. X, 337 pages. 2008.

Vol. 4929: M. Helmert, Understanding Planning Tasks. XIV, 270 pages. 2008.

Vol. 4924: D. Riaño (Ed.), Knowledge Management for Health Care Procedures. X, 161 pages. 2008.

Vol. 4923: S.B. Yahia, E.M. Nguifo, R. Belohlavek (Eds.), Concept Lattices and Their Applications. XII, 283 pages. 2008.

Vol. 4914: K. Satoh, A. Inokuchi, K. Nagao, T. Kawamura (Eds.), New Frontiers in Artificial Intelligence. X, 404 pages. 2008.

Vol. 4911: L. De Raedt, P. Frasconi, K. Kersting, S. Muggleton (Eds.), Probabilistic Inductive Logic Programming. VIII, 341 pages. 2008.

Vol. 4908: M. Dastani, A. El Fallah Segrouchni, A. Ricci, M. Winikoff (Eds.), Programming Multi-Agent Systems. XII, 267 pages. 2008.

Vol. 4898: M. Kolp, B. Henderson-Sellers, H. Mouratidis, A. Garcia, A.K. Ghose, P. Bresciani (Eds.), Agent-Oriented Information Systems IV. X, 292 pages. 2008.

Vol. 4897: M. Baldoni, T.C. Son, M.B. van Riemsdijk, M. Winikoff (Eds.), Declarative Agent Languages and Technologies V. X, 245 pages. 2008.

Vol. 4894: H. Blockeel, J. Ramon, J. Shavlik, P. Tadepalli (Eds.), Inductive Logic Programming. XI, 307 pages. 2008.

Vol. 4885: M. Chetouani, A. Hussain, B. Gas, M. Milgram, J.-L. Zarader (Eds.), Advances in Nonlinear Speech Processing. XI, 284 pages. 2007.

Vol. 4874: J. Neves, M.F. Santos, J.M. Machado (Eds.), Progress in Artificial Intelligence. XVIII, 704 pages. 2007.

Vol. 4870: J.S. Sichman, J. Padget, S. Ossowski, P. Noriega (Eds.), Coordination, Organizations, Institutions, and Norms in Agent Systems III. XII, 331 pages. 2008.

Vol. 4869: F. Botana, T. Recio (Eds.), Automated Deduction in Geometry. X, 213 pages. 2007.

Vol. 4865: K. Tuyls, A. Nowe, Z. Guessoum, D. Kudenko (Eds.), Adaptive Agents and Multi-Agent Systems III. VIII, 255 pages. 2008.

Vol. 4850: M. Lungarella, F. Iida, J.C. Bongard, R. Pfeifer (Eds.), 50 Years of Artificial Intelligence. X, 399 pages. 2007.

Vol. 4845: N. Zhong, J. Liu, Y. Yao, J. Wu, S. Lu, K. Li (Eds.), Web Intelligence Meets Brain Informatics. XI, 516 pages. 2007.

Vol. 4840: L. Paletta, E. Rome (Eds.), Attention in Cognitive Systems. XI, 497 pages. 2007.

Vol. 4830: M.A. Orgun, J. Thornton (Eds.), AI 2007: Advances in Artificial Intelligence. XIX, 841 pages. 2007.

Vol. 4828: M. Randall, H.A. Abbass, J. Wiles (Eds.), Progress in Artificial Life. XII, 402 pages. 2007.

Vol. 4827: A. Gelbukh, Á.F. Kuri Morales (Eds.), MICAI 2007: Advances in Artificial Intelligence. XXIV, 1234 pages. 2007.

Vol. 4826: P. Perner, O. Salvetti (Eds.), Advances in Mass Data Analysis of Signals and Images in Medicine, Biotechnology and Chemistry. X, 183 pages. 2007.

Vol. 4819: T. Washio, Z.-H. Zhou, J.Z. Huang, X. Hu, J. Li, C. Xie, J. He, D. Zou, K.-C. Li, M.M. Freire (Eds.), Emerging Technologies in Knowledge Discovery and Data Mining. XIV, 675 pages. 2007.

Vol. 4811: O. Nasraoui, M. Spiliopoulou, J. Srivastava, B. Mobasher, B. Masand (Eds.), Advances in Web Mining and Web Usage Analysis. XII, 247 pages. 2007.

Vol. 4798: Z. Zhang, J.H. Siekmann (Eds.), Knowledge Science, Engineering and Management. XVI, 669 pages. 2007.

Vol. 4795: F. Schilder, G. Katz, J. Pustejovsky (Eds.), Annotating, Extracting and Reasoning about Time and Events. VII, 141 pages. 2007.

Vol. 4790: N. Dershowitz, A. Voronkov (Eds.), Logic for Programming, Artificial Intelligence, and Reasoning. XIII, 562 pages. 2007.

Vol. 4788: D. Borrajo, L. Castillo, J.M. Corchado (Eds.), Current Topics in Artificial Intelligence. XI, 280 pages. 2007.

Vol. 4775: A. Esposito, M. Faundez-Zanuy, E. Keller, M. Marinaro (Eds.), Verbal and Nonverbal Communication Behaviours. XII, 325 pages. 2007.

Vol. 4772: H. Prade, V.S. Subrahmanian (Eds.), Scalable Uncertainty Management. X, 277 pages. 2007.

Vol. 4766: N. Maudet, S. Parsons, I. Rahwan (Eds.), Argumentation in Multi-Agent Systems. XII, 211 pages. 2007.

Vol. 4760: E. Rome, J. Hertzberg, G. Dorffner (Eds.), Towards Affordance-Based Robot Control. IX, 211 pages. 2008.

Vol. 4755: V. Corruble, M. Takeda, E. Suzuki (Eds.), Discovery Science. XI, 298 pages. 2007.

Vol. 4754: M. Hutter, R.A. Servedio, E. Takimoto (Eds.), Algorithmic Learning Theory. XI, 403 pages. 2007.

Vol. 4737: B. Berendt, A. Hotho, D. Mladenic, G. Semeraro (Eds.), From Web to Social Web: Discovering and Deploying User and Content Profiles. XI, 161 pages. 2007.

Vol. 4733: R. Basili, M.T. Pazienza (Eds.), AI*IA 2007: Artificial Intelligence and Human-Oriented Computing. XVII, 858 pages. 2007.

Vol. 4724: K. Mellouli (Ed.), Symbolic and Quantitative Approaches to Reasoning with Uncertainty. XV, 914 pages. 2007.

Vol. 4722: C. Pelachaud, J.-C. Martin, E. André, G. Chollet, K. Karpouzis, D. Pelé (Eds.), Intelligent Virtual Agents. XV, 425 pages. 2007.

Vol. 4720: B. Konev, F. Wolter (Eds.), Frontiers of Combining Systems. X, 283 pages. 2007.

Vol. 4702: J.N. Kok, J. Koronacki, R. Lopez de Mantaras, S. Matwin, D. Mladenič, A. Skowron (Eds.), Knowledge Discovery in Databases: PKDD 2007. XXIV, 640 pages. 2007.

Vol. 4701: J.N. Kok, J. Koronacki, R. Lopez de Mantaras, S. Matwin, D. Mladenič, A. Skowron (Eds.), Machine Learning: ECML 2007. XXII, 809 pages. 2007.

Vol. 4696: H.-D. Burkhard, G. Lindemann, R. Verbrugge, L.Z. Varga (Eds.), Multi-Agent Systems and Applications V. XIII, 350 pages. 2007.

Vol. 4694: B. Apolloni, R.J. Howlett, L. Jain (Eds.), Knowledge-Based Intelligent Information and Engineering Systems, Part III. XXIX, 1126 pages. 2007.

Vol. 4693: B. Apolloni, R.J. Howlett, L. Jain (Eds.), Knowledge-Based Intelligent Information and Engineering Systems, Part II. XXXII, 1380 pages. 2007.

Vol. 4692: B. Apolloni, R.J. Howlett, L. Jain (Eds.), Knowledge-Based Intelligent Information and Engineering Systems, Part I. LV, 882 pages. 2007.

Vol. 4687: P. Petta, J.P. Müller, M. Klusch, M. Georgeff (Eds.), Multiagent System Technologies. X, 207 pages. 2007.

Vol. 4682: D.-S. Huang, L. Heutte, M. Loog (Eds.), Advanced Intelligent Computing Theories and Applications. XXVII, 1373 pages. 2007.

Vol. 4676: M. Klusch, K.V. Hindriks, M.P. Papazoglou, L. Sterling (Eds.), Cooperative Information Agents XI. XI, 361 pages. 2007.

Vol. 4667: J. Hertzberg, M. Beetz, R. Englert (Eds.), KI 2007: Advances in Artificial Intelligence. IX, 516 pages. 2007.

Vol. 4660: S. Džeroski, L. Todorovski (Eds.), Computational Discovery of Scientific Knowledge. X, 327 pages. 2007.

Vol. 4659: V. Mařík, V. Vyatkin, A.W. Colombo (Eds.), Holonic and Multi-Agent Systems for Manufacturing. VIII, 456 pages. 2007.

Vol. 4651: F. Azevedo, P. Barahona, F. Fages, F. Rossi (Eds.), Recent Advances in Constraints. VIII, 185 pages. 2007.

Vol. 4648: F. Almeida e Costa, L.M. Rocha, E. Costa, I. Harvey, A. Coutinho (Eds.), Advances in Artificial Life. XVIII, 1215 pages. 2007.

Vol. 4635: B. Kokinov, D.C. Richardson, T.R. Roth-Berghofer, L. Vieu (Eds.), Modeling and Using Context. XIV, 574 pages. 2007.

Vol. 4632: R. Alhajj, H. Gao, X. Li, J. Li, O.R. Zaïane (Eds.), Advanced Data Mining and Applications. XV, 634 pages. 2007.

Vol. 4629: V. Matoušek, P. Mautner (Eds.), Text, Speech and Dialogue. XVII, 663 pages. 2007.

Vol. 4626: R.O. Weber, M.M. Richter (Eds.), Case-Based Reasoning Research and Development. XIII, 534 pages. 2007.

Vol. 4617: V. Torra, Y. Narukawa, Y. Yoshida (Eds.), Modeling Decisions for Artificial Intelligence. XII, 502 pages. 2007.

Vol. 4612: I. Miguel, W. Ruml (Eds.), Abstraction, Reformulation, and Approximation. XI, 418 pages. 2007.

Vol. 4604: U. Priss, S. Polovina, R. Hill (Eds.), Conceptual Structures: Knowledge Architectures for Smart Applications. XII, 514 pages. 2007.

Vol. 4603: F. Pfenning (Ed.), Automated Deduction – CADE-21. XII, 522 pages. 2007.

Vol. 4597: P. Perner (Ed.), Advances in Data Mining. XI, 353 pages. 2007.

Vol. 4594: R. Bellazzi, A. Abu-Hanna, J. Hunter (Eds.), Artificial Intelligence in Medicine. XVI, 509 pages. 2007.

Vol. 4585: M. Kryszkiewicz, J.F. Peters, H. Rybinski, A. Skowron (Eds.), Rough Sets and Intelligent Systems Paradigms. XIX, 836 pages. 2007.

Vol. 4578: F. Masulli, S. Mitra, G. Pasi (Eds.), Applications of Fuzzy Sets Theory. XVIII, 693 pages. 2007.

Vol. 4573: M. Kauers, M. Kerber, R. Miner, W. Windsteiger (Eds.), Towards Mechanized Mathematical Assistants. XIII, 407 pages. 2007.

Vol. 4571: P. Perner (Ed.), Machine Learning and Data Mining in Pattern Recognition. XIV, 913 pages. 2007.

Vol. 4570: H.G. Okuno, M. Ali (Eds.), New Trends in Applied Artificial Intelligence. XXI, 1194 pages. 2007.

Vol. 4565: D.D. Schmorrow, L.M. Reeves (Eds.), Foundations of Augmented Cognition. XIX, 450 pages. 2007.

Vol. 4562: D. Harris (Ed.), Engineering Psychology and Cognitive Ergonomics. XXIII, 879 pages. 2007.

Vol. 4548: N. Olivetti (Ed.), Automated Reasoning with Analytic Tableaux and Related Methods. X, 245 pages. 2007.